**Polysaccharides and Polyamides
in the Food Industry**

Properties, Production, and Patents

Edited by A. Steinbüchel and S. K. Rhee

Related Titles

A. Steinbüchel
Biopolymers. 10 Volumes + Index
2003
ISBN 3-527-30290-5

Single "Biopolymers" Volumes:

M. Hofrichter, A. Steinbüchel
Vol. 1: Lignin, Humic Substances and Coal
2001
ISBN 3-527-30220-4

T. Koyama, A. Steinbüchel
Vol. 2: Polyisoprenoids
2001
ISBN 3-527-30221-2

Y. Doi, A. Steinbüchel
Vol. 3a: Polyesters I – Biological Systems and Biotechnological Production
2002
ISBN 3-527-30224-7

Y. Doi, A. Steinbüchel
Vol. 3b: Polyesters II – Properties and Chemical Synthesis
2002
ISBN 3-527-30219-0

Y. Doi, A. Steinbüchel
Vol. 4: Polyesters III – Applications and Commercial Products
2002
ISBN 3-527-30225-5

E. J. Vandamme, S. De Baets, A. Steinbüchel
Vol. 5: Polysaccharides I – Polysaccharides from Prokaryotes
2002
ISBN 3-527-30226-3

S. De Baets, E. J. Vandamme, A. Steinbüchel
Vol. 6: Polysaccharides II – Polysaccharides from Eukaryotes
2002
ISBN 3-527-30227-1

S. R. Fahnestock, A. Steinbüchel
Vol. 7: Polyamides and Complex Proteinaceous Materials I
2003
ISBN 3-527-30222-0

S. R. Fahnestock, A. Steinbüchel
Vol. 8: Polyamides and Complex Proteinaceous Materials II
2003
ISBN 3-527-30223-9

S. Matsumura, A. Steinbüchel
Vol. 9: Miscellaneous Biopolymers and Biodegradation of Polymers
2003
ISBN 3-527-30228-X

A. Steinbüchel
Vol. 10: General Aspects and Special Applications
2003
ISBN 3-527-30229-8

A. Steinbüchel
Cumulative Index
2003
ISBN 3-527-30230-1

H. F. Mark
Encyclopedia of Polymer Science and Technology
Part 1: Volumes 1–4, **2003**, ISBN 0-471-28824-1
Part 2: Volumes 5–8, **2003**, ISBN 0-471-28781-4
Part 3: Volumes 9–12, **2004**, ISBN 0-471-28780-6

Polysaccharides and Polyamides in the Food Industry

Properties, Production, and Patents

Edited by A. Steinbüchel and S. K. Rhee

Volume 1
Polysaccharides

WILEY-VCH Verlag GmbH & Co. KGaA

Editors:

Prof. Dr. Alexander Steinbüchel
Institut für Mikrobiologie
Westfälische Wilhelms-Universität
Corrensstrasse 3
48149 Münster
Germany

Dr. Sang Ki Rhee
Korea Institute of Industrial Technology
Evaluation and Planning (ITEP)
701-7, Yeoksam-Dong
Gangnam-Gu
135-080 Seoul
Korea

■ All books published by Wiley-VCH are carefully produced. Nevertheless, authors, editors, and publisher do not warrant the information contained in these books, including this book, to be free of errors. Readers are advised to keep in mind that statements, data, illustrations, procedural details or other items may inadvertently be inaccurate.

Library of Congress Card No: applied for

British Library Cataloguing-in-Publication Data: A catalogue record for this book is available from the British Library

Bibliographic information published by Die Deutsche Bibliothek
Die Deutsche Bibliothek lists this publication in the Deutsche Nationalbibliografie; detailed bibliographic data is available in the Internet at <http://dnb.ddb.de>.

© 2005 Wiley-VCH Verlag
GmbH & Co. KGaA, Weinheim
All rights reserved (including those of translation into other languages). No part of this book may be reproduced in any form – nor transmitted or translated into machine language without written permission from the publishers. Registered names, trademarks, etc. used in this book, even when not specifically marked as such, are not to be considered unprotected by law.

Printed in the Federal Republic of Germany
Printed on acid-free paper.

Composition: Konrad Triltsch
Print und digitale Medien GmbH
Ochsenfurt-Hohestadt
Printing: Strauss GmbH, Mörlenbach
Bookbinding: J. Schäffer GmbH, Grünstadt

ISBN-13 978-3-527-31345-7
ISBN-10 3-527-31345-1

Introductory Preface

General Aspects

Living organisms are able to synthesize an overwhelming variety of polymers which can be distinguished into eight major classes according to their chemical structure: (1) nucleic acids, (2) proteins and other polyamides, (3) polysaccharides, (4) polyoxoesters (polyhydroxyalkanoic acids), (5) polythioesters, (6) polyanhydrides (polyphosphate), (7) polyisoprenoids, and (8) lignin. These biopolymers accomplish quite different essential or beneficial functions for the organisms. Microorganisms are capable of synthesizing biopolymers belonging to classes 1–6, whereas eukaryotes synthesize mainly biopolymers belonging to classes 1–3 and 6–8. Among the biopolymers produced are many used for various applications in industry. Biotechnological production of polymers is at present mostly achieved by fermentation of microorganisms in stirred tank bioreactors, and the biopolymers can be obtained as extracellular or intracellular compounds. Alternatively, biopolymers can also be produced by enzymatic *in vitro* processes. However, by far the largest amounts of biopolymers are still extracted from plant and animal sources or from algae.

Biopolymers exhibit fascinating properties and play a major role in the food industry for processing food and modifying food texture and properties. Among the various biopolymers, polysaccharides and polyamides are the most important in the food industry.

Polysaccharides

Polysaccharides are *per se* renewable resources or are produced from renewable resources, and they offer a wide variety of potentially useful products to human life. They comprise a distinct class of biopolymers, produced universally by living organisms including microorganisms, plants, and animals. They exhibit a large variety of unique and in most cases rather complex chemical structures, different physiological functions, and a wide range of potential applications, particularly for foodstuff. For instance, a number of plant polysaccharides such as starch have been widely used in food for a very long time. More recently, other plant or microbial polysaccharides such as levan, inulin, curdlan, and pullulan have also found use in the food industry. Similarly, marine algae have yielded such useful products as agar, alginate, and carrageenan. Cellulose and its derivatives have also found many uses as food additives because of their easy availability and cheapness. While cellulose, chitin, and some other polysaccharides are insoluble, many other polysaccharides are water-soluble and are capable of significantly altering the rheology of aqueous-based solutions and find a wide range of application in the food industry.

Proteins, Proteinaceous Materials, and Poly(Amino Acids)
Proteins are biopolymers composed of amino acids and are essential components of every biological system. They are catalysts of diverse capability, agents of molecular recognition, mediators of self-assembly, transducers of energy and information, media of communication, producers of directed motion, and librarians of the genetic program. The specific amino acids used and the sequence of amino acids in a protein polymer chain as well as the length of the polymer chain are determined by the corresponding DNA and RNA templates, respectively. Therefore, ribosomal protein biosynthesis yields principally monodisperse products which a clearly defined chemical structure. Proteinaceous materials provided by nature have been exploited by human technology for millennia. Silks, furs, leather, bone, horn, feathers, all have been—and many still remain, despite modern chemistry—essential materials for all human cultures. In addition, organisms and in particular microorganisms are capable of synthesizing other biopolymers also consisting of amino acids; they are referred to as poly(amino acids) or polyamides. Polyamides are in contrast to proteins synthesized by soluble synthetases, which use free amino acids as substrates in ATP-dependent reactions. Since synthesis is not directed by a template, the products are polydisperse. In addition, the specificities of these synthetases are also not restricted to their "natural" substrates, and they are therefore not strictly specific. Moreover, R-stereoisomers of amino acids and linkage types not found in proteins occur in poly(amino acids). One of these poly(amino acids) is cyanophycin, which is a copolymer of aspartic acid and arginine. In addition to cyanophycin, microorganisms synthesize poly(glutamic acid) and poly(lysine) in a similar way by template-independent mechanisms. Whereas cyanophycin is a storage compound for nitrogen, carbon, and energy, and occurs as insoluble cytoplasmic inclusions in many bacteria, poly(glutamic acid) and poly(lysine) are only synthesized by a few microorganisms and the polymers occur as extracellular polymers.

Many proteins and poly(amino acids) are of commercial interest because of their catalytic or physicochemical properties. The advances in recombinant DNA technology have presented new approaches and opportunities for design and biosynthesis of protein materials. In addition, many traditional proteins prepared by extraction of animal (e.g., collagen) or plant (e.g., soy or zein from corn) tissue are being chemically or physically modified for new applications in biotechnology and in the food industry.

Scope of this Book
Taking the importance and the applications of biopolymers in the food industry as our guide, we carefully selected from Volumes 5–8 of the ten-volume work *Biopolymers* 19 chapters all dealing with polysaccharides, proteins, proteinaceous materials, and poly(amino acids). The present two-volume spin-off product covers 13 different types of polysaccharides in Volume 1 and six different polyamides in Volume 2. The polymers from these selected chapters are successfully being used as key food additives, e.g., for gelling, viscosifying, stabilizing, and stiffening of food products. They include bacterial cellulose, curdlan, xanthan, dextran, levan, exopolysaccharides of lactic acid bacteria, pullulan, alginates from algae, carrageenan, pectins, starch, inulin, chitin, and chitosan from animal sources of polysaccharides. Enzymes for technical applications, collagen and gelatins, sweet-tasting proteins, and the seed storage proteins vicilin and legumin are also included. Poly(ε-1-lysine) and poly(γ-glutamic acid) represent the poly(amino acids) relevant for the subject of this volume. Each

polymeric substance is treated similarly, covering properties, production, patents, and applications, which range from traditional uses, e.g., of starch and pectins, to novel applications such as sweet-tasting proteins.

In compiling this handbook, it has been our intention to provide the scientific and industrial community with a comprehensive view of the current state of knowledge on polysaccharides and polyamides. This handbook attempts to review what is currently known about these fascinating biopolymers in terms of their discovery, occurrence, chemical and physical properties, analysis, biosynthesis, molecular genetics, physiological role, fermentative production, isolation, purification, and applications. With the title more focused and at a price more affordable than that of the complete *Biopolymers* series, this two-volume handbook will be of interest in particular to medium-sized laboratories that are interested or active in this area, and to libraries.

We consider it a strength of this collection that the individual chapters are diverse in style and in purpose. Some paint an area in broad, conceptual strokes, others in fine technical detail. Some present information, others arguments or interpretation. Some summarize past accomplishments; others point to future possibilities. We are convinced that each is in some way appropriate to its topic. As broad as this field is, there is of course no hope of completeness. We have attempted to sample broadly, but key omissions are inevitable and we readily acknowledge them. We will feel this handbook has been successful if some of these chapters stimulate readers to become interested in and solve specific problems, or make the field more accessible to newcomers.

Acknowledgments

Whatever is accomplished is of course the achievement of the authors. We are most grateful to all of them for devoting so much of their valuable time to this endeavor and for sharing their knowledge and insights so generously. We are also particularly grateful to the authors of the selected chapters for allowing the contents of their *Biopolymers* contributions to be included in this new title.

Last but not least, we would like to thank Wiley-VCH for publishing this new handbook with their customary professionalism and excellence and for their outstanding help throughout the gestation and birth of this handbook. Special thanks are due to Mrs. Karin Dembowsky, who initiated the *Biopolymers* series, to Dr. Andreas Sendtko, who continued and finalized it, and to many others at Wiley-VCH for their initiative, constant efforts, helpful suggestions, constructive criticism, and wonderful ideas.

Alexander Steinbuechel Münster and Seoul
Sang Ki Rhee May 2005

Contents

Volume 1

Introductory Preface V

I Polysaccharides

1 Alginates from Algae 1
  Kurt Ingar Draget, Olav Smidsrød, Gudmund Skjåk-Bræk

2 Bacterial Cellulose 31
  Stanislaw Bielecki, Alina Krystynowicz, Marianna Turkiewicz, Halina Kalinowska

3 Carrageenan 85
  Fred van de Velde, Gerhard A. De Ruiter

4 Chitin and Chitosan from Animal Sources 115
  Martin G. Peter

5 Curdlan 209
  In-Young Lee

6 Dextran 233
  Timothy D. Leathers

7 Exopolysaccharides of Lactic Acid Bacteria 257
  Isabel Hallemeersch, Sophie De Baets, Erick J. Vandamme

8 Inulin 281
  Anne Franck, Leen De Leenheer

9 Levan 323
  Sang-Ki Rhee, Ki-Bang Song, Chul-Ho Kim, Buem-Seek Park, Eun-Kyung Jang, Ki-Hyo Jang

10 Pectins 351
  Marie-Christine Ralet, Estelle Bonnin, Jean-François Thibault

11 Pullulan 387
  Timothy D. Leathers

12	Starch *Richard Frank Tester, John Karkalas*	423
13	Xanthan *Karin Born, Virginie Langendorff, Patrick Boulenguer*	481

Volume 2

II Polyamides

14	Collagens and Gelatins *Barbara Brodsky, Jerome A. Werkmeister, John A. M. Ramshaw*	521
15	Enzymes for Technical Applications *Thomas Schäfer, Ole Kirk, Torben Vedel Borchert, Claus Crone Fuglsang, Sven Pedersen, Sonja Salmon, Hans Sejr Olsen, Randy Deinhammer, Henrik Lund*	557
16	Poly-γ-glutamic Acid *Makoto Ashiuchi, Haruo Misono*	619
17	ε-Poly-L-Lysine *Toyokazu Yoshida, Jun Hiraki, Toru Nagasawa*	671
18	Sweet-tasting Proteins *Ignacio Faus and Ms. Heidi Sisniega*	687
19	Structure, Function, and Evolution of Vicilin and Legumin Seed Storage Proteins *James Martin Dunwell*	707
20	Index	739

I
Polysaccharides

1
Alginates from Algae

Dr. Kurt Ingar Draget[1], Prof. Dr. Olav Smidsrød[2], Prof. Dr. Gudmund Skjåk-Bræk[3]
[1] Norwegian Biopolymer Laboratory, Department of Biotechnology, Norwegian University of Science and Technology, Sem Saelands vei 6-8, N-7491 Trondheim, Norway; Tel.: +47-73598260; Fax: +47-73591283;
E-mail: Kurt.I.Draget@chembio.ntnu.no
[2] Norwegian Biopolymer Laboratory, Department of Biotechnology, Norwegian University of Science and Technology, Sem Saelands vei 6-8, N-7491 Trondheim, Norway; Tel.: +47-735-98260; Fax: +47-735-93337;
E-mail: Olav.Smidsroed@chembio.ntnu.no
[3] Norwegian Biopolymer Laboratory, Department of Biotechnology, Norwegian University of Science and Technology, Sem Saelands vei 6-8, N-7491 Trondheim, Norway. Tel.: +47-735-98260; Fax: +47-735-93340;
E-mail: Gudmund.Skjaak-Braek@chembio.ntnu.no

1	Introduction	2
2	Historical Outline	3
3	Chemical Structure	4
4	Conformation	4
5	Occurrence and Source Dependence	5
6	Physiological Function	6
7	Chemical Analysis and Detection	6
7.1	Chemical Composition and Sequence	6
7.2	Molecular Mass	7
7.3	Detection and Quantification	7
8	Biosynthesis and Biodegradation	7

Polysaccharides and Polyamides in the Food Industry. Properties, Production, and Patents.
Edited by A. Steinbüchel and S. K. Rhee
Copyright © 2005 WILEY-VCH Verlag GmbH & Co. KGaA, Weinheim
ISBN: 3-527-31345-1

9	**Production: Biotechnological and Traditional**	8
9.1	Isolation from Natural Sources / Fermentative Production	8
9.2	Molecular Genetics and *in vitro* Modification	8
9.3	Current and Expected World Market and Costs	9
9.4	Alginate Manufacturers	11
10	**Properties**	11
10.1	Physical Properties	11
10.1.1	Solubility	11
10.1.2	Selective Ion Binding	13
10.1.3	Gel Formation and Ionic Cross-linking	14
10.1.4	Gel Formation and Alginic Acid Gels	15
10.2	Material Properties	15
10.2.1	Stability	15
10.2.2	Ionically Cross-linked Gels	16
10.2.3	Alginic Acid Gels	19
10.3	"Biological" Properties	20
11	**Applications**	20
11.1	Technical Utilization	21
11.2	Medicine and Pharmacy	21
11.3	Foods	22
12	**Relevant Patents**	23
13	**Outlook and Perspectives**	24
14	**References**	26

DP	degree of polymerization
EDTA	etylenediamine tetraacetic acid
G	α-L-guluronic acid
GDL	D-glucono-δ-lactone
M	β-D-mannuronic acid (M)
$N_{G>1}$	average G-block length larger than 1
NMR	nuclear magnetic resonance spectroscopy
PGA	propylene glycol alginate
pK_a	dissociation constants for the uronic acid monomers

1 Introduction

Alginates are quite abundant in nature since they occur both as a structural component in marine brown algae (*Phaeophyceae*), comprising up to 40% of the dry matter, and as capsular polysaccharides in soil bacteria (see Chapter 8 on bacterial alginates in Volume 5 of this series). Although present research and results point toward a possible production by microbial fermentation and also by post-polymerization modification of the alginate molecule, all commercial alginates are at present still extracted from algal sources. The industrial applications of alginates are linked to its ability to retain water, and its gelling, viscosifying, and stabilizing properties. Upcoming biotechnological applications, on the other hand, are based either on specific biological effects of the alginate molecule itself or on its unique, gentle, and almost temperature-independent sol/gel transition in the presence of multivalent cations (e.g., Ca^{2+}), which makes alginate highly suitable as an immobilization matrix for living cells.

Traditional exploitation of alginates in technical applications has been based to a large extent on empirical knowledge. However, since alginates now enter into more knowledge-demanding areas such as pharmacy and biotechnology, new research functions as a locomotive for a detailed further investigation of structure–function relationships. New scientific breakthroughs are made, which in turn may benefit the traditional technical applications.

2
Historical Outline

The British chemist E. C. C. Stanford first described alginate (the preparation of "algic acid" from brown algae) with a patent dated 12 January 1881 (Stanford, 1881). After the patent, his discovery was further discussed in papers from 1883 (Standford, 1883a,b). Stanford believed that alginic acid contained nitrogen and contributed much to the elucidation of its chemical structure.

In 1926, some groups working independently (Atsuki and Tomoda, 1926; Schmidt and Vocke, 1926) discovered that uronic acid was a constituent of alginic acid. The nature of the uronic acids present was investigated by three different groups shortly afterwards (Nelson and Cretcher, 1929, 1930, 1932; Bird and Haas, 1931; Miwa, 1930), which all found D-mannuronic acid in the hydrolysate of alginate. The nature of the bonds between the uronic acid residues in the alginate molecule was determined to be β1,4, as in cellulose (Hirst et al., 1939)

This very simple and satisfactory picture of the constitution of alginic acid was, however, destroyed by the work of Fischer and Dörfel (1955). In a paper chromatographic study of uronic acids and polyuronides, they discovered the presence of a uronic acid different from mannuronic acid in the hydrolysates of alginic acid. This new uronic acid was identified as L-guluronic acid. The quantity of L-guluronic acid was considerable, and a method for quantitative determination of mannuronic and guluronic acid was developed.

Alginate then had to be regarded as a binary copolymer composed of α-L-guluronic and β-D-mannuronic residues. As long as alginic acid was regarded as a polymer containing only D-mannuronic acid linked together with β-1,4 links, it was reasonable to assume that alginates from different raw materials were chemically identical and that any given sample of alginic acid was chemically homogeneous. From a practical and a scientific point of view, the uronic acid composition of alginate from different sources had to be examined, and methods for chemical fractionation of alginates had to be developed. These tasks were undertaken mainly by Haug and coworkers (Haug, 1964), as described in Section 3 below. The discovery of alginate as a block-copolymer, the correlation between physical properties and block structure, and the discovery of a set of epimerases converting mannuronic to guluronic acid in a sequence-dependent manner also are discussed further in later sections.

3
Chemical Structure

Being a family of unbranched binary copolymers, alginates consist of $(1 \rightarrow 4)$ linked β-D-mannuronic acid (M) and α-L-guluronic acid (G) residues (see Figure 1a and b) of widely varying composition and sequence. By partial acid hydrolysis (Haug, 1964; Haug et al., 1966; Haug and Larsen, 1966; Haug et al., 1967a; Haug and Smidsrød, 1965), alginate was separated into three fractions. Two of these contained almost homopolymeric molecules of G and M, respectively, while a third fraction consisted of nearly equal proportions of both monomers and was shown to contain a large number of MG dimer residues. It was concluded that alginate could be regarded as a true block copolymer composed of homopolymeric regions of M and G, termed M- and G-blocks, respectively, interspersed with regions of alternating structure (MG-blocks; see Figure 1c). It was further shown (Painter et al., 1968; Larsen et al., 1970; Smidsrød and Whittington, 1969) that alginates have no regular repeating unit and that the distribution of the monomers along the polymer chain could not be described by Bernoullian statistics. Knowledge of the monomeric composition is hence not sufficient to determine the sequential structure of alginates. It was suggested (Larsen et al., 1970) that a second-order Markov model would be required for a general approximate description of the monomer sequence in alginates. The main difference at the molecular level between algal and bacterial alginates is the presence of O-acetyl groups at C2 and/or C3 in the bacterial alginates (Skjåk-Bræk et al., 1986).

4
Conformation

Knowledge of the monomer ring conformations is necessary to understand the polymer properties of alginates. X-ray diffraction studies of mannuronate-rich and guluronate-rich alginates showed that the guluronate residues in homopoly-meric blocks were in the $^1C_4$ conformation (Atkins et al., 1970), while the mannuronate residues had

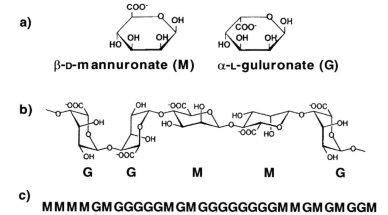

Fig. 1 Structural characteristics of alginates: (a) alginate monomers, (b) chain conformation, (c) block distribution.

the $^4C_1$ conformation (see Figure 1a). Viscosity data of alginate solutions indicated that the stiffness of the chain blocks increased in the order MG < MM < GG. This series could be reproduced only by statistical mechanical calculations when the guluronate residues were set in the $^1C_4$ conformation (Smidsrød et al., 1973) and was later confirmed by ^{13}C-NMR (Grasdalen et al., 1977). Hence, alginate contains all four possible glycosidic linkages: diequatorial (MM), diaxial (GG), equatorial-axial (MG), and axial-equatorial (GM) (see Figure 1b).

The diaxial linkage in G-blocks results in a large, hindered rotation around the glycosidic linkage, which may account for the stiff and extended nature of the alginate chain (Smidsrød et al., 1973). Additionally, taking the polyelectrolyte nature of alginate into consideration, the electrostatic repulsion between the charged groups on the polymer chain also will increase the chain extension and hence the intrinsic viscosity. Extrapolation of dimensions both to infinite ionic strength and to θ-conditions (Smidsrød, 1970) yielded relative dimensions for the neutral, unperturbed alginate chain being much higher than for amylose derivatives and even slightly higher than for some cellulose derivatives.

Another parameter reflecting chain stiffness and extension is the exponent in the Mark-Houwink-Sakurada equation,

$$[\eta] = K \cdot M^a$$

where M is the molecular weight of the polymer, $[\eta]$ is the intrinsic viscosity, and the exponent a generally increases with increasing chain extension. Some measurements on alginates (Martinsen et al., 1991; Smidsrød and Haug, 1968a, Mackie et al., 1980), yielded a-values ranging from 0.73 to 1.31, depending on ionic strength and alginate composition. Low and high a-values are related to large fractions of the flexible MG-blocks and the stiff and extended G-blocks, respectively (Moe et al., 1995).

5
Occurrence and Source Dependence

Commercial alginates are produced mainly from Laminaria hyperborea, Macrocystis pyrifera, Laminaria digitata, Ascophyllum nodosum, Laminaria japonica, Eclonia maxima, Lessonia nigrescens, Durvillea antarctica, and Sargassum spp. Table 1 gives some sequential parameters (determined by high-field NMR-spectroscopy) for samples of these

Tab. 1 Composition and sequence parameters of algal alginates (Smidsrød and Draget 1996)

Source	F_G	F_M	F_{GG}	F_{MM}	$F_{GM,MG}$
Laminaria japonica	0.35	0.65	0.18	0.48	0.17
Laminaria digitata	0.41	0.59	0.25	0.43	0.16
Laminaria hyperborea, blade	0.55	0.45	0.38	0.28	0.17
Laminaria hyperborea, stipe	0.68	0.32	0.56	0.20	0.12
Laminaria hyperborea, outer cortex	0.75	0.25	0.66	0.16	0.09
Lessonia nigrescens[a]	0.38	0.62	0.19	0.43	0.19
Ecklonia maxima	0.45	0.55	0.22	0.32	0.32
Macrocystis pyrifera	0.39	0.61	0.16	0.38	0.23
Durvillea antarctica	0.29	0.71	0.15	0.57	0.14
Ascophyllum nodosum, fruiting body	0.10	0.90	0.04	0.84	0.06
Ascophyllum nodosum, old tissue	0.36	0.64	0.16	0.44	0.20

[a] Data provided by Bjørn Larsen

alginates. The composition and sequential structure may, however, vary according to seasonal and growth conditions (Haug, 1964; Indergaard and Skjåk-Bræk, 1987). High contents of G generally are found in alginates prepared from stipes of old *Laminaria hyperborea* plants, whereas alginates from *A. nodosum*, *L. japonica*, and *Macrocystis pyrifera* are characterized by low content of G-blocks and low gel strength.

Alginates with more extreme compositions containing up to 100% mannuronate can be isolated from bacteria (Valla et al., 1996). Alginates with a very high content of guluronic acid can be prepared from special algal tissues such as the outer cortex of old stipes of *L. hyperborea* (see Table 1), by chemical fractionation (Haug and Smidsrød, 1965; Rivera-Carro, 1984) or by enzymatic modification *in vitro* using mannuronan C-5 epimerases from *A. vinelandii* (Valla et al., 1996; see Section 9.2). This family of enzymes is able to epimerize M-units into G-units in different patterns from almost strictly alternating to very long G-blocks. The epimerases from *A. vinelandii* have been cloned and expressed, and they represent at present a powerful new tool for the tailoring of alginates. It is also obvious that commercial alginates with less molecular heterogeneity, with respect to chemical composition and sequence, can be obtained by a treatment with one of the C-5 epimerases (Valla et al., 1996).

6
Physiological Function

The biological function of alginate in brown algae generally is believed to be as a structure-forming component. The intercellular alginate gel matrix gives the plants both mechanical strength and flexibility (Andresen et al., 1977). Simply speaking, alginates in marine brown algae may be regarded as having physiological properties similar to those of cellulose in terrestrial plants. This relation between structure and function is reflected in the compositional difference of alginates in different algae or even between different tissues from the same plant (see Table 1). In *L. hyperborea*, an alga that grows in very exposed coastal areas, the stipe and holdfast have a very high content of guluronic acid, giving high mechanical rigidity. The leaves of the same algae, which float in the streaming water, have an alginate characterized by a lower G-content, giving it a more flexible texture. The physiological function of alginates in bacteria will be covered elsewhere in this series.

7
Chemical Analysis and Detection

Since alginates are block copolymers, and because of the fact that their physical properties rely heavily on the sequence of these blocks, it obvious that the development of techniques enabling a sequence quantification is of the utmost importance. Additionally, molecular mass and its distribution (polydispersity) is a significant parameter in some applications.

7.1
Chemical Composition and Sequence

Detailed information about the structure of alginates became available by introduction of high-resolution ^{1}H and ^{13}C NMR-spectroscopy (Grasdalen et al., 1977, 1979; Penman and Sanderson, 1972; Grasdalen, 1983) in the sequential analysis of alginate. These powerful techniques make it possible to determine the monad frequencies F_M and F_G; the four nearest neighboring (diad) frequencies F_{GG}, F_{MG}, F_{GM}, and F_{MM}; and the eight next nearest neighboring (triad) frequencies. Knowledge of these frequencies enables, for example, the calculation of the average G-block length larger than 1:

$$N_{G>1} = (F_G - F_{MGM}) / F_{GGM}.$$

This value has been shown to correlate well with gelling properties. It is important to realize that in an alginate chain population, neither the composition nor the sequence of each chain will be alike. This results in a composition distribution of a certain width.

7.2
Molecular Mass

Alginates, like polysaccharides in general, are polydisperse with respect to molecular weight. In this aspect they resemble synthetic polymers rather than other biopolymers such as proteins and nucleic acids. Because of this polydispersity, the "molecular weight" of an alginate is an average over the whole distribution of molecular weights.

In a population of molecules where N_i is the number of molecules and w_i is the weight of molecules having a specific molecular weight M_i, the number and the weight average are defined respectively as:

$$\overline{M_n} = \frac{\Sigma_i N_i M_i}{\Sigma_i N_i}$$

$$\overline{M_w} = \frac{\Sigma_i w_i M_i}{\Sigma_i w_i} = \frac{\Sigma_i N w_i M_i^2}{\Sigma_i N_i M_i}$$

For a randomly degraded polymer, we have $\overline{M_w} \approx 2\overline{M_n}$ (Tanford, 1961). The fraction $\overline{M_w}/\overline{M_n}$ is called the polydispersity index. Polydispersity index values between 1.4 and 6.0 have been reported for alginates and have been related to different types of preparation and purification processes (Martinsen et al., 1991; Smidsrød and Haug, 1968a; Mackie et al., 1980; Moe et al., 1995).

The molecular-weight distribution can have implications for the uses of alginates, as low-molecular-weight fragments containing only short G-blocks may not take part in gel-network formation and consequently do not contribute to the gel strength. Furthermore, in some high-tech applications, the leakage of mannuronate-rich fragments from alginate gels may cause problems (Stokke et al., 1991; Otterlei et al., 1991) and a narrow molecular-weight distribution therefore is recommended.

7.3
Detection and Quantification

Detection and quantification of alginates in the presence of other biopolymers, such as proteins, are not straightforward mainly because of interference. Once isolated, a number of colorimetric methods can be applied to quantify alginate. The oldest and most common is the general procedure for carbohydrates, the phenol/sulfuric acid method (Dubois et al., 1956), but there are also two slightly refined formulas specially designed for uronic acids (Blumenkrantz and Asboe-Hansen, 1973; Filisetti-Cozzi and Carpita, 1991).

8
Biosynthesis and Biodegradation

Our knowledge of the alginate biosynthesis mainly comes from studying alginate-producing bacteria. Figure 2 shows the principal enzymes involved in alginate biosynthesis, and the activity of all enzymes (1–7) has been identified in brown algae. During the last decade, the genes responsible for alginate synthesis in *Pseudomonas* and *Azotobacter* have been identified, sequenced, and cloned. For further information on alginate biosynthesis, please see Chapter 8 on bacte-

THE BIOSYNTHETIC PATHWAY OF ALGINATE

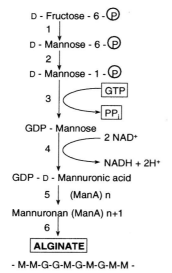

Fig. 2 Biosynthetic pathway of alginates.

rial alginates in Volume 5 of this series. Because of their potential use in alginate modification, the only enzymes we will comment on here are the alginate lyases and the mannuronan C-5 epimerases.

Alginates are not degraded in the human gastric-intestinal tract, and hence do not give metabolic energy. Some lower organisms have, however, developed lyases that degrade alginates down to single components, resulting in alginates that function as a carbon source. Alginate lyases catalyze the depolymerization of alginate by splitting the 1–4 glycosidic linkage in a β-elimination reaction, leaving an unsaturated uronic acid on the non-reducing end of the molecules. Alginate lyases are widely distributed in nature, including in organisms growing on alginate as a carbon source such as marine gastropods, prokaryotic and eukaryotic microorganisms, and bacteriophages. They also are found in the bacterial species producing alginate such as *Azotobacter vinelandii* and *Pseudomonas aeruginosa*. All of them are endolyases and may exhibit specificity to either M or G. Since the aglycon residue will be identical for both M and G, the use of lyases for structural work is limited. Table 2 lists a range of lyases and their specificities.

9
Production: Biotechnological and Traditional

There has been significant progress in the understanding of alginate biosynthesis over the last 10 years. The fact that the alginate molecule enzymatically undergoes a post-polymerization modification with respect to chemical composition and sequence opens up the possibility for *in vitro* modification and tailoring of commercially available alginates.

9.1
Isolation from Natural Sources / Fermentative Production

As already described, all commercial alginates today are produced from marine brown algae (Table 1). Alginates with more extreme compositions can be isolated from the bacterium *Azotobacter vinelandii*, which, in contrast to *Pseudomonas* species, produces polymers containing G-blocks. Production by fermentation therefore is technically possible but is not economically feasible at the moment.

9.2
Molecular Genetics and *in vitro* Modification

Alginate with a high content of guluronic acid can be prepared from special algal tissues by chemical fractionation or by *in vitro* enzymatic modification of the alginate *in vitro* using mannuronan C-5 epimerases from *A. vinelandii* (Ertesvåg et al., 1994, 1995, 1998b; Høydal et al., 1999). These epimerases, which convert M to G in the

Tab. 2 Substrate specificity and biochemical properties of some alginate lyases (Gacesa, 1992; Wong et al., 2000)

Source	Localization	Sequence specifity	Major end-product	pHopt	Mw (kDa)	pI	Reference
K. aerogenes	Extracellular	G↓X	Trimer	7.0	31.4	8.9	Boyd and Turvey, 1978
Enterobacter cloacae	Extracellular	G	–	7.8	32–38	8.9	Nibu et al., 1995
P. aeruginosa	Intracellular	G↓G	Dimer/pentamer	7.5	31–39	8.9	Shimokawa et al., 1997
(AlgL)	Periplasmic	M-X	Trimer	7.0	39	9.0	Boyd and Turvey, 1977
A. vinelandii	Periplasmic (AlgL)	M↓X M$_{Ac}$↓X	Trimer/tetramer	8.1–8.4	39	5.1	Ertesvåg et al., 1998a
	Extracellular (AlgE7)	G↓X	Tetramer-septamer	6.3–7.3	90.4	–	Ertesvåg et al., 1998a
P. alginovora	Extracellular	G↓G	–	7.5	28	5.5	Boyen et al., 1990a
	Intracellular	M↓M	–	–	24	5.8	Boyen et al., 1990a
Haliotis tuberculata	Hepato-pancreas	M↓X, G↓M	Trimer/dimer	8	34	–	Boyen et al., 1990b
Sphingomonas sp. ALYI-III	Cytoplasmic	M↓X M$_{Ac}$↓X	–	5.6–7.8	38	10.16	Murata et al., 1993
Littorina sp.	Hepato-pancreas	M↓M	Trimer	5.6	~40	–	Elyakova and Favorov, 1974
A. vinelandii phage	Extracellular	M↓X M$_{Ac}$↓X	Trimer	7.7	30–35	–	Davidson et al., 1977

polymer chain, recently have allowed for the production of highly programmed alginates with respect to chemical composition and sequence. A. vinelandii encodes a family of 7 exocellular isoenzymes with the capacity to epimerize all sorts of alginates and other mannuronate-containing polymers, as shown in Figure 3, where the mode of action of AlgE4 (giving alternating introduction of G) is presented. Although the genes have a high degree of homology, the enzymes they encode exhibit different specificities. Different epimerases may give alginates with different distribution of M and G, and thus alginates with tailored physical and chemical properties can be made as illustrated in Figure 4. None of the enzymatically modified polymers, however, are commercially available at present. Table 3 lists the modular structure of the mannuronan C-5 epimerase family and its specific action.

9.3
Current and Expected World Market and Costs

Industrial production of alginate is roughly 30,000 metric tons annually, which is probably less than 10% of the annually biosynthesized material in the standing macroalgae crops. Because macroalgae also may be cultivated (e.g., in mainland China where 5 to 7 million metric tons of wet *Laminaria japonica* are produced annually) and because production by fermentation is technically possible, the sources for industrial production of alginate may be regarded as unlimited even for a steadily growing industry.

It is expected that future growth in the alginate market most likely will be of a qualitative rather than a quantitative nature. Predictions suggest that manufacturers will move away from commodity alginate production toward more refined products, e.g., for the pharmaceutical industry.

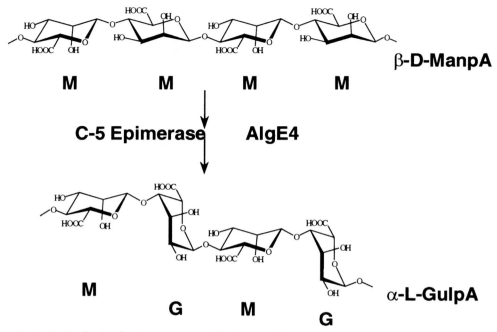

Fig. 3 Mode of action for the mannuronan C5-epimarase AlgE4.

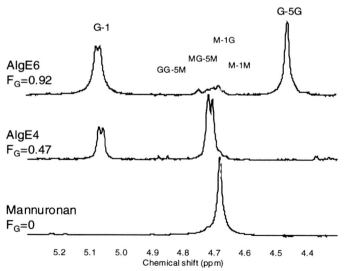

Fig. 4 Resulting chemical composition and sequence after treating mannuronan with different C5-epimerases.

Tab. 3 The seven AlgE epimerases from *A. vinelandi*[a]

Type	[kDa]	Modular structure	Products
AlgE1	147.2	A1 R1 R2 R3 A2 R4	Bi-functional G-blocks + MG-blocks
AlgE2	103.1	A1 R1 R2 R3 R4	G-blocks (short)
AlgE3	191	A1 R1 R2 R3 A2 R4 R5 R6 R7	Bi-functional G-block + MG-blocks
AlgE4	57.7	A1 R1	MG-blocks
AlgE5	103.7	A1 R1 R2 R3 R4	G-blocks (medium)
AlgE6	90.2	A1 R1 R2 R3	G-blocks (long)
AlgE7	90.4	A1 R1 R2 R3	Lyase activity + G-blocks + MG-blocks

[a] A – 385 amino acids; R – 155 amino acids

The cost of alginates can differ extremely depending on the degree of purity. Technical grade, low-purity alginate (containing a substantial amount of algae debris) can be obtained from around 1 USD per kilogram, and ordinary purified-grade alginate can be obtained from approximately 10 USD per kilo, whereas ultra-pure (low in endotoxins) alginate specially designed for immobilization purposes typically costs around 5 USD per gram.

9.4
Alginate Manufacturers

The alginate producers members list of the Marinalg hydrocolloid association includes six different companies. These are China Seaweed Industrial Association, Danisco Cultor (Denmark), Degussa Texturant Systems (Germany), FMC BioPolymer (USA), ISP Alginates Ltd. (UK), and Kimitsu Chemical Industries Co., Ltd. (Japan). In addition to these, Pronova Biomedical A/S (Norway) now commercially manufactures ultra-pure alginates that are highly compatible with mammalian biological systems following the increased popularity of alginate as an immobilization matrix. These qualities are low in pyrogens and facilitate sterilization of the alginate solution by filtration due to low content of aggregates.

10
Properties

The physical properties of the alginate molecule were revealed mainly in the 1960s and 1970s. The last couple of decades have exposed some new knowledge on alginate gel formation

10.1
Physical Properties

Compared with other gelling polysaccharides, the most striking features of alginate's physical properties are the selective binding of multivalent cations, being the basis for gel formation, and the fact that the sol/gel transition of alginates is not particularly influenced by temperature.

10.1.1
Solubility

There are three essential parameters determining and limiting the solubility of alginates in water. The pH of the solvent is important because it will determine the presence of electrostatic charges on the uronic acid residues. Total ionic strength of the solute also plays an important role (salting-out effects of non-gelling cations), and, obviously, the content of gelling ions in the solvent limits the solubility. In the latter case, the "hardness" of the water (i.e., the content of Ca^{2+} ions) is most likely to be the main problem.

Potentiometric titration (Haug, 1964) revealed that the dissociation constants for mannuronic and guluronic acid monomers were 3.38 and 3.65, respectively. The pK_a value of the alginate polymer differs only slightly from those of the monomeric residues. An abrupt decrease in pH below the pK_a value causes a precipitation of alginic acid molecules, whereas a slow and controlled release of protons may result in the formation of an "alginic acid gel". Precipitation of alginic acid has been studied extensively (Haug, 1964; Haug and Larsen, 1963; Myklestad and Haug, 1966; Haug et al., 1967c), and addition of acid to an alginate solution leads to a precipitation within a relatively narrow pH range. This range depends not only on the molecular weight of the alginate but also on the chemical composition and sequence. Alginates containing more of the "alternating" structure (MG-blocks) will precipitate at lower pH values compared with the alginates containing more homogeneous block structures (poly-M and poly-G). The presence of homopolymeric blocks seems to favor precipitation by the formation of crystalline regions stabilized by hydrogen bonds. By increasing the degree of alternating "disorder" in the alginate chain, as in alginates isolated from *Ascophyllum nodosum* (see Table 1), the formation of these crystalline regions is not formed as easily. A certain alginate fraction from *A. nodosum* is soluble at a pH as low as 1.4 (Myklestad and Haug, 1966). Because of this relatively limited solubility of alginates at low pH, the esterified propylene glycol alginate (PGA) is applied as a food stabilizer under acidic conditions (see Section 10.3).

Any change of ionic strength in an alginate solution generally will have a profound effect, especially on polymer chain extension and solution viscosity. At high ionic strengths, the solubility also will be affected. Alginate may be precipitated and fractionated to give a precipitate enriched with mannuronate residues by high concentrations of inorganic salts like potassium chloride (Haug and Smidsrød, 1967; Haug, 1959a). Salting-out effects like this exhibit large hysteresis in the sense that less than 0.1 M salt is necessary to slow down the kinetics of the dissolution process and limit the solubility (Haug, 1959b). The gradient in the chemical potential of water between the bulk solvent and the solvent in the alginate particle, resulting from a very high counterion concentration in the particle, is most probably the drive of the dissolution process of alginate in water. This drive becomes severely reduced when attempts are made to dissolve alginate in an aqueous solvent already containing ions. If alginates are to be applied at high salt concentrations, the polymer should first be fully hydrated in pure water followed by addition of salt under shear.

For the swelling behavior of dry alginate powder in aqueous media with different concentrations of Ca^{2+}, there seems to be a limit at approximately 3 mM free calcium ions (unpublished results). Alginate can be solubilized at $[Ca^{2+}]$ above 3 mM by the addition of complexing agents, such as polyphosphates or citrate, before addition of the alginate powder.

10.1.2
Selective Ion Binding

The basis for the gelling properties of alginates is their specific ion-binding characteristics (Haug, 1964; Smidsrød and Haug, 1968b; Haug and Smidsrød, 1970; Smidsrød, 1973, 1974). Experiments involving equilibrium dialysis of alginate have shown that the selective binding of certain alkaline earth metals ions (e.g., strong and cooperative binding of Ca^{2+} relative to Mg^{2+}) increased markedly with increasing content of α-L-guluronate residues in the chains. Poly-mannuronate blocks and alternating blocks were almost without selectivity. This is illustrated in Figures 5 and 6, where a marked hysteresis in the binding of Ca^{2+} ions to G-blocks also is seen.

The high selectivity between similar ions such as those from the alkaline earth metals indicates that some chelation caused by structural features in the G-blocks takes place. Attempts were made to explain this phenomenon by the so-called "egg-box" model (Grant et al., 1973), based upon the linkage conformations of the guluronate residues (see Figure 1b). NMR studies (Kvam et al., 1986) of lanthanide complexes of related compounds suggested a possible binding site for Ca^{2+} ions in a single alginate chain, as given in Figure 7 (Kvam 1987).

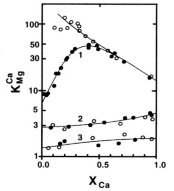

Fig. 6 Selectivity coefficients, K_{mg}^{Ca} as a function of ionic composition (X_{Ca}) for different alginate fragments. Curve 1: Fragments with 90% guluronate residues. Curve 2: Alternating fragment with 38% guluronate residues. Curve 3: Fragment with 90% mannuronate residues. ●: Dialysis of the fragments in their Na^+ form. ○: Dialysis first against 0.2 M $CaCl_2$, then against mixtures of $CaCl_2$ and $MgCl_2$.

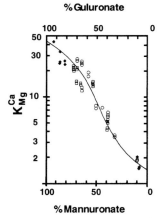

Fig. 5 Selectivity coefficients, K_{mg}^{Ca}, for alginates and alginate fragments as a function of monomer composition. The experimental points are obtained at $X_{Ca} = X_{Mg} = 0.5$. The curve is calculated using $K_{mg\ guluronate}^{Ca} = 40$ and $K_{mg\ mannuronate}^{Ca} = 1.8$.

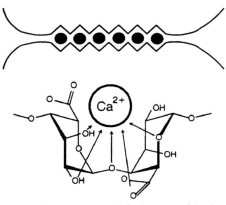

Fig. 7 The egg-box model for binding of divalent cations to homopolymeric blocks of α-L-guluronate residues, and a probably binding site in a GG-sequence.

Although more accurate steric arrangements have been suggested, as supported by x-ray diffraction (Mackie et al., 1983) and NMR spectroscopy (Steginsky et al., 1992), the simple "egg-box" model still persists, as it is principally correct and gives an intuitive understanding of the characteristic chelate-type of ion-binding properties of alginates. The simple dimerization in the "egg-box" model is at present questionable, as data from small-angle x-ray scattering on alginate gels suggest lateral association far beyond a pure dimerization with increasing [Ca^{2+}] and G-content of the alginate (Stokke et al., 2000). In addition, the fact that isolated and purified G-blocks (totally lacking elastic segments; typically DP = 20) are able to act as gelling modulators when mixed with a gelling alginate suggests higher-order junction zones (Draget et al., 1997).

The selectivity of alginates for multivalent cations is also dependent on the ionic composition of the alginate gel, as the affinity toward a specific ion increases with increasing content of the ion in the gel (Skjåk-Bræk et al., 1989b) (see Figure 6). Thus, a Ca-alginate gel has a markedly higher affinity toward Ca^{2+} ions than has the Na-alginate solution. This has been explained theoretically (Smidsrød, 1970; Skjåk-Bræk et al., 1989b) by a near-neighbor, auto-cooperative process (Ising model) and can be explained physically by the entropically unfavorable binding of the first divalent ion between two G-blocks and the more favorable binding of the next ions in the one-dimensional "egg-box" (zipper mechanism).

10.1.3
Gel Formation and Ionic Cross-linking

A very rapid and irreversible binding reaction of multivalent cations is typical for alginates; a direct mixing of these two components therefore rarely produces homogeneous gels. The result of such mixing is likely to be a dispersion of gel lumps ("fish-eyes"). The only possible exception is the mixing of a low-molecular-weight alginate with low amounts of cross-linking ion at high shear. The ability to control the introduction of the cross-linking ions hence becomes essential.

A controlled introduction of cross-linking ions is made possible by the two fundamental methods for preparing an alginate gel: the diffusion method and the internal setting method. The diffusion method is characterized by allowing a cross-linking ion (e.g., Ca^{2+}) to diffuse from a large outer reservoir into an alginate solution (Figure 8a). Diffusion setting is characterized by rapid gelling kinetics and is utilized for immobilization purposes where each droplet of alginate solution makes one single gel bead with entrapped (bio-) active agent (Smidsrød and Skjåk-Bræk, 1990). High-speed setting is also beneficial, e.g., in restructuring of foods when a given size and shape of the final product is desirable. The molecular-weight dependence in this system is negligible as long as the weight average molecular weight of the alginate is above 100 kDa (Smidsrød, 1974).

The internal setting method differs from the diffusion method in that the Ca^{2+} ions are released in a controlled fashion from an inert calcium source within the alginate solution (Figure 8b). Controlled release usually is obtained by a change in pH, by a limited solubility of the calcium salt source, and/or the by presence of chelating agents. The main difference between internal and diffusion setting is the gelling kinetics, which is not diffusion-controlled in the former case. With internal setting, the tailor-making of an alginate gelling system toward a given manufacturing process is possible because of the controlled, internal release of cross-linking ions (Draget et al., 1991). Internally set gels generally show a

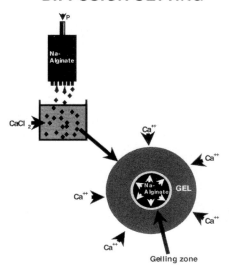

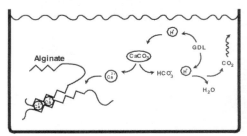

Fig. 8 Principal differences between the diffusion method exemplified by the immobilization technique and the internal setting method exemplified by the CaCO$_3$/GDL technique.

10.1.4
Gel Formation and Alginic Acid Gels

It is well known that alginates may form acid gels at pH values below the pK$_a$ values of the uronic residues, but these alginic acid gels traditionally have not been as extensively studied as their ionically cross-linked counterparts. With the exception of some pharmaceutical uses, the number of applications so far is also rather limited. The preparation of an alginic acid gel has to be performed with care. Direct addition of acid to, e.g., a Na-alginate solution leads to an instantaneous precipitation rather than a gel. The pH must therefore be lowered in a controlled fashion, and this is most conveniently carried out by the addition of slowly hydrolyzing lactones like D-glucono-δ-lactone (GDL).

10.2
Material Properties

Since alginates are traditionally used for their gelling, viscosifying, and stabilizing properties, the features of alginate based materials are of utmost importance for a given application. Recently some quite unique biological effects of the alginate molecule itself have been revealed.

10.2.1
Stability

Alginate, being a single-stranded polymer, is susceptible to a variety of depolymerization processes. The glycosidic linkages are cleaved by both acid and alkaline degradation mechanisms and by oxidation with free radicals. As a function of pH, degradation of alginate is at its minimum nearly neutral and increases in both directions (Haug and Larsen, 1963) (Figure 9). The increased instability at pH values less than 5 is explained by a proton-catalyzed hydrolysis, whereas the reaction responsible for the

more pronounced molecular weight dependence compared with diffusion set gels. It has been reported that the internally set gels depend on molecular weight even at 300 kDa (Draget et al., 1993). This could be due to the fact that internally set gels are more calcium-limited compared with the gels made by diffusion, implying that the non-elastic fractions (sol and loose ends) at a given molecular weight will be higher in the internally set gels.

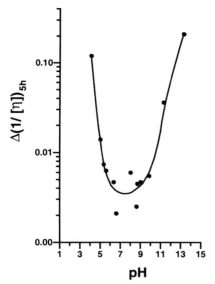

Fig. 9 Degradation of alginate isolated from *Laminaria digitata* measured as the change (Δ) in intrinsic viscosity ([η]) after 5 h at different pH and at 68°C.

degradation at pH 10 and above is the β-alkoxy elimination (Haug et al., 1963, 1967b). Free radicals degrade alginate mainly by oxidative-reductive depolymerization reactions (Smidsrød et al., 1967; Smidsrød et al., 1963a,b) caused by contamination of reducing agents like polyphenols from the brown algae. Since all of these depolymerization reactions increase with temperature, autoclaving generally is not recommended for the sterilization of an alginate solution. Since alginate is soluble in water at room temperature, sterile filtering rather than autoclaving has been recommended as a sterilization method for immobilization purposes in order to reduce polymer breakdown and to maintain the mechanical properties of the final gel (Draget et al., 1988).

Sterilization of dry alginate powder is also troublesome. The effect of γ-irradiation is often disastrous and leads to irreversible damage. It is generally believed that, under these conditions, O_2 is depleted rapidly with formation of the very reactive OH· free radical. A short-term exposure in an electron accelerator could be an alternative to long-term exposure from a traditional ^{60}Co source. It has been shown that sterilization doses applied by ^{60}Co irradiation reduce the molecular weight to the extent that the gelling capacity is almost completely lost (Leo et al., 1990).

10.2.2
Ionically Cross-linked Gels

In contrast to most gelling polysaccharides, alginate gels are cold-setting, implying that alginate gels set more or less independent of temperature. The kinetics of the gelling process, however, can be strongly modified by a change in temperature, but a sol/gel transition will always occur if gelling is favored (e.g., by the presence of cross-linking ions). It is also important to realize that the properties of the final gel most likely will change if gelling occurs at different temperatures. This is due to alginates being non-equilibrium gels and thus being dependent upon the history of formation (Smidsrød, 1973).

Alginate gels can be heated without melting. This is the reason that alginates are used in baking creams. It should be kept in mind that alginates, as described earlier, are subjected to chemical degrading processes. A prolonged heat treatment at low or high pH might thus destabilize the gel because of an increased reaction rate of depolymerizing processes such as proton-catalyzed hydrolysis and the β-elimination reaction (Moe et al., 1995).

Because the selective binding of ions is a prerequisite for alginate gel formation, the alginate monomer composition and sequence also have a profound impact on the final properties of calcium alginate gels. Figure 10 shows gel strength as a function of

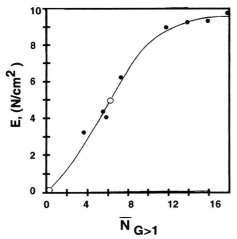

Fig. 10 Elastic properties of alginate gels as function of average G-block length.

strength when $N_{G>1}$ changes from 5 to 15. This coincides with the range of G-block lengths found in commercial alginates.

The polyelectrolyte nature of the alginate molecule is also important for its function, especially in mixed systems where, under favorable conditions, alginates may interact electrostatically with other charged polymers (e.g., proteins), resulting in a phase transition and altering the rheological behavior. Generally, it can be stated that if the purpose is to avoid such electrostatic interactions, the mixing of alginate and protein should take place at a relatively high pH, where most proteins have a net negative charge (Figure 11). These types of interactions also can be utilized to stabilize mixtures and to increase the gel strength of some restructured foods. In studies involving gelling of bovine serum albumin and alginate in both the sodium and the calcium form, a consid-

the average length of G-blocks larger than one unit ($N_{G>1}$). This empirical correlation shows that there is a profound effect on gel

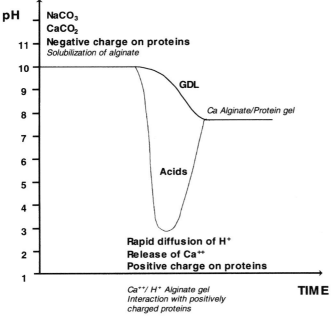

Fig. 11 Alginate/protein mixed gels exemplified by the internal gelation with $CaCO_3$. Release of Ca^{2+} is achieved by either a slow pH-lowering agent (GDL) or by a fast lowering with acids.

erable increase in Young's modulus was found within some range of pH and ionic strength (Neiser et al., 1998, 1999). These results suggest that electrostatic interactions are the main driving force for the observed strengthening effects.

An important feature of gels made by the diffusion-setting method is that the final gel often exhibits an inhomogeneous distribution of alginate, the highest concentration being at the surface and gradually decreasing towards the center of the gel. Extreme alginate distributions have been reported (Skjåk-Bræk et al., 1989a), with a five-fold increase at the surface (as calculated from the concentration in the original alginate solution) and virtually zero concentration in the center (Figure 12). This result has been explained by the fact that the diffusion of gelling ions will create a sharp gelling zone that moves from the surface toward the center of the gel. The activity of alginate (and of the gelling ion) will equal zero in this zone, and alginate molecules will diffuse from the internal, non-gelled part of the gelling body toward the zero-activity region

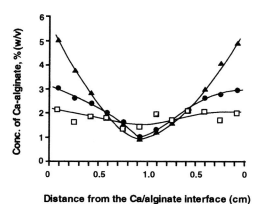

Distance from the Ca/alginate interface (cm)

Fig. 12 Polymer concentration profiles of alginate gel cylinders formed by dialyzing a 2% (w/v) solution of Na-alginate from *Laminaria hyperborea* against 0.05 M $CaCl_2$ in the presence of NaCl. ▫: 0.2 M; ●: 0.05 M; ▲: no NaCl.

(Skjåk-Bræk et al., 1989a, Mikkelsen and Elgsæter, 1995). Inhomogeneous alginate distribution may or may not be beneficial in the final product. It is therefore important to know that the degree of homogeneity can be controlled and to know which parameters govern the final alginate distribution. Maximum inhomogeneity is reached by placing a high-G, low-molecular-weight alginate gel in a solution containing a low concentration of the gelling ion and an absence of non-gelling ions. Maximum homogeneity is reached by gelling a high-molecular-weight alginate with high concentrations of both gelling and non-gelling ions (Skjåk-Bræk et al., 1989b).

The presence of non-gelling ions in alginate-gelling systems also affects the stability of the gels. It has been shown that alginate gels start to swell markedly when the ratio between non-gelling and gelling ions becomes too high and that the observed destabilization increases with decreasing F_G (Martinsen et al., 1989).

Swelling of alginate gels can be increased dramatically by a covalent cross-linking of preformed Ca-alginate gels with epichlorohydrin, followed by subsequent removal of Ca^{2+} ions by etylenediamine tetraacetic acid (EDTA) (Skjåk-Bræk and Moe, 1992). These Na-alginate gels can be dried, and they exhibit unique swelling properties when re-hydrated. The forces affecting the swelling of a polymer network can be split into three terms. Two of these terms favor swelling and can be said to constitute what might be called "swelling pressure": (1) the mixing term (Π_{mix} = the osmotic pressure generated by polymer/solvent mixing) and (2) the ionic term (Π_{ion} = the osmotic effect of an unequal distribution of the polymer counter-ions between the inside and the outside of the gel; the Donnan equilibrium). Of these two terms, the ionic part has been shown to contribute approximately 90% of the swelling pressure, even at 1 molar ionic strength, for highly ionic gels like Na-alginate (Moe et al., 1993). The third term (Π_{el} = the reduction in osmotic pressure due to the elastic response of the polymer network) balances the "swelling pressure" so that the total of these three terms equals zero at equilibrium.

These Na-alginate gels would function well as water absorbents in hygiene and pharmaceutical applications. However, Π_{ion} depends upon the ionic strength of the solute; with increasing ionic strength, the difference in chemical potential is reduced because of a more even distribution of the mobile ions between the inside and the outside of the gel. Therefore, reduced swelling will be observed at physiological ionic conditions compared with deionized water, but this reduction will be less pronounced than that for other water-absorbing materials, such as cross-linked acrylates, as a result of the inherent stiffness of the alginate molecule itself (Skjåk-Bræk and Moe, 1992) (Figure 13).

10.2.3
Alginic Acid Gels

It has been shown (Draget et al., 1994) that the gel strength of acid gels becomes independent of pH below a pH of 2.5, which equals 0.8 M GDL in a 1.0% (w/v) solution. Table 4 shows the Young's moduli of acid gels prepared (1) by a direct addition of GDL and (2) by converting an ionic cross-linked gel to the acid form by mineral acid. The modulus seems to be rather independent of the history of formation. Therefore, a most important feature of the acid gels compared with the ionic cross-linked gels seems to be that the former reaches equilibrium in the gel state.

Figure 14 shows the observed elastic moduli of acid gels made from alginates with different chemical composition, together with expected values for ionically cross-linked gels. From these data, it can be concluded that acid gels resemble ionic gels in the sense that high contents of guluronate (long stretches of G-blocks) give the strongest gels. However, it is also seen that poly-mannuronate sequences support alginic acid gel formation, whereas poly-alternating sequences seem to perturb this transition. The obvious demand for homopolymeric sequences in acid gel formation suggests

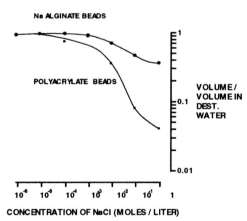

Fig. 13 Salt tolerance of covalently cross-linked Na-alginate and polyacrylate gel beads measured as swelling at different ionic strengths.

Tab. 4 E_{app} (kPa) for gels made from three different high-G alginates at 2% (w/v) concentration

Ca-alginate gel	Ca-gel to alginic acid gel	Syneresis correction	Direct addition of GDL
105 ± 4.6	52 ± 4.3	15.5 ± 0.3	15 ± 1.1
116 ± 11	64 ± 8.1	17.1 ± 1.8	17.8 ± 1.4
127 ± 6.4	79 ± 5.8	19.8 ± 1.3	20.4 ± 0.7

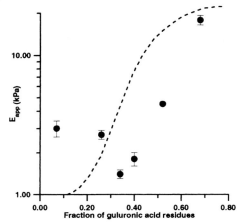

Fig. 14 Young's modulus E_{app} of alginic acid gels at apparent equilibrium as function of guluronic acid content. Dashed line refers to expected results for Ca^{2+} cross-linked alginate gels.

that cooperative processes are involved just as in the case of ionic gels (Draget et al., 1994). High molecular weight dependence has been observed, and this dependence becomes more pronounced with increasing content of guluronic acid residues.

A study of the swelling and partial solubilization of alginic acid gels at pH 4 has confirmed the equilibrium properties of the gels (Draget et al., 1996). By comparing the chemical composition and molecular weight of the alginate material leaching out from the acid gels with the same data for the whole alginate, an enrichment in mannuronic acid residues and a reduction in the average length of G-blocks were found together with a lowering of the molecular weight.

10.3
"Biological" Properties

Through a series of papers, it has been established that the alginate molecule itself has different effects on biological systems. This is more or less to be expected because of the large variety of possible chemical compositions and molecular weights of alginate preparations. A biological effect of alginate initially was hinted at in the first animal transplantation trials of encapsulated Langerhans islets for diabetes control. Overgrowth of alginate capsules by phagocytes and fibroblasts, resembling a foreign body/inflammatory reaction, was reported (Soon-Shiong et al., 1991). In bioassays, induction of tumor necrosis factor and interleukin 1 showed that the inducibility depended upon the content of mannuronate in the alginate sample (Soon-Shiong et al., 1993). This result directly explains the observed capsule overgrowth; mannuronate-rich fragments, which do not take part in the gel network, will leach out of the capsules and directly trigger an immune response (Stokke et al., 1993). This observed immunologic response can be linked in part to $(1 \rightarrow 4)$ glycosidic linkages, as other homopolymeric di-equatorial polyuronates, like D-glucuronic acid (C6-oxidised cellulose), also exhibit this feature (Espevik and Skjåk-Bræk, 1996). The immunologic potential of polymannuronates have now been observed in *in vivo* animal models in such diverse areas as for protection against lethal bacterial infections and irradiation and for increasing non-specific immunity (Espevik and Skjåk-Bræk, 1996).

11
Applications

Given the large number of different applications, alginates must be regarded as one of the most versatile polysaccharides. These applications span from traditional technical utilization, to foods, to biomedicine.

11.1
Technical Utilization

The quantitatively most important technical application of alginates is as a shear-thinning viscosifyer in textile printing, in which alginate has gained a high popularity because of the resulting color yield, brightness, and print levelness. Alginates also are used for paper coating to obtain surface uniformity and as binding agents in the production of welding rods. In the latter case, alginate gives stability in the wet stage and functions as a plasticizer during the extrusion process. As a last example of technical applications, ammonium alginate frequently is used for can sealing. The ammonium form is used because of its very low ash content (Onsøyen, 1996).

11.2
Medicine and Pharmacy

Alginates have been used for decades as helping agents in various human-health applications. Some examples include use in traditional wound dressings, in dental impression material, and in some formulations preventing gastric reflux. Alginate's increasing popularity as an immobilization matrix in various biotechnological processes, however, demonstrates that alginate will move into other and more advanced technical domains in addition to its traditional applications. Entrapment of cells within Ca-alginate spheres has become the most widely used technique for the immobilization of living cells (Smidsrød and Skjåk-Bræk, 1990). This immobilization procedure can be carried out in a single-step process under very mild conditions and is therefore compatible with most cells. The cell suspension is mixed with a sodium alginate solution, and the mixture is dripped into a solution containing multivalent cations (usually Ca^{2+}). The droplets then instantaneously form gel-spheres entrapping the cells in a three-dimensional lattice of ionically cross-linked alginate. The possible uses for such systems in industry, medicine, and agriculture are numerous, ranging from production of ethanol by yeast, to production of monoclonal antibodies by hybridoma cells, to mass production of artificial seed by entrapment of plant embryos (Smidsrød and Skjåk-Bræk, 1990).

Perhaps the most exciting prospect for alginate gel immobilized cells is their potential use in cell transplantation. Here, the main purpose of the gel is to act as a barrier between the transplant and the immune system of the host. Different cells have been suggested for gel immobilization, including parathyroid cells for treatment of hypocalcemia and dopamine-producing adrenal chromaffin cells for treatment of Parkinson's disease (Aebisher et al., 1993). However, major interest has been focused on insulin-producing cells for the treatment of Type I diabetes. Alginate/poly-L-lysine capsules containing pancreatic Langerhans islets have been shown to reverse diabetes in large animals and currently are being clinically tested in humans (Soon-Shiong et al., 1993, 1994). Table 5 lists some biomedical applications of alginate-encapsulated cells.

Tab. 5 Some potential biomedical application of alginate-encapsulated cells

Cell type	Treatment of
Adrenal chromaffin cells	Parkinson's disease[a]
Hepatocytes	Liver failure[a]
Paratyroid cells	Hypocalcemia[a]
Langerhans islets (β-cells)	Diabetes[b]
Genetically altered cells	Cancer[c]

[a] Aebischer et al., 1993 [b] Soon-Shiong et al., 1993, 1994 [c] Read et al., 2000

11.3
Foods

Alginates are used as food additives to improve, modify, and stabilize the texture of foods. This is valid for such properties as viscosity enhancement; gel-forming ability; and stabilization of aqueous mixtures, dispersions, and emulsions. Some of these properties stem from the inherent physical properties of alginates themselves, as outlined above, but they also may result from interactions with other components of the food product, e.g., proteins, fat, or fibers. As an example, alginates interact readily with positively charged amino acid residues of denatured proteins, which are utilized in pet foods and reformed meat. Cottrell and Kovacs (1980), Sime (1990), and Littlecott (1982) have given numerous descriptions and formulations on alginates in food applications. A general review on this topic is given by McHugh (1987).

Special focus perhaps should be placed on restructured food based on Ca-alginate gels because of its simplicity (gelling being independent upon temperature) and because it is a steadily growing alginate application. Restructuring of foods is based on binding together a flaked, sectioned, chunked, or milled foodstuff to make it resemble the original. Many alginate-based restructured products are already on the market (see Figure 15), as is exemplified by meat products (both for human consumption and as pet food), onion rings, pimento olive fillings, crabsticks, and cocktail berries.

For applications in jams, jellies, fruit fillings, etc., the synergetic gelling between alginates high in guluronate and highly esterified pectins may be utilized (Toft et al., 1986). The alginate/pectin system can give thermoreversible gels in contrast to the purely ionically cross-linked alginate gels. This gel structure is almost independent of

Fig. 15 An example of alginate used in the restructuring of foods: the pimiento fillings of olives. (picture kindly provided by FMC BioPolymer).

sugar content, in contrast to pectin gels, and therefore may be used in low calorie products.

The only alginate derivative used in food is propylene glycol alginate (PGA). Steiner (1947) first prepared PGA, and Steiner and McNeely improved the process (1950). PGA is produced by a partial esterification of the carboxylic groups on the uronic acid residues by reaction with propylene oxide. The main product gives stable solutions under acidic conditions where the unmodified alginate would precipitate. It is now used to stabilize acid emulsions (such as in French dressings), acid fruit drinks, and juices. PGA also is used to stabilize beer foam.

As for the regulatory status, the safety of alginic acid and its ammonium, calcium, potassium, and sodium salts was last evaluated by the Joint FAO/WHO Expert Committee on Food Additives (JECFA) at its 39th meeting in 1992. An ADI "not specified" was allocated. JECFA allocated an ADI of 0 to 25 mg/kg bw to propylene glycol alginate at its 17th meeting.

In the U.S, ammonium, calcium, potassium, and sodium alginate are included in a list of stabilizers that are generally recognized as safe (GRAS). Propylene glycol alginate is approved as a food additive (used as an emulsifier, stabilizer, or thickener) and in several industrial applications (used as a coating for fresh citrus fruit, as an inert pesticide adjuvant, and as a component of paper and paperboard in contact with aqueous and fatty foods). In Europe, alginic acid and its salts and propylene glycol are all listed as EC-approved additives other than colors and sweeteners.

Alginates are inscribed in Annex I of the Directive 95/2 of 1995, and as such can be used in all foodstuffs (except those cited in Annex II and those described in Part II of the Directive) under the Quantum Satis principle of the EU.

12
Relevant Patents

A search in one of the international databases for patents and patent applications yielded well above 2000 hits on alginate. It is outside the scope of this chapter, not to mention beyond the capabilities of its authors, to systematically discuss all this literature, which covers inventions for improved utilization of alginates in the technical, pharmaceutical, food, and agricultural areas. We have therefore limited this section to only a handful of *prior art* inventions (Table 6), with the present authors as co-inventors, that point toward a production of alginate and alginate fractions with novel structures and some new biomedical applications based on the physical and biological properties of certain types of alginates with specified chemical structures.

US Patent 5,459,054 may represent a large number of patents dealing with the immobilization of living cells in immuno-protective alginate capsules for implantation purposes as discussed in Section 10.2. US Patents 5,169,840 and 6,087,342 cover the use of alginates enriched with mannuronate for the stimulation of cytokine production in monocytes, which could be of future importance in the treatment of microbial infections, cancer, and immune deficiency and autoimmune diseases. This stimulating effect has been discussed and connected to the use of alginate fibers in wound-healing dressings. A closer look at these specific effects is presented in Section 9.3.

When the calcium ions in alginate gels are exchanged by covalent cross links, the resulting gel with monovalent cations as counter-ions has the ability to swell several hundred times its own weight in water or salt solutions at low ionic strength, as shown in US Patent 5,144,016. This super-absorbent system, further elaborated in Section 9.2.2, still has not found any commercial uses, mainly because of competition from similar materials based on starch and cellulose derivatives, but certain biomedical applications may be foreseen.

A patent on the genes encoding the different C5-epimerases (US Patent 5,939,289) points to the possibility of producing alginates with a large number of different predetermined compositions and sequences and opens up the possibility for the tailor-making of different alginates, as discussed in Sections 8.2 and 12.

An alternative way of manufacturing alginate fractions with extreme compositions is by using selective extraction techniques, as disclosed in Patent WO 98/51710. One possible use of such fractions as gelling modifiers is revealed in Patent WO 98/02488, where it is suggested that these purified low-molecular-weight guluronate blocks give a gel enforcement at high concentrations of calcium ions by connecting and shortening elastic segments that otherwise would be topologically restricted. In conclusion, it may be argued that this relative high rate of patent filing suggests

Tab. 6 Summary of prior art patents on alginates pointing toward specialty applications

Patent number	Holder	Inventors	Patent title	Public date
U.S. 5,459,054	Neocrine	G. Skjåk-Bræk O. Smidsrød T. Espevik M. Otterlei P. Soon-Shiong	Cells encapsulated in alginate containing a high content of α-L-guluropnic acid	2 July 1993
U.S. 5,169,840	Nobipol, Protan Biopolymer	M. Otterlei T. Espevik O. Smidsrød G. Skjåk-Bræk	Diequatorially bound β-1,4-polyuronates and use of same for cytokine stimulation	27 March 1991
U.S. 6,087,342	FMC BioPolymer	T. Espevik G. Skjåk-Bræk	Substrates having bound polysaccharides and bacterial nucleic acids	15 May 1998
U.S. 5,144,016	Protan Biopolymer	G. Skjåk-Bræk S. Moe	Alginate gels	29 May 1991
U.S. 5,939,289	Pronova Biopolymer, Nobipol	H. Ertesvåg S. Valla G. Skjåk-Bræk B. Larsen	DNA compounds comprising sequences encoding mannuroran C-5-epimerase	9 May 1995
WO 98/51710	FMC BioPolymer	M. K. Simensen O. Smidsrød K. I. Draget F. Hjelland	Procedure for producing uronic acid blocks from alginate	11 November 1998
WO 98/02488	FMC BioPolymer	M. K. Simensen K. I. Draget E. Onsøyen O. Smidsrød T. Fjæreide	Use of G-block polysaccharides	22 January 1998

that new alginate-based products are being developed and that there is continuous stable demand for alginates and their products.

13
Outlook and Perspectives

From a chemical point of view, the alginate molecule may look very simple, as it contains only the two monomer units M and G linked by the same 1,4 linkages. This simplification of its chemical structure may lead potential commercial users of alginate to treat it as a commodity like many of the cellulose derivatives. In this chapter, we have shown that alginate represents a very high diversity with respect to chemical composition and monomer sequence, giving the alginate family of molecules a large variety of physical and biological properties. This may represent a challenge to the unskilled users of alginate, but it may be an advantage for the producers and new users of alginate who are interested in developing research-based, high-value applications. When microbial alginate and epimerase-modified alginate enter into the marked in the future, the possibility of alginate being tailor-made to diverse applications will be increased even further.

We therefore see a future trend, which has already started, in which the exploitation of alginate gradually shifts from low-tech applications with increasing competition from cheap alternatives to more advanced, knowledge-based applications in the food, pharmaceutical, and biomedical areas. We then foresee continuous, high research activity in industry and academia to describe, understand, and utilize alginate-containing products to the benefit of society.

Acknowledgements

The authors would like to thank Anne Bremnes and Hanne Devle for most skillful assistance in preparing graphic illustrations and Nadra J. Nilsen for collecting the data on alginate lyases.

14
References

Aebischer, P., Goddard, M., Tresco, P. A. (1993) Cell encapsulation for the nervous system, in: *Fundamentals of Animal Cell Encapsulation and Immobilization* (Goosen, M. F. A., Ed.), Boca Raton, FL: CRC Press, 197–224.

Andresen, I.-L., Skipnes, O., Smidsrød, O., Østgaard, K., Hemmer, P. C. (1977) Some biological functions of matrix components in benthic algae in relation to their chemistry and the composition of seawater, *ACS Symp. Ser.* **48**, 361–381.

Atkins, E. D. T., Mackie, W., Smolko, E. E. (1970) Crystalline structures of alginic acids, *Nature* **225**, 626–628.

Atsuki, K., Tomoda, Y. (1926) Studies on seaweeds of Japan I. The chemical constituents of Laminaria, *J. Soc. Chem. Ind. Japan* **29**, 509–517.

Bird, G. M., Haas, P. (1931) XLVII. On the constituent nature of the cell wall constituents of *Laminaria* spp. Mannuronic acid, *Biochem. J.* **25**, 26–30.

Boyd, J., Turvey, J. R. (1977) Isolation of a poly-"-L-guluronate lyase from *Klebsiella aerogenes*, *Carbohydr. Res.* **57**, 163–171.

Boyd, J., Turvey, J. R. (1978) Structural studies of alginic acid, using a bacterial poly-L-guluronate lyase, *Carbohydr. Res.* **66**, 187–194.

Boyen, C., Bertheau, Y., Barbeyron, T., Kloareg, B. (1990a) Preparation of guluronate lyase from *Pseudomonas alginovora* for protoplast isolation in Laminaria, *Enzyme Microb. Technol.* **12**, 885–890.

Boyen, C., Kloareg, B., Polne-Fuller, M., Gibor, A. (1990b) Preparation of alginate lyases from marine molluscs for protoplast isolation in brown algae, *Phycologia* **29**, 173–181.

Blumenkrantz, N., Asboe-Hansen, G. (1973) New method for quantitative determination of uronic acids, *Anal. Biochem.* **54**, 484–489.

Cottrell, I.W., Kovacs, P. (1980). Alginates, in: *Handbook of Water-Soluble Gums and Resins* (Crawford, H.B., Williams, J., Eds.), Auckland, New Zealand: McGraw-Hill, 21–43.

Davidson I.W., Lawson, C.J., Sutherland, I.W. (1977) An alginate lyase from *Azotobacter vinelandii* phage, *J. Gen. Microbiol.* **98**, 223–229.

Draget, K. I., Myhre, S., Skjåk-Bræk, G., Østgaard, K. (1988) Regeneration, cultivation and differentiation of plant protoplasts immobilized in Ca-alginate beads, *J. Plant Physiol.* **132**, 552–556.

Draget, K.I., Østgaard, K., Smidsrød, O. (1991) Homogeneous alginate gels; a technical approach, *Carbohydr. Polym.* **14**, 159–178.

Draget, K.I., Simensen, M.K., Onsøyen, E., Smidsrød, O. (1993) Gel strength of Ca-limited gels made *in situ*, *Hydrobiologia*, **260/261**, 563–565.

Draget, K.I., Skjåk-Bræk, G., Smidsrød, O. (1994) Alginic acid gels: the effect of alginate chemical composition and molecular weight, *Carbohydr. Polym.* **25**, 31–38.

Draget, K. I., Skjåk-Bræk, G., Christensen, B. E., Gåserød, O., Smidsrød, O. (1996) Swelling and partial solubilization of alginic acid gel beads in acidic buffer, *Carbohydr. Polym.* **29**, 209–215.

Draget, K.I., Onsøyen, E., Fjæreide, T., Simensen M. K., Smidsrød O. (1997) Use of G-block polysaccharides', *Intl. Pat. Appl.#* PCT/NO97/00176.

Dubois, M., Gilles, K.A., Hamilton, J.K., Rebers, P.A., Smith, F. (1956) Colorimetric method for determination of sugars and related substances, *Anal. Chem.* **28**, 350–356.

Elyakova, L. A., Favorov, V. V. (1974) Isolation and certain properties of alginate lyase VI from the mollusk *Littorina* sp., *Biochim. Biophys. Acta* **358**, 341–354.

Ertesvåg, H., Larsen, B., Skjåk-Bræk, G., Valla, S. (1994) Cloning and expression of an *Azotobacter vinelandii* mannuronan-C-5 epimerase gene, *J. Bacteriol.* **176**, 2846–2853.

14 References

Ertesvåg, H., Høidal, H.K., Hals, I.K., Rian, A., Doseth, B., Valla, S. (1995) A family of moddular type mannuronana C-5 epimerase genes controls the alginate structure in *Azotobacter vinelandii*, *Mol. Microbiol* **16**, 719–731.

Ertesvåg, H., Erlien, F., Skjåk-Bræk, G., Rehm, B.H.A., Valla, S. (1998a) Biochemical properties and substrate specificities of a recombinantly produced *Azotobacter vinelandii* alginate lyase, *J. Bacteriol.* **180**, 3779–3784.

ErtesvågH., Høydal, H., Skjåk-Bræk, G., Valla, S. (1998b) The *Azotobacter vinelandii* mannuronan C-5 epimerase AlgE1 consists of two separate catalytic domains, *J. Biol. Chem.* **273**, 30927–30938.

Espevik, T., Skjåk-Bræk, G. (1996) Application of alginate gels in biotechnology and biomedicine, *Carbohydr. Eur.* **14**, 19–25.

Filisetti-Cozzi, T. M. C. C., Carpita, N. C. (1991) Measurement of uronic acids without interference from neutral sugars, *Anal. Biochem.* **197**, 157–162.

Fisher, F. G., Dörfel, H. (1955) Die Polyuronsäuren der Braunalgen (Kohlenhydrate der Algen), *Z. Physiol. Che.,* **302**, 186–203.

Gacesa P., (1992) Enzymic degradation of alginates, *Int. J. Biochem.* **24**, 545–552.

Grant, G. T., Morris, E. R., Rees, D. A., Smith, P. J. C., Thom, D. (1973) Biological interactions between polysaccharides and divalent cations: The egg-box model, *FEBS Lett.* **32**, 195–198.

Grasdalen, H., Larsen, B., Smidsrød O. (1977) ^{13}C-NMR studies of alginate, *Carbohydr. Res.* **56**, C11–C15.

Grasdalen, H., Larsen, B., Smidsrød, O. (1979) A PMR study of the composition and sequence of uronate residues in alginate, *Carbohydr. Res.* **68**, 23–31.

Grasdalen, H. (1983) High-field ^{1}H-nmr spectroscopy of alginate: Sequential structure and linkage conformations, *Carbohydr. Res.* **118**, 255–260.

Haug, A. (1959a) Fractionation of alginic acid, *Acta Chem. Scand.* **13**, 601–603.

Haug, A. (1959b) Ion exchange properties of alginate fractions, *Acta Chem. Scand.* **13**, 1250–1251.

Haug, A., Larsen B. (1963) The solubility of alginate at low pH, *Acta Chem. Scand.* **17**, 1653–1662.

Haug, A., Larsen, B., Smidsrød, O. (1963) The degradation of alginates at different pH values, *Acta Chem. Scand.* **17**, 1466–1468.

Haug, A. (1964) Composition and properties of alginates, Thesis, Norwegian Institute of Technology, Trondheim.

Haug, A., Smidsrød O. (1965) Fractionation of alginates by precipitation with calcium and magnesium ions, *Acta Chem. Scand.* **19**, 1221–1226.

Haug, A., Larsen B. (1966) A study on the constitution of alginic acid by partial acid hydrolysis, *Proc. Int. Seaweed Symp.* **5**, 271–277.

Haug, A., Larsen B., Smidsrød O. (1966) A study of the constitution of alginic acid by partial hydrolysis, *Acta Chem. Scand.* **20**, 183–190.

Haug, A., Smidsrød O. (1967) Precipitation of acidic polysaccharides by salts in ethanol-water mixtures, *J. Polym. Sci.* **16**, 1587–1598.

Haug, A., Larsen B., Smidsrød O. (1967a) Studies on the sequence of uronic acid residues in alginic acid, *Acta Chem. Scand.* **21**, 691–704.

Haug, A., Larsen, B., Smidsrød, O. (1967b) Alkaline degradation of alginate, *Acta Chem. Scand.* **21**, 2859–2870.

Haug, A., Myklestad, S., Larsen, B., Smidsrød, O. (1967c) Correlation between chemical structure and physical properties of alginate, *Acta Chem. Scand.* **21**, 768–778.

Haug, A., Smidsrød, O. (1970) Selectivity of some anionic polymers for divalent metal ions, *Acta Chem. Scand.* **24**, 843–854.

Hirst, E. L., Jones, J. K. N., Jones, W. O., (1939) The structure of alginic acid. Part I, *J. Chem. Soc.* 1880–1885.

Høydal, H., Ertesvåg, H., Stokke, B. T., Skjåk-Bræk, G., Valla, S. (1999) Biochemical properties and mechanism of action of the recombinant *Azotobacter vinelandii* mannuronan C-5 epimerase, *J. Biol. Chem.* **274**, 12316–12322.

Indergaard, M., Skjåk-Bræk, G. (1987) Characteristics of alginate from *Laminaria digitata* cultivated in a high phosphate environment, *Hydrobiologia* **151/152**, 541–549.

Kvam, B. J., Grasdalen, H., Smidsrød, O., Anthonsen, T. (1986) NMR studies of the interaction of metal ions with poly-(1,4-hexuronates). VI. Lanthanide(III) complexes of sodium (methyl ∀-D-galactopyranosid)uronate and sodium (phenylmethyl ∀-D-galactopyranosid)uronate, *Acta Chem. Scand.* **B40**, 735–739.

Kvam, B. (1987) Conformational conditions and ionic interactions of charged polysaccharides. Application of NMR techniques and the Poisson-Boltzmann equation, Thesis, Norwegian Institute of Technology, Trondheim.

Larsen, B., Smidsrød O., Painter T. J., Haug A. (1970) Calculation of the nearest-neighbour frequencies in fragments of alginate from the yields of free monomers after partial hydrolysis, *Acta Chem. Scand.* **24**, 726–728.

Leo, W. J., McLoughlin, A. J., Malone, D. M. (1990) Effects of terilization treatments on some properties of alginate solutions and gels, *Biotechnol. Prog.* **6**, 51–53.

Littlecott, G. W. (1982). Food gels–the role of alginates, *Food Technol. Aust.* **34**, 412–418.

Mackie, W., Noy, R., Sellen, D. B. (1980) Solution properties of sodium alginate, *Biopolymers* **19**, 1839–1860.

Mackie, W., Perez, S., Rizzo, R., Taravel, F., Vignon, M. (1983) Aspects of the conformation of polyguluronate in the solid state and in solution, *Int. J. Biol. Macromol.* **5**, 329–341.

Martinsen, A., Skjåk-Bræk, G., Smidsrød, O. (1989) Alginate as immobilization material: I. Correlation between chemical and physical properties of alginate gel beads, *Biotechnol. Bioeng.* **33**, 79–89.

Martinsen, A., Skjåk-Bræk, G., Smidsrød, O., Zanetti, F., Paoletti, S. (1991) Comparison of different methods for determination of molecular weight and molecular weight distribution of alginates, *Carbohydr. Polym.* **15**, 171–193.

McHugh, D.J. (1987). Production, properties and uses of alginates, in: Production and Utilization of Products from Commercial Seaweeds, FAO Fisheries Technical Paper No. 288 (McHugh, D. J., Ed.), Rome: FAO, 58–115.

Miawa, T. (1930) Alginic acid, *J. Chem. Soc. Japan* **51**, 738–745.

Mikkelsen, A., Elgsæter, A. (1995) Density distribution of calcium-induced alginate gels, *Biopolymers* **36**, 17–41.

Moe, S. T., Skjåk-Bræk, G., Elgsæter, A., Smidsrød, O. (1993) Swelling of covalently cross-linked ionic polysaccharide gels: Influence of ionic solutes and nonpolar solvents, *Macromolecules* **26**, 3589–3597.

Moe, S., Draget, K., Skjåk-Bræk, G., Smidsrød, O. (1995) Alginates, in: *Food Polysaccharides and Their Applications* (Stephen, A. M., Ed.), New York: Marcel Dekker, 245–286.

Murata, K., Inose, T., Hisano, T., Abe, S., Yonemoto, Y. (1993) Bacterial alginate lyase: enzymology, genetics and application, *J. Ferment. Bioeng.* **76**, 427–437.

Myklestad. S., Haug A. (1966) Studies on the solubility of alginic acid *from Ascophyllum nodosum* at low pH, *Proc. Int. Seaweed Symp.* **5**, 297–303.

Neiser, S., Draget, K. I., Smidsrød O. (1998) Gel formation in heat-treated bovine serum albumin–sodium alginate systems, *Food Hydrocolloids* **12**, 127–132.

Neiser, S., Draget, K. I., Smidsrød O. (1999) Interactions in bovine serum albumin–calcium alginate gel systems, *Food Hydrocolloids* **13**, 445–458.

Nelson, W. L., Cretcher, L. H. (1929) The alginic acid from *Macrocystis pyrifera*, *J. Am. Chem. Soc.* **51**, 1914–1918.

Nelson, W. L., Cretcher, L. H. (1930) The isolation and identification of *d*-mannuronic acid lactone from the *Macrocystis pyrifera*, *J. Am. Chem. Soc.* **52**, 2130–2134.

Nelson, W. L., Cretcher, L. H. (1932) The properties of *d*-mannuronic acid lactone, *J. Am. Chem. Soc.* **54**, 3409–3406.

Nibu, Y., Satoh, T., Nishi, Y., Takeuchi, T., Murata, K., Kusakabe, I. (1995) Purification and characterization of extracellular alginate lyase from Enterobacter cloacae M-1, *Biosci. Biotechnol. Biochem.* **59**, 632–637.

Onsøyen, E. (1996) Commercial applications of alginates, *Carbohydr. Eur.* **14**, 26–31.

Otterlei, M., Østgaard, K., Skjåk-Bræk, G., Smidsrød, O., Soon-Shiong, P., Espevik, T. (1991). Induction of cytokine production from human monocytes stimulated with alginate, *J. Immunother.* **10**, 286–291.

Painter, T. J., Smidsrød O., Haug A. (1968) A computer study of the changes in composition-distribution occurring during random depolymerisation of a binary linear heteropolysaccharide, *Acta Chem. Scand.* **22**, 1637–1648.

Penman, A., Sanderson G. R. (1972) A method for the determination of uronic acid sequence in alginates, *Carbohydr. Res.* **25**, 273–282.

Read, T.-A., Sorensen, D.R., Mahesparan, R., Enger, P.Ø., Timpl, R., Olsen, B.R., Hjelstuen, M.H.B., Haraldseth, O., Bjerkevig, R. (2000) Local endostatin treatment of gliomas administered by microencapsulated producer cells, *Nature Biotechnol.* **19**, 29–34.

Rivera-Carro H. D. (1984) Block structure and uronic acid sequence in alginates, Thesis, Norwegian Institute of Technology, Trondheim.

Schmidt, E., Vocke, F. (1926) Zur Kenntnis der Polyglykuronsäuren, *Chem. Ber.* **59**, 1585–1588.

Shimokawa, T., Yoshida, S., Kusakabe, I., Takeuchi, T., Murata, K., Kobayashi, H. (1997) Some properties and action mode of "-L-guluronan lyase from *Enterobacter cloacae* M-1. *Carbohydr. Res.* **304**, 125–132.

Sime, W. (1990). Alginates, in: *Food Gels* (Harris, P., Ed.), London: Elsevier, 53–78.

Skjåk-Bræk, G., Larsen B., Grasdalen H. (1986) Monomer sequence and acetylation pattern in some bacterial alginates, *Carbohydr. Res.* **154**, 239–250.

Skjåk-Bræk, G., Grasdalen, H., Smidsrød, O. (1989a) Inhomogeneous polysaccharide ionic gels, *Carbohydr. Polym.* **10**, 31–54.

Skjåk-Bræk, G., Grasdalen, H., Draget, K. I., Smidsrød, O. (1989b). Inhomogeneous calcium alginate beads, in: Biomedical and Biotechnological Advances in Industrial Polysaccharides (Crescenzi, V., Dea, I. C. M., Paoletti, S., Stivala, S. S., Sutherland, I. W., Eds.), New York: Gordon and Breach, 345–363.

Skjåk-Bræk, G., Moe, S. T. (1992) Alginate gels, US Patent 5,144,016.

Smidsrød, O., Haug, A., Larsen, B. (1963a) The influence of reducing substances on the rate of degradation of alginates, *Acta Chem. Scand.* **17**, 1473–1474.

Smidsrød, O., Haug, A., Larsen, B. (1963b) Degradation of alginate in the presence of reducing compounds, *Acta Chem. Scand.* **17**, 2628–2637.

Smidsrød, O., Haug, A., Larsen, B. (1967) Oxidative-reductive depolymerization: a note on the comparison of degradation rates of different polymers by viscosity measurements, *Carbohydr. Res.* **5**, 482–485.

Smidsrød, O., Haug, A. (1968a) A light scattering study of alginate, *Acta Chem. Scand.* **22**, 797–810.

Smidsrød, O., Haug, A. (1968b) Dependence upon uronic acid composition of some ion-exchange properties of alginates, *Acta Chem. Scand.* **22**, 1989–1997.

Smidsrød, O., Whittington S. G. (1969) Monte Carlo investigation of chemical inhomogeneity in copolymers, *Macromolecules* **2**, 42–44.

Smidsrød, O. (1970) Solution properties of alginate, *Carbohydr. Res.* **13**, 359–372.

Smidsrød, O. (1973). Some physical properties of alginates in solution and in the gel state, Thesis, Norwegian Institute of Technology, Trondheim.

Smidsrød, O., Glover, R. M., Whittington, S. G. (1973) The relative extension of alginates having different chemical composition, *Carbohydr. Res.* **27**, 107–118.

Smidsrød, O. (1974) Molecular basis for some physical properties of alginates in the gel state, *J. Chem. Soc. Farad. Trans* **57**, 263–274.

Smidsrød, O., Skjåk-Bræk, G. (1990) Alginate as immobilization matrix for cells, *Trends Biotechnol.* **8**, 71–78.

Smidsrød, O., Draget, K. I. (1996) Alginates: chemistry and physical properties, *Carbohydr. Eur.* **14**, 6–13.

Soon-Shiong, P., Otterlei, M., Skjåk-Bræk, G., Smidsrød, O., Heintz, R., Lanza, R. P., Espevik, T. (1991) An immunologic basis for the fibrotic reaction to implanted microcapsules, *Transplant Proc.* **23**, 758–759.

Soon-Shiong, P., Feldman, E., Nelson, R., Heints, R., Yao, Q., Yao, T., Zheng, N., Merideth, G., Skjåk-Bræk, G., Espevik, T., Smidsrød, O., Sandford P. (1993) Long-term reversal of diabetes by the injection of immunoprotected islets, *Proc. Natl. Acad. Sci. USA* **90**, 5843–5847.

Soon-Shiong, P., Heintz, R. E., Merideth, N., Yao, Q. X., Yao, Z. W., Zheng, T. L., Murphy, M., Moloney, M. K., Schmehl, M., Harris, M., Mendez, R., Mendez, R., Sandford, P. A. (1994) Insulin independence in a type 1 diabetic patient after encapsulated islet transplantation, *Lancet* **343**, 950–951.

Stanford, E. C. C. (1881). British Patent 142.

Stanford, E. C. C. (1883a) New substance obtained from some of the commoner species of marine algæ; Algin, *Chem. News* **47**, 254–257.

Stanford, E. C. C. (1883b) New substance obtained from some of the commoner species of marine algæ; Algin, *Chem. News* **47**, 267–269.

Steginsky, C. A., Beale, J. M., Floss, H. G., Mayer, R. M. (1992) Structural determination of alginic acid and the effects of calcium binding as determined by high-field n.m.r., *Carbohydr. Res.* **225**, 11–26.

Steiner, A. B. (1947). Manufacture of glycol alginates, US Patent 2,426,215.

Steiner, A. B., McNeilly, W. H. (1950). High-stability glycol alginates and their manufacture, US Patent 2,494,911.

Stokke, B. T., Smidsrød, O., Bruheim, P., Skjåk-Bræk, G. (1991). Distribution of uronate residues in alginate chains in relation to alginate gelling properties, *Macromolecules* **24**, 4637–4645.

Stokke, B. T., Smidsrød, O., Zanetti, F., Strand, W., Skjåk-Bræk G. (1993) Distribution of uronate residues in alginate chains in relation to gelling properties 2:Enrichment of -D-mannuronic acid and depletion of "-L-guluronic acid in the sol fraction, *Carbohydr. Polym.* **21**, 39–46.

Stokke B. T., Draget K. I., Yuguchi Y., Urakawa H., Kajiwara K. (2000) Small angle X-ray scattering and rheological characterization of alginate gels. 1 Ca-alginate gels, *Macromolecules* **33**, 1853–1863.

Tanford, C. (1961). *Physical Chemistry of Macromolecules.* New York: John Wiley & Sons, Inc.

Toft, K., Grasdalen, H., Smidsrød, O. (1986). Synergistic gelation of alginates and pectins, *ACS Symp. Ser.* **310**, 117–132.

Valla, S., Ertesvåg, H., Skjåk-Bræk, G. (1996) Genetics and biosynthesis of alginates, *Carbohydr. Eur.* **14**, 14–18.

Wong, T. Y., Preston, L. A., Schiller, N. L. (2000) Alginate lyase: review of major sources and enzyme characteristics, structure-function analysis, biological roles, and applications, *Ann. Rev. Microbiol.* **54**, 289–340.

2
Bacterial Cellulose

Prof. Dr. Eng. Stanislaw Bielecki[1], Dr. Eng. Alina Krystynowicz[2], Prof. Dr. Marianna Turkiewicz[3], Dr. Eng. Halina Kalinowska[4]

[1] Institute of Technical Biochemistry, Technical University of Lódz, Stefanowskiego 4/10, 90-924 Lódz, Poland; Tel: +48-4263-13442; Fax: +48-4263-402; E-mail: biochem@ck-sg.p.lodz.pl

[2] Institute of Technical Biochemistry, Technical University of Lódz, Stefanowskiego 4/10, 90-924 Lódz, Poland; Tel: +48-4263-13442; Fax: +48-4263-402; E-mail: biochem@ck-sg.p.lodz.pl

[3] Institute of Technical Biochemistry, Technical University of Lódz, Stefanowskiego 4/10, 90-924 Lódz, Poland; Tel: +48-4263-13442; Fax: +48-4263-402; E-mail: biochem@ck-sg.p.lodz.pl

[4] Institute of Technical Biochemistry, Technical University of Lódz, Stefanowskiego 4/10, 90-924 Lódz, Poland; Tel: +48-4263-13442; Fax: +48-4263-402; E-mail: biochem@ck-sg.p.lodz.pl

1	Introduction	34
2	Historical Outline	34
3	Structure of BC	34
4	Chemical Analysis and Detection	37
5	Occurrence	38
6	Physiological Function	38
7	**Biosynthesis of BC**	39
7.1	Synthesis of the Cellulose Precursor	39
7.2	Cellulose Synthase	41
7.3	Mechanism of Biosynthesis	42
7.3.1	Mechanism of 1,4-β-Glucan Polymerization	42
7.3.2	Assembly and Crystallization of Cellulose Chains	46

Polysaccharides and Polyamides in the Food Industry. Properties, Production, and Patents.
Edited by A. Steinbüchel and S. K. Rhee
Copyright © 2005 WILEY-VCH Verlag GmbH & Co. KGaA, Weinheim
ISBN: 3-527-31345-1

7.4	Genetic Basis of Cellulose Biosynthesis	48
7.5	Regulation of Bacterial Cellulose Synthesis	49
7.6	Soluble Polysaccharides Synthesized by *A. xylinum*	50
7.7	Role of Endo- and Exocellulases Synthesized by *A. xylinum*	51

| 8 | **Biodegradation of BC** | 52 |

9	**Biotechnological Production**	53
9.1	Isolation from Natural Sources and Improvement of BC-producing Strains	54
9.1.1	Improvement of Cellulose-producing Strains by Genetic Engineering	54
9.2	Fermentation Production	55
9.2.1	Carbon and Nitrogen Sources	55
9.2.2	Effect of pH and Temperature	57
9.2.3	Static and Agitated Cultures; Fermentor Types	57
9.2.4	Continuous Cultivation	59
9.3	*In vitro* Biosynthesis	60
9.4	Chemo-enzymatic Synthesis	60
9.5	Production Processes Expected to be Applied in the Future	61
9.6	Recovery and Purification	61

| 10 | **Properties** | 62 |

11	**Applications**	62
11.1	Technical Applications	63
11.2	Medical Applications	65
11.3	Food Applications	66
11.4	Miscellaneous Uses	66

| 12 | **Patents** | 67 |

| 13 | **Outlook and Perspectives** | 67 |

| 14 | **References** | 79 |

A-BC	bacterial cellulose from agitated culture
ATP	adenosine triphosphate
BC	bacterial cellulose
CBH	cellobiohydrolase
c-di-GMP	cyclic diguanosine monophosphate
Cel$^-$	cellulose-negative mutant
Cel6A, Cel7A	cellobiohydrolases belonging to 6A and 7A families, respectively
CM	carboxymethyl-
CS	cellulose synthase
CSL	corn steep liquor
D	aspartic acid

DMSO	dimethyl sulfoxide
DP	degree of polymerization
E	glutamic acid
FBP	fructose-1,6-biphosphate phosphatase
FK	fructokinase
Fru-bi-P	fructose-1,6-biphosphate
Fru-6-P	fructose-6-phosphate
G	guanine
GK	glucokinase
Glc	glucose
Glc-6(1)-P	glucose-6(1)-phosphate
G6PDH	glucose-6-phosphate dehydrogenase
H-S medium	Hestrin and Schramm medium (1954)
IS	insertion sequence
Lip	lipid
LP	UDPGPT: lipid pyrophosphate: UDPGlc- phosphotransferase
LPP	lipid pyrophosphate phosphohydrolase
Man	mannose
NMR	nuclear magnetic resonance
PC	plant cellulose
PDEA	phosphodiesterase A
PDEB	phosphodiesterase B
Pel$^-$	pellicle non-forming
1PFK	fructose-1-phosphate kinase
PGA	phosphogluconic acid
PGI	phosphoglucoisomerase
PMG	phosphoglucomutase
PTS	system of phosphotransferases
Q	glutamine
R	arginine
Rha	rhamnose
Rib	D-ribose
S	serine
S-BC	bacterial cellulose from static culture
TC	terminal complexe
U	uridine
UDP	uridine diphosphate
UDPGlc	uridine diphosphoglucose
UGP	pyrophosphorylase uridine diphosphoglucose
UMP	uridine monophosphate
v.v.m.	volume per volume per minute
W	tryptophan

1
Introduction

Cellulose is the most abundant biopolymer on earth, recognized as the major component of plant biomass, but also a representative of microbial extracellular polymers. Bacterial cellulose (BC) belongs to specific products of primary metabolism and is mainly a protective coating, whereas plant cellulose (PC) plays a structural role.

Cellulose is synthesized by bacteria belonging to the genera *Acetobacter, Rhizobium, Agrobacterium,* and *Sarcina* (Jonas and Farah, 1998). Its most efficient producers are Gram-negative, acetic acid bacteria *Acetobacter xylinum* (reclassified as *Gluconacetobacter xylinus,* Yamada et al., 1997; Yamada, 2000), which have been applied as model microorganisms for basic and applied studies on cellulose (Cannon and Anderson, 1991). Investigations have been focused on the mechanism of biopolymer synthesis, as well as on its structure and properties, which determine practical use (Legge, 1990; Ross et al., 1991). One of the most important features of BC is its chemical purity, which distinguishes this cellulose from that from plants, usually associated with hemicelluloses and lignin, removal of which is inherently difficult.

Because of the unique properties, resulting from the ultrafine reticulated structure, BC has found a multitude of applications in paper, textile, and food industries, and as a biomaterial in cosmetics and medicine (Ring et al., 1986). Wider application of this polysaccharide is obviously dependent on the scale of production and its cost. Therefore, basic studies run together with intensive research on strain improvement and production process development.

2
Historical Outline

Although synthesis of an extracellular gelatinous mat by *A. xylinum* was reported for the first time in 1886 by A. J. Brown, BC attracted more attention in the second half of the 20th century. Intensive studies on BC synthesis, using *A. xylinum* as a model bacterium, were started by Hestrin et al. (1947, 1954), who proved that resting and lyophilized *Acetobacter* cells synthesized cellulose in the presence of glucose and oxygen. Next, Colvin (1957) detected cellulose synthesis in samples containing cell-free extract of *A. xylinum,* glucose, and ATP. Further milestones in studies on BC synthesis, presented in this review, contributed to the elucidation of mechanisms governing not only the biogenesis of the bacterial polymer, but also that of plants, thus leading to the understanding of one of the most important processes in nature. The true historical outline is presented throughout all the paragraphs below, including the references.

3
Structure of BC

Cellulose is an unbranched polymer of β-1,4-linked glucopyranose residues. Extensive research on BC revealed that it is chemically identical to PC, but its macromolecular structure and properties differ from the latter (Figure 1). Nascent chains of BC aggregate to form subfibrils, which have a width of approximately 1.5 nm and belong to the thinnest naturally occurring fibers, comparable only to subelemental fibers of cellulose detected in the cambium of some plants and in quinee mucous (Kudlicka, 1989). BC subfibrils are crystallized into microfibrils (Jonas and Farah, 1998), these into bundles, and the latter into ribbons (Yamanaka et al.,

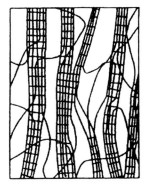 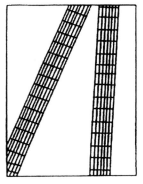

Fig. 1 Schematic model of BC microfibrils (right) drawn in comparison with the 'fringed micelles'; of PC fibrils (Iguchi et al., 2000; with kind permission).

2000). Dimensions of the ribbons are 3–4 (thickness) × 70–80 nm (width), according to Zaar (1977), 3.2 × 133 nm, according to Brown et al. (1976), or 4.1 × 117 nm, according to Yamanaka et al. (2000), whereas the width of cellulose fibers produced by pulping of birch or pine wood is two orders of magnitude larger (1.4–4.0 × 10^{-2} and 3.0–7.5 × 10^{-2} mm, respectively). The ultrafine ribbons of microbial cellulose, the length of which ranges from 1 to 9 μm, form a dense reticulated structure (Figure 2), stabilized by extensive hydrogen bonding (Figure 3). BC is also distinguished from its plant counterpart by a high crystallinity index (above 60%) and different degree of polymerization (DP), usually between 2000 and 6000 (Jonas and Farah, 1998), but in some cases reaching even 16,000 or 20,000 (Watanabe et al., 1998b), whereas the average DP of plant polymer varies from 13,000 to 14,000 (Teeri, 1997).

Macroscopic morphology of BC strictly depends on culture conditions (Watanabe et al., 1998a; Yamanaka et al., 2000). In static conditions (Figure 4), bacteria accumulate cellulose mats (S-BC) on the surface of nutrient broth, at the oxygen-rich air–liquid interface. The subfibrils of cellulose are continuously extruded from linearly ordered pores at the surface of the bacterial cell, crystallized into microfibrils, and forced deeper into the growth medium. Therefore, the leather-like pellicle, supporting the population of A. xylinum cells, consists of overlapping and intertwisted cellulose rib-

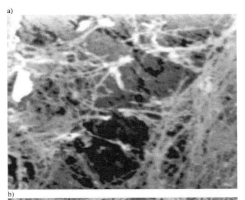
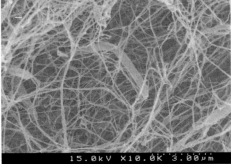

Fig. 2 Scanning electron microscopy images of BC membrane from static culture of A. xylinum (a) and bacterial cell with attached cellulose ribbons (b).

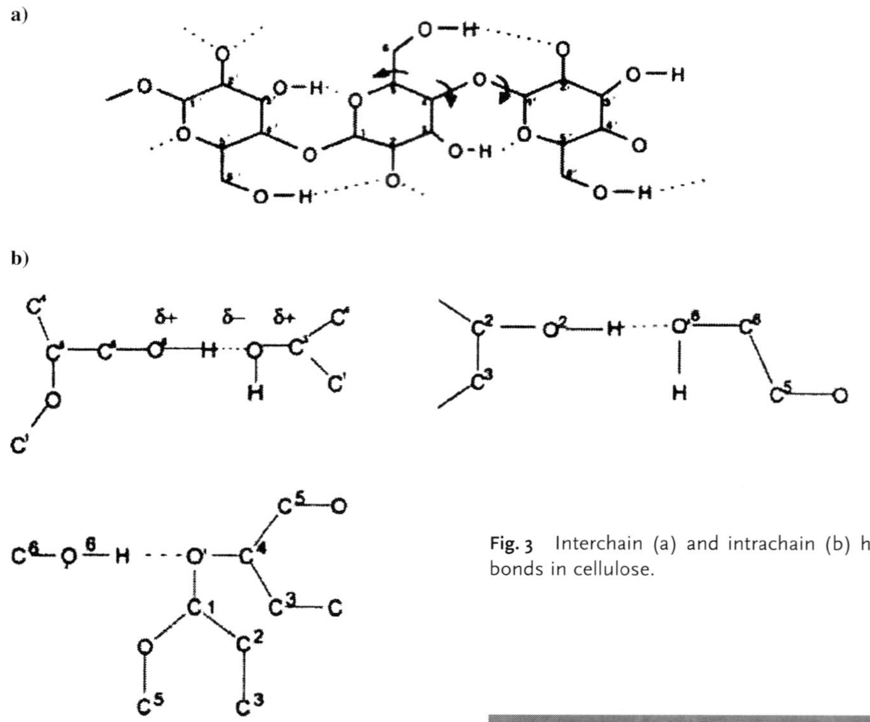

a)

b)

Fig. 3 Interchain (a) and intrachain (b) hydrogen bonds in cellulose.

bons, forming parallel but disorganized planes (Jonas and Farah, 1998). The adjacent S-BC strands branch and interconnect less frequently than these in BC produced in agitated culture (A-BC), in a form of irregular granules, stellate and fibrous strands, well-dispersed in culture broth (Figure 5) (Vandamme et al., 1998). The strands of reticulated A-BC interconnect to form a grid-like pattern, and have both roughly perpendicular and roughly parallel orientations (Watanabe et al., 1998a).

Differences in three-dimensional structure of A-BC and S-BC are noticeable in their scanning electron micrographs. The S-BC fibrils are more extended and piled above one another in a criss-crossing manner. Strands of A-BC are entangled and curved (Johnson and Neogi, 1989). Besides, they

Fig. 4 BC pellicle formed in static culture.

Fig. 5 BC pellets formed in agitated culture.

have a larger cross-sectional width (0.1–0.2 µm) than S-BC fibrils (usually 0.05–0.10 µm). Morphological differences between S-BC and A-BC contribute to varying degrees of crystallinity, different crystallite size and I_α cellulose content.

Two common crystalline forms of cellulose, designated as I and II, are distinguishable by X-ray, nuclear magnetic resonance (NMR), Raman spectroscopy, and infrared analysis (Johnson and Neogi, 1989). It is known that in the metastable cellulose I, which is synthesized by the majority of plants and also by *A. xylinum* in static culture, parallel β-1,4-glucan chains are arranged uniaxially, whereas β-1,4-glucan chains of cellulose II are arranged in a random manner. They are mostly antiparallel and linked with a larger number of hydrogen bonds that results in higher thermodynamic stability of the cellulose II.

A-BC has a lower crystallinity index and a smaller crystallite size than S-BC (Watanabe et al., 1998a). It was also observed that a significant portion of cellulose II occurred in BC synthesized in agitated culture. In nature, cellulose II is synthesized by only a few organisms (some algae, molds, and bacteria, such as *Sarcina ventriculi*) (Jonas and Farah, 1998); the industrial production of this kind of cellulose is based on chemical conversion of PC.

Using CP/MAS ^{13}C-NMR it is possible to reveal the presence of cellulose I_α and I_β – two distinct forms of cellulose (Watanabe et al., 1998a). These forms occur in algae-, bacteria-, and plant-derived cellulose. The latter one contains less I_α cellulose than BC (Johnson and Neogi, 1989). The irreversible crystal transformation from cellulose I_α to I_β shifts the X-ray and CP/MAS ^{13}C-NMR spectra because of the difference in the unit cell. S-BC contains more cellulose I_α than A-BC. It was reported that the difference in cellulose I_α content between A-BC and S-BC exceeded that in crystallinity index (Watanabe et al., 1998a), and the mass fraction of cellulose I_α was closely related to the crystallite size.

4
Chemical Analysis and Detection

For the detection of either crystalline or amorphous cellulose, several direct dyes, specific for the linear β-1,4-glucan, are used (Mondal and Kai, 2000). All of them are fluorescent brightening agents and form dye–cellulose complexes, stabilized by van der Waals and/or hydrogen bonding. One of these dyes is the fluorescent brightener Calcofluor. The direct dyes do not only enable visualization of cellulose chains, but also have been intensively applied for studies on nascent cellulose chains association and crystallization (see Section 7.3.2).

The weight-average DP of cellulose and the DP distribution are determined by high-performance gel-permeation chromatography (Watanabe et al., 1998a) of nitrated cellulose samples.

The differences between the reticulated structure of microbial cellulose, produced under agitated culture conditions, and the disorganized layered structure of cellulose pellicle, formed in static culture, are noticeable in scanning electron microscopy images (Johnson and Neogi, 1989).

To distinguish the parallel chain crystalline lattice of cellulose I from the antiparallel one of cellulose II, X-ray diffraction, Raman spectroscopy, infrared analysis, and NMR are applied (Johnson and Neogi, 1989). The crystallinity index and crystallite size are calculated based on X-ray diffraction measurements (Watanabe et al., 1998a).

Two distinct forms of cellulose I, i.e. cellulose I_α and I_β, are not distinguishable by X-ray diffraction, and therefore CP/MAS ^{13}C-NMR analysis, carried out on freeze-dried cellulose samples, has to be performed to determine their mass fractions (Watanabe et al., 1998a).

The physicochemical properties of cellulose such as water holding capacity, viscosity of disintegrated cellulose suspension, and the Young's modulus of dried sheets are determined using conventional methods (Watanabe et al., 1998a, Iguchi et al., 2000).

5
Occurrence

BC is synthesized by several bacterial genera, of which *Acetobacter* strains are best known. An overview of BC producers is presented in Table 1 (Jonas and Farah, 1998). The polymer structure depends on the organism, although the pathway of biosynthesis and mechanism of its regulation are probably common for the majority of BC-producing bacteria (Ross et al., 1991; Jonas and Farah, 1998).

Tab. 1 BC producers (Jonas and Farah, 1998, modified)

Genus	Cellulose structure
Acetobacter	extracellular pellicle composed of ribbons
Achromobacter	fibrils
Aerobacter	fibrils
Agrobacterium	short fibrils
Alcaligenes	fibrils
Pseudomonas	no distinct fibrils
Rhizobium	short fibrils
Sarcina	amorphous cellulose
Zoogloea	not well defined

A. xylinum (synonyms *A. aceti* ssp. *xylinum*, *A. xylinus*), which is the most efficient producer of cellulose, has been recently reclassified and included within the novel genus *Gluconacetobacter*, as *G. xylinus* (Yamada et al., 1998, 2000) together with some other species (*G. hansenii*, *G. europaeus*, *G. oboediens*, and *G. intermedius*).

6
Physiological Function

In natural habitats, the majority of bacteria synthesize extracellular polysaccharides, which form envelopes around the cells (Costeron, 1999). BC is an example of such a substance. Cells of cellulose-producing bacteria are entrapped in the polymer network, frequently supporting the population at the liquid–air interface (Wiliams and Cannon, 1989). Therefore, BC-forming strains can inhabit sewage (Jonas and Farah, 1998). The polymer matrix takes part in adhesion of the cells onto any accessible surface and facilitates nutrient supply, since their concentration in the polymer lattice is markedly enhanced due to its adsorptive properties, in comparison to the surrounding aqueous environment (Jonas and Farah, 1998; Costeron, 1999). Some authors suppose that cellulose synthesized by *A. xylinum* also plays a storage role and can be utilized by the starving microorganisms. Its decomposition would be then catalyzed by exo- and endo-glucanases, the co-presence of which was detected in the culture broth of some cellulose-producing *A. xylinum* strains (Okamoto et al., 1994).

Because of the viscosity and hydrophilic properties of the cellulose layer, the resistance of producing bacterial cells against unfavorable changes (a decrease in water content, variations in pH, appearance of toxic substances, pathogenic organisms, etc.) in an habitat is increased, and they can further grow and develop inside the envelope. It was also found that cellulose placed over bacterial cells protects them from ultraviolet radiation. As much as 23% of the acetic acid bacteria cells covered with BC survived a 1h treatment with ultraviolet irradiation. Removal of the protective polysaccharide brought about a drastic decrease in their viability (3% only) (Ross et al., 1991).

7
Biosynthesis of BC

Synthesis of BC is a precisely and specifically regulated multi-step process, involving a large number of both individual enzymes and complexes of catalytic and regulatory proteins, whose supramolecular structure has not yet been well defined. The process includes the synthesis of uridine diphosphoglucose (UDPGlc), which is the cellulose precursor, followed by glucose polymerization into the β-1,4-glucan chain, and nascent chain association into characteristic ribbon-like structure, formed by hundreds or even thousands of individual cellulose chains. Pathways and mechanisms of UDPGlc synthesis are relatively well known, whereas molecular mechanisms of glucose polymerization into long and unbranched chains, their extrusion outside the cell, and self-assembly into fibrils require further elucidation.

Moreover, studies on BC synthesis may contribute to better understanding of PC biogenesis.

7.1
Synthesis of the Cellulose Precursor

Cellulose synthesized by *A. xylinum* is a final product of carbon metabolism, which depending on the physiological state of the cell involves either the pentose phosphate cycle or the Krebs cycle, coupled with gluconeogenesis (Figure 6) (Ross et al., 1991; Tonouchi et al., 1996). Glycolysis does not operate in acetic acid bacteria since they do not synthesize the crucial enzyme of this pathway – phosphofructose kinase (EC 2.7.1.56) (Ross et al., 1991). In *A. xylinum*, cellulose synthesis is tightly associated with catabolic processes of oxidation and consumes as much as 10% of energy derived from catabolic reactions (Weinhouse, 1977). BC production does not interfere with other anabolic processes, including protein synthesis (Ross et al., 1991).

A. xylinum converts various carbon compounds, such as hexoses, glycerol, dihydroxyacetone, pyruvate, and dicarboxylic acids, into cellulose, usually with about 50% efficiency. The latter compounds enter the Krebs cycle and due to oxalacetate decarboxylation to pyruvate undergo conversion to hexoses via gluconeogenesis, similarly to glycerol, dihydroxyacetone, and intermediates of the pentose phosphate cycle (Figure 6).

The direct cellulose precursor is UDPGlc, which is a product of a conventional pathway, common of many organisms, including plants, and involving glucose phosphoryla-

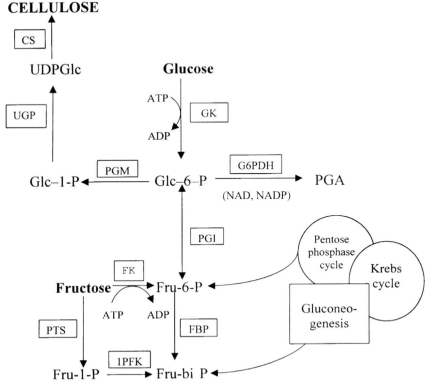

Fig. 6 Pathways of carbon metabolism in A. *xylinum*. CS, cellulose synthase (EC 2.4.2.12); FBP, fructose-1,6-biphosphate phosphatase (EC 3.1.3.11); FK, glucokinase (EC 2.7.1.2); G6PDH, glucose-6-phosphate dehydrogenase (EC 1.1.1.49); 1PFK, fructose-1-phosphate kinase (EC 2.7.1.56); PGI, phosphoglucoisomerase; PMG, phosphoglucomutase (EC 5.3.1.9); PTS, system of phosphotransferases; UGP, pyrophosphorylase UDPGlc (EC 2.7.7.9); Fru-bi-P, fructose-1,6-bi-phosphate; Fru-6-P, fructose-6-phosphate; Glc-6(1)-P, glucose-6(1)-phosphate; PGA, phosphogluconic acid; UDPGlc, uridine diphosphoglucose.

tion to glucose-6-phosphate (Glc-6-P), catalyzed by glucokinase, followed by isomerization of this intermediate to Glc-α-1-P, catalyzed by phosphoglucomutase, and conversion of the latter metabolite to UDPGlc by UDPGlc pyrophosphorylase. This last enzyme seems to be the crucial one involved in cellulose synthesis, since some phenotypic cellulose-negative mutants (Cel⁻) are specifically deficient in this enzyme (Valla et al., 1989), though they display cellulose synthase (CS) activity, this was confirmed *in vitro* by means of observation of cellulose synthesis, catalyzed by cell-free extracts of Cel⁻ strains (Saxena et al., 1989). Furthermore, the pyrophosphorylase activity varies between different A. *xylinum* strains and the highest activity was detected in the most effective cellulose producers, such as A. *xylinum* ssp. sucrofermentans BPR2001. The latter strain prefers fructose as a carbon source, displays high activity of phosphoglucoisomerase, and possesses a system of phosphotransferases, dependent on phosphoenolpyruvate. The system catalyzes conversion of fructose to fructose-1-phosphate and further to fructose-1,6-biphosphate (Figure 6).

7.2
Cellulose Synthase

Cellulose biosynthesis both in plants and in prokaryotes is catalyzed by the uridine diphosphate (UDP)-forming CS which is basically a processing 4-β-glycosyltransferase (EC 2.4.1.12, UDPGlc: 1,4-β-glucan 4-β-D-glucosyltransferase), since it transfers consecutive glucopyranose residues from UDPGlc to the newly formed polysaccharide chain and is all the time linked with this chain. Oligomeric CS complexes are frequently called terminal complexes (TCs). It is presumed that TCs are responsible first for β-1,4-glucan chains synthesis, extrusion through the outer membrane (if the globular, catalytic domain of CS is localized on the cytoplasmic side of the cell membrane), as well as association and crystallization into defined supramolecular structures, that follow the first two processes.

Cellulose synthase of *A. xylinum* is a typical membrane-anchored protein, having a molecular mass of 400–500 kDa (Lin and Brown, 1989), and is tightly bound to the cytoplasmic membrane. Because of this localization, purification of CS was extremely difficult, and isolation of the membrane fraction was necessary before CS solubilization and purification. Furthermore, *A. xylinum* CS appeared to be a very unstable protein (Lin and Brown, 1989). CS isolation from membranes was carried out using digitonin (Lin and Brown, 1989), or detergents (Triton X-100) and treatment with trypsin (Saxena et al., 1989), followed by CS entrapment on cellulose. According to Lin and Brown (1989), the purified CS preparation contained three different types of subunits, having molecular masses of 90, 67, and 54 kDa. Saxena et al. (1989) found only two types of polypeptides (83 and 93 kDa). The latter result seems to be more probable, since the mass of both subunits corresponds to the size of two genes *cesA* and *cesB*, detected in the CS operon by Saxena et al. (1990b, 1991), and the genes *bcsA* and *bcsB*, reported by Wong et al. (1990) and Ben-Bassat et al. (1993) for the same operon (see Section 7.4).

Photolabeling affinity studies indicate that the 83-kDa polypeptide is a catalytic subunit, displaying high affinity towards UDPGlc (Lin et al., 1990). According to the gene sequence, it contains 723 amino acid residues, of which the majority are hydrophobic. This subunit is probably synthesized as a proprotein, which contains a signal sequence, composed of 24 amino acid residues, which is cut off during the maturation process (Wong et al., 1990). The mature protein is anchored in the cell membrane by means of a transmembrane helix, close to the N-terminus of the polypeptide chain. The protein contains five more transmembrane helices, which can interact both with each other and with a large, catalytic site-comprising globular domain submerged in cytoplasma. Brown and Saxena (2000) claim that the catalytic subunit of *A. xylinum* CS operates in the same manner as processing glycosyltransferases of the second family, i.e. it catalyzes the direct glycoside bond synthesis in cellulose, assisted by simultaneous inversion of the configuration on the anomeric carbon, from the α-configuration in the UDPGlc, which is the monomer donor, to the β-configuration in the polysaccharide. A catalytically active aspartic acid residue (D), and short sequences DXD and QXXRW were found in the globular fragment of the numerous processing glycosyltransferases (Saxena and Brown, 1997). The same motives were detected in *A. xylinum* CS. This globular fragment has a two-domain character (i.e., comprises two domains A and B) in many of glycoside synthases (Saxena et al., 1995). The domain A includes the D residue, crucial for catalysis, and the DXD motive, which probably binds the nucleotide–sugar substrate. Participation of this motive in UDPGlc binding by *A. xylinum* CS was recently confirmed by Brown and Saxena (2000), who investigated enzyme mutants obtained using site-directed mutagenesis. It is not yet clear if the globular fragment of *A. xylinum* CS comprises one or two domains, although this second possibility is more probable. However, it is known that the catalytic subunit of *A. xylinum* CS is glycosylated. Two potential sites of glycosylation were found in its primary structure, deduced based on the gene sequence (Saxena et al., 1990a).

The 93-kDa polypeptide is tightly bound to the catalytic subunit. It contains the signal sequence close to its N-terminus and one transmembrane helix close to the C-terminus. The helix enables anchoring in the cell membrane. The polypeptide is probably a regulatory subunit, which interacts with the CS activator – cyclic diguanosine monophospahte (c-di-GMP) (see Section 7.5). The 93-kDa polypeptide does not combine with the 83-kDa subunit antibodies. However, it is not yet clear if CS is composed of these two subunits only, since this oligomeric complex plays multiple roles, i.e. β-1,4-glucan polymerization, extrusion of the subfibrils outside the cell, as well as self-assembly and crystallization of cellulose ribbons, composed of microfibrils. Therefore, the CS complex is probably comprised of other polypeptides involved in transmembrane pore formation. The analysis of microscopic images of the purified CS indicates that the enzyme tends to form tetrameric or octameric aggregates (Lin et al., 1989).

7.3
Mechanism of Biosynthesis

Synthesis of the metastable cellulose I allomorph, in *A. xylinum* and other cellulose-producing organisms, including plants, involves at least two intermediary steps, i.e. (1) polymerization of glucose molecules to the linear 1,4-β-glucan, and (2) assembly and crystallization of individual nascent polymer chains into supramolecular structures, characteristic for each cellulose-producing organism.

Assuming that the CS globular domain is localized on the cytoplasmic side of the cell membrane (see Section 7.2), the transfer of 1,4-β-glucan chains through the membrane, outside the cell is also required. All three steps are tightly coupled, and the rate of polymerization is limited by the rate of assembly and crystallization.

7.3.1
Mechanism of 1,4-β-Glucan Polymerization

Formation of BC is catalyzed by the CS complexes aligned linearly in the *A. xylinum* cytoplasmic membrane. Since these complexes can polymerize up to 200,000 molecules of Glc s^{-1} into the β-1,4-glucan chain (Ross et al., 1991), the process must run with high intensity. The mechanism of the reaction has not yet been definitely clarified. Currently, two different hypotheses for this mechanism in *A. xylinum* have been proposed.

The first of them, developed in Brown's laboratory (Brown, 1996; Brown and Saxena, 2000), assumes that 1,4-β-glucan polymerization does not involve a lipid intermediate, which transfers glucose from UDPGlc to the newly synthesized polymer chain. This hypothesis agrees with the fact that glycosyltransferases responsible for synthesis of unbranched homopolysaccharides are processing enzymes. The scheme of 1,4-β-glucan polymerization, proposed by Brown and his colleagues, is presented in Figure 7.

According to this model, there are three catalytic sites in the globular fragment of the CS catalytic subunit, similar to other proc-

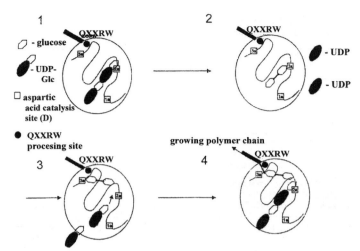

Fig. 7 Generalized concept for the polymerization reactions leading to β-1,4-glucan chain biosynthesis (Brown and Saxena, 2000; with kind permission).

essing glycosyltransferases. One of the catalytic sites (2a), comprising the DXD motive, binds two UDPGlc molecules (see Figure 7.1). The second catalytic site (1a), containing the crucial acidic aspartic acid (D) residue, catalyzes formation of the β-1,4-linkage between the two Glc residues docked in the pocket (see Figure 7.2), accompanied by releasing two UDP molecules. The third catalytic site (3a), containing the QXXRW motif, pulls the reducing end of the synthesized cellobiose (see Figure 7.3). The disaccharide leaves the area 1a–2a, which may bind two subsequent UDPGlc molecules (see Figure 7.4), and forms the second cellobiose molecule. In the next step, involving the QXXRW motif, the reducing end of the first cellobiose molecule is forced to the site of extrusion. At the same moment, the second cellobiose molecule is linked (with the aid of the D residue in the site 1a) to the nonreducing end of the first one, thus giving cellotetraose. The emptied 1a–2a area may bind two subsequent UDPGlc molecules, to repeat the cycle of reactions, attaching two more glucose residues to the chain. The simplified scheme of the proposed model of polymerization is depicted in Figure 8. According to this model, extrusion of the newly synthesized β-1,4-glucan chain starts from its reducing end.

Studies by Koyama et al. (1997) confirm this model. The authors proved that glucose residues are added to the nonreducing end of the polysaccharide and that reducing ends of nascent polymer chains are situated away from the cells. Furthermore, the torsion angle between two adjacent glucose residues in cellulose molecule is 180°, thus adding cellobiose residues (rather than single glucose moieties) to the growing chain favors maintaining the 2-fold screw axis of the β-1,4-glucan (Brown, 1996). Also Kuga and Brown (1988), who applied silver labeling, proved that cellulose chains were extruded outside the cell, starting from their reducing ends.

Han and Robyt (1998) who studied BC synthesis by the *A. xylinum* ATCC 10821 strain (resting cells and membrane preparations) using the ^{14}C pulse and chase reaction with D-glucose and UDPGlc, respectively,

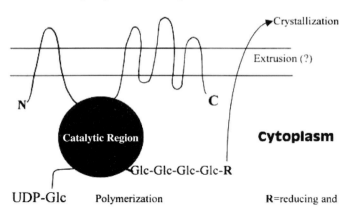

Fig. 8 Simplified scheme of BC biosynthesis according to the Brown's model (Brown and Saxena, 2000; with kind permission).

proposed a second, different molecular mechanism of this process. They believe that the consecutive residues of the activated glucopyranoses (from UDPGlc) are linked to the reducing end of the growing cellulose chain, because the ratio of D-[^{14}C]glucitol obtained from the extruded reducing end of the cellulose chain to D-[^{14}C]glucose was decreasing with time. Han and Robyt assumed that a lipid pyrophosphate with a polyisoprenoid character takes part in the biosynthesis. According to them, formation of the lipid intermediate during BC synthesis was confirmed by an extraction of as much as 33% of the pulsed ^{14}C label with a mixture of chloroform and methanol. Its quantitative extraction was impossible, since when the polysaccharide chain length exceeded 8–10 sugar moieties, the complex with lipid became insoluble in the extraction mixture.

Han and Robyt proposed that BC biosynthesis involved three enzymes embedded in the cytoplasmic membrane: CS, lipid pyrophosphate: UDPGlc phosphotransferase (LipPP: UDPGlc-PT), and lipid pyrophosphate phosphohydrolase (LPP). The reaction mechanism, called by the authors the insertion reaction, is presented in the Figure 9.

The first enzyme transfers Glc-α-1-P from UDPGlc onto the lipid monophosphate (Lip-P), thus giving the lipid pyrophosphate-α-D-Glc (LipPP-α-Glc) (reaction 1, Figure 9). The α configuration on the anomeric carbon, involved in the Glc phosphoester bond, remains the same as in the substrate molecule. The second product of this reaction is UMP (according to Brown's model, UDP is released).

In the next reaction (reaction 2, Figure 9), catalyzed by CS, the glucose residue is transferred from one LipPP-α-Glc molecule onto another one and the β-1,4-glycosidic linkage between the two glucose residues is formed, due to the attack of the C-4 hydroxyl group of one of them onto C-1 hydroxyl group of the second Glc (from the second LipPP-α-Glc).

In the third reaction (reaction 3, Figure 9), hydrolysis of the lipid pyrophosphate formed in the previous step occurs and

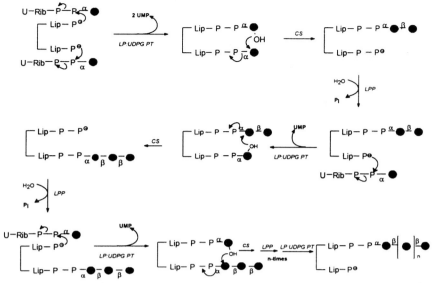

Fig. 9 Mechanism of BC biosynthesis involving lipid intermediates (Han and Robyt, 1998; with kind permission; modified). Lip, lipid; P, phosphoryl; Rib, D-ribose; U, uridine, ●, glucosyl residue; LP: UDPGPT, lipid pyrophosphate: UDPGlc-phosphotransferase; CS; cellulose synthase; LPP, lipid pyrophosphate phosphohydrolase.

another Glc-α-1P from UDPGlc can be attached to the LipP, released in this reaction. The cycle of these three reactions is continued to give the β-1,4-glucan chain of an appropriate length. The mechanism includes the inversion of a configuration on the anomeric carbon in the Glc residue transferred either from one LipPP-α-Glc molecule to another, or to cellobiose, -triose, -tetraose, and subsequently to n-ose. An acceptor of the elongated chain of β-1,4-glucan is always a single α-D-Glc residue, linked with one of two LipPP molecules present in the polymerizing system. It means that the chain elongation occurs from its reducing end and nonreducing ends of nascent cellulose chains are situated away from the cells; this is contradictory to Brown's model. According to Han and Robyt's model, both the two cooperating LipPP molecules and the CS complex, which is not a processing glycosyltransferase, are embedded in the cytoplasmic membrane of the bacterial cell.

Participation of a lipid intermediate in cellulose biosynthesis was confirmed for *Agrobacterium tumefaciens* (Matthysse et al., 1995), as well as for synthesis of *Salmonella* O-antigen polysaccharide (Bray and Robbins, 1967) and *Xanthomonas campestris* xanthan (Ielpi et al., 1981). Furthermore, the lipid intermediate is probably also involved in the synthesis of acetan (De Lannino et al., 1988), which is a soluble polysaccharide produced together with cellulose by numerous *A. xylinum* strains (see Section 7.6). Further studies have to be carried out to elucidate whether the lipid intermediate plays in cellulose formation, the role postulated by Han and Robyt (1998). On the contrary, Brown's hypothesis (1996, 2000) is confirmed by cellulose synthesis *in vitro*, catalyzed by *A. xylinum* membrane preparations solubilized with digitonin (Lin et al., 1985; Bureau and Brown, 1987), and by polymer synthesis by purified CS subunits (Lin and Brown, 1989), which did not contain the lipid component.

According to both proposed hypotheses, BC biosynthesis does not require any primer, in agreement with earlier deductions (Canale-Parda and Wolte, 1964).

7.3.2
Assembly and Crystallization of Cellulose Chains

Polymerization of glucopyranose molecules to the β-1,4-glucan, whatever the mechanism, is the least complicated stage of cellulose synthesis, also in bacteria. The unique structure and properties of cellulose, which are dependent on its origin, result from the course of further stages: it means extrusion of the chains and their assembly outside the cell, giving supramolecular structures. Most important in these processes is the specific organization of multimeric complexes of CS, which are anchored in cytoplasmic membrane. In *A. xylinum* cells these complexes are ordered linearly along the long axis of the cell and one cell contains 50–80 TCs (Brown et al., 1976), whereas in vascular plants TCs form characteristic, 6-fold symmetrical rosettes (Brown and Saxena, 2000).

Stronger aeration, e.g., during agitated *A. xylinum* cultures, or the presence of certain substances, that cannot penetrate inside the cells, but can form competitive hydrogen bonds with the β-1,4-glucan chains (e.g., carboxymethyl-(CM)-cellulose, fluorescent brightener Calcofluor white), bring about significant changes in the supramolecular organization of cellulose chains, and instead of the ribbon-like polymer, i.e. the metastable cellulose I (Figure 10), which fibrils are colinear to CS complexes and gathered in a row, the second, thermodynamically more stable allomorph amorphous cellulose II is formed. This phenomenon is accompanied by disorganization of the TC' linear order. Discussing this problem, Brown and Saxena (2000) state that the conditions-dependent form of the final product may somehow determine the TC' arrangement in the cytoplasmic membrane, though it is believed that in *A. xylinum* these complexes have a stationary character, as opposed to some algae (Kudlicka, 1989). Worth mentioning is the well-known phenomenon of *A. xylinum* cell motion during BC synthesis, in the direction opposite to cellulose chain extrusion (Brown et al., 1976). Reversal movement of *A. xylinum* cells enabled calculation

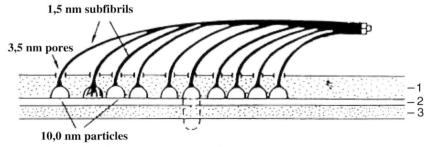

Fig. 10 Model of *A. xylinum* cellulose subfibrils formation: 1, lipopolysaccharide layer; 2, periplasmic space; 3, plasmalemma; 10 nm particles, CS subunits (Kudlicka, 1989; with kind permission).

of the typical chain elongation rate, which is equal to 2 µm min^{-1}, and corresponds to a polymerization of more than 10^8 glucose molecules into the β-1,4-glucan per hour and per single bacterial cell. Thus if forces associated with subfibrils assembly to bundles and ribbons, are strong enough to shift the motion of whole A. xylinum cells, these forces may as well influence the spatial arrangement of CS subunits, localized in the semi-liquid lipid bilayer of the membrane.

The spatial assembly of the β-1,4-glucan chains has a hierarchical character (Figure 10). In the first stage, 10–15 nascent β-1,4-glucan chains form a subfibril (also called a protomicrofibril), 1.5 nm in diameter, which is not a rod-like structure, but a left-handed triple helix (Ross et al., 1991). The subfibrils gather to form (also twisted) microfibrils composed of numerous parallel chains. The next structure in hierarchy is a bundle of microfibrils, followed by loosely wound ribbon, comprising about 1000 individual chains of the β-1,4-glucan (Haigler, 1985). In the presence of Calcofluor, which penetrates the outer membrane of the cell envelope, and forms complexes with protomicrofibrils, their aggregation to microfibrils is stopped. In the presence of much larger molecules such as CM-cellulose and xyloglucan (Yamamoto et al., 1996), which are too large to traverse the membrane, assembly of the bundles of microfibrils and ribbons is disrupted, thus indicating that formation of these two latter structures occurs exocellularly. Ross et al. (1991), who analyzed the mechanism of A. xylinum cellulose chain association, emphasized the important role of specific sites of adhesion in the inner and outer membranes of cell envelope, whose presence in bacterial cells was first reported by Bayer (1979). Their occurrence would enable an export of cellulose microfibrils without any interactions with the peptidoglycan gel, which fills the periplasmic space. Such interactions would prevent formation of correct hydrogen bonding between the protomicrofibrils.

The molecular organization of pores, through which A. xylinum cellulose microfibrils migrate outside the cell, remains unknown. The participation of expression products of the bcsC and bcsD genes, belonging to CS operon (see Section 7.4), in their formation cannot be excluded (Wong et al., 1990), all the more since these expression products appeared to be essential for cellulose synthesis in vivo (Ben-Bassat et al., 1993). Also the function of three proteins: CSAP20, 54, and 59 (CS-associated proteins), has to be scrutinized. They bind to cellulose together with the CS catalytic subunit, as revealed by experiments on CS purification (Benziman and Tal, 1995). It seems probable that isolation and purification of other components of the complex machinery responsible for cellulose synthesis, in concert with direct mutations of genes from A. xylinum CS operon, may help to elucidate the sequence of events during this process on the molecular level.

The process of assembly of nascent cellulose chains, not separated from parental A. xylinum cells, displays a unique and remarkable dynamics. Extrusion of these chains outside and their assembly into ordered structures is assisted by reversal of the motion of the cells. Throughout the whole process of the β-1,4-glucan synthesis and secretion, A. xylinum cell simultaneously turns around its own axis (uncoiling motion) (Brown et al., 1976). Also, this movement is driven by forming of twists of the ribbons, which are anchored in extrusion loci, localized on the side surface of the elongated cell. These twists are noticeable under the electron microscope.

7.4
Genetic Basis of Cellulose Biosynthesis

Cellulose biosynthesis involves several enzymes. Their function starts from the synthesis of the direct cellulose precursor UDPGlc, followed by polymerization of the monomer. It was found that the first reactions catalyzed by glucokinase, phosphoglucomutase and UDPGlc pyrophosphorylase, do not limit the final rate of BC synthesis, since A. xylinum synthesizes an excess of UDPGlc (Ben-Basat et al., 1993). The rate of BC synthesis is limited by diguanylate cyclase and oligomeric, regulatory complexes of CS. Therefore construction of more efficient cellulose-producing A. xylinum mutants, requires determination of organization of the genes coding for these enzymes. This has been studied by Wong et al. (1990) and Ben-Bassat et al. (1993), who proved that BC biosynthesis involves four coupled genes: bcsA (2261 bp), bcsB (2405 bp), bcsC (3956 bp), and bcsD (467 bp), forming the CS operon, which is 9217 bp in length, and is transcribed simultaneously as polycistron mRNA.

Although the function of the proteins encoded by each of these genes has not yet been precisely determined, some information on their role was derived from genetic complementation experiments (Wong et al., 1990). Results of these studies indicate that CS-deficient strains can be complemented by the gene bcsB, which codes for the 85-kDa protein (802 amino acid residues), and that A. xylinum mutants defective in both CS and diguanylate cyclase are complemented by the gene bcsA, coding for the 84-kDa protein (754 amino acid residues) (Ben-Bassat et al., 1993). Based on different mutations it was concluded that protein A takes part in the interaction of the CS complex with c-di-GMP, which is an allosteric activator of CS, earlier postulated by Saxena et al. (1991).

The protein B is capable of binding UDPGlc and synthesizing β-1,4-glucan chains. A gene coding for the catalytic subunit (cesA) and analogous to the gene bcsB (Wong et al., 1990; Ben-Bassat et al., 1993) was also described by Saxena et al. (1990a), who one year later identified the gene coding for the regulatory subunit (cesB). However, Saxena et al. (1990a, 1991) suppose that the position of the first two genes in the CS operon is opposite and that the first gene codes for the CS catalytic subunit.

The role of the expression products of both bcsC and bcsD genes has not been determined so precisely. These genes probably also code for proteins (molecular mass of 141 and 17 kDa, respectively), anchored in the membrane, and both essential for cellulose synthesis in vivo and for protein secretion from the cells. Ben-Bassat et al. (1993) reported that splitting or deletion of the bcsD gene markedly reduced synthesis of cellulose. The presence of genes, which govern the assembly of cellulose chains, in the CS operon was also predicted by Saxena et al. (1990a) and Ross et al. (1991).

It is not clear if plasmid DNA participates in BC synthesis and, if so, what its role is. However it was proven that the composition and size of plasmids detected in 60% of the non-reverting Cel⁻ mutants of the A. xylinum ATCC 10245 (currently 17005) strain, that were obtained by means of mutagenesis with N-methyl-N'-nitro-N-nitrosoguanidine, were markedly different from that of the wild strain. This observation suggests the plausible relationship between plasmid DNA structure and BC synthesis.

Mobile DNA fragments, including insertion sequences (IS), particularly widespread among prokaryotes, and 750–2500 bp in length (Galas and Chandler, 1989), are important factors of regulation of many genes. The ISs may activate or inactivate genes, and start DNA rearrangement such as deletions, inversions, and cointegrations. It was revealed that ISs participated in regulation of extracellular polysaccharides synthesis in *Pseudomonas atlantica* (Bartlett and Silverman, 1989), *Xanthomonas campestris* (Hotte et al., 1990), and *Zoogloea ramigera* (Easson et al., 1987). Their participation in BC synthesis regulation is also probable.

The best recognized insertion sequence in *A. xylinum* is the IS 1031 (950 bp, with known nucleotide sequence). Alteration in the IS 1031 profile was detected in the majority of Cel$^-$ mutants, in comparison to the wild strain. Some recombinants contained two or more IS 1031 fragments. Coucheron (1991) observed even more significant changes in the IS pattern, pointing to DNA rearrangement and tried to obtain revertants by splitting off the additional ISs from the inactivated gene and obtained pseudorevertants, which produced a wax-like substance on the surface of nutrient broth, but the capability of cellulose synthesis was not restored. These experiments indicate that the presence of ISs contributes to the genetic instability of *A. xylinum*.

7.5 Regulation of Bacterial Cellulose Synthesis

Cyclo-3,6′:3′6 diguanosine monophosphate (c-di-GMP, see Figure 11) is a reversible allosteric activator of *A. xylinum* CS and plays a crucial role in regulation of the whole β-1,4-glucan biogenesis (Ross et al., 1987).

This compound binds to the enzyme regulatory subunit, and induces conformational changes that facilitate association of the CS protomers and lead to the enhancement of its reactivity (Ross et al., 1987).

The concentration of c-di-GMP in *A. xylinum* cells is regulated by 3 enzymes: diguanyl cyclase (CDG), phosphodiesterase A (PDEA) and phosphodiesterase B (PDEB)

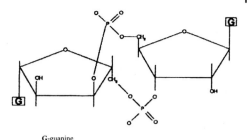

G-guanine

Fig. 11 Structure of c-di-GMP – the allosteric activator of *A. xylinum* CS.

(Figure 12) (Ross et al., 1987). PDEA and PDEB are anchored in the cytoplasmic membrane, and CDG has two forms. One of them is anchored in the cytoplasmic membrane and the second one operates in the cytoplasma (Ross et al., 1991). Recently, Weinhouse et al. (1997) reported on a new c-di-GMP-binding protein, which probably also participates in the intracellular regulation of free c-di-GMP concentration. Tal et al. (1998) proved that the cellular turnover of c-di-GMP in *A. xylinum* is controlled by three operons.

CDG, the key regulatory enzyme in the *A. xylinum* cellulose synthesis system, is composed of two polypeptide chains and encoded by two genes (Nichols and Singletary, 1998). CDG is activated by Mg^{2+} ions and specifically inhibited by saponin (Ohana et al., 1998), a glycosylated terpenoid. CDG converts two GTP molecules at first into the linear (pppGpG) and then into the cyclic (c-di-GMP) diguanosine monophosphate, which activates CS. PDEA splits active c-di-GMP into a linear inactive dimer (pGpG, di-GMP), further decomposed by PDEB into two molecules of 5′GMP (Figure 12). PDEA is inhibited by Ca^{2+} ions, which do not influence PDEB. Therefore, the Ca^{2+} concentration indirectly affects the rate of cellulose synthesis. High concentration of these ions enhances this rate, since the

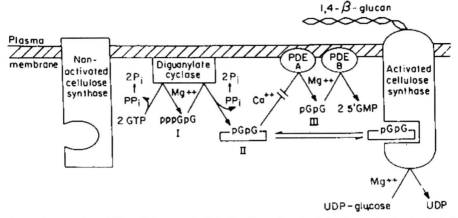

Fig. 12 Proposed model for cellulose synthesis in *A. xylinum*. For simplicity, the synthesis of a single β-1,4-glucan chain is depicted, although a more complex form of CS, polymerizing several chains simultaneously, might be the active enzyme unit in cellulose biogenesis (Ross et al., 1991; with kind permission). PDEA, phosphodiesterase A; PDEB, phosphodiesterase B; pppGpG, diphospho-di-GMP; pGpG (II), c-di-GMP (cyclic-di-GMP); pGpG, di-GMP (linear di-GMP).

conversion of c-di-GMP to the inactive linear form is inhibited.

It is believed that both the molecular background of the CS activation by c-di-GMP as well as the regulation of its synthesis, involving a positive influence of CDG, and a negative effect of PDEA and PDEB, have a unique character. To date, such a mechanism of cellulose synthesis regulation has only been detected in *A. xylinum*.

7.6
Soluble Polysaccharides Synthesized by *A. xylinum*

Apart from cellulose, some related soluble polysaccharides are also synthesized by *A. xylinum*. In 1970s, a soluble β-homoglucan was detected. Its main backbone was composed of β-1,4-linked glucose residues and every third glucose moiety of this chain was substituted with another glucose via β-1,2 linkage (Colvin and Leppard, 1977; Colvin et al., 1979). *A. xylinum* cellulose pellicle non-forming (Pel⁻) strains synthesize α-glucan linked with the cell envelope (Dekker et al., 1977). Valla and Kjosbakken (1981) isolated a soluble polysaccharide containing Glc, Man, Rha, and glucuronic acid residues (molar ratio of 3:1:1:1, respectively) from the culture broth of the Cel⁻ strain. The related polymer, composed of the same moieties (altered molar ratio of 6:2:1:1, respectively), synthesized by another Cel⁻ strain, was detected by Minakami et al. (1984). Probably some of the residues in these polymers are acetylated (Tayama et al., 1985). Polysaccharides of this type have received the common name acetan (De Lannino et al., 1988). Moreover, some wild *A. xylinum* strains synthesize a similar soluble polysaccharide (Glc, Man, Rha, and glucuronic acid; 6:1:1:1, respectively) together with cellulose (Savidge and Colvin, 1985).

The question is whether cellulose-assisting soluble polysaccharides play any role in its biogenesis. It seems that they do not play any direct role, although kinetics of their biosynthesis are the same as that of cellulose (Marx-Figini and Pion, 1974) and, moreover,

UDPGlc was preferentially used for production of these soluble saccharides instead of the production of cellulose (Delmer, 1982). Some authors state that *in vivo* nascent cellulose microfibrils are coated with amorphous material (Leppard et al., 1975), which may be composed of the above-mentioned soluble polysaccharides. The absence of these soluble polymers in the liquid fraction of static culture broths and their presence in the cellulose mat confirms that assumption (Valla and Kjosbakken, 1981).

Much earlier, Ben-Hayyim and Ohad revealed in 1965 that CM-cellulose present in the medium, used for BC synthesis, resulted in incorporation of this soluble derivative of cellulose into microfibrils of the polymer. However, incorporation of the *A. xylinum* soluble exopolysaccharides into cellulose microfibrils has not been observed.

7.7
Role of Endo- and Exocellulases Synthesized by *A. xylinum*

The first reports on *A. xylinum* cellulases, detected independently by Okamoto et al. (1994) and Standal et al. (1994), included description of their genes. The first authors selected from *A. xylinum* IFO 3288 DNA gene libraries one gene coding for a 24-kDa protein (218 amino acid residues), which displayed CM cellulase activity. Standal et al. (1994) found another endocellulase gene in the *A. xylinum* ATCC 23769 Cel⁻ mutant, localized upstream the CS operon. They revealed that the loss of cellulase synthesis capability resulted from gene splitting. The conclusions of Tonouchi et al. (1997), who investigated the localization of the *A. xylinum* BPR 2001 endo-1,4-β-glucanase gene, were similar. The latter gene was localized upstream the CS operon, whereas the gene of the second cellulolytic enzyme, produced by this strain, i.e. an exo-1,4-β-glucosidase, was found downstream this operon.

A. xylinum BPR 2001 endo- and exocellulase were purified and characterized (Oikawa et al., 1997; Tahara et al., 1998). The studies of Tahara et al. (1997) also revealed that at pH 5.0, optimal for growth and BC synthesis, the activity of both cellulases is several times higher than at pH 4.0, at which the BC production is only slightly declined. It was also observed that at pH 5.0, the average DP was decreasing from 16,800 to 11,000 with the time of cultivation, whereas at pH 4.0 these changes were negligible. The cellulose obtained at pH 5.0 had inferior physical properties, i.e. a lower tensile strength (a lower Young's modulus value). These results and colocalization of genes of cellulases and CS operon, suggest that the endo-1,4-β-glucanases and exo-1,4-β-glucosidase are involved in *A. xylinum* cellulose biogenesis (Tahara, 1998).

More particularly, it relates to possible degradation of the nascent β-glucan chains, synthesized in the later phase of growth. Furthermore, the *A. xylinum* BPR 2001 exo-1,4-β-glucosidase displays both hydrolytic and transglycosylating activity (Tahara et al., 1998), similarly to other glycosidases, which do not cause inversion of configuration on the anomeric carbon. The latter activity may be responsible *in vivo* for changes in the average DP of the polymer; however, further research is necessary to explain this hypothesis. Tentative evidence for such a possibility might be the behavior of soybean cells cultured *in vitro*. The activity of their β-glucosidase, which also has transglycosylating activity, increases together with the length of the cells, since this enzyme probably participates in the transfer of glycosidic residues of hemicelluloses precursors to the growing terminus of the cell wall, in which intensive synthesis of these polysaccharides is taking place (Nari et al., 1983).

Whether *A. xylinum* cellulases are involved in releasing the energy stored in the form of cellulose, in periods of starvation, is still a question to answer, although some studies confirm this thesis (Okamoto et al., 1994). It is known that some *Sclerotium* endo-1,3-β-glucanases play such a role (Jones et al., 1974).

Current observations indicate that modification of the physical properties and DP of BC can be achieved either by using compounds that influence its biosynthesis, or by exploiting the activity and specificity of *A. xylinum* cellulases.

8
Biodegradation of BC

Independent of its origin, cellulose undergoes total biodegradation in nature. However, in comparison to PC, associated both with polymers susceptible to degradation, like hemicelluloses and pectin, and with lignin, which is the most resistant plant polymer, BC is relatively pure and *a priori* more susceptible to attack by cellulolytic enzymes, which are produced mainly by fungi and numerous bacterial species. In this respect, commercial application of BC is friendly for the environment.

Furthermore, the structure and properties of BC (large accessible surface area) make it the superior model substrate for studies on cellulases (1,4-β-glucan 4-glucanohydrolases, formerly endocellulases, EC 3.2.1.4) and cellobiohydrolases (cellulose 1,4-β-cellobiosidases, EC 3.2.1.91), that are the main components of cellulosomes, the specialized multienzymatic particles from cellulolytic bacteria, as well as fungal systems for cellulose decomposition.

Recently, Boisset et al. (1999) proved that *Clostridium thermocellum* cellulosomes completely decompose *A. xylinum* cellulose microfibrils faster than microcrystalline *Valonia ventricosa* cellulose. Ultrastructural observations of the hydrolysis process, using transmission electron microscopy, infrared spectroscopy, and X-ray diffraction analysis, indicated that the rapid hydrolysis of BC resulted from very efficient synergistic action of the various enzymic components, present in the cellulosome scaffolding structure. Further studies of Boisset et al. (2000) revealed that *Humicola insolvens* cellobiohydrolase Cel7A (previously CBH I) brought about thinning of dispersed BC ribbons, whereas the mixture of Cel6A (CBH II) and Cel7A (in the ratio of 2:1, respectively) cut the ribbons to shorter pieces, thus suggesting the partly endo-manner of CBH II attack. The phenomenon of low inherent endoactivity of some exoglucanases has been known for several years and explains more efficient synergistic action of cellobiohydrolases in comparison to endoglucanases (Teeri, 1997). According to current opinions, the system of multiple exo- and endocellulases, both in bacterial cellulosomes, and in fungal cellulolytic complexes, represents the perfectly balanced continuum of activity, providing efficient degradation of various cellulosic substrates.

BC susceptibility to cellulases was also observed by Samejima et al. (1997), who found that the mixture of *Trichoderma viride* CBH I and endoglucanase II, drastically disintegrated the twisted and bent ribbon-like structure of microfibril bundles to linear needle-like microcrystallites, and also caused rapid polymer fragmentation. Transformation of the coiled structure of BC ribbons to the needle-like one is driven by the remarkable twisting motion of the substrate, which is probably a result of a tension – released when the microfibrils are being decomposed to shorter fragments by cellulases. Samejima et al. (1997) also detected that the acid-treated BC, containing many microfibril aggregates, was not susceptible to attack of both enzymes.

Similar results were achieved by Srisodsuk et al. (1998), who digested microcrystalline BC with *Trichoderma reesei* CBH I and observed rapid solubilization of the polymer, but slow decrease in its DP. The endocellulase I from this organism attacked the substrate in the opposite manner. Both enzymes hydrolyzed cotton cellulose more slowly than BC, though cotton cellulose exhibits relatively high purity as compared to other plants.

One of the reasons of the susceptibility of BC to the attack of cellulases is its large accessible surface area, that even increases throughout hydrolysis and facilitates effective binding of cellulolytic enzymes, the crucial step in the process of degradation. It was concluded based on studies of Palonen et al. (1999), who found that the number of *T. reesei* CBH I and CBH II molecules linked to BC, increased during hydrolysis. Furthermore, the CBH II binding was markedly stronger (the adsorption/desorption of the enzyme from its substrate was assisted by 60–70% hysteresis), pointing to a distinctly different processing character of this enzyme.

Stalbrandt et al. (1998) compared the mode of attack of four *Cellulomonas fimi* endo- and exocellulases against microcrystalline BC and acid-swollen Avicel cellulose. The latter was decomposed more effectively by all the enzymes. In most cases 45–65% solubilization and a decrease in DP were observed, except for Cbh B, which caused 27% solubilization. A high degree of bacterial polymer solubilization (67%) was achieved for Cbh A, known as the processing enzyme. Results of Stalbrandt et al. (1998) proved that only the external surface of BC fibrils is accessible for *C. fimi* cellulases. Since Cbh A is the processing enzyme, it can remove external fibrils much faster than the other three cellulases and attack the deeper internal surface of the polymer. Therefore, the other enzymes (two nonprocessing endocellulases and Cbh B, which displays weaker processing properties) could not solubilize BC so effectively, whereas amorphous soluble cellulose was equally available to all of them.

Current results indicate that complete digestion of the highly crystalline bacterial polymer will not cause any problems, all the more since organisms responsible for BC biodegradation in natural environments produce a multitude of various endo- and exocellulases, whose activities complement each other. Furthermore, as opposed to cellulose of plant origin, which requires pretreatment to provide easier enzymatic conversion to sugars, BC can be directly digested by cellulases.

9
Biotechnological Production

To achieve high productivity and yields of BC and to reduce cost of its production, special emphasis should be given to the following aspects:

- development of screening methods providing selection of *A. xylinum* strains, which can efficiently produce cellulose from various inexpensive waste carbon sources – wild strains derived from the screening could be improved using genetic engineering methods;
- optimization of *A. xylinum* culture conditions (static or agitated culture, fermentor type), determining both a form (a pellicle or an amorphous gel) and properties (resilience, elasticity, mechanical strength, absorbency) of BC, which have to be tailored to the further polymer application;

- optimization of culture broth composition (carbon and nitrogen sources, biosynthesis stimulating compounds, microelements, etc.) and process conditions (pH, temperature, aeration).

These are discussed in the next sections.

9.1
Isolation from Natural Sources and Improvement of BC-producing Strains

One of the methods enabling selection of proper *A. xylinum* strains, is the screening for strains, which cannot oxidize glucose via gluconic acid to 2-, 5-, or 2,5-ketogluconate (Winkelman and Clark, 1984; Johnson and Neogi, 1989; De Wulf et al., 1996; Vandamme, 1998). In this respect, De Wulf et al. (1996) applied an agar nutrient medium containing Br^- and BrO_3^- ions, that combine at acidic pH to release molecular bromine, toxic to *A. xylinum* cells. Only those mutants, which do not convert glucose into gluconate or its derivatives, can survive in this medium. Vandamme et al. (1998) successfully used this method to isolate *A. xylinum* KJ33 strain, producing 3.3 g cellulose L^{-1}. *A. xylinum* strains, which do not metabolize glucose to gluconic acids, were also selected on $CaCO_3$ containing medium (Johnson and Neogi, 1989). Another approach to avoid conversion of glucose to organic acids, was screening for mutants defective in glucose-6-phosphate dehydrogenase. The mentioned increase in final concentration, volumetric productivity and total cellulose yield, was achieved in agitated cultures. For the same purpose, i.e. selection of the best cellulose producers, preparations of cellulases, added into the growth medium, have also been used (Brown, 1989c).

Mutation with chemical compounds, such as *N*-methyl-*N'*-nitro-*N*-nitrosoguanidine or ethylmethanesulphate, as well as with ultraviolet irradiation gave mutants that displayed higher cellulose productivity, reduced synthesis of the soluble polysaccharide acetane, and a much lower degree of glucose conversion into organic acids (Johnson and Noegi, 1989).

Screening of *A. xylinum* strains was also aimed at isolation of spontaneous or induced cellulose II-producing mutants. This cellulose is synthesized by bacteria that form smooth colonies and do not produce a pellicle on the surface of the nutrient broth. Instead, the polymer is dispersed in the whole volume of the medium.

Selection of the best BC producers has been also performed traditionally, by determination of the synthetic activity of a plethora of individual monocultures. For example, Toyosaki et al. (1995) isolated 2096 *Acetobacter* strains from a large number of plant samples (fruits, flowers) and 412 strains forming a mat on the surface of nutrient medium were subjected to further investigations. The procedure yielded *A. xylinum* ssp. sucrofermentans BPR2001 – the strain which efficiently synthesized BC in agitated culture.

9.1.1
Improvement of Cellulose-producing Strains by Genetic Engineering

Expression of genes coding for various carbohydrases from organisms other than *A. xylinum* enhances the scope of carbon sources available, that may increase BC yield and decrease the nutrient broth price. An example is the expression of sucrose phosphorylase (EC 2.4.1.7) gene from *Leuconostoc mesenteroides* in *Acetobacter* sp. (Tonouchi et al., 1998), that resulted in utilization of sucrose as a carbon source and increased cellulose production. In addition, Nakai et al. (1999) obtained double cellulose yield by means of expression of the mutant sucrose synthase (EC 2.4.1.13) gene from mung bean (*Vigna radiata*, Wilczek) in *A. xylinum* strain. In the mutant enzyme, the eleventh serine residue was replaced by glutamic acid (S11E); this caused higher affinity of the engineered enzyme toward sucrose and favored cleavage of this sugar for the synthesis of UDPGlc. Introduction of the mutant gene into *A. xylinum* not only changed sucrose metabolism in the recombinant strain, but also created a new metabolic pathway of direct UDPGlc synthesis (Figure 13). The energy driving this process is derived only from the cleavage of the glycosidic bond in sucrose without any other energy input required, which is the case when UDPGlc is synthesized via the conventional pathway. Furthermore, the UDP molecules resulting from glucose polymerization process can be directly recycled by the sucrose synthase activity, which increases the rate of the coupled reactions and a higher rate of cellulose synthesis is observed. It should be emphasized that higher plants have both systems of UDPGlc synthesis.

9.2
Fermentation Production

Growth on the surface of liquid media and synthesis of the gelatinous, leather-like mat, are natural properties of *A. xylinum* strains. Therefore static conditions seemed to be optimal for BC synthesis. However, the main drawbacks of static culture, such as the synthesis of the polymer only in the form of a sheet and relatively low productivity, have contributed to the development of new fermentation processes.

9.2.1
Carbon and Nitrogen Sources

In studies on factors affecting the BC production yield, much attention has been paid to carbon sources. Numerous mono-, di-, and polysaccharides, alcohols, organic acids, and other compounds were compared by Jonas and Farah (1998), who found out that the preferred carbon sources were D-arabitol and D-mannitol, which resulted in a 6.2- or 3.8-fold higher cellulose production, respectively, in comparison to glucose. Both sugar alcohols provided stabilization of the pH throughout the culture period, since they were not converted to gluconic acids.

Tonouchi et al. (1996), who used a strain of *A. xylinum* to obtain cellulose from glucose and fructose, found that fructose stimulated the activities of phosphoglucose isomerase and UDPGlc pyrophosphorylase, thus enhancing cellulose yield.

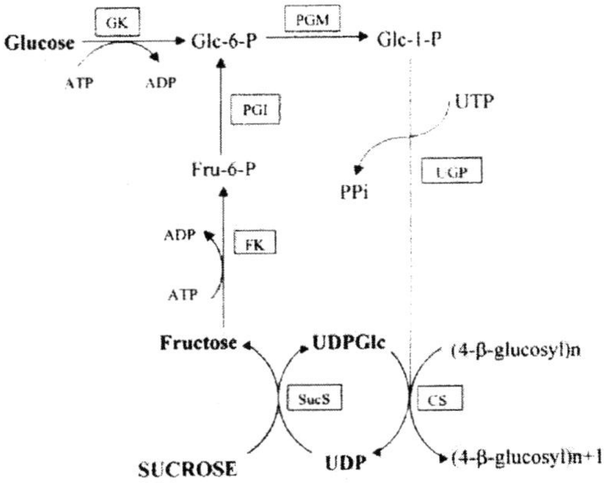

Fig. 13 UDPGlc biosynthesis pathways in a recombinant strain of *A. xylinum* containing the sucrose synthase (SucS) gene. Other abbreviations are the same as in Figure 6 (see Section 7.1).

When maltose was the sole carbon source in the nutrient medium (Masaoka, 1993) BC production was even 10 times lower than in glucose-containing medium and the polymer was markedly shorter (DP = 4000–5000) than in the presence of glucose (DP = 11,500).

Matsuoka et al. (1996) investigated BC synthesis by *A. xylinum* ssp. sucrofermentous BPR2001 in agitated culture and found out that the presence of lactate in the growth medium stimulated bacterial growth and enhanced the cellulose yield 4–5 times. The source of lactate is usually corn steep liquor (CSL), which is one of the main medium components applied for BC production, especially in agitated cultures. Contrary to glucose and fructose, lactate is not converted to UDPGlc, but is metabolized via pyruvate and oxalacetate in the Krebs cycle, being a source of energy for cellulose production. To provide lactate in the growth medium, cultivation of mixed cultures of lactic acid and acetic acid bacteria has been carried out. The best results gave strains of *Lactobacillus*, *Leuconostoc*, *Pediococcus*, and *Streptococcus*. The lactic and acetic acid bacteria were grown also in the presence of sucrose-hydrolyzing *Saccharomyces* yeast (β-fructofuranosidase producers). This principle provided a high BC yield, since after 14 days of agitated culture, as much as 8.1 g of cellulose L^{-1} was obtained, when in the absence of *Lactobacillus* strain, only 6.4 g L^{-1} (Seto et al., 1997).

One of the cellulose synthesis-stimulating compounds appeared to be ethanol (Naritomi et al., 1998). Continuous culture of *A. xylinum* in fructose-containing medium, revealed that 10 g ethanol L^{-1} enhanced a BC yield, but a concentration of 15 g L^{-1} prevented polymer synthesis. These results suggest that similarly to lactate, ethanol is also a source of energy (accumulated as ATP) and not a substrate for cellulose synthesis. ATP activates fructose kinase and inhibits glucose-6-phosphate dehydrogenase, thus halting conversion of 6-P-glucose to 6-phosphogluconate. It can be concluded that – due to these coupled reactions – the amount of fructose further isomerized to glucose-6-phosphate is increased.

Each cellulose-producing strain requires a specific complex nitrogen source, providing not only amino acids, but vitamins and mineral salts as well. These requirements are met with yeast extract, CSL as well as hydrolyzates of casein and other proteins. The preferred nitrogen sources are yeast extract and peptone, which are basic components of the model medium developed by Hestrin and Schramm (1954; H-S medium), applied in numerous studies on BC synthesis. However the most recommended nitrogen source for agitated cultures is CSL (Johnson and Noegi, 1989).

It was also found that a significant part of the expensive medium components, i.e., yeast extract and bactopeptone, can be replaced with CSL or even white cabbage juice. Also waste plant materials such as sugar beet molasses, spent liquors after glucose separation from starch hydrolyzates, as well as whey and some fermentation industry wastes (e.g., spent liquors after dextran precipitation with ethanol) were appropriate medium components (Krystynowicz et al., 2000).

Studies on the influence of vitamins on BC synthesis (Ishikawa et al., 1995, 1996b) revealed that the most stimulating ones were pyridoxine, nicotinic acid, *p*-aminobenzoic acid and biotin, and even CSL media should be fortified with these substances. Some other compounds, strongly stimulating cellulose production by *A. xylinum* strains, such as derivatives of choline, betaine, and fatty acids (salts and esters), were also identified (Hikawu et al., 1996).

High BC production (up to 25 g L^{-1}) can be achieved using optimum growth medium composition, designed by mathematical methods and computer analysis (Joris et al., 1990; Embuscado et al., 1994; Vandamme et al., 1998; Galas et al., 1999).

9.2.2
Effect of pH and Temperature

Analysis of the influence of pH on *A. xylinum* cellulose yield and properties, indicates that optimum pH depends on the strain, and usually varies between 4.0 and 7.0 (Johnson and Neogi, 1989; Galas et al., 1999). For instance, Ishikawa et al. (1996a) and Tahara et al. (1997), who applied two different *A. xylinum* strains, observed the highest polymer yield at pH 5.0. The same pH was found to be optimal by Krystynowicz et al. (1997). Comparison of adsorptive properties of the polymer obtained at different pH, learned that cellulose, accumulated in the S-H medium pH 4.8–6.0, displayed the highest water-binding capacity (Wlochowicz, 2001).

In addition to the pH of the nutrient broth, also the temperature influences BC yield and properties. In majority of reported experiments, the temperature ranged from 28 to 30 °C, and its variations caused changes of cellulose DP and water-binding capacity. For instance, BC synthesized at 30 °C had a lower DP (approximately 10,000) and a higher water-binding capacity (164%) in comparison to that produced at 25 and 35 °C (Wlochowicz, 2001).

9.2.3
Static and Agitated Cultures; Fermentor Types

Synthesis of BC is run either in static culture or in submerged conditions, providing proper agitation and aeration, necessary for medium homogeneity and effective mass transfer. The choice of culture conditions strictly depends on the polymer use and destination.

BC yield in static cultures is mostly dependent on the surface/volume ratio. Optimum surface/volume ratio protects from either too high (unnecessary) or too low aeration (cell growth and BC synthesis termination). Reported values of surface:volume ratio vary from 2.2 cm^{-1} (Joris et al., 1990) to 0.7 cm^{-1} (Krystynowicz et al., 1997). BC synthesis in static conditions can be achieved either in a one-step (medium inoculated with 5–10% cell suspension) or a two-step procedure. The latter one starts from agitated fermentation, followed by the static culture (Okiyama et al., 1992). Krystynowicz et al. (1997) modified the two-step procedure and applied two consecutive static cultures. Pellicles containing entrapped *A. xylinum* E$_{25}$ cells, obtained in the first step (24 h), were cut to uniform pieces and used as an inoculum to start the next culture, run for 4–5 days. The method provided uniform bacterial growth and production of homogeneous pellicles.

The control of BC synthesis in static culture is very difficult since the pellicle limits access to the liquid medium. A particularly important parameter, which requires continuous control, is the pH. Accumulation of keto-gluconic acids in the culture broth brings about a decrease in pH far below the value that is optimum for growth and polysaccharide synthesis. Because conventional methods of pH adjustment cannot be used in static cultures, Vandamme et al. (1998) applied an *in situ* pH control via an optimized fermentation medium design, based on introducing acetic acid as an additional substrate for *Acetobacter* sp. LMG1518. Products of acetic acid catabolism counteracted the pH decrease caused by keto-gluconate formation, and provided a constant pH of 5.5 of the growth medium, throughout the whole process.

Fig. 14 *A. xylinum* cellulose formed on the surface of a roller in horizontal fermentor.

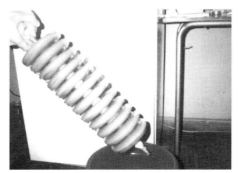

Fig. 15 *A. xylinum* cellulose deposited on disks in rotating disk fermentor.

Better control of BC synthesis can be achieved in special fermentors. Cellulose production in horizontal fermentors provides a combination of stationary and submerged cultures. The polymer is deposited on the surface of rollers or disks, rotating around the long axis (Figure 14). Part of their surface temporarily dips in the liquid medium or is above its surface (in the air). The advantages of this method include a larger polymer surface, synthesis of cellulose in a form of hollow fibers, different in diameter, as well as good process control, easy scale up, appropriate accessible surface for adhesion of bacteria and product deposition, better substrate utilization, higher rate of cellulose production, possibility of nutrient supply, additional aeration during the process, etc. (Sattler and Fiedler, 1990). Bungay and Serafica (1997) produced BC in a 1-L disk (12 cm in diameter) fermentor and revealed that optimum sugar (sucrose or glucose) concentration was 10 g L^{-1}, disks rotation rate 12 r.p.m., and constant pH 5.0. Krystynowicz et al. (1997) successfully applied a 11-L disk reactor containing 3 l of the H-S medium for BC synthesis by the *A. xylinum* E_{25} strain (Figure 15). The optimum rate of rotation of its 11 disks (12 cm in diameter each) appeared to be 3 r.p.m. As much as 4.2 g L^{-1} of dry cellulose was obtained after 7 d of growth.

BC production in submerged cultures usually requires common fermentors, equipped with some static parts (baffles, blades, etc.), enabling cell adhesion, since cellulose synthesis in the free aqueous phase usually drops. In analogy, various water-insoluble microparticles, such as sand, diatomaceous earth, or glass beads, added to the culture medium, enhanced BC productivity, since the biofilm formed by bacteria on the particles probably limited oxygen transfer and stopped glucose oxidation to gluconic acids (Vandamme et al., 1998).

Large-scale cellulose production, in fermentors with continuous agitation and aeration, encounters many problems, including spontaneous appearance of Cel$^-$ mutants, which contribute to a decline in cellulose production. Optimized agitation and aeration prevent turbulence, which negatively effects cellulose polymerization and crystallization, thus reducing the polysaccharide yield. For instance, a rate of agitation equal to 60 r.p.m. and aeration of 0.6 volume per volume per minute (v.v.m.) were optimum for the *A. aceti* ssp. xylinum ATCC 2178 strain cultured in a 300-L fermentor for 45 h at 30 °C, and $10 \text{ g BC L}^{-1} \text{ d}^{-1}$ was obtained (Laboureur, 1988). Ben-Bassat et al. (1989) investigated BC production by the *A. xylinum* 1306–21 strain in 14-L Chemap fermentor, cultured in CSL medium containing glucose and CSL, 2% of each. The rate of agitation was equal to 900 r.p.m. and the dissolved oxygen concentration corresponded to 30% air saturation. The polymer yield amounted to 5.1 g $L^{-1} d^{-1}$. Chao et al. (2000) applied a 50-L internal loop airlift reactor for BC synthesis with the A. xylinum sp. BPR2001 strain. Aeration with oxygen-enriched air enlarged cellulose yield from 3.8 to 8 g L^{-1} after 67 h.

Production of BC in fermentors encounters similar agitation problems as cultivation of fungi or streptomycetes, since most of these organisms grow in pellet or filament form and their culture broths become non-Newtonian fluids. A high concentration of mycelium in the form of a dense suspension of diversely shaped particles limits agitation and gas transfer. Increasing the rate of agitation in order to improve aeration leads to damage to the product structure due to shearing forces. Bauer et al. (1992) took this aspect into consideration and introduced a polyacrylamide-protecting agent into the growth medium, applied for BC synthesis in fermentors. The protector reduced the shear damage and affected the specific productivity at high densities negatively since the cell growth rate was lower.

Accumulation of some metabolites can also affect production of cellulose. For instance, studies of Kouda et al. (1998) revealed that a high partial pressure of CO_2 negatively affected A. xylinum growth and reduced BC yield.

Laboureur (1998) developed a method for BC production by A. xylinum sp. ATCC 21780 strain in 300- and 500-L fermentors in a medium containing 5% sucrose, 0.05% yeast extract, 0.2% citric acid, nitrogen salts, Mg^{2+}, and phosphates. The medium was inoculated with 12% of inoculum, cultivated for 160 h. The process was run for 45 h at 30 °C and pH 4.8, and an aeration rate of 0.6 v.v.m. BC synthesis yield reached 18 g L^{-1} (10 g $L^{-1} d^{-1}$). Another strain, Acetobacter sp. ATCC 8303, was grown at 28 °C and pH 4.6, in a modified medium, containing 0.28% glucose, 0.07% maltose, 0.03% CSL, and 0.03% yeast extract. The aeration was 1 v.v.m. for the first 33 h and 0.5 v.v.m. for the last 12 h of the process, and the BC yield was 13 g $L^{-1} d^{-1}$.

Preparing an inoculum of appropriate cell density for large-volume fermentors is also a problem, mainly because the cells are entrapped in cellulose. To liberate the cells and increase their density, Brown (1989c) applied preparations of cellulases for partial cellulose hydrolysis. In the presence of these enzymes, the cell density reached 10^8 mL^{-1}, whereas in their absence it was only $1.12 \cdot 10^7$ mL^{-1}, after 30 h of growth.

In addition to the above-mentioned methods, the production of cellulose in the form of hollow fibrils of various diameters has also been tried (Yamanaka et al., 1990). Such material could be especially useful for the production of small-diameter blood vessels. The hollow BC fibril is obtained by growing the BC-producing bacteria on the inner and/ or outer surface of an oxygen-permeable hollow carrier, produced from, for example, cellophane, Teflon, silicone, ceramics, etc.

9.2.4
Continuous Cultivation

Microbial cellulose can be also produced in a static continuous culture (Sakair et al., 1998). A. xylinum strain was cultured on trays, in the S-H medium, and after 2 d the pellicle produced on the surface was picked up, passed through a sodium dodecylsulfate bath to kill the cells, and set on a winding roller. The process was continued for a couple of weeks at a winding rate of 35 mm h^{-1} and fresh S-H medium was added into the trays every 8–12 h to keep its optimum level. A cellulose filament of more than 5 m was collected using this method, indicating its industrial application potential.

9.3
In vitro Biosynthesis

In vitro synthesis of cellulose has been one of the most difficult topics in the area of cellulose research. For the first time, cellulose was synthesized utilizing the 'reversed action' of cellulase as a catalyst, by Kobayashi et al. (1991), even though chemical synthetic methods have been applied before. Several monomers and catalysts have been used (Nakatsubo et al., 1989), but none of them gave the desired product, the stereoregular polymer of β-1,4-linked glucopyranoses. The idea of BC synthesis in vitro originated from experiments using *A. xylinum* cell-free extracts (Glaser, 1958), raw preparations of membranes (Colvin, 1980; Swissa et al., 1980; Aloni et al., 1982), and membranes solubilized with digitonin (Aloni et al., 1983). Such investigations proved that UDPGlc is the substrate for *A. xylinum* cellulose synthesis, c-di-GMP is the specific activator (Ross et al., 1987), and Ca^{2+} and Mg^{2+} ions are essential for this process (Swissa et al., 1980). Synthesis in vitro, carried out using *A. xylinum* membranes solubilized with digitonin (Lin et al., 1985), gave cellulose fibers 17 ± 2 Å in diameter. Their morphology and size resembled *A. xylinum* cellulose fibrils formed in the presence of factors disturbing crystallization of the nascent polymer (see Section 7.3.2). Further research, including X-ray diffraction analysis (Bureau and Brown, 1987) of cellulose produced by the *A. xylinum* membrane fractions, revealed that synthesis in vitro led to the amorphous cellulose II allomorph, whereas in vivo the highly crystalline cellulose I was obtained. Lin and Brown (1989) proved that purified CS also catalyzed cellulose synthesis in vitro. All these experiments were run on the microscale level and it is hard to believe today that commercial in vitro production of microbial polymer is possible, all the more since in the 1990s no progress in this area has been made.

9.4
Chemo-enzymatic Synthesis

Experiments on the chemical synthesis of cellulose have not given the expected results. Branched and low-molecular-weight glucans (Husemann and Muller, 1966) or polymers of β-1,4- and α-1,4-linked glucopyranoses (Micheel and Brodde, 1974) were obtained.

Successful experiments of Kobayashi et al. (1991) indicate that using coupled chemical and enzymatic methods may lead to the development of commercial in vitro synthesis of cellulose. The authors applied β-D-cellobiosyl fluoride, obtained by means of chemical synthesis, as a substrate and *T. reesei* cellulase as a catalyst in an aqueous–organic solvent system (a mixture of acetic buffer pH 5.0 and acetonitrile, 1:5 v/v) and achieved water-insoluble 'synthetic cellulose' having a DP $>$ 22. X-ray and ^{13}C-NMR analyzes revealed that the product was the cellulose II. The authors stated that the DP of the product depends on reaction conditions, especially on the aqueous–organic solvent composition. They found that in the presence of higher concentrations of the substrate or acetonitrile, the cellulase – which acts as a glycosyltransferase – produces mainly soluble cellooligosaccharides (DP $\leq$ 8), which may also find numerous uses. Although these studies have not been continued, the method proposed by Kobayashi might be a promising alternative for total enzymatic in vitro cellulose production.

9.5
Production Processes Expected to be Applied in the Future

Recently, an original method of BC production on a large-scale was proposed by Nichols and Singletary (1998), who patented the idea of the construction of transgenic plants capable of this polymer synthesis. The authors intend to express three *A. xylinum* CS genes (*bcsA*, *bcsB*, and *bcsC*, see Section 7.4) and two *A. xylinum* diguanyl cyclase genes in storage tissues (roots, tubers, grains) of crops (potato, maize, oat, sorghum, millet, wheat, rice, sugarcane, etc.). They suppose to obtain masses of the pure polymer, easy and cheap to recover. According to their method, its production would be less expensive than by means of fermentation. The authors also emphasize the ecological aspect of their procedure, since an additional benefit of breeding of pure cellulose-producing transgenic plants would be a more economical use of forest resources.

9.6
Recovery and Purification

Microbial cellulose obtained via stationary or agitated culture is not completely pure and contains some impurities, such as culture broth components and whole *A. xylinum* cells. Prior to use in medicine, food production, or even the papermaking industry, all these impurities must be removed.

One of the most widely used purification methods is based on treatment of BC with a solution of hydroxides (mainly sodium and potassium), sodium chlorate or hypochlorate, H_2O_2, diluted acids, organic solvents, or hot water. The reagents can be used alone or in combination (Yamanaka et al., 1990). Immersing BC in such solutions (for 14–18 h, in some cases up to 24 h) at elevated temperature (55–65 °C) markedly reduces the number of cells and coloration degree.

Boiling in 2% NaOH solution after preliminary rinsing with running tap water was also reported (Yamanaka et al., 1989). Watanabe et al. (1998a) immersed the polymer in 0.1 NaOH at 80 °C for 20 min and then washed it with distilled water. Takai et al. (1997) treated BC with distilled water and 2% NaOH, and neutralized the mat with 2% acetic acid.

Krystynowicz et al. (1997) developed a procedure starting from washing crude BC in running tap water (overnight), followed by boiling in 1% NaOH solution for 2 h, washing in tap water to accomplish NaOH removal (1 d), neutralization with 5% acetic acid, and its removal with tap water. The final BC preparations obtained contained less than 3% protein, and were suitable for certain food and medical purposes.

Medical application of BC requires special procedures to remove bacterial cells and toxins, which can cause pyrogenic reactions. One of the most effective protocols begins with gentle pressing of the cellulose pellicle between absorbent sheets to expel about 80% of the liquid phase and then immersing the mat in 3% NaOH for 12 h. This procedure is repeated 3 times, and after that the pellicle is incubated in 3% HCl solution, pressed, and thoroughly washed in distilled water. The purified pellicle is sterilized in an autoclave or by ^{60}Co irradiation. It performs excellently as a wound dressing since it contains only 1–50 ng of lipopolysaccharide endotoxins, whereas BC purified using conventional methods usually contains 30 µg or more of these substances (Ring et al., 1986).

10
Properties

BC is an extremely insoluble, resilient, and elastic polymer, having high tensile strength. It has a reticulated structure, in which numerous ribbon-shaped fibrils are composed of highly crystalline and highly uniaxially oriented cellulose subfibrils. This three-dimensional structure, not found in plant-originating cellulose, brings about a higher crystallinity index (60–70%) of BC. Furthermore, fibers of the plant polymer are about 100 times thicker than BC microfibrils and therefore the bacterial polymer has an about 200 times larger accessible surface. In comparison to *A. xylinum* cellulose from agitated cultures, the polymer synthesized under static conditions has a higher DP (14,400 and 10,900, respectively), crystallinity index (71 and 63%, respectively), tensile strength (Young's modulus of 33.3 and 28.3, respectively), but lower water holding-capacity (45 and 170 g water g BC^{-1}, respectively) and suspension viscosity (0.04 and 0.52 Pa·s, respectively) in its disintegrated form (Watanabe et al., 1998a).

Microbial cellulose appears to be gelatinous, since its liquid component (usually water), present in voids among very fine ribbons, amounts to at least 95% by weight. The bacterial polysaccharide has a high water-binding capacity, but the majority of the water is not bound to the polymer and it can be squeezed out by gentle pressing. Drying of BC leads to paper-like sheets, having a thickness of 0.01–0.5 mm (Yamanaka et al., 1990; Krystynowicz et al., 1995, 1997) and good absorptive properties. In addition to a high Young's modulus, BC also displays high sonic velocity and, because of these unique mechanical properties, it can be applied as acoustic membranes (Vandamme et al., 1998).

The properties of BC can be modified either during its synthesis or when the culture is completed. Some compounds, like cellulose derivatives, sulfonic acids, alkylphosphates, or other polysaccharides (starch, dextran), introduced into the nutrient broth alter the macroscopic morphology, the tensile strength, the optical density, and the absorptive properties of the final product (Yamanaka et al., 2000). BC can also be combined with other substances, added either to wet or dried cellulose, thus giving composites of desired physicochemical properties (Yamanaka, 1990). The auxiliary materials used for this purpose comprise granules and fibers of various inorganic and organic compounds, such as alumina, glass, agar, alginates, carrageenan, pullulan, dextran, polyacrylamide, heparin, polyhydroxylalcohols, gelatin, collagen, etc. They are combined with BC sheets by impregnation, lamination, or adsorption, or mixed with the disintegrated polymer. The composites are further subjected to a shaping treatment, thus giving various products.

Recently, Kim et al. (1999) reported an enzymatic method of BC modification using *L. mesenteroides* dextransucrase and alternansucrase. In their presence *A. xylinum* ATCC 10821 synthesized 'soluble cellulose', which was a glucan composed of 1,4-, 1,6-, and 1,3-linked monomers.

The basic BC properties are summarized in Table 2.

11
Applications

BC belongs to the generally recognized as safe (GRAS) polysaccharides and therefore it has already been put to a multitude of different uses. Commercial application of this polymer results from its unique properties and developments in effective technologies of production, based on growth of improved bacterial strains on cheap waste materials. The advantage of BC is its chem-

11 Applications

Tab. 2 Properties of BC

High purity
High degree of crystallinity
Greater surface area than that of conventional wood pulp
Sheet density from 300 to 900 kg m^{-3}
High tensile strength even at low sheet density (below 500 kg m^{-3})
High absorbency
High water-binding capacity
High elasticity, resilience, and durability
Nontoxicity
Metabolic inertness
Biocompatibility
Susceptibility to biodegradation
Good shape retention
Easy tailoring of physicochemical properties

ical purity and the absence of substances usually occurring in the plant polysaccharide, which requires laborious purification. In addition to the shape of BC sheets, their area and thickness can be tailored by means of culture conditions. Relatively easy BC modification during its biosynthesis enables regulation of such properties as molecular mass, elasticity, resilience, water-holding capacity, crystallinity index, etc. BC microfibrils may bind both low-molecular-weight substances and polymers, added for instance to the growth medium, thus giving novel commodities. BC can be also a raw material for further chemical modifications.

Based on the assumption that as much as 10,000 kg of the bacterial polymer can be obtained per year from static culture having 1 hectare surface area and only 600 kg of cotton is harvested in the same period of time from a field of the same area, the perspectives for a wider use of BC become more apparent (Kudlicka, 1989).

The main potential BC applications are summarized in the Table 3.

11.1
Technical Applications

Compared to PC sheets, BC has satisfactory tensile strength even at low sheet density (300–500 kg m^{-3}) (Johnson and Neogi,

Tab. 3 Applications of bacterial cellulose

Sector	Application
Cosmetics	Stabilizer of emulsions such as creams, tonics, nail conditioners, and polishes; component of artificial nails
Textile industry	Artificial skin and textiles; highly adsorptive materials
Tourism and sports	Sport clothes, tents, and camping equipment
Mining and refinery waste treatment	Spilt oils collecting sponges, materials for toxins adsorption, and recycling of minerals and oils
Sewage purification	Municipal sewage purification and water ultrafiltration
Broadcasting	Sensitive diaphragms for microphones and stereo headphones
Forestry	Artificial wood replacer, multi-layer plywood and heavy-duty containers
Paper industry	Specialty papers, archival documents repair, more durable banknotes, diapers, and napkins;
Machine industry	Car bodies, airplane parts, and sealing of cracks in rocket casings
Food production	Edible cellulose and 'nata de coco'
Medicine	Temporary artificial skin for therapy of decubitus, burns and ulcers; component of dental and arterial implants
Laboratory/research	Immobilization of proteins, chromatographic techniques, and medium component of *in vitro* tissue cultures

1989). Therefore, BC is an excellent component of papers, providing better mechanical properties. Microfibrils of the bacterial polymer form a great number of hydrogen bonds when the paper is subjected to drying, thus giving improved chemical adhesion and tensile strength. BC-containing paper not only shows better retention of solid additives, such as fillers and pigments, but is also more elastic, air permeable, resistant to tearing and bursting forces, and binds more water (Iguchi et al., 2000).

A beneficial effect such as an improved aging resistance was achieved by adding small amounts of BC to cotton fibers to obtain hand-made paper, used for old document repair. The paper displayed appropriate ink receptivity and specific snap (Krystynowicz et al., 1997). Paper containing 1% of BC meets the international standard ISO 9706:1994 for information and documentation papers, and also has a specific snap comparable to that of rag paper. It is possible to use BC for pressboard making, as well as production of paperboard used as an electroinsulating material and for bookbinding.

BC can also be used as a surface coating for specialty papers. For this purpose, a filtered suspension of BC (homogenized and mixed with other components) is added using special applicators, either within wet sheet formation, or to a partially or completely dried sheet. The coating improves properties such as gloss, brightness, smoothness, porosity, ink receptivity, and tensile strength. Other additives like starch, organic polymers, including CM-cellulose, organic, or inorganic pigments may also be used. Johnson and Neogi (1989) claim that paper coated with 3% of BC (solid matter) displays gloss properties and surface strength similar to rotogravure paper having 20% traditional coating. The authors also state that coating with a mixture of BC and CM-cellulose gives even better properties, since they act synergistically.

BC is also a valuable component of synthetic paper (Iguchi et al., 2000) since nonpolar polypropylene and polyethylene fibers, providing insulation, heat resistance, and fire-retarding properties, cannot form hydrogen bonds. The amount of wood pulp in this type of paper is usually from 20 to 50% to achieve good quality. Using BC enables us to decrease the amount of the additives without any effect on the synthetic paper properties.

BC also appeared to be a good binder in nonwoven fabric-like products (Yamanaka et al., 1990) commonly used in surgical drapes and gowns, and containing various hydrophilic and hydrophobic, natural and synthetic fibers, such as cellulose esters, polyolefin, nylon, acrylic glass, or metal fibers. Even small amounts of BC improve tensile and tear strength of the fabric, e.g. 10% of the bacterial polymer is equivalent to 20–30% of latex binder.

The scope of BC uses can be even wider since it is modifiable during synthesis (Brown, 1989a,b). For this purpose, CM-cellulose or copolymers of saccharides and dicarboxylic acids are added directly to the culture medium (Yamanaka et al., 1989). Cellulose obtained in the presence of CM-cellulose and dried with organic solvents has a resilient and elastic character, as well as higher water binding capacity (adsorbs faster more water). The optimum CM-cellulose concentrations range from 0.1 to 5% (w/v). The polarity of the organic solvent also influences the BC features. After treatment with acetone the polymer is elastic and rubber-like, whereas after drying with absolute ethanol, BC resembles leather.

Since BC is susceptible to enzymatic digestion, it is modified to obtain various composites of satisfactory biodegradability and strength. Their production is based either on chemical reaction between cellulose and the copolymer or on culturing the BC-producing strain in the copolymer-containing medium.

11.2
Medical Applications

A cellulose pad from a static culture is a ready-to-use, naturally 'woven' wound dressing material that meets the standards of modern wound dressings (Figure 16). It is sterilizable, biocompatible, porous, elastic, easy to handle and store, adsorbs exudation, provides optimum humidity, which is essential for fast wound healing, protects from secondary infection and mechanical injury, does not stick to the newly regenerated tissue, and alleviates pain by heat adsorption from burns. BC sheets are also excellent carriers for immobilization of medicinal preparations, which speed up the healing process.

Since BC pellicles can have various dimensions, it is relatively easy to produce dressings for extensive wounds. Because of recent problems with products of animal origin, collagen dressings can be replaced with BC ones. This thesis is additionally supported by undoubtedly positive results of clinical tests. For instance, the BC preparation, Prima Cel™, produced by Xylos Corp., according to the Rensselaer Polytechnic Institute (USA), has been applied as a wound dressing in clinical tests to heal ulcers. The results obtained were satisfactory, since after 8 weeks 54% of the patients recovered and the remaining ulcers were almost healed (Jonas and Farah, 1998). Other commercial preparations of *A. xylinum* cellulose, such as Biofil® and Bioprocess®, appeared to be excellent as skin transplants, and in the treatment of third-degree burns, ulcers, and decubitus. Another preparation, Gengiflex®, found application in recovering periodontal tissues (Jonas and Farah, 1998). Investigations on *A. xylinum* cellulose pads prepared by Krystynowicz et al. (2000) (Figure 17) indicate that general use as wound dressings is possible.

Results of investigations on BC hollow fibers use as artificial blood vessels and ureters are also promising (Yamanaka et al., 1990). Antithrombic BC property (blood

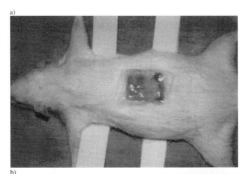

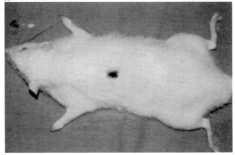

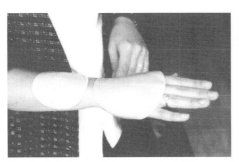

Fig. 16 Cellulose pellicle as a material for wound dressings.

Fig. 17 Burns treated with a BC dressing, before (a) and after (b) healing (Krystynowicz et al., 2000).

compatibility) was evaluated by a test based on replacement of adult dog blood vessels (parts of the descending aorta and jugular vein) with the artificial counterpart made of BC. One month later the artificial vessel was removed and the state of adhesion of thrombi on its inner surface was examined. A good open state of the BC blood vessel was maintained.

Another successful experiment showing the application of biosynthetic cellulose was substitution of the dog dura matter in the brain (Jonas and Farah, 1998).

Because of its high tensile strength, elasticity, and permeability to liquids and gases, dried BC was applied as an additional membrane to protect immobilized glucose oxidase in biosensors used for assays of blood glucose levels. This BC membrane enhanced the electrode stability in 10 times diluted human blood solution, up to 200 h. Other commercial protecting membranes, such as cuprophan (AKZO, England), provided an electrode stability up to 30 h only. In undiluted human blood the biosensor coated with cuprophan was stable for 3–4 h, whereas the BC membrane prolonged its stability up to 24 h.

11.3
Food Applications

Chemically pure and metabolically inert BC has been applied as a noncaloric bulking and stabilizing agent in processed food. Similar to pure preparations of the plant polysaccharide and its derivatives, it is used for stabilization of foams, pectin and starch gels, emulsions, e.g. canned chocolate drinks and soups, texture modification, e.g. improvement of the consistency of pulps, enhancement of adhesion, replacement of lipids, including oils, and dietary fiber supplements (Ang and Miller, 1991; Kent et al., 1991; Krystynowicz et al., 1999).

The first successful commercial application of BC in food production is 'nata de coco' (Sutherland, 1998). It is a traditional dessert from the Philippines, prepared from coconut milk or coconut water with sucrose, which serves as a growth medium for BC-producing bacteria. Consumption of the pellicle is believed to protect against bowel cancer, artheriosclerosis, and coronary thrombosis, and prevents a sudden rise of glucose in the urine. Therefore 'nata de coco' is becoming increasingly popular, not only in Asia.

Another popular BC-containing food product is Chinese Kombucha (teakvass or tea-fungus), obtained by growing yeast and acetic acid bacteria on tea and sugar extract. The pellicle formed on the surface contains both cellulose and enzymes healthy for humans. Their abiotic activity is especially stimulating for the large bowel and the whole alimentary tract. Kombucha is believed to protect against certain cancers (Iguchi et al., 2000).

Results of studies carried out by the authors, who applied BC pellicles synthesized by *A. xylinum* E_{25} for wine and juice filtration, and for immobilization of polyphenols, are promising (Krystynowicz et al., 1999). Preparations of bioactive anthocyanins, enriched in dietary fiber, are excellent for functional food production. BC also appeared to be an attractive component of bakery products, since it plays a role of dietary fiber, is taste and odorless, and prolongs the shelf-life.

11.4
Miscellaneous Uses

The large accessible surface area, high durability, and superior adsorptive properties as well as possibility of modification by means of physical or chemical methods, means that BC can be applied as a carrier for immobilization of biocatalysts. Cellulose gels containing immobilized animal cells were used for their culture to produce interferon, interleukin-1, cytostatics, and monoclonal antibodies (Iguchi et al., 2000).

BC was also applied for adsorption of cells of *Gluconobacter oxydans*, *Acetobacter methanolyticus* and *Saccharomyces cerevisiae*. The immobilized strains appeared to be effective producers of gluconate (84–92% yield), dihydroxyacetone (90–98% yield), and ethanol (88–92% yield), respectively, and displayed better operational and thermal stability.

Purified BC can be a raw material for synthesis of cellulose acetate, nitrocellulose, CM-cellulose, hydroxymethylcellulose, methylcellulose, and hydroxycellulose (Yamanaka, 1990). If BC is produced in the presence of compounds that interfere with regular fibril assembly and which influence the β-1,4-glucan structure, such as CM-cellulose or other cellulose derivatives, and other carbohydrates (starch, dextran), sulfonates, and alkylphosphates, the resulting microbial polymer has novel, additive-dependent, and useful properties, including optical transparency or higher water-binding capacity, even after repeated soaking and drying.

The potential application of BC in the chemical, paper, and textile industries depends on its price and accessibility. To meet these demands, BC has to be produced by highly efficient strains, growing on cheap substrates, in sophisticated surface, solid-state, or submerged fermentors (Vandamme et al., 1998).

12
Patents

Growing rapidly, the interest in BC is reflected in a number of patents (annually about 20 since the 1980s) and publications (20–40 per year for the last 10 years) devoted to this unique polymer (Iguchi et al., 2000). Some basic patents concerning the biosynthesis, properties, and applications of BC are presented in Table 4. These patents that are cited in chapter are listed in the references.

13
Outlook and Perspectives

The first scientific paper reporting on an unusual substance formed by acetic acid bacteria and known for ages in many countries as 'vinegar plant' (Yamanaka et al., 1989) was published 115 years ago (Brown, 1886). Further studies revealed that the substance is a super-pure cellulose. Metabolic pathways and the complicated molecular machinery of the polysaccharide biosynthesis, as well as the intriguing dynamics of its nascent chains association into a structure with unique properties, has been elucidated. Although progress and limited commercialization of BC have undoubtedly taken place, the relevant biotechnology, competitive to modern industrial technologies for PC production, has not yet been developed.

So, what to do, to achieve success and accomplish commercialization of BC production? First of all, stable overproducer strains of the polymer have to be constructed, using recent achievements in molecular genetics and biology. These strains have to assimilate a wide range of carbon sources, as well as display lower tendency towards spontaneous mutation to Cell$^-$ mutants and effectively synthesize cellulose (10–15 g L^{-1} d^{-1}) under agitated culture conditions. Construction of new bioreactors for both stationary and submerged culture is necessary. The considerable reduction of BC production cost could be attained by replacing expensive nutrient media components

Tab. 4 Selected patents concerning BC

Patent number	Patent holder	Inventors	Title	Date of publication	Major claims
WO 0125470	Novozymes A/S, Denmark	Herbert, W., Chanzy, H. D., Ernst, S., Schulein, M., Husum, T. L., Kongsbak, L.	Cellulose films for screening	2001	BC microfibril films containing fluorescein-labeled hemoglobin or galactomannan can be used to detect proteases and mannanases, respectively
WO 0105838	Pharmacia Corp., USA	Yang, Z. F., Sharma, S., Mohan, C., Kobzeff, J.	Process for drying reticulated bacterial cellulose without co-agents	2001	The reticulated *Acetobacter* cellulose subjected to dispersing in a solvent, e.g. hydrocarbon (hexane), aliphatic alcohol, and/or alkyl sulfoxide (DMSO), separation from the solvent, and drying, may be rehydrated to provide uniform dispersions having high viscosity
JP 11255806 A2	Bio-Polymer Research Co. Ltd, Japan	Watanabe, O.	Freeze drying method for microfibrous cellulose concentrate	1999	Freeze drying of *Acetobacter* cellulose gives dry microfibrous cellulose with good retention of its original properties after redispersing in water
WO 9940153	Monsanto	Smith, B. A., Colegrove, G. T., Rakitsky, W. G.	Acid-stable, cationic compatible cellulose compositions useful as rheology modifiers	1999	Cationic co-agents, acids, and/or cationic surfactants are used to form acid-stable *Acetobacter* cellulose compositions, which are useful as rheology modifiers
WO 9943748	Sony	Uryu, M., Tokura, K.	Biodegradable composite polymer material	1999	A new composite material has been obtained by drying the *A. xylinum* cellulose, its pulverization, and blending with a biodegradable polymer material
JP 0056669 A2	Canon Co., Japan	Minami, M., Mihara, C., Takeda, T., Kikuchi, Y.	Composite, for use in thermoformed articles, comprises cellulose and saccharide ester derivative	1999	A composite comprising BC and a saccharide ester can be used for production of biodegradable, thermoformed articles, which have improved processability, mechanical strength and flexibility
JP 11172115 A2	Ajinomoto Co. Inc., Japan	Suzuki, O., Kitamura, N., Matsumoto, R.	Bacterial cellulose-containing composite absorbents with high liquid absorption	1999	Dispersing highly water-absorbing polymer particles, dissolved in an organic solvent, in an aqueous solution of defibrillated BC, followed by organic solvent removal and partial drying, gives excellent and stable absorbents

13 Outlook and Perspectives

Tab. 4 (cont.)

Patent number	Patent holder	Inventors	Title	Date of publi-cation	Major claims
JP 11246602 A2	Bio-Polymer Research Co. Ltd, Japan	Tahara, N., Hagamida, T., Miyashita, H., Watanabe, O.	Preservation of wet bacterial cellulose	1999	Wet BC can be preserved with alkyl sulfate salts or NaOH and/or KOH
JP 11187896 A2	Bio-Polymer Research Co. Ltd, Japan	Tabata, T., Toyosaki, H., Tsuchida, T., Yoshinaga, F.	A method for screening cellulose-producing bacteria using cellulase	1999	A rapid and convenient method for screening a large number of cellulose membrane-producing strains is presented; the metabolic peculiarities of the strains and conditions needed for them to produce cellulose pellicle are explained
JP 11092502 A2	Bio-Polymer Research Co. Ltd, Japan	Shoda, M., Kanno, Y., Koda, T., Yoshinaga, F.	Manufacture of bacterial cellulose under oxygen-enriched conditions	1999	Passing more than 21% oxygen-containing air through A. xylinum ssp. sucrofermentans culture in an air-lift fermentor, enables accumulation of 6.5 g L^{-1} of BC after 75 h
JP 11117120 A2	Toray Industries Inc., Japan	Hara, T., Amano, J.	Fibers made from blends of bacterial cellulose and polymers having flexible main backbone	1999	The blends contain polyvinyl alcohol-type polymers at weight ratio of 2–50%
JP 11181001 A2	Bio-Polymer Research Co. Ltd, Japan	Matsuoka, M., Toyosaki, H., Matsumura, T., Ougiya, H., Tsuchida, T., Yoshinaga, F.	Production of bacterial cellulose	1999	BC, useful for filler retention aids for paper-making, can be produced in agitated cultures of Acetobacter strains, in the presence of water-soluble polysaccharides, e.g. CM-cellulose
JP 11137163 A2	Shikishima Seipan Co. Ltd, Japan	Kondo, M., Yamada, M., Inoue, S.	Manufacture of bread from dough containing bacterial cellulose	1999	Addition of BC increases water absorption of dough, thus giving bread with high water-holding capacity
JP 11221072 A2	Bio-Polymer Research Co. Ltd, Japan	Ishikawa, A., Tsuchida, T., Yoshinaga, F.	Bacterial cellulose production enhancement	1999	An A. xylinum mutant having higher cellular levels of UDP, UTP, and UDPGlc as well as higher carbamoyl phosphate synthetase activity is an excellent cellulose producer

Tab. 4 (cont.)

Patent number	Patent holder	Inventors	Title	Date of publication	Major claims
JP 11269797 A2	Toppan Printing Co. Ltd, Nakano Vinegar Co. Ltd, Japan	Yamawaki, K., Tomita, T., Harasawa, A., Kaminaga, J., Kawasaki, K., Matsuo, R., Sasaki, N., Fukagai, M., Tsukamoto, Y.	Impregnated paper with good water resistance and stiffness	1999	Paper impregnated with silane coupler-grafted BC derivatives is moisture resistant
PP 299907	Technical University of Lodz, Poland	Krystynowicz, A., Czaja, W., Bielecki, S.	Biosynthesis and application of bacterial cellulose	1999	Production of BC by wild and mutant *A. xylinum* strains, cultured in static or agitated cultures, and under various conditions as well as an influence of culture conditions on the cellulose properties are described
JP 11018758 A2	Bio-Polymer Research Co. Ltd, Japan	Yamamoto, T., Yano, H., Yoshinaga, F.	Horizontal type spinner culture vessel, having high oxygen-supplying efficiency	1999	BC can be produced in the spinner culture tank, equipped with mixing impellers, and providing high efficiency of oxygen supply
JP 10298204 A2	Ajinomoto Co. Inc., Japan	Ishikara, M., Yamanaka, S.	Bacterial cellulose with ribbon-like microfibril shape	1998	Cellulose fibrils having a short axis 10–100 nm and a long axis 160–1000 nm are produced extracellularly by cellulose-generating bacteria, e.g. *Acetobacter pasteurianus*, in a culture containing cell division inhibitor
US 005723764	Pioneer Hi-Bred International Inc., USA	Nichols, S. E., Singletary, G. W.	Cellulose synthesis in the storage tissue of transgenic plants	1998	Introducing the genes for cellulose biosynthesis from the species *A. xylinum* into a given plant provides a method of synthesizing cellulose in the storage tissue of transgenic plants
JP 10077302	Bio-Polymer Research Co. Ltd, Japan	Tabuchi, M., Watanabe, K., Morinaga, Y.	Solubilized bacterial cellulose and its compositions or composites for moldings and coatings.	1998	Cellulose synthesized by *A. xylinum* in stationary culture was solubilized by stirring in a mixture of DMSO and paraformaldehyde (25:5) at 100 °C for 3 h

Tab. 4 (cont.)

Patent number	Patent holder	Inventors	Title	Date of publication	Major claims
WO 97/05271	Rensselaer Polytechnic Institute, USA	Bungay, H. R., Serafica, G.	Production of microbial cellulose using a rotating disc film bioreactor	1997	BC can be deposited by A. xylinum, cultured in a liquid medium inside the horizontal fermentor, on a plurality of disks, which rotate around the long axis of the fermentor and are partly submerged in the culture medium
JP 09025302 A2	Bio-Polymer Research Co. Ltd, Japan	Hioki, S., Watanabe, K., Ogya, H., Morinaga, Y.	Preparation of disintegrated bacterial cellulose for improved additive retention in paper manufacture	1997	BC which helps additive retention and causes no harm to freeness during paper manufacturing can be obtained by disintegration using a self-excited ultrasonic pulverizer
JP 09107892 A	Nakano Vinegar Co. Ltd, Japan	Furukawa, H., Maruyama, Y., Fukaya, M., Tsukamoto, Y., Kawamura, K.	Composition for stabilizing dispersion used in food	1997	A fine particulate (210 µm average particle diameter) BC, obtained by hydrolyzing A. xylinum cellulose with a mineral acid, is of low viscosity and is a sufficient dispersion stability agent in food products
WO 9744477	Bio-Polymer Research Co. Ltd, Japan	Naritomi, T., Kouda, T., Naritomi, M., Yano, H., Yoshinaga, F.	Continuous preparation of bacterial cellulose having a high production rate and yield	1997	BC obtained in continuous culture of A. xylinum (production rate at least $0.4~g~L^{-1}~h^{-1}$) in a medium containing a substance enhancing the apparent substrate affinity to sugar (e.g. lactic acid) is useful as a thickener, humectant or stabilizer for production of food, cosmetics or paints, etc.
WO 9740135	Bio-Polymer Research Co. Ltd, Japan	Tsuchida, T., Tonouchi, N., Seto, A., Kojima, Y., Matsuoka, M., Yoshinaga, F.	Novel cellulose-producing bacteria and a process of producing it	1997	Production of cellulose with a new A. xylinum ssp. nonactoxidans, which lacks an ability to oxidize acetates and lactates, yields odor- and taste-less products, having excellent dispersability in water
WO 9712987	Bio-Polymer Research Co. Ltd, Japan	Kouda, T., Naritomi, T., Yano, H., Yoshinaga, F.	Production process for bacterial cellulose which is useful as material in various fields	1997	A new process for BC production involving maintaining a certain pressure inside a fermentor, reduces power required for agitation, and elevates production rate and yield
JP 09056392 A	Kikkoman Corp., Japan	Fukazawa, K., Imai, H., Kijima, T., Kikuchi, T.	Production of microorganism cellulose	1997	A cellulose pellicle synthesized by A. xylinum strain precultured in stationary conditions can be homogenized and used to inoculate the fresh culture medium

Tab. 4 (cont.)

Patent number	Patent holder	Inventors	Title	Date of publication	Major claims
PL 171952 B1	Technical University of Lodz, Poland	Galas, E., Krystynowicz, A.	Method of obtaining bacterial cellulose	1997	A method of BC production in static culture using *A. xylinum* P23 strain is described
US 0824096	Bio-Polymer Research Co. Ltd, Japan	Kouda, T., Nagata, Y., Yano, H., Yoshinaga, F.	Production of bacterial cellulose through cultivation of cellulose-producing bacteria under specified conditions in aerated and agitated culture	1997	Cellulose produced by species of *Acetobacter*, *Agrobacterium*, *Rhizobium*, *Sarcina*, *Pseudomonas*, *Achromobacter*, *Alcaligenes*, *Aerobacter*, *Azotobacter*, and *Zoogloea*, in aerated and agitated cultures can be applied in production of food and cosmetics
JP 97–21905	Bio-Polymer Research Co. Ltd, Japan	Seto, H., Tsuchida, T., Yoshinaga, F.	Manufacture of bacterial cellulose by mixed culture of microorganisms	1997	To provide lactate and split sucrose, cellulose-producing *A. xylinum* strain can be grown together with lactic acid bacteria and *Saccharomyces* yeast; their presence in the culture broth markedly enhances BC yield
JP 08127601 A	Bio-Polymer Research Co. Ltd, Japan	Hioki, S., Watabe, O., Hori, S., Morinaga, Y., Yoshinaga, F.	Freeness regulating agent	1996	Production of a freeness regulating agent, comprising a defiberized BC, synthesized by *A. xylinum* ssp. sucrofermentans is described
JP 96316922	Bio-Polymer Research Co. Ltd, Japan	Hikawu, S., Hiroshi, T., Takayasu, T., Yoshinaga, F.	Manufacture of bacterial cellulose by addition of cellulose formation stimulators	1996	Compounds such as derivatives of choline, betaine, and fatty acids (salts and esters) appeared to stimulate cellulose production by *A. xylinum* strains
JP 08056689 A	Bio-Polymer Research Co. Ltd, Japan	Seto, H., Tsuchida, T., Yoshinaga, F.	Production of bacterial cellulose	1996	To obtain an edible BC, excellent in aqueous dispersibility, useful for retaining the viscosity of foods, cosmetics, coatings, etc., and enrichment of foods, *Acetobacter* strains can be cultured in saponin-containing medium
JP 08033494 A	Bio-Polymer Research Co. Ltd, Japan	Tawara, N., Koda, T., Hagamida, T., Morinaga, Y., Yano, H.	Method for circulating continuous production and separation of bacterial cellulose	1996	Circulating a culture liquid containing *A. xylinum* cells between a culture and a separation apparatus enables separation of the produced cellulose; also, a flotation separator or an edge filter can be applied for this purpose

Tab. 4 (cont.)

Patent number	Patent holder	Inventors	Title	Date of publication	Major claims
JP 08034802 A	Gun Ei Chemical Industries Co. Ltd, Japan	Hirooka, S., Hamano, T., Miyashita, Y., Hanaue, K., Yamazaki, K., Shiichi, K., Shiichi, F.	Bacterial cellulose, production thereof and processed product made therefrom	1996	*A. xylinum* can synthesize cellulose in culture broths containing difficult to ferment, branched oligosaccharides
JP 08000260 A	Bio-Polymer Research Co. Ltd, Japan	Ishikawa, A., Tsuchida, T., Yoshinaga, F.	Production of bacterial cellulose with pyrimidine analogue-resistant strain	1996	A method of production of BC, in a high yield and at a low cost, using a pyrimidine analog-resistant *Acetobacter* mutant is presented
JP 08325301 A	Bio-Polymer Research Co. Ltd, Japan	Hori, S., Watabe, O., Morinaga, Y., Yoshinaga, F.	Cellulose having high dispersibility and its production	1996	BC synthesized by *A. xylinum* ssp. sucrofermentans, subjected after harvesting to partial hydrolysis with HCl, yields a fraction exhibiting high birefringence
JP 08276126	Bio-Polymer Research Co. Ltd, Japan	Ogiya, H., Watabe, O., Shibata, A., Hioki, S., Morinaga, Y., Yoshinaga, F.	Emulsification stabilizer	1996	The method of production of the emulsification stabilizer containing BC obtained from agitated culture of *A. xylinum* ssp. sucrofermentans and having low index of crystallization, is presented
JP 08009965 A	Bio-Polymer Research Co. Ltd, Japan	Ishikawa, A., Tsuchida, T., Yoshinaga, F.	Production of bacterial cellulose using microbial strain resistant to inhibitor of DHO-dehydrogenase	1996	A method of efficient production of BC, using *Acetobacter* strains resistant either to an inhibitor of DHO-dehydrogenase or to DNP is presented
US 005382656	Weyerhauser Co., USA	Benziman, M., Tal, R.	Cellulose synthase associated proteins	1995	CS-associated proteins, which have molecular weights of 20, 54, and 59 kDa are not encoded by CS operon genes *bcsA, B, C,* or *D*
JP 07313181 A	Bio-Polymer Research Co. Ltd, Japan	Takemura, H., Tsuchida, T., Yoshinaga, F., Matsuoka, M.	Production of bacterial cellulose using PQQ-unproductive strain	1995	Pyrroloquinolinequinone-unproductive *Acetobacter* mutant strain enables high-yield production of BC
JP 07184675 A	Bio-Polymer Research Co. Ltd, Japan	Matsuoka, M., Tsuchida, T., Yoshinaga, F.	Production of bacterial cellulose	1995	To obtain BC useful for retaining the viscosity of foods, cosmetics, or coatings, *Acetobacter* strains can be cultured in a methionine-containing medium

Tab. 4 (cont.)

Patent number	Patent holder	Inventors	Title	Date of publi-cation	Major claims
JP 07268128	Fujitsuko Co. Ltd., Japan	Kiriyama, S., Fukui, H., Toda, T., Yamagishi, H.	Dried material of cellulose derived from microorganism and its production	1995	Water-soluble stabilizer (preferably glucose or gelatin) added to cellulose gel obtained by culturing A. *xylinum* strain before drying of the material provides excellent and stable physical properties
WO 95/32279	Bio-Polymer Research Co. Ltd, Japan	Tonouchi, N., Tsuchida, T., Yoshinaga, F., Horinouchi, S., Beppu, T.	Cellulose-producing bacterium transformed with gene coding for enzyme related to sucrose metabolism	1995	A. *xylinum* transformant with a gene coding for invertase accumulates cellulose in a sucrose-containing medium
JP 07184677 A	Bio-Polymer Research Co. Ltd, Japan	Seto, H., Tsuchida, T., Yoshinaga, F.	Production of bacterial cellulose	1995	High-yield production of an edible BC useful for retaining the viscosity of cosmetics or coatings can be achieved by culturing an *Acetobacter* strain in an invertase-added medium containing sucrose as a carbon source
JP 07039386 A	Bio-Polymer Research Co. Ltd, Japan	Matsuoka, M., Takemura, H., Tsuchida, T., Yoshinaga, F.	Production of bacterial cellulose	1995	BC, useful for retaining the viscosity of a food, a cosmetic, a coating, etc., and usable as a food additive, an emulsion stabilizer, etc., can be obtained by culturing A. *xylinum* in a culture medium containing a carboxylic acid (e.g. lactic acid) salt
JP 07274987 A	Bio-Polymer Research Co. Ltd, Japan	Toda K., Asakura, T.	Production of bacterial cellulose	1995	To obtain BC in high yield, the cellulose-producing *Acetobacter* strain is cultured in a cylindrical container and oxygen is fed through an oxygen-permeable membrane at the bottom
JP 06125780	Nakano Vinegar Co. Ltd, Japan	Fukaya, M., Okumura, H., Kawamura, K.	Production of cellulosic substance of microorganism	1994	A method used to improve production efficiency of cellulose by an *Acetobacter* strain is presented; the method is based on adding a protein having affinity to the cellulose to the culture medium
JP 06248594 A	Mitsubishi Paper Mills Ltd, Japan	Katsura, T., Okafuro, K.	Low-density paper having high smoothness	1994	Excellent, low-density paper having high smoothness contains BC and the broad-leaved pulp in the ratio of 1/99:1/1

Tab. 4 (cont.)

Patent number	Patent holder	Inventors	Title	Date of publication	Major claims
JP 06206904	Shin Etsu Chemical Co. Ltd, Japan	Horii, F., Yamamoto, H.	Bacterial cellulose, its production and method for controlling crystal structure thereof	1994	Xanthan gum or sodium CM-cellulose added to the culture broth of A. xylinum enable control of the crystal structure of BC
JP 06113873 A	Nippon Paper Industries Co. Ltd, Japan	Samejima, K., Mamoto, K.	Production of microbial cellulose	1994	Adding a sulfite pulp waste liquor and/or its permeate from ultrafiltration into a culture medium of a cellulose-producing microorganism enhances the cellulose yield and lowers the cost of its production
WO 94/20626	Bio-Polymer Research Co. Ltd, Japan	Beppu, T., Tonouchi, N., Horinouchi, S., Tsuchida, T.	*Acetobacter*, plasmid originating therein, and shuttle vector constructed from said plasmid	1994	Genetic recombination of cellulose-producing *Acetobacter* strains is performed using an *Acetobacter* strain, its endogenous plasmid, and a shuttle vector constructed from the latter plasmid and a plasmid from *Escherichia coli*
JP 06001647 A	Shimizu Corp., Japan	Yano, H., Narutomi, T., Okamura, K., Kawai, T., Minami, S.	Concrete and coating material	1994	The concrete or coating material containing disaggregated BC displays better dispersibility of cement or pigment particles and an enhanced fluidity
US 005268274 A	Cetus Corp., USA	Ben-Bassat, A., Calhoon, R. D., Fear, A. L., Gelfand, D. H., Meade, J. H., Tal, R., Wong, H., Benziman, M.	Methods and nucleic acid sequences for the expression of the cellulose synthase operon	1993	Nucleic acid sequences encoding the BC synthase from *A. xylinum*, and methods for isolating the genes and their expression in hosts are presented
EP 0396344 A2	Ajinomoto Co. Inc., Japan	Yamanaka, S., Ono, E., Watanabe, K., Kusakabe, M., Suzuki, Y.	Hollow microbial cellulose, process for preparation thereof, and artificial blood vessel formed of said cellulose	1990	BC prepared by culturing a cellulose-producing strain on one or both surfaces of an oxygen-permeable hollow carrier is useful as a substitute for a blood vessel or another internal hollow organ; the cellulose can be impregnated with a medium, cured, and cut if necessary

Tab. 4 (cont.)

Patent number	Patent holder	Inventors	Title	Date of publi-cation	Major claims
US 004950597	University of Texas, USA	Saxena, I. M., Roberts, E. M., Brown, R. M.	Modification of cellulose normally synthesized by cellulose-producing micro-organisms	1990	Mutants of *A. xylinum* that do not form a pellicle in liquid culture and synthesize cellulose almost exclusively as the allomorph cellulose II, that arise spontaneously or by nitrosoguanidine mutagenesis are described and the cellulose they produce is characterized
JP 02182194 A	Asahi Chemical Industries Co. Ltd, Japan	Matsuda, Y., Kamiide, K.	Production of cellulose with acetic acid bacterium	1990	Acetic acid bacteria having a synchronized cell cycle enable efficient production of BC, excellent in water holding properties, tensile strength, and purity, and displaying a relatively low DP
WO 89/12107	Brown, R. M.	Brown, R. M.	Microbial cellulose as a building block resource for specialty products and processes thereof	1989	A novel process for manufacturing BC using different bacterial species belonging to *Acetobacter*, *Rhizobium*, *Agrobacterium*, and *Pseudomonas*, and production of various articles from this polymer are described
EP 0258038 A3	Brown, R. M.	Brown, R. M.	Use of cellulase preparations in the cultivation and use of cellulose-producing microorganisms	1989	To prepare an inoculum of appropriate cell density for large-volume fermentors, *A. xylinum* cells entrapped in cellulose can be liberated with cellulase preparation, which causes a partial cellulase hydrolysis
US 004863565	Weyerhauser Co., USA	Johnson, D. C., Neogi, A. M.	Sheeted products formed from reticulated microbial cellulose	1989	Strains of *Acetobacter* that are stable under agitated culture conditions and that exhibit reduced gluconic acids production, synthesize unique reticulated cellulose sheets, characterized by resistance to densification and great tensile strength
WO 89/11783	Brown, R. M.	Brown, R. M.	Microbial cellulose composites and processes for producing same	1989	Methods enabling production of various objects utilizing BC produced *in situ* or applied as a film are presented; a process for manufacturing currency from BC is described

Tab. 4 (cont.)

Patent number	Patent holder	Inventors	Title	Date of publication	Major claims
EP 0323717 A3	ICI Plc, UK	Byrom, D.	Process for the production of microbial cellulose	1988	A process for the production of BC using a novel strain of the genus *Acetobacter* is described
EP 0 289993 A3	Weyerhaeuser Co., USA	Johnson, D. C., LeBlanc, H. A., Neogi, A. N.	Bacterial cellulose as surface treatment for fibrous web	1988	BC applied at relatively low concentrations, singularly or in combination, to at least one surface of a fibrous web gives excellent properties of gloss, smoothness, ink receptivity and holdout, and surface strength
WO 88/09381	Financial Union for Agricultural Development, France	Labourer, P. F.	Process for producing bacterial cellulose from material of plant origin	1988	Culturing of *A. xylinum* strain in a plant polysaccharide-containing medium enables efficient cellulose production
US 004655758	Johnson & Johnson products, Inc., USA	Ring, D. F., Nashed, W., Dow, T.	Microbial polysaccharide articles and methods of production	1987	After removal of excess liquid and bacterial cells, the cellulose pellicle can be impregnated and used for various purposes
WO 86/02095	Bio-Fill Industria e Comercio de Produtos Medico Hospitalares, Ltd, Brazil	Farah, L. F. X.	Process for the preparation of cellulose film, cellulose film produced thereby, artificial skin graft and its use	1986	BC film preparation, including optimal conditions of *A. xylinum* culturing, and methods of removing the formed film are described; the film appeared to be suitable for use as an artificial skin graft, a separating membrane, or artificial leather
EP 0200409 A3	Ajinomoto Co. Inc., Japan	Iguchi, M., Mitsuhashi, S., Ichimura, K., Nishi, Y., Uryu, M., Yamanake, S., Watanabe, K.	Molded material comprising bacteria-produced cellulose	1986	BC is an excellent component of molded materials having high dynamic strength as compared to conventional molded materials

with industrial wastes, rich in proper carbon sources, such as spent liquors from crystalline glucose production, etc. Simultaneously, a significant environmental benefit would be obtained. However, the major advantage of mass production of BC would be the protection of forests, which are presently disappearing at an alarming rate, thus leading to soil eutrophication and global climate changes.

BC has already been put to numerous uses, presented in this review. The polysaccharide can not only be replaced by some animal polymers (collagen), but also carriers of substances having a positive impact on human health (e.g. antioxidants and prebiotics). The usefulness of the bacterial polymer in medicine (wound, burn and ulcer dressing materials, component of implants) is not longer questioned. Moreover, cellulose granulates can be an excellent matrix for the immobilization of medicinal preparations. For example, if specific substances (receptors) are adsorbed on BC, the resulting molecules can be scavengers of either toxins or of the pathogenic microflora inhabiting the alimentary tract. Recently, unique nanocrystals ($30 \times 600-800$ nm) of BC have been obtained, derived from its commercial preparation Prima Cel™ (Xylos). Selective modification (trimethyl silylation) of the surface of these nanocrystals while leaving their core intact has been achieved. Such modified crystals have great potential in several advanced technologies.

Deciphering of all the riddles of cellulose biosynthesis will lead to improvement and tailoring of BC supramolecular structure and properties; as a consequence, novel concepts for both its inexpensive production, and for its bulk and specialty applications will be developed.

14 References

Aloni, Y., Benziman, M. (1982) Intermediates of cellulose synthesis in *Acetobacter*, in: *Cellulose and Other Natural Polymers System* (Brown, R. M., Jr., Ed.), New York: Plenum Press, 341–361.

Aloni, Y., Cohen, Y., Benziman, M., Delmer, D. P. (1983) Solubilization of UDP-glucose: 1,4-β-glucan 4-β-D-glucosyl transferase (cellulose synthase) from *Acetobacter xylinum*, *J. Biol. Chem.* **258**, 4419–4423.

Ang, J. F., Miller, W. B. (1991) Multiple functions of powdered cellulose as a food ingredient, *J. Am. Ass. Cereal Chem.* **36**, 558–564.

Bartlett, D. H., Silverman, M. (1989) Nucleotide sequences of IS492, a novel insertion sequence causing variation of extracellular production in the marine bacterium *Pseudomonas atlantica*, *J. Bacteriol.* **171**, 1763–66.

Bauer, K., Codolington, K., Ben-Bassat, A. (1992) Methods for improving production of bacterial cellulose, *Abstr. Paper Am. Chem. Soc.*, **203** Meet., Pt 1, Biot. 94.

Bayer, M. E. (1979) The fusion sites between outer membrane and cytoplasmic membrane in bacteria: their role in membrane assembly and virus infection, in: *Bacterial Outer Membranes* (Inouye, M., Ed.), New York: John Wiley & Sons, 167–202.

Ben-Bassat, A., Bruner, R., Wong, H., Shoemaker, S., Aloni, Y. (1989) Production of bacterial cellulose by *Acetobacter*, *Abstr. Pap. Am. Chem. Soc.*, **198** Meet., MBTD20.

Ben-Bassat, A., Calhoon, R. D., Fear, A. L., Gelfand, D. H., Mead, J. H., Tal, R., Wong, H., Benziman, M. (1993) Methods and nucleic acid sequences for expression of the cellulose synthase operon, US patent 5 268 274.

Ben-Hayyim, G., Ohad, I. (1965) Synthesis of cellulose by *Acetobacter xylinum*; VIII. On the formation and orientation of bacterial cellulose fibrils in the presence of acidic polysaccharides, *J. Cell Biol.* **25**, 191–207.

Benziman, M., Tal, R. (1995) Cellulose synthase associated proteins. US patent 5 382 656.

Boisset, C., Chanzy, H., Henrissat, B., Lamed, R., Shoham, Y., Bayer, E. A. (1999) Digestion of crystalline cellulose substrates by the *Clostridium thermocellum* cellulosome: structural and morphological aspects, *Biochem. J.* **340**, 829–835.

Boisset, C., Fraschini, C., Schulein, M., Henrissat, B., Chanzy, H. (2000) Imaging the enzymatic digestion of bacterial cellulose ribbons reveals the endo character of the cellobiohydrolase Cel6A from *Humicola insolens* and its mode of synergy with cellobiohydrolase Cel7A, *Appl. Environ. Microbiol.* **66**, 1444–1452.

Brown, A. J. (1886) An acetic ferment which forms cellulose, *J. Chem. Soc.* **49**, 432–439.

Brown R. M., Jr. (1989a) Microbial cellulose composites and processes for producing same. WO 89/11783.

Brown, R. M., Jr. (1989b) Microbial cellulose as a building block resource for specialty products and processes thereof, WO 89/12107.

Brown, R. M., Jr. (1989c) Use of cellulase preparations in the cultivation and use of cellulose-producing microorganisms, European patent 0258038A3.

Brown, R. M., Jr. (1996) The biosynthesis of cellulose, *Pure Appl. Chem.* **A33**, 1345–1373.

Brown, R. M., Jr., Saxena, I. M. (2000) Cellulose biosynthesis: a model for understanding the assembly of biopolymers, *Plant Physiol. Biochem.* **38**, 57–60.

Brown, R. M., Jr., Willison, J. H. M., Richardson, C. L. (1976) Cellulose biosynthesis in *Acetobacter xylinum*: visualisation of the site of synthesis and direct measurement of the *in vivo* process, *Proc. Natl. Acad. Sci. USA* **73**, 4565–4569.

Bungay, H. R., Serafica, G. (1997) Production of microbial cellulose using a rotating disc film bioreactor, WO 97/05271.

Bureau, T. E., Brown, R. M., Jr. (1987) In vitro synthesis of cellulose II from a cytoplasmic membrane fraction of *Acetobacter xylinum*, Proc. Natl. Acad. Sci. USA **84**, 6985–6989.

Canale-Parda, E., Wolfe, R. S. (1964) Synthesis of cellulose by *Sarcina ventriculi*, Biochim. Biophys. Acta **82**, 403–405.

Cannon, R. E., Anderson, S. M. (1991) Biogenesis of bacterial cellulose, Crit. Rev. Microbiol. **17**, 435–439.

Chao, Y., Ishida, T., Sugano, Y., Shoda, M. (2000) Bacterial cellulose production by *Acetobacter xylinum* in a 50 L internal-loop airlift reactor, Biotechnol. Bioeng. **68**, 345–352.

Colvin, J. R. (1957) Formation of cellulose microfibrils in a homogenate of *Acetobacter xylinum*, Arch. Biochem. Biophys. **70**, 294–295.

Colvin, J. R. (1980) The biosynthesis of cellulose, in: *Plant Biochemistry* (Priess, J., Ed.), New York: Academic Press, 543–570, Vol. 3

Colvin, J. R., Leppard, G. G. (1977) The biosynthesis of cellulose by *Acetobacter xylinum* and *Acetobacter acetigenus*, Can. J. Microbiol. **23**, 701–709.

Colvin, J. R., Sowden, L. C., Daoust, V., Perry, M. (1979) Additional properties of a soluble polymer of glucose from cultures of *Acetobacter xylinum*, Can. J. Biochem. **57**, 1284–1288.

Costeron, J. W. (1999) The role of bacterial exopolysaccharides in nature and disease, J. Ind. Microbiol. Biotechnol. **22**, 551–563.

Coucheron, D. H. (1991) An *Acetobacter xylinum* insertion sequence element associated with inactivation of cellulose production, J. Bacteriol. **173**, 5723–2731.

Dekker, R. F. H., Rietschel, E. T., Sandermann, H. (1977) Isolation of α-glucan and lipopolysaccharide fractions from *Acetobacter xylinum*, Arch. Microbiol. **115**, 353–357.

De Lannino, N., Cuoso, R. O., Dankert, M. A. (1988) Lipid-linked intermediates and the synthesis of acetan in *A. xylinum*, J. Gen. Microbiol. **134**, 1731–1736.

Delmer, D. P. (1982) Biosynthesis of cellulose, Adv. Carbohydr. Chem. Biochem. **41**, 105–153.

De Wulf, P., Joris, K., Vandamme, E. J. (1996) Improved cellulose formation by an *Acetobacter xylinum* mutant limited in (keto)gluconate synthesis, J. Chem. Tech. Biotechnol. **67**, 376–380.

Easson, D. D., Jr., Sinskey, A. J., Peoples, O. P. (1987) Isolation of *Zoogloea ramigera* I-16 M exopolysaccharides biosynthetic genes and evidence for instability within this region, J. Bacteriol. **169**, 4518–4524.

Embuscado, M. E., Marks, J. S., Miller, J. N. (1994) Bacterial cellulose. II. Optimization of cellulose production by *Acetobacter xylinum* through response surface methodology, Food Hydrocolloids **8**, 419–430.

Galas, D. J., Chandler, M. (1989) Bacterial insertion sequences, in: *Mobile DNA* (Berg, D. E., Howe, M. M., Eds.) Washington, DC: ASM, 109–162.

Galas, E., Krystynowicz, A., Tarabasz-Szymanska, L., Pankiewicz, T., Rzyska, M. (1999) Optimization of the production of bacterial cellulose using multivariable linear regression analysis, Acta Biotechnol. **19**, 251–260.

Glaser, L. (1958) The synthesis of cellulose in cell-free extracts of *Acetobacter xylinum*, J. Biol. Chem. **232**, 627–636.

Haigler, C. H. (1985) The function and biogenesis of native cellulose, in: *Cellulose Chemistry and its Applications* (Nevel, R. P., Zeronian, S. H., Eds.), Chichester: Ellis Horwood, 30–83.

Han, N. S., Robyt, J. F. (1998) The mechanism of *Acetobacter xylinum* cellulose biosynthesis: direction of chain elongation and the role of lipid pyrophosphate intermediates in the cell membrane, Carbohydr. Res. **313**, 125–133.

Hestrin, S., Aschner, M., Mager J. (1947) Synthesis of cellulose by resting cells of *Acetobacter xylinum*, Nature **159**, 64–65.

Hestrin, S., Schramm, M. (1954) Synthesis of cellulose by *Acetobacter xylinum*, II. Preparation of freeze-dried cells capable of polymerizing glucose to cellulose, Biochem. J. **58**, 345–352.

Hikawu, S., Hiroshi, T., Takayasu, T., Yoshinaga, F. (1996) Manufacture of bacterial cellulose by addition of cellulose formation stimulators, Japanese patent 96316922.

Hotte, B., Roth-Arnold, I., Puhler, A., Simon, R. (1990) Cloning and analysis of a 35.3 kb DNA region involved in exopolysaccharide production by *Xanthomonas campestris*, J. Bacteriol. **172**, 2804–2807.

Husemann, E., Muller, G. J. M. (1966) Synthesis of unbranched polysaccharides, Macromol. Chem. **91**, 212–230.

Ielpi, L., Couso, R., Dankert, M. A. (1981) Lipid-linked intermediates in the biosynthesis of xanthan gum, FEBS Lett. **130**, 253–256.

Iguchi, M., Yamanaka, S., Budhioko, A. (2000) Bacterial cellulose – a masterpiece of nature's arts, J. Mater. Sci. **35**, 261–270.

Ishikawa, A., Matsuoka, M., Tsuchida, T., Yoshinaga, F. (1995) Increasing of bacterial cellulose production by sulfoguanidine-resistant mutants derived from *Acetobacter xylinum* subsp. sucrofermentans BPR2001, *Biosci. Biotechnol. Biochem.* **59**, 2259–2263.

Ishikawa, A., Tsuchida, T., Yoshinaga, F. (1996a) Production of bacterial cellulose using microbial strain resistant to inhibitor of DHO-dehydrogenase, Japanese patent 08009965A.

Ishikawa, A., Tsuchida, T., Yoshinaga, F. (1996b) Production of bacterial cellulose with pyrimidine analogue-resistant strain, Japanese patent 08000260.

Johnson, D. C., Neogi, A. N. (1989) Sheeted products formed from reticulated microbial cellulose, US patent 4 863 565.

Jonas, R., Farah, L. F. (1998) Production and application of microbial cellulose, *Polym. Degrad. Stabil.* **59**, 101–106.

Jones, D., Gordon, A. H., Bacon, J. S. D. (1974) β-1,3-Glucanases from *Sclerotium rolfsii*, *Biochem. J.* **140**, 47–55.

Joris, K., Billiet, F., Drieghe, S., Brachx, D., Vandamme, E. (1990) Microbial production of β-1,4-glucan, *Meded. Fac. Landbouwwet Rijksuniv. Gent* **55**, 1563–1566.

Kenji, K., Yukiko, M., Hidehi, L., Kunihiko, O. (1990) Effect of culture conditions of acetic acid bacteria on cellulose biosynthesis, *Br. Polym. J.* **22**, 167–171.

Kent, R. A., Stephens, R. S., Westland, J. A. (1991) Bacterial cellulose fiber provides an alternative for thickening and coating, *Food Technol.* **45**, 108.

Kim, D., Kim, Y. M., Park, D. H. (1999) Modification of *Acetobacter xylinum* bacterial cellulose using dextransucrase and alternansucrase, *J. Microbiol. Biotechnol.* **9**, 704–708.

Kobayashi, S., Kashiwa, K., Kawasaki, T., Shoda, S. (1991) Novel method for polysaccharide synthesis using an enzyme: the first *in vitro* synthesis of cellulose via nonbiosynthetic path utilising cellulase as catalyst, *J. Am. Chem. Soc.* **113**, 3079–3084.

Kouda, T., Naritomi, T., Yano, H., Yoshinaga, F. (1998) Inhibitory effect of carbon dioxide on bacterial cellulose production by *Acetobacter* in agitated culture, *J. Ferment. Bioeng.* **85**, 318–321.

Koyama, M., Helbert, W., Imai, T., Sugiyama, J., Henrissat, B. (1997) Parallel-up structure evidence the molecular directionality during biosynthesis of bacterial cellulose, *Proc. Natl. Acad. Sci. USA* **94**, 9091–9095.

Krystynowicz, A., Turkiewicz, M., Drynska, E., Galas, E. (1995) Bacterial cellulose – biosynthesis and application, *Biotechnologia* **30**, 120–132.

Krystynowicz, A., Galas, E., Pawlak, E. (1997) Method of bacterial cellulose production. Polish patent P-299907.

Krystynowicz, A., Czaja, W., Bielecki, S. (1999) Biosynthesis and application of bacterial cellulose, *Zywnosc* **3**, 22–33.

Krystynowicz, A., Czaja, W., Pomorski, L., Kolodziejczyk, M., Bielecki, S. (2000) The evaluation of usefulness of microbial cellulose as a wound dressing material, 14th Forum for Applied Biotechnology, Gent, Belgium, *Meded. Fac. Landbouwwet Rijksuniv. Gent*, Proceedings Part I, 213–220.

Kudlicka K. (1989) Terminal complexes in cellulose synthesis, *Postepy biologii komórki* **16**, 197–212 (abstract in English).

Kuga, G., Brown, R. M., Jr. (1988) Silver labeling of the reducing ends of bacterial cellulose, *Carbohydr. Res.* **180**, 345–350.

Laboureur, P. (1988) Process for producing bacterial cellulose from material of plant origin, WO 88/09381.

Legge R. L. (1990) Microbial cellulose as a specialty chemical, *Biotechnol. Adv.* **8**, 303–319.

Leppard G. G., Sowden, L. C., Ross, C. J. (1975) Nascent stage of cellulose biosynthesis, *Science* **189**, 1094–1095.

Lin, F. C., Brown, R. M., Jr. (1989) Purification of cellulose synthase from *Acetobacter xylinum*, in: *Cellulose and Wood Chemistry and Technology* (Schmerck, C., Ed.), New York: John Wiley & Sons, 473–492.

Lin, F. C., Brown, R. M., Jr., Cooper, J. B., Delmer, D. P. (1985) Synthesis of fibrils *in vitro* by a solubilized cellulose synthase from *Acetobacter xylinum*, *Science* **230**, 822–825.

Lin, F. C., Brown, R. M., Jr., Drake, R. R., Jr., Haley. B. E. (1990) Identification of the uridine-5′-diphosphoglucose (UDPGlc) binding subunit of cellulose synthase in *Acetobacter xylinum* using the photoaffinity probe 5-azido-UDPGlc, *J. Biol. Chem.* **265**, 4782–4784.

Marx-Figini, M., Pion, B. G. (1974) Kinetic investigations on biosynthesis of cellulose by *Acetobacter xylinum*, *Biochim. Biophys. Acta* **338**, 382–393.

Masaoka, S., Ohe, T., Sakota, N. (1993) Production of cellulose from glucose by *Acetobacter xylinum*, *J. Ferment. Bioeng.* **75**, 18–22.

Matsuoka, M., Tsuchida, T., Matsushita, K., Adachi, O., Yoshinaga, F. (1996) A synthetic medium for bacterial cellulose production by *Acetobacter xylinum* subsp. sucrofermentans, *Biosci. Biotechnol. Biochem.* **60**, 575–579.

Matthyse, A., Thomas, D. I., White, A. R. (1995) Mechanism of cellulose synthesis in *Agrobacterium tumefaciens*, *J. Bacteriol.* **177**, 1076–1081.

Micheel, F., Brodde, O. E. (1974) Polymerization of 1,4-anhydro-2,3,6-tri-*O*-benzyl-α-D-glucopyranose, *Liebigs Ann. Chem.* **124**, 702–708.

Minakami, H., Entani, K., Tayama, S., Fujiyama, S., Masai, H. (1984) Isolation and characterization of a new polysaccharide-producing *Acetobacter* sp. *Agric. Biol. Chem.* **48**, 2405–2414.

Mondal, I. H., Kai, A. (2000) Control of the crystal structure of microbial cellulose during nascent stage, *J. Appl. Polym. Sci.* **79**, 1726–1734.

Nakai, T., Tonouchi, N., Konishi, T., Kojima, Y., Tsuchida, T., Yoshinaga, F., Sakai, F., Hayashi, T. (1999) Enhancement of cellulose production by expression of sucrose synthase in *Acetobacter xylinum*, *Proc. Natl. Acad. Sci. USA* **96**, 14–18.

Nakatsubo, F., Takano, T., Kawada, T., Murakami, K. (1989) Toward the synthesis of cellulose: synthesis of cellooligosaccharides, in: *Cellulose: Structural and Functional Aspects* (Kennedy, J. F., Philips, G. O., Williams, P. A., Eds.), New York: Ellis Horwood, 201–206.

Nari, J., Noat, G., Richard, J., Franchini, E., Monstacas, A. M. (1983) Catalytic properties and tentative function of a cell wall β-glucosyltransferase from soybean cells cultured *in vitro*, *Plant Sci. Lett.* **28**, 313–320.

Naritomi, T., Kouda, T., Yano, H., Yoshinaga, F. (1998) Effect of ethanol on bacterial cellulose production from fructose in continuous culture, *J. Ferment Bioeng.* **85**, 598–603.

Nichols, S. E., Singletary, G. W. (1998) Cellulose synthesis in the storage tissue of transgenic plants, US patent 5 723 764.

Ohana, P., Delmer, D. P., Volman, G., Benziman, M. (1998) Glycosylated triterpenoid saponin: a specific inhibitor of diguanylate cyclase from *Acetobacter xylinum*, *Plant Cell Physiol.* **39**, 153–159.

Oikawa, T., Kamatani, N., Kaimura, T., Ameyama, M., Soda, K. (1997) Endo-β-glucanase from *A. xylinum* – purification and characterization, *Curr. Microbiol.* **34**, 309–313.

Okamoto, T., Yamano, S., Ikeaga, H., Nakamura, K. (1994) Cloning of the *A. xylinum* cellulase gene and its expression in *E. coli* and *Zymomonas mobilis*, *Appl. Microbiol. Biotechnol.* **42**, 563–568.

Okiyama, A., Shirae, H., Kano, H., Yamanaka, S. (1992) Two-stage fermentation process for cellulose production by *Acetobacter aceti*, *Food Hydrocolloids* **6**, 471–477.

Palonen, H., Tenkanen, M., Linder, M. (1999) Dynamic interaction of *Trichoderma reesei* cellobiohydrolases Cel6A and Cel7A and cellulose at equilibrium and during hydrolysis, *Appl. Environ. Microbiol.* **65**, 5229–5233.

Ring, D. F., Nashed, W., Dow, T. (1986) Liquid loaded pad for medical applications, US patent 4 588 400.

Ross, P., Weinhouse, H., Aloni, Y., Michaeli, D., Ohana, P., Mayer, R., Braun, S., de Vroom, E., van der Marel, G. A., van Boom, J. H., Benziman, M. (1987) Regulation of cellulose synthesis in *Acetobacter xylinum* by cyclic diguanylic acid, *Nature* **325**, 279–281.

Ross, P., Mayer, R., Benziman, M. (1991) Cellulose biosynthesis and function in bacteria, *Microbiol. Rev.* **55**, 35–58.

Sakair, N., Asamo, H., Ogawa, M., Nishi, N., Tokura, S. (1998) A method for direct harvest of bacterial cellulose filaments during continuous cultivation of *Acetobacter xylinum*, *Carbohydr. Polym.* **35**, 233–237.

Samejima, M., Sugiyama, J., Igarashi, K., Eriksson, K. E. L. (1997) Enzymatic hydrolysis of bacterial cellulose, *Carbohydr. Res.* **305**, 281–288.

Sattler, K., Fiedler, S. (1990) Production and application of bacterial cellulose. II. Cultivation in a rotating drum fermentor, *Zbl. Microbiol.* **145**, 247–252.

Savidge, R. A., Colvin, J. R. (1985) Production of cellulose and soluble polysaccharides by *Acetobacter xylinum*, *Can. J. Microbiol.* **31**, 1019–1025.

Saxena, I. M., Brown, R. M., Jr. (1989) Cellulose biosynthesis in *Acetobacter xylinum*: a genetic approach, in: *Cellulose and Wood Chemistry and Technology* (Schnerck, C., Ed.). New York: John Wiley & Sons, 537–557.

Saxena, I. M., Brown, R. M., Jr. (1997) Identification of cellulose synthase(s) in higher plants: Sequence analysis of processive β-glycosyltransferases with common motif 'D, D, D 35Q (RQ)XRW', *Cellulose* **4**, 33–49.

Saxena, I. M., Brown, R. M., Jr. (2000) Cellulose synthases and related proteins, *Curr. Opin. Plant Biol.* **3**, 523–531.

Saxena, I. M., Lin, F. C., Brown, R. M., Jr. (1990a) Cloning and sequencing of the cellulase synthase catalytic subunit gene of *Acetobacter xylinum*, *Plant Mol. Biol.* **15**, 673–683.

Saxena, I. M., Roberts, E. M., Brown, R. M., Jr. (1990b) Modification of cellulose normally synthesised by cellulose-producing microorganisms, US patent 4 950 597.

Saxena, I. M., Lin, F. C., Brown, R. M., Jr. (1991) Identification of a new gene in an operon for cellulose biosynthesis in *Acetobacter xylinum*, *Plant Mol. Biol.* **16**, 947–954.

Saxena, I. M., Brown, R. M., Jr., Fevre, M., Geremia, R. A., Henrissat, B. (1995) Multidomain architecture of β-glycosyl transferase: Implications for mechanism of action, *J. Bacteriol.* **177**, 1419–1424.

Seto, H., Tsuchida, T., Yoshinaga, F., Beppu, T., Horinouchi, S. (1996) Characterization of the biosynthetic pathway of cellulose from glucose and fructose in *Acetobacter xylinum*, *Biosci. Biotechnol. Biochem.* **60**, 1377–1379.

Seto, H., Tsuchida, T., Yoshinaga, F. (1997) Manufacture of bacterial cellulose by mixed culture of microorganisms, Japanese patent 9721905.

Srisodsuk, M., Kleman-Leyer, K., Keranen, S., Kirk, T. K., Teeri, T. T. (1998) Modes of action on cotton and bacterial cellulose of a homologous endoglucanase-exoglucanase pair from *Trichoderma reesei*, *Eur. J. Biochem.* **62**, 185–187.

Stalbrandt, H., Mansfield, S. D., Saddler, J. N., Kilburn, D. G., Warren, R. A. J., Gilkes, N. R. (1998) Analysis of molecular size of cellulose by recombinant *Cellulomonas fimi* β-1,4-glucanase, *Appl. Environ. Microbiol.* **64**, 2374–2379.

Standal, R., Iversen, T. G., Coucheron, D. H., Fjaervik, E., Blatny, J. M., Valla, S. (1994) A new gene required for cellulose production and gene encoding cellulolytic activity in *A. xylinum* are localized with *bcs* operon, *J. Bacteriol.* **176**, 665–672.

Sutherland, I. W. (1998) Novel and established applications of microbial polysaccharides, *TIBTECH* **16**, 41–46.

Swissa, M., Aloni, Y., Weinhouse, H., Benziman, M. (1980) Intermediary steps in cellulose synthesis in *Acetobacter xylinum*: studies with whole cells and cell-free preparation of the wild type and a cellulose-less mutant, *J. Bacteriol.* **143**, 1142–1150.

Tahara, N., Tabuchi, M., Watanabe, K., Yano, H., Morinaga, Y., Yoshinaga, F. (1997) Degree of polymerisation of cellulose from *A. xylinum* BPR 2001 decreased by cellulase producing strain, *Biosci. Biotechnol. Biochem.* **61**, 1862–1865.

Tahara, N., Tonouchi, M., Yano, H., Yoshinaga, F. (1998) Purification and characterization of exo-1,4-β-glucosidase from *A. xylinum* BPR 2001, *J. Ferment. Bioeng.* **85**, 589–594.

Takai, M., Tsuta, Y., Watanabe, S. (1997) Biosynthesis of cellulose by *Acetobacter xylinum* and characterization of bacterial cellulose, *Polym. J.* **7**, 137–146.

Tal, R., Wong, H. C., Calhon, R., Gelfand, D., Fear, A. L., Volman, G., Mayer, R., Ross, P., Amikam, D., Weinhouse, H., Cohen, A., Sapir, S., Ohana, P., Benziman, M. (1998) Three *cdg* operons control cellular turnover of c-di-GMP in *Acetobacter xylinum*: genetic organization and occurrence of conserved domain in isozymes, *J. Bacteriol.* **180**, 4416–4425.

Tayama, K., Minakami, H., Entani, E., Fujiyama, S., Masai, H. (1985) Structure of an acidic polysaccharide from *Acetobacter*, sp. NBI 1022, *Agric. Biol. Chem.* **49**, 959–966.

Teeri, T. T. (1997) Crystalline cellulose degradation: new insight into the function of cellobiohydrolases, *TIBTECH* **15**, 160–167.

Tonouchi, N., Tsuchida, T., Yoshinaga, F., Beppu, T. (1996) Characterization of the biosynthetic pathway of cellulose from glucose and fructose in *Acetobacter xylinum*, *Biosci. Biotechnol. Biochem.* **60**, 1377–1379.

Tonouchi, N., Tahara, N., Kojima, Y., Nakai, T., Sakai, F., Hayashi, T., Tsuchida, T., Yoshinaga, F. (1997) A β-glucosidase gene downstream of the cellulase synthase operon in cellulase producing *Acetobacter*, *Biosci. Biotechnol. Biochem.* **61**, 1789–1790.

Tonouchi, N., Hirinouchi, S., Tsuchida, T., Yoshinaga, F. (1998) Increased cellulose production by *Acetobacter* after introducing the sucrose phosphorylase gene, *Biosci. Biotechnol. Biochem.* **62**, 1778–1780.

Toyosaki, H., Naritomi, T., Seto, A., Matsuoka, M., Tsuchida, T., Yoshinga, F. (1995) Screening of bacterial cellulose-producing *Acetobacter* strains suitable for agitated culture, *Biosci. Biotechnol. Biochem.* **59**, 1498–1502.

Valla, S., Kjosbakken, J. (1981) Isolation and characterization of a new extracellular polysaccharide from a cellulose-negative strain of *Acetobacter xylinum*, *Can. J. Microbiol.* **27**, 599–603.

Valla, S., Coucheron, D. H., Fjaervik, E., Kjosbakken, J., Weinhose, H., Ross, P., Amikam, D., Benziman, M. (1989) Cloning of a gene involved in cellulose biosynthesis in *Acetobacter xylinum*: complementation of cellulose-negative mutant by the UDPG pyrophosphorylase structure gene, *Mol. Gen. Genet.* **217**, 26–30.

Vandamme, E. J., De Baets, S., Vanbaelen, A., Joris, K., De Wulf P. (1998) Improved production of bacterial cellulose and its application potential, *J. Polymer Degrad. Stabil.* **59**, 93–99.

Watanabe, K., Tabuchi, M., Morinaga, Y., Yoshinaga, F. (1998a) Structural features and properties of bacterial cellulose produced in agitated culture, *Cellulose* **5**, 187–200.

Watanabe, K., Tabuchi, M., Ischikawa, M., Takemura, H., Tsuchida, T., Morinaga, Y. (1998b)

Acetobacter xylinum mutant with high cellulose productivity and ordered structure, *Biosci. Biotechnol. Biochem.* **62**, 1290–1292.

Weinhouse, R. (1977) Regulation of carbohydrate metabolism in *Acetobacter xylinum*, PhD thesis, Hebrew University of Jerusalem. Jerusalem, Israel.

Weinhouse, H., Sapir, S., Amikam, D., Shiko, Y., Volman, G., Ohana, P., Benziman, M. (1997) c-di-GMP-binding protein, a new factor regulating cellulose synthesis in *Acetobacter xylinum*, *FEBS Lett.* **416**, 207–211.

Wiliams, W. S., Cannon, R. E. (1989) Alternative environmental roles for cellulose produced by *Acetobacter xylinum*, *Appl. Environ. Microbiol.* **55**, 2448–2452.

Winkelman, J. W., Clark, D. P. (1984) Proton suicide: general method for direct selection of sugar transport and fermentation-defective mutants, *J. Bacteriol.* **11**, 687–690.

Wlochowicz, A. (2001) Personal communication.

Wong, H. C., Fear, A. L., Calhoon, R. D., Eichinger, G. M., Mayer, R., Amikam, D., Benziman, M., Gelfand, D. H., Meade, J. H., Emerick, A. W., Bruner, R., BenBassat, A., Tal, R. (1990) Genetic organization of cellulose synthase operon in *A. xylinum*, *Proc. Natl. Acad. Sci. USA* **87**, 8130–8134.

Yamada, Y., Hoshino, K., Ishikawa, T. (1997) The phylogeny of acetic acid bacteria based on the partial sequences of 16 S ribosomal RNA: the elevation of the subgenus *Gluconacetobacter* to the generic level, *Biosci. Biotechnol. Biochem.* **61**, 1244–51.

Yamada, Y., Hoshino, K., Ishikawa, T. (1998) *Gluconacetobacter corrig. (Gluconoacetobacter* [sic]), in: Validation of publication of new names and new combinations previously effectively published outside the IJSB, List no 64, *Int. J. Syst. Bacteriol.* **48**, 327–328.

Yamada, Y. (2000) Transfer *Acetobacter oboediens* and *A. intermedius* to the genus *Gluconacetobacter* as *G. oboediens* comb. nov. and *G. intermedius* comb. nov., *Int. J. System. Evolut. Microbiol.* **50**, 2225–2227.

Yamamoto, H., Horii, F., Hirai, A. (1996) *In situ* crystallization of bacterial cellulose. 2. Influence of different polymeric additives on the formation of celluloses I_α and I_β at the early stage of incubation, *Cellulose* **3**, 229–242.

Yamanaka, S., Watanabe, K., Kitamura, N., Iguchi, M., Mitsuhashi, S., Nishi, Y., Uryu, M. (1989) The structure and mechanical properties of sheets prepared from bacterial cellulose, *J. Mater. Sci.* **24**, 3141–3145.

Yamanaka, S., Watanabe, K., Suzuki, Y. (1990) Hollow microbial cellulose, process for preparation thereof, and artificial blood vessel formed of said cellulose, European patent 0396344A2.

Yamanaka, S., Ishihara, M., Sugiyama, J. (2000) Structural modification of bacterial cellulose, *Cellulose* **7**, 213–225.

Zaar, K. (1977) The biogenesis of cellulose by *Acetobacter xylinum*, *Cytobiologie* **16**, 1–15.

3
Carrageenan

Dr. Ir. Fred van de Velde[1], Dr. Gerhard A. De Ruiter[2]

[1] Wageningen Centre for Food Sciences and TNO Nutrition and Food Research Institute, Carbohydrate Technology Department; present address NIZO food research, PO Box 20, 6710 BA Ede, The Netherlands; Tel.: +31-318-659-582; Fax: +31-318-650-400; E-mail: Fvelde@nizo.nl

[2] NIZO food research, Product Technology Department, PO Box 20, 6710 BA Ede, The Netherlands; Tel.: +31-318-659-636; Fax: +31-318-650-400; E-mail: DeRuiter@nizo.nl

1	**Introduction**	87
2	**Historical Outline**	87
3	**Chemical Structure**	88
3.1	General Description	88
3.2	Molecular Structure	89
4	**Occurrence**	90
5	**Physiological Function**	92
6	**Chemical Analysis**	92
6.1	Isolation and Fractionation	92
6.1.1	Isolation	92
6.1.2	Fractionation	93
6.1.3	Separation of Low-molecular-mass Fractions	93
6.2	Infrared Spectroscopy	93
6.3	Nuclear Magnetic Resonance Spectroscopy	94
6.4	Chromatographic Analysis	96
6.4.1	Molecular Mass Determination	96
6.4.2	Sulfate Content	96
6.4.3	Monosaccharide Composition	97
6.4.4	Glycosidic Linkage Analysis	97

Polysaccharides and Polyamides in the Food Industry. Properties, Production, and Patents.
Edited by A. Steinbüchel and S. K. Rhee
Copyright © 2005 WILEY-VCH Verlag GmbH & Co. KGaA, Weinheim
ISBN: 3-527-31345-1

7	**Biosynthesis**	97
8	**Extracellular Biodegradation**	98
8.1	Enzymology of Degradation	98
8.2	Genetic Basis of Degradation	99
9	**Production**	99
9.1	Seaweed Harvesting	99
9.2	Seaweed Farming	99
9.3	Manufacturing	100
9.4	Modified Carrageenan Functionalities	101
9.5	Current World Market	102
9.6	Companies Producing Carrageenan	102
10	**Properties**	102
10.1	Physical Properties	102
10.1.1	Solubility	102
10.1.2	Coil-helix Transitions	103
10.1.3	Viscosity	104
10.1.4	Gelation	104
10.1.5	Synergism with Gums	105
10.1.6	Interaction with Proteins	105
10.2	Chemical Properties	105
10.3	Safety	106
11	**Applications**	106
11.1	Technical Applications	106
11.2	Medical Applications	106
11.3	Excipient Applications in Drugs	106
11.4	Personal Care and Household	107
11.5	Agriculture	107
11.6	Food Application	107
11.7	Other Applications	109
12	**Relevant Patents**	109
13	**Current Problems and Limitations**	109
14	**Outlook and Perspectives**	110
15	**References**	111

AMF	alkali-modified flour
ARC	alternatively refined carrageenan
ERF	enzyme-resistant fractions

GC	gas chromatography
GPC	gel permeation chromatography
GRAS	generally recognized as safe
HPAEC	high-performance anion-exchange chromatography
HPLC	high-performance liquid chromatography
IR	infrared
MALLS	multi-angle laser-light scattering
MMB	methylmorpholine-borane
NMR	nuclear magnetic resonance
PES	processed *Eucheuma* seaweed
PNG	Philippines natural grade
SEC	size exclusion chromatography
SRC	semi-refined carrageenan

1
Introduction

Carrageenan is the generic name for a family of gel-forming, viscosifying polysaccharides that are obtained commercially by extraction of certain species of red seaweeds (*Rhodophyceae*). The main species responsible for most of today's carrageenan production belong to the following genera:

- *Gigartina* (Argentina/Chile, France, Morocco),
- *Chondrus* (France, North Atlantic),
- *Iridaea* (Chile), and
- *Eucheuma* (Philippines/Indonesia).

Carrageenans are composed of a linear galactose backbone with a varying degree of sulfatation (between 15% and 40%). Different carrageenan types differ in composition and conformation, resulting in a wide range of rheological and functional properties. Carrageenans are used in a variety of commercial applications as gelling, thickening, and stabilizing agents, especially in food products such as frozen desserts, chocolate milk, cottage cheese, whipped cream, instant products, yogurt, jellies, pet foods, and sauces. Aside from these functions, carrageenans are used in pharmaceutical formulations, cosmetics, and industrial applications such as mining.

2
Historical Outline

For several hundred years, carrageenan has been used as a thickening and stabilizing agent in food in Europe and the Far East. In Europe the use of carrageenan started more than 600 years ago in Ireland. In the village of Carraghen on the south Irish coast, flans were made by cooking the so-called Irish moss (red seaweed species *Chondrus crispus*) in milk. The name carrageenin, the old name for carrageenan, was first used in 1862 for the extract from *C. crispus* and was dedicated to this village (Tseng, 1945). Schmidt described the extraction procedure in 1844.

Since the 19th century, Irish moss also has been used for industrial beer clarification and textile sizing. The commercial production began in the 1930s in the U.S. During that time, the trading shifted from dried seaweed meal to refined carrageenan (Therkelsen, 1993). After the Second World War, a general increase in the standard of living forced an increase in carrageenan production.

Fractionation of crude carrageenan extracts started in the early 1950s (Smith et al., 1955), resulting in the characterization of the different carrageenan types. A Greek prefix was introduced to identify the different carrageenans. In the same period, the molecular structure of carrageenans was determined (O'Neill 1955a,b). The structure of 3,6-anhydro-D-galactose in κ-carrageenan, as well as the type of linkages between galactose and anhydrogalactose rings, was determined.

Today, the industrial manufacture of carrageenan is no longer limited to extraction from Irish moss, and numerous red seaweed species are used. Traditionally, these seaweeds have been harvested from naturally occurring populations. Seaweed farming to increase the production started almost 200 years ago in Japan. Scientific information about the seaweed life cycles allowed artificial seeding in the 1950s. Today, nearly a dozen seaweed taxa are cultivated commercially, lowering the pressure on naturally occurring populations.

During the past few years, the total carrageenan market has shown a growth rate of 3% per year, reaching estimated worldwide sales of 310 million US$ in 2000. At the end of the 20th century, a few large corporations that account for over 80% of the supply dominate the carrageenan market, including:

- FMC Corporation (USA),
- CP Kelco (USA),
- Degussa (Germany),
- Danisco (Denmark),
- Ceamsa (Spain), and
- Quest International (The Netherlands).

3
Chemical Structure

3.1
General Description

Carrageenan is a high molecular mass material with a high degree of polydispersity. The molecular mass distribution varies from sample to sample, depending upon the sample history, e.g., age of the harvested seaweed, season of harvesting, way of extracting, and duration of heat treatment. Commercial (food-grade) carrageenans have a weight average molecular mass (M_w) ranging from 400–600 kDa with a minimum of 100 kDa. This minimum is set in response to reports of cecal and colonic ulceration induced by highly degraded carrageenan. In 1976 the U.S. Food and Drug Administration defined food-grade carrageenan as having a water viscosity of no less than 5 mPa·s (5 cP) at 1.5% concentration and 75°C, which corresponds to the above-mentioned 100 kDa.

Besides the traditionally extracted carrageenan, called refined carrageenan in trade, a new type of carrageenan product is promoted by a group of Philippine producers (Seaweed Industry Association of the Philippines). This product is marketed under the name Philippines natural grade (PNG). Other synonyms for this type of carrageenan product are processed *Eucheuma* seaweed (PES is the regulatory name), semi-refined carrageenan (SRC), alternatively refined carrageenan (ARC), and alkali-modified flour (AMF). These *Eucheuma* seaweeds (*E. cottonii* and *E. spinosum*) are harvested around Indonesia and the Philippines and treated with a more cost-effective process that avoids extraction of carrageenan in dilute solutions (see Section 9.3). The above-mentioned carrageenan differs from the traditionally refined carrageenan in that it contains 8% to 15% acid-insoluble matter compared with 2% in extracted carrageenan. The acid-insoluble matter consists mainly of cellulose, which is normally present in algae cell walls. Also, the heavy metal content of processed *Eucheuma* seaweed is higher than that of traditionally refined carrageenan (Imeson, 2000). The water-soluble component in PES is κ-carrageenan and is almost indistinguishable from the refined carrageenan. The molecular mass of κ-carrageenan present in PES can be slightly higher than that of refined carrageenan (Hoffmann et al., 1996).

3.2
Molecular Structure

Carrageenan is not a single biopolymer but a mixture of water-soluble, linear, sulfated galactans. They are composed of alternating 3-linked β-D-galactopyranose (G-units) and 4-linked α-D-galactopyranose (D-units) or 4-linked 3,6-anhydrogalactose (DA-units), forming the "ideal" disaccharide-repeating unit of carrageenans (see Figure 1). The sulfated galactans are classified according to the presence of the 3,6-anhydrogalactose on the 4-linked residue and the position and number of sulfate groups. For commercial carrageenan, the sulfate content falls within the range of 22% to 38% (w/w). Besides galactose and sulfate, other carbohydrate residues (e.g., xylose, glucose, and uronic acids) and substituents (e.g., methyl ethers and pyruvate groups) are present in carrageenans. Since natural carrageenan is a mixture of nonhomologous polysaccharides, the term disaccharide-repeating unit refers to the idealized structure. To describe more complex structures, a letter-code-based nomenclature for red algae galactans has been developed (Knutsen et al., 1994).

The most common types of carrageenan are traditionally identified by a Greek prefix. The three commercially most important carrageenans are called ι-, κ-, and λ-carrageenan. The corresponding IUPAC (International Union of Pure and Applied Chemistry) names and letter codes are carrageenose 2,4′-sulfate (G4S-DA2S), carrageenose 4′-sulfate (G4S-DA), and carrageenan 2,6,2′-trisulfate (G2S-D2S,6S). ι- and κ-carrageenan are gel-forming carrageenans, whereas λ-carrageenan is a thickener/viscosity builder.

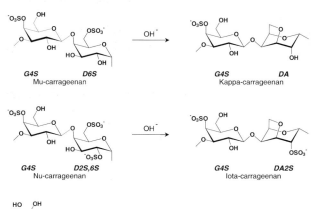

Fig. 1 Schematic representation of the different idealized repeating units of carrageenans. The letter codes refer to the alternative nomenclature (Knutsen et al., 1994).

The difference in rheological behavior between ι- and κ-carrageenan on the one hand and λ-carrageenan on the other results from the fact that the DA-units of the gelling ones have the $^1C_4$-conformation that results from the 3,6-anhydro bridges and λ-carrageenan does not. The natural precursors of ι- and κ-carrageenan are called ν- and μ-carrageenan (letter code G4S-D2S,6S and G4S-D6S, respectively) and are also non-gelling carrageenans with the D-units in the $^4C_1$-conformation as a consequence of the absence of the 3,6-anhydro bridge.

The 3,6-anhydro bridges are formed by the elimination of the sulfate from the C-6 sulfate ester of the precursors and the concomitant formation of the 3,6-anhydro bridge. *In vivo*, ι- and κ-carrageenan are formed enzymatically from their precursors, by a sulfohydrolase (see also Section 7). In industrial processing, the cyclization reaction is carried out with OH⁻ as a catalyst. The $^1C_4$-conformation of the 3,6-anhydro-D-galactopyranosyl units in ι- and κ-carrageenan allows for a helical tertiary structure, which is essential for the gel-forming properties. Occurrence of disaccharide units without the 3,6-anhydro ring and, as a consequence, with a $^4C_1$-conformation causes "kinks" in the regular chain and prevents the formation of helical strands, thus, preventing the gelation of the carrageenan.

4
Occurrence

All of the seaweeds that produce carrageenan as their main cell-wall material belong to the class of the red algae, or Rhodophyceae (Figure 2). Different seaweed species produce different types of carrageenans. The

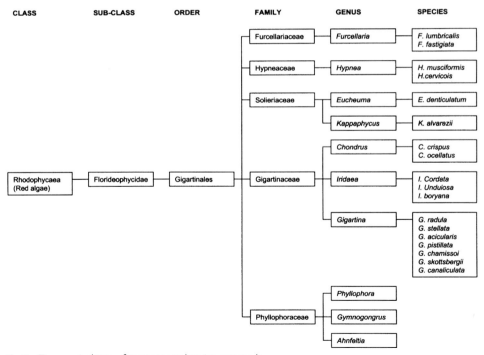

Fig. 2 Taxonomical tree of carrageenan-bearing seaweeds.

tropical seaweed *Kappaphycus alvarezii*, known in trade as *Eucheuma cottonii* (or simply cottonii), yields a relatively homogeneous κ-carrageenan after alkali treatment. This seaweed accounts for the largest production worldwide (De Ruiter and Rudolph, 1997; Rudolph, 2000). Another important species is *Eucheuma denticulatum* (trade name Spinosum), which yields ι-carrageenan upon alkali treatment. λ-Carrageenan is obtained from different species from the *Gigartina* and *Chondrus* genera. These seaweeds have a life cycle with alternating gametophytic (male and female; 1n: one set of chromosomes each) and sporophytic (2n: two sets of chromosomes) phases. The sporophytic plants produce λ-carrageenan, whereas the gametophytic plants produce a κ/ι-hybrid type of carrageenan. These κ/ι-hybrid carrageenans consist of a mixed chain containing both κ- and ι-units (VandeVelde et al., 2001) and range from almost pure ι-carrageenan to almost pure κ-carrageenan (Bixler, 1996).

Kappaphycus alvarezii (Figure 3a) has a natural distribution in the Philippines, Indonesia, and East Africa. It grows on the inner side of coral reefs in the upper subtidal zone. It has a bushy thallus consisting of numerous round branches. The surface can be both rough and smooth with a green to brown color. Its normal size is around 20–30 cm in diameter, but large ones can grow to a size of 1 m.

Eucheuma denticulatum (Figure 3b) has the same natural distribution as *K. alvarezii*. Their morphologies are also the same, except that the branches of *E. denticulatum* have spines of 3–4 mm all over the surface of the thallus and are smaller and more slender. The average size is around 75 cm in diameter.

Chondrus crispus (Figure 3c) is distributed from Norway to Morocco and from Newfoundland (Canada) to Cape Cod (Massachusetts). It grows from a basal disc that is adhered to rocks or stones. *Chondrus* is found from the upper subtidal zone down to a depth of 5 to 6 m. It has a dark violet thallus, with numerous dichotomous branches and has a size of around 25 cm (Rudolph, 2000).

Besides the above-mentioned three important species, several other species are used commercially. *Gracilaria* is harvested in many regions, especially in Chile, Indonesia, Namibia, Japan, Thailand, Taiwan, and Vietnam. *Gelidium* species are found all over the world, but the commercial harvesting is restricted to Morocco, Senegal, New Zealand, Japan, Chile, Venezuela, Spain, and Portugal. *Gigartina radula* is harvested in Chile and Argentina. *Hypnea* is found in the Caribbean Sea, Indian Ocean, and the tropical Pacific Ocean, but commercial harvesting is focused in Brazil.

Fig. 3 Photographs of the red seaweeds *Kappaphycus alvarezii* (Philippines), *Eucheuma denticulatum* (Philippines), and *Chondrus crispus* (Canada); copyright Dr. B. Rudolph (CP Kelco, Lille Skensved, Denmark).

5
Physiological Function

The majority of the red seaweeds have a cell-wall structure that is analogous to that of the higher plants: cellulosic microfibrils embedded in matrix polysaccharides. Because red algae inhabit aquatic environments, the composition and organization of their intercellular matrices differ from those of land plants. Land plants require a rigid structure to support them against the pull of gravity, making cellulose an important part of the cell wall of these plants. The crystalline structure of cellulose offers a rigid structure. In red algal cell walls, cellulose is only a minor component, varying from 1% to 8% of the dry weight of the seaweed. Marine plants require a more flexible structure to accommodate the varying stresses of tidal and wave motion. An extracellular matrix composed of mainly carrageenan gives the seaweeds the required flexible and gelatinous structure. However, the physiological significance of the composition of the intercellular matrix of red seaweeds in mechanical, hydration, and osmotic regulation in marine environments is still a matter of debate (Kloareg and Quantrano, 1988).

Attempts have been made to visualize the distribution of carrageenans over the cells, cell walls, and intercellular matrix of different red algae using different techniques, such as gold-labeled monoclonal antibodies against either κ- or ι-carrageenan or gold-labeled κ-carrageenases. The different types of carrageenan are distributed in different ways, and distributions differ from species to species. For example, in *K. alvarezii*, κ-carrageenan is located in the intercellular matrix, but not in the cell walls. In *Agardhiella subulata*, ι-carrageenan is found in both the cell walls and the intercellular matrix (Gretz et al., 1997).

6
Chemical Analysis

Chemical analysis of carrageenans is done mostly on isolated and purified carrageenan samples to reveal their (detailed) molecular structure. In the beginning, chemical modification and degradation methods were time-consuming and tedious analytical techniques. In the mid-1970s, a real boost was given by the introduction of nuclear magnetic resonance (NMR) spectroscopy, which was followed by introduction of reductive hydrolysis to preserve the 3,6-anhydro bridge. The most important analytical techniques for carrageenan determination will be discussed below. At present, there is still lack of adequate analytical techniques to determine the amounts, the polydispersity, and the purity of carrageenans in raw materials and food products. The different techniques and approaches used for this type of analysis, e.g., colorimetry, immunoassays, combined with high-performance chromatography (HPLC), and electrophoresis (Roberts and Quemener, 1999), will not be discussed in this section.

6.1
Isolation and Fractionation

6.1.1
Isolation

Commercial extraction of carrageenans is normally performed under strong alkaline conditions at elevated temperatures (see Section 9.3). However, in order to obtain specific carrageenan fractions or carrageenan with a high content of precursor units for analytical and research purposes, several mild extraction methods have been developed. Hot water extraction is probably the oldest and most applied method for the extraction of carrageenan from seaweed. Prior to carrageenan extraction, seaweeds can be extracted with organic solvents, such as acetone, alcohol, or diethyl ether, to remove undesired compounds. By choosing the appropriate extraction time and temperature, different carrageenan fractions are obtained. Salts, such as sodium chloride or sodium bicarbonate, are added to the extraction medium to fine-tune the extraction procedure. Amyloglucosidase can be added to a cooled extraction mixture to digest any floridean starch present. Dialysis may be applied to remove low-molecular-mass impurities, such as salts and/or sugars.

Centrifugation or filtration is applied to remove the undissolved cell-wall material and to obtain a clear solution. As in industrial processing, carrageenans are obtained by precipitation with ethanol or isopropanol followed by a drying step. Lyophilization is an elegant method to obtain carrageenans on a small scale. Dialysis or ion-exchange resins are applied to obtain carrageenans with specific counter ions.

6.1.2
Fractionation

A widely used separation technique is the KCl fractionation for the separation of κ-carrageenan from other types of carrageenans (Smith and Cook, 1953). The method is based on the unique property of κ-carrageenan helices, which cluster together in the presence of potassium ions forming gel structures. Both ι- carrageenan and λ-carrageenan do not have this gel-forming mechanism. The KCl fractionation was developed to separate κ- and λ-carrageenan from a hot water extract of *Chondrus crispus*. The addition of potassium ions to a sufficiently diluted carrageenan solution permits a sharp separation into a κ-carrageenan containing gel phase and a solution containing λ-carrageenan. This technique also has been applied to distinguish mixtures of κ- and ι-carrageenan from κ/ι-hybrid carrageenans (van deVelde et al., 2001).

The leaching method is a modified KCl-fractionation method for the extraction of λ-carrageenan (Stancioff and Stanley, 1969). Dried, powdered crude carrageenan extracts are stirred in a KCl solution, in which the κ-carrageenan fraction remains in swollen gel particles but the λ-carrageenan fraction leaches out. The use of stepwise, increasing KCl concentrations in both methods leads to a series of carrageenan fractions, precipitating at different salt concentrations.

6.1.3
Separation of Low-molecular-mass Fractions

Low-molecular-mass fractions of carrageenans are obtained either by unspecific acid hydrolysis or by specific enzymatic hydrolysis. Enzyme hydrolysates are important for the determination of carrageenan fine structures (Knutsen and Grasdalen, 1992). The first step is the ethanol precipitation of the higher molecular mass or enzyme-resistant fractions (ERF). In a second step, different fractions are obtained by gel filtration and reversed-phase HPLC. Improvements in the performance of chromatographic materials resulted in the development of a rapid analysis method that allowed the separation of oligosaccharides with a degree of polymerization (DP) of 2 to approximately 12 in about 20 min (Knutsen et al. 2001).

6.2
Infrared Spectroscopy

Infrared (IR) spectroscopy has been one of the most frequently used technologies. Infrared absorption results from the stretching and bending vibrations of molecular bonds and thus reflects the molecular structure of the material under study. An overview of the absorption bands of the carrageenan structural elements is given in Table 1. The broad

Tab. 1 Identification of carrageenan types by infrared spectroscopy

Vibrations	Bond(s)/Group(s)	Carrageenan dimeric units				
Wavenumbers (cm^{-1})^a		ι	κ	λ	ν	μ
1240	S=O of sulfate esters	+	+	+	+	+
930	C-O of 3,6-anhydrogalacote (DA)	+	+	−	−	−
845	C-O-S on C-4 of G4S	+	+	−	+	+
830	C-O-S on C-2 of D2S,6S	−	−	+	+	−
820	C-O-S on C-6 of D2S, or D6s	−	−	+	+	+
805	C-O-S on C-2 of DA2S	+	−	−	−	−

^{a} Wavenumbers from Chopin and Whalen (1993) and Stancioff and Stanley (1969) and letter codes from Knutsen et al. (1994).

band at 1240 cm^{-1} is common for all sulfated polysaccharides, as it results for the S=O vibration of the sulfate groups. This band correlates with the sulfate content of the sample. Another important absorption band is the 903 cm^{-1} band, which corresponds with the presence of 3,6-anhydrogalactose bridges in the sample. The 805 cm^{-1} band is restricted to ι-carrageenan and is used to determine the ι-to-κ ratio in κ/ι-hybrid carrageenans.

Infrared spectroscopy has been used to study the composition of carrageenan samples from different origins as well as their fractions obtained by precipitation and fractionation (Correa-Diaz et al., 1990; Stancioff and Stanley 1969). This technique has been limited by the fact that only soluble carrageenans could be measured. The technique is broadened to dried, ground algal material by the introduction of FT IR diffuse reflectance spectroscopy (Chopin and Whalen, 1993). Because its accuracy is low, IR spectroscopy normally is used to quantify the different carrageenan types present in a sample without quantifying their amounts.

6.3
Nuclear Magnetic Resonance Spectroscopy

Today, NMR spectroscopy (both ^{1}H- and ^{13}C-NMR) is one of the standard tools for the determination of the chemical structure of carrageenan samples (Usov, 1998). Prior to recording the spectra, the carrageenan samples are sonicated to reduce the viscosity of the solution (high viscosity results in line broadening). Because of the low natural abundance of the ^{13}C isotope, samples for ^{13}C-NMR are prepared at relatively high concentrations (5% to 10% w/w) compared with ^{1}H-NMR samples (0.5% to 1.0% w/w).

^{13}C-NMR spectroscopy is used mostly to reveal the molecular composition of carrageenan samples, e.g., the proportion of κ-, ι-, λ-, μ-, and ν-repeating units. Quantification of ^{13}C-NMR spectra, based on the intensities of the resonances of the anomeric carbons, is feasible, provided that sufficiently long interpulse delays are used (above 1.5 s is appropriate for the most important carrageenans) (van deVelde et al., 2001). The major drawback of ^{13}C-NMR spectroscopy is its low sensitivity, which results in a lack of information about minor components (components present in amounts below 5%).

The ^{13}C-NMR spectra of κ-, ι-, μ-, and ν-carrageenan have been fully elucidated for many years and used for the determination of carrageenan samples and fractions (Ciancia et al., 1993; Usov et al., 1980). The ^{13}C-NMR spectrum of λ-carrageenan has been a matter of debate; the calculated chemical shifts of λ-carrageenan, together with all

Tab. 2 ^{13}C-NMR spectral assignment of different carrageenans

Carrageenan	Unit	Chemical shift (ppm)[a]					
		C-1	C-2	C-3	C-4	C-5	C-6
ι	G4S	103.1	69.9	77.3	72.5	75.3	61.8
	DA2S	92.4	75.3	78.7	78.9	77.5	70.3
κ	G4S	103.1	70.2	79.1	74.4	75.3	61.8
	DA	95.5	70.2	79.7	78.9	77.3	70.0
λ	G2S	103.9	77.4	75.8	64.6	74.2	61.7
	D2S,6S	92.0	74.8	69.5	80.3	68.7	68.1
ν	G4S	105.3	70.8	80.4	74.4	75.3	61.8
	D2S,6S	98.7	76.8	68.7	79.7	68.7	68.3
μ	G4S	105.3	70.8	78.9	74.4	75.3	61.8
	D6S	98.4	69.1	70.9	79.7	68.7	68.3

[a] Chemical shifts from Ciancia et al. (1993), Falshow and Furneaux (1994) and Stortz et al. (1994) and letter codes from Knutzsen et al. (1994).

other possible carrageenan structures, were published in 1992 (Stortz and Cerezo, 1992). Two years later, the chemical shifts of λ-carrageenan were published; however, some of the observed resonances were not assigned (Stortz et al., 1994). Table 2 gives an overview of the chemical-shift data of ^{13}C-NMR spectra of the major carrageenan types. Pyruvate acetal substitution of the carrageenan backbone results in the appearance of resonances at 25.5 ppm (methyl), 101.6 ppm (acetal), and 175.7 ppm (carboxyl) (Chiovitti et al., 1999). Floridean starch is a reserve polysaccharide in red seaweeds and is identified in carrageenan samples by resonances with chemical shifts of 72.9 and 102.5 ppm (Turquois et al., 1996). Commercial carrageenan samples are often blended with sucrose. This limits the composition determination, as the resonance of the C-1 carbon of α-D-glucose overlaps with that of C-1 of ι-carrageenan (DA2S at 92.4 ppm) (Turquois et al., 1996). Small molecules, such as sugar, can be removed by dialysis prior to NMR analysis.

^{1}H-NMR spectroscopy is much more sensitive than ^{13}C-NMR spectroscopy but shows significantly less chemical shift dispersion. Therefore, ^{1}H-NMR is mainly used for two different purposes. For the first purpose, the high sensitivity is used for a fast analysis of low concentrated samples. The quantification of the different carrageenan types in a sample is based on the fact that the α-anomeric protons give five resolved signals in the region from 5.1–5.7 ppm (see Table 3). The signals for the β-anomeric protons cannot be used for either identification or quantification of the diads. ^{1}H-NMR spectroscopy can in this way be used for routine analysis of commercial carrageenan samples, which are

Tab. 3 ^{1}H-NMR chemical shifts of the α-anomeric protons of carrageenans

Carregeenan	Unit	Chemical shift (ppm)[a]
ι	DA2S	5.32
κ	DA	5.11
λ	D2S,6S	5.59
ν	D2S,6S	5.52
μ	D6S	5.26

[a] Chemical shifts from Chiancia et al. (1993) and Stortz et al. (1994) and letter codes from Knutsen et al. (1994).

often blended with sugars and salts, for their total carrageenan content and molar ratio of the different carrageenan types.

In the second approach, oligosaccharides are prepared by enzymatic hydrolysis of the carrageenan samples by either κ- or ι-specific carrageenases (see Section 8.2). ^{1}H-NMR spectroscopy is used for its high sensitivity and possibility to resolve the detailed molecular structure of these carrageenan oligosaccharides. The higher- and lower-molecular-mass fractions were separated by precipitation methods and further fractionated by gel permeation chromatography (GPC) (see Section 6.1). The obtained fractions are then investigated by high field ^{1}H-NMR spectroscopy. The detailed information about the molecular structure of the different di-, tetra-, and oligosaccharides in the different fractions is used to deduce the structure of the original carrageenan sample (Knutsen, 1992; Knutsen and Grasdalen, 1992).

6.4
Chromatographic Analysis

Chromatographic techniques are applied to study different structural aspects of carrageenans, such as molecular mass distribution, sulfate ester content, monosaccharide composition, and glycosidic linkage analysis.

6.4.1
Molecular Mass Determination

The molecular mass of carrageenans is an important parameter of their rheological behavior (e.g., yield stress and gel strength) and, consequently, their application. Molecular mass distributions are normally determined by size exclusion chromatography (SEC) or GPC with multi-angle laser-light scattering (MALLS) detection (Hoffmann et al., 1996). The chromatographic system separates the carrageenans according to their hydrodynamic volume, giving the molecular mass distribution, whereas the light-scattering detector reveals the absolute molecular mass of each individual peak. However, several problems are associated with SEC or GPC, such as adsorption of the sample onto the column packing, degradation of the polymers due to the high shear forces within the column, and difficulties with separation of the sample with a high-molecular-mass or aggregated material, resulting from the fact that only dissolved material can be analyzed. The above problems are less severe in asymmetrical flow field-flow fractionation. Recently, field-flow fractionation with MALLS detection was used successfully for the molecular mass determination of κ-carrageenans (Viebke and Williams, 2000).

6.4.2
Sulfate Content

Sulfate groups are such an important structural substituent of the galactose backbone of carrageenans that a rapid and reproducible method for sulfate ester content is important. The officially used method for determining the sulfate content in carrageenans is based on the selective hydrolysis of the sulfate ester by acid and subsequent selective precipitation of the sulfate ions as barium sulfate. Barium sulfate is then measured by weighing or turbidimetry as described by the FAO/WHO Joint Expert Committee on Food Additives and most recently updated in 1992 (FAO, 1992). This method has disadvantages: it is laborious and requires gram quantities of sample. A promising development in this area is the use of HPLC to measure free sulfate that is liberated from carrageenans by acid hydrolysis. Typically, this analysis is performed on an HPLC system using an anion-exchange column with conductivity detection (Jol et al., 1999).

6.4.3
Monosaccharide Composition

The monosaccharide composition of carrageenans generally is measured to determine the ratio between galactose and anhydrogalactose. Normal acid hydrolysis of polysaccharides to obtain monosaccharides results in the conversion of anhydrogalactose to either galactose or degradation products, such as 5-hydroxymethylfurfural. The introduction of the reductive hydrolysis remedies this problem. Acid hydrolysis in the presence of a methylmorpholine-borane (MMB) complex produces the alditols of monosaccharides without destroying the 3,6-anhydro moiety. Acetylation of the alditols is necessary for the analysis with gas chromatography (GC) However, this derivatizing step is time-consuming and results in loss of accuracy. Recently, high-performance anion-exchange chromatography (HPAEC) was described as an accurate method for the direct analysis of alditols, obtained by reductive hydrolysis with MMB and trifluoroacetic acid (Jol et al., 1999). This method also reveals the presence of small amounts (below 5 mol%) of glucose and xylose in commercial carrageenan samples. The reductive hydrolysis also can be applied to seaweeds samples and, thus, facilitate chemotaxonomic studies.

6.4.4
Glycosidic Linkage Analysis

The glycosidic linkages in carrageenans can be revealed in principle by traditional methylation analysis. Methylation of carrageenans prior to hydrolysis modifies the unsubstituted hydroxyl groups. The major problem encountered with methylation analysis is the incomplete methylation of hydroxyl groups of sulfated polysaccharides. This problem can be eliminated by conversion of sulfate ester groups into the triethylammonium salts (Stevenson and Furneaux, 1991). Reductive hydrolysis of the methylated carrageenans results in the formation of the corresponding monosaccharide alditols that can be analyzed directly by HPAEC or by GC after acetylation. Information is obtained about the number and distribution of unsubstituted hydroxyl groups. Removal of the sulfate ester groups prior to methylation and hydrolysis reveals the number and position of hydroxyl groups that are involved in linkages between the different monosaccharide units.

7
Biosynthesis

Very little is known about the biosynthesis of carrageenans. The synthesis of the galactan backbone is believed to take place within the Golgi bodies in the cells (Bellion and Hamer, 1981). The sulfate esterification of the galactan backbone is catalyzed by sulfate transferases that occur in the Golgi dictyosomes, not in the cell wall (Craigie, 1990). The mechanisms and enzymes responsible for the introduction of other substituents, such as pyruvate and methoxy groups, await elucidation.

The (biological) precursors of the gel forming κ- and ι-carrageenan are called μ- and ν-carrageenan, respectively, and lack the 3,6-anhydro bridge. The formation of the 3,6-anhydro bridges is catalyzed by sulfohydrolases that are thought to act in the cell wall (Wong and Craigie, 1978). For ι-carrageenan the relationship between precursor and product is clear, but for κ-carrageenan this remains ambiguous because several κ/ι-hybrid carrageenan samples contain only ν-carrageenan as biological precursor (Amimi et al., 2001; VandeVelde et al., 2001). Bellion et al. (1983) report the presence of ν-carrageenan in samples of κ-carrageenan from *K. alvarazii*, known for its production of almost pure κ-carrageenan. Suggestions are made that ν-carrageenan is the common precursor for both κ- and ι-carrageenan. Further research is necessary to elucidate the different stages in the biosynthesis of the different carrageenans.

Recently, two sulfohydrolases from *Chondrus crispus* were purified to electrophoretic homogeneity and cloned (De Ruiter et al., 2000; Genicot et al., 2000). The two sulfohydrolases, sulfohydrolase I and sulfohydrolase II, exhibit the same substrate specificity and are indicated as ν-carrageenan 6-O-sulfohydrolases. Sulfohydrolase I releases sulfate without modifying the viscosity of the polymer. It is therefore likely that this enzyme acts randomly. In contrast, sulfohydrolase II is thought to remove sulfate from the ν-carrageenan processively, resulting in long uninterrupted chains of ι-carrageenan dimers. This results in a significant increase of the polysaccharide gelling behavior, indicating the formation of a helical structure.

8
Extracellular Biodegradation

Carrageenans are degraded by enzymes called carrageenases. Several marine bacteria produce κ- and/or ι-carrageenases that degrade κ- and ι-carrageenan, respectively. The structural biology and molecular evolution of carrageenases have been reviewed recently (Barbeyron et al., 2001), as well as their role in carrageenan biotechnology (De Ruiter and Rudolph, 1997).

8.1
Enzymology of Degradation

As far as we are aware, the first enzyme able to modify carrageenans was described in 1943 by the Japanese researcher Mori (1943), followed 12 years later by a Canadian group (Yaphe and Baxter, 1955). After that time, many carrageenan-modifying enzymes have been described, as listed in Table 1 of the review about carrageenan biotechnology (De Ruiter and Rudolph, 1997). A considerable number of those enzymes have been purified from marine bacteria such as *Alteromonas carrageenovora* (previously called *Pseudomonas carrageenovora*) and from different algae (e.g., *Gigartina* spp.).

The best-characterized enzyme is the κ-carrageenase isolated from *Alteromonas carrageenovora* (Bellion et al., 1982; Knutsen, 1992; Potin et al., 1995). This enzyme acts as an endo-hydrolase specifically hydrolyzing the β-glycosidic linkage between 3,6-anhydro-D-galactose and D-galactose, resulting in the formation of neocarrabiose oligosaccharide DA-G4S with 3-linked β-D-galactopyranose 4-sulfate as reducing end. The lytic mechanism and specificity were revealed by detailed analysis of the reaction products by ^{1}H-NMR (Knutsen and Grasdalen, 1992). ι-Carrageenases from *Alteromonas fortis* and *Zobellia galactanovorans* hydrolyze the β-1,4 linkages of ι-carrageenan (Barbeyron et al., 2001). The hydrolysis is a one-step nucleophilic substitution reaction and results in inversion of the anomeric bond configuration. In contrast to κ-carrageenases (Potin et al., 1995), ι-carrageenases do not show transglycosylation activity (Barbeyron et al., 2001). To our knowledge, no enzymes have been reported that are able to cleave the glycosidic linkage of the resulting neocarrabiose dimers or that cleave the 3,6-anhydro-α-D-galactose linkage, resulting in a reducing end of 3,6-anhydro-galactose residues.

The crystallization and preliminary X-ray analysis are reported for the κ-carrageenase from *P. carrageenovora* and the ι-carrageenase from *A. fortis*. The structure of the κ-carrageenase has been resolved to 1.54 Å resolution (Michel et al., 2001). This was the first three-dimensional structure of a carrageenase. The active site of this enzyme is tunnel-shaped, indicating that the enzyme is able to degrade solid substrates.

8.2
Genetic Basis of Degradation

The gene encoding for the κ-carrageenase from *P. carrageenova* has been cloned from a genomic library of this bacterium (Barbeyron et al., 1994). Based on the amino acid sequence, the enzyme was classified in family 16 of the glycosyl hydrolases of the system designed by Dr. Bernard Henrissat (Potin et al., 1995). The family-16 glycoside hydrolases share two conserved glutamic acid residues and one aspartic acid residue in their active sites. The κ-carrageenase from *Zobellia galactanovorans*, whose gene has been cloned (Barbeyron et al., 1998a), belongs also to this family 16 of glycoside hydrolases. In contrast, the ι-carrageenases do not belong to the family 16 of glycoside hydrolases (Barbeyron et al., 2001). They are totally unrelated to the κ-carrageenases and belong to a new family of glycoside hydrolases.

9
Production

9.1
Seaweed Harvesting

The harvesting of carrageenan-containing seaweeds is labor-intensive, as the largest part is manually collected (Rudolph, 2000; Therkelsen, 1993). This is done in seaweed farming areas for seaweeds that have been thrown ashore after a storm and at low tides for *Iridaea* and *Gigartina* species. Hand- or drag-rakes are used to harvest *C. crispus*, which adheres to rocks. This raking from boats was developed in Canada to harvest fresh seaweed of a good quality. Floating seaweeds, e.g., *Furcellaria*, are easily harvested by trawling. Collection of seaweeds from the shore has its limitations. The amount that will reach the shore is limited, and seaweeds on the shore decompose fast when exposed to rain and sunlight.

9.2
Seaweed Farming

Harvesting of seaweeds from natural populations always contains the risk of overproduction and loss of the total population; therefore, much effort has been put into seaweed farming. Seaweed farming started almost 200 years ago in Japan. Fisherman stacked brush on the seashore to expand the settling grounds of *Porphyra*. Domestication of seaweed cultivars is important to remove the harvest pressure on target species in sensitive ecosystems. Scientific information about the seaweed life cycles allowed artificial seeding in the 1950s. Today, nearly a dozen seaweed taxa are cultivated commercially, which lowers the pressure on naturally occurring populations. An example of an important success in this field is in the Philippines, where Maxwell S. Doty pioneered the marine culture of *Eucheuma* species (Santelices, 1999). Seedlings of a variety that grows rapidly and that has a good resistance to diseases are tied to lines. The lines are placed at the inside of coral reefs, where the seedlings can grow without floating away. Plants are harvested when they reach a size of 0.8 to 1.2 kg (Rudolph, 2000). Today's production of *Eucheuma* and *Kappaphycus* cultivars counts for most of this country's carrageenan production without the need of harvesting natural resources.

Another approach to seaweed farming is the co-cultivation of seaweeds with intensive fish farming. Intensive fish and shrimp farming results in the release of nutrients that add to costal eutrophication. Seaweeds are successfully used as biofilters to remove these nutrients. Co-cultivation in tank systems or cultivation on ropes near a fish farm increases the seaweed production yield and reduces the amount of nutrients released by the intensive fish and shrimp farming. Studies on co-cultivation started in the mid-1970s and gained new interest at the end of the 20th century (Troell et al., 1999).

9.3 Manufacturing

Harvested seaweed is washed to remove sand and stones, followed by drying to preserve the quality of carrageenan and to reduce weight. In tropical regions, sun drying is mainly used, but in colder climates fuel-fired rotary air dryers are applied. After drying, seaweed is shipped to the production plants or warehouses where it is stored before use. For manufacturing plants that are located near the harvesting area, the use of wet seaweed is economically attractive because it reduces the drying costs.

To produce a certain carrageenan product with constant properties, different batches are combined. The quality of the final product is assured by detailed analysis of the raw material. The manufacturing processes for both refined and semi-refined carrageenans are drawn schematically in Figure 4.

In the refined carrageenan process, the seaweed is extracted under alkaline conditions. In this way, the 3,6-anhydro bridges are formed during the extraction step and a good quality of κ- and ι-carrageenan is obtained. The extraction is performed for several hours at a temperature near the boiling point of the alkaline solution

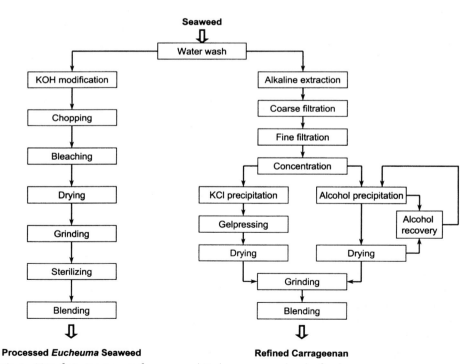

Fig. 4 Manufacturing process for processed *Eucheuma* seaweed and refined carrageenan.

(≥ 110°C). Different types of alkali can be used to manipulate the salt type of the carrageenan produced. Numerous patents describe different conditions to obtain carrageenans with special properties. For example, extraction at lower pH values or lower temperatures, resulting in an increasing amount of precursor units, is applied to obtain special carrageenan preparations (Hansen et al., 2000), which are typically used as an alternative to gelatin in specific applications.

The extract is subsequently filtered by multistage filtration. A coarse filter removes the residual seaweed material (mostly cellulose), and a fine filter is used to obtain a clear liquid. The diluted carrageenan solution is concentrated by multistage vacuum evaporators to a final carrageenan concentration of 3% to increase the yield in the precipitation step. The carrageenan is precipitated by the addition of isopropanol. The gelled fibrous material is pressed to remove the excess of solvent. The alcohol water mixture is distilled and the alcohol reused to reduce process costs. For κ-carrageenan an alternative precipitation protocol, also referred to as "gel press technology", based on the specific gelation of κ-carrageenan in the presence of potassium ions followed by removing the liquid phase, is possible. In this process the carrageenan solution is extruded into a concentrated KCl solution, in which the carrageen precipitates. The gelled material is subsequently pressed followed by a freeze/thaw cycle to remove the water.

The pressed fibers derived from both processes are dried and finally ground to an appropriate particle size. To obtain a product of consistent quality, extracted carrageenan from several batches are blended and/or mixed with salts or sugars. Different additives are introduced to obtain specific properties and to meet customer requirements.

In the manufacture of semi-refined carrageenan, the seaweed is treated with a potassium hydroxide solution at elevated temperature (70°C to 80°C). This temperature is high enough to catalyze the formation of the 3,6-anhydro bridges in both κ- and ι-carrageenan, but is not high enough to allow the carrageenan to be extracted. Most of the low-molecular-mass material is removed by the alkaline treatment. The modified seaweed is chopped and bleached to reduce the color of the final product. After washing, the material is dried and ground to the required particle size. Before blending and standardization, this product is sometimes sterilized to improve the microbiological quality of the final product.

9.4
Modified Carrageenan Functionalities

A continuous challenge for manufacturers of carrageenans is to make new, high value-added products and to widen the applications of carrageenan. An attractive market opportunity occurred in the 1990s when, as a result of "mad cow disease", the food industry started a search for gelatin alternatives for food applications. Although carrageenan cannot replace gelatin in all applications, some very good gelatin mimics are on the market based on ι-carrageenan for use in confectionery, e.g., in wine gums.

A promising development in this area is the production of engineered carrageenans with the help of polysaccharide-modifying enzymes. In collaboration with CP Kelco, Dr. Bernard Kloareg and coworkers (CNRS Station Biologique, Roscoff, France) cloned two sulfohydrolases from *Chondrus crispus* (De Ruiter et al., 2000). These enzymes provide a new tool to produce ν/ι-hybrid carrageenans with special properties, which can be tailor-made for application in food industry.

Sulfated oligogalactans are recognized as valuable, active biocompounds for non-food applications, such as agriculture, cosmetics, or pharmaceuticals. For this reason, a variety of carrageenan modifying enzymes have been purified and cloned. Several carrageenases have been developed in Dr. Kloareg's lab for the selective hydrolysis of κ- or ι-carrageenan (Barbeyron et al., 2001). These enzymes can be used to produce carrageenan oligosaccharides with specific structures and properties. Moreover, the κ-carrageenase from *Pseudoalteromona carrageenovora* was shown to display endo-transglycosylation activity (Potin et al., 1995). This activity may be used to transfer various substituents onto κ-carrageenan oligosaccharides. The patent on this group of enzymes belongs to Goëmar Laboratories (Barbeyron et al., 1998b).

9.5
Current World Market

Carrageenan is the third most important hydrocolloid in the food area worldwide, after gelatin and starch. The total world production of seaweeds for carrageenan extraction is currently around 140 million kg dry weight ("as sold"), giving approximately 26 million kg of carrageenans with an estimated value of 310 million USD in 2000. The total carrageenan market showed a grow rate of 3% per year during the last few years.

9.6
Companies Producing Carrageenan

Today, six companies manufacture over 80% of the supply of carrageenans. Two types of companies can be distinguished: the traditional biopolymer-producing companies and the total ingredient suppliers, which manufacture their own carrageenan for the use in blends. The biopolymer producing companies include:

- FMC Corporation (USA; the former Marine Colloids Inc. known from several innovative patents in the sixties; www.fmcbiopolymer.com);
- CP Kelco (USA; formed in 2000 by merging the Copenhagen Pectin/Food Gums division of Hercules Incorporated and the Kelco Biopolymers group of Monsanto/Pharmacia Corporation; www.cpkelco.com); and
- Ceamsa (Spain; www.ceamsa.com).

The worldwide operating companies that, besides carrageenans, also manufacture a complete range of food ingredients, such as emulsifiers, flavors and so forth, include:

- Degussa (Germany; formerly SKW Biosystems; www.degussa.com or www.texturantsystems.com);
- Danisco (Denmark; www.danisco.com); and
- Quest International (Netherlands; part of ICI; www.questintl.com).

Because of the price erosion of carrageenan during the last 10 years, the use of carrageenan in blends increased at the expense of the traditional biopolymer companies.

10
Properties

10.1
Physical Properties

10.1.1
Solubility

Carrageenans are water-soluble, but their solubility depends on the type of carrageenan, temperature, pH, type, and concentration of counter ions and other solutes present. The most soluble carrageenan is λ-carrageenan, which lacks the more hydrophobic 3,6-anhydrogalactose and has three hydrophilic sulfate ester groups. λ-Carrageenan is readily water-soluble under most conditions. κ-Carrageenan is the most difficult carrageenan to dissolve. The sodium salt of κ-carrageenan is soluble in cold water, whereas the potassium salt of κ-carrageenan is soluble only in hot solutions. ι-Carrageenan has an intermediate solubility. In many practical systems, dispersions of carrageenans are used to add carrageenan to the process.

10.1.2
Coil-helix Transitions

κ-Carrageenan and ι-carrageenan can occur in two forms: the unstructured random coil conformation typically at elevated temperatures and the structured double helices, which are usually formed upon cooling. By cooling hot solutions of the carrageenans, the double helices are formed from the random coils at a certain temperature, called coil-helix transition temperature. This coil-helix transition temperature is an important parameter for the functional properties of carrageenans and depends heavily on intrinsic factors of the carrageenans, such as polydispersity, distribution of sulfate ester groups, amount of precursor units, and extrinsic conditions such as concentration and type of salts, pH, and cooling speed.

At temperatures above 60°C, κ-carrageenan is always in the random coil conformation, and, depending on the detailed chemical structure of the κ-carrageenan and the system, a typical coil-helix transition temperature in water and milk is approximately 38°C. In the presence of potassium ions, the double helices of κ-carrageenan cluster together into bigger units that are called "fibers" (Snoeren and Payens, 1976), "helix aggregates" (Morris et al., 1980), "superstrands" (Hermansson et al., 1991), or "superhelical rods" (Piculell, 1995). Electron microscopy revealed that the maximum size of the aggregates is about 20 nm (Hermansson et al., 1991). Potassium acts as intramolecular glue, forming electrostatic interactions with the sulfate esters and anhydrooxygen atom of κ-carrageenan (Figure 5) (Tako and Nakamura, 1986). The level of aggregation heavily determines the physical properties such as gel strength, syneresis, etc.

ι-Carrageenan also is always in the random coil conformation at temperatures above 60°C. Upon cooling, helices are formed, but these are very different from those of κ-carrageenan and are independent of the type of cation that is present. As far as is known in the literature, ι-carrageenan does not form big aggregates, as does κ-carrageenan, but under some conditions some aggregation of the helices has been observed (Morris et al., 1980). The coil-helix transition temperature of ι-carrageenan is also very dependent on intrinsic and extrinsic factors and in typical water or milk systems is approximately 45°C. The single helices of ι-carrageenan can be bridged by intermolecular calcium (Figure 6) (Tako et al., 1987).

λ-Carrageenan does not form helices and always is present in the random coil conformation. Therefore, λ-carrageenan cannot form gels.

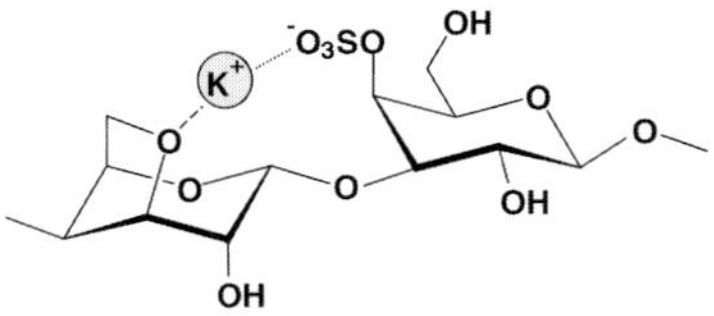

Fig. 5 Intramolecular binding of potassium ions to κ-carrageenan according to Tako and Nakamura (1986).

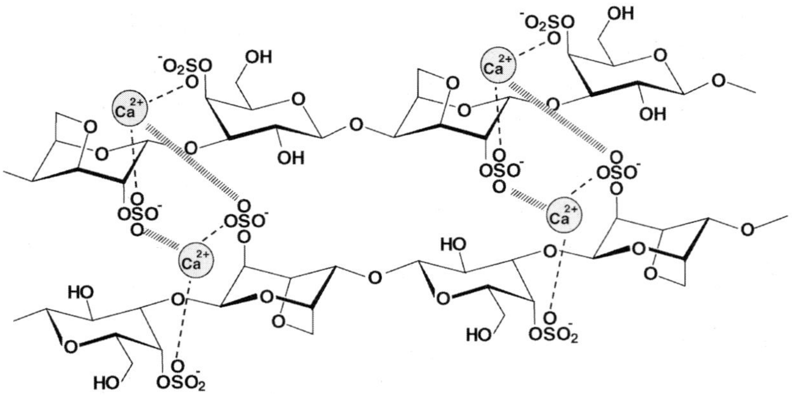

Fig. 6 Schematic representation of the intermolecular Ca^{2+}-bridges between ι-carrageenan helices as proposed by Tako et al. (1987).

10.1.3
Viscosity

Carrageenans above their coil-helix transition temperature typically form highly viscous solutions. In this situation, carrageenan molecules are random coils and are often highly extended because of the electrostatic repulsion of the negatively charged sulfate ester groups along the linear galactan chain. The viscosity increases almost exponentially with concentration. Addition of salts to a carrageenan solution reduces the viscosity because of shielding of the charges on the polymer. Viscosity increases with the molecular mass of the carrageenan, following the so-called Mark-Houwink equation,

$$[\eta] = K (M_w)^a \qquad (1)$$

where $[\eta]$ is the intrinsic viscosity and M_w is the weight-average molecular mass (strictly speaking, the viscosity-average molecular mass). The values for the constants K and a are dependent of the carrageenan type. The exponent a is close to unity, which indicates rigid, rod-like molecules.

10.1.4
Gelation

Gel formation is the most important feature of carrageenans. Only κ- and ι-carrageenan are able to form a gel, as the other carrageenans lack the essential $^1C_4$-conformation that results from the 3,6-anhydrobridge (see Section 3.2). Gels are formed upon cooling of a hot carrageenan solution. The gelling behavior of carrageenans is described in general in several textbooks (Guiseley et al., 1980; Imeson, 2000; Therkelsen, 1993) and in detail in review articles (Lahaye, 2001; Piculell, 1995).

In general terms, κ-carrageenan gels are hard, strong, and brittle gels that are freeze/thaw instable, whereas ι-carrageenan forms soft and weak gels that are freeze/thaw stable. The strength of both types of gels is controlled by the concentration of gelling cations. Both types of gels are thermoreversible, which means that the gels will melt when heated and form a gel again upon cooling. ι-Carrageenan forms gels in the presence of calcium, and the strength is related to the amount of calcium. The gelation of κ-carrageenan is specifically promoted by some monovalent cations

(K^+, Rb^+, Cs^+, and NH_4^+). These cations promote the aggregation of κ-carrageenan double helices to form so-called aggregated "domains" or "superhelical rods" (see Section 10.1.2). This association of helical strains explains the hysteresis observed in optical rotation and rheological measurements of κ-carrageenan. The coil-to-helix transition of ι-carrageenan appears not to be cation-specific. The thermal conformational transition of ι-carrageenan shows no hysteresis, which suggests that there is little or no interhelical aggregation. Therefore, the gel forming in ι-carrageenan gels is assumed to take place at the helical level. The ι-carrageenan gels are thixotropic, which means that a gel, whose structure has been destroyed, will reform if left for an appropriate period of time without disturbance.

10.1.5
Synergism with Gums
κ-Carrageenans show synergism with galactomannan gums, especially with locust bean gum, less with konjac gum and guar gum. A synergistic effect means that the minimum gelling polymer concentration can be reduced even though locust bean gum is non-gelling alone. The mechanism of the interaction is assumed to proceed via alignment of the non-branched mannan backbone regions of the galactomannans with the aggregates of the double helices of κ-carrageenan (Dea and Morrison, 1975; Harding et al., 1995). Addition of galactomannans to κ-carrageenan gels changes the morphology from brittle, rigid gels to strong elastic gels with low syneresis.

Some starches are known to have synergism with ι-carrageenan; however, the detailed mechanism has not been revealed so far.

10.1.6
Interaction with Proteins
In the dairy industry, carrageenans are used for their specific synergistic interactions with milk proteins. In milk systems, five times lower carrageenan concentrations are necessary to obtain a gel compared to water systems.

In milk systems, carrageenan concentrations one-fifth of those required for water systems are necessary to obtain a gel. The synergistic effect is related to the interactions of the negatively charged carrageenans with the positively charged amino acids at the surface of casein micelles, especially between κ-carrageenan and κ-casein. Very low levels of 0.015% to 0.025% w/w of carrageenan are sufficient to prevent whey separation and to stabilize cocoa particles in chocolate milk.

10.2
Chemical Properties

Carrageenans are very sensitive to acid and oxidative breakdown. Cleavage of the glycosidic linkages increases with temperature and time. Acid-catalyzed hydrolysis occurs mainly at the $(1 \rightarrow 3)$ glycosidic linkage. The 3,6-anhydrogalactose-ring system favors the hydrolysis, whereas the presence of a sulfate ester group at the 2-position reduces hydrolysis (Therkelsen, 1993). The acid hydrolysis of κ-carrageenan is faster than that of ι-carrageenan. Depolymerization of carrageenans also could occur by acidification of the sulfate groups (releasing sulfuric acid) as a result of exhaustive dialysis against deionized water. Autohydrolysis also could occur during a lyophilization step normally following the dialysis or when carrageenans are stored for a long time at dry conditions. Carrageenans in the gelled state retain their associated cations and are protected from autohydrolysis (Hoffmann et al., 1996).

10.3
Safety

Seaweed products in general and carrageenans in particular have been used as food ingredient for centuries. Food-grade carrageenans, with a typical weight-average molecular mass above 100 kDa, have been demonstrated to be completely safe and nontoxic (see also Section 11.6) and therefore can be used in food in unlimited amounts. Food-grade carrageenan is not absorbed and there is no evidence that carrageenan is ulcerogenic in humans (Guiseley et al., 1980; Weiner 1991).

11
Applications

Overviews of carrageenan applications are given in almost every textbook dealing with carrageenans (Guiseley et al., 1980; Imeson, 2000; Rudolph, 2000; Therkelsen, 1993). The most important fields of carrageenan application are listed below.

11.1
Technical Applications

Water-based paints and inks can be thickened and stabilized with κ- or ι-carrageenan (0.15% to 0.25% w/w) to prevent settling of the pigment particles. Addition of carrageenan also improves the flow behavior of the paint. ι-Carrageenan (0.25% to 0.8% w/w) is used to suspend and stabilize insolutes in abrasive suspensions and ceramic glazes.

Beverage clarification is performed with all types of carrageenans, κ-carrageenan being the preferred one. Carrageenan added to beer or wine serves as a fining agent by complexation with proteins. Filtration or centrifugation removes insoluble particles. Semi-refined carrageenan or bleached and ground seaweed is preferred over refined carrageenans.

In oil-well drilling, carrageenans and other polysaccharides are used to increase the viscosity of drilling fluid. Increased viscosity enhances the carrying capacity of the fluid.

11.2
Medical Applications

In medical research, carrageenans are used in the so-called rat-paw edema assay for the screening of new drugs (Renn, 1997). Administration of carrageenan fractions directly into the blood system causes a systemic effect on the immune system. Carrageenan injections are used in animal models of inflammation. The rat-paw edema assay has led to the development of several anti-inflammatory drugs.

Today, carrageenans are recognized as agents that can prevent or inhibit development of sexually transferred viral infections (Neushul, 1990; Schaeffer and Krylov, 2000). Thus, carrageenans are considered promising new anti-HIV drugs. The mode of action of carrageenans (and other sulfated polysaccharides) is believed to be inhibition of the virus reverse transcriptase (Nakashima et al., 1987). Seaweed fucans are more active than carrageenans because of their higher degree of sulfatation and, therefore, are preferred for application in drugs.

11.3
Excipient Applications in Drugs

In the pharmaceutical industry, carrageenans are used as excipients because of their physical properties. ι-Carrageenan (0.1% to 0.5% w/w) is used to stabilize emulsions and suspensions of mineral oil and insoluble drug preparation. Complexes between carrageenan and drug molecules can be used for controlled release of pharmaceutical compounds. In tablet manufacture, carrageenan is used as ingredient as well as coating agent.

11.4
Personal Care and Household

One of the largest non-food applications of carrageenan is as viscosifier in high-quality toothpastes. Modern toothpaste is a complex mixture of inorganic and organic compounds suspended in a continuous phase, stabilized by a hydrocolloid. ι-Carrageenan (0.8% to 1.2% w/w) is added to prevent separation of the liquid phase and the abrasive. Addition of carrageenan offers texture, good stand-up, and thixotropic flow. The functional properties of carrageenan are influenced by interactions with the other ingredients, such as pigments, calcium carbonate, dicalcium phosphate, and silica. As carrageenans cannot be degraded easily by enzymes, the use of carrageenan relative to cellulose derivatives, such as carboxymethylcellulose (CMC) improves the stability of the paste during storage and use in areas where high temperature and humidity prevail.

λ-Carrageenan (0.1% to 1.0% w/w) is used in hand lotions and creams to provide slip and improve rubout. Both natural and low-molecular-mass carrageenans are used for the stabilization and thickening of shampoos.

Air-freshener gels are prepared from κ-carrageenans, combined with other gums and gelling salts (up to 2.5% w/w of gums). Fragrance oils and odor-absorbing compounds that are incorporated within the gel release uniformly from the gel surface as the gel dries down.

11.5
Agriculture

Carrageenan oligosaccharides, prepared by either acid hydrolysis or enzymatic degradation, are marketed by Goëmar Laboratories (Saint-Malo, France) as fertilizers and growth biostimulants. When applied at the instant of flowering, these compounds stimulate the nutrition and reproduction of numerous crops, resulting in better flower fertilization and fruit formation. Sulfated oligosaccharides are recognized to stimulate the defense mechanism of plants (Bouarab et al., 1999; Potin et al., 1999). Therefore, oligocarrageenans can be used as a natural growth-enhancing agent.

11.6
Food Application

In the U.S., carrageenan is generally recognized as safe (GRAS) by experts of the Food and Drug Administration (21 CFR 182.7255) and is approved as a food additive (21 CFR 172.620). In the EU, a difference is made between carrageenan and processed *Eucheuma* seaweed. They are approved as E407 and E407a, respectively, in the list of permitted emulsifiers, stabilizers, and thickening and gelling agents. The World Health Organization Joint Expert Committee of Food Additives has concluded that it is not necessary to specify an acceptable daily intake limit for carrageenan. These registrations made carrageenans applicable in food products. The use of carrageenan in food applications has recently been reviewed in the Handbook of Hydrocolloids (Imeson, 2000). The most important food applications of carrageenan are summarized in Table 4.

Tab. 4 Typical food applications for carrageenan[a]

Use	Carrageenan type	Function	Use level (% w/w)
Water dessert gels	Kappa + iota	Gelation	0.6–0.9
	Kappa + iota + LBG[b]		
Low-calorie gels	Kappa + iota	Gelation	0.5–1.0
Cooked flans	Kappa, kappa + iota	Gelation, mouthfeel	0.2–0.3
Cold prepared custards	Kappa, iota, lambda	Thickening, gelation	0.2–0.3
Instant breakfasts	Lambda	Suspension, bodying	0.1–0.2
Chocolate milk	Kappa, lambda	Suspension, mouthfeel	0.01–0.03
Filled milk	Iota, lambda	Emulsifying	0.05
Dairy creams	Kappa, iota	Stabilization of the emulsion	0.01–0.05
Ice cream	Kappa + GG, LBG, XG	Whey prevention, emulsion stabilization, control meltdown	0.01–0.02
Whipped cream	Lambda	Stabilize overrun	0.5–0.15
Soy milk	Kappa + iota	Suspension, mouthfeel	0.02–0.04
Processed cheese	Kappa	Improve slicing, control melting	0.05–3.0
Canned and processed meats	Kappa	Moisture retention, slicing properties	0.6
Salad dressings	Iota	Stabilization of suspended herbs	0.3
Sauces	Kappa	Bodying	0.2–0.5
Pie fillings	Kappa	Reduced starch, lower burn-on	0.1–0.2
Tart glazing	Kappa + LBG	Gelation	0.7–1.0
Syrups	Kappa, lambda	Suspension, bodying	0.3–0.5
Beer or wine fining	Kappa	Complexation of proteins	?
Wine gums	Iota	Creating structure	?

[a] Data from (Guiseley et al. (1980), Imeson (2000), and Therkelsen (1993). [b] LBG = locust bean gum; GG = guar gum; XG = xanthan gum.

11.7
Other Applications

Chibata and coworkers pioneered the use of carrageenans for the immobilization of enzymes and microorganism (Chibata et al., 1979). Especially κ-carrageenan is used for the immobilization of microorganisms, plant cells, and enzymes (Iborra et al., 1997). For this application, the potassium-induced gelation of κ-carrageenan is used. Cells or enzymes are mixed with an aqueous solution of carrageenan, and beads are prepared by dripping this solution in a (cold) 0.3 M KCl solution, in which the κ-carrageenan solution immediately forms a gel. When the biomaterial is thermostable, the thermal reversibility of carrageenan can be used to prepare gels by cooling a hot solution. In this way, various shapes, e.g., cubes, beads, or membranes, of immobilized biocatalysts can be tailored for a particular application.

λ-Carrageenan is used as a chiral selector in capillary electrophoresis separations (Beck et al., 2000). With λ-carrageenan as the chiral selector, various racemic beta-blockers and tryptophan derivatives are successfully separated into their enantiomers.

12
Relevant Patents

Numerous patents describe the extraction/processing procedures for carrageenan manufacture as well as the specific application of carrageenans in particular areas. The patent of Goëmar Laboratories (France) describes the production of glycosylhydrolases, such as κ- and ι-carrageenase, for the biodegradation of carrageenans (Barbeyron et al., 1998b). The patent of CP Kelco (USA) describes the production of two sulfohydrolases from *Chondrus crispus* for the enzyme-catalyzed gelation of ν/ι-mixed carrageenans (De Ruiter et al., 2000). In addition, there is the development of a mild alkaline extraction process to prepare ι-carrageenan that contains a small fraction of ν-repeating units (around 5%) to obtain a ι-carrageenan preparation with enhanced functional properties (Hansen et al., 2000).

13
Current Problems and Limitations

Carrageenans have been used as healthy and natural products in food for centuries. However, reports of cecal and colonal ulceration induced by a highly degraded carrageenan resulted in a negative image of carrageenans. Unfortunately, the carrageenans used in these studies were significantly chemically modified compared to the regular carrageenan polymers used in food for so long; therefore, these results, scientifically speaking, cannot be extrapolated to food carrageenans. Intensive investigation into carrageenan's safety by the FDA resulted in the GRAS-status for food-grade carrageenan in 1976. Unfortunately, this could not remove the public opinion that carrageenan consumption has a negative association.

Further development of carrageenan technology is hampered by the fact that the whole industry has a slight overcapacity, which during the last 10 years resulted in price erosion. This limits the willingness of the traditional biopolymer industry to invest in research and development. While the total annual sales of pectin and carrageenan are almost equal, the number of patents and scientific articles on pectin is estimated to be at least a factor of 10 higher than on carrageenan.

14
Outlook and Perspectives

In order to improve process efficiency, a further increase in the production of semi-refined carrageenan (such as processed *Eucheuma* seaweed) in the country of harvesting will take place. Production in the country of harvesting reduces the transportation costs and the loss of material due to degradation on storage and drying. Development of better technologies for the drying and storage of carrageenan will increase the seaweed quality and will reduce the need for raw materials (Kapraun, 1999).

Because the exploitation of natural populations results in a decrease of resources, the domestication of seaweed cultivars is of greatest importance. The marine culture of commercial seaweed species will result in less environmental impact. The integration of seaweed production with other processes will become more important for the cultivation of seaweeds for carrageenan production. Co-cultivation of seaweeds with intensive marine culture system, e.g., salmon farming, reduces both the nutrient effluent of the salmon farm and the harvest of natural seaweed populations. In an integrated approach of carrageenan production and agriculture, the seaweed residues can be used as fertilizer.

15 References

Amimi, A., Mouradi, A., Givernaud, T., Chiadmi, N., Lahaye, M. (2001) Structural analysis of *Gigartina pistillata* carrageenans (Gigartinaceae, Rhodophyta), *Carbohydr. Res.* **333**, 271–279.

Barbeyron, T., Henrissat, B., Kloareg, B. (1994) The gene encoding the kappa-carrageenase of Alteromonas carrageenovora is related to beta-1,3-1,4-glucanases, *Gene* **139**, 105–109.

Barbeyron, T., Gerard, A., Potin, P., Henrissat, B., Kloareg, B. (1998a) The kappa-carrageenase of the marine bacterium *Cytophaga drobachiensis*. Structural and phylogenetic relationships within family-16 glycoside hydrolases, *Mol. Biol. Evol.* **15**, 528–537.

Barbeyron, T., Henrissat, B., Kloareg, B., Potin, P., Richard, C., Yvin, J. C. (1998b) Glycosylhydrolase genes and their use for producing enzymes for the biodegradation of carrageenans, World patent application: WO9815617.

Barbeyron, T., Flament, D., Michel, G., Potin, P., Kloareg, B. (2001) The sulphated-galactan hydrolases, agarases and carrageenases: structural biology and molecular evolution, *Cah. Biol. Mar.* **42**, 169–183.

Beck, G. M., Neau, S. H., Holder, A. J., Hemenway, J. N. (2000) Evaluation of quantitative structure property relationships necessary for enantioresolution with lambda- and sulfobutylether lambda-carrageenan in capillary electrophoresis, *Chirality* **12**, 688–696.

Bellion, C., Hamer, G. K. (1981) Analysis of kappa-iota hybrid carrageenans with kappa-carrageenase, iota-carrageenase and ^{13}C NMR, in: *Proceedings of the Xth Int. Seaweed Symp.* (Levring, T., Ed.), Berlin: de Gruyter, 379–384.

Bellion, C., Hamer, G. K., Yaphe, W. (1982) The degradation of *Eucheuma spinosum* and *Eucheuma cottonii* carrageenans by ι-carrageenases and κ-carrageenases from marine bacteria, *Can. J. Microbiol.* **28**, 874–880.

Bellion, C., Brigand, G., Prome, J.-C., Bociek, D. W. S. (1983) Identification et caractérisation des précurseurs diologiques des carraghénanes par spectroscopie de RMN-13C, *Carbohdr. Res.* **119**, 31–48.

Bixler, H. J. (1996) Recent developments in manufacturing and marketing carrageenan, *Hydrobiologia* **326/327**, 35–57.

Bouarab, K., Potin, P., Correa, J., Kloareg, B. (1999) Sulfated oligosaccharides mediate the interaction between a marine red alga and its green algal pathogenic endophyte, *Plant Cell* **11**, 1635–1650.

Chibata, I., Tosa, T., Takata, I. (1979) Immobilized catalytically active substance and method of preparing the same, US Patent No. 4,138,292.

Chiovitti, A., Bacic, A., Kraft, G. T., Craik, D. J., Liao, M.-L. (1999) Pyruvated carrageenans from *Solieria robusta* and its adelphoparasite *Tikvahiella candida*, *Hydrobiologia* **398/399**, 401–409.

Chopin, T., Whalen, E. (1993) A new and rapid method for carrageenan identification by FT IR diffuse reflectance spectroscopy directly on dried, ground algal material, *Carbohydr. Res.* **246**, 51–59.

Ciancia, M., Matulewicz, M. C., Finch, P., Cerezo, A. S. (1993) Determination of the structures of cystocarpic carrageenans from *Gigartina skottsbergii* by methylation analysis and NMR spectroscopy, *Carbohydr. Res.* **238**, 241–248.

Correa-Diaz, F., Aguilar-Rosas, R., Aguilar-Rosas, L. E. (1990) Infrared analysis of eleven carrageenophytes form Baja California, Mexico, *Hydrobiologia* **204/205**, 609–614.

Craigie, J. S. (1990) Cell walls, in: *Biology of the Red Algae* (Cole, K. M., Sheath, R. G., Ed.), Cambridge: Cambridge University Press, 221–257.

Dea, I. C. M., Morrison, A. (1975) Chemistry and interactions of seed galactomannans, in: *Ad-*

vances in Carabohydrate Chemistry and Biochemistry (Tipson, R. S., Horton, D., Ed.), New York: Academic Press, 241–312.

De Ruiter, G. A., Rudolph, B. (1997) Carrageenan biotechnology, Trends Food Sci. Technol. **8**, 389–395.

De Ruiter, G., Richard, O., Rudolph, B., Genicot, S., Kloareg, B., Penninkhof, B., Potin, P. (2000) Sulfohydrolases, corresponding amino acid and nucleotide sequences, sulfohydrolase preparations, processes, and products thereof, World patent application: WO0068395.

Falshaw, R., Furneaux, R. (1994) Carrageenan from the tetrasporic stage of *Gigartina decipiens* (Gigartinaceae, Rhodophyta), Carbohydr. Res. **252**, 171–182.

FAO (1992) *Compendium of Food Additive Specifications*. Rome: FAO/WHO Joint Expert Committee on Food Additives, Food and Agricultural Organization of the United Nations.

Genicot, S., Richard, O., Crépineau, F., Boyen, C., Penninkhof, B., Potin, P., Rousvoal, S., Rudolph, B., Kloareg, B. (2000) Purification and cloning of polysaccharide-modifying enzymes from marine algae, in: *Polymerix 2000* (Conanec, R., Ed.), Centre de Biotechnologies en Bretagne, 157–163.

Gretz, M. R., Mollet, J.-C., Falshaw, R. (1997) Analysis of red algal extracellular matrix polysaccharides, in: *Techniques in Glycobiology* (Townsend, R. R., Hotchkiss, A. T., Ed.), New York: Marcel Dekker, 613–628.

Guiseley, K. B., Stanley, N. F., Whitehouse, P. A. (1980) Carrageenan, in: *Handbook of Water-Soluble Gums and Resins* (Davidson, R. L., Ed.), New York: McGraw-Hill, Chapter 5.

Hansen, J. H., Larsen, H., Groendal, J. (2000) Carrageenan compositions and methods for their production, U.S. patent application: US 6,063,915.

Harding, S. E., Hill, S. E., Mitchell, J. R. (1995) Biopolymer mixtures, Nottingham University Press, Nottingham.

Hermansson, A. M., Eriksson, E., Jordansson, E. (1991) Effects of potassium, sodium and calcium on the microstructure and rheological behavior of kappa-carrageenan gels, Carbohydr. Polym. **16**, 297–320.

Hoffmann, R. A., Russell, A. R., Gidley, M. J. (1996) Molecular weight distribution of carrageenans: characterisation of commercial stabilisers and effect of cation depletion on depolymerisation, in: *Gums and Stabilisers for the Food Industry 8* (Phillips, G. O., Williams, P. J., Wedlock, D. J., Ed.), Oxford: Oxford University Press, 137–148.

Iborra, J. L., Manjón, A., Cánovas, M. (1997) Immobilization in carrageenans, in: *Methods in Biotechnology* (Bickerstaff, G. F., Ed.), Totowa, NJ: Humana Press Inc, 53–60.

Imeson, A. P. (2000) Carrageenan, in: *Handbook of hydrocolloids* (Phillips, G. O., Williams, P. A., Ed.), Cambridge: Woodhead Publishing, 87–102.

Jol, C. N., Neiss, T. G., Penninkhof, B., Rudolph, B., De Ruiter, G. A. (1999) A novel high-performance anion-exchange chromatographic method for the analysis of carrageenans and agars containing 3,6-anhydrogalactose, Anal. Biochem. **268**, 213–222.

Kapraun, D. F. (1999) Red algal polysaccharides industry: economics and research status at the turn of the century, Hydrobiologia **398/399**, 7–14.

Kloareg, B., Quantrano, R. S. (1988) Structure of the cell walls of marine algae and ecophysiological functions of the matrix polysaccharides, Oceanogr. Mar. Biol. Annu. Rev. **26**, 259–315.

Knutsen S. H. (1992) Isolation and analysis of red algal galactans, University of Trondheim, Norway.

Knutsen, S. H., Grasdalen, H. (1992) Analysis of carrageenans by enzymic degradation gel, filtration and ^{1}H NMR spectroscopy, Carbohydr. Polym. **19**, 199–210.

Knutsen, S. H., Myslabodski, D. E., Larsen, B., Usov, A. I. (1994) A modified system of nomenclature for red algal galactans, Bot. Mar. **37**, 163–169.

Knutsen, S. H., Sletmoen, M., Kristensen, T., Barbeyron, T., Kloareg, B., Potin, P. (2001) A rapid method for the separation and analysis of carrageenan oligosaccharides released by *iota-* and *kappa-*carrageenase, Carbohydr. Res. **331**, 101–106.

Lahaye, M. (2001) Chemistry and physico-chemistry of phycocolloids, Cah. Biol. Mar. **42**, 137–157.

Michel, G., Chantalat, L., Duee, E., Barbeyron, T., Henrissat, B., Kloareg, B., Dideberg, O. (2001) The kappa-carrageenase of *P. carrageenovora* features a tunnel-shaped active site: A novel insight in the evolution of clan-B glycoside hydrolases, Structure **9**, 513–525.

Mori, T. (1943) The enzyme catalyzing the decomposition of mucilage of *Chondrus ocellatus* III. Purification, unit determination, and distribution of the enzyme, J. Agric. Chem. Soc. Jpn. **19**, 740–742.

Morris, E. R., Rees, D. A., Robinson, G. (1980) Cation-specific aggregation of carrageenan helices: domain model of polymer gel structure, J. Mol. Biol. **138**, 349–362.

Nakashima, H., Kido, Y., Kobayashi, N., Motoki, Y., Neushul, M., Yamamoto, N. (1987) Purification and characterization of an avian myeloblastosis and human immunodeficiency virus reverse transcriptase inhibitor, sulfated polysaccharides extracted from sea algae, *Antimicrob. Agents Chemother.* **31**, 1524–1528.

Neushul, M. (1990) Antiviral carbohydrates from marine red algae, *Hydrobiologica* **204/205**, 99–104.

O'Neill, A. N. (1955a) 3,6-Anhydro-D-galactose as a constituent of kappa-carrageenin, *J. Am. Chem. Soc.* **77**, 2837–2839.

O'Neill, A. N. (1955b) Derivatives of 4-O-beta-D-galactopyranosyl-3,6-anhydro-D-galactose form kappa-carrageenin, *J. Am. Chem. Soc.* **77**, 6324–6326.

Piculell, L. (1995) Gelling carrageenans, in: *Food Polysaccharides and Their Applications* (Stephan, A. M., Ed.), New York: Marcel Dekker, 205–244.

Potin, P., Richard, C., Barbeyron, T., Henrissat, B., Gey, C., Petillot, Y., Forest, E., Dideberg, O., Rochas, C., Kloareg, B. (1995) Processing and hydrolytic mechanism of the cgkA-encoded kappa-carrageenase of *Alteromonas carrageenovora*, *Eur. J. Biochem.* **228**, 971–975.

Potin, P., Bouarab, K., Kupper, F., Kloareg, B. (1999) Oligosaccharide recognition signals and defence reactions in marine plant-microbe interactions, *Curr. Opin. Microbiol.* **2**, 276–283.

Renn, D. (1997) Biotechnology and the red seaweed polysaccharide industry: status, needs and prospects, *Trends Biotechnol.* **15**, 9–14.

Roberts, M. A., Quemener, B. (1999) Measurement of carrageenans in food: challenges, progress, and trends in analysis, *Trends Food Sci. Technol.* **10**, 169–181.

Rudolph, B. (2000) Seaweed product: red algae of economic significance, in: *Marine and Freshwater Products Handbook* (Martin, R. E., Carter, E. P., Davis, L. M., Flich, G. J., Ed.), Boca Raton, FL: CRC Press, 515–529.

Santelices, B. (1999) A conceptual framework for marine agronomy, *Hydrobiologia* **398/399**, 15–23.

Schaeffer, D. J., Krylov, V. S. (2000) Anti-HIV activity of extracts and compounds from algae and cyanobacteria, *Ecotoxicol. Environ. Saf.* **45**, 208–227.

Smith, D. B., Cook, W. H. (1953) Fractionation of carrageenin, *Arch. Biochem. Biophys.* **232**, 232–233.

Smith, D. B., O'Neill, A. N., Perlin, A. S. (1955) Studies on the heterogeneity of carrageenin, *Can. J. Chem.* **32**, 1352–1360.

Snoeren, T. H. M., Payens, T. A. J. (1976) On the sol-gel transition in solutions of kappa-carrageenan, *Biochim. Biophys. Acta* **437**, 264–272.

Stancioff, D. J., Stanley, N. F. (1969) Infrared and chemical studies on algal polysaccharides, *Proc. Intl. Seaweed Symp.* **6**, 595–609.

Stevenson, T. T., Furneaux, R. H. (1991) Chemical methods for the analysis of sulphated galactans from red algae, *Carbohydr. Res.* **210**, 277–298.

Stortz, C. A., Cerezo, A. S. (1992) The ^{13}C NMR spectroscopy of carrageenans: calculation of chemical shifts and computer-aided structural determination, *Carbohydr. Polym.* **18**, 237–242.

Stortz, C. A., Bacon, C. E., Cherniak, R., Cerezo, A. S. (1994) High-field NMR spectroscopy of cystocarpic and tetrasporic carrageenans form *Iridaea undulosa*, *Carbohydr. Res.* **261**, 317–326.

Tako, M., Nakamura, S. (1986) Indicative evidence for a conformational transition in kappa-carrageenan from studies of viscosity-shear rate dependence, *Carbohydr. Res.* **155**, 200–205.

Tako, M., Nakamura, S., Kohda, Y. (1987) Indicative evidence for a conformational transitions in iota-carrageenan, *Carbohydr. Res.* **161**, 247–255.

Therkelsen, G. H. (1993) Carrageenan, in: *Industrial Gums: Polysaccharides and Their Derivatives* (Whistler, R. L., BeMiller, J. N., Ed.), San Diego, CA: Academic Press, 145–180.

Troell, M., Rönnbäck, P., Halling, C., Kautsky, N., Buschmann, A. (1999) Ecological engineering in aquaculture: use of seaweeds for removing nutrients from intensive marineculture, *J. Appl. Phycol.* **11**, 89–97.

Tseng, C. K. (1945) The terminology of seaweed colloids, *Science* **101**, 597–602.

Turquois, T., Acquistapace, S., Arce-Vera, F., Welti, D. H. (1996) Composition of carrageenan blends inferred from ^{13}C-NMR and infrared spectroscopy, *Carbohydr. Polym.* **31**, 269–278.

Usov, A. I., Yarotsky, S. V., Shashkov, A. S. (1980) ^{13}C-NMR spectroscopy of red algal galactans, *Biopolymers* **19**, 977–990.

Usov, A. I. (1998) Structural analysis of red seaweed galactans of agar and carrageenan groups, *Food Hydrocolloids* **12**, 301–308.

VandeVelde, F., Peppelman, H. A., Rollema, H. S., Tromp, R. H. (2001) On the structure of kappa/iota-hybrid carrageenans, *Carbohydr. Res.* **331**, 271–283.

Viebke, C., Williams, P. A. (2000) Determination of molecular mass distribution of kappa-carrageenan and xanthan using asymmetrical flow field-flow fractionation, *Food Hydrocolloids* **14**, 265–270.

Weiner, M. L. (1991) Toxicological properties of carrageenan, *Agents Actions* **32**, 46–51.

Wong, K. F., Craigie, J. S. (1978) Sulfohydrolase activity and carrageenan biosynthesis in *Chondrus crispus* (Rhodophyceae), *Plant Physiol.* **61**, 663–666.

Yaphe, W., Baxter, B. (1955) The enzymatic hydrolysis of carrageenan, *Appl. Microbiol.* **3**, 380–383.

4
Chitin and Chitosan from Animal Sources

Prof. Dr. Martin G. Peter
University of Potsdam, Institute of Organic Chemistry and Structure Analysis, and Interdisciplinary Research Center for Biopolymers, Karl-Liebknecht-Str. 25, D-14476 Golm, Germany; Tel.: +49-331-977-5401; Fax: +49-331-977-5300; E-mail: peter@serv.chem.uni-potsdam.de

1	**Introduction**	119
2	**Historical Outline**	119
3	**Structure of Chitin and Chitosan**	120
3.1	Conformation in Solution	120
3.2	Crystal Structures	121
4	**Occurrence**	121
5	**Physiological Function**	122
6	**Detection of Chitin in Animals and Analysis of Chitin and Chitosan**	123
6.1	Detection of Chitin in Biological Samples	123
6.2	Determination of F_A	124
6.2.1	IR Spectroscopy	124
6.2.2	NMR Spectroscopy	124
6.2.3	Titration Methods	126
6.3	Mass Spectrometry of Chitin and Chitosan Oligosaccharides	126
6.4	Macromolecular Characterization of Chitin and Chitosan	127
6.4.1	Viscosimetry	127
6.4.2	Chromatography	128
7	**Biosynthesis of Chitin in Animals**	128
7.1	Synthesis of Substrates for the Polymerizing Enzyme	128
7.2	Enzymology of Chitin Synthase	129
7.2.1	Assays of Chitin Biosynthesis	129

Polysaccharides and Polyamides in the Food Industry. Properties, Production, and Patents.
Edited by A. Steinbüchel and S. K. Rhee
Copyright © 2005 WILEY-VCH Verlag GmbH & Co. KGaA, Weinheim
ISBN: 3-527-31345-1

7.2.2	Polymerization of GlcNAc	130
7.2.3	Translocation and Finishing of the Polymer	130
7.2.4	Inhibition of Chitin Synthesis	131
7.3	Genetic Basis of Chitin Synthesis	131
7.3.1	Chitin Synthase-like Genes in Bacteria and in Vertebrates	131
7.4	Regulation of Chitin Synthesis	132
7.4.1	At the Enzymatic Level	132
7.4.2	At the Translational Level	132
7.4.3	At the Transcriptional Level	133

8	**Biodegradation**	133
8.1	Enzymology of Chitin Degradation	133
8.1.1	Chitinases: An Overview	133
8.1.2	Occurrence and Functions of Chitinases	138
8.1.3	Structures and Mechanisms of Chitinases	139
8.1.4	Inhibition of Chitinases	141
8.1.5	Lysozymes	141
8.1.6	Chitin-binding Proteins and Lectins	141
8.1.7	Transmembrane Transport and Intracellular Degradation of Chitooligosaccharides	142
8.2	Chitinase Genes	143
8.3	Regulation of Degradation	145

9	**Production of Chitin and Chitosan**	146
9.1	Isolation of Chitin and Chitosan from Shellfish Waste	146
9.1.1	Resources	146
9.1.2	Chemical Processes	147
9.1.3	Fermentation Processes	147
9.1.4	Enzymatic Deacetylation of Chitin	148
9.2	Preparation of Low Molecular-weight Chitin, Chitosan, and of Chitooligosaccharides	148
9.2.1	Depolymerization of Chitin and Chitosan	148
9.2.2	Synthesis of Chitooligosaccharides	149
9.2.3	Abiotic Synthesis of Chitin	149
9.3	Current World Market and Economics	150
9.4	Companies Producing the Polymer	151

10	**Properties of Chitin and Chitosan**	152
10.1	Physico-chemical Properties	152
10.2	Materials Produced from Chitin or Chitosan	152
10.2.1	Films, Membranes, and Fibers	153
10.2.2	Polyelectrolyte Complexes	153
10.2.3	Lipid-binding Properties	153
10.3	Chemistry of Chitin and Chitosan	153
10.3.1	Reactivity	154

10.3.2	Derivatives	154
10.3.3	Hybrid Polymers	155
10.4	Biological Properties	155
10.4.1	Biological Activities in Mammalian Systems	155
10.4.2	Antimicrobial Activity	156
10.4.3	Elicitor Activity in Plants	156
11	**Applications of Chitin and Chitosan**	**157**
11.1	Technical Applications	157
11.1.1	Waste-water Engineering	157
11.1.2	Fibers, Textiles, and Nonwoven Fabrics	158
11.1.3	Paper Technology	159
11.1.4	Biotechnology	159
11.2	Medicine and Healthcare	160
11.2.1	Treatment of Obesity and Hyperlipidemia	160
11.2.2	Bone Regeneration and Prosthetic Implants	161
11.2.3	Vascular Medicine and Surgery	161
11.2.4	Wound Care and Artificial Skin	162
11.2.5	Other Medical Applications	162
11.2.6	Toxicology of Chitin and Chitosan	162
11.2.7	Regulatory Aspects	163
11.3	Pharmaceutical Applications	163
11.3.1	Transmucosal Drug Delivery	164
11.3.2	Sustained-release Formulations	164
11.4	Cosmetics	164
11.5	Agriculture	164
11.6	Food	164
12	**Current Problems and Limits**	**165**
13	**Outlook and Perspectives**	**165**
14	**Relevant Patents for Isolation, Production and Applications of Chitin and Chitosan**	**166**
15	**References**	**191**

CBP	chitin-binding protein
CDA	chitin deacetylase
CP/MAS NMR	cross-polarization/magic angle spinning nuclear magnetic resonance (solid-state)
CSA	10-camphorsulfonic acid
DA	degree of acetylation
DD	degree of deacetylation
DEAc	N,N-diethylacetamide

DMAc	N,N-dimethylacetamide
DMSO	dimethylsulfoxide
DP	degree of polymerization
DS	degree of substitution
ELISA	enzyme-linked immunosorbent assay
ESI MS	electrospray ionization mass spectrometry
F_A	mole fraction of N-acetylglucosamine residues
FAB MS	fast atom bombardment mass spectrometry
FT-IR	Fourier transformation infrared spectroscopy
Gal	D-galactopyranose
Glc	D-glucopyranose
GlcN	2-amino-2-deoxy-D-glucopyranose, β-(1-4)-linked in chitin/chitosan
GlcNAc	2-acetamido-2-deoxy-D-glucopyranose, β-(1-4)-linked in chitin/chitosan
GlcNAc-MU	4-methylumbelliferyl N-acetylglucosaminide
GlcNAc-pNP	p-nitrophenyl N-acetylglucosaminide
GlcNAc-oNP	o-nitrophenyl N-acetylglucosaminide
GlcNase	glucosaminidase
GPC	gel-permeation chromatography
HMW	high molecular weight
HP-GPC	high-performance gel-permeation chromatography
HPLC	high-performance liquid chromatography
HP-SEC	high-performance size-exclusion chromatography
HTS	high-throughput screening
IUB	International Union of Biochemistry
LHRH	luteinizing hormone releasing hormone
LMW	low molecular weight
Man	D-mannopyranose
MALDI TOF MS	matrix-assisted laser desorption ionization time-of-flight mass spectrometry
MALLS	multiple-angle laser light-scattering
MCCh	microcrystalline chitosan
M	molecular mass (daltons)
M_n	number average of molecular mass
M_v	viscosity average molecular mass
M_w	mass average of molecular mass
MS	mass spectrometry
MU	4-methylumbelliferyl
NMR	nuclear magnetic resonance
20-OH-E	20-hydroxyecdysone
PAL	phenylalanine ammonia lyase
PD MS	plasma desorption mass spectrometry
PEG	polyethylene glycol
$\langle R_g \rangle z$	radius of gyration
SEC	size-exclusion chromatography
SLS	static light scattering

TGF   transforming growth factor
THF   tetrahydrofuran
TIM   triose phosphate isomerase
p-TosOH   *p*-toluenesulfonic acid
UDP-GlcNAc   uridine diphospho-*N*-acetylglucosamine
WGA   wheat germ agglutinin

1
Introduction

Chitin and chitosan are aminoglucopyranans composed of *N*-acetylglucosamine (GlcNAc) and glucosamine (GlcN) residues. These polysaccharides are renewable resources which are currently being explored intensively by an increasing number of academic and industrial research groups. Indeed, an impressive number of applications – especially of chitosan and its derivatives – have been suggested in the literature, and several of those have been commercialized.

The chemistry and biochemistry of chitin is reviewed in this chapter. Although chitosan is a biopolymer that occurs naturally in several fungi, its technical production is based mostly on processing of chitinous resources from animals. Hence, the chemistry and applications of chitosan are included here, while the compound's biochemistry is discussed in more detail in Chapter 5 of this volume. Due to space limitations, only a small fraction of the extensive literature (over 14,000 references during the past decade, including approximately 3500 patents) will be considered. (Only in exceptional cases is literature available in Chinese or Japanese language cited in this review.)

2
Historical Outline

Chitin is a rather "old" molecule, with chemically detectable remains having been found in fossil insects from the Oligocene period (24.7 million years ago) (Stankiewicz et al., 1997). The history of chitin research began in 1811, when the French Professor of Natural History, H. Braconnot, published a paper entitled "Sur la nature des champignons" in *Ann. Chim. Phys.* (Paris). Prof. Braconnot described an alkali-insoluble material from higher fungi which he named "fungine" (for references, see Muzzarelli, 1977a; Roberts, 1992). Twelve years later, A. Odier isolated a similar fraction from insect exoskeletons which he named chitin (Greek: χιτων = tunic, armor). In 1859, C. Rouget showed that saponification of chitin produces chitosan, whilst in 1878 G. Ledderhose found that hydrolysis of chitin yields "glycosamine" and acetic acid. F. Tiemann demonstrated in 1886 that the phenylosazone of glucosamine was identical to the phenylosazones prepared two years earlier by E. Fischer from "dextrose" (= glucose) and laevulose (= fructose). F. Hoppe-Seyler reported in 1894 on the occurrence of chitin in crabs, scorpions, and spiders, whilst in 1903, E. Fischer and H. Leuchs obtained glucosamine and mannosamine from arabinose. W. V. Haworth and coworkers finally presented proof of the absolute configuration by synthesis of D-glucosamine in 1939. The first mention of a chitinase appeared in 1929, when P. Karrer, A. Hofmann, and G.

von Francois used "Schneckensaft" (an extract from *Helix pomatia*) to show that chitin from lobsters, *Homarus vulgaris*, showed essentially the same behavior as chitin from the edible mushroom, *Boletus edulis*, both yielding *N*-acetylglucosamine (GlcNAc). The first patents on chitin, together with several applications, were filed in 1935 by G. W. Rigby, and in 1936 by Du Pont de Nemours & Co.

The first textbook on chitin appeared in 1977 (Muzzarelli, 1977). (The original book is out of print; a facsimile copy is available from Franklin Book Company, Inc., 7804 Montogomery Avenue, Eliks Park, PA 19117, USA, FAX: +1-225-635-6155.) This was followed by a comprehensive presentation of chitin chemistry (Roberts, 1992). *Chitin Handbook*s are available in Japanese (Society for the Study of Chitin and Chitosan, 1991; Japanese Society for Chitin and Chitosan, 1995) and in English (Muzzarelli and Peter, 1997). Besides a growing number of publications in scientific journals, important reference sources are available in a series of conference proceedings, most of which are referred to throughout this chapter.

3
Structure of Chitin and Chitosan

It is usually understood that chitin (Chemical Abstracts Registry (CAS) 1398–61-4) is the polymer of β-1,4-linked *N*-acetylglucosamine (2-acetamido-2-deoxy-β-D-glucopyranose, GlcNAc) whereas chitosan (CAS 9012–76-4) is the corresponding polymer of glucosamine (GlcN). However, neither chitin nor chitosan are homopolymers, as both contain varying fractions of GlcNAc and GlcN residues (Figure 1). Roberts (1997) suggested indicating the mole fraction of GlcNAc residues by the "F_A value". Thus, the homopolymer of GlcNAc is chitin (F_A 1.0) and the homopolymer of GlcN is chitosan (F_A 0.0). Other (sometimes confusing) terms are used throughout the literature, such as degree of acetylation (DA), degree of deacetylation (DD or DDA), or residual degree of acetyl groups. The polymers may be distinguished by their solubility in 1% aqueous acetic acid. Chitin, containing ca. >40% GlcNAc residues (F_A >0.4) is insoluble, whereas soluble polymers are named chitosan.

Chitin from animals is highly acetylated (F_A >0.9); the average molecular weight of chitin in insect cuticles is estimated at 1–2×10^6, corresponding to a degree of polymerization (DP) of ca. 5000–10,000 (Hackman, 1987).

Chitosan is prepared from suitable chitinous raw materials, mostly by a sequence of deproteinization, demineralization, and chemical deacetylation procedures. The molecular weight of chitosan depends on the source of the biological material, as well as on the conditions of the deacetylation process (see Section 9).

3.1
Conformation in Solution

Light scattering of low-viscosity solutions of depolymerized chitin (DP ca. 40–50) in DMAc/5% LiCl ([η] <10 dL g^{-1}) indicate that the polysaccharide forms stable aggregates of rod-shaped, semirigid chains with a

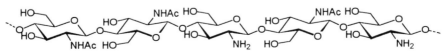

Fig. 1 Structure of chitin (F_A 0.60).

persistence length of 21.4 ± 1 nm, whereas higher-viscosity solutions contain molecularly dispersed chitin (Terbojevich et al., 1996).

It is generally accepted that the worm-like chain model applies to the conformation of chitosan, independent of F_A (Beri et al., 1993; Berth et al., 1998; Terbojevich et al., 1988). Due to the polyelectrolyte nature of chitosan, however, the data derived for conformational analysis are sensitive to M_w, polydispersity, ionic strength, polymer concentration, and the presence of hydrogen bond-breaking reagents such as urea. Thus, other models describing rod-shaped chains and increasing stiffness with increasing F_A, as well as random coil conformations have been suggested (Errington et al., 1993; Vårum et al., 1994; Tsaih and Chen, 1997, 1999; Chen and Tsaih, 2000; Signini et al., 2000).

3.2
Crystal Structures

Similar to cellulose, chitin occurs in three polymorphic forms, named α-, β-, and γ-chitin, which differ in the orientation of the polysaccharide chains. They are antiparallel in α-chitin, where intrachain hydrogen bonds are donated from C(3′)-OH to O(5), from C(6′)-OH to the acetamido carbonyl O(7), and from C(6)-OH to O(6′). Interchain hydrogen bonds exist between NH and acetamido carbonyl O(7′) and between C(6′)-OH and O(6). The unit cell dimensions of α-chitin are $a = 4.74$, $b = 10.32$, and $c = 18.86$ Å. In β-chitin, the chains are parallel whereas γ-chitin possesses two parallel chains in association with one antiparallel (for a comprehensive discussion and references, see Roberts, 1992; Ebert, 1993; Chanzy, 1997).

Crystalline chitosan exists in several polymorphs, all having an extended two-fold helical structure, but differing in packing density and water content. Intrachain hydrogen bonding occurs between C(3)-OH and O(5′). Complexes with acids or metal salts are grouped into type I, showing a 10.3 Å axial repeat, while type II has a 40 Å axial repeat (Ogawa et al., 2000; Okuyama et al., 2000).

4
Occurrence

Chitin is widely distributed in living organisms; its occurrence has been detected in fungi, algae, Protozoa, Cnidaria, Aschelmintes, Endoprocta, Bryozoa, Phoronida, Brachiopoda, Echiurida, Annelida, Mollusca, Onychophora, Arthropoda, Chaetognatha, Pognophora, and Tunicata (Muzzarelli, 1977; Roberts, 1992). Chitooligosaccharides, i.e. $(GlcNAc)_n$, $n > 2$, are synthesized in numerous organisms, including bacteria of the family Rhizobiaceae (Denarie et al., 1996), filaria (Nematoda) (Haslam et al., 1999), teleost fish (Jeuniaux et al., 1996), and even humans (Bakkers et al., 1999) (see also Section 7). N,N'-Diacetylchitobiose $(GlcNAc)_2$ is a normal constituent of the core structure of N-glycans of glycoproteins.

Chitin occurs mostly in the thermodynamically stable α-modification, whereas the metastable β-chitin is present in the shell of Inarticulata (Brachiopoda), the chaetae and gizzard cuticle of some Annelidae, the pen of *Loglio* and *Octopus* (Cephalopoda), and in the tubes of Pognophora. γ-Chitin is observed rarely, e.g. in the stalk cuticle of Inarticulata (Brachiopoda) and in the stomach cuticle of cephalopods (Muzzarelli, 1977; Roberts, 1992).

It is frequently stated that chitin is, after cellulose, the second most abundant polysaccharide on earth. Actually, chitin represents only a small fraction of the estimated 2.7×10^{11} tons of organic carbon in the biosphere, about 99% of which (i.e., 2.63×10^{11} tons) is located in plants, with 40% of that (i.e., ca. 1.1×10^{11} tons) being bound in cellulose (Ebert, 1993). The annual regeneration of cellulose by photosynthesis is estimated at 1.3×10^9 tons (an "average" tree synthesizes about 14 g of cellulose per day) (Falbe and Regitz, 1989). The synthesis of chitin by Crustacea (including the suborders Entomostraca and Malacostraca) in the hydrosphere alone is estimated at ca. 2.3×10^9 tons per year (Jeuniaux et al., 1993; Cauchie, 1997), some authors even citing a production of 10^{11} tons per annum. The chitin released into the seas by molting processes and dead chitinous organisms is called "marine snow" (for references, see Keyhani and Roseman, 1999). Cellulose is most abundant in terms of static occurrence, but chitin is most prominent in terms of metabolism dynamics, exceeding the annual regeneration of cellulose probably by roughly one order of magnitude!

The relative amounts of chitin vary considerably within the organisms and their chitinous organs or structures, ranging from traces to 80% of the organic fraction. Particularly rich sources of chitin are fungi, which contain up to 45% chitin (see Chapter 5, this volume), and arthropods. The organic fraction of insect cuticle contains 20–60% chitin, while that of decalcified crustacean cuticle contains up to 80% (Hackman, 1987).

5
Physiological Function

Chitin is often compared with cellulose: both are extracellular polysaccharides forming fibrous elements in biological composite materials. Thus, the most prominent role of chitin is apparent in the construction of the insect and crustacean exoskeleton which, depending on the degree of sclerotization (or mineralization, respectively), provides mechanical rigidity for the maintenance of morphology. There is a fundamental difference however in the function of extracellular structures in plants and animals in that the exoskeleton of the invertebrates also serves as attachments for tendons and muscles and, of course, this is the premise for mobility.

Chitin fibers are always embedded in a matrix of proteins which presumably are covalently linked to the polysaccharide. Although many proteins have been isolated and characterized, particularly from insect cuticles (Andersen et al., 1995), the extent of cross-linking as well as the exact chemical structure of the polysaccharide–polypeptide linkages are essentially unknown. With the possible exception of the chitin synthase complex (see Section 7), neither a glycosyltransferase nor a peptidyltransferase catalyzing the formation of covalent chitin–protein bonds in arthropods has yet been described. X-ray analysis of chitin–protein complexes from the ovipositor of the fly *Megarhyssa lunator* showed that the proteins form a sheath of a 6_1 helix of subunits around the chitin microfiber (Blackwell and Weigh, 1984).

The mechanical properties of the biological structures are clearly correlated with the packing of the chitin microfibrils, forming smectic, nematic, or cholesteric mesophases (Hackman, 1987). In most cuticles of insects, plywood-like stacking of the fibers yields isotropic materials. The chitin occurs in the form of microfibrils with diameters varying from 2.5 to 25 nm, and a length of ca. 0.36 µm (for references, see Vincent, 1990). The stiffness of the chitin fibers is estimated at 70–90 GPa, and that of the protein matrix at 120 MPa. Anisotropic materials such as the tendon of locusts show a parallel fiber orientation. The value of Young's modulus is 11 GPa in the longitudinal direction, and 0.15 GPa in the transverse direction. A helicoidal packing of chitin fibrils around the tracheae and other discontinuities in the cuticle yields materials that are isotropic in the plane (Vincent, 1990).

In addition to fiber orientation, the nature of the protein also determines the functional properties. Thus, the wing hinge of the locust and the intersegmental membranes of the integument of insects contain the rubber-like protein resilin (for a classic text on structure–function relationships in insects, see Chapman, 1971; for some recent articles on resilin-containing structures in insects, see Frazier et al., 1999; Haas et al., 2000; Neff et al., 2000).

The polysaccharide–protein complex of insects is further stabilized in soft cuticles with low molecular-weight phenolic compounds, in particular catecholamine derivatives, which are enzymatically oxidized and polymerized in sclerotized cuticles (Peter, 1993; Andersen et al., 1996). Evidence for the covalent coupling of polyphenols to chitin has been obtained by solid-state NMR spectroscopy (Kramer et al., 1995) and ESI mass spectrometry (Kerwin et al., 1999). In crustaceans, hardening of the exoskeleton occurs by mineralization, i.e., by the incorporation predominantly of calcium carbonate.

Besides occurring in the exoskeleton of arthropods, chitin in combination with proteins and further components serves as a fibrous element in many other biological structures, such as in the peritrophic membrane that forms the intestinal lining in insects (Tellam and Eisemann, 2000). The ovipositor of many insects and the mandibles of herbivores are examples of particularly hard structures, being reinforced by up to 10% of their dry weight with zinc or manganese ions (Vincent, 1990). The eggshells of nematodes and the cysts of the brine shrimp, *Artemia salina*, are protected against dehydration and mechanical damage by a chitin-containing envelope.

6
Detection of Chitin in Animals and Analysis of Chitin and Chitosan

Methods used to detect chitin in samples from animals are outlined below. Additional procedures are available for the analysis of chitin and chitosan in fungi (see Chapter 5, this volume). General methods of determining structural parameters and macromolecular properties of chitin are described in this section.

6.1
Detection of Chitin in Biological Samples

Earlier gravimetric methods as well as the determination of chitin by ion-exchange chromatography, using instrumentation for amino acid analysis, are not considered here.

Chitin can be detected with high sensitivity (nmol equivalents of GlcNAc or GlcN) by using an enzyme-linked immunosorbent assay (ELISA) which is based on the immobilization of chitin, chitosan or oligosaccharides on the surface of microtiter plates. Quantification is achieved either by measuring the binding of a polyclonal chitin antibody by means of IgG-peroxidase, or by competitive displacement of a known amount of antibody with the analytical sample (Buss et al., 1996). A similar procedure was developed for the quantification of oligomers of GlcNAc and GlcN, using a polyclonal oligosaccharide antibody (Kim et al., 2000).

Degradation of chitin by chitinase yields oligomers, mainly $(GlcNAc)_2$ which are further cleaved to GlcNAc by N-acetylhexosaminidase (see Section 8), and determination of chitin is based on the quantification of GlcNAc by chromatographic or colorimetric methods. Digestion of crude biological samples with chitinase yields GlcNAc from the fraction of the polysaccharide that is accessible to the enzyme ("free chitin"), whereas digestion of purified chitin from the same specimen indicates the total chitin. The difference is then calculated as "bound chitin" (Jeuniaux and Voss-Foucard, 1997).

The lectin wheatgerm agglutinin (WGA) possesses a high-affinity binding site for oligomers of GlcNAc (DP > 3) and chitin. Application of WGA-gold labeling in combination with electron microscopy is a powerful tool for the ultrastructural localization of chitin in various organisms (Neuhaus et al., 1997; Peters and Latka, 1997), including fungi (see Chapter 5, this volume). Other methods which are mostly used to detect chitin in fungi include staining by means of WGA-fluorescein or Calcofluor White.

6.2
Determination of F_A

Many methods for the determination of F_A of chitin and chitosan are described in the literature, including infrared (IR), CP/MAS ^{13}C-NMR, first-derivative-UV, and CD spectroscopy, potentiometric and dye adsorption titration, pyrolysis-gas chromatography, quantification of acetic acid in hydrolyzates by liquid chromatography, thermal and elemental analysis (Roberts, 1992; Muzzarelli et al., 1997b). Only the most commonly applied methods are considered here. It should be mentioned that each method has its advantages or disadvantages, depending on F_A, but calibration with an independent method is recommended.

6.2.1
IR Spectroscopy

The variables, i.e., the intensities of the amide absorptions at 1660 and 1630 cm^{-1} ($\nu_{C=O}$, amide I, two types of H-bonds) and at 1550 (δ_{NH}; amide II) cm^{-1} relative to the invariables, i.e., absorption at 3450 (ν_{OH}), 2950–2880 (ν_{CH}), or 1150 and 1020–1100 (ν_{C-O}, asymm. and symm.) cm^{-1} (KBr) are calculated by various methods, differing in selection of the absorption band and definition of the baseline of the IR spectrum (Roberts, 1992; Shigemasa et al., 1996a,b; Muzzarelli et al., 1997b; Ratajska et al., 1997; Struszczyk et al., 1997; Duarte et al., 2000; Struszczyk et al., 2000; Brugnerotto et al., 2001). In the author's laboratory, the intensities of the amide I band and ν_{OH} are used for calculation of F_A from A_{1650}/A_{3450} (Figure 2) (Struszczyk et al., 1997).

The absorption bands selected may be obscured by ν_{OH} at ca. 3480, ν_{NH} at 3269, δ_{OH} at ca. 1640, and, particularly in low F_A samples, by NH$_2$ at 1597 cm^{-1}. Absorptions in the fingerprint regions are very sensitive to differences in structure (Duarte et al., 2000).

6.2.2
NMR Spectroscopy

Solid-state NMR is the method of choice for analysis of insoluble samples, usually requiring amounts of about 200–300 mg. Soluble chitosans of low to medium DP (M ≤ 25,000 Daltons) may also be analyzed in aqueous systems using D$_2$O as the solvent.

Solid-state ^{13}C-NMR
CP/MAS ^{13}C-NMR (natural ^{13}C abundance) is useful to determine the degree of acetylation of chitin of $F_A > 0.05$. It is also suitable for the analysis of chitin in highly complex biological matrices such as sclerotized insect cuticle which contains also proteins and polyphenols (Grün et al., 1984; Kramer et al., 1995; Zhang et al., 2000). The principles and application of CP/MAS ^{13}C-NMR for chitin analysis are explained by Ebert and Fink (1997). Representative spectra are depicted in Figure 3 and Table 1.

Besides determination of F_A, high-resolution CP/MAS ^{13}C-NMR is used for analysis

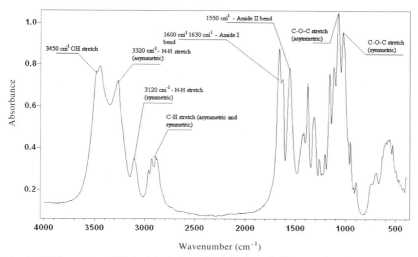

Fig. 2 FT-IR spectrum (KBr) of chitin from *Pandalus borealis* (Struszczyk et al., 1997).

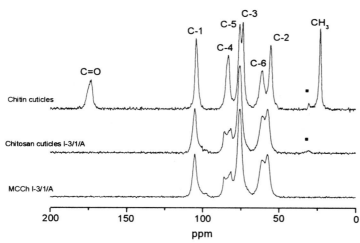

Fig. 3 ^{13}C-NMR spectra of chitin and chitosan from *Calliphora erythrocephala* and of MCCh. ■: Signals of impurities. Chitin cuticles: deproteinized insect integuments; chitosan cuticles I-3/1/A: chitin cuticles treated with 50% NaOH at 100 °C for 3 h (Struszczyk et al., 2000).

of chitin polymorphs. The signals of C(3) and C(5) are well resolved in the spectra of α-chitin, but not in β-chitin (Takai et al., 1989; Vincendon et al., 1989). The duplication of the C(4) signal observed in CP/MAS ^{13}C-NMR spectra of chitosan is explained by the presence of type I and II polymorphs (Takai et al., 1989).

^{1}H- and ^{13}C-NMR in Solution

In oligosaccharides, the chemical shifts of the ^{1}H and ^{13}C nuclei are influenced significantly by the next-neighbor sugar residue. High-resolution ^{1}H- and ^{13}C-NMR spectroscopy of partially acetylated chitosan (F_A < 0.7) yields statistical information about the occurrence of GlcNAc and GlcN sequen-

Tab. 1 ^{13}C Chemical shifts of solid chitin and chitosan. (External standard: adamantane or benzene; for references, see Ebert and Fink, 1997.)

	C=O	C-1	C-4	C-5	C-3	C-6	C-2	CH$_3$
α-Chitin								
Crab shell	173.7	104.5	83.2	76.0	73.6	61.0	55.4	23.1
Shrimp shell	173.7	104.4	83.3	75.8	73.6	61.4	55.5	23.1
β-Chitin								
Tevnia tube (dried)	175.5	105.3	84.4	75.4	73.1	59.8	55.2	22.7
	176.4						56.0	
Squid pen	174.8	104.8	83.7	75.4	74.5	59.3	56.0	23.3
						61.0		24.1
Chitosan								
Crab shell		105.0	85.6	75.5		60.3	56.4	
			81.1					
Shrimp shell		105.2	86.1	75.6		61.0	57.2	
			81.2					

ces in diads and triads, as well as about the nature of the sugar at the reducing and nonreducing ends. HMW chitosan must be depolymerized partially, e.g., by acid-catalyzed limited hydrolysis or by treatment with nitrous acid (cf. Section 10) (Vårum et al., 1991a, b). The ^1H chemical shifts of (GlcNAc)$_2$, (GlcNAc)$_3$, and the O-acetates, as well as the ^{13}C shifts of the sugar carbon atoms in diads and triads of different GlcNAc and GlcN sequences were tabulated (Schanzenbach and Peter, 1997).

6.2.3
Titration Methods

Titration of chitosan is useful for analysis of low F$_A$ samples. The pK_a of chitosan falls between 6.2 and 6.4. Potentiometry or conductometry is used for titration of chitosan hydrochloride with NaOH (Raymond et al., 1993; Koetz and Kosmella, 1997) or of chitosan acetate with perchloric acid in glacial acetic acid/anhydrous 1,4-dioxan (Bodek, 1994, 1995). The F$_A$ is calculated from the inflexion points of the potentiometric respectively from the section indicating dissociation of the ammonium groups in the conductometric titration curve.

For polyelectrolyte titration, sodium polystyrenesulfonate is the polyanion of choice. The end-point is determined by turbidity measurements (Koetz and Kosmella, 1997) or by toluidine blue adsorption (Park et al., 1995). The latter method requires a minimum of four sequential GlcN residues (Hattori et al., 1999).

6.3
Mass Spectrometry of Chitin and Chitosan Oligosaccharides

Although various polymers are conveniently analyzed using soft ionization techniques, few reports are available on the investigation of polysaccharides (Mohr et al., 1995). Chitin and chitosan are difficult to analyze by MS, due to the insolubility of chitin, and with regard to the polycationic nature of chitosan. The highest oligomer identified in a mixture of oligosaccharides prepared from fully deacetylated chitin (F$_A$ < 0.02) is (GlcN)$_{49}$ (calculated mass: 7932 [M + Na]$^+$) (Figure 4) (Letzel et al., 2000).

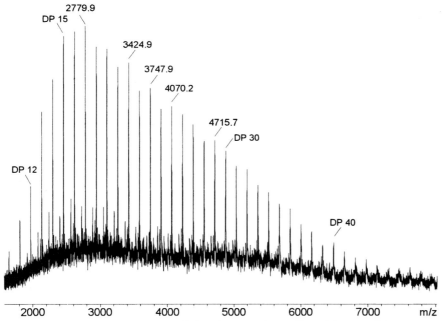

Fig. 4 MALDI-TOF MS (matrix: THAP) of GlcN oligomers obtained by HCl cleavage of chitosan ($F_A < 0.02$). The labeled peaks refer to $[M+Na]^+$ ions of the $(GlcN)_n$ series (Letzel et al., 2000).

Mass spectrometry of chitooligosaccharides is frequently used to determine the products of enzymatic degradation or deacetylation of chitin and chitosan (for a survey of MS methods, see Haebel et al., 1997). Fully acetylated GlcNAc oligomers of DP ≤ 7 were analyzed by FAB (Bosso and Domard, 1992), PD (Lopatin et al., 1995), or MALDI TOF MS (Akiyama et al., 1995a). The fragmentation of GlcN oligomers of DP ≤ 9 was investigated by positive and negative ion mode FAB MS (Bosso and Domard, 1992). MALDI TOF mass spectra of partially acetylated chitooligosaccharides containing GlcNAc and GlcN were measured for pure compounds of DP ≤ 6 (Akiyama et al., 1995b) and of mixtures containing up to DP 17 which were obtained by depolymerization of chitosan ($F_A = 0.24$) with an enzyme cocktail (Zhang et al., 1999) (see Section 8). The spectra of higher partially acetylated oligomers become increasingly complex due to the chemical inhomogeneity of the sample.

6.4
Macromolecular Characterization of Chitin and Chitosan

The most commonly used methods for determination of M_v, M_w, and M_n are viscosimetry, light scattering (SLS and MALLS), and GPC (SEC) (Terbojevich and Cosani, 1997).

6.4.1
Viscosimetry

The relationship between M_v and intrinsic viscosity $[\eta]$ is expressed by the Mark-Houwink-Kuhn-Sakurada equation, $[\eta] = KM^a$, where the viscosity parameters K and a depend on the polymer, the temperature,

Tab. 2 Viscosity parameters of chitosan (for references and discussion, see Roberts, 1992)

M_v	Solvent	K	a
113,000–492,000	0.2 M HOAc, 0.1 M NaCl, 4 M urea	8.93×10^{-4}	0.71
90,000–1,140,000	0.1 M HOAc, 0.2 M NaCl	1.81×10^{-3}	0.93
13,000–135,000	0.33 M HOAc, 0.3 M NaCl	3.41×10^{-3}	1.02

the solvent, and the salt concentration. The values found for chitin in DMAc-5% LiCl solution are $K = 2.2 \times 10^{-4}$ dL g^{-1}, a = 0.88 (Terbojevich et al., 1996). The viscosity parameters used widely for chitosan are listed in Table 2 (Roberts, 1992).

6.4.2
Chromatography

Separation of GlcN oligomers of low DP (< 50) is achieved by size-exclusion chromatography (SEC) on polyacrylamide gels, such as Biogel P (Domard and Cartier, 1991). However, the method is not universally applicable to chitosan separation because higher oligomers (DP > 50) cannot be separated, even by the use of larger pore-size gels. The apparent pK_a of chitosan decreases with increasing DP, and strong interactions can cause strong adsorption of the polymer onto the stationary phase (Domard and Cartier, 1991). Irreversible adsorption may also occur with lower GlcN oligomers on some lots of Biogel (M. G. Peter, unpublished observations). The supplier does not guarantee the specific applicability of these materials, and it is recommended that the suitability of each lot of Biogel be tested before setting up columns for preparative separations, even of low-DP GlcN oligomers.

HP-SEC is conveniently performed with a series of TSK-60 and/or TSK-50 columns (Beri et al., 1993; Terbojevich and Cosani, 1997). Calibration of the chromatographic system with dextrans may provide erroneous data, as the M_w of chitosan is overestimated, especially when small quantities of the polymers are injected. HP-SEC in combination with MALLS provides more reliable values of polydispersity.

Lower oligosaccharides of mixed acetylation patterns are separated by ion exchange on CM Sephadex C-25 columns (Mitsutomi et al., 1996). For chitosan, Sepharose CL-2B is suitable for determination of polydispersity in open column systems (Berth et al., 1998).

7
Biosynthesis of Chitin in Animals

All chitin-containing organisms must synthesize the polysaccharide. In addition, chitin synthase homologous genes occur in bacteria and in vertebrates, including humans. This section focuses on chitin synthesis, mostly in insects; chitin synthase-like activities in bacteria and vertebrates are also briefly reviewed.

7.1
Synthesis of Substrates for the Polymerizing Enzyme

An overview of the enzymatic transformations involved in chitin biosynthesis is given in Figure 5. Trehalose [α-D-Glc(1-1)α-D-Glc] is the primary storage form of glucose in many organisms, e.g., fungi, lichens, algae, bacteria, mosses, and arthropods. Enzymatic hydrolysis by the α-glycosidase trehalase [EC 3.2.1.28] gives glucose (Figure 5, step 1). Apparently, trehalose does not occur in vertebrates. Trehalase has only rarely been

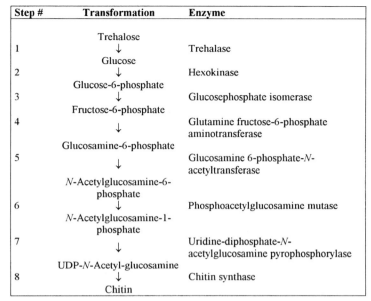

Fig. 5 Biochemical pathway of chitin biosynthesis.

found in higher organisms, and therefore, the enzyme appears as an attractive target for the development of pesticides (for a review on the chemistry and biological activity of trehalase inhibitors and analogues, see Berecibar et al., 1999). Steps 2–7 in Figure 5, i.e., the transformation of Glc into UDP-GlcNAc are most common biochemical reactions.

7.2
Enzymology of Chitin Synthase

The enzymology of chitin synthesis in arthropods is fairly well understood. However, the ultrastructural organization as well as the mechanism of chitin synthase [chitin-(UDP-GlcNAc)-transferase, EC 2.4.1.16] in animals are essentially unknown. Chitin synthase is located in membrane fractions, most likely as a component of a multiprotein complex, and activity is lost during attempts to purify the enzyme (for reviews on chitin synthesis in arthropods, covering also aspects of chitin degradation and regulation of synthesis, see Kramer and Koga, 1986; Spindler and Spindler-Barth, 1996; Londershausen et al., 1997).

In contrast to the large number of studies on arthropods, the investigation of chitin synthesis in lower organisms is limited to incorporation studies of radiolabeled precursors and/or histochemical evidence in the ciliated protozoon *Eufolliculina uhligi* (Schermuly et al., 1996; Mulisch and Schermuly, 1997) and the hydrothermal vent worm *Riftia pachytila* (Vestimentifera) (Shillito et al., 1993). Both organisms produce β-chitin.

7.2.1
Assays of Chitin Biosynthesis

Chitin synthesis is studied in tissue or cell homogenates, or in microsomal fractions. The incorporation of radiolabeled GlcNAc, GlcN, or UDP-GlcNAc into the alkali-resistant fraction yields the amount of chitin synthesized (Spindler-Barth, 1997). The presence of chitin is verified by acid hydrolysis and/or by digestion with chitinase, followed by thin-layer chromatography (TLC) analysis of the products. Also, ELISA or staining with WGA or Calcofluor White is used to assay for the presence of chitin (see Section 6). Chitin synthesis may be enhanced in crude homogenates when a chitinase inhibitor is added to the assay mixture (Peter and Schweikart, 1990).

7.2.2
Polymerization of GlcNAc

Assembly of the chitin chain occurs by transfer of GlcNAc from UDP-GlcNAc, most likely to the reducing end of the growing chain. The pH optimum of chitin synthesis in cell-free systems is ca. 7.0–8.0, and the K_m values for UDP-GlcNAc are in the range of 0.1 to 1 mM (Kramer and Koga, 1986). Chitin synthesis shows an apparent K_m of 4.1 mM for GlcNAc in homogenates of embryonic tissues of the tick, *Boophilus microplus* (Londershausen et al., 1993). Reports on the initialization of the polymerization reaction by transfer of GlcNAc oligomers from dolichol-P-(GlcNAc)$_n$ to "chitinoproteins", involving the Golgi apparatus, are limited to the crustacean species *Artemia salina* (Horst, 1986; Horst and Walker, 1993), and await confirmation by analogous studies in other species as well as rigorous identification of the intermediary lipid-linked chitooligosaccharides. Tunicamycin, which inhibits (UDP-GlcNAc):dolichyl-P-GlcNAc-1-P transferase, apparently interferes strongly with chitin synthesis in intact tissues, but much less in subcellular microsomal fractions (for discussion and references, see Kramer and Koga, 1986; Londershausen et al., 1997).

Reports on the occurrence of sulfated "chitinoproteins" in *Drosophila* and other insect species, yielding (GlcNAc)$_2$ and (GlcNAc)$_3$ upon chitinase digestion (Kramerov et al., 1986, 1990), should be critically re-evaluated by application of appropriate methods of structure analysis, especially as O-linked GlcNAc occur in many nuclear and cytoplasmic proteins (for a review, see Haltiwanger, 2000). On the other hand, chitooligosaccharides appear to be involved in hyaluronan biosynthesis (see below).

7.2.3
Translocation and Finishing of the Polymer

Translocation of the polymer may occur by exocytosis and/or extrusion, the exocytosis being mediated by GTP binding proteins. Indirect supportive evidence for exocytosis is provided by the finding that an analogue of GTP, γ-S-GTP, stimulates chitin synthesis (Londershausen et al., 1997). On the other hand, extrusion of the polysaccharide chain ensues when the chitin synthase complex is located in the plasma membrane. It is of interest to note that fungal chitin synthase genes contain a myosin motor-like sequence (see Chapter 5, this volume).

With respect to the ultimate steps of chitin synthesis, two intriguing questions remain unanswered to date:

- The mechanism of crystallization of α-chitin is not understood, though several models have been proposed. When the polymer grows into the lumen of a vesicle, the chains could be aligned in an antiparallel fashion. However, crystallization could also occur at overlapping ends, and additional processes would be required to achieve higher degrees of crystallinity.
- The mechanism accounting for the presence of a small but significant number of GlcN residues in the polymer is unknown. The most simple hypothesis would be that GlcN residues in animal chitin are artifacts, resulting from the isolation process. The occurrence of a chitin deacetylase (an earlier report on the enzymatic deacetyla-

tion of chitin in arthropods was apparently not followed up by further studies) or of a chitin-polypeptide acyltransferase in animals has not yet been described. However, as GlcN is incorporated into chitin via GlcNAc, a heteropolymer could result when the enzymes transforming GlcN into UDP-GlcNAc and catalyzing transfer of UDP-GlcNAc to chitin are not absolutely substrate-specific.

7.2.4
Inhibition of Chitin Synthesis

A large variety of chitin synthesis inhibitors have been described as potential pesticides (for reviews, see Wright and Retnakaran, 1987; Spindler et al., 1990; Cohen, 1993; Palli and Retnakaran, 1999). The most important, commercially exploited inhibitors of chitin synthesis are benzoylphenyl ureas (Wright and Retnakaran, 1987; Palli and Retnakaran, 1999). The mechanism of inhibition remains unknown, though several sites of action have been proposed (Londershausen et al., 1997). Analogues of UDP-GlcNAc, i.e., polyoxins and nikkomycins, are effective in insects (Cohen, 1987; Tellam et al., 2000), but have not found any practical application. Several compounds inhibit chitin synthesis nonspecifically by disrupting membrane organization (Londershausen et al., 1997). As mentioned above, data on the inhibition of chitin synthesis by tunicamycin are at variance. Other natural products possessing general toxicity sometimes inhibit chitin synthesis quite efficiently, e.g., the naphthoquinone plumbagin which shows $IC_{50} = 10$ μM in *Chiromomus tentans* cell line preparations and $IC_{50} = 6$ μM in tick embryonic tissue homogenates (Londershausen et al., 1993).

7.3
Genetic Basis of Chitin Synthesis

Various chitin synthase and chitin synthase-like genes have been identified in fungi (see Chapter 5, this volume), bacteria, and vertebrates (see below). In contrast, only two putative chitin synthase genes of arthropods have been cloned to date, i.e., chitin synthase encoding cDNA from the mosquito *Aedes aegypti* (Ibrahim et al., 2000) and chitin synthase-like protein, LcCS-1, from larvae of the fly *Lucilia cuprina* (Tellam et al., 2000). Chitin synthase is a single copy gene, at least in *A. aegypti*. The sequences show high homology to related genes from the nematode *Caenorhabditis elegans* and lower (but significant) similarity to yeast, *Saccharomyces cerevisiae* chitin synthases, with stronger conservation centered in the active sites of the yeast enzymes. The cDNA of *A. aegypti* is 3.5 kb in length with an open reading frame of 2.6 kb, encoding a protein of 865 amino acids with a predicted $M = 99,500$ daltons. The cDNA of *L. cuprina* is 5757 bp in length, coding for a protein of 1592 amino acids, and $M = 180,717$ Daltons. LcCS-1 contains 15 to 18 potential transmembrane segments, indicative of an integral membrane protein. Highly related genomic sequences were demonstrated in three insect orders, in one arachnid and in the nematode *C. elegans* (Tellam et al., 2000).

7.3.1
Chitin Synthase-like Genes in Bacteria and in Vertebrates

The products of chitin synthase activity in Rhizobia and in vertebrates are chitin oligosaccharides of DP 4–6, which in the bacteria are modified via a variety of reactions to yield lipochitooligosaccharides (Nod factors) (Figure 6) (for reviews, see Denarie et al., 1996; Bakkers et al., 1999).

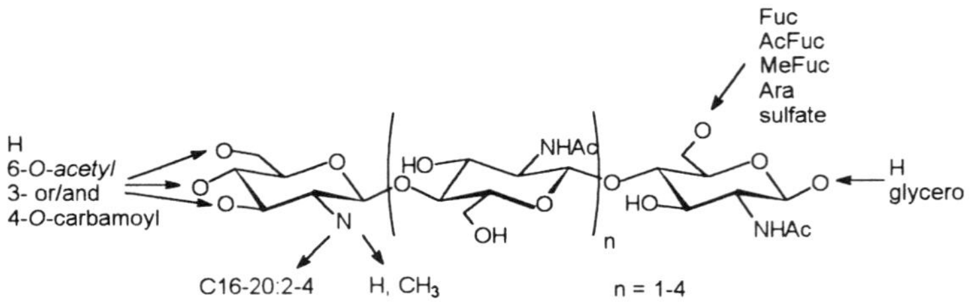

Fig. 6 Structures of *Rhizobium* Nod factors.

Nod factors stimulate nodule formation in leguminous plants, their synthesis being initiated by a flavonoid signal released by the plant. Assembly of the chitooligosaccharide is catalyzed by NodC, transferring GlcNAc from UDP-GlcNAc to the nonreducing end of free GlcNAc and the growing oligosaccharide chain (Kamst et al., 1999).

A developmentally regulated gene of *Xenopus*, DG42, which is expressed between the late midblastula and neurulation stages of embryonic development, was shown to synthesize chitooligosaccharides (Semino and Robbins, 1995). Analogous genes occur also in zebrafish and mouse (Semino et al., 1996). Interference with chitooligosaccharide synthesis or their modification by the fucosyltransferase NodZ in developing embryos of zebrafish leads to severe defects in trunk and tail development (Bakkers et al., 1997). On the other hand, in the presence of UDP-GlcNAc and UDP-GlcA, DG42 synthesizes hyaluronan and it was suggested that chitooligosaccharides act as primers for hyaluronan biosynthesis (Meyer and Kreil, 1996) (for comments, see Varki, 1996). Indeed, recombinant mouse hyaluronan synthase *has*1 incorporates UDP-GlcNAc into chitooligosaccharides (Yoshida et al., 2000). DG42 and various hyaluronan synthase genes share >70% sequence homology and significant homology with the chitin synthases NodC from Rhizobia and *chs*3 from *S. cerevisiae* (Bakkers et al., 1999; Spicer and McDonald, 1998).

7.4
Regulation of Chitin Synthesis

A wealth of information is available on metabolic, translational, and transcriptional regulation of chitin synthesis in arthropods. Apparently, regulation of chitin synthesis has not been investigated in lower chitinous organisms.

7.4.1
At the Enzymatic Level

Proteolytic activation of chitin synthesis seems to be required for efficient *in vitro* incorporation of sugar precursors into the polysaccharide (Kramer and Koga, 1986; Ludwig et al., 1991).

In contrast to fungal chitin synthase (see Chapter 5, this volume), the enzyme from an insect cell line of *Chironomus tentans* is not allosterically regulated by GlcN or GlcNAc, but is stimulated by Mg^{2+} and inhibited by UDP, UMP, and UTP (Ludwig et al., 1991).

7.4.2
At the Translational Level

In-situ hybridization studies of midgut samples of *A. aegypti* revealed that chitin synthase mRNA increases following a blood meal. Localization at the periphery of the epithelial cells facing the midgut lumen underlines the role of these cells in formation of the chitin-containing peritrophic membrane (Ibrahim et al., 2000).

7.4.3
At the Transcriptional Level

In arthropod cuticle, the activity of chitin synthesis is correlated with molting (ecdysis) and differentiation. Both processes are strictly regulated by an array of hormones, *inter alia* prothoracicotrophic hormone, juvenile hormones and ecdysteroids, the latter regulating also chitin synthesis. Hormones in general induce expression of transcription factors which act in various ways on other functions (for reviews of ecdysteroid regulation of chitin synthesis, see Marks and Ward, 1987; Palli and Retnakaran, 1999).

Molting in insects is initiated by a rise in the titer of 20-hydroxyecdysone (20-OH-E) which also induces or suppresses chitin synthesis, depending critically on hormone concentration as well as on changes in hormone titers (Spindler and Spindler-Barth, 1996). In the pupal-adult development of holometabolous insects, high concentrations of 20-OH-E inhibit chitin synthesis during differentiation of organs in imaginal discs. A decline in hormone titer induces formation of the epicuticle, which is devoid of chitin, whilst formation of the cuticle requires low basal titers of 20-OH-E. In larval molting, high ecdysterone titers induce chitin synthesis. A high permanent ecdysteroid titer also inhibits chitin synthesis in an insect cell line from *C. tentans* (Spindler-Barth, 1993).

8
Biodegradation

The enormous amounts of chitin synthesized by living organisms do not accumulate in the biosphere, because the polysaccharide is degraded by microorganisms which use chitin as an energy source. In addition, chitin degradation plays an essential role during the metamorphosis of arthropods, as well as in a variety of developmentally regulated processes in many organisms, including vertebrates.

8.1
Enzymology of Chitin Degradation

A wide variety of enzymes is involved in chitin and chitosan degradation, including endo- and exochitinases, chitosanases, chitodextrinases, chitotriosidases, chitobiosidases, chitobiases, N-acetylhexosaminidases, glucosaminidases, and, finally, enzymes of the terminal metabolism of GlcNAc and GlcN. This section focuses on chitinases, excluding enzymes from fungi (see Chapter 5, this volume). An overview of the enzymes involved in chitin degradation is given in Table 3. Lysozymes and enzymes of the terminal metabolism of $(GlcNAc)_2$ are mentioned briefly.

8.1.1
Chitinases: An Overview

There is a wide diversity of chitinases that differ in structure, catalytic mechanism, and substrate specificity. Chitinases, besides being of academic interest, possess a great potential for many applications, such as plant protection in agriculture, and processing in the food industry, as well as for the biotechnological production of chitooligosaccharides, which show intriguing biological activities (see Section 10). Various aspects of these enzymes have been reviewed (Beintema, 1994; Schrempf, 1996; Kramer and Muthukrishnan, 1997; Keyhani and Roseman, 1999; Koga et al., 1999; Robertus and Monzingo, 1999; Patil et al., 2000).

Nomenclature, Substrate Specificities and Cleavage Patterns of Chitinolytic Enzymes

Chitinases [EC 3.1.14.1] hydrolyze glycosidic bonds of $(GlcNAc)_n$, $n > 2$. Based on sequence similarities, chitinases are classified

Tab. 3 Overview of enzymes of chitin and chitosan degradation

Enzyme	EC no.	Family[a]	Substrate	Occurrence
Chitinase	3.2.1.14	18	Chitin	Widely
Chitinase	3.2.1.14	19	Chitin	Plants, *Streptomyces*
Chitodextrinase	3.2.1.–	18	$(GlcNAc)_n$	*Vibrio furnissii*
Lysozyme	3.2.1.17	22–25		Widely
di-N-Acetylchitobiase[b]	3.2.1.52	20	$(GlcNAc)_2$	Widely
β-N-Acetylglucosaminidase (exo)	3.2.1.52	20	$(GlcNAc)_n$	Widely
Chitin deacetylase	3.5.1.41	–	Chitin	Fungi
Chitosanase	3.2.1.132	46	Chitosan	Bacteria, fungi
β-Glucosaminidase	–	–	4′-terminal GlcN	Fungi
GlcNAc-6-P-deacetylase	3.5.1.25	–		Bacteria, viruses

[a] For classification based on sequence similarities of carbohydrate active enzymes, see Henrissat (1999) and a URL (Coutinho and Henrissat, 1999) which also provides links to additional relevant databases. [b] Former EC 3.2.1.30 was deleted and chitobiase is now hexosaminidase, EC 3.2.1.52.

as family 18 or family 19 glycosidases (Coutinho and Henrissat, 1999; Henrissat, 1999). Family 18 plant chitinases are further divided into classes III and V, and family 19 chitinases are grouped into classes I, II, IV, VI enzymes, according to domain structures and cellular location (for reviews, see Terwisscha van Scheltinga, 1997; Patil et al., 2000). A revised nomenclature dividing chitinase genes into four families has been suggested (Neuhaus et al., 1996). In addition, chitinases are assigned as endo- or exo-enzymes, as chitobiosidases, chitotriosidases, chitodextrinases, acidic and basic enzymes, or by molecular mass. In fact, the situation is becoming increasingly complex (if not confusing) as an increasing number of genes and enzymes has become available, while standard rules for nomenclature are lacking.

The substrate specificities of chitinases differ considerably, as illustrated by a few recent examples in Table 4. In most cases, the smallest substrate cleaved is $(GlcNAc)_3$ and the end-products of chitin hydrolysis are GlcNAc and $(GlcNAc)_2$. *Plasmodium* chitinases do not cleave $(GlcNAc)_3$ (Vinetz et al., 1999, 2000), and the smallest substrate cleaved by the plant chitinase hevamine is $(GlcNAc)_5$ (Terwisscha van Scheltinga et al., 1995; Bokma et al., 2000).

Chitodextrinase is different from chitinase, as it cleaves chitooligosacharides but not $(GlcNAc)_2$ or chitin (Bassler et al., 1991).

The substrate-binding characteristics of chitinases and cleavage patterns of partially acetylated chitosans are determined by NMR spectroscopy (Stokke et al., 1995; Vårum and Smidsrød, 1997) or by enzymatic sequencing of the products (Mitsutomi et al., 1996).

Chitinase Assays

A great variety of assays for chitinase activity has been described in the literature, using different experimental set-ups and substrates (for a description of the most common methods, see Spindler, 1997). Selection of the appropriate substrate is crucial when discrimination between chitinase, chitodextrinase, and N-acetylglucosaminidase is desired, and additional investigations may be necessary when the function of a particular enzyme is investigated. Different kinetics, pH and temperature optima for oligosaccharide versus polymeric substrates have been reported for various chitinases (Brurberg et al., 1996; for additional references, see Keyhani and Roseman, 1999):

Tab. 4 Substrates and cleavage patterns of chitinases

Organism	Substrate(s)	Results	References
Human			
Homo sapiens, granulocyte-rich homogenates	[^{3}H] chitin	Release of 7.2 nmol GlcNAc min^{-1} mg protein^{-1}	Escott and Adams (1995)
Insects			
Bombyx mori, larvae 65 and 88 kDa chitinases	(GlcNAc)$_3 \rightarrow$ (GlcNAc)$_4 \rightarrow$ (GlcNAc)$_5 \rightarrow$ (GlcNAc)$_6 \rightarrow$	GlcNAc plus (GlcNAc)$_2$ 2 (GlcNAc)$_2$ (GlcNAc)$_2$ plus (GlcNAc)$_3$ (GlcNAc)$_2$ plus (GlcNAc)$_4$ (65%), and 2 (GlcNAc)$_3$ (35%)	Koga et al. (1997)
Molluscs			
Todarodes pacificus (squid)	glycol chitin		
Endochitinase	(GlcNAc)$_4 \rightarrow$ (GlcNAc)$_5 \rightarrow$ (GlcNAc)$_6 \rightarrow$	2 (GlcNAc)$_2$ (GlcNAc)$_2$ plus (GlcNAc)$_3$ (GlcNAc)$_2$ plus (GlcNAc)$_4$ (84%) and 2 (GlcNAc)$_3$ (16%)	Matsumiya and Mochizuki (1997)
Plants			
Dioscorea opposita (yam) chitinase A	glycol chitin; (GlcNAc)$_n$, n = 2–6	Endo/random, released the monosaccharide from all hydrolyzed oligosaccharides; prefers oligosaccharides with longer chain lengths	Arakane and Koga (1999)
Rehmannia glutinosa, leaves: chitinase P2	(GlcNAc)$_{4-6}$ chitin	(GlcNAc)$_2$	Lee et al. (1999a)
Hordeum vulgare	(GlcNAc)$_6$	2 (GlcNAc)$_3$, and (GlcNAc)$_4$ plus (GlcNAc)$_2$; (GlcNAc)$_3$-MU: K_m 33 μM, k_{cat} 0.33 min^{-1}; (GlcNAc)$_4$: K_m 3 μM, k_{cat} 35 min^{-1}; K_d for (GlcNAc)$_n$: n = 2: 43, n = 3: 19, n = 4: 6 μM; Bi-bi kinetics, with (GlcNAc)$_4$ and H$_2$O as substrates and (GlcNAc)$_2$ as product	Hollis et al. (1997)
Bacteria			
Serratia marcescens ChiA; endo/exochitinase	(GlcNAc)$_n$; chitin (GlcNAc)$_3$-MU $\rightarrow$	GlcNAc plus (GlcNAc)$_2$; shows higher specific activity towards chitin than ChiB GlcNAc plus (GlcNAc)$_2$ plus MU	Brurberg et al. (1996)
S. marcescens ChiB Endo/exochitinase	(GlcNAc)$_n$; chitin (GlcNAc)$_3$-MU $\rightarrow$	(GlcNAc)$_2$ plus GlcNAc; synergistic effects upon combining ChiA and ChiB GlcNAc-MU plus (GlcNAc)$_2$	Brurberg et al. (1996)
Streptomyces griseus HUT 6037, 49 kDa chitinase	chitosan F_A 0.46	(GlcNAc/GlcN)$_n$, always GlcNAc at reducing end; partially N-acetylated chitosan is hydrolyzed more easily than colloidal chitin	Tanabe et al. (2000)

Tab. 4 (cont.)

Organism	Substrate(s)	Results	References
Protozoa			
Plasmodium falciparum PfCHT1; endochitinase	glycol chitin (GlcNAc)$_3$ (GlcNAc)$_4$ → (GlcNAc)$_5$ → (GlcNAc)$_6$ →	Is not hydrolyzed 2 (GlcNAc)$_2$ (GlcNAc)$_2$ plus (GlcNAc)$_3$ (GlcNAc)$_3$ (major), and (GlcNAc)$_2$ plus (GlcNAc)$_4$ (minor)	Vinetz et al. (1999)
Plasmodium gallinaceum PgCHT1; endochitinase		Similar to PfCHT1	Vinetz et al. (2000)

- Radiolabeled colloidal chitin is prepared by acetylation of chitosan, preferentially with [^{3}H]acetic anhydride. Dye-modified chitin derivatives (Chitin Red or Chitin Blue), suitable for photometric assays, are available commercially. In a recently described plate assay, microorganisms are grown on plates prepared from a suspension of colloidal chitin dyed with Remazol Brilliant Red in agar. The presence of chitinase is detected through observation of a clearing zone (Draborg et al., 1997). Other frequently used polymer substrates are 6-O-hydroxyethylchitin or 6-O-acetoxyethylchitin. Differences in the hydrolysis rates were observed, depending on the degree of substitution and source of the chitin used for derivatization (Koga and Kramer, 1983; Inui et al., 1996).
- The use of chromogenic or fluorogenic oligosaccharide derivatives, i.e., (GlcNAc)$_n$-pNP, (GlcNAc)$_n$-oNP, or (GlcNAc)$_n$-MU, respectively, is probably the most widely used assay for quantification of hydrolytic activities of enzymes specified by EC 3.2.1.14, EC 3.2.1.52, and EC 3.2.1.17. Advantages are high sensitivity, especially with fluorogenic 4-methylumbelliferyl (MU) derivatives, the simple protocol, and suitability for high-throughput screening (HTS) on microtiter plates. The assay is also useful to analyze cleavage patterns of chitinases and lysozymes, though care must be taken to differentiate chitinases with respect to substrate specificity and function. Uncritical use of these substrate my lead to invalid conclusions and contribute further to the confusion about the nomenclature of chitinases (Keyhani and Roseman, 1999).
- HPLC assays are used to detect (GlcNAc)$_n$ of DP ≤ 7. Amide-80 columns allow the analysis of substrate cleavage patterns as well as the anomeric ratios of free oligosaccharides (Terwissscha van Scheltinga et al., 1995; Koga et al., 1998). Partially acetylated oligomers are not eluted from the column. Fully acetylated oligomers – but not anomers – may also be separated on amino phases (Mitsutomi et al., 1996). Reductive amination with p-aminoethylbenzoate of an oligosaccharide mixture followed by HPLC on a C$_{18}$ column was used to determine chitinase kinetics. This method is not suitable for analysis of anomers, and the derivatives of (GlcNAc)$_4$ and (GlcNAc)$_5$ are not separated (Bokma et al., 2000).

Tab. 5 Properties of chitinases

Organism	M [kDa]	pI	pH opt.	Temperature optimum [°C]	Comments	References
Human						
Homo sapiens, granulocyte-rich homogenates	48 and 56				Partial inhibition by allosamidin at 9 µM; low levels in serum	Escott and Adams (1995)
Homo sapiens, plasma, spleen	39 and 50	8.0; 7.2			Both enzymes have identical N-terminal sequence	Renkema et al. (1995)
Insects						
Bombyx mori, larvae	65 and 88		5.5 6.5		pH optima 5.5 and 6.5 with (GlcNAc)$_5$ and pH 4–10 with glycolchitin; substrate inhibition with small (GlcNAc)$_n$; 88 kDa chitinase is four-fold times more active with glycol chitin than 65 kDa enzyme	Koga et al. (1997)
Molluscs						
Todarodes pacificus (squid), liver	38	8.3	1.5 and 8.5	50		Matsumiya and Mochizuki (1997)
Plants						
Dioscorea opposita (yam) chitinase A	28	3.6	4.0	60	Inhibited 76% by 55 µM allosamidin	Arakane and Koga (1999)
Rehmannia glutinosa, leaves, chitinase P2	28.6	8.46	5.0	60	Basic exochitinase	Lee et al. (1999a)
Hordeum vulgare endochitinase					pK_a 3.9 (Glu89) and 6.9 (Glu67)	Hollis et al. (1997)
Bacteria						
Serratia marcescens ChiA	58.5		5.0–6.0	50–60		Brurberg et al. (1996)
Serratia marcescens ChiB	55.4		5.0–6.0	50–60		Brurberg et al. (1995, 1996)
Streptomyces erythraeus	30.9				290 AA residues, two disulfide bridges	Kamei et al. (1989)
Streptomyces griseus HUT 6037	49	7.3			N-terminal amino acid sequence fibronectin type III-like sequence	Tanabe et al. (2000)
Streptomyces thermoviolaceus OPC-520	30	3.8	4.0	60	Also 25 and 40 kDa chitinases found	Tsujibo et al. (2000a)
Arthrobacter sp. NHB-10	30	6.8	5.0	45	Internal sequences AGPQLLTGYY and IGGVMT identified	Okazaki et al. (1999)

Tab. 5 (cont.)

Organism	M [kDa]	pI	pH opt.	Temperature optimum [°C]	Comments	References
Protozoa						
Plasmodium falciparum PfCHT1	39		5.0		Inhibited by allosamidin, IC$_{50}$ 0.04 µM (pH > 6.0)	Vinetz et al. (1999)
Plasmodium gallinaceum PgCHT1	60		5.0		Inhibited by allosamidin, IC$_{50}$ 12 µM (pH > 6.0)	Vinetz et al. (2000)

8.1.2
Occurrence and Functions of Chitinases

The CAZy database (Coutinho and Henrissat, 1999) contains (at the time of writing) sequences or fragments with 247 entries for family 18 chitinases, and 220 entries for family 19 chitinases. In addition, numerous related sequences of unknown or undetermined enzymatic activities, including also lectins, are listed. Family 18 chitinases are widely distributed in mammals, arthropods, plants, fungi, bacteria, and viruses. The family 19 enzymes occur typically in plants, but have also been observed in bacteria of the genus *Streptomyces* (Watanabe et al., 1999; Tsujibo et al., 2000c; for reviews, see Koga et al., 1999; Robertus and Monzingo, 1999). Some recent data on the occurrence and properties of chitinases are listed in Table 5.

Chitinase activity has also been discovered in human serum (Overdijk and Van Steijn, 1994), and the enzyme as well as its production by cloning and use as an antifungal has been patented (Aerts, 1996). Guinea pigs infected with *Aspergillus fumigatus* show elevated levels of the enzyme, and it is thought that chitinase is part of the mammalian defense against fungi (Overdijk et al., 1996, 1999). However, in ca. 6% of Caucasian people, the enzyme shows pseudodeficiency (i.e., very low activity without apparent symptoms) (Eiberg and Dentandt, 1997). Apparently, there are other, yet unknown functions of chitinases. A several hundred-fold increase of chitinase (chitotriosidase) was found in lipid-loaden macrophages and in plasma of symptomatic Gaucher disease patients (Renkema et al., 1995; Aerts et al., 1996). Furthermore, chitotriosidase enzyme activity is elevated up to 55-fold in extracts of atherosclerotic tissue, and it is postulated that chitinase plays a role in cell migration and tissue remodeling during atherogenesis (Renkema et al., 1998; Bleau et al., 1999; Boot et al., 1999). The binding domain of human chitinase was shown to recognize specifically chitin and several yeasts and fungi (Tjoelker et al., 2000).

Other occurrences of chitinases in vertebrates include the intestine of cod, *Gadus morhua* (Danulat, 1986), and a variety of other fishes (Matsumiya and Mochizuki, 1996), their function clearly being to digest chitinous food components.

Chitinases have important functions in the regulation of growth and development of many organisms which often contain several synergistically acting chitinases.

Insects digest part of the cuticle during molting, and therefore inhibition of chitinases is one of many the strategies for the development of new pesticides (for reviews, see Kramer and Koga, 1986; Kramer and Muthukrishnan, 1997). Two chitinase isoenzymes were purified from fifth-instar larvae of *Bombyx mori*, showing similar characteristics as the chitinases from *Manduca sexta* (Koga et al., 1997). They are most likely involved in the initial and intermediate stages of chitin degradation during molting.

In plants, the most important function of chitinases is to act as phytoalexins in defense against fungal infections. A great variety of chitinase isoenzymes are known. In yam tubers (*Dioscorea opposita*), chitinases H1 and H2 are family 18 enzymes, whereas chitinases E and F belong to family 19, as revealed by immunological properties of the isoenzymes (Arakane et al., 2000). In contrast to yam chitinase G, chitinases E, F, and H1 show high lytic activity against the plant pathogen, *Fusarium oxysporum*. Chitinases E, F and H1 show two optimum pH ranges of 3–4 and 7.5–9 towards glycolchitin.

Many bacteria metabolize chitin ultimately to CO_2, NH_4^+, and H_2O (Keyhani and Roseman, 1999). Chitinase-producing soil bacteria are active against fungi and nematodes (Cohen-Kupiec and Chet, 1998). The marine bacterium, *Vibrio furnissii*, shows chemotaxis to GlcNAc oligomers and produces at least two extracellular chitinases – one to provide the nutrient $(GlcNAc)_2$, and the other to release oligosaccharides for chemotaxis from a remote source of chitin. *Serratia marcescens* produces at least three family 18 extracellular chitinases (ChiA, ChiB, ChiC), but ChiB is also located in the periplasm (Brurberg et al., 1996; Watanabe et al., 1997). *Streptomyces* produce both, family 18 and family 19 chitinases (Schrempf et al., 1993; Saito et al., 1999; Watanabe et al., 1999).

A number of serious human and animal infectious diseases are caused by protozoan and metazoan parasites, and evidence is accumulating rapidly that chitinases play crucial roles in these infectious mechanisms. Thus, during the development of the malarial parasite, *Plasmodium*, ookinetes migrate from the intestine of the mosquito host into the hemolymph, traversing the chitinous peritrophic membrane (Shahabuddin and Kaslow, 1994; Ramasamy et al., 1997), and family 18 chitinases occur in the human malarial parasite *Plasmodium falciparum* (Vinetz et al., 1999) and in *P. gallinaceum* (Vinetz et al., 2000). Other parasites producing chitinases include *Leishmania*, *Trypanosoma*, Filaria, and Entamoeba (for references, see Wu et al., 1996; De la Vega et al., 1997; Shakarian and Dwyer, 1998; Shahabuddin and Vinetz, 1999; Shakarian and Dwyer, 2000).

8.1.3
Structures and Mechanisms of Chitinases

Crystal structures are presently known of five family 18 chitinases (Table 6). The common motif of the catalytic domain is a $(\beta\alpha)_8$-barrel (TIM-barrel; Figure 7). The solution structure of the catalytic domain of *Bacillus circulans* WL-12 chitinase A1 was also resolved by NMR spectroscopy (Ikegami et al., 2000).

In addition to the catalytic domain, many chitinases contain chitin-binding and/or FnIII type domains which are important for recognition of the polysaccharidic substrate and possibly also assist in substrate binding (see also Section 8.2). Deletion of

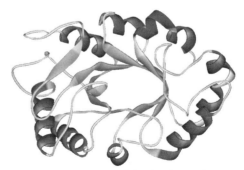

Fig. 7 Ribbon display of the family 18 chitinase hevamine (Terwisscha van Scheltinga et al., 1996). PDB code 2HVM; display by WebLab viewer.

Tab. 6 Known crystal structures of family 18 chitinases

Genus	Enzyme	M [kDa]	Catalytic residue	Reference
Plants	*Hevea brasiliensis*, hevamine	29	Glu127	Terwissscha van Scheltinga et al. (1996)
Bacteria	*Serratia marcescens*, ChiA	58.7	Glu315	Perrakis et al. (1994)
Bacteria	*Serratia marcescens*, ChiB	55.5	Glu144	van Aalten et al. (2000)
Bacteria	*Bacillus circulans* WL-12, ChiA1	74	Glu204	Matsumoto et al. (1999)
Fungi	*Coccidioides immitis*, CiX1	47.4	Glu171	Hollis et al. (2000)

the chitin-binding domain does not abolish catalytic activity towards smaller oligosaccharides, but does diminish or abolish the hydrolysis of chitin (Vorgias, 1997; Hashimoto et al., 2000; Tjoelker et al., 2000).

Serratia marcescens ChiA is a family 18 endo/exo-chitinase that possesses a deep active site groove composed of six sugar-binding subsites. The chitin-binding FnIII type domain of ChiA extends the substrate binding cleft towards the nonreducing side of chitin. Release of chitobiose must therefore occur from the reducing end of the substrate. In contrast, *S. marcescens* ChiB possesses a relatively short binding pocket, and the chitin-binding domain is directed towards the reducing end of the substrate. Therefore, chitobiose/chitotriose are released from the nonreducing end of chitin (van Aalten et al., 2000). Whereas both ChiA and ChiB from *S. marcescens* each have considerable exo-activity, the plant chitinase hevamine has a much more open active site and appears to be an endochitinase.

Family 18 chitinases hydrolyze the substrate with retention of configuration at the anomeric center, with the acetamido group providing anchimeric assistance. Binding of the substrate and protonation of the glycosidic bond leads to distortion of the sugar ring at the −1 binding site to a boat-like conformation, and to an approximately perpendicular orientation of the planes of sugar rings at the − and the + binding sites as well as a co-linear orientation of the leaving glycosidic oxygen and the incoming acetamido carbonyl oxygen, leading to an oxazolinium ion intermediate (Tews et al., 1997; Brameld and Goddard, 1998). Substrate-assisted catalysis is also evident from kinetic data. Chitinase A1 from *Bacillus circulans* WL-12 cleaves MU from (GlcNAc)$_2$-MU, GlcN-GlcNAc-MU, and GlcNAc-GlcN-MU, showing k_{cat}/K_m values of 145.3, 8.3, and 0.1 s^{-1}M^{-1}, respectively. The fully deacetylated disaccharide glycoside (GlcN)$_2$-MU is not hydrolyzed at all (Honda et al., 2000b).

Family 19 chitinases follow the classical mechanism of inverting glycosidases, cleaving the glycosidic bond by direct displacement of the leaving glycosidic oxygen with a water molecule. To date, two crystal structures of plant family 19 chitinases, i.e., from jack bean (Hahn et al., 2000) and from *Hordeum vulgare* (Hart et al., 1995), are known. The X-ray structure of a chitinase from barley, *H. vulgare*, shows a lysozyme type α+β fold, which is located within the structure, composed of a β-sheet, 10 α-helices, and three disulfide bonds. The catalytic site contains two glutamic acid residues where Glu67 acts as proton donor and Glu89 acts as a base coordinating the attacking water molecule (Robertus and Monzingo, 1999). Jack bean chitinase shows analogous structural features (Hahn et al., 2000).

Glycosidases are further distinguished with respect to the direction of the laterally occurring protonation as *syn* or *anti* protonators (for a review, see Heightman and Vasella, 1999).

8.1.4
Inhibition of Chitinases

The most potent, well-known chitinase inhibitors are allosamidin and congeners, including several synthetic and semi-synthetic analogues (for reviews, see Berecibar et al., 1999; Koga et al., 1999; Spindler and Spindler-Barth, 1999; Rast et al., 2000). Although not all chitinases which are inhibited by allosamidin are classified, it is thought that the compound is specific for family 18 chitinases. Remarkable species selectivity is observed, with IC_{50} values of 0.32 µM for insect (*Lucilia cuprina*), 0.69 µM for tick (*Boophilus microplus*), and 0.048 µM for nematode (*Haemonchus contortus*) chitinases (Londershausen et al., 1996). Allosamidin inhibits the family 18 plant chitinase hevamine with a rather high K_i of 3.1 µM (Bokma et al., 2000), and a family 19 plant class I chitinase from *Phaseolus vulgaris* with IC_{50} 1 µM (Londershausen et al., 1996). A number of chitooligosaccharide-derived glycosylamines were found to yield different inhibition of ChiA and ChiB from *S. marcescens* (Rottmann et al., 1999, 2000). Although the IC_{50} values are rather high (in the range of 0.2 to 0.5 mM), the results suggest that the number of sugar residues is crucial for binding in the case of ChiA, whereas the hydrophobicity of the aglycone contributes to affinity towards ChiB. A mixture of partially acetylated chitooligosaccharides strongly inhibits ChiB of *S. marcescens* (Letzel et al., 2000).

Several nonsugar chitinase inhibitors have been described recently. The fungal cyclic peptides, argifin and argadin, inhibit the insect *Lucilia cuprina* chitinase with IC_{50} 3.7 µM and 150 nM respectively at 37 °C and 0.1 µM and 3.4 nM respectively at 10 °C (Arai et al., 2000; Shiomi et al., 2000). Styloguanidins were isolated from a marine sponge, *Stylotella aurantium*, and shown to inhibit the molting of cyprid larvae of barnacles (Kato et al., 1995).

8.1.5
Lysozymes

Lysozymes [EC 3.2.1.17] hydrolyze *N*-acetylmuramic acid. Although not classified as chitinases, theses enzymes show also chitinase [EC 3.2.1.14] activity. Hydrolysis of depolymerized chitosan of DP 30 and F_A 0.68, with hen egg-white lysozyme revealed that substrate binding requires at least three GlcNAc residues in sequence, and that cleavage occurs exclusively between two GlcNAc residues (Vårum et al., 1996; Vårum and Smidsrød, 1997). Chitosan of $F_A < 0.16$ is not hydrolyzed by lysozyme (Tokura et al., 1984). Many plant chitinases also show lysozyme activity (Beintema and Terwisscha van Scheltinga, 1996); however, it was proven that, at least for hevamine, hydrolysis of *N*-acetylmuramic acid occurs with release of GlcNAc as the reducing sugar and therefore, the enzyme is not a lysozyme (Bokma et al., 1997).

Interactions of lysozyme with partially deacetylated chitosan were studied by NMR spectroscopy (Kristiansen et al., 1998), with $(GlcNAc)_6$ by mass spectrometry (He et al., 1999), and with $(GlcNAc)_n$ or MU derivatives by X-ray crystallography (Harata and Muraki, 1995; Karlsen and Hough, 1995; Kopacek et al., 1999).

8.1.6
Chitin-binding Proteins and Lectins

A large number of chitin-binding proteins (CBPs) and lectins are known that show no enzymatic activity. A characteristic feature of chitin-binding domains and of CBPs is a tryptophan-rich sequence. The CBPs from *Vibrio* species are expressed constitutively or are induced by chitin or higher chitooligosaccharides. Their function is to mediate adhesion of the bacteria to chitin and to immobilize larger chitin fragments on the cell membrane (Keyhani and Roseman, 1999). The CBPs of *Streptomyces olivaceoviridis* are small proteins which specifically bind α-chitin, but not β-chitin (Schrempf, 1999).

Chitin-specific plant lectins were reviewed recently (Beintema, 1994; Van Damme et al., 1998). Sequence homologies and X-ray structures of narbonin (Hennig et al., 1992) and concanavalin B (Hennig et al., 1995) revealed a catalytically inactive $(\beta\alpha)_8$ barrel where a glutamine is in the position of the catalytic glutamic acid residue of a homologous chitinase. The three-dimensional (3D) structure of the plant lectin hevein from *Hevea brasiliensis* is also related to family 18 chitinases, as shown by NMR spectroscopy (Asensio et al., 1995, 1998; Poveda et al., 1997). A lectin related to hevein occurs in the stinging nettle, *Urtica dioica* and 3D structures of complexes with NAG oligomers were determined by X-ray crystallography (Harata and Muraki, 2000; Saul et al., 2000).

8.1.7
Transmembrane Transport and Intracellular Degradation of Chitooligosaccharides

The catabolism of chitin by bacteria which use the polysaccharide as an energy source depends ultimately on the uptake of low molecular-weight precursors. Thus, a variety of proteins mediating transport and hydrolysis of oligosaccharides as well as metabolizing GlcNAc and GlcN are required.

Chitoporins

In Gram-negative bacteria, the oligosaccharides generated from chitin by extracellular chitinases are transported via porins into the periplasm where they are further hydrolyzed by chitodextrinase and β-N-acetylglucosaminidase to GlcNAc and $(GlcNAc)_2$ (for a review, see Keyhani and Roseman, 1999). A specific chitoporin, M = 40 kDa, which is inducible by $(GlcNAc)_n$, n = 2–6, occurs in the outer membrane of *Vibrio furnissii*. The protein was cloned and sequenced recently (Keyhani et al., 2000).

The Final Steps

The final steps in the degradation of chitin in Gram-negative bacteria encompass transport of GlcNAc and $(GlcNAc)_2$ into the cytoplasm (Keyhani and Roseman, 1999). A periplasmic N-acetylglucosaminidase cleaves oligosaccharides specifically from the nonreducing terminal. Translocation of GlcNAc is driven by the phosphoenolpyruvate:glycose phosphotransferase system, yielding GlcNAc-6-P as the product, whereas $(GlcNAc)_2$ is transported unchanged or may also become phosphorylated (Park et al., 2000). Additional enzymes of yet unknown function include a chitobiase which is incapable of hydrolyzing GlcNAc-pNP-or GlcNAc-MU but which hydrolyzes $(GlcNAc)_2$, and a β-N-acetylhexosaminidase, which hydrolyzes GlcNAc-pNP and GlcNAc-MU, but not chitin or chitin oligosaccharides. A detailed discussion is beyond the scope of this chapter, and the reader is referred to the review by Keyhani and Roseman (1999).

The chitobiase (hexosaminidase) of *S. marcescens* is a family 20 glycosidase observing a retaining mechanism of hydrolysis via the oxazoline intermediate (Drouillard et al., 1997). The crystal structure of this enzyme has been resolved (Tews et al., 1996).

N-Acetyl-β-D-hexosaminidases of arthropods occur in large varieties (Kramer and Koga, 1986). Three enzymes were characterized from the brine shrimp, *Artemia salina*, showing K_m 0.16–0.72 mM for GlcNAc-pNP, and M = 83, 110, and 56 kDa respectively (Spindler and Funke-Höpfner, 1991). The N-acetyl-β-D-hexosaminidases from the liver of the prawn *Penaeus japonicus* are acidic enzymes with $pI < 3.2$, optimum pH 5.0–5.5, and optimum temperature 50 °C, $M = 64$ and 110 kDa, respectively, preferring shorter oligosaccharides from which GlcNAc is cleaved from the nonreducing end (Koga et al., 1996). *Alteromonas* contains an outer-membrane β-N-acetylglucosaminidase which was characterized by gene cloning (Tsujibo et al., 2000b). Finally, deacetylation of substrates may occur on the level of GlcNAc, GlcNAc-6-P, and/or $(GlcNAc)_2$-P (Yamano et al., 1996).

8.2
Chitinase Genes

Chitinase genes from a variety of organisms have been cloned and sequenced, as reviewed on several occasions (Graham and Sticklen, 1994; Kramer and Muthukrishnan, 1997; Neuhaus, 1999; Saito et al., 1999; Gokul et al., 2000; Patil et al., 2000), and a few examples are listed in Table 7.

Many chitinases are composed of a catalytic domain and one or more chitin binding and/or FnIII type domains (Figure 8). The catalytic domain of family 18 chitinases contains the highly conserved motif DXXDXDXE (Synstad et al., 2000). Analysis of gene structures often reveals the presence of a signal sequence of 21–23 amino acid residues.

Human chitinase (chitotriosidase) contains a C-terminal chitin-binding domain of at least 49 amino acid residues, including six cysteines accounting for three disulfide bridges which are essential for chitin binding (Tjoelker et al., 2000). The gene is located on chromosome 1q (Eiberg and Dentandt, 1997). Cloning of a cDNA encoding human macrophage chitotriosidase confirmed assignment to family 18 chitinases. Expression is strongly regulated as the mRNA occurs at a late stage of differentiation of monocytes to activated macrophages (Boot et al., 1995). Several mammalian proteins of yet unknown function, including human articular cartilage chondrocyte protein YKL-39, YKL-40, oviductal glycoprotein, and macrophage YM-1, share significant sequence homologies with $(\beta\alpha)_8$ family 18 chitinases (Hu et al., 1996; Jin et al., 1998; Renkema et al., 1998; Bleau et al., 1999).

Insect chitinase genes have been cloned in a few cases, i.e., the tobacco hornworm, *Manduca sexta* (Kramer et al., 1993; Choi et al., 1997), the silkworm *Bombyx mori* (Kim et al., 1998), and *Aedes*, *Anopheles*, and *Drosophila* species (De la Vega et al., 1998). Typically, these genes code for N-terminal signal sequences and for chitin-

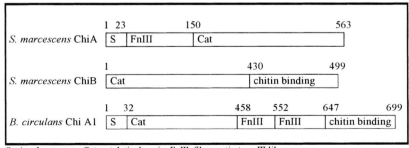

S: signal sequence, Cat: catalytic domain, FnIII: fibronectin type III like sequence

Fig. 8 Examples of typical domain organizations of family 18 chitinases. The numbers indicate the positions of amino acids in the sequence.

Tab. 7 Chitinase genes

Organism	No. of amino acids	Comments	References
Human			
Homo sapiens	466	N-terminal signal sequence	Aerts et al. (1996)
Insects			
Bombyx mori	565	75% homology with M. sexta chitinase; contains signal peptide; N-glycosylated	Kim et al. (1998)
Hyphantria cunea	553	77–80% homology with M. sexta chitinase; contains signal peptide; N-glycosylated	Kim et al. (1998)
Bacteria			
Serratia marcescens chitinase A	563	N-terminal signal peptide of 23 residues; domain structure: see Figure 8	Brurberg et al. (1994)
Serratia marcescens chitinase B	499	Does not contain a signal peptide; domain structure: see Figure 8	Brurberg et al. (1995)
Streptomyces thermoviolaceus; thermostable chitinase Chi30	347	Catalytic domain only; similarity with ChiA from S. coelicolor and ChiA from S. lividans; 12-bp direct repeat sequence involved probably in induction by chitin and repression by Glc	Tsujibo et al. (2000a)
Protozoa			
Plasmodium falciparum PfCHT1	378	N-terminal signal sequence; lacks proenzyme and chitin-binding domains	Vinetz et al. (1999)
Plasmodium gallinaceum PgCHT1	587	N-terminal signal sequence; C-terminal chitin binding domain; a proenzyme form was identified	Vinetz et al. (2000)

binding domains, besides the catalytic domain of typically family 18 chitinases. An exception is the chitinase from a cell line from *Chironomus tentans* which is devoid of a chitin-binding domain (Feix et al., 2000).

In plants, chitinases belong to the PR-genes, and degrade chitin as a component of fungal cell walls. Rapid evolution in plant chitinases shows that nonsynonymous substitution rates in plant class I chitinase often exceed synonymous rates in the plant genus (Bishop et al., 2000). The chitinases from bulbs of genus *Tulipa* TBC-1 and TBC-2 each consist of 275 amino acid residues with M = 30,825 and 30,863 daltons, respectively (Yamagami and Ishiguro, 1998). The enzymes share 90% and 63% identity respectively with *Gladiolus* chitinase A. The chitinase A (M = 22,391 daltons) from leaves of pokeweed (*Phytolacca americana*) does not contain a chitin-binding domain (Yamagami et al., 1998).

The gene encoding ChiB from *S. marcescens* does not contain a signal peptide, and the enzyme is exported to the periplasm without processing (Brurberg et al., 1995). Also, family 18 ChiC1 of *S. marcescens* lacks a signal sequence (Suzuki et al., 1999). Deletion of the C-terminal domain reduces the hydrolytic activity towards powdered chitin and regenerated chitin, but not towards colloidal chitin and glycol chitin, illustrating the importance of the chitin-binding domain for efficient hydrolysis of crystalline chitin. ChiC2 corresponds to the catalytic domain of ChiC1, and is probably generated by proteolytic removal of the FnIII-like and chitin-binding domains. Phylogenetic analysis shows that family 18 chitinases can be clustered in three subfamilies which have diverged at an early stage of bacterial chitinase evolution (Suzuki et al., 1999). *S. marcescens* chitinase C1 is found in one subfamily, whereas chitinases A and B of the same bacterium belong to another subfamily. Chitinase ChiA1 of *Bacillus circulans* WL-12 contains an N-terminal catalytic domain, two FnIII type domains, and a C-terminal chitin-binding domain (Hashimoto et al., 2000).

A high multiplicity of family 18 and 19 chitinase genes exists in *Streptomyces*, and proteolytic cleavage of the primary gene products contributes to the diversity of chitinases in these bacteria (Saito et al., 1999). The family 19 chitinase genes, chi35 and chi25 of *Streptomyces thermoviolaceus* OPC-520, are arranged in tandem (Tsujibo et al., 2000c). Alignment of the deduced amino acid sequences reveals that Chi35 has a N-terminal polysaccharide-binding domain, similar to xylanase but not yet observed in chitinase, containing three reiterated amino acid sequences starting from C-L-D and ending with W, and a catalytic domain. Chi25 encodes only a catalytic domain.

The chitinases of *Plasmodium* differ in domain structure. In contrast to PgCHT1, PfCHT1 lacks proenzyme and chitin-binding domains. (Vinetz et al., 1999, 2000).

8.3
Regulation of Degradation

In insects, chitinase digests chitin in nonsclerotized parts of the cuticle during molting. Induction of the integumental enzyme by 20-OH-E was observed in fifth instar larvae of the tobacco hornworm, *Manduca sexta* (Fukamizo and Kramer, 1987), in the silkworm, *Bombyx mori* (Koga et al., 1991), and in insect cell lines of *Chironomus tentans* (Spindler-Barth, 1993). Gene expression is up-regulated during the molting process, larval–pupal transformation and pupal–adult transformation.

In plants, different ethylene-inducible genes are involved in ripening of fruit, senescence of flower petals, or during pathogen infection (Deikman, 1997; Ohme-Takagi et al., 2000). The ethylene-responsive element was identified as the promoter GCC box (AGCCGCC), and transcription factors respond to extracellular signals to modulate GCC box-mediated gene expression. Transcription of class I chitinase genes is enhanced in response to subnanomolar concentrations of $(GlcNAc)_n$, $n > 5$. The elicitor signal evokes also phytoalexin production and the expression of elicitor-responsive genes is probably transmitted through a 75 kDa high-affinity binding plasma membrane protein. However, the induction of chitinase expression by N-acetylchitooligosaccharides may require protein phosphorylation, but not *de novo* protein synthesis (Nishizawa et al., 1999).

In bacteria, chitinase synthesis is induced by chitin and repressed by the presence of glucose (for a review, see Keyhani and Roseman, 1999). The promoter region of most of the *Streptomyces* chitinase genes contains a pair of 12 bp direct repeat sequences, which apparently is involved in both chitin induction and glucose repression (Saito et al., 1999; Tsujibo et al., 2000a). A pleiotropic regulatory gene, *reg1*, was identified in *Streptomyces lividans*, which controlled both amylase and chitinase expression (Nguyen et al., 1997).

9 Production of Chitin and Chitosan

Chemical as well as biotechnological procedures are currently under investigation for chitin and chitosan production. In practical terms, chemical processing of the waste fraction of the shellfish industry is the most important source of these polysaccharides, but biotechnological procedures for deproteinization, demineralization, and deacetylation by fermentation or enzymatic reactions are being evaluated in a number of research laboratories. To date, the investigation of additional sources such as insects (Struszczyk et al., 2000; Zhang et al., 2000) and fungi (see Chapter 5, this volume) has not provided any viable technical process for chitin and chitosan production.

9.1 Isolation of Chitin and Chitosan from Shellfish Waste

The primary sources of α-chitin and chitosan are waste fractions from crabs, prawns, shrimps, and freshwater crayfish which are harvested for human food consumption, but never for the sole purpose of producing chitin or chitosan. Chitin and chitosan are produced by chemical processes suitable for removing proteins, minerals, pigments, and varying fractions of the N-acetyl groups, respectively. As a rule, the yield of chitin from crustaceans is in the order of 1–2% of the raw material wet weight.

9.1.1 Resources

The waste generated globally from processing of crustaceans is estimated at 1.44×10^6 tons per annum on a dry matter basis, and the potential for chitin production is approximately 2×10^5 tons (for references, see Roberts, 1992; Shahidi and Synowiecki, 1992). Some data collected from recent publications are listed in Table 8.

The estimated resources of the Northern Atlantic deep-water shrimp, *Pandalus borealis*, in the Barents Sea and the Spitzbergen area alone are ca. 161,000 tons, 25% of which are landed in Norway. Additional resources of *Pandalus*, utilized predominantly in Canada, Greenland, and Iceland, occur in the North-West Atlantic. Shrimp shells and heads yield about 40% of the processed material, most of which is either disposed as waste (thereby causing serious pollution

Tab. 8 Potential resources for chitin and chitosan production

Country/year	Source	Annual catch [tons]	References
India, 1997	Various shellfish	80,000 (waste)	See Kumar (2000)
Japan, 1996	Krill, *Euphausia superba*	60,500	Krasavtsev (2000)
Mexico, 1998	Prawn, *Penaeus monodon*	80,000	Cira et al. (2000)
Norway, 1993	Shrimp, *Pandalus borealis*	40,000	Stenberg and Wachter (1996)
Panama, 1996	Krill, *Euphausia superba*	495	Krasavtsev (2000)
Poland, 1996	Krill, *Euphausia superba*	20,600	Krasavtsev (2000)
Russia, 1993	Shrimp, *Pandalus borealis*	20,000	Stenberg and Wachter (1996)
Tasmania, 1992	Lobster, *Jasus edwardsii* and *J. verreouxi*;	1907	Kow (1994)
	Crab, *Pseudocarcinus gigas*;	76	
	Squid, *Nototodarus gouldi*	14	
Thailand, 1997	Shrimp, *Penaeus monodon*	200,000 (waste)	Rao and Stevens (1997)
Ukraine, 1996	Krill, *Euphausia superba*	13,400	Krasavtsev (2000)

problems) or processed to shrimp meal to be used as a feed additive for aquacultured salmon. The production of chitosan from *Pandalus* waste has been established in some countries along the North-East Atlantic coast line, including Norway and Iceland (Stenberg and Wachter, 1996).

The primary source for chitin and chitosan production in Japan are shells from limbs of the Red Crab, *Chionoecetes japonicas*, which yield chitosan at about 20–30% of dry matter (Hirano, 1989).

The Antarctic krill, *Euphausia superba*, is caught on a large scale by fishing fleets, mainly of Japan, Panama, Poland, and Ukraine. Estimates of the resources vary between 50×10^6 and 1000×10^6 tons annually (Peter and Köhler, 1996; Krasavtsev, 2000). The actual world catch of Antarctic krill is in the order of 100,000 tons, which is $<1\%$ of the permissible amount of $15-16 \times 10^6$ tons. Approximately 35,000 tons are harvested by Polish and Ukrainian fleets (Krasavtsev, 2000). However, krill is apparently not used for the production of chitin or chitosan on a technical scale.

9.1.2
Chemical Processes

The technology of chitin and chitosan production has been reviewed (Muzzarelli, 1977; Roberts, 1992; No and Meyers, 1997); however, only the basic principles of the process are outlined in this section.

The procedure for isolation of chitin from crustacean waste consists principally of four steps (Table 9). The concentrations of reagents, temperatures and reaction times vary widely, depending on the source of the starting material and the desired specifications of the final product, i.e., mainly M_v and F_A of chitosan. There is an inverse correlation between M_v and F_A versus reagent concentration as well as reaction temperature and reaction time during deacetylation (Wojtasz-Pajak et al., 1994; Roberts, 1997; Struszczyk et al., 1998; Ng et al., 2000; Roberts and Wood, 2000; Struszczyk et al., 2000).

A modified process consisting of an initial demineralization of crustacean shells, followed by deproteinization and a second demineralization, before deacetylation yields chitosan of M_v 900,000–1,000,000 and F_A 0.2–0.12, with an ash content $<0.1\%$ (Wachter et al., 1995). Mechanical shear does not significantly affect M_v and F_A (Rege and Block, 1999).

Insect chitin has a lower crystallinity than crustacean chitin, and relatively mild conditions for deacetylation yield chitosans of high M_v and low F_A (Struszczyk et al., 2000; Zhang et al., 2000). β-Chitin was isolated from pens of the squid, *Loglio vulgaris*, and used as starting material for the isolation of chitosan of $M = 500,000$ daltons (Tolaimate et al., 2000).

9.1.3
Fermentation Processes

Fermentation of prawn waste with *Lactobacillus* is performed under cofermentation

Tab. 9 Principal steps of chitin and chitosan production (for details, see No and Meyers, 1997; Roberts, 1992)

Step	Reagent	Temperature [°C]	Time [h]
Deproteinization	0.5–15 % NaOH	25–100 °C	0.5–72
Demineralization	2.5–8%	15–30 °C	0.5–48
Decoloration	Various organic solvents; NaOCl, H_2O_2	20–30 °C	Washing, 1.0
Deacetylation	39–60% NaOH	60–150 °C	0.5–144

with additional carbon sources, such as sugar cane, lactose and whey powder (Rao and Stevens, 1997; Cira et al., 2000). The process removes up to 90% of the protein and 80% of the calcium carbonate and calcium phosphate. The crude chitin from silage is treated with acid and alkali to complete the removal of minerals and proteins.

Acetic acid fermentation using *Lactococcus lactis* or *Clostridium formicoaceticum* and a chitinous fraction of crawfish, *Programbarum clarkii*, containing ca. 18% protein and 52% ash, affords chitin of ca. 95% purity, containing ca. 1% protein and 2% ash (Bautista et al., 2000).

Screening of proteolytic microorganisms including *Pseudomonas maltophilia*, *Bacillus subtilis*, *Streptococcus faecium*, *Pediococcus pentosaseus*, and *Aspergillus oryzae* resulted in 82% deproteinization of demineralized prawn shell when an inoculum containing four strains (*B. subtilis*, *S. faecium*, *P. pentosaseus*, and *A. oryzae*) is used (Bustos and Healy, 1994).

9.1.4
Enzymatic Deacetylation of Chitin

Chitin deacetylases (CDA) occur in bacteria and fungi, and the biochemistry of the enzyme is described in Chapter 5, this volume. In this section, approaches towards the utilization of CDA for preparation of chitosan are briefly mentioned, though a review of CDA was published recently (Tsigos et al., 2000). The extracellular CDAs of *Mucor rouxii*, *Absidia glauca* and *A. coerulea* species are effective on amorphous chitin of medium F_A, whereas data on the specificity of CDA of *Colletotrichum* are at variance (Martinou et al., 1997; Win et al., 2000). The *Mucor* enzyme deacetylates chitin of F_A 0.42 or 0.28 to yield, within 6 to 13 h, chitosans of F_A 0.02 or 0.03, respectively. Crystalline α-chitin from shrimps or crabs reacts very slowly, with < 1% deacetylation occurring, while amorphous crustacean chitin undergoes ca. 10% deacetylation (Martinou et al., 1997). Although the process has been demonstrated on a laboratory scale, it is not yet technically feasible, though the optimization of enzyme production is currently under investigation (Win et al., 2000).

9.2
Preparation of Low Molecular-weight Chitin, Chitosan, and of Chitooligosaccharides

Oligomers of GlcNAc, GlcN, as well as chitooligosaccharides composed of both monosaccharides are of increasing interest because of their remarkable biological activities (see Section 10). Principally, these compounds can be obtained by enzymatic or chemical depolymerization of chitin of any F_A, or by enzymatic, chemical or chemoenzymatic synthesis.

9.2.1
Depolymerization of Chitin and Chitosan

The traditional procedures of hydrolysis of chitin by means of hydrochloric acid, hydrogen fluoride, or sulfuric acid/acetic anhydride are discussed elsewhere (Defaye et al., 1989; Schanzenbach et al., 1997; Scheel and Thiem, 1997). Other methods for depolymerization of chitin or chitosan include γ-irradiation (Lim et al., 1998), ultrasonic irradiation (Takiguchi et al., 1999), or reaction with hydroxyl radicals (Tanioka et al., 1996). Depolymerization of chitosan by diazotization of amino groups with nitrous acid results in the formation of 2,5-anhydro-D-mannose units at the cleaved glycosidic bond (Allan and Peyron, 1997). Treatment with sodium borohydride then yields the corresponding anhydromannitol (Figure 9).

Low molecular-weight chitosan of M_v 11,000 to 436,000 and polydispersities between 2.01 and 4.16 is obtained by acid and alkali treatment of β-chitin (Shimojoh et al., 1998).

Enzymatic depolymerization of partially acetylated chitin is achieved by chitinases, chitosanases, and by lysozymes (cf. Section 8). Chitooligosaccharides of DP 3–6 are produced by degradation of chitosan (F_A 0.11) in an ultrafiltration membrane reactor using chitosanase from *Bacillus pumilus* (Jeon and Kim, 2000).

Besides lysozymes and chitinases, hemicellulase, cellulase, α-amylase, papain, tannase, and lipases can also be used for depolymerization of chitin and for the preparation of chitooligosaccharides (Pantaleone and Yalpani, 1993; Muzzarelli et al., 1994, 1995; Yalpani and Pantaleone, 1994; Muzzarelli, 1997a; Zhang et al., 1999). Generally, high enzyme concentrations and rather long reaction times are required, however.

9.2.2
Synthesis of Chitooligosaccharides

The synthesis of β-(1-6)-linked GlcN or GlcNAc as well as attachment of GlcNAc to glycans containing for example Man or Gal residues is not considered here. An impressive number of outstanding chemical investigations have been published on the synthesis of $(GlcN)_n$, $n \leq 12$ (Kuyama and Ogawa, 1997) and on lipochitooligosaccharides (Nod factors) (for a recent reference, see Demont-Caulet et al., 1999). Polycondensation of O-benzyl-protected ethylthio-N-acetylglucosaminide affords O-benzyl oligomers of GlcNAc of DP ≤ 12 (Hashimoto et al., 1989).

Oligosaccharides of β-(1-4)-linked GlcN and GlcNAc are most conveniently synthesized by chemoenzymatic approaches using enzymatic transglycosylations and selective N-acetylation or N-deacetylation reactions. A summary of products and methods is provided in Table 10.

9.2.3
Abiotic Synthesis of Chitin

The synthesis of chitin of M_v 46,000 has recently been achieved using an elegant chemoenzymatic approach with the intermediate of the enzymatic hydrolysis of chitin, i.e., the oxazoline which is readily prepared from $(GlcNAc)_2$ (Figure 10) (Kobayashi et al., 1996). X-ray diffraction and ^{13}C-CP/MAS NMR revealed the product to be α-chitin (Kiyosada et al., 1998).

A hyperbranched chitin analogous dendrimer of M_w 4.5×10^5 daltons, consisting of β-(1-3/1-4)-linked GlcNAc, is obtained by camphorsulfonic acid-catalyzed polycondensation of the 1,2-oxazoline derivative of 6-O-tosyl-GlcNAc (Kadokawa et al., 1998).

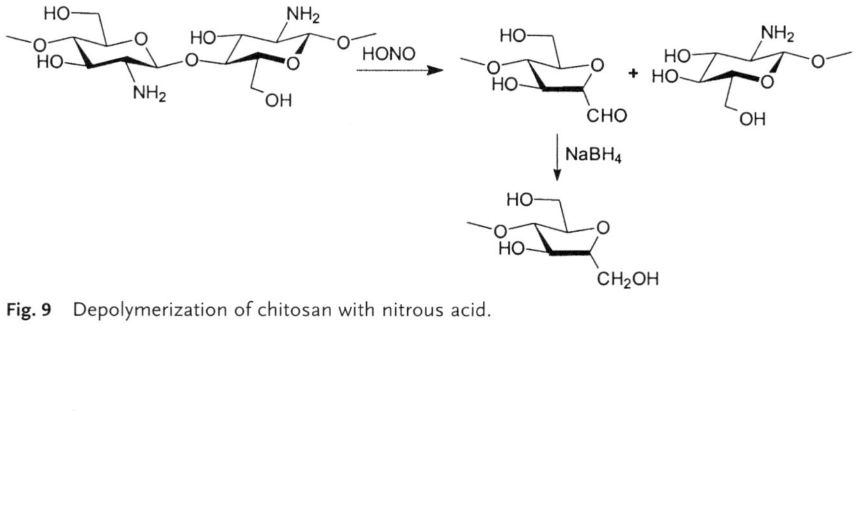

Fig. 9 Depolymerization of chitosan with nitrous acid.

Tab. 10 Chemoenzymatic preparation of chitin and chitosan oligosaccharides.

Product	Method	Reference
$(GlcNAc/GlcN)_n$, $n = 4-12$	Transglycosylation with lysozyme; substrates: $(GlcNAc)_3$ and/or $(GlcN\text{-}COCH_2Cl)_3$	Akiyama et al. (1995a)
$(GlcNAc)_3$- and $(GlcNAc)_4$-6-O-sulfate	NodST and Arylsulfotransferase IV as PAPS regenerating system	Burkart et al. (1999)
GlcNAc-GlcN	N-Acetylation of $(GlcN)_2$ by reverse hydrolysis with chitin deacetylase from *C. lindemuthianum*	Tokuyasu et al. (1999a)
GlcNAc-GlcNAc-GlcNAc-GlcN	N-Acetylation of $(GlcN)_4$ by reverse hydrolysis with chitin deacetylase from *C. lindemuthianum*	Tokuyasu et al. (2000)
$(GlcNAc)_5$; $(GlcN)(GlcNAc)_4$	Synthesis with NodC, N-terminal deacetylation with NodBC, expressed in recombinant *E. coli*	Samain et al. (1997)
GlcN-GlcNAc-MU; GlcNAc-GlcN-MU; $(GlcN)_2$-MU	N-Deacetylation with chitin deacetylase from *C. lindemuthianum*	Honda et al. (2000a)
$(GlcNAc)_2$	Enzymatic transfer of GlcNAc-oxazoline to GlcNAc	Kobayashi et al. (1997)
$(GlcNAc)_2$-pNP	Transglycosylation of GlcNAc-pNP with hexosaminidase	Kubisch et al. (1999)
GlcN-GlcNAc-pNP	Enzymatic deacetylation of $(GlcNAc)_2$-pNP	Tokuyasu et al. (1999b)
$(GlcNAc)_n$, $n = 3-6$	N-Acetylation of oligomers produced by chitosanase digestion of chitosan	Izume et al. (1992)
$(GlcNAc)_n$, $n = 3-8$	Transglycosylation of $(GlcNAc)_n$, $n = 2-7$ with hexosaminidase	Dvorakova et al. (2001)

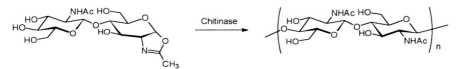

Fig. 10 Chemoenzymatic synthesis of chitin (Kobayashi et al., 1996).

9.3 Current World Market and Economics

The commercial aspects of chitin and chitosan have been detailed recently (Technical-Insights, 1998). Bulk prices for chitosan vary between approximately US$ 10 and US$ 150 per kg, depending primarily on the product specifications, though clearly the cost of highly pure, pyrogen-free chitosan suitable for high-value applications is significantly higher.

Reliable data relating to the current world production of chitin and chitosan are not available. In 1988, the total capacity of chitin production by 15 Japanese companies was reported as 2000 tons, but with an actual output of 700 tons. This was used in the production of chitosan (500 tons), glucosamine and oligosaccharides (60 tons), and for other purposes (40 tons), leaving an overproduction of 100 tons of chitin (Hirano, 1989). Since then, the demand for chitin and chitosan has grown remarkably, and production sites for both compounds have been established in many countries worldwide. Based on information received by the author from various sources, some companies are planning to increase production capacities for chitosan to several 1000 tons per annum, and so it is probably safe to estimate that the global annual production of

Tab. 11 Some companies offering chitin, chitosan, and related products (alphabetical listing)

Company	Product	Contact
BioPrawn AS, Tromsø, Norway / Henkel KGaA, Düsseldorf, Germany	Chitosan for use in cosmetics and pharmaceuticals	
Chito-Bios, Ancona, Italy	N-Carboxyisobutyl chitosan	Via Torresi, 8; I-60128 Ancona, Italy. Ph.: +39 (0) 71 89 39 58 FAX: +1-902-466 6889
ChitoGenics, Halifax, Canada	Chitin, chitosan, N,O-carboxymethylchitosan	
Genis hf, Iceland	Chitosan, Oligosaccharides	http://www.genis.is
Heppe GmbH, Halle-Queis, Germany	Chitosan; R&D contracts	http://www.biolog-heppe.de
Katakura Chikkarin Co., Ltd., Tokyo, Japan	Chitin, chitosan	
Kate International	Chitin, chitosan	http://www.kateinternational.com
Kitto Life Co., Seoul, Korea	Chitosan, oligosaccharides, glucosamine	http://www.kittolife.co.kr
Korea Chitosan CO., Ltd.; Youngdeok Chitosan Co., Ltd., Korea	Chitin / chitosan fiber for medical / textile use, water soluble chitosan, derivatives of chitin / chitosan,	http://www.koreachitosan.com
Micromod Gmbh, Germany	Chitosan micro- and nanocapsules	http://www.micromod.de/contact.htm
Primex Ingredients ASA, Norway	Chitosan for food and cosmetics	http://www.primex.no
Pronova, Norway	Various chitosans of low to high viscosity / Mw; drug master files are expected	www.pronova.com
Seafresh Chitosan, Thailand	Chitosan	seafresh@samarth.co.th
SONAT, JSC, Russia	Chitosan, intended for use in food industry, as a food ingredient and emulgator	chitosan@col.ru

chitin and chitosan currently ranges from 3000–10,000 tons. (A recent report of >100 billion tons of chitosan being produced in Japan alone is clearly incorrect.) It appears that approximately 1% of the shellfish waste produced globally is actually utilized for chitin and chitosan production.

It has been suggested that chitin and derivatives might become competitors of cellulose or synthetic polymers, but this appears to be unrealistic for reasons of cost and availability. Cellulose is available in amounts exceeding those of chitin by several orders of magnitude. However, the unique properties of chitosan might lead to a considerable number of high-added value applications, in addition to the bulk use of cheap chitosan in some technical processes (see Section 11).

9.4
Companies Producing the Polymer

Companies which produce chitin, chitosan, and/or related products are listed in Table 11, this list having been compiled from various sources in recent years. The list is most likely incomplete, as not all producers are listed on the Internet, nor do they attend chitin conferences. However, some companies may recently have ceased production. Additional addresses of companies and research institutions are listed in a commercial reference source (Technical-Insights, 1998; the Table of Contents is available at the URL cited).

10
Properties of Chitin and Chitosan

Some of the remarkable physico-chemical and biological properties of chitin and chitosan have been referred to in previous sections of this chapter. Here, attention will be focused on those properties that are especially important with respect to the development of practical applications of the polysaccharides (see Section 11).

10.1
Physico-chemical Properties

The latest comprehensive compilation of data on the physical and chemical properties of chitin and chitosan appeared some ten years ago (Roberts, 1992; see also Muzzarelli, 1977). Some of the fundamental parameters are summarized in Table 12.

Water-soluble chitin is obtained by steeping α-chitin in concentrated aqueous NaOH, followed by the addition of crushed ice, thereby reducing the NaOH concentration to 10%; the solution is then maintained at room temperature for 3 h. The resulting water-soluble product is actually chitin of F_A 0.45–0.55. β-Chitin swells readily in water, and is also soluble in formic acid. A recently described organic solvent for chitosan consists of DMSO and p-toluenesulfonic acid (p-TosOH) or 10-camphorsulfonic acid (CSA) (Sashiwa et al., 2000).

Microcrystalline chitosan (MCCh) is prepared by limited hydrolysis of chitosan, followed by precipitation with alkali under high shear (Struszczyk, 1987). This material possesses interesting properties, for example a high water retention value (Table 13) (Struszczyk and Kivekäs, 1992).

10.2
Materials Produced from Chitin or Chitosan

A variety of materials can be prepared from chitosan, including films, semipermeable membranes, fibers, micro- and nanocapsules, as well as composites with inorganic components.

Tab. 12 Chemical properties of chitin and chitosan (Roberts, 1992)

Parameter	α-Chitin	Chitosan
M	$1-2 \times 10^6$ Da (estimated)	Varying widely; c.f. Section 9
Solubility	5% LiCl in DMAc[a]; hexafluoroacetone; hexafluoro-2-propanol; 12 M cold HCl[b]	Dilute aqueous organic or mineral acids
pK_a	–	6.5, decreasing with increasing DP

[a] Similar solvents are 5% LiCl in DEAc or N-methyl-2-pyrrolidone.
[b] Chitin is slowly depolymerized in cold concentrated HCl.

Tab. 13 Properties of microcrystalline chitosan (MCCh) (Struszczyk and Kivekäs, 1992)

Parameter	Hydrogel	Dry powder
M_w	10^4–10^6	10^4–10^6
Water retention (%)	500–5000	200–800
F_A	<0.65	<0.65
Crystallinity (%)	–	≤95
Polymer content (%)	0.1–10.0	85–95
pH	≥7	6.5–7.5

10.2.1
Films, Membranes, and Fibers

Chitosan forms films when diluted aqueous solutions of the polymer (<2%) in for example 1% acetic or lactic acid are allowed to dry, using either solvent-casting or surface-coating techniques. The tensile strengths of chitosan films fall in the range of 38 to 66 MPa, which is approximately twice the tensile strength of polyethylene (Remunan-Lopez and Bodmeier, 1996; Park et al., 1999). Trimethylsilyl chitin provides films from solutions in organic solvents (Kurita et al., 1999). Membranes formed from chitosan are less permeable to oxygen, nitrogen and carbon dioxide than cellulose acetate membranes. Many blends have been developed in order to modify the properties of films and other materials prepared from chitosan. Blend films of bacterial poly(3-hydroxybutyric acid) (PHB) with chitin or chitosan show enhanced biodegradability as compared with materials which contain a single component (Ikejima and Inoue, 2000).

Fibers are manufactured by wet spinning of appropriate solutions, applying principally the technologies used in the cellulose industry (Rathke and Hudson, 1994). Treatment of materials prepared from chitosan with acetic anhydride affords the corresponding insoluble chitin materials.

10.2.2
Polyelectrolyte Complexes

Chitosan, as a polycation, forms polyelectrolyte complexes with poylanions such as dextran sulfate (Sakiyama et al., 1999), chondroitin sulfate, and hyaluronan (Denuziere et al., 1996). Microcapsules derived from chitin as the only precursor are prepared from chitosan and 6-oxychitin (Muzzarelli, 2000a; Muzzarelli et al., 2000a) or chitosan sulfate (Holme and Perlin, 1997; Voigt et al., 1999). The mechanical strength and permeability of the capsules is optimized by cross-linking, using glutaraldehyde or other bifunctional reagents.

10.2.3
Lipid-binding Properties

Chitosan is well known for its lipid-binding properties, including fatty acids, di- and triglycerides, and steroids. The mechanism of adsorption is not fully understood, though the electrostatic interactions with carboxylates and emulsifying properties are well documented. Chitosan interacts with fatty acids, leading to flocculation or dispersion of the system (Domard and Demarger-Andre, 1994). Columns of chitosan, when percolated with olive oil, retain the oil at up to twice the weight of the chitosan, and the retained fraction is enriched in phytosteroids (Muzzarelli et al., 2000b). Stable emulsions with sunflower oil are formed in 0.2–2% solutions of chitosan (F_A 0.05–0.25) in diluted hydrochloric acid at weight ratios of ca. 4–5 parts chitosan to 1 part oil (Rodriguez et al., 2000).

10.3
Chemistry of Chitin and Chitosan

Chitosan possesses unique chemical functionalities, as the biopolymer contains primary and secondary hydroxy groups as well as primary amino groups of comparatively low pK_a. Thus, numerous derivatives and

Tab. 14 Derivatives of chitin and chitosan

Derivative	R^1; R^2	R^3	R^4	References
Triphenylsilylchitin	H, Ac	Ph_3Si	Ph_3Si	Vincendon (1997)
N-Acylchitosan	H, acyl	H	H	Hirano (1997)
N-Alkylchitosan	H, alkyl	H	H	Muzzarelli and Tanfani (1985)
Quaternary chitosan salt	$(R_3)^+$	H	H	Lang et al. (1997b)
N-Alkylidenechitosan	=CH-alkyl	H	H	Hirano (1997)
N-Arylidenechitosan	=CH-aryl	H	H	Hirano (1997)
N- or O-Carboxymethylchitosan	CH_2COO^-	H, CH_2COO^-	H, CH_2COO^-	Muzzarelli (1988)
N,N-Dicarboxymethylchitosan	CH_2COO^-	H	H	Muzzarelli et al. (1997c)
N-or O-Sulfonyl chitosan	H, SO_3^-	H	H	Holme and Perlin (1997); Nishimura et al. (1998)
N- or O-Hydroxyalkylchitosan	alkyl-OH	H	H or alkyl-OH	Lang et al. (1997a)

procedures for their preparation have been described in the literature.

10.3.1
Reactivity

α-Chitin, because of its insolubility, is rarely subjected to chemical reactions, except for the preparation of chitosan by deacetylation, though techniques utilizing highly reactive regenerated chitin by precipitation from phosphorous acid have been described (Vincendon, 1997). β-Chitin has relatively high reactivity (Kurita et al., 1994a).

10.3.2
Derivatives

The derivatives prepared from chitin or chitosan are numerous, and only a very limited selection of recently produced examples is considered here (Table 14) (for reviews, see Hirano, 1989; Lang, 1995; Kurita, 1997; Kurita et al., 2000).

Triphenylsilylchitin of DS 1.5 containing predominantly 6-O-silyl groups is soluble in dioxan, tetrahydrofuran (THF), chloroform, and toluene, forming low-viscosity solutions. The polymer is stable in water and thermostable up to 250 °C (Vincendon, 1997). Likewise, 3,6-di-O-trimethylsilyl chitin of DS 2.0, prepared from β-chitin is soluble in organic solvents and can be modified further to give for example 6-O-tosylated or 6-O-glycosylated chitin, containing α-Man or β-GlcNAc branches (Kurita et al., 1994b, 1999). 6-O-Tosylchitin is used for further modifications, such as preparation of 6-mercaptochitin (Kurita et al., 1997). N-Phthaloyl-chitosan is another organosoluble precursor for the regioselective introduction of various branches into chitin (Kurita, 1997; Kurita et al., 2000).

N-Acyl derivatives include various fatty acid and amino acid amides. The reaction of solid N-α-chloropropionyl or N-chloroacetyl chitosan with ammonia or aromatic amines affords the corresponding N-alanyl and N-(N'-aryl)glycyl derivatives of chitosan, respectively (Hirano et al., 1992).

Carboxymethyl chitins and chitosans have long been known (Muzzarelli, 1988). Substitution occurs preferentially at the 6-OH position, yielding betaines which are soluble at acidic or basic pH values, but which form gels at pH ca. 7. Depending on the reaction conditions, N-, 6-O- or 3,6-di-O-, or N,N-dicarboxymethylchitosans are obtained.

The oxidation of chitin or fungal chitin–glucan complexes by means of NaOCl in the presence of 2,2,6,6-tetramethylpiperidine-N-oxide and NaBr gives 6-oxychitin (Figure 11). This was suggested as a substitutive for hyaluronan (Muzzarelli et al., 1999) and, indeed, possesses interesting biological properties (see below). Bromination of N-acetylchitosan followed by reduction with $NaBH_4$ affords 6-deoxychitin (see Figure 11) (Zhang et al., 1997).

10.3.3
Hybrid Polymers

Copolymers containing alternating chitooligosacharides and poly(propylene glycol) are resistant to chitinase digestion (Kadokawa et al., 1995). Reaction of chitosan with PEG-aldehyde followed by N-acetylation produces a polymer which is soluble in water and in phosphate buffer (Sugimoto et al., 1998). N-Ethyl-N'-(dimethylaminopropyl)carbodiimide is used to prepare chitosan esters at neutral pH, whereas dialkylpyrocarbonate and cyanuryl dichloride are preferably used under weakly basic conditions (Aiba, 1993). Graft copolymerization of 2-methyl-2-oxazo-

Fig. 11 Structures of 6-oxychitin and 6-deoxychitin.

line onto tosyl- or iodo-chitins produces poly(N-acetylethyleneimine) side chains (Kurita et al., 1996a). Likewise, mercaptochitin is used for graft copolymerization of methyl methacrylate or styrene onto chitin (Kurita et al., 1996b,c, 2000).

Chitosan is a suitable carrier for peptide synthesis, as shown by preparation of protected precursors of the putative *Ras* protein farnesyl transferase inhibitors, i.e., Fmoc-Cys-Aib-Aib-Met-NH_2, Fmoc-Cys-Aib-Aib-Met-OH, and Fmoc-Cys-Val-Gly-Met-NH_2. Different cleavable linkers such as 4-(2',4'-dimethoxyphenyl-Fmoc-aminomethyl)-phenoxy-acetic acid, 4-hydroxy-methyl-3-methoxyphenoxy-acetic acid, or 4-methylene-3-methoxy-phenol were grafted onto chitosan prior to peptide synthesis (Neugebauer et al., 2000).

10.4
Biological Properties

Although the outstanding biological activities of chitin, chitosan and chitooligosaccharides were discovered some 15 years ago, some novel aspects are reviewed in this section.

10.4.1
Biological Activities in Mammalian Systems

The most spectacular results were observed when chitosan or derivatives were assayed in vertebrates for acceleration of bone regeneration and wound healing. Other biological effects include antimetastatic and antiviral activities of both chitooligosaccharides and chitosan derivatives.

Bone Regeneration and Wound Healing

Chitosan, N,N-dicarboxymethylchitosan, and 6-oxychitin show a remarkable acceleration effect on bone regeneration (Muzzarelli, 2000a) and wound healing (Stone et al., 2000). The molecular mechanisms of these effects are unknown, but chitosan-coated surfaces promote the growth of human osteoblasts and chondrocytes and propagate the expression of extracellular matrix proteins (Klokkevold et al., 1997b; Lahiji et al., 2000). Chitosan films promote the growth of keratinocytes, but not of fibroblasts which adhere to the surface of the films almost twice as strongly as do keratinocytes (Chatelet et al., 2001). Moreover, as discussed in Section 7.3, chitooligosaccharides ultimately are primers for hyaluronan biosynthesis (Varki, 1996; Muzzarelli, 1997b; Muzzarelli et al., 1997a), whilst GlcN is both substrate for, and regulator of, glycosaminoglycan biosynthesis (Rubin et al., 2000). Chitosan also inhibits the production of nitric oxide (NO) in activated macrophages, thus reducing cytotoxicity in cell proliferation during inflammation processes in wound healing (Hwang et al., 2000).

Activation of complement C3 and C5, as well as stimulation of fibroblasts to produce interleukin-8 are important events in wound healing acceleration by chitin and chitosan (Minami et al., 1997). Complement C3 activation by chitosan decreases with decreasing F_A, but is not seen with low-M_w chitosans (Suzuki et al., 2000). An investigation of the effects of chitin and derivatives on proliferation of fibroblasts *in vitro* revealed that the biological response is mediated through additional factors, including the composition of the culture medium (Mori et al., 1997).

Other Effects

Although (GlcNAc)$_6$ shows significant antimetastatic effects in mice bearing Lewis lung carcinoma (Suzuki et al., 1986, 1989), this issue does not appear to have been investigated with appropriate clinical studies.

Sulfation of 6-O-tritylchitosan at N-2 and O-3 affords potent antiretroviral agents which inhibit the infection of HIV-1 to T lymphocytes at 0.28 μg mL^{-1}, without significant cytotoxicity, whilst 6-O-sulfochitin strongly inhibits blood coagulation (Nishimura et al., 1998).

10.4.2
Antimicrobial Activity

GlcN oligomers of DP ≥ 30 possess antimicrobial activity against a number of Gram-negative, Gram-positive and lactic acid bacteria, whereas low-DP oligomers are ineffective (Ueno et al., 1997; Jeon et al., 2001). Apparently, the smaller chitosan oligomers serve as nutrients for bacteria, whereas the higher oligomers are toxic by virtue of their charge-mediated adhesion to the cell membrane, which in turn prevents the uptake of nutrients through the cell wall (Tokura et al., 1997).

10.4.3
Elicitor Activity in Plants

Besides induction of cell division in the cortex of Leguminosae by Nod factors (see Section 7), the most prominent activity of chitin and/or chitosan in plants is the elicitation of defense reactions, i.e., the induction of chitinases or of metabolites of the shikimate pathway, including lignin.

Partially N-acetylated, polymeric chitosans elicit phenylalanine-ammonia-lyase (PAL) and peroxidase activities in wheat (*Triticum aestivum*) leaves. Maximum activity is observed with chitosans of intermediate F_A values, whereas (GlcNAc)$_n$, $n \geq 7$, elicits peroxidase. GlcN oligomers are not active as elicitors. All chitosans (but not the chitin oligomers) induce the deposition of lignin (Vander et al., 1998). Partially acetylated (F_A 0.33–0.52) chitooligosaccharides of DP 5–7 elicit pisatin in pea epicotyls (Akiyama et al., 1995b).

Chitosan stimulates the production of secondary metabolites in plant cell cultures, as was shown for anthraquinone by *Morinda citrifolia* and amaranthin by *Chenopodium rubrum*, presumably by elicitation and membrane permeabilization (Doernenburg and Knorr, 1996).

Chitin-oligosaccharides induce an extracellular class III chitinase and PAL in a rice suspension culture. The maximum activity is observed with $(GlcNAc)_6$ at $1~\mu g~mL^{-1}$, whereas lower GlcNAc and GlcN oligosaccharides are less active (Inui et al., 1997). The induction of anti-plant pathogen-specific chitinases by various elicitors, including chitin, chitosan and chitooligosaccharides has been demonstrated in grass and yam (Koga, 1996).

At the physiological level, GlcNAc oligomers of DP ≥ 4 cause a transient alkalinization of the culture medium of suspension-cultured tomato cells at 1.4 nM for whole cells and 23 nM for microsomal membranes, whereas chitosan oligomers are inactive (Baureithel et al., 1994).

11
Applications of Chitin and Chitosan

The great potential for applications of chitin and chitosan is reflected by the coexistence of approximately 3500 patents or patent applications, in addition to a much higher number of scientific articles which have appeared in the literature during the past decade. Many of the claims suggest uses for chitin and chitosan where these biopolymers actually replace existing synthetic or natural polymers, irrespective of economical aspects (for general reviews, see Lang, 1995; Peter, 1995; Daly and Macossay, 1997; Kurita, 1997; Kumar, 2000). The most frequently cited advantages of chitosan are seen in the unique physico-chemical and biological properties of the polymer, including antimicrobial activity and biodegradability. A survey of applications suggested over the years is listed in Table 15. The commercial exploration of chitosan has been particularly successful in water engineering, the manufacture of textiles, and as ingredients in cosmetics and dietary health food products.

11.1
Technical Applications

Numerous technical applications of chitosan have been suggested, the most intensively studied topics being the removal of proteins, metal ions, and dyes from various factory effluents, and applications in the textile and paper industry.

11.1.1
Waste-water Engineering

As chitosan interacts with negatively charged proteins, it is not surprising that applications for the flocculation of proteins from waste waters have been reported (for example, Guerrero et al., 1998; Savant and Torres, 2000). The recovered chitin–protein complexes can then be used for animal feed.

The metal ion-complexing properties of chitosan and derivatives are also well documented (Guibal et al., 1997; Domard and Piron, 2000), and a compilation of data is given in Table 16 (Peter, 1995). Possible applications are seen for example in the galvanizing industry, or in the purification of effluents from nuclear power plants.

Chitosan possesses a high affinity for a wide range of dyes, in particular negatively charged compounds. The sorption characteristics are influenced primarily by the pH of the solution, and also by the presence of other contaminants. Applications are envisaged in the purification of waste water from the textile industry (Kumar, 2000).

Chitosan pervaporation membranes are used to separate ethanol or DMSO from water (Kurita, 1997).

Tab. 15 Summary of applications of chitosan

Application	Properties of chitosan utilized
Technical	
Water engineering: adsorption of metal ions and dyes; flocculation of proteins; separation of organic solvents	Polycation; metal ion complexation; biodegradability
Textiles, fibers, nonwoven fabrics, leather	Polycation; film formation; antibacterial properties
Paper coating	Complex formation with polyanions and polysaccharides
Biotechnology: enzyme immobilization; plant culture medium supplement; cell encapsulation; protein purification	Chemical functionality; polyelectrolyte complexes
Medicine and healthcare	
Lowering of serum lipids	Polycation; lipid complexation
Bone regeneration; treatment of rheumatoid diseases	Osteoconductivity; GAG synthesis regulation
Vascular medicine and surgery	Tissue adhesion; hemostatic
Wound care; artificial skin; hemostasis	Antibacterial; biological activity on cells
Cosmetics	
Skin moisturizing ingredient; hair shampoos; hair styling; dentrifices	Gel and film formation; antibacterial
Agriculture	
Plant growth regulators; elicitors of plant defense; seed conservation; soil fertilizer; anti-fungals and anti-nematodals	Plant growth regulator; regulation of resistance proteins; stimulation of chitinase-producing soil bacteria
Food	
Health ingredient, dietary fiber; foam stabilizer; preservative; clarification of beverages; packaging materials	Viscosity; polycation; bacteriostatic activity; film formation

11.1.2

Fibers, Textiles, and Nonwoven Fabrics

A comprehensive review on fibers and films made from chitin and chitosan was published by Rathke and Hudson (1994) (see also Struszczyk, 1997a; Kumar, 2000). Solvents used for the spinning of chitin fibers are chosen according to the solubility of the polymer in LiCl/DMAc or halogenated organic solvents. Alternatively, an aqueous alkaline solution of sodium xanthate can be used, and N-propionylchitosan fibers are obtained in the same way. N-Acylchitosan fibers containing higher fatty acid residues are obtained by N-acylation of chitosan fibers; these may be prepared by spinning a 3% solution of chitosan in 2% acetic acid into an aqueous 10% NaOH solution containing 30% sodium acetate (Hirano et al., 2000). The mechanical properties of chitosan fibers (Table 17) are modulated by stretching or by immersing the fibers in phosphate- and phthalate-containing solutions (Knaul et al., 1999).

Chitosan–cellulose blend fibers are generated from mixtures of the corresponding xanthates or of chitosan salts and cellulose xanthate (Hirano and Midorikawa, 1998). Conventional textile fibers may also be coated with chitosan (Struszczyk, 1997b) and cross-linked, for example by periodate activation of the cellulose (Liu et al., 2001).

Chitosan-containing textiles, which are manufactured by several companies, possess

Tab. 16 Binding capacities of polymers for metal ions (mEq g⁻¹) (Peter, 1995)

Polymer	Method	Mg²⁺	Ca²⁺	Sr²⁺	Ba²⁺	Mn²⁺	Ni²⁺	Cu²⁺	Cd²⁺	Pb²⁺	Zn²⁺	Hg²⁺	Co²⁺	Cr²⁺	Fe²⁺	Ag⁺	Au³⁺	Pt⁴⁺	Pd²⁺
Chitosan (F$_A$ 0.45)	A	0.3				1.1	3.5	5.3	6.5		5.5						5.84	4.52	6.28
Chitosan (F$_A$ 0.03)	A	0.5				0.5	2.3	4.8	4.9		3.2							3.82	0.87
Chitosan (F$_A$ not specified)	B		0.8	1.5	1.1	1.44	3.15	3.12	2.78	3.97	3.70	5.60	2.47	0.46	1.18	3.26	5.75		
p-Aminostyrene	B		0.4	0.6	0.8		0.02	1.31	0.10	0.17	0.52	5.70	0.07	0.03	0.05	1.98			
6-O-CM-chitin	C	2.08	2.50	2.12	2.15	2.22	2.22		2.42	2.13									
3,6-di-O-CM-chitin	C	4.0	3.3*	2.8	10.1⁺		3.4	4.1⁺	3.1⁺	9.2⁺	8.5⁺								

Method A: 0.04 M metal ion; pH 7.4; 20 h; 30 °C. Method B: 1 g polymer; 0.02–0.4 mM metal chloride or nitrate; 24 h. Method C: not specified. 6-O-CM-chitin = 6-carboxymethylchitin; 3,6-di-O-CM-chitin = 3,6-di-O-carboxymethylchitin. *Gel formation or precipitation. ⁺Ca²⁺ can be desorbed by means of EDTA.

Tab. 17 Some properties of chitosan fibers (Struszczyk, 1997a)

Property	Value
Titer [dtex]	1.5–3.0
Tenacity in standard conditions [cN/tex]	10–15
Tenacity in wet conditions [cN/tex]	3–7
Loop tenacity [cN/tex]	3–7
Elongation in standard conditions [%]	10–20
Water retention [%]	≤250

antimicrobial properties and so are especially useful in the manufacture of underwear and sanitary products. Nonwoven fabrics may also be used as wound dressings (see below).

11.1.3
Paper Technology

Paper coated with chitosan shows improved breaking strength, burst resistance or tearing strength as compared with noncoated paper (Wieczorek and Mucha, 1997). Techniques for the manufacture of this include casting after addition of chitosan to paper pulp, as well as coating of paper sheets with chitosan.

11.1.4
Biotechnology

Chitosan is used as a stationary phase for the chromatography of various biomolecules, e.g., proteins and saccharides. Porous beads are manufactured from chitosan by cross-linking, and are available commercially in Japan as Chitopearl® with a range of specifications, including chemically modified materials, e.g., for chiral recognition, ion-exchange, gel-permeation, or affinity chromatography, as well as for the immobilization of cells, enzymes, antibodies, or other polypeptides.

Microbial and mammalian cells are also immobilized by encapsulation, often using alginates as the negatively charged polymer component, with possible applications in biotechnology, medicine, and pharmacy (Zielinski and Aebischer, 1994; Vogt et al., 1996; Quong et al., 1997; Bartkowiak and Hunkeler, 2000).

11.2
Medicine and Healthcare

The low toxicity, antimicrobial activity and physico-chemical functionality of chitosan offer a great potential for medical applications, as documented in several reviews (Muzzarelli, 1997b; Muzzarelli et al., 1997a; Paul and Sharma, 2000; see also Muzzarelli, 2000b). Chitosan is currently not registered as a drug for the treatment of diseases, but GlcN – either as the hydrochloride or the sulfate – is used widely for the relief of pain in musculoskeletal rheumatoid diseases, including arthrosis, arthritis, or osteoporosis. GlcN is available in some European countries by prescription, but is sold elsewhere (e.g., the USA) as an over-the-counter (OTC) pharmaceutical. GlcN sulfate is prepared by the hydrolysis of chitin with hydrochloric acid, followed by ion exchange. The starting material for GlcN production is chitin, and significant amounts of the available resources of this polymer are utilized to satisfy the huge market resulting from applications in human and veterinary medicine.

11.2.1
Treatment of Obesity and Hyperlipidemia

Although reports on the hypocholesterolemic and hypolipidemic activity of chitosan in animals first appeared some 25 years ago, a comprehensive discussion of the hypolipidemic action (including possible mechanisms) and drug formulations was published only recently (Furda, 2000).

It has been suggested that chitosan binds to bile acids, thereby removing them from the enterohepatic circulation, and showing similar effects as other polycations such as cholestyramine or cholestipol. Oral administration of chitosan to mice showed that fatty and bile acids, cholesterol and triglycerides which bind to chitosan and are not absorbed, resulting in hypolipidemic and anti-atherogenic effects (Muzzarelli, 1998; Ormrod et al., 1998; Han et al., 1999). In rats, cholesterol resorption in the intestine was reduced, while fat and bile acid excretion was enhanced after feeding chitosan (Gallaher et al., 2000). On the other hand, a hypocholesterolemic action of chitosan was absent in rabbits with high serum cholesterol levels, though enhanced excretion of sterols and bile acids were noted (Hirano and Akiyama, 1995)

Chitosan at a dose of 3–6 g per day to humans was reported to lower total serum cholesterol, increase high-density lipoprotein (HDL), and enhance the excretion of primary bile acids, cholic acid and chenodeoxycholic acid (Maezaki et al., 1993). Derivatives of chitosan bearing a tertiary or quaternary amino group show enhanced binding capacity for glycocholate (Table 18) (Lee et al., 1999b).

A clinical study showed that 66% of a moderately obese population responded to a dose of 2.25 g per day chitosan over 12 weeks, with a mean weight reduction of ca. 4 kg (Thom and Wadstein, 2000). Chitosan

Tab. 18 Binding capacity of polymers for glycocholic acid (Lee et al., 1999b)

Polymer	Binding capacity [mmol g^{-1}]
Chitosan	1.42
DEAE-chitosan	3.12
Quaternized DEAE-chitosan	4.06
Cholestyramine	2.78

in combination with a hypocaloric diet, with or without additional fibers and micronutrients, was reported also to reduce body weight, serum triglycerides, total and low-density lipoprotein (LDL) cholesterol, but to increase in HDL cholesterol (Colombo and Sciutto, 1996; Girola et al., 1996; Macchi, 1996; Veneroni et al., 1996a,b). Chitooligosaccharides appeared to have similar effects, in addition to lowering blood pressure (Kawasaki et al., 1998).

Some critical issues were raised, however. For example, Ernst and Pittler (1998) and other investigators reported no significant differences in body weight reduction or serum lipid levels when a normal diet was administered with or without chitosan (Pittler et al., 1999; Stern et al., 2000), including microcrystalline chitosan (MCCh) (Wuolijoki et al., 1999). Possible side effects, such as impairment of the resorption of minerals or lipid-soluble vitamins were not observed in all studies. Vitamin K levels were increased from 0.33 (placebo) to 0.56 ng mL^{-1} (Pittler et al., 1999).

When administered orally to dogs, β-chitin was not degraded during passage through the intestine, whereas chitosan (M = 4 × 10^5; particle size ca. 0.5–3 mm) underwent transformation into a gel in the stomach, thereby losing ca. 15% of its weight. Further degradation of chitosan was found to take place in the large intestine, and < 10% was recovered in the feces (Okamoto et al., 2001). The DP of the degradation products was not determined. Similar results were obtained in rabbits and chickens, whereas the reverse situation appears to exist in sheep (for references, see Okamoto et al., 2001). Feeding of [^{14}C]chitosan to rats resulted in the distribution of radioactivity among all tissues (Nishimura et al., 1997).

11.2.2
Bone Regeneration and Prosthetic Implants

The osteoconductive properties of chitosan and some derivatives are described in Section 10.3.1. 6-Oxychitin was shown to promote reconstruction of the osteo-architectural morphology in surgical lesions in rat condylus, even though healing was slower when compared with N,N-dicarboxymethylchitosan (Mattioli-Belmonte et al., 1999). Complete healing was observed with N,N-dicarboxymethylchitosan within 3 weeks. A complex of N,N-dicarboxymethylchitosan with calcium acetate and Na$_2$HPO$_4$ promoted bone mineralization in osteogenesis, leading to complete healing of otherwise nonhealing surgical defects in sheep (Muzzarelli et al., 1998).

A mixture of chitosan, hydroxyapatite granules, ZnO, and CaO, pH 7.4, was used to prepare a self-hardening paste that produced much less heat on setting as compared with conventional poly(methyl methacrylate) bone cement (Maruyama and Ito, 1996).

The coating of prosthetic materials made from ceramics or titanium with chitosan showed improved biocompatibility and facilitated the attachment of fibroblasts and osteoblasts, thus promoting bone healing and integration of the implant (Muzzarelli, 2000a; Pittermann et al., 1997b).

11.2.3
Vascular Medicine and Surgery

N-Acylchitosans are blood-compatible and show anticoagulative effects, especially in the case of N-hexanoylchitosan of DS 0.2–0.5, which is biodegradable by lysozyme (Lee et al., 1995). In contrast, chitosans of F_A 0.3–0.57 show hemostatic activity (Klokkevold et al., 1997a; Sugamori et al., 2000). MCCh was found to seal arterial puncture sites when applied after arterial catheterization via an arterial sheath (Hoekstra et al., 1998).

Photocross-linking by UV irradiation of chitosan containing *p*-azidobenzamide and lactobionic acid amide side chains produces an insoluble hydrogel within 60 s. The binding strength of the noncytotoxic hydrogel is similar to that of fibrin glue, and the material seals against air leakage from pinholes on isolated small intestine, aorta, and also from incisions on isolated trachea (Ono et al., 2000).

11.2.4
Wound Care and Artificial Skin

Spectacular results on wound healing have been observed when chitosan was applied to damaged skin in the form of bandages made from nonwoven tissues, or sponges, films and membranes (Muzzarelli, 1997b; Muzzarelli et al., 1997a). MCCh was shown to absorb and retain blood, to induce the release of substances involved in platelet activation, such as β-thromboglobulin and platelet factor 4, and to possess excellent hemoagglutinative activity in dogs (Sugamori et al., 2000). Chitosan, as a semipermeable biological dressing, maintained a sterile wound exudate beneath a dry scab, preventing dehydration and contamination of the wound, facilitating rapid wound re-epithelialization, including regeneration of nerves within a vascular dermis (Stone et al., 2000). Various formulations of chitosan with other biopolymers, such as collagen, hyaluronan, or chondroitin sulfate have been described (Denuziere et al., 1996; Choi et al., 2001).

11.2.5
Other Medical Applications

Chitosan–iron complexes have been investigated for the treatment of hyperphosphatemia (Baxter et al., 2000). Chitosan immobilized on poly(L-lysine) binds bilirubin, and is suggested as an adsorbent for hematoperfusion in hyperbilirubinemia (Chandy and Sharma, 1992). Contrast agents used in X-ray, ultrasound or NMR imaging diagnostics have been described (mostly in the patent literature) in considerable variety as formulations containing chitosan or derivatives.

11.2.6
Toxicology of Chitin and Chitosan

Chitin is a normal constituent of seafood and mushrooms, both of which are considered to be nontoxic and safe for human consumption, except for a very small percentage of the population that is allergic to seafood products due to the presence of proteinaceous impurities. However, hypersensitization is most likely due to protein components rather than to the polysaccharide. Intravenous or peritoneal application of chitin is not possible, and has actually never been suggested.

Earlier studies published some 20 years ago indicated a very low toxicity of chitosan in mice (for references, see Hwang and Damodaran, 1995). A recent extensive toxicological study with PROTASAN™ UP (ultrapure; F_A 0.17), including various application routes as well as hypersensitization, mutagenicity, and cytotoxicity assays,

Tab. 19 Toxicology of chitosan glutamate (PROTASAN™ UP) in rats (Dornish et al., 1997a)

Application	No toxic effects seen at	Application for
Oral	up to 600 mg kg^{-1} per day	13 weeks
Intravenous	up to 25 mg kg^{-1}	Once
Intraperitoneal	bolus injection, up to 500 mg kg^{-1}	7 days
Nasal mucosa	up to 3 mg per rat per day	7 days

showed that chitosan glutamate is essentially nontoxic in rats (Table 19) (Dornish et al., 1997a). The testing of cosmetic preparations of Hydagen® CMF on the skin of volunteers revealed favorable properties of high-M_v chitosan when erythema and squamations were evaluated (Pittermann et al., 1997a).

Data on the toxicological properties of carboxymethylchitosans are rare. The LD_{50} was reported as >5 g kg^{-1} in rats (Liu et al., 1997).

11.2.7
Regulatory Aspects

As mentioned earlier, chitosan has not been registered for drug use, for a number of reasons, including:

- Lack of standards: apparently, no obligatory protocol for defining the chemical, physico-chemical, and purity specifications as well as other parameters has been established (Table 20).
- Scarcity of data on efficacy: although numerous spectacular results have been detailed in the literature, some are apparently at variance, and the results of appropriate clinical studies are extremely rare. However, this situation may soon change as some companies are clearly investigating drug master files.
- Lack of long-term toxicological and safety data. The use of chitosan as a pharmaceutical excipient is currently hampered by a lack of appropriate preclinical testing and assessment of the compound's safety (Baldrick, 2000).

11.3
Pharmaceutical Applications

The most intensively studied topics in pharmaceutical applications are enhanced transmucosal drug delivery, sustained drug release, and drug targeting. Again, the polycationic nature of chitosan as well as its biodegradability are the most important features of chitosan with regard to interactions with negatively charged tissue surfaces, as well as the development of slow-release formulations, including micro- and nanocapsular systems.

Tab. 20 Specifications of chitosan for pharmaceutical-grade products

Parameter	Specification	Reference[a]
Viscosity	Should be specified	A, B, C
F_A	Should be specified	A, B, C
Acid content	Should be specified	A
Moisture	Should be specified	A, B, C
Solubility, turbidity	$>99.9\%$	A, B, C
Heavy metals	<27 ppm	A, B
Endotoxin	<625 EU g^{-1}	A
Microbiology	<1 c.f.u. g^{-1}	A
Color	Colorless, grayish, cream, or pink	B, C
Ash	$<3\%$	B, C
Appearance	Powder, flakes	B
Grain size	Less than 25% >2 mm	B
Chromatographic analysis	Suggested	C
Spectroscopic analysis	Suggested	C
Chemical analysis; I_2, $KMnO_4$	Suggested	C
Protein/amino acids	Suggested	C

[a] A = Dornish et al. (1997); B = Wojtasz-Pajak et al. (1994); C = Stevens (2001).

11.3.1
Transmucosal Drug Delivery

Chitosan is strongly mucoadhesive (Deacon et al., 1999; Harding et al., 1999). Furthermore, chitosans of DP > 50 and $F_A < 0.4$ have been shown to induce the opening of tight junctions in confluent monolayers of Caco-2 cells (Holme et al., 2000). An increased transport of polar drugs across epithelial surfaces was observed when insulin was applied with chitosan glutamate (PROTASAN™) to the nasal epithelia of rats, with little or no toxicity to the nasal mucosa being observed (Dornish et al., 1997b). Hydrocortisone or transforming growth factor β (TGF-β) was applied with chitosan lactate to porcine buccal mucosa, and this resulted in a six-fold enhancement of tissue permeability (Senel et al., 2000a,b). Peptide delivery through the intestinal epithelia has also been demonstrated in Caco-2 cells with the luteinizing hormone-releasing hormone (LHRH) analogue buserelin, using trimethylchitosonium chloride (DS 0.4 or 0.6) (Thanou et al., 2000).

By contrast, the screening of cationic compounds as absorption enhancers for nasal drug delivery has led to the conclusion that poly(L-arginine) is superior to chitosan in terms of efficiency and side effects when fluorescein isothiocyanate-labeled dextran was used as a model drug in rats (Natsume et al., 1999).

11.3.2
Sustained-release Formulations

In addition to peptides (for a review, see Bernkop-Schnurch, 2000), a variety of other drugs, e.g., nifedipine, diclofenac, and propranolol, have been investigated for sustained oral, buccal, transdermal, or colon-specific delivery from beads, granules, membranes, films, tablets, or microcapsules, manufactured from chitosan, often in combination with other polymers (for some recent reviews, see Zecchi et al., 1997; Felt et al., 1998; Illum, 1998; Davis, 2000; Gupta and Kumar, 2000; Kumar, 2000; Paul and Sharma, 2000).

$(GlcNAc)_n$ has been investigated as a slow-release form in the treatment of osteoarthritis (Rubin et al., 2000).

11.4
Cosmetics

Applications of chitosan are based on the film-forming properties on negatively charged surfaces, and on the formation of hydrogels. Hair conditioners, shampoos, and styling products containing either chitosan or derivatives such as carboxymethyl- and hydroxyalkylchitosans as well as quaternary ammonium salts have long been used (Lang, 1995), while numerous other uses of chitosan have been suggested in skin lotions and creams, soaps, and dentifrices. A chitosan-containing soap is currently marketed in Korea.

11.5
Agriculture

Chitosan can be used as an antimicrobial film for fruit and seed coating, resulting in higher crop yields. Application of chitosan to the soil stimulates growth of the chitinase-producing bacteria that mediate antifungal and anti-nematodal effects (Gooday, 1999; Herrera-Estrella and Chet, 1999). The elicitor activities for plant defense reactions and the stimulation of nodulation were discussed in Sections 8 and 9.

11.6
Food

Chitosan, though not registered as a drug, is successfully marketed as a health food ingredient and dietary fiber by several companies, predominantly in the Far East, North America, and in some European countries (Muzzarelli, 1996; Yalpani, 1997; Shahidi et al., 1999).

Applications of chitosan in food processing technology include uses as a foam stabilizer and emulsifier, as a preservative agent, and as a flocculent of tannins and proteins in beverages. A review of these topics is beyond the scope of this chapter, however.

12
Current Problems and Limits

The appearance of the book *Chitin* almost a quarter of a century ago (Muzzarelli, 1977) marked the beginning of a world-wide systematic exploitation of chitin and chitosan. Whilst it is likely that all major chemical and pharmaceutical companies have filed patents on the production and uses of biopolymers, the question remains as to why chitin and chitosan have not yet acquired a position in the large, high value-added markets. It is possible that chitosan has remained underinvestigated, even though the routes of production of both chitin and chitosan are well established and the chemistry of the polymers is reasonably well understood.

In addition to the relatively high price of good quality chitosan, practical utilization is slowed by regulatory matters (see Section 11). Other problems are sometimes thought to arise from limitations of the availability of raw materials, which depends not only on the population dynamics of marine organisms, but also on political and ecological issues. However, in view of the vast amounts of shellfish waste produced globally, it is envisaged that huge resources are always available, even when the catch of a particular crustacean species might be restricted. To secure the resources initially is a matter of developing suitable logistics. The fishing and seafood processing industries are most likely not aware of the potential value of shellfish waste, and much of this is simply returned either to the sea or to ecologically questionable terrestrial landfill sites. Another major problem arises from the nature of the biological material itself, as it rapidly decomposes and must therefore be processed with minimal delay after peeling of the crustacean catch. Nonetheless, it appears that these problems can be solved, as demonstrated by the fact that several companies now produce these polymers successfully.

The comment has often been made that the production of chitosan leads to environmental problems, mainly because of the harsh conditions involved such as high temperatures, aggressive acid and alkali solutions, and the large volumes of fresh water required. However, similar concerns might be expressed with regard to the production of cellulose.

13
Outlook and Perspectives

It is expected that the global demand for chitosan will increase dramatically, once the polysaccharide has been registered for use as a food additive and pharmaceutical excipient, and for certain medical and agricultural applications. Moreover, it is possible that this might happen in the near future. Subsequently, the production of chitin and chitosan will depend for many years essentially on marine and freshwater crustacean resources, though biotechnological processes to remove most of the protein and mineral fractions show great promise. However, it is unlikely that the development of alternative routes such as enzymatic deacetylation, or the use of either insects, fungi, or synthetic approaches will become practicable in the near future.

Often, when the subject of chitin is raised, it is the exoskeleton of insects and crabs that comes first to mind, with chitosan remaining something of an unknown quantity. However, there is an increasing awareness of the fascinating properties of chitosan and its vast potential in high value-added applications. Consequently, as our knowledge of chitosan develops, then this biopolymer – which is clearly much more than simply an "interesting skeletal material" – is likely to be exploited to a much greater extent.

14
Relevant Patents for Isolation, Production and Applications of Chitin and Chitosan

A patent search in *Chemical Abstracts* on chitin or chitosan related citations yielded a total of 3527 references containing relevant terms either in the title and/or in the abstract (date of patent search: May 25, 2001). The number of references is rapidly growing, averaging presently ca. 250 *per annum* (Figure 12).

A summary of the topics covered in the patent literature is given in Table 21.

Tables 22–34 refer to details. The citations are sorted alphabetically by the first inventor.

Obviously, it is neither practical nor useful to evaluate all patents and thus, a rather narrow selection is made from the International (World) patents or applications which were published after the year 1990. Original texts were reviewed in a few cases only, and most of the information is retrieved from the titles and abstracts. Patents describing uses of these polysaccharides as substitutes for other biopolymers are generally excluded.

This survey should not be used as a support for future patent applications. It cannot not replace a literature search which must always be done when a new patent should be considered. Indeed, the selection is very incomplete, covering less than 10% of the relevant literature. However, we believe that it is representatively illustrating the various activities going on in the exciting field of research and applications of chitin and chitosan.

Acknowledgements
The investigations performed in the author's laboratory were supported by the Deutsche Forschungsgemeinschaft, by EU grant no. BIO4-CT-960670, and by the Fonds der Chemischen Industrie. The author's thanks are also extended to H.-M. Cauchie (Luxembourg), V. G. H Eijsink (Ås), J.-P. Thomé (Liège), and J. F. V. Vincent (Bath) for their valuable comments on different sections of the manuscript.

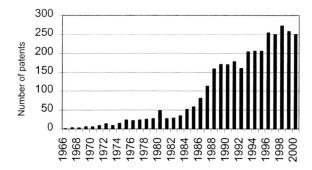

Fig. 12 Annual number of chitin or chitosan related patents or patent applications, 1966–2000.

Tab. 21 Summary of the main areas and topics of chitin or chitosan relevant patents cited in Tables 22–34

Field	Table No.	Number of patents cited	Selected topics
Agrochemical bio-regulators	22	14	Antimicrobial chitinolytic enzymes and chitin binding proteins Chitosan as plant growth regulator Chitosan in release formulations for antimicrobial agents
Biochemical genetics	23	16	Chitinases as antifreeze proteins in plants Construction of transgenic plants expressing chitinases Genes and cloning of chitinases, including human chitinase Uses of cloned chitinases and chitin binding proteins as antifungal agents Yeast chitin synthase
Biochemical methods	24	6	Chitinases or chitin binding proteins as diagnostic or analytical tools Chitosan materials for cell culture
Enzymes	25	8	Chitin deacetylases, including c-DANN's Enzymes as antimicrobial agents, as therapeutic and analytical tools Enzymes for production of chitooligosaccharides Uses of chitosan for immobilization of enzymes
Cosmetics and ingredients	26	11	Liquid cleansers Products for skin and hair care, including microcapsule formulations Sunscreen formulations Toothpastes and oral hygiene
Industrial biotechnology	27	4	Cell adhesive surfaces Chitosan beads as carriers for proteins Manufacture of chitin and chitosan using fungal cultures Production of N-acetylglucosamine from chitin Use of chitosan for improved production of natural products in plant tissue cell culture
Food and feed chemistry	28	5	Chitosan for use in weight loss programmes Gels and emulsions for food processing Ingredient in food preservatives Protective coating of fruits, vegetables and other food Stabilization and formulation of vitamins
Immunochemistry	29	3	Chitinase or chitin binding proteins for diagnostics to detect fungal infections Chitosan as adjuvant in influenza vaccines Immunstimulation and -potentiation by chitosan
Polysaccharide chemistry and processes	30	38	Applications in drug delivery or targeting systems, wound dressings, magnetic supports, coatings Carboxymethylchitosans, chitin-glucan-complexes Chitin or derivatives in form of hydrogels, films, fibers, foams, beads, porous particles, micro- and nanocapsules Cyclodextrin derivatives Manufacture of chitooligosaccharides Methylpyrrolidinone chitosans N-Halochitosans for use as flocculants Processes and technologies for production of chitosan Processes for crosslinking of chitosan Quaternary ammonium salts and their hypocholesterolemic properties

Tab. 21 (cont.)

Field	Table No.	Number of patents cited	Selected topics
Pharmaceuticals and pharmacology	31	100	Various Derivatives for uses in cosmetics, transfection of cells, synthetic bone and teeth Adjuvants for vaccines Antimicrobial non-woven fabrics Antiviral, antifungal, or antiinflammatory formulations Artificial tears for treatment of dry-eye syndrome Biodegradable cationic bioadhesives Bone substitutes N,O-Carboxymethylchitosan for prevention of postsurgical tissue adhesion Chitosan for prevention of infections Chitosan-drug conjugates Compositions and formulations for treatment of rheumatoid diseases Detection and monitoring of fungal infections Doxorubicin-chitooligosaccharide conjugates Drug targeting and enhancement of resorption in the intestinal tract Enhancers for transmucosal drug delivery Foam scaffolds for cell entrappment and growth supports for encapsulated cells Formulations for reducing skin irritation Hygiene products Imageing materials Immunostimulants for tumor therapy Implantable polymers, encapsulation of cells for implantation Liquid and odour absorbent materials for medical and pharmaceutical uses Manufacture of UV radiation absorbers Nanoparticles for drug delivery Osteoinductive materials pH Controlled gel formation Promotion of tissue repair and tissue engineering Reduction of fat absorption from food Sustained drug release formulations Thromboresistant surfaces Treatment of fungal infections with chitin synthesis inhibitors Treatment of periodontic diseases Wound dressings
Textiles and Paper	32	11	Modified fibers Pulp and paper additives Textile finishing
Water purification	33	6	Adsorption of anions, dyes, metal ions, and organochemicals
Various applications	34		Anticorrosives Filtration media Liquid absorbers Metal ion adsorbers Packaging materials Semipermeable membranes, also for pervaporation

14 Relevant Patents for Isolation, Production and Applications of Chitin and Chitosan | 169

Tab. 22 Agrochemical bioregulators

Patent number	Patent holder	Inventors	Title of patent	Publication date (YYYYMMDD)
WO 0032041	Safescience, Inc., USA	Ben-Shalom, N., Pinto, R.	Chitosan metal complexes as agents for controlling microbial diseases in crops	20000608
WO 9942594	Cornell Research Foundation, Inc., USA	Broadway, R. M., Harman, G. E.	Fungus and insect control with chitinolytic enzymes from *Streptomyces albidoflavus*	19990826
WO 9411511	Zeneca Ltd., UK	Broekaert, W. F., Cammue, B. P. A., Osborn, R. W., Rees, S. B.	Antimicrobial plant proteins and their purification	19940526
WO 0051428	Virbac S.A., France	Broussaud, O., Derrieu, G., Jouan Daniel, R., Pougnas, J.-I.	Microparticulate or nanoparticulate vectors for active agents in aquatic media	20000908
WO 9732973	Cornell Research Foundation, Inc., USA	Harman, G. E., Lorito, M., Di Pietro, A., Hayes, C. K., Scala, F., Kubicek, C. P.	Combinations of fungal cell wall degrading enzyme synergistic use with fungicides, transgenic plants, and agricultural and therapeutic uses	19970912
WO 9832335	DCV, Inc., USA.	Heinsohn, G. E., Bjornson, A. S.	Chitosan salts as crop yield enhancers	19980730
WO 9639824	Chitogenics, Inc., USA	Henderson, S. E., Curran, D. T., Elson, C. M.	Negatively-charged chitosan derivative semiochemical delivery system	19961219
WO 0024260	Université Pierre et Marie Curie (Paris VI), France	Khoury, C., Minier, M., Le Goffic, F., Van Huynh, N.	Glycolytic enzyme-based fungicidal composition	20000504
WO 9708944	Kansas State University Research Foundation, USA	Kramer, K. J., Muthukrishnan, S., Choi, H. K., Corpuz, L., Gopalakrishnan, B.	A chitinase of *Manduca sexta* and a cDNA encoding it and their use in the preparation of insect resistant plants	19970313
WO 9942488	Vanson, Inc. USA	Nichols, E. J.	Composition and method for reducing transpiration in plants	19990826
WO 9403062	Novasso OY, Finland	Struszczyk, H., Kivekaes, O.	Seed encrusting with microcrystalline chitosan	19940217
WO 0119187	Instytut Wlokien Chemicznych, Poland; Instytut Ochrony Roslin; Instytut Sadownictwa i Kwiaciarstwa	Struszczyk, H., Niekraszewicz, A., Wisniewska-Wrona, M., Urbanowski, A., Pospieszny, H., Orlikowski, L., Wojdyla, A., Skrzypczak, C.	Chitosan gel for protecting plants against diseases	20010322
WO 9945784	Oji Paper Co., Ltd., Japan	Takahashi, T.	Antimicrobial agents containing Eucalyptus leaf extracts and chitosan	19990916
WO 9709879	Bioestimulantes Organicos, Lda., Portugal	Villanueva, J., Valenzuela, P.	Chitosan formulation to increase resistance of plants to pathogenic agents and environmental stress	19970320

Tab. 23 Biochemical genetics

Patent number	Patent holder	Inventors	Title of patent	Publication date (YYYYMMDD)
WO 0009729	Pioneer Hi-Bred International, Inc., USA	Dhugga, K. S., Anderson, P. C., Nichols, S. E.	Expression of chitin synthase and chitin deacetylase genes in plants to alter the cell wall for industrial uses and improved disease resistance	20000224
WO 9201792	SANOFI S. A., France; Societe Nationale Elf Aquitaine	Dubois, M., Grison, R., Leguay, J. J., Pignard, A., Toppan, A.	Chimeric endochitinase gene and transgenic, pathogen-resistant plants expressing it	19920206
WO 0024874	Johns Hopkins University, USA	Fomenkov, A., Keyhani, N. O., Roseman, S.	Polynucleotide and polypeptide of *Vibrio furnissii* gene chiE extracellular chitinase, sequences and biological uses thereof	20000504
WO 9747752	Icos Corp., USA	Gray, P. W.	Human chitinase, its cDNA sequence and cloning, and its use in antifungal compositions	19971218
WO 9946390	Icos Corp., USA	Gray, P. W., Tjoelker, L. W.	Chitinase chitin-binding fragments and their use in chitin determination and as anti-fungal agents	19990916
WO 9424288	Cornell Research Foundation, Inc., USA	Harman, G. E., Tronsmo, A., Hayes, C. K., Lorito, M.	Gene encoding for endochitinase	19941027
WO 9906565	Ice Biotech Inc., Canada	Hew, C., Xiong, F., Moffatt, B., Griffith, M.	Cloning expression systems and cDNA sequences encoding antifreeze proteins from winter rye	19990211
WO 9736917	Smith-Kline Beecham Corp., USA; Human Genome Sciences, Inc.	Kirkpatrick, R. B., Thotakura, N. R., Ni, J., Gentz, R., Chopra, A.	Human chitotriosidase and its cDNA, detection of altered chitotriosidase expression, and diagnosis of tissue remodeling disorder	19971009
WO 9716540	Chemgenics Pharmaceuticals, Inc., USA	Koltin, Y., Riggle, P., Gavrias, V., Bulawa, C., Winter, K.	Yeast chitin synthase 1 gene sequence, recombinant vector and auxotrophic host cell, and method for determining growth-associated proteins and genetic elements	19970509
WO 0116353	United States of America, as Represented by the Secretary of Agriculture, USA; Novo Nordisk Biotech, Inc.	Okubara, P. A., Blechl, A. E., Hohm, T. M., Berka, R. M.	Nucleic acid sequences encoding cell wall-degrading enzymes and their use to engineer plant resistance to *Fusarium* and other pathogens	20010308
WO 0109283	The Regents of the University of California, USA; The Li, J.	Ronald, P., He, Z., Chory, J., Lamb, C.,	DNA constructs encoding chimeric plant RRK receptors (Bri1::Xa21 and Hevein::Xa21), and	20010208

Tab. 23 (cont.)

Patent number	Patent holder	Inventors	Title of patent	Publication date (YYYYMMDD)
	Salk Institute for Biological Studies		their use in production of transgenic plants	
WO 9625424	Johns Hopkins University, USA	Roseman, S., Bassler, B., Keyhani, N., Chitlaru, E., Rowe, C., Yu, C.	Bacterial catabolism of chitin	19960822
WO 0056908	Pioneer Hi-Bred International, Inc., USA	Simmons, C. R., Yalpani, N.	Maize chitinases and cDNAs and method of modulating chitinase activity in plants	20000928
WO 0102577	Provalis UK Ltd., UK	Smith, C. J., Thompson, S. E., Smith, M. W., Peek, K., Sizer, P. J. H., Wilkinson, M. C.	*Pseudomonas aeruginosa* antigens related to chitinase and groEL genes and their use in vaccine preparation	20010111
WO 9829542	Japan Tobacco Inc., Japan	Takakura, Y., Inoue, T., Saito, H., Ito, T.	Cloning of gene for chitinase of rice, characterization of floral organ-specific promoter of the monocotyledonous plant, and use for breeding pathogen-resistant plant	19980709
WO 9961635	Mycogen Corp., USA	Yenofsky, R. L., Fine, M., Rangan, T. S., Anderson, D. M.	Cotton cells, plants, and seeds genetically engineered to express insecticidal and fungicidal chitin binding proteins (lectins)	19991202

Tab. 24 Biochemical methods

Patent number	Patent holder	Inventors	Title of patent	Publication date (YYYYMMDD)
WO 9925390	Albert Einstein College of Medicine of Yeshiva University, USA	Burns, E. R., Wittner, M., Faskowitz, F.	A method for detecting chitin-containing organisms	19990527
WO 0047716	Advanced Medical Solutions Ltd., UK	Hamilton, D. W., Ives, C. L., Middleton, I. P., Rosetto, C.	Fibers for culturing eukaryotic cells	20000817
WO 9735037	Smith-Kline Beecham Corp., USA	Kirkpatrick, R. B.	Tissue remodeling proteins	19970925
WO 9802742	Board of Supervisors of Louisiana State University, USA	Laine, R. A., Lo, W. C. J.	Diagnosis of fungal infections with a chitinase	19980122
WO 9851712	National Institute of Sericultural and Entomological Science Ministry, Japan	Tsukada, M., Shirata, A., Hayasaka, S.	Chitin beads, chitosan beads, processes for producing these beads, carriers made of these beads, and processes for producing microsporidian spores	19981119
WO 9217786	SRI International, USA	Tuse, D., Dousman, L.	Assay device and enzymic-spectrophotometric method of detecting chitin	19921015

Tab. 25 Enzymes

Patent number	Patent holder	Inventors	Title of patent	Publication date (YYYYMMDD)
WO 9640940	Universiteit Van Amsterdam, The Netherlands	Aerts, J. M. F.	A human macrophage chitinase for use in the treatment or prophylaxis of infection by chitin-containing pathogens and manufacture of the enzyme by expression of the cloned gene	19961219
WO 9402598	Cornell Research Foundation, Inc., USA	Harman, G. E., Broadway, R. M., Tronsmo, A., Lorito, M., Hayes, C. K., Di, P. A.	Chitobiase and endochitinase of *Trichoderma harzianum* and their use in control of plant pathogenic fungi	19940203
WO 9958650	Biomarin Pharmaceuticals, USA	Klock, J. C., Mishra, C., Starr, C. M.	Lytic enzymes from *Trichoderma useful* for treating fungal infections	19991118
WO 9854304	Institut für Pflanzengenetik und Kulturpflanzenfor-schung, Germany	Schlesier, B., Wolf, A., Koch, G., Horstmann, C., Konig, S., Gillandt, R.-m.	Chitinase of *Canavalia ensiformis* seed and its use in preparation of N-acetyl-D-glucosamine from chitin	19981203
WO 9413815	Institute for Molecular Biology and Biotechnology/Forth, Greece	Thireos, G., Kafetzopoulos, D.	Microbial chitin deacetylases and the genes encoding them	19940623
WO 9853053	Foundation for Research and Technology, Hellas, Greece	Thireos, G., Kafetzopoulos, D.	Cloning of cDNA for an arthropod chitin synthase and its use for developing chitin synthase inhibitors	19981126
WO 9806859	Human Genome Sciences, Inc., USA	Thotakura, N. R., Chopra, A., Choi, G. H., Genzt, R. L., Rosen, C. A.	Human chitinase α and chitinase α-2 and their cDNA sequences and therapeutic applications	19980219
WO 0073488	Board of Regents of the University of Texas System, USA	Vinetz, J. M.	*Plasmodium* gene CHT1 chitinases and cDNAs and methods for preventing malaria transmission by mosquitoes	20001207

Tab. 26 Cosmetics and ingredients

Patent number	Patent holder	Inventors	Title of patent	Publication date (YYYYMMDD)
WO 0060038	Laboratoire Medidom S.A., Switzerland	Cantoro, A.	Viscosity-enhanced ophthalmic solutions having detergent action and their use on contact lenses	20001012
WO 9219218	Procter and Gamble Co., USA	Cooke, R. J., Date, R. F., Barrington, N. G., Spengler, E. G.	Carboxymethylchitin-containing cosmetic composition for skin and hair	19921112
WO 9822210	Merck S.A., France	Grandmontagne, B., Marchio, F.	Chitin or chitin derivative microcapsules containing a hydrophobic sunscreen	19980528
WO 0054738	Cognis Deutschland GmbH, Germany; Biotec ASA	Griesbach, U., Wachter, R., Ansmann, A., Fabry, B., Eisfeld, W., Engstad, R. E.	Cosmetic preparations containing chitosans and β-(1-3)-glucans	20000921
WO 9525752	Kao Corp., Japan	Hasebe, Y., Sawada, M., Furukawa, M., Nakayama, T., Kodama, K., Ito, Y., Nakamura, G., Fukumoto, Y.	Fine porous polysaccharide particles and their use in cosmetics	19950928
WO 9817245	Henkel KGaA, Germany	Heilemann, A., Holzer, J., Horlacher, P., Sander, A., Wachter, R.	Collagen-free cosmetic preparations obtained from crosslinked chitosan hydrogels	19980430
WO 0047177	Cognis Deutschland GmbH, Germany	Kropf, C., Fabry, B., Foerster, T., Wachter, R., Reil, S., Panzer, C.	Use of nanoscale chitosans and/or chitosan derivatives	20000817
WO 9530403	Larm, O., Sweden	Larm, O.	Dentifrices containing chitosan and sulfated polysaccharides for improvement of oral hygiene	19951116
WO 0025734	Cognis Deutschland GmbH, Germany	Panzer, C., Tesmann, H., Wachter, R.	Ethanolic cosmetic preparations containing chitosan	20000511
WO 9413774	Allergan, Inc., USA	Powell, C. H., Karageozian, H., Currie, J. P.	Compositions and methods for cleaning hydrophilic contact lenses	19940623
WO 9629979	Unilever Plc, UK	Tsaur, L. S., Shen, S., Jobling, M., Aronson, M. P.	Liquid cleansers comprising hydrogel dispersions	19961003

Tab. 27 Industrial biotechnology

Patent number	Patent holder	Inventors	Title of patent	Publication date (YYYYMMDD)
WO 9317121	Phyton Catalytic, Inc., USA; Prince, C. L.; Schubmehl, B. F.; Kane, E. J.; Roach, B.	Bringi, V., Kadkade, P. G.	Increased yields of taxol and taxanes from cell cultures of Taxus species	19930902
WO 9731121	University of British Columbia, Canada	Haynes, C. A., Aloise, P., Creagh, A. L.	Process for producing N-acetyl-D-glucosamine	19970828
WO 9409148	Toray Industries, Inc., Japan	Miwa, K., Fukuyama, M., Uchiyama, T.	Process for producing major histocompatibility antigen class II (MHC) protein and immobilized MHC II or fragments	19940428
WO 9314216	Nakamura, Tomotaka, Japan	Nakamura, K., Yamada, C.	Manufacture of chitin and chitosan with dermatophyte	19930722

Tab. 28 Food and feed chemistry.

Patent number	Patent holder	Inventors	Title of patent	Publication date (YYYYMMDD)
WO 0004086	Kemestrie Inc., Canada	Chornet, E., Dumitriu, S.	Polyionic hydrogels based on xanthan and chitosan for stabilization and controlled release of vitamins	20000127
WO 0005974	E-Nutriceuticals, Inc., USA	Hardinge-Lyme, N.	Fat-absorbing chitosan-containing liquid compositions for use in weight-loss programs	20000210
WO 0103511	CH20 Incorp., USA	Iverson, C. E., Ager, S. P.	Method of coating food products, and a coating composition	20010118
WO 9748291	Fuji Oil Co., Ltd., Japan	Nakamura, A., Hattori, M., Maeda, H.	Food preservatives	19971224
WO 9837775	Nestle S.A., Switzerland	Popper, G., Ekstedt, S., Walkenstrom, P., Hermansson, A.-M.	Gelled emulsion products containing chitosan	19980903

Tab. 29 Immunochemistry.

Patent number	Patent holder	Inventors	Title of patent	Publication date (YYYYMMDD)
WO 0123430	Icos Corp., USA	Allison, D. S., Dietsch, G. N., Gray, P. W., Shaw, K. D., Steiner, B. H.	Chitinase immunoglobulin fusion products	20010405
WO 9716208	Medeva Holdings B.V., The Netherlands	Chatfield, S. N.	Influenza vaccine compositions	19970509
WO 9609805	Zonagen, Inc., USA	Podolski, J. S., Hsu, K., Bhogal, B., Singh, G.	Chitosan induced immunopotentiation	19960404

Tab. 30 Polysacchride chemistry and processes.

Patent number	Patent holder	Inventors	Title of patent	Publication date (YYYYMMDD)
WO 9964470	Bioeffect A/S, Norway	Arntsen, D.	An integrated plant and method for automatic multistage production of chitosan	19991216
WO 0024490	Cognis Deutschland Gmbh, Germany	Blum, S., Horlacher, P., Trius, A., Wagemans, P., Weitkemper, N., Albiez, W.	Continuous extraction of natural substances and production of chitin and chitosan	20000504
WO 9626965	Akzo Nobel N.V., The Netherlands	Bredehorst, R., Pomato, N., Scheel, O., Thiem, J.	High-yield manufacture of water-soluble dimeric to decameric chitin oligomers	19960906
WO 9206136	Pfizer Inc., USA	Chandler, C. E., Curatolo, W. J.	Hypocholesterolemic quaternary ammonium salts derived from chitosan	19920416
WO 9208742	Du Pont de Nemours, E. I., and Co., USA	Dingilian, E. O., Heinsohn, G. E.	*N*-halochitosans and their preparation and uses	19920529
WO 9641818	Drohan, W. N., et al., USA	Drohan, W. N., Macphee, M. J., Miekka, S. I., Singh, M., Elson, C., Taylor, J.	Manufacture and use of supplemented chitin hydrogels	19961227
WO 9400512	Albany International Corp., USA	Eagles, D. B., Bakis, G., Jeffery, A. B., Mermingis, C., Hagoort, T. H.	Method of producing polysaccharide foams	19940106
WO 9806489	Wolff Walsrode AG, Germany	Engelhardt, J., Koch, W., Pannek, J.-B., Vorlop, K.-D., Patel, A. V.	Polymeric immobilizate for encapsulation of useful materials	19980219

Tab. 30 (cont.)

Patent number	Patent holder	Inventors	Title of patent	Publication date (YYYYMMDD)
WO 9921648	Everest-Todd Research & Development Ltd., UK	Everest-Todd, S.	Formulation of oils	19990506
WO 9730092	Innovative Technologies Ltd., UK	Gilding, D. K., Al-Lamee, K. G.	Biocompatible hydrogels made from crosslinked chitosan and galactomannan-type or/and pectic substances and their manufacture and uses	19970821
WO 9745453	Virginia Tech Intellectual Properties, Inc., USA	Glasser, W. G., Jain, R. K.	Manufacture of ester-crosslinked chitosan hydrogels as support material and quaternized, cyclodextrin-modified chitosan	19971204
WO 9900452	Virginia Tech Intellectual Properties, Inc., USA	Glasser, W. G., Jain, R. K.	Method of making magnetic, crosslinked chitosan support materials and products thereof	19990107
WO 8606082	Matcon Raadgwende Ingenioerfirma A/S, Denmark	Joensen, J. O.	Recovering chitin from materials in which chitin occurs together with or connected to proteinaceous substances	19861023
WO 9938893	Matcon Radgivende Ingeniorfirma A/S, Denmark	Joensen, J. O.	Chitosan production by deacetylation of chitin with reuse of the deacetylating hydroxide solution	19990805
WO 9808877	Wakunaga Seiyaku Kabushiki Kaisha, Japan	Kakimoto, M., Ushijima, M., Kasuga, S., Shiraishi, S., Itakura, Y.	Chitosan derivatives, their manufacture and uses in treatment of constipation	19980305
WO 9742226	Nippon Suisan Kaisha, Ltd., Japan	Kawahara, H., Jinno, S., Okita, Y.	Chitin derivatives having lower carboxyalkyl groups as hydrophilic substituents and hydrophobic substituents, and high-molecular micellar supports and aqueous compositions thereof	19971113
WO 9817693	Transgene S.A., France	Kolbe, H.	Chitosan compositions for transferring therapeutic agents into host cells	19980430
WO 0132751	Cognis Deutschland GmbH, Germany	Kropf, C., Reil, S.	Method for producing nanoparticulate chitosans or chitosan derivatives	20010510
WO 0011038	Kumar, G., Payne, G. F., USA	Kumar, G., Payne, G. F.	Modified chitosan polymers and enzymic methods for their production	20000302

Tab. 30 (cont.)

Patent number	Patent holder	Inventors	Title of patent	Publication date (YYYYMMDD)
WO 9948809	Teknimed S.A., France	Lacout, J.-L., Hatim, Z., Frache-Botton, M.	Preparation and *in situ* hardening of hydroxyapatite formulations for biomaterial for synthetic bone and teeth	19990930
WO 9961482	Trustees of Tufts College, USA; University of Massachusetts Lowell; United States Dept. of the Army	Lee, J. W., Yeomans, W. G., Allen, A. L., Deng, F., Gross, R. A., Kaplan, D. L.	Production of chitosan- and chitin-like polymers	19991202
WO 9209635	Merck Patent GmbH, Germany	Muzzarelli, R.	Methylpyrrolidinone-substituted chitosans, production and uses thereof	19920611
WO 0123398	Carbion Oy, Finland	Natunen, J.	Preparation of acetamidodeoxy fucosylated oligosaccharides via enzymic glycosidation reaction	20010405
WO 9729132	Merck Patent GmbH, Germany	Nies, B.	Crosslinked products of amino-group-containing biopolymers	19970814
WO 9533773	Vitaphore Corp., USA	Prior, J. J., Yamamoto, R. K., Brode, G. L.	Lysozyme-degradable biomaterials and their uses	19951214
WO 9306136	Fidia S.P.A., Italy	Romeo, A., Toffano, G., Callegaro, L.	Chitosan-based carboxylic polysaccharide derivatives	19930401
WO 0104207	Cognis Deutschland GmbH, Germany	Schafer, G., Holzer, J., Sander, A., Heilemann, A.	Chitosan preparations free from crosslinking agents	20010118
WO 9105808	Firextra Oy, Finland	Struszczyk, H., Kivekaes, O.	Method and apparatus for manufacture of chitosan and other products from shells of organisms, especially marine organisms	19910502
WO 9100298	Firextra Oy, Finland; Instytut Wlokien Chemicznych	Struszczyk, H., Niekraszewicz, A., Wrzesniewska-Tosik, K., Koch, S., Kivekas, O.	Method for continuous manufacture of microcrystalline chitosan	19910110
WO 9901480	National Starch and Chemical Investment Holding Corp., USA	Sugimoto, M., Shigemasa, Y.	Hydrophilic chitin derivatives and manufacture methods	19990114
WO 9901479	National Starch and Chemical Investment Holding Corp., USA	Sugimoto, M., Shigemasa, Y.	Chitosan derivatives and manufacture methods	19990114

Tab. 30 (cont.)

Patent number	Patent holder	Inventors	Title of patent	Publication date (YYYYMMDD)
WO 9625437	Abion Beteiligungs- und Verwaltungsgesellschaft mbH, Germany	Teslenko, A.	Preparation of chitosan-glucan complexes and their uses	19960822
WO 0014155	University of Strathclyde, UK	Uchegbu, I. F.	Erodible solid hydrogels for delivery of biologically active materials	20000316
WO 9944710	Virginia Tech Intellectual Properties, Inc., USA	Velander, W. H., Van Cott, K. E., Van Tassell, R.	Inside-out crosslinked and commercial-scale hydrogels, and sub-macromolecular selective purification using the hydrogels	19990910
WO 9803523	Villalva Basabe, J., Spain	Villalva Basabe, J.	Process for obtaining [2,1-d]-2-oxazolines of amino sugars from chitin or its derivatives	19980129
WO 0068273	Cognis Deutschland GmbH, Germany	Wachter, R., Griesbach, U., Horlacher, P.	Collagen-free compositions for cosmetics	20001116
WO 9840411	Henkel KGaA, Germany	Wachter, R., Priebe, C., Ritter, W., Skwiercz, M.	Crosslinking of chitosans	19980917
WO 9707139	Ciba-Geigy A.-G., Switzerland	Yen, S.-F., Sou, M.	Process for sterilizing solutions of chitosan or its derivatives	19970227

Tab. 31 Pharmaceuticals and pharmacology

Patent number	Patent holder	Inventors	Title of patent	Publication date (YYYMMDD)
WO 9407999	Brown University Research Foundation, USA	Aebischer, P., Zielinski, B. A.	Chitosan matrixes for encapsulated cells for implantation	19940414
WO 9804244	Universidade de Santiago de Compostela, Spain	Alonso Fernandez, M. J., Calvo Salve, P., Remunan Lopes, C., Vila Jato, J. L.	Application of nanoparticles based on hydrophilic polymers as pharmaceutical forms	19980205
WO 0056362	The Secretary of State for Defence, UK	Alpar, H. O., Eyles, J. E., Somavarapu, S., Williamson, E. D., Baillie, L. W. J.	Immunostimulants comprising polycationic carbohydrates	20000928
WO 0050090	DCV, Inc., USA	Angerer, J. D., Cyron, D. M., Iyer, S., Jerrell, T. A.	Dry acid-chitosan complexes	20000831
WO 9919003	Battelle Memorial Institute, USA	Armstrong, B. L., Campbell, A. A., Gutowska, A., Song, L.	Implantable polymer/ceramic composites	19990422
WO 9620730	Astra AB, Sweden	Artursson, P., Schipper, N.	Chitosan polymer with specific degree of acetylation as drug absorption enhancer	19960711
WO 9211844	Enzytech, Inc., USA	Auer, H. E., Brown, L. R., Gross, A.	Stabilization of proteins by cationic biopolymers	19920723
WO 0009123	F. Hoffmann-La Roche A.-G., Switzerland	Bailly, J., Fleury, A., Hadvary, P., Lengsfeld, H., Steffen, H.	Pharmaceutical compositions containing lipase inhibitors and chitosan	20000224
WO 9313137	Howmedica Inc., USA	Barry, J. J., Higham, P. A., Mann, N. N.	Process for insolubilizing N-carboxyalkyl derivatives of chitosan	19930708
WO 0001373	Ecole Polytechnique Federale De Lausanne, Switzerland; Kiddle, Simon	Bartkowiak, A., Hunkeler, D.	Materials and methods relating to encapsulation by formation of complex reaction between oppositely charged polymers	20000113
WO 9948480	Aventis Research & Technologies GmbH & Co. KG, Germany	Bayer, U.	Method for the production of microcapsules	19990930
WO 9948479	Aventis Research & Technologies GmbH & Co. KG, Germany	Bayer, U., Hahn, B., Majeres, A.	Slow release microcapsules	19990930

Tab. 31 (cont.)

Patent number	Patent holder	Inventors	Title of patent	Publication date (YYYYMMDD)
WO 9503786	Fidia Advanced Biopolymers SRL, Italy	Benedetti, L., Callegaro, L.	Pharmaceutical compositions for topical use containing hyaluronic acid and its derivatives	19950209
WO 0100246	Shearwater Polymers, Incorp., USA	Bentley, M. D., Zhao, X.	Hydrogels derived from chitosan and poly(ethylene glycol) or related polymers	20010104
WO 0122973	Medicarb AB, Sweden	Bergstrom, T., Trybala, E., Back, M., Larm, O.	The use of a positively charged carbohydrate polymer such as chitosan for the prevention of infection	20010405
WO 0057931	Paul Hartmann A.-G., Germany	Bitterhof, A.	Breast-milk absorbent pad containing chitosan to improve performance	20001005
WO 9961079	Kimberly-Clark Worldwide, Inc., USA	Boney, L. C., Borders, R. A., Di Luccio, R. C., Kepner, E. S., Yahiaoui, A.	Enhanced odor absorption by natural and synthetic polymers	19991202
WO 0126635	Elan Pharma International Ltd., Ireland; McGurk, S. L.	Bosch, H. W., Cooper, E. R.	Bioadhesive nanoparticulate compositions having cationic surface stabilizers	20010419
WO 9209636	Baker Cummins Dermatologicals, Inc., USA	Brode, G. L., Adams, N. W., Figueroa, R.	Skin protective compositions and method of inhibiting skin irritation	19920611
WO 9523235	Myco Pharmaceuticals Inc., USA	Bulawa, C. E.	Methods for identifying inhibitors of fungal pathogenicity and use in pharmacologically and agriculturally useful fungicide analysis and insecticide analysis	19950831
WO 9610421	Medeva Holdings B.V., The Netherlands	Chatfield, S. N.	Vaccine compositions	19960411
WO 9907416	Bio-Syntech Ltd., Canada	Chenite, A., Chaput, C., Combes, C., Jalal, F., Selmani, A.	Temperature-controlled pH-dependant formation of ionic polysaccharide gels	19990218
WO 9613253	Dong Kook Pharmaceutical Co., Ltd., S. Korea	Chung, C. P., Lee, S. J.	Biodegradable sustained release preparation for treating periodontitis	19960509
WO 9901498	Danbiosyst UK Ltd., UK	Davis, S. S., Lin, W., Bignotti, F., Ferruti, P.	Conjugate of polyethylene glycol and chitosan	19990114

Tab. 31 (cont.)

Patent number	Patent holder	Inventors	Title of patent	Publication date (YYYYMMDD)
WO 9426788	Development Biotechnological Processes S.N.C., Italy; IMS-International Medical Servica S.R.L	De Rosa, A., Rossi, A., Affaitati, P.	Process for the preparation of iodinated biopolymers having disinfectant and cicatrizing activity, and the iodinated biopolymers obtainable thereby	19941124
WO 0056275	Virbac, France	Derrieu, G., Pougnas, J.-I.	Pharmaceutical compositions based on chondroitin sulphate and chitosan for preventing or treating rheumatic disorders by systemic administration	20000928
WO 9909962	Reckitt & Colman Products Ltd., UK	Dettmar, P. W., Jolliffe, I. G., Skaugrud, O.	In situ formation of bioadhesive polymeric material	19990304
WO 0107486	Polygene Ltd., Israel	Domb, A. J.	A biodegradable polycation composition for delivery of an anionic macromolecule in gene therapy	20010201
WO 9932697	Kimberly-Clark Worldwide, Inc., USA	Dutkiewicz, J., Bevernitz, K. J., Huard, L. S., Qin, J., Sun, T., Wallajapet, P. R. R.	Antimicrobial structures based on polymer nonwoven fabrics for diapers	19990701
WO 9635433	Chitogenics, Inc., USA	Elson, C. M.	N,O-carboxymethylchitosan for prevention of surgical adhesions	19961114
WO 0071136	Chitogenics, Inc., USA	Elson, C. M. Lee, T. D. G.	Adhesive N,O-carboxymethyl chitosan coatings which inhibit attachment of substrate-dependent cells and proteins	20001130
WO 0015192	Zonagen, Inc., USA	Fontenot, G. K.	Methods and materials related to bioadhesive contraceptive gels	20000323
WO 9528158	Janssen Pharmaceutica N.V., Belgium	Francois, M. K. J., Embrechts, R. C. A., Illum, L.	Intranasal antimigraine composition	19951026
WO 9729760	Furda, Ivan, USA	Furda, I.	Multifunctional fat absorption and blood cholesterol reducing formulation containing chitosan and nicotinic acid	19970821
WO 9902252	Norsk Hydro ASA, Norway	Gaserod, O., Skaugrud, O., Dettmar, P., Sjak-Braek, G., Jolliffe, I.	Preparation of capsules based on ionic polysaccharides	19990121
WO 9602260	Medicarb AB, Sweden	Gouda, I., Larm, O.	Wound healing agents comprising polysaccharides conjugates with chitosan	19960201

Tab. 31 (cont.)

Patent number	Patent holder	Inventors	Title of patent	Publication date (YYYYMMDD)
WO 0030609	Laboratoire Medidom S.A., Switzerland	Gurny, R., Felt, O.	Aqueous ophthalmic formulations comprising chitosan	20000602
WO 0125321	Alvito Biotechnologie GmbH, Germany	Hahnemann, B., Pahmeier, A., Loetzbeyer, T.	Three-dimensional matrix for producing cell transplants	20010412
WO 9608149	USA	Haimovich, B., Freeman, A., Greco, R.	Thromboresistant surface treatment for biomaterials	19960321
WO 9725994	Hansson, H.-A., Sweden	Hansson, H.-A.	Control of healing process	19970724
WO 9602258	Astra AB, Sweden	Hansson, H.-a., Johansson-Ruden, G., Larm, O.	Immobilized polysaccharide as anti-adhesion agent in wound healing	19960201
WO 9602259	Astra AB, Sweden	Hansson, H.-A., Johansson-Ruden, G., Larm, O.	Hard tissue stimulating agent comprising chitosan and a polysaccharide immobilized thereto	19960201
WO 9707802	Shaman Pharmaceuticals, Inc., USA	Hector, R. F., Sabouni, A.	Methods and compositions for treating fungal infections in mammals	19970306
WO 9855149	Shionogi & Co., Ltd., Japan	Horie, K., Masuda, K., Sakagami, M.	Use of drug carriers for producing lymph node migrating drugs	19981210
WO 9531184	Vaccine Technologies Pty. Ltd., Australia	Husband, A., Kingston, D.	Film-coated microparticles for bioactive molecule delivery	19951123
WO 9315737	Danbiosyst UK Ltd., UK	Illum, L.	Compositions for nasal administration containing polar metabolites of opioid analgesics	19930819
WO 9720576	Danbiosyst UK Ltd., UK	Illum, L.	Vaccine compositions for intranasal administration containing chitosan as adjuvant	19970612
WO 9801160	Danbiosyst UK Ltd., UK	Illum, L.	Chitosan compositions suitable for delivery of genes to epithelial cells	19980115
WO 9203480	Drug Delivery System Institute, Ltd., Japan	Inoue, K., Ito, T., Okuno, S., Aono, K.	Preparation of N-acetyl-O-(carboxymethyl)chitosan derivatives for drug targeting	19920305
WO 9833521	Korea Institute of Science and Technology, S. Korea	Jeong, S. Y., Kwon, I.-C., Kim, Y.-H., Seong, S.-Y.	Pneumococcal polysaccharide conjugate with cholera toxin B subunit for use in vaccines	19980806

Tab. 31 (cont.)

Patent number	Patent holder	Inventors	Title of patent	Publication date (YYYYMMDD)
WO 9826788	Noviscens AB, Sweden	Johansson, B., Niklasson, B.	Medical composition and use thereof for the manufacture of a topical barrier formulation, a UV-radiation absorbing formulation, or an antiviral, antifungal, or antiinflammatory formulation	19980625
WO 9916478	Metagen, Llc, USA	Johnson, J. R., Marx, J. G., Johnson, W. D.	Bone substitutes comprising rigid framework filled with osteoconductive materials and resilient polymers	19990408
WO 9822114	Dumex-Alpharma A/S, Denmark	Jorgensen, T., Moss, J., Nicolajsen, H. V., Nielsen, L. S.	Method and carbohydrate composition for promoting tissue repair	19980528
WO 0050507	Cognis Corporation, USA	Kraechter, H. U., Wachter, R.	Latex products with reduced hypersensitivity	20000831
WO 0074720	Mochida Pharmaceutical Co., Ltd., Japan	Kudo, Y., Ueshima, H., Sakai, K.	System for release in lower digestive tract	20001214
WO 9805341	Medicarb AB, Sweden	Larm, O., Back, M., Bergstrom, T.	The use of heparin or heparan sulfate in combination with chitosan for the prevention or treatment of infections caused by herpes virus	19980212
WO 9819718	Challenge Bioproducts Co., Ltd., Taiwan; Lee, H. L, Lin, C.-k., Sung, H.-w.	Lee, T. C.-j., Sung, H.-w.	Chemical modification of biomedical materials with genipin	19980514
WO 9805304	Cytotherapeutics Inc., USA	Li, R., Hazlett, T. F.	Biocompatible devices with foam scaffolds	19980212
WO 0006210	Minnesota Mining and Manufacturing Company, USA	Lyon, K. R., Rock, M. M., Jr.	Method for the manufacture of antimicrobial articles	20000210
WO 9927960	Medeva Europe Ltd., UK	Makin, J. C., Bacon, A. D.	Vaccine compositions for mucosal administration comprising chitosan	19990610
WO 0100792	Marchosky, J. A., USA	Marchosky, J. A.	Compositions containing growth factors and methods for forming and strengthening bone	20010104
WO 9922739	Taisho Pharmaceutical Co., Ltd., Japan	Miwa, A., Yamahira, T., Ishibe, A., Itai, S.	Medicinal steroid compound composition	19990514

14 Relevant Patents for Isolation, Production and Applications of Chitin and Chitosan

Tab. 31 (cont.)

Patent number	Patent holder	Inventors	Title of patent	Publication date (YYYYMMDD)
WO 9742975	Genemedicine, Inc., USA	Mumper, R. J., Rolland, A.	Chitosan-related compositions and methods for delivery of nucleic acids into a cell	19971120
WO 9919366	Abiogen Pharma SRL, Italy	Muzzarelli, R., Rosini, S., Trasciatti, S.	Chelates of chitosans and alkaline-earth metal insoluble salts and the use thereof as medicaments useful in osteogenesis	19990422
WO 9921566	Rexall Sundown, Inc., USA	Myers, A. E., Priddy, M. R.	Dietary compositions with lipid binding properties for weight management and serum lipid reduction	19990506
WO 0124840	Coloplast A/S, Denmark	Nielsen, B.	Wound care device based on chitosan fibers	20010412
WO 9947162	Wound Healing of Oklahoma, Inc., USA	Nordquist, R. E., Chen, W. R., Carubelli, R.	Laser/sensitizer-assisted immunotherapy of cancer	19990923
WO 0115716	Medex Scientific (UK) Ltd., UK	Oben, J. E.	Plant extracts, chitosan derivatives and antioxidant vitamins for treatment of weight-related disorders	20010308
WO 9747323	Zonagen, Inc., USA	Podolski, J. S., Hsu, K. T., Singh, G.	Chitosan drug delivery system	19971218
WO 9842374	Zonagen, Inc., USA	Podolski, J. S., Martinez, M. L.	Chitosan-induced immunopotentiation	19981001
WO 9613284	Innovative Technologies Ltd., UK	Qin, Y., Gilding, K. D.	Wound treatment composition	19960509
WO 9613282	Innovative Technologies Ltd., UK	Qin, Y., Gilding, K. D.	Absorbent wound dressing	19960509
WO 9706782	Ciba-Geigy A.-G., Switzerland	Reed, K. W., Yen, S.-F.	Ophthalmic pharmaceuticals containing O-carboxyalkyl chitosan	19970227
WO 0124795	SKW Trostberg A.-G., Germany	Schuhbauer, H., Pischel, I., Bernkop-Schnuerch, A.	Delayed-release dosage form containing α-lipoic acid (derivatives)	20010412
WO 9965521	Zonagen, Inc., USA	Seid, C. A., Singh, G.	Prostate-associated antigen composition with chitosan metal chelate for the treatment of prostatic carcinoma	19991223
WO 0021567	Societe De Conseils De Recherches Et D'applications Scientifiques S.A., France	Shalaby, S. W., Jackson, S. A., Ignatious, F. X., Moreau, J.-P., Russell, R. M.	Ionic molecular conjugates of N-acylated derivatives of poly(2-amino-2-deoxy-D-glucose) and polypeptides	20000420
WO 0110467	Dainippon Pharmaceutical Co., Ltd., Japan	Shimono, N., Mori, M., Higashi, Y.	Solid preparations containing chitosan powder and process for producing the same	20010215

Tab. 31 (cont.)

Patent number	Patent holder	Inventors	Title of patent	Publication date (YYYYMMDD)
WO 0015766	Oncopharmaceutical, Inc., USA	Singh, S. S.	Treatment of oncologic tumors with an injectable formulation of a Golgi apparatus disturbing agent	20000323
WO 0030658	The University of Akron, USA	Smith, D. J., Serhatkulu, S.	Chitosan-based nitric oxide donor compositions	20000602
WO 0016817	Korea Atomic Energy Research Institute, S. Korea	Son, Y.-S., Youn, Y.-H., Hong, S.-I., Lee, S.-H., Gin, Y.-J.	Dermal scaffold using neutralized chitosan sponge or neutralized chitosan/collagen mixed sponge	20000330
WO 9324112	Clover Consolidated, Ltd., Switzerland	Soon, S. P., Heintz, R. A.	Microencapsulation of cells for transplantation	19931209
WO 0024785	Novasso Oy, Finland	Struszczyk, H., Pomoell, H., Wulff, M., Saynatjoki, E., Ylitalo, P., Wuolijoki, E., Hakli, H.	Chitosan-based pharmaceuticals for reduction of cholesterol and lipid contents	20000504
WO 9947186	University of Pittsburgh, USA; Wayne State University	Suh, J.-K., Matthew, H., Fu, F. H., Sechriest, F. V., Manson, T. T.	Chitosan-based composite materials containing glycosaminoglycan for cartilage repair	19990923
WO 9702125	FMC Corp., USA	Thomas, W. R., Pfeffer, H. A., Guiliano, B. A., Sewall, C. J., Tomko, S.	Process for making biopolymer gel microbeads	19970123
WO 9811903	Chicura Cooperatie U.A., The Netherlands	Timmermans, C.	Chitosan derivatives, and preparation thereof, for treating malignant tumors	19980326
WO 9945900	Kabushiki Kaisha Soken, Japan	Tokuyama, T., Jo, M.	Topical compositions for improving skin conditions	19990916
WO 9842755	University of Strathclyde, UK	Uchegbu, I. F.	Particulate drug carriers	19981001
WO 9205791	Dana-Farber Cancer Institute, Inc., USA	Vercellotti, S. V., Ruprecht, R. M., Gama Sosa, M. A.	Inhibition of viral replication with chitosan derivatives	19920416
WO 9843609	Henkel KGaA, Germany	Wachter, R., Horlacher, P.	Chitosan microspheres	19981008
WO 9930751	Kimberly-Clark Worldwide, Inc., USA	Wallajapet, P. R. R., Romans-Hess, A. Y., Ta, E. T., Qin, J.	Absorbent structure having balanced pH profile	19990624

Tab. 31 (cont.)

Patent number	Patent holder	Inventors	Title of patent	Publication date (YYYYMMDD)
WO 9312875	Vanderbilt University, USA	Wang, T. G. J.	Method and apparatus for producing uniform polymeric spheres	19930708
WO 9936075	Danbiosyst UK Ltd., UK	Watts, P. J., Davis, S. S.	Oral oil sorbing compositions for treatment of obesity	19990722
WO 9605810	Danbiosyst UK Ltd., UK	Watts, P. J., Illum, L.	Drug delivery composition containing chitosan or its derivative having a defined zeta-potential	19960229
WO 9830207	Danbiosyst UK Ltd., UK	Watts, P. J., Illum, L.	Chitosan-gelatin A microparticles	19980716
WO 9603142	Danbiosyst UK Ltd., UK	Watts, P.J., Illum, L.	Drug delivery composition for the nasal administration of antiviral agents	19960208
WO 9823275	Henkel KGaA,	Weitkemper, N., Fabry, B.	Use of mixtures of phytostanols and tocopherols for the production of hypocholesteremic agents	19980604
WO 9404135	Teikoku Seiyaku Kabushiki Kaisha, Japan	Yamada, A., Wato, T., Uchida, N., Kadoriku, M., Takama, S., Inamoto, Y.	Oral preparation for release in lower digestive tract	19940303
WO 9741894	Taisho Pharmaceutical Co., Ltd., Japan	Yamahira, T., Miwa, A., Ishibe, A., Itai, S.	Micellar aqueous composition and method for solubilizing hydrophobic drug	19971113
WO 9610426	Yokota, S., et al., Japan	Yokota, S., Shimokawa, S., Sonohara, R., Okada, A., Takahashi, K.	Osteoplastic graft	19960411
WO 0027889	Netech Inc., Japan	Yura, H., Saito, Y., Ishihara, M., Ono, K., Saeki, S.	Functional chitosan derivatives	20000518

Tab. 32 Textiles and paper

Patent number	Patent holder	Inventors	Title of patent	Publication date (YYYYMMDD)
WO 0049219	Foxwood Research Limited, UK	Blowes, P. C., Taylor, A. J., Roberts, G., Wood, F.	Substrates with biocidal properties and process for making them	20000824
WO 9851709	Hong, Y. K., S. Korea	Hong, Y. K.	Aqueous cellulose solution containing chitosan or alginic acid or derivatives and rayon fiber produced therefrom	19981119
WO 9904083	Consejo Superior De Investigaciones Cientificas, Spain	Julia Ferres, R., Erra Serrabasa, P., Diez Tascon, J. M.	Shrinkproofing process for wool	19990128
WO 0134897	Moritoshi Co., Ltd., Japan	Mori, T.	Fiber-treating agent and fiber	20010517
WO 9109163	Kemira Oy, Finland	Nousiainen, P., Struszczyk, H.	Microcrystalline chitosan-modified viscose fibers and method for their manufacture	19910627
WO 9812369	Mitsubishi Rayon Co., Ltd., Japan	Ohnishi, H., Nishihara, Y., Hosokawa, H., Oishi, S., Iwamoto, M., Fujii, Y., Itoh, H., Ohsuga, N.	Chitosan-containing acrylic fibers and process for preparing the same	19980326
WO 0046441	Gunze Limited, Japan	Ozawa, N., Akieda, S., Mukai, H.	Deodorizing fibers and production process	20000810
WO 9511925	Kimberly-Clark Corp., USA	Qin, J., Gross, J. R., Mui, W. J., Ning, X., Schroeder, W. Z., Sun, T.	Crosslinked polysaccharides having enhanced absorbency and process for their preparation	19950504
WO 9830752	Du Pont De Nemours and Co., USA	Ramachandran, S.	Chitosan-coated pulp, a paper using the pulp, and a process for making them	19980716
WO 9409192	Novasso OY, Finland	Struszczyk, H., Nousiainen, P., Kivekaes, O., Niekraszewicz, A.	Modified viscose fibers and their manufacture	19940428
WO 9842818	Procter & Gamble, USA	Surutzidis, A., Heist, B. M., Leblanc, M. J.	Laundry additive particle having multiple surface coatings	19981001

Tab. 33 Water purification

Patent number	Patent holder	Inventors	Title of patent	Publication date (YYYYMMDD)
WO 9937584	Fountainhead Technologies, Inc., USA	Denkewicz, R. P., Jr., Senderov, E. E., Grenier, J. W.	Biocidal compositions for treating water	19990729
WO 9202460	Archaeus Technology Group Ltd., UK	Harris, R., Jacques, A. M., Buchan, A., Brown, M.	Decolorization of water	19920220
WO 9533690	Vanson L.P., USA	Heinsohn, G. E.	Removal of phosphates from aqueous media	19951214
WO 9817386	I.N.P. - Industrial Natural Products SRL, Italy	Pifferi, P., Spagna, G., Manenti, I.	Method for removing pesticides and/or phytodrugs from liquids using cellulose, chitosan and pectolignincellulosic material derivatives	19980430
WO 9002708	Firextra Oy, Finland	Struszczyk, H., Kivekaes, O., Lindberg, J. J., Riekkola, M. L., Holopainen, K.	Method for purification of waste aqueous medium	19900322
WO 9737008	Yissum Research Development Company of the Hebrew University of Jerusalem, Israel	Van Rijn, J., Nussinovitch, A., Tal, J.	Bacterial immobilization in polymeric beads for nitrate removal from aquarium water	19971009

Tab. 34 Various applications

Patent number	Patent holder	Inventors	Title of patent	Publication date (YYYYMMDD)
WO 9723390	Tetra Laval Holdings & Finance SA, Switzerland	Berlin, M.	A laminated packaging material, its production, and packaging containers having gas barrier properties even in humid environments	19970703
WO 9325716	EMC Services, France	Durecu, S., Berthelin, J., Thauront, J.	Detoxication of combustion residues by removing soluble toxic compounds and by binding and concentrating the compounds obtained from the treatment solutions	19931223
WO 9702077	Kimberly-Clark Corp., USA	Gillberg-La, F. G. E., Turkevich, L. A., Kiick-Fischer, K. L.	Surface-modified fibrous material as filtration medium	19970123
WO 9822540	Merck S.A., France	Grandmontagne, B., Brousse, B.	Coating of mineral pigment microparticles with chitin	19980528
WO 0053295	University of Waterloo, Canada	Huang, R. Y. M., Pal, R., Moon, G. Y.	Two-layer composite pervaporation and reverse-osmosis membranes consisting of alginic acid derivative and chitosan	20000914
WO 9962974	Kao Corp., Japan	Kuwahara, K., Hasebe, Y., Akaogi, A., Yasumura, T., Takahashi, T., Tachizawa, O., Terada, E.	Core-shell polymer emulsions and process their manufacture	19991209
WO 9824832	Kimberly-Clark Worldwide, Inc., USA	Qin, J., Wallajapet, P. R. R.	Absorbent composition for disposable absorbent sheets	19980611
WO 9114499	Toray Industries, Inc., Japan	Shiro, K., Himeshima, Y., Yamada, S., Watanabe, T., Uemura, T., Kurihara, M.	Polythiophenylene-polysulfone composite membranes with active layers	19911003
WO 0121854	Potchefstroom University for Christian Higher Education, S. Africa	Vorster, S. W., Waanders, F. B., Geldenhuys, A. J.	Chitosan-based corrosion inhibitor for steel in aqueous acidic media	20010329

15 References

Aerts, J. M. F. (1996) Universiteit Van Amsterdam, Netherlands; Aerts, Johannes Maria Franciscus Gerardus, A human macrophage chitinase for use in the treatment or prophylaxis of infection by chitin-containing pathogens and manufacture of the enzyme by expression of the cloned gene, WO 9640940 [*Chem. Abstr.* **126**, 128715].

Aerts, J. M. F. G., Boot, R. G., Renkema, G. H., van Weely, S., Hollak, C. E. M., Donker-Koopman, W., Strijland, A., Verhoek, M. (1996) Chitotriosidase: a human macrophage chitinase that is a marker for Gaucher disease manifestation, in: *Chitin Enzymology*, (Muzzarelli, R. A. A., Ed.), Grottammare: Atec, 3–10, vol. 2.

Aiba, S. (1993) Studies on chitosan. 5. Reactivity of partially N-acetylated chitosan in aqueous media, *Makromol. Chem.* **194**, 65–75.

Akiyama, K., Kawazu, K., Kobayashi, A. (1995a) A novel method for chemo-enzymic synthesis of elicitor-active chitosan oligomers and partially N-deacetylated chitin oligomers using N-acylated chitotrioses as substrates in a lysozyme-catalyzed transglycosylation reaction system, *Carbohydr. Res.* **279**, 151–160.

Akiyama, K., Kawazu, K., Kobayashi, A. (1995b) Partially N-deacetylated chitin oligomers (pentamer to heptamer) are potential elicitors for (+)-pisatin induction in pea epicotyls, *Z. Naturforsch.* **50C**, 391–397.

Allan, G. G., Peyron, M. (1997) Depolymerization of chitosan by means of nitrous acid, in: *Chitin Handbook* (Muzzarelli, R. A. A., Peter, M. G., Eds.), pp. 175–180. Grottammare: Atec.

Andersen, S. O., Hojrup, P., Roepstorff, P. (1995) Insect cuticular proteins, *Insect Biochem. Mol. Biol.* **25**, 153–176.

Andersen, S. O., Peter, M. G., Roepstorff, P. (1996) Cuticular sclerotization in insects, *Comp. Biochem. Physiol.* **B113**, 689–705.

Arai, N., Shiomi, K., Yamaguchi, Y., Masuma, R., Iwai, Y., Turberg, A., Kolbl, H., Omura, S. (2000) Argadin, a new chitinase inhibitor, produced by *Clonostachys* sp FO-7314, *Chem. Pharm. Bull.* **48**, 1442–1446.

Arakane, Y., Koga, D. (1999) Purification and characterization of a novel chitinase isozyme from yam tuber, *Biosci. Biotechnol. Biochem.* **63**, 1895–1901.

Arakane, Y., Hoshika, H., Kawashima, N., Fujiya-Tsujimoto, C., Sasaki, Y., Koga, D. (2000) Comparison of chitinase isozymes from yam tuber – Enzymatic factor controlling the lytic activity of chitinases, *Biosci. Biotechnol. Biochem.* **64**, 723–730.

Asensio, J. L., Canada, F. J., Bruix, M., Rodriguez-Romero, A., Jimenez-Barbero, J. (1995) The interaction of hevein with N-acetylglucosamine-containing oligosaccharides. Solution structure of hevein complexed to chitobiose, *Eur. J. Biochem.* **230**, 621–633.

Asensio, J. L., Canada, F. J., Bruix, M., Gonzalez, C., Khiar, N., Rodriguez-Romero, A., Jimenez-Barbero, J. (1998) NMR investigations of protein-carbohydrate interactions: refined three-dimensional structure of the complex between hevein and methyl beta-chitobioside, *Glycobiology* **8**, 569–577.

Bakkers, J., Semino, C. E., Stroband, H., Kijne, J. W., Robbins, P. W., Spaink, H. P. (1997) An important developmental role for oligosaccharides during early embryogenesis of cyprinid fish, *Proc. Natl. Acad. Sci. USA* **94**, 7982–7986.

Bakkers, J., Kijne, J. W., Spaink, H. P. (1999) Function of chitin oligosaccharides in plant and animal development, *EXS* **87** (Chitin and Chitinases), 71–83.

Baldrick, P. (2000) Pharmaceutical excipient development: The need for preclinical guidance, *Regul. Toxicol. Pharmacol.* **32**, 210–218.

Bartkowiak, A., Hunkeler, D. (2000) Alginate-oligochitosan microcapsules. II. Control of mechanical resistance and permeability of the membrane, *Chem. Mater.* **12**, 206–212.

Bassler, B. L., Yu, C., Lee, Y. C., Roseman, S. (1991) Chitin utilization by marine bacteria, *J. Biol. Chem.* **266**, 24276–24286.

Baureithel, K., Felix, G., Boller, T. (1994) Specific, high affinity binding of chitin fragments to tomato cells and membranes. Competitive inhibition of binding by derivatives of chitooligosaccharides and a NOD factor of *Rhizobium*, *J. Biol. Chem.* **269**, 17931–17938.

Bautista, J., Cremades, O., Corpas, R., Ramos, R., Iglesias, F., Vega, J., Fontiveros, E., Perales, J., Parrados, J., Millan, F. (2000) Preparation of chitin by acetic acid fermentation, *Adv. Chitin Sci.* **4**, 28–33.

Baxter, J., Shimizu, F., Takiguchi, Y., Wada, M., Yamaguchi, T. (2000) Effect of iron(III) chitosan intake on the reduction of serum phosphorus in rats, *J. Pharm. Pharmacol.* **52**, 863–874.

Beintema, J. J. (1994) Structural features of plant chitinases and chitin-binding proteins, *FEBS Lett.* **350**, 159–163.

Beintema, J. J., Terwisscha van Scheltinga, A. C. (1996) Plant lysozymes, *EXS* **75** (Lysozymes: Model Enzymes in Biochemistry and Biology), 75–86.

Berecibar, A., Grandjean, C., Siriwardena, A. (1999) Synthesis and biological activity of natural aminocyclopentitol glycosidase inhibitors: mannostatins, trehazolin, allosamidins and their analogues, *Chem. Rev.* **99**, 779–844.

Beri, R. G., Walker, J., Reese, E. T., Rollings, J. E. (1993) Characterization of chitosans via coupled size-exclusion chromatography and multiple-angle laser light-scattering technique, *Carbohydr. Res.* **238**, 11–26.

Bernkop-Schnurch, A. (2000) Chitosan and its derivatives: potential excipients for peroral peptide delivery systems, *Int. J. Pharm.* **194**, 1–13.

Berth, G., Dautzenberg, H., Peter, M. G. (1998) Physicochemical characterization of chitosans varying in degree of acetylation, *Carbohydr. Polym.* **36**, 205–216.

Bishop, J. G., Dean, A. M., Mitchell-Olds, T. (2000) Rapid evolution in plant chitinases: Molecular targets of selection in plant-pathogen coevolution, *Proc. Natl. Acad. Sci. USA* **97**, 5322–5327.

Blackwell, J., Weigh, M. A. (1984) The structure of chitin–protein complexes, in: *Chitin, Chitosan, and Related Enzymes* (Zikakis, J. P., Ed.), New York: Academic Press, 257–272.

Bleau, G., Massicotte, F., Merlen, Y., Boisvert, C. (1999) Mammalian chitinase-like proteins, *EXS* **87** (Chitin and Chitinases), 211–221.

Bodek, K. H. (1994) Potentiometric method for determination of the degree of deacetylation of chitosan, in: *Chitin World* (Karnicki, Z. S., Wojtasz-Pajak, A., Brzeski, M. M., Bykowski, P. J., Eds.), Bremerhaven: Wirtschaftsverlag NW, 456–461.

Bodek, K. H. (1995) Evaluation of the potentiometric titration in anhydrous medium for determination of the chitosan deacylation degree, *Acta Pol. Pharm.* **52**, 337–343.

Bokma, E., van Koningsveld, G. A., Jeronimus-Stratingh, M., Beintema, J. J. (1997) Hevamine, a chitinase from the rubber tree Hevea brasiliensis, cleaves peptidoglycan between the C-1 of N-acetylglucosamine and C-4 of N-acetylmuramic acid and therefore is not a lysozyme, *FEBS Lett.* **411**, 161–163.

Bokma, E., Barends, T., van Scheltinga, A. C. T., Dijkstra, B. W., Beintema, J. J. (2000) Enzyme kinetics of hevamine, a chitinase from the rubber tree *Hevea brasiliensis*, *FEBS Lett.* **478**, 119–122.

Boot, R. G., Renkema, G. H., Strijland, A., van Zonneveld, A. J., Aerts, J. M. F. G. (1995) Cloning of a cDNA encoding chitotriosidase, a human chitinase produced by macrophages, *J. Biol. Chem.* **270**, 26252–26256.

Boot, R. G., Van Achterberg, T. A. E., Van Aken, B. E., Renkema, G. H., Jacobs, M. J. H. M., Aerts, J. M. F. G., De Vries, C. J. M. (1999) Strong induction of members of the chitinase family of proteins in atherosclerosis: chitotriosidase and human cartilage gp-39 expressed in lesion macrophages, *Arterioscler. Thromb. Vasc. Biol.* **19**, 687–694.

Bosso, C., Domard, A. (1992) Characterization of glucosamine and N-acetylglucosamine oligomers by fast atom bombardment mass spectrometry, *Org. Mass Spectrom.* **27**, 799–806.

Brameld, K. A., Goddard, W. A. (1998) Substrate distortion to a boat conformation at subsite-1 is critical in the mechanism of family 18 chitinases, *J. Am. Chem. Soc.* **120**, 3571–3580.

Brugnerotto, J., Lizardi, J., Goycoolea, F. M., Arguelles-Monal, W., Desbrieres, J., Rinaudo, M. (2001) An infrared investigation in relation with chitin and chitosan characterization, *Polymer* **42**, 3569–3580.

Brurberg, M. B., Eijsink, V. G. H., Nes, I. F. (1994) Characterization of a chitinase gene (chiA) from *Serratia marcescens* BJL200 and one-step purification of the gene product, *FEMS Microbiol. Lett.* **124**, 399–404.

Brurberg, M. B., Eijsink, V. G. H., Haandrikman, A. J., Venema, G., Nes, I. F. (1995) Chitinase B from *Serratia marcescens* BJL200 is exported to the periplasm without processing, *Microbiology* **141**, 123–131.

Brurberg, M. B., Nes, I. F., Eijsink, V. G. H. (1996) Comparative studies of chitinases A and B from *Serratia marcescens*, *Microbiology* **142**, 1581–1589.

Burkart, M. D., Izumi, M., Wong, C. H. (1999) Enzymatic regeneration of 3′-phosphoadenosine-5′-phosphosulfate using aryl sulfotransferase for the preparative enzymatic synthesis of sulfated carbohydrates, *Angew. Chem. Int. Edn.* **38**, 2747–2750.

Buss, U., Vårum, K. M., Peter, M. G., Spindler-Barth, M. (1996) ELISA for quantitation of chitin, chitosan and related compounds, *Adv. Chitin Sci.* **1**, 254–261.

Bustos, R. O., Healy, M. G. (1994) Microbial extraction of chitin from prawn waste, in: *Chitin World* (Karnicki, Z. S., Wojtasz-Pajak, A., Brzeski, M. M., Bykowski, P. J., Eds.), Bremerhaven: Wirtschaftsverlag NW, 15–25.

Cauchie, H.-M. (1997) An attempt to estimate crustacean chitin production in the hydrosphere, *Adv. Chitin Sci.* **2**, 32–39.

Chandy, T., Sharma, P. (1992) Polylysine-immobilized chitosan beads as adsorbents for bilirubin, *Artif. Organs* **16**, 568–576.

Chanzy, H. (1997) Chitin crystals, *Adv. Chitin Sci.* **2**, 11–21.

Chapman, R. F. (1971) *The Insects, Structure and Function*, London: The English Universities Press.

Chatelet, C., Damour, O., Domard, A. (2001) Influence of the degree of acetylation on some biological properties of chitosan films, *Biomaterials* **22**, 261–268.

Chen, R. H., Tsaih, M. L. (2000) Urea-induced conformational changes of chitosan molecules and the shift of break point of Mark-Houwink equation by increasing urea concentration, *J. Appl. Polym. Sci.* **75**, 452–457.

Choi, H. K., Choi, K. H., Kramer, K. J., Muthukrishnan, S. (1997) Isolation and characterization of a genomic clone for the gene for an insect molting enzyme, chitinase, *Insect Biochem. Mol. Biol.* **27**, 37–47.

Choi, Y. S., Lee, S. B., Hong, S. R., Lee, Y. M., Song, K. W., Park, M. H. (2001) Studies on gelatin-based sponges. Part III: A comparative study of cross-linked gelatin/alginate, gelatin/hyaluronate and chitosan/hyaluronate sponges and their application as a wound dressing in full-thickness skin defect of rat, *J. Mater. Sci., Mater. Med.* **12**, 67–73.

Cira, L. A., Huerta, S., Guerrero, I., Rosas, R., Hall, G. M., Shirai, K. (2000) Scaling up of lactic acid fermentation of prawn wastes in packed-bed column reactor for chitin recovery, *Adv. Chitin Sci.* **4**, 21–27.

Cohen, E. (1987) Interference with chitin biosynthesis in insects, in: *Chitin and Benzoylphenyl Ureas* (Wright, J. E., Retnakaran, A., Eds.), Dordrecht: W. Junk, 43–73.

Cohen, E. (1993) Chitin synthesis and degradation as targets for pesticide, *Arch. Insect Biochem. Physiol.* **22**, 245–261.

Cohen-Kupiec, R., Chet, I. (1998) The molecular biology of chitin digestion, *Curr. Opin. Biotechnol.* **9**, 270–277.

Colombo, P., Sciutto, A. M. (1996) Nutritional aspects of chitosan employment in hypocaloric diet, *Acta Toxicol. Ther.* **17**, 287–302.

Coutinho, P. M., Henrissat, B. (1999) Carbohydrate-Active Enzymes server at URL: http://afmb.cnrs-mrs.fr/~pedro/CAZY/db.html.

Daly, W. H., Macossay, J. (1997) An overview of chitin and derivatives for biodegradable materials applications, *Fib. Text. East. Eur.* **5**, 22–27.

Danulat, E. (1986) The effects of various diets on chitinase and beta-glucosidase activities and the condition of cod, *Gadus morhua* (L.), *J. Fish Biol.* **28**, 191–197.

Davis, S. S. (2000) Drug delivery systems, *Interdiscipl. Sci. Rev.* **25**, 175–183.

De la Vega, H., Specht, C. A., Semino, C. E., Robbins, P. W., Eichinger, D., Caplivski, D., Ghosh, S., Samuelson, J. (1997) Cloning and expression of chitinases of Entamoebae, *Mol. Biochem. Parasitol.* **85**, 139–147.

De la Vega, H., Specht, C. A., Liu, Y., Robbins, P. W. (1998) Chitinases are a multi-gene family in *Aedes*, *Anopheles* and *Drosophila*, *Insect Mol. Biol.* **7**, 233–239.

Deacon, M. P., Davis, S. S., White, R. J., Nordman, H., Carlstedt, I., Errington, N., Rowe, A. J., Harding, S. E. (1999) Are chitosan–mucin interactions specific to different regions of the stomach? Velocity ultracentrifugation offers a clue, *Carbohydr. Polym.* **38**, 235–238.

Defaye, J., Gadelle, A., Pedersen, C. (1989) Chitin and chitosan oligosaccharides, in: *Chitin and Chitosan* (Skjåk-Bræk, G., Anthonsen, T., Sandford, P., Eds.), London: Elsevier Applied Science, 415–429.

Deikman, J. (1997) Molecular mechanisms of ethylene regulation of gene transcription, *Physiol. Plantarum* **100**, 561–566.

Demont-Caulet, N., Maillet, F., Tailler, D., Jacquinet, J.-C., Prome, J.-C., Nicolaou, K. C., Truchet, G., Beau, J.-M., Denarie, J. (1999) Nodule-inducing activity of synthetic *Sinorhizobium meliloti* nodulation factors and related lipo-chitooligosaccharides on alfalfa. Importance of the acyl chain structure, *Plant Physiol.* **120**, 83–92.

Denarie, J., Debelle, F., Prome, J.-C. (1996) Rhizobium lipo-chitooligosaccharide nodulation factors: signaling molecules mediating recognition and morphogenesis, *Annu. Rev. Biochem.* **65**, 503–535.

Denuziere, A., Ferrier, D., Domard, A. (1996) Chitosan–chondroitin sulfate and chitosan–hyaluronate polyelectrolyte complexes. Physicochemical aspects, *Carbohydr. Polym.* **29**, 317–323.

Doernenburg, H., Knorr, D. (1996) Chitosan for elicitation, permeabilization and immobilization of plant cell cultures, *Bioforum* **19**, 52–62.

Domard, A., Cartier, N. (1991) Glucosamine oligomers: 3. Study of the elution process on polyacrylamide gels, application to other polyamine oligomer series, *Polym. Commun.* **32**, 116–119.

Domard, A., Demarger-Andre, S. (1994) Chitosan behaviors in fatty acid dispersions, *Macromol. Rep.* **A31**, 849–856.

Domard, A., Piron, E. (2000) Recent approach of metal binding by chitosan and derivatives, *Adv. Chitin Sci.* **4**, 295–301.

Dornish, M., Hagen, A., Hansson, E., Pecheur, C., Verdier, F., Skaugrud, O. (1997a) Safety of Protasan: ultrapure chitosan salts for biomedical and pharmaceutical use, *Adv. Chitin Sci.* **2**, 664–670.

Dornish, M., Skaugrud, O., Illum, L., Davis, S. S. (1997b) Nasal drug delivery with Protasan, *Adv. Chitin Sci.* **2**, 694–697.

Draborg, H., Kauppinen, S., Christgau, S., Dalboge, H. (1997) A sensitive plate assay for the detection and expression cloning of endochitinase from *Trichoderma harzianum*, in: *Chitin Handbook* (Muzzarelli, R. A. A., Peter, M. G., Eds.), Grottammare: Atec, 261–266.

Drouillard, S., Armand, S., Davies, G. J., Vorgias, C. E., Henrissat, B. (1997) *Serratia marcescens* chitobiase is a retaining glycosidase utilizing substrate acetamido group participation, *Biochem. J.* **328**, 945–949.

Duarte, M. L., M.C. Ferreira, M. C., Marvão, M. R. (2000) A statistical evaluation of IR spectroscopic methods to determine the degree of acetylation of α-chitin and chitosan, *Adv. Chitin Sci.* **4**, 367–374.

Dvorakova, J., Schmidt, D., Hunkova, Z., Thiem, J., Kren, V. (2001) Enzymatic rearrangements of chitin hydrolysates with β-N-acetylhexosaminidase from *Aspergillus oryzae*, *J. Mol. Catal. B: Enzymatic* **11**, 225–232.

Ebert, A., Fink, H.-P. (1997) Solid-state NMR spectroscopy of chitin and chitosan, in: *Chitin Handbook* (Muzzarelli, R. A. A., Peter, M. G., Eds.), Grottammare: Atec, 137–143.

Ebert, G. (1993) *Biopolymere*, Stuttgart: Teubner.

Eiberg, H., Dentandt, W. R. (1997) Assignment of human plasma methylumbelliferyl-tetra-N-acetylchitotetraose hydrolase or chitinase to chromosome 1q by a linkage study, *Hum. Genet.* **101**, 205–207.

Ernst, E., Pittler, M. H. (1998) Chitosan as a treatment for body weight reduction? A meta-analysis, *Perfusion* **11**, 461–462.

Errington, N., Harding, S. E., Vårum, K. M., Illum, L. (1993) Hydrodynamic characterization of chitosans varying in degree of acetylation, *Int. J. Biol. Macromol.* **15**, 113–117.

Escott, G. M., Adams, D. J. (1995) Chitinase activity in human serum and leukocytes, *Infect. Immun.* **63**, 4770–4773.

Falbe, J., Regitz, M. (Eds.) (1989) *Römpp Chemie Lexikon*, Stuttgart: Thieme, 613.

Feix, M., Gloggler, S., Londershausen, M., Weidemann, W., Spindler, K. D., Spindler-Barth, M. (2000) A cDNA encoding a chitinase from the epithelial cell line of *Chironomus tentans* (Insecta, Diptera) and its functional expression, *Arch. Insect Biochem. Physiol.* **45**, 24–36.

Felt, O., Buri, P., Gurny, R. (1998) Chitosan: a unique polysaccharide for drug delivery, *Drug Dev. Ind. Pharm.* **24**, 979–993.

Frazier, S. F., Larsen, G. S., Neff, D., Quimby, L., Carney, M., DiCaprio, R. A., Zill, S. N. (1999) Elasticity and movements of the cockroach tarsus in walking, *J. Comp. Physiol.* **A185**, 157–172.

Fukamizo, T., Kramer, K. J. (1987) Effect of 20-hydroxyecdysone on chitinase and β-N-acetylglucosaminidase during the larval-pupal transformation of *Manduca sexta*, *Insect Biochem. Mol. Biol.* **17**, 547–550.

Furda, I. (2000) Reduction of absorption of dietary lipids and cholesterol by chitosan and its derivatives and special formulations, in: *Chitosan per os, from Dietary Supplement to Drug Carrier* (Muzzarelli, R. A. A., Ed.), Grottamare: Atec, 41–63.

Gallaher, C. M., Munion, J., Hesslink, R., Wise, J., Gallaher, D. D. (2000) Cholesterol reduction by glucomannan and chitosan is mediated by changes in cholesterol absorption and bile acid and fat excretion in rats, *J. Nutr.* **130**, 2753–2759.

Girola, M., De Bernardi, M., Contos, S., Tripodi, S., Ventura, P., Guarino, C., Marletta, M. (1996) Dose effect in lipid-lowering activity of a new dietary integrator (chitosan, *Garcinia cambogia* extract and chrome), *Acta Toxicol. Ther.* **17**, 25–40.

Gokul, B., Lee, J. H., Song, K. B., Rhee, S. K., Kim, C. H., Panda, T. (2000) Characterization and applications of chitinases from *Trichoderma harzianum* – A review, *Bioproc. Eng.* **23**, 691–694.

Gooday, G. W. (1999) Aggressive and defensive roles for chitinases, *EXS* **87** (Chitin and Chitinases), 157–169.

Graham, L. S., Sticklen, M. B. (1994) Plant chitinases, *Can. J. Bot.* **72**, 1057–1083.

Grün, L., Förster, H., Peter, M. G. (1984) CP/MAS-^{13}C-NMR spectra of sclerotized insect cuticle and of chitin, *Angew. Chem. Int. Edn.* **23**, 638–639.

Guerrero, L., Omil, F., Mendez, R., Lema, J. M. (1998) Protein recovery during the overall treatment of wastewaters from fish-meal factories, *Bioresour. Technol.* **63**, 221–229.

Guibal, E., Milot, C., Roussy, J. (1997) Chitosan gel beads for metal ion recovery, in: *Chitin Handbook* (Muzzarelli, R. A. A., Peter, M. G., Eds.), Grottammare: Atec, 423–429.

Gupta, K. C., Kumar, M. N. V. R. (2000) Trends in controlled drug release formulations using chitin and chitosan, *J. Sci. Ind. Res.* **59**, 201–213.

Haas, F., Gorb, S., Blickhan, R. (2000) The function of resilin in beetle wings, *Proc. R. Soc. London, Ser. B, Biol. Sci.* **267**, 1375–1381.

Hackman, R. H. (1987) Chitin and the fine structure of insect cuticles, in: *Chitin and Benzoylphenyl Ureas* (Wright, J. E., Retnakaran, A., Eds.), Dordrecht: W. Junk, 1–32.

Haebel, S., Peter-Katalinic, J., Peter, M. G. (1997) Mass spectrometry of chito-oligosaccharides, in: *Chitin Handbook* (Muzzarelli, R. A. A., Peter, M. G., Eds.), Grottammare: Atec, 205–214.

Hahn, M., Hennig, M., Schlesier, B., Höhne, W. (2000) Structure of jack bean chitinase, *Acta Crystallogr.* **D56**, 1096–1099.

Haltiwanger, R. S. (2000) Structures and function of nuclear and cytoplasmic glycoproteins, in: *Carbohydrates in Chemistry and Biology* (Ernst, B., Hart, G. W., Sinay, P., Eds.), Weinheim: Wiley-VCH, 651–667, vol. 4.

Han, L. K., Kimura, Y., Okuda, H. (1999) Reduction in fat storage during chitin-chitosan treatment in mice fed a high-fat diet, *Int. J. Obes.* **23**, 174–179.

Harata, K., Muraki, M. (1995) X-ray structure of turkey egg lysozyme complex with di-*N*-acetylchitobiose. Recognition and binding of α-anomeric form, *Acta Crystallogr.* **D51**, 718–724.

Harata, K., Muraki, M. (2000) Crystal structures of *Urtica dioica* agglutinin and its complex with tri-*N*-acetylchitotriose, *J. Mol. Biol.* **297**, 673–681.

Harding, S. E., Davis, S. S., Deacon, M. P., Fiebrig, I. (1999) Biopolymer mucoadhesives, *Biotechnol. Genet. Engin. Rev.* **16**, 41–86.

Hart, P. J., Pfluger, H. D., Monzingo, A. F., Hollis, T., Robertus, J. D. (1995) The refined crystal structure of an endochitinase from *Hordeum vulgare* L. seeds at 1.8 Å resolution, *J. Mol. Biol.* **248**, 402–413.

Hashimoto, H., Abe, Y., Horito, S., Yoshimura, J. (1989) Synthesis of chitooligosaccharide derivatives by condensation polymerization, *J. Carbohydr. Chem.* **8**, 307–311.

Hashimoto, M., Ikegami, T., Seino, S., Ohuchi, N., Fukada, H., Sugiyama, J., Shirakawa, M., Watanabe, T. (2000) Expression and characterization of the chitin-binding domain of chitinase A1 from *Bacillus circulans* WL-12, *J. Bacteriol.* **182**, 3045–3054.

Haslam, S. M., Houston, K. M., Harnett, W., Reason, A. J., Morris, H. R., Dell, A. (1999) Structural studies of *N*-glycans of filarial parasites. Conservation of phosphorylcholine-substituted glycans among species and discovery of novel chito-oligomers, *J. Biol. Chem.* **274**, 20953–20960.

Hattori, T., Katai, K., Kato, M., Izume, M., Mizuta, Y. (1999) Colloidal titration of chitosan and critical unit of chitosan to the potentiometric colloidal titration with poly(vinyl sulfate) using Toluidine Blue as indicator, *Bull. Chem. Soc. Jpn.* **72**, 37–41.

He, F., Ramirez, J., Lebrilla, C. B. (1999) Evidence for enzymatic activity in the absence of solvent in gas-phase complexes of lysozyme and oligosaccharides, *Int. J. Mass Spectrometry* **193**, 103–114.

Heightman, T. D., Vasella, A. T. (1999) Recent insights into inhibition, structure, and mechanism of configuration-retaining glycosidases, *Angew. Chem. Int. Edn.* **38**, 750–770.

Hennig, M., Schlesier, B., Dauter, Z., Pfeffer, S., Betzel, C., Höhne, W. E., Wilson, W. S. (1992) A TIM barrel protein without enzymatic activity? Crystal-structure of narbonin at 1.8 Å resolution, *FEBS Lett.* **306**, 80–84.

Hennig, M., Jansonius, J. N., Terwisscha van Scheltinga, A. C., Dijkstra, B. W., Schlesier, B. (1995) Crystal structure of concanavalin B at 1.65 Å resolution. An inactivated chitinase from seeds of *Canavalia ensiformis*, *J. Mol. Biol.* **254**, 237–246.

Henrissat, B. (1999) Classification of chitinases modules, *EXS* **87** (Chitin and Chitinases), 137–156.

Herrera-Estrella, A., Chet, I. (1999) Chitinases in biological control, *EXS* **87** (Chitin and Chitinases), 171–184.

Hirano, S. (1989) Production and application of chitin and chitosan in Japan, in: *Chitin and Chitosan* (Skjåk-Bræk, G., Anthonsen, T., Sandford, P., Eds.), London: Elsevier Applied Science, 37–43.

Hirano, S. (1997) N-Acyl-, N-arylidene- and N-alkylidene chitosans and their hydrogels, in: *Chitin Handbook* (Muzzarelli, R. A. A., Peter, M. G., Eds.), Grottammare: Atec, 71–75.

Hirano, S., Akiyama, Y. (1995) Absence of a hypocholesterolemic action of chitosan in high-serum-cholesterol rabbits, *J. Sci. Food Agric.* **69**, 91–94.

Hirano, S., Midorikawa, T. (1998) Novel method for the preparation of N-acylchitosan fiber and N-acylchitosan-cellulose fiber, *Biomaterials* **19**, 293–297.

Hirano, S., Sakaguchi, T., Kuramitsu, K. (1992) N-Alanyl and some N-(N'-aryl)glycyl derivatives of chitosan, *Carbohydr. Polym.* **19**, 135–138.

Hirano, S., Zhang, M., Chung, B., Kim, S. (2000) The N-acylation of chitosan fibre and the N-deacetylation of chitin fibre and chitin-cellulose blended fibre at a solid state, *Carbohydr. Polym.* **41**, 175–179.

Hoekstra, A., Struszczyk, H., Kiveka̋s, O. (1998) Percutaneous microcrystalline chitosan application for sealing arterial puncture sites, *Biomaterials* **19**, 1467–1471.

Hollis, T., Honda, Y., Fukamizo, T., Marcotte, E., Day, P. J., Robertus, J. D. (1997) Kinetic analysis of barley chitinase, *Arch. Biochem. Biophys.* **344**, 335–342.

Hollis, T., Monzingo, A. F., Bortone, K., Ernst, S., Cox, R., Robertus, J. D. (2000) The X-ray structure of a chitinase from the pathogenic fungus *Coccidioides immitis*, *Protein Sci.* **9**, 544–551.

Holme, H. K., Hagen, A., Dornish, M. (2000) Influence of chitosans on permeability of human intestinal epithelial (Caco-2) cells: the effect of molecular weight, degree of deacetylation and exposure time, *Adv. Chitin Sci.* **4**, 259–265.

Holme, K. R., Perlin, A. S. (1997) Chitosan N-sulfate. A water-soluble polyelectrolyte, *Carbohydr. Res.* **302**, 7–12.

Honda, Y., Tanimori, S., Kirihata, M., Kaneko, S., Tokuyasu, K., Hashimoto, M., Watanabe, T. (2000a) Chemo- and enzymatic synthesis of partially and fully N-deacetylated 4-methylumbelliferyl chitobiosides: fluorogenic substrates for chitinase, *Bioorg. Med. Chem. Lett.* **10**, 827–829.

Honda, Y., Tanimori, S., Kirihata, M., Kaneko, S., Tokuyasu, K., Hashimoto, M., Watanabe, T., Fukamizo, T. (2000b) Kinetic analysis of the reaction catalyzed by chitinase A1 from *Bacillus circulans* WL-12 toward the novel substrates, partially N-deacetylated 4-methylumbelliferyl chitobiosides, *FEBS Lett.* **476**, 194–197.

Horst, M. N. (1986) Lipid-linked intermediates in crustacean chitin synthesis, in: *Chitin in Nature and Technology* (Muzzarelli, R. A. A., Jeuniaux, C., Gooday, G. W., Eds.), New York: Plenum Press, 45–52.

Horst, M. N., Walker, A. N. (1993) Crustacean chitin synthesis and the role of the Golgi apparatus, in: *Chitin Enzymology* (Muzzarelli, R. A. A., Ed.), Lyon: European Chitin Society, 109–118.

Hu, B., Trinh, K., Figueira, W. F., Price, P. A. (1996) Isolation and sequence of a novel human chondrocyte protein related to mammalian members of the chitinase protein family, *J. Biol. Chem.* **271**, 19415–19420.

Hwang, D.-C., Damodaran, S. (1995) Selective precipitation and removal of lipids from cheese whey using chitosan, *J. Agric. Food Chem.* **43**, 33–37.

Hwang, S. M., Chen, C. Y., Chen, S. S., Chen, J. C. (2000) Chitinous materials inhibit nitric oxide production by activated RAW 264.7 macrophages, *Biochem. Biophys. Res. Commun.* **271**, 229–233.

Ibrahim, G. H., Smartt, C. T., Kiley, L. M., Christensen, B. M. (2000) Cloning and characterization of a chitin synthase cDNA from the mosquito *Aedes aegypti*, *Insect Biochem. Mol. Biol.* **30**, 1213–1222.

Ikegami, T., Okada, T., Hashimoto, M., Seino, S., Watanabe, T., Shirakawa, M. (2000) Solution structure of the chitin-binding domain of *Bacillus circulans* WL-12 chitinase A1, *J. Biol. Chem.* **275**, 13654–13661.

Ikejima, T., Inoue, Y. (2000) Crystallization behavior and environmental biodegradability of the blend films of poly(3-hydroxybutyric acid) with chitin and chitosan, *Carbohydr. Polym.* **41**, 351–356.

Illum, L. (1998) Chitosan and its use as a pharmaceutical excipient, *Pharm. Res.* **15**, 1326–1331.

Inui, H., Yoshida, M., Hirano, S. (1996) Effects of a 6-O-hydroxyethyl group on the hydrolysis of 6-O-hydroxyethylchitin (glycolchitin) by chitinase, *Biosci. Biotechnol. Biochem.* **60**, 1886–1887.

Inui, H., Yamaguchi, Y., Hirano, S. (1997) Elicitor actions of N-acetylchitooligosaccharides and laminarioligosaccharides for chitinase and L-phenylalanine ammonia-lyase induction in rice suspension culture, *Biosci. Biotechnol. Biochem.* **61**, 975–978.

Izume, M., Nagae, S., Kawagishi, H., Ohtakara, A. (1992) Preparation of normal-acetylchitooligosaccharides from enzymatic hydrolyzates of chitosan, *Biosci. Biotechnol. Biochem.* **56**, 1327–1328.

Japanese Society for Chitin and Chitosan (1995) *Kichin, Kitosan Handobukku (Chitin and Chitosan Handbook)*, Tokyo: Gihodo Shuppan. [*Chem. Abstr.* **122**, 291440].

Jeon, Y. J., Kim, S. K. (2000) Production of chitooligosaccharides using an ultrafiltration membrane reactor and their antibacterial activity, *Carbohydr. Polym.* **41**, 133–141.

Jeon, Y. J., Park, P. J., Kim, S. K. (2001) Antimicrobial effect of chitooligosaccharides produced by bioreactor, *Carbohydr. Polym.* **44**, 71–76.

Jeuniaux, C., Voss-Foucart, M. F. (1997) A specific enzymatic method for the quantitative estimation of chitin, in: *Chitin Handbook* (Muzzarelli, R. A. A., Peter, M. G., Eds.), Grottammare: Atec, 3–7.

Jeuniaux, C., Voss-Foucart, M. F., Bussers, J.-C. (1993) La production de chitine par les crustacés dans les écosystèmes marins, *Aquat. Living Resour.* **6**, 331–341.

Jeuniaux, C., Compere, P., Toussaint, C., Decloux, N., Voss-Foucart, M.-F. (1996) Confirmation of the presence of chitin in fin cuticles of some Blennidae (Teleostei) by enzymic and cytochemical methods, *Adv. Chitin Sci.* **1**, 18–25.

Jin, H. M., Copeland, N. G., Gilbert, D. J., Jenkins, N. A., Kirkpatrick, R. B., Rosenberg, M. (1998) Genetic characterization of the murine Ym1 gene and identification of a cluster of highly homologous genes, *Genomics* **54**, 316–322.

Kadokawa, J.-I., Yamashita, K., Karasu, M., Tagaya, H., Chiba, K. (1995) Preparation and enzymic hydrolysis of block copolymer consisting of oligochitin and poly(propylene glycol), *J. Macromol. Sci., Pure Appl. Chem.* **A32**, 1273–1280.

Kadokawa, J., Sato, M., Karasu, M., Tagaya, H., Chiba, K. (1998) Synthesis of hyperbranched aminopolysaccharides, *Angew. Chem. Int. Edn.* **37**, 2373–2376.

Kamei, K., Yamamura, Y., Hara, S., Ikenaka, T. (1989) Amino acid sequence of chitinase from *Streptomyces erythraeus, J. Biochem.* **105**, 979–985.

Kamst, E., Bakkers, J., Quaedvlieg, N. E. M., Pilling, J., Kijne, J. W., Lugtenberg, B. J. J., Spaink, H. P. (1999) Chitin oligosaccharide synthesis by *Rhizobia* and zebrafish embryos starts by glycosyl transfer to O-4 of the reducing-terminal residue, *Biochemistry* **38**, 4045–4052.

Karlsen, S., Hough, E. (1995) Crystal structures of three complexes between chito-oligosaccharides and lysozyme from the rainbow trout. How distorted is the NAG sugar in site D? *Acta Crystallogr.* **D51**, 962–978.

Kato, T., Shizuri, Y., Izumida, H., Yokoyama, A., Endo, M. (1995) Styloguanidines, new chitinase inhibitors from the marine sponge *Stylotella aurantium, Tetrahedron Lett.* **36**, 2133–2136.

Kawasaki, T., Kawasaki, M., Notomi, A., Itoh, K., Ikeyama, N. (1998) Effects of chitin-oligosaccharides on blood pressure, lipid and glucose metabolism in clinically healthy subjects – a randomized single-blind placebo-controlled crossover trial, *Kichin, Kitosan Kenkyu* **4**, 316–324 [*Chem. Abstr.* **130**, 60903].

Kerwin, J. L., Whitney, D. L., Sheikh, A. (1999) Mass spectrometric profiling of glucosamine, glucosamine polymers and their catecholamine adducts. Model reactions and cuticular hydrolysates of *Toxorhynchites amboinensis* (Culicidae) pupae, *Insect Biochem. Mol. Biol.* **29**, 599–607.

Keyhani, N. O., Roseman, S. (1999) Physiological aspects of chitin catabolism in marine bacteria, *Biochim. Biophys. Acta* **1473**, 108–122.

Keyhani, N. O., Li, X. B., Roseman, S. (2000) Chitin catabolism in the marine bacterium *Vibrio furnissii* – Identification and molecular cloning of a chitoporin, *J. Biol. Chem.* **275**, 33068–33076.

Kim, M. G., Shin, S. W., Bae, K. S., Kim, S. C., Park, H. Y. (1998) Molecular cloning of chitinase cDNAs from the silkworm, *Bombyx mori* and the fall webworm, *Hyphantria cunea, Insect Biochem. Mol. Biol.* **28**, 163–171.

Kim, S. Y., Shon, D. H., Lee, K. H. (2000) Enzyme-linked immunosorbent assay for detection of chitooligosaccharides, *Biosci. Biotechnol. Biochem.* **64**, 696–701.

Kiyosada, T., Shoda, S., Kobayashi, S. (1998) Enzymic ring-opening polyaddition for chitin synthesis. A cationic mechanism in basic solution? *Macromol. Symp.* **132**, 415–420.

Klokkevold, P., Fukayama, H., Sung, E. (1997a) The effect of chitosan on hemostasis: current work and review of the literature, *Adv. Chitin Sci.* **2**, 698–704.

Klokkevold, P., Redd, M., Salamati, A., Kim, J., Nishimura, R. (1997b) The effect of chitosan on guided bone regeneration: a pilot study in the rabbit, *Adv. Chitin Sci.* **2**, 656–663.

Knaul, J. Z., Hudson, S. M., Creber, K. A. M. (1999) Improved mechanical properties of chitosan fibers, *J. Appl. Polym. Sci.* **72**, 1721–1732.

Kobayashi, S., Kiyosada, T., Shoda, S. (1996) Synthesis of artificial chitin: irreversible catalytic behavior of a glycosyl hydrolase through a transition state analogue substrate, *J. Am. Chem. Soc.* **118**, 13113–13114.

Kobayashi, S., Kiyosada, T., Shoda, S. (1997) A novel method for synthesis of chitobiose via enzymic glycosylation using a sugar oxazoline as glycosyl donor, *Tetrahedron Lett.* **38**, 2111–2112.

Koetz, J., Kosmella, S. (1997) Polyelectrolyte complex formation with chitosan, *Adv. Chitin Sci.* **2**, 476–483.

Koga, D. (1996) Induction of chitinase in fine bentgrass and yam by various elicitors, *Front. Biomed. Biotechnol.* **3**, 231–241.

Koga, D., Kramer, K. J. (1983) Hydrolysis of glycol chitin by chitinolytic enzymes, *Comp. Biochem. Physiol.* **76B**, 291–293.

Koga, D., Funakoshi, T., Fujimoto, H., Kuwano, E., Eto, M., Ide, A. (1991) Effects of 20-hydroxyecdysone and KK-42 on chitinase and β-N-acetylglucosaminidase during the larval-pupal transformation of *Bombyx mori*, *Insect Biochem.* **21**, 277–284.

Koga, D., Hoshika, H., Matsushita, M., Tanaka, A., Ide, A., Kono, M. (1996) Purification and characterization of β-N-acetylhexosaminidase from the liver of a prawn, *Penaeus japonicus*, *Biosci. Biotechnol. Biochem.* **60**, 194–199.

Koga, D., Sasaki, Y., Uchiumi, Y., Hirai, N., Arakane, Y., Nagamatsu, Y. (1997) Purification and characterization of *Bombyx mori* chitinases, *Insect Biochem. Mol. Biol.* **27**, 757–767.

Koga, D., Yoshioka, T., Arakane, Y. (1998) HPLC analysis of anomeric formation and cleavage pattern by chitinolytic enzymes, *Biosci. Biotechnol. Biochem.* **62**, 1643–1646.

Koga, D., Mitsutomi, M., Kono, M., Matsumiya, M. (1999) Biochemistry of chitinases, *EXS* **87** (Chitin and Chitinases), 111–123.

Kopacek, P., Vogt, R., Jindrak, L., Weise, C., Safarik, I. (1999) Purification and characterization of the lysozyme from the gut of the soft tick *Ornithodoros moubata*, *Insect Biochem. Mol. Biol.* **29**, 989–997.

Kow, F. (1994) Seafood offal available as raw material for chitin and chitosan production in Tasmania, in: *Chitin World* (Karnicki, Z. S., Wojtasz-Pajak, A., Brzeski, M. M., Bykowski, P. J., Eds.), Bremerhaven: Wirtschaftsverlag NW, 26–29.

Kramer, K. J., Koga, D. (1986) Insect chitin: physical state, synthesis, degradation, and metabolic regulation, *Insect Biochem.* **16**, 851–877.

Kramer, K. J., Muthukrishnan, S. (1997) Insect chitinases: molecular biology and potential use as biopesticides, *Insect Biochem. Mol. Biol.* **27**, 887–900.

Kramer, K. J., Corpuz, L., Choi, H. K., Muthukrishnan, S. (1993) Sequence of a cDNA and expression of the gene encoding epidermal and gut chitinases of *Manduca sexta*, *Insect Biochem. Mol. Biol.* **23**, 691–701.

Kramer, K. J., Hopkins, T. L., Schaefer, J. (1995) Applications of solid-state NMR to the analysis of insect sclerotized structures, *Insect Biochem. Mol. Biol.* **25**, 1067–1080.

Kramerov, A. A., Mukha, D. V., Metakovsky, E. V., Gvozdev, V. A. (1986) Glycoproteins containing sulfated chitin-like carbohydrate moiety, *Insect Biochem.* **16**, 417–432.

Kramerov, A. A., Rozovsky, Y. M., Baikova, N. A., Gvozdev, V. A. (1990) Cognate chitinoproteins are detected during *Drosophila melanogaster* development and in cell cultures from different insect species, *Insect Biochem.* **20**, 769–775.

Krasavtsev, V. E. (2000) Krill as a promising material for the production of chitin in Europe, *Adv. Chitin Sci.* **4**, 1–3.

Kristiansen, A., Vårum, K. M., Grasdalen, H. (1998) Quantitative studies of the non-productive binding of lysozyme to partially N-acetylated chitosans. Binding of large ligands to a one-dimensional binary lattice studied by a modified McGhee and Von Hippel model, *Biochim. Biophys. Acta* **1425**, 137–150.

Kubisch, J., Weignerova, L., Kotter, S., Lindhorst, T. K., Sedmera, P., Kren, V. (1999) Enzymatic synthesis of *p*-nitrophenyl β-chitobioside, *J. Carbohydr. Chem.* **18**, 975–984.

Kumar, M. N. V. R. (2000) A review of chitin and chitosan applications, *React. Funct. Polym.* **46**, 1–27.

Kurita, K. (1997) Chitin and chitosan derivatives, in: *Desk Reference of Functional Polymers, Synthesis and Applications* (Arshady, R., Ed.), Washington, DC: American Chemical Society, 239–259.

Kurita, K., Ishii, S., Tomita, K., Nishimura, S. I., Shimoda, K. (1994a) Reactivity characteristics of squid β-chitin as compared with those of shrimp chitin: high potentials of squid chitin as a starting material for facile chemical modifications, *J. Polym. Sci., Part A: Polym. Chem.* **32**, 1027–1032.

Kurita, K., Kobayashi, M., Munakata, T., Ishii, S., Nishimura, S.-I. (1994b) Synthesis of non-natural branched polysaccharides. Regioselective introduction of α-mannoside branches into chitin, *Chem. Lett.*, 2063–2066.

Kurita, K., Hashimoto, S., Ishi, S., Mori, T., Nishimura, S.-I. (1996a) Efficient graft copolymerization of 2-methyl-2-oxazoline onto tosyl- and iodo-chitins in solution, *Polym. J.* **28**, 686–689.

Kurita, K., Hashimoto, S., Ishii, S., Mori, T. (1996b) Chitin/poly(methyl methacrylate) hybrid materials. Efficient graft copolymerization of methyl methacrylate onto mercapto-chitin, *Polym. Bull.* **36**, 681–686.

Kurita, K., Hashimoto, S., Yoshino, H., Ishii, S., Nishimura, S.-I. (1996c) Preparation of chitin/polystyrene hybrid materials by efficient graft copolymerization based on mercaptochitin, *Macromolecules* **29**, 1939–1942.

Kurita, K., Yoshino, H., Nishimura, S.-I., Ishii, S., Mori, T., Nishiyama, Y. (1997) Mercapto-chitins: a new type of supports for effective immobilization of acid phosphatase, *Carbohydr. Polym.* **32**, 171–175.

Kurita, K., Hirakawa, M., Nishiyama, Y. (1999) Silylated chitin: a new organo-soluble precursor for facile modifications and film casting, *Chem. Lett.*, 771–772.

Kurita, K., Inoue, M., Nishiyama, Y. (2000) Graft copolymerization of methyl methacrylate onto mercapto-chitin, *Adv. Chitin Sci.* **4**, 417–421.

Kuyama, H., Ogawa, T. (1997) Synthesis of chitosan oligomers, in: *Chitin Handbook* (Muzzarelli, R. A. A., Peter, M. G., Eds.), Grottammare: Atec, 181–189.

Lahiji, A., Sohrabi, A., Hungerford, D. S., Frondoza, C. G. (2000) Chitosan supports the expression of extracellular matrix proteins in human osteoblasts and chondrocytes, *J. Biomed. Mater. Res.* **51**, 586–595.

Lang, G. (1995) Chitosan derivatives – preparation and potential uses, in: *Chitin Chitosan* (Zakaria, M. B., Wan-Muda, W. M., Pauzi, A., Eds.), Bangi: Penerbit University Kebangsaan Malaysia, 109–118.

Lang, G., Maresch, G., Birkel, S. (1997a) Hydroxyalkyl chitosans, in: *Chitin Handbook* (Muzzarelli, R. A. A., Peter, M. G., Eds.), Grottammare: Atec, 61–66.

Lang, G., Maresch, G., Birkel, S. (1997b) Quaternary chitosan salts, in: *Chitin Handbook* (Muzzarelli, R. A. A., Peter, M. G., Eds.), Grottammare: Atec, 67–70.

Lee, E. A., Pan, C. H., Son, J. M., Kim, S. I. (1999a) Isolation and characterization of basic exochitinase from leaf extract of *Rehmannia glutinosa*, *Biosci. Biotechnol. Biochem.* **63**, 1781–1783.

Lee, J. K., Kim, S. U., Kim, J. H. (1999b) Modification of chitosan to improve its hypocholesterolemic capacity, *Biosci. Biotechnol. Biochem.* **63**, 833–839.

Lee, K. Y., Ha, W. S., Park, W. H. (1995) Blood compatibility and biodegradability of partially N-acylated chitosan derivatives, *Biomaterials* **16**, 1211–1216.

Letzel, M. C., Synstad, B., Eijsink, V. G. H., Peter-Katalinic, J., Peter, M. G. (2000) Libraries of chitooligosaccharides of mixed acetylation patterns and their interactions with chitinases, *Adv. Chitin Sci.* **4**, 545–552.

Lim, L.-Y., Khor, E., Koo, O. (1998) γ-Irradiation of chitosan, *J. Biomed. Mater. Res.* **43**, 282–290.

Liu, W., Chen, X., Zhang, X., Liu, C., Cong, R., Jin, L., Lang, G., Wang, S. (1997) Study on toxicity of CM-chitosan, *Zhongguo Haiyang Yaowu* **16**, 17–19 [*Chem. Abstr.* **129**, 50606].

Liu, X. D., Nishi, N., Tokura, S., Sakairi, N. (2001) Chitosan coated cotton fiber: preparation and physical properties, *Carbohydr. Polym.* **44**, 233–238.

Londershausen, M., Tuberg, A., Buss, U., Spindler-Barth, M., Spindler, K. D. (1993) Comparison of chitin synthesis from an insect cell line and embryonic tick tissues, in: *Chitin Enzymology* (Muzzarelli, R. A. A., Ed.), Lyon: European Chitin Society, 101–108.

Londershausen, M., Turberg, A., Bieseler, B., Lennartz, M., Peter, M. G. (1996) Characterization and inhibitor studies of chitinases from a parasitic blowfly (*Lucilia cuprina*), a tick (*Boophilus microplus*), an intestinal nematode (*Haemonchus contortus*) and a bean (*Phaseolus vulgaris*), *Pest. Sci.* **48**, 305–314.

Londershausen, M., Turberg, A., Ludwig, M., Hirsch, B., Spindler-Barth, M. (1997) Properties of insect chitin synthase: effects of inhibitors, γ-S-GTP, and compounds influencing membrane lipids, in: *Applications of Chitin and Chitosan* (Goosen, M. F. A., Ed.), Lancaster: Technomic, 155–170.

Lopatin, S. A., Ilyin, M. M., Pustobaev, V. N., Bezchetnikova, Z. A., Varlamov, V. P., Davankov, V. A. (1995) Mass-spectrometric analysis of N-acetylchitooligosaccharides prepared through enzymic hydrolysis of chitosan, *Anal. Biochem.* **227**, 285–288.

Ludwig, M., Spindler-Barth, M., Spindler, K. D. (1991) Properties of chitin synthase in homogenates from *Chironomus* cells, *Arch. Insect Biochem. Physiol.* **18**, 251–263.

Macchi, G. (1996) A new approach to the treatment of obesity: chitosan's effects on body weight reduction and plasma cholesterol's levels, *Acta Toxicol. Ther.* **17**, 303–320.

Maezaki, Y., Tsuji, K., Nakagawa, Y., Kawai, Y., Akimoto, M., Tsugita, T., Takekawa, W., Terada, A., Hara, H., Mitsuoka, T. (1993) Hypocholesterolemic effect of chitosan in adult males, *Biosci. Biotechnol. Biochem.* **57**, 1439–1444.

Marks, E. P., Ward, G. B. (1987) Regulation of chitin synthesis: mechanisms and methods, in: *Chitin and Benzoylphenyl Ureas* (Wright, J. E., Retnakaran, A., Eds.), Dordrecht: W. Junk, 33–42.

Martinou, A., Tsigos, I., Bouriotis, V. (1997) Preparation of chitosan by enzymatic deacetylation, in: *Chitin Handbook* (Muzzarelli, R. A. A., Peter, M. G., Eds.), pp. 501–505. Grottammare: Atec.

Maruyama, M., Ito, M. (1996) In vitro properties of a chitosan-bonded self-hardening paste with hydroxyapatite granules, *J. Biomed. Mater. Res.* **32**, 527–532.

Matsumiya, M., Mochizuki, A. (1996) Distribution of chitinase and β-N-acetylhexosaminidase in the organs of several fishes, *Fish Sci.* **62**, 150–151.

Matsumiya, M., Mochizuki, A. (1997) Purification and characterization of chitinase from the liver of Japanese common squid *Todarodes pacificus*, *Fish Sci.* **63**, 409–413.

Matsumoto, T., Nonaka, T., Hashimoto, M., Watanabe, T., Mitsui, Y. (1999) Three-dimensional structure of the catalytic domain of chitinase A1 from *Bacillus circulans* WL-12 at a very high resolution, *Proc. Jpn. Acad., Ser. B, Phys. Biol. Sci.* **75**, 269–274.

Mattioli-Belmonte, M., Nicoli-Aldini, N., De Benedittis, A., Sgarbi, G., Amati, S., Fini, M., Biagini, G., Muzzarelli, R. A. A. (1999) Morphological study of bone regeneration in the presence of 6-oxychitin, *Carbohydr. Polym.* **40**, 23–27.

Meyer, M. F., Kreil, G. (1996) Cells expressing the DG42 gene from early *Xenopus* embryos synthesize hyaluronan, *Proc. Natl. Acad. Sci. USA* **93**, 4543–4547.

Minami, S., Okamoto, Y., Mori, T., Fujinaga, T., Shigemasa, Y. (1997) Mechanism of wound healing acceleration by chitin and chitosan, *Adv. Chitin Sci.* **2**, 633–639.

Mitsutomi, M., Ueda, M., Arai, M., Ando, A., Watanabe, T. (1996) Action patterns of microbial chitinases and chitosanases on partially N-acetylated chitosan, in: *Chitin Enzymology*, (Muzzarelli, R. A. A., Ed.), Grottammare: Atec, 272–284, vol. 2.

Mohr, M. D., Boernsen, K. O., Widmer, H. M. (1995) Matrix-assisted laser desorption/ionization mass spectrometry: improved matrix for oligosaccharides, *Rapid Commun. Mass Spectrom.* **9**, 809–814.

Mori, T., Okumura, M., Matsuura, M., Ueno, K., Tokura, S., Okamoto, Y., Minami, S., Fujinaga, T. (1997) Effects of chitin and its derivatives on the proliferation and cytokine production of fibroblasts in vitro, *Biomaterials* **18**, 947–951.

Mulisch, M., Schermuly, G. (1997) Chitin synthesis in lower organisms, in: *Chitin Handbook* (Muzzarelli, R. A. A., Peter, M. G., Eds.), Grottammare: Atec, 337–344.

Muzzarelli, R. A. A. (1977) *Chitin*, Oxford: Pergamon Press.

Muzzarelli, R. A. A. (1988) Carboxymethylated chitins and chitosans, *Carbohydr. Polym.* **8**, 1–21.

Muzzarelli, R. A. A. (1996) Chitosan-based dietary foods, *Carbohydr. Polym.* **29**, 309–316.

Muzzarelli, R. A. A. (1997a) Depolymerization of chitins and chitosans with hemicellulase, lysozyme, papain and lipases, in: *Chitin Handbook* (Muzzarelli, R. A. A., Peter, M. G., Eds.), Grottammare: Atec, 153–163.

Muzzarelli, R. A. A. (1997b) Human enzymic activities related to the therapeutic administration of chitin derivatives, *Cell. Mol. Life Sci.* **53**, 131–140.

Muzzarelli, R. A. A. (1998) Management of hypercholesterolemia and overweight by oral administration of chitosans, *Biomed. Health Res.* **16**, 135–142.

Muzzarelli, R. A. A. (2000a) Chemical and preclinical studies on 6-oxychitin, *Adv. Chitin Sci.* **4**, 171–175.

Muzzarelli, R. A. A. (Ed.) (2000b) *Chitosan per os, from Dietary Supplement to Drug Carrier*, Grottammare: Atec.

Muzzarelli, R. A. A., Peter, M. G. (Eds.) (1997) *Chitin Handbook*, Grottammare: Atec.

Muzzarelli, R. A. A., Tanfani, F. (1985) The N-permethylation of chitosan and the preparation of N-trimethyl chitosan iodide, *Carbohydr. Polym.* **5**, 297–307.

Muzzarelli, R. A. A., Tomasetti, M., Ilari, P. (1994) Depolymerization of chitosan with the aid of papain, *Enzyme Microb. Technol.* **16**, 110–114.

Muzzarelli, R. A. A., Xia, W., Tomasetti, M., Ilari, P. (1995) Depolymerization of chitosan and substituted chitosans with the aid of a wheat germ lipase preparation, *Enzyme Microb. Technol.* **17**, 541–545.

Muzzarelli, R. A. A., Mattioli-Belmonte, M., Muzzarelli, B., Mattei, G., Fini, M., Biagini, G. (1997a) Medical and veterinary applications of chitin and chitosan, *Adv. Chitin Sci.* **2**, 580–589.

Muzzarelli, R. A. A., Rochetti, R., Stanic, V., Weckx, M. (1997b) Methods for the determination of the degree of acetylation of chitin and chitosan, in: *Chitin Handbook* (Muzzarelli, R. A. A., Peter, M. G., Eds.), Grottammare: Atec, 109–119.

Muzzarelli, R. A. A., Rosini, S., Trasciatti, S. (1997c) Abiogen Pharma S.R.L., Italy, Chelates of chitosans and alkaline-earth metal insoluble salts and the use thereof as medicaments useful in osteogenesis, WO 9919366 [*Chem. Abstr.* **130**, 301742].

Muzzarelli, R. A. A., Ramos, V., Stanic, V., Dubini, B., Mattioli-Belmonte, M., Tosi, G., Giardino, R. (1998) Osteogenesis promoted by calcium phosphate-N,N-dicarboxymethyl chitosan, *Carbohydr. Polym.* **36**, 267–276.

Muzzarelli, R. A. A., Muzzarelli, C., Cosani, A., Terbojevich, M. (1999) 6-Oxychitins, novel hyaluronan-like regiospecifically carboxylated chitins, *Carbohydr. Polym.* **39**, 361–367.

Muzzarelli, R. A. A., Miliani, M., Cartolari, M., Genta, I., Perugini, P., Modena, T., Pavanetto, F., Conti, B. (2000a) Oxychitin-chitosan microcapsules for pharmaceutical use, *STP Pharm. Sci.* **10**, 51–56.

Muzzarelli, R. A. A., Frega, N., Miliani, M., Cartolari, M. (2000b) Interactions of chitin, chitosan, N-laurylchitosan, and N-dimethylaminopropyl chitosan with olive oil, *Adv. Chitin Sci.* **4**, 275–279.

Natsume, H., Iwata, S., Ohtake, K., Miyamoto, M., Yamaguchi, M., Hosoya, K.-I., Kobayashi, D., Sugibayashi, K., Morimoto, Y. (1999) Screening of cationic compounds as an absorption enhancer for nasal drug delivery, *Int. J. Pharm.* **185**, 1–12.

Neff, D., Frazier, S. F., Quimby, L., Wang, R. T., Zill, S. (2000) Identification of resilin in the leg of cockroach, *Periplaneta americana*: confirmation by a simple method using pH dependence of UV fluorescence, *Arthropod Struct. Devel.* **29**, 75–83.

Neugebauer, W. A., D'Orléans-Juste, P., Bkaily, G. (2000) Peptide synthesis on chitosan/chitin, *Adv. Chitin Sci.* **4**, 411–416.

Neuhaus, B., Bresciani, J., Peters, W. (1997) Ultrastructure of the pharyngeal cuticle and lectin labelling with wheat germ agglutinin–gold conjugate indicating chitin in the pharyngeal cuticle of *Oesophagostomum dentatum* (Strongylida, Nematoda), *Acta Zool.* **78**, 205–213.

Neuhaus, J.-M. (1999) Plant chitinases (PR-3, PR-4, PR-8, PR-11), in: *Pathogen-related Proteins in Plants* (Datta, S. K., Muthukrishnan, S., Eds.), Boca Raton, FL: CRC Press, 77–105.

Neuhaus, J.-M., Fritig, B., Linthorst, H. J. M., Meins, F., Jr., Mikkelsen, J. D., Ryals, J. (1996) A revised nomenclature for chitinase genes, *Plant Mol. Biol. Rep.* **14**, 102–104.

Ng, C.-H., Chandrkrachang, S., Stevens, W. F. (2000) Effect of the rate of deacetylation on the physico-chemical properties of cuttlefish chitin, *Adv. Chitin Sci.* **4**, 50–54.

Nguyen, J., Francou, F., Virolle, M.-J., Guerineau, M. (1997) Amylase and chitinase genes in *Streptomyces lividans* are regulated by reg1, a pleiotropic regulatory gene, *J. Bacteriol.* **179**, 6383–6390.

Nishimura, S., Kai, H., Shinada, K., Yoshida, T., Tokura, S., Kurita, K., Nakashima, H., Yamamoto, N., Uryu, T. (1998) Regioselective syntheses of sulfated polysaccharides: specific anti-HIV-1 activity of novel chitin sulfates, *Carbohydr. Res.* **306**, 427–433.

Nishimura, Y., Watanabe, Y., Hong, J. M., Takeda, H., Wada, M., Yukawa, M. (1997) Intestinal absorption of ^{14}C-chitosan in rats, *Kichin, Kitosan Kenkyu* **3**, 55–61 [*Chem. Abstr.* **126**, 340532].

Nishizawa, Y., Kawakami, A., Hibi, T., He, D.-Y., Shibuya, N., Minami, E. (1999) Regulation of the chitinase gene expression in suspension-cultured rice cells by N-acetylchitooligosaccharides: differences in the signal transduction pathways leading to the activation of elicitor-responsive genes, *Plant Mol. Biol.* **39**, 907–914.

No, K. H., Meyers, S. P. (1997) Preparation of chitin and chitosan, in: *Chitin Handbook* (Muzzarelli, R. A. A., Peter, M. G., Eds.), Grottammare: Atec, 475–489.

Ogawa, K., Kawada, J., Yui, T., Okuyama, K. (2000) Crystalline behavior of chitosan, *Adv. Chitin Sci.* **4**, 324–329.

Ohme-Takagi, M., Suzuki, K., Shinshi, H. (2000) Regulation of ethylene-induced transcription of defense genes, *Plant Cell Physiol.* **41**, 1187–1192.

Okamoto, Y., Nose, M., Miyatake, K., Sekine, J., Oura, R., Shigemasa, Y., Minami, S. (2001) Physical changes of chitin and chitosan in canine gastrointestinal tract, *Carbohydr. Polym.* **44**, 211–215.

Okazaki, K., Kawabata, T., Nakano, M., Hayakawa, S. (1999) Purification and properties of chitinase from *Arthrobacter* sp. NHB-10, *Biosci. Biotechnol. Biochem.* **63**, 1644–1646.

Okuyama, K., Noguchi, K., Kanenari, M., Egawa, T., Osawa, K., Ogawa, K. (2000) Structural diversity of chitosan and its complexes, *Carbohydr. Polym.* **41**, 237–247.

Ono, K., Saito, Y., Yura, H., Ishikawa, K., Kurita, A., Akaike, T., Ishihara, M. (2000) Photocrosslinkable chitosan as a biological adhesive, *J. Biomed. Mater. Res.* **49**, 289–295.

Ormrod, D. J., Holmes, C. C., Miller, T. E. (1998) Dietary chitosan inhibits hypercholesterolemia and atherogenesis in the apolipoprotein E-deficient mouse model of atherosclerosis, *Atherosclerosis* **138**, 329–334.

Overdijk, B., Van Steijn, G. J. (1994) Human serum contains a chitinase: identification of an enzyme, formerly described as 4-methylumbelliferyl-tetra-N-acetylchitotetraoside (MU-TACT hydrolase), *Glycobiology* **4**, 797–803.

Overdijk, B., Van Steijn, G. J., Odds, F. C. (1996) Chitinase levels in guinea pig blood are increased after systemic infection with *Aspergillus fumigatus*, *Glycobiology* **6**, 627–634.

Overdijk, B., Van Steijn, G. J., Odds, F. C. (1999) Distribution of chitinase in guinea pig tissues and increases in levels of this enzyme after systemic infection with *Aspergillus fumigatus*, *Microbiology* **145**, 259–269.

Palli, S. R., Retnakaran, A. (1999) Molecular and biochemical aspects of chitin synthesis inhibition, *EXS* **87** (Chitin and Chitinases), 85–98.

Pantaleone, D., Yalpani, M. (1993) Unusual susceptibility of aminoglycans to enzymic hydrolysis, *Front. Biomed. Biotechnol.* **1**, 44–51.

Park, H. J., Jung, S. T., Song, J. J., Kang, S. G., Vergano, P. J., Testin, R. F. (1999) Mechanical and barrier properties of chitosan-based biopolymer film, *Kichin, Kitosan Kenkyu* **5**, 19–26 [*Chem. Abstr.* **131**, 20503].

Park, J. K., Keyhani, N. O., Roseman, S. (2000) Chitin catabolism in the marine bacterium *Vibrio furnissii* – identification, molecular cloning, and characterization of a N,N'-diacetylchitobiose phosphorylase, *J. Biol. Chem.* **275**, 33077–33083.

Park, R.-D., Cho, Y.-Y., Kim, K.-Y., Bom, H.-S., Oh, C.-S., Lee, H.-C. (1995) Adsorption of Toluidine Blue O onto chitosan, *Han'guk Nonghwa Hakhoechi* **38**, 447–451 [*Chem. Abstr.* **124**, 225723].

Patil, R. S., Ghormade, V., Deshpande, M. V. (2000) Chitinolytic enzymes: an exploration, *Enzyme Microb. Technol.* **26**, 473–483.

Paul, W., Sharma, C. P. (2000) Chitosan, a drug carrier for the 21st century: a review, *STP Pharm. Sci.* **10**, 5–22.

Perrakis, A., Tews, I., Dauter, Z., Oppenheim, A. B., Chet, I., Wilson, K. S., Vorgias, C. E. (1994) Crystal structure of a bacterial chitinase at 2.3 Å resolution, *Structure* **2**, 1169–1180.

Peter, M. G. (1993) Die molekulare Architektur des Exoskeletts von Insekten, *Chem. uns. Zeit* **27**, 189–197.

Peter, M. G. (1995) Applications and environmental aspects of chitin and chitosan, *J. Macromol. Sci., Pure Appl. Chem.* **A32**, 629–640.

Peter, M. G., Köhler, L. (1996) Chitin und Chitosan: Nachwachsende Rohstoffe aus dem Meer, in: *Perspektiven nachwachsender Rohstoffe in der Chemie* (Eierdanz, H., Ed.), Weinheim: VCH, 328–331.

Peter, M. G., Schweikart, F. (1990) Chitin biosynthesis enhancement by the endochitinase inhibitor allosamidin, *Biol. Chem. Hoppe Seyler* **371**, 471–473.

Peters, W., Latka, I. (1997) Wheat Germ Agglutinin-gold labelling, in: *Chitin Handbook* (Muzzarelli, R. A. A., Peter, M. G., Eds.), Grottammare: Atec, 33–40.

Pittermann, W., Hörner, V., Wachter, R. (1997a) Efficiency of high molecular weight chitosan in skin care applications, in: *Chitin Handbook* (Muzzarelli, R. A. A., Peter, M. G., Eds.), Grottammare: Atec, 361–372.

Pittermann, W., Wachter, R., Hörner, V. (1997b) Henkel K.G.a.A., Germany, Use of chitosan and/or chitosan derivatives for surface coating of implants and medical instruments, DE 19724869 [*Chem. Abstr.* **130**, 71621].

Pittler, M. H., Abbot, N. C., Harkness, E. F., Ernst, E. (1999) Randomized, double-blind trial of chitosan for body weight reduction, *Eur. J. Clin. Nutr.* **53**, 379–381.

Poveda, A., Asensio, J. L., Espinosa, J. F., Martin Pastor, M., Canada, J., Jimenez Barbero, J. (1997) Applications of nuclear magnetic resonance spectroscopy and molecular modeling to the study of protein–carbohydrate interactions, *J. Mol. Graph. Model.* **15**, 9–17.

Quong, D., Groboillot, A., Darling, G. D., Poncelet, D., Neufeld, R. J. (1997) Microencapsulation with cross-linked chitosan membranes, in: *Chitin Handbook* (Muzzarelli, R. A. A., Peter, M. G., Eds.), Grottammare: Atec, 405–410.

Ramasamy, M. S., Kulasekera, R., Wanniarachchi, I. C., Srikrishnaraj, K. A., Ramasamy, R. (1997) Interactions of human malaria parasites, *Plasmodium vivax* and *P. falciparum*, with the midgut of *Anopheles* mosquitoes, *Med. Vet. Entomol.* **11**, 290–296.

Rao, M. S., Stevens, W. F. (1997) Processing parameters in scale-up of *Lactobacillus* fermentation of shrimp biowaste, *Adv. Chitin Sci.* **2**, 88–93.

Rast, D. M., Merz, R. A., Jeanguenat, A., Mösinger, E. (2000) Enzymes of chitin metabolism for the design of antifungals, *Adv. Chitin Sci.* **4**, 479–505.

Ratajska, M., Struszczyk, M. H., Boryniec, S., Peter, M. G., Loth, F. (1997) The degree of deacetylation of chitosan: optimization of the IR method, *Polimery* **42**, 572–575.

Rathke, T. D., Hudson, S. M. (1994) Review of chitin and chitosan as fiber and film formers, *J. Macromol. Sci., Rev. Macromol. Chem. Phys.* **C34**, 375–437.

Raymond, L., Morin, F. G., Marchessault, R. H. (1993) Degree of deacetylation of chitosan using conductometric titration and solid-state NMR, *Carbohydr. Res.* **246**, 331–336.

Rege, P. R., Block, L. H. (1999) Chitosan processing: influence of process parameters during acidic and alkaline hydrolysis and effect of the processing sequence on the resultant chitosan's properties, *Carbohydr. Res.* **321**, 235–245.

Remunan-Lopez, C., Bodmeier, R. (1996) Mechanical and water vapor transmission properties of polysaccharide films, *Drug Dev. Ind. Pharm.* **22**, 1201–1209.

Renkema, G. H., Boot, R. G., Muijsers, A. O., Donker-Koopman, W. E., Aerts, J. M. F. G. (1995) Purification and characterization of human chitotriosidase, a novel member of the chitinase family of proteins, *J. Biol. Chem.* **270**, 2198–2202.

Renkema, G. H., Boot, R. G., Au, F. L., Donker-Koopman, W. E., Strijland, A., Muijsers, A. O., Hrebicek, M., Aerts, J. M. F. G. (1998) Chitotriosidase, a chitinase, and the 39-kDa human cartilage glycoprotein, a chitin-binding lectin, are homologs of family 18 glycosyl hydrolases secreted by human macrophages, *Eur. J. Biochem.* **251**, 504–509.

Roberts, G. A. F. (1992) *Chitin Chemistry*, Houndmills: Macmillan.

Roberts, G. A. F. (1997) Chitosan production routes and their role in determining the structure and properties of the product, *Adv. Chitin Sci.* **2**, 22–31.

Roberts, G. A. F., Wood, F. A. (2000) Inter-source reproducibility of the chitin deacetylation process, *Adv. Chitin Sci.* **4**, 34–39.

Robertus, J. D., Monzingo, A. F. (1999) The structure and action of chitinases, *EXS* **87** (Chitin and Chitinases), 125–135.

Rodriguez, M. S., Albertengo, L. A., Agulló, E. (2000) Chitosan emulsification properties, *Adv. Chitin Sci.* **4**, 382–388.

Rottmann, A., Synstad, B., Eijsink, V. G. H., Peter, M. G. (1999) Synthesis of N-acetylglucosaminyl and diacetylchitobiosyl amides of heterocyclic carboxylic acids as potential chitinase inhibitors, *Eur. J. Org. Chem.*, 2293–2297.

Rottmann, A., Synstad, B., Thiele, G., Schanzenbach, D., Eijsink, V. G. H., Peter, M. G. (2000) Approaches towards the design of new chitinase inhibitors, *Adv. Chitin Sci.* **4**, 553–557.

Rubin, B. R., Talent, J. M., Pertusi, R. M., Forman, M. D., Gracy, R. W. (2000) Oral polymeric N-acetyl-D-glucosamine as potential treatment for patients with osteoarthritis, *Adv. Chitin Sci.* **4**, 266–269.

Saito, A., Fujii, T., Miyashita, K. (1999) Chitinase system in *Streptomyces, Actinomycetologica* **13**, 1–10.

Sakiyama, T., Takata, H., Kikuchi, M., Nakanishi, K. (1999) Polyelectrolyte complex gel with high pH-sensitivity prepared from dextran sulfate and chitosan, *J. Appl. Polym. Sci.* **73**, 2227–2233.

Samain, E., Drouillard, S., Heyraud, A., Driguez, H., Geremia, R. A. (1997) Gram-scale synthesis of recombinant chitooligosaccharides in *Escherichia coli*, *Carbohydr. Res.* **302**, 35–42.

Sashiwa, H., Shigemasa, Y., Roy, R. (2000) Dissolution of chitosan in dimethyl sulfoxide by salt formation, *Chem. Lett.*, 596–597.

Saul, F. A., Rovira, P., Boulot, G., Van Damme, E. J. M., Peumans, W. J., Truffa-Bachi, P., Bentley, G. A. (2000) Crystal structure of *Urtica dioica* agglutinin, a superantigen presented by MHC molecules of class I and class II, *Struct. Fold. Design* **8**, 593–603.

Savant, V. D., Torres, J. A. (2000) Chitosan-based coagulating agents for treatment of cheddar cheese whey, *Biotechnol. Progr.* **16**, 1091–1097.

Schanzenbach, D., Matern, C., Peter, M. G. (1997) Cleavage of chitin by means of sulfuric acid/acetic anhydride, in: *Chitin Handbook* (Muzzarelli, R. A. A., Peter, M. G., Eds.), Grottammare: Atec, 171–174.

Schanzenbach, D., Peter, M. G. (1997) NMR spectroscopy of chito-oligosaccharides, in: *Chitin Handbook* (Muzzarelli, R. A. A., Peter, M. G., Eds.), Grottammare: Atec, 199–204.

Scheel, O., Thiem, J. (1997) Cleavage of chitin by means of aqueous hydrochloric acid and isolation of chito-oligosaccharides, in: *Chitin Handbook* (Muzzarelli, R. A. A., Peter, M. G., Eds.), Grottammare: Atec, 165–170.

Schermuly, G., Markmann-Mulisch, U., Mulisch, M. (1996) In vivo studies on the pathway of chitin synthesis in the ciliated protozoon *Eufolliculina uhligi*, *Adv. Chitin Sci.* **1**, 10–17.

Schrempf, H. (1996) The chitinolytic system of *Streptomycetes*, *Adv. Chitin Sci.* **1**, 123–128.

Schrempf, H. (1999) Characteristics of chitin-binding proteins from *Streptomycetes*, *EXS* **87** (Chitin and Chitinases), 99–108.

Schrempf, H., Blaak, H., Schnellmann, J., Stock, S. (1993) Chitinases from *Streptomyces olivaceoviridis*, *DECHEMA Monogr.* **129**, 261–271.

Semino, C. E., Robbins, P. W. (1995) Synthesis of "Nod"-like chitin oligosaccharides by the *Xenopus* developmental protein DG42, *Proc. Natl. Acad. Sci. USA* **92**, 3498–3501.

Semino, C. E., Specht, C. A., Raimondi, A., Robbins, P. W. (1996) Homologs of the *Xenopus* developmental gene DG42 are present in zebrafish and mouse and are involved in the synthesis of Nod-like chitin oligosaccharides during early embryogenesis, *Proc. Natl. Acad. Sci. USA* **93**, 4548–4553.

Senel, S., Kremer, M. J., Kas, S., Wertz, P. W., Hincal, A. A., Squier, C. A. (2000a) Effect of chitosan in enhancing drug delivery across buccal mucosa, *Adv. Chitin Sci.* **4**, 254–258.

Senel, S., Kremer, M. J., Kas, S., Wertz, P. W., Hincal, A. A., Squier, C. A. (2000b) Enhancing effect of chitosan on peptide drug delivery across buccal mucosa, *Biomaterials* **21**, 2067–2071.

Shahabuddin, M., Kaslow, D. C. (1994) *Plasmodium*: parasite chitinase and its role in malaria transmission, *Exp. Parasitol.* **79**, 85–88.

Shahabuddin, M., Vinetz, J. M. (1999) Chitinases of human parasites and their implications as antiparasitic targets, *EXS* **87** (Chitin and Chitinases), 223–234.

Shahidi, F., Synowiecki, J. (1992) Quality and compositional characteristics of Newfoundland shellfish processing discards, in: *Advances in Chitin and Chitosan* (Brine, C. J., Sandford, P. A., Zikakis, J. P., Eds.), London: Elsevier Applied Science, 617–626.

Shahidi, F., Arachchi, J. K. V., Jeon, Y.-J. (1999) Food applications of chitin and chitosans, *Trends Food Sci. Technol.* **10**, 37–51.

Shakarian, A. M., Dwyer, D. M. (1998) The Ld Cht1 gene encodes the secretory chitinase of the human pathogen *Leishmania donovani*, *Gene* **208**, 315–322.

Shakarian, A. M., Dwyer, D. M. (2000) Pathogenic *Leishmania* secrete antigenically related chitinases which are encoded by a highly conserved gene locus, *Exp. Parasitol.* **94**, 238–242.

Shigemasa, Y., Matsuura, H., Sashiwa, H., Saimoto, H. (1996a) An improved IR spectrometric determination of degree of deacetylation of chitin, *Adv. Chitin Sci.* **1**, 204–209.

Shigemasa, Y., Matsuura, H., Sashiwa, H., Saimoto, H. (1996b) Evaluation of different absorbance ratios from infrared spectroscopy for analyzing the degree of deacetylation in chitin, *Int. J. Biol. Macromol.* **18**, 237–242.

Shillito, B., Lechaire, J. P., Goffinet, G., Gaill, F. (1993) The chitin secreting microvilli of hydrothermal vent worms, in: *Chitin Enzymology* (Muzzarelli, R. A. A., Ed.), Lyon: European Chitin Society, 129–136.

Shimojoh, M., Fukushima, K., Kurita, K. (1998) Low-molecular-weight chitosans derived from β-chitin: preparation, molecular characteristics and aggregation activity, *Carbohydr. Polym.* **35**, 223–231.

Shiomi, K., Arai, N., Iwai, Y., Turberg, A., Kolbl, H., Omura, S. (2000) Structure of argifin, a new chitinase inhibitor produced by *Gliocladium* sp., *Tetrahedron Lett.* **41**, 2141–2143.

Signini, R., Desbrieres, J., Campana, S. P. (2000) On the stiffness of chitosan hydrochloride in acid-free aqueous solutions, *Carbohydr. Polym.* **43**, 351–357.

Society for the Study of Chitin and Chitosan (1991) *Kichin, Kitosan Jikken Manyuaru (Manual for Chitin and Chitosan Experiments)*, Tokyo: Gihodo Shuppan. [*Chem. Abstr.* **117**, 107775].

Spicer, A. P., McDonald, J. A. (1998) Characterization and molecular evolution of a vertebrate hyaluronan synthase gene family, *J. Biol. Chem.* **273**, 1923–1932.

Spindler, K. D. (1997) Chitinase and chitosanase assays, in: *Chitin Handbook* (Muzzarelli, R. A. A., Peter, M. G., Eds.), Grottammare: Atec, 229–235.

Spindler, K. D., Funke-Höpfner, B. (1991) N-Acetyl-β-D-hexosaminidases of the brine shrimp *Artemia*: partial purification and characterization, *Z. Naturforsch.* **46C**, 781–788.

Spindler, K.-D., Spindler-Barth, M. (1996) Chitin degradation and synthesis in arthropods, in: *Chitin in Life Sciences* (Giraud-Guille, M. M., Ed.), Lyon: André, 41–52.

Spindler, K.-D., Spindler-Barth, M. (1999) Inhibitors of chitinases, *EXS* **87** (Chitin and Chitinases), 201–209.

Spindler, K. D., Spindler-Barth, M., Londershausen, M. (1990) Chitin metabolism: a target for drugs against parasites, *Parasitol. Res.* **76**, 283–288.

Spindler-Barth, M. (1993) Hormonal regulation of chitin metabolism in insect cell lines, in: *Chitin Enzymology* (Muzzarelli, R. A. A., Ed.), Lyon: European Chitin Society, 75–82.

Spindler-Barth, M. (1997) Quantitative determination of chitin biosynthesis, in: *Chitin Handbook* (Muzzarelli, R. A. A., Peter, M. G., Eds.), Grottammare: Atec, 331–335.

Stankiewicz, B. A., Briggs, D. E. G., Evershed, R. P., Flannery, M. B., Wuttke, M. (1997) Preservation of chitin in 25-million-year-old fossils, *Science* **276**, 1541–1543.

Stenberg, E., Wachter, R. (1996) Potential of the northern shrimp (*Pandalus borealis*) as a source of chitosan in Norway, *Adv. Chitin Sci.* **1**, 166–172.

Stern, J. S., Gades, M. D., Halsted, C. H. (2000) Chitosan does not block fat absorption in men fed a high fat diet, *Obesity Res.* **8**, B94.

Stevens, W. F. (2001) Proceedings, 8th ICCC, Yamaguchi, Japan, September 20–23, 2000, in press, and personal communication.

Stokke, B. T., Vårum, K. M., Holme, H. K., Hjerde, R. J. N., Smidsrød, O. (1995) Sequence specificities for lysozyme depolymerization of partially N-acetylated chitosans, *Can. J. Chem.* **73**, 1972–1981.

Stone, C. A., Wright, H., Clarke, T., Powell, R., Devaraj, V. S. (2000) Healing at skin graft donor sites dressed with chitosan, *Br. J. Plast. Surgery* **53**, 601–606.

Struszczyk, H. (1987) Microcrystalline chitosan. I. Preparation and properties of microcrystalline chitosan, *J. Appl. Polym. Sci.* **33**, 177–189.

Struszczyk, H. (1997a) Preparation of chitosan fibers, in: *Chitin Handbook* (Muzzarelli, R. A. A., Peter, M. G., Eds.), Grottammare: Atec, 437–440.

Struszczyk, H. (1997b) Coating of textile fibers with chitosan, in: *Chitin Handbook* (Muzzarelli, R. A. A., Peter, M. G., Eds.), Grottammare: Atec, 441–444.

Struszczyk, H., Kivekäs, O. (1992) Recent developments in microcrystalline chitosan applications, in: *Advances in Chitin and Chitosan* (Brine, C. J., Sandford, P. A., Zikakis, J. P., Eds.), London: Elsevier Applied Science, 549–555.

Struszczyk, M. H., Loth, F., Peter, M. G. (1997) Analysis of degree of deacetylation in chitosans from various sources, *Adv. Chitin Sci.* **2**, 71–77.

Struszczyk, M. H., Loth, F., Köhler, L. A., Peter, M. G. (1998) Characterization of chitosan, in: *Carbohydrates as Organic Raw Materials*, (Praznik, W., Huber, A., Eds.), Vienna: WUV Universitätsverlag, 110–117, vol. 4.

Struszczyk, M. H., Hahlweg, R., Peter, M. G. (2000) Comparative analysis of chitosans from insects and Crustacea, *Adv. Chitin Sci.* **4**, 40–49.

Sugamori, T., Iwase, H., Maeda, M., Inoue, Y., Kurosawa, H. (2000) Local hemostatic effects of microcrystalline partially deacetylated chitin hydrochloride, *J. Biomed. Mater. Res.* **49**, 225–232.

Sugimoto, M., Morimoto, M., Sashiwa, H., Saimoto, H., Shigemasa, Y. (1998) Preparation and characterization of water-soluble chitin and chitosan derivatives, *Carbohydr. Polym.* **36**, 49–59.

Suzuki, K., Mikami, T., Okawa, Y., Tokoro, A., Suzuki, S., Suzuki, M. (1986) Antitumor effect of hexa-N-acetylchitohexaose and chitohexaose, *Carbohydr. Res.* **151**, 403–408.

Suzuki, K., Taiyoji, M., Sugawara, N., Nikaidou, N., Henrissat, B., Watanabe, T. (1999) The third chitinase gene (ChiC) of *Serratia marcescens* 2170 and the relationship of its product to other bacterial chitinases, *Biochem. J.* **343**, 587–596.

Suzuki, S., Matsumoto, T., Tsukada, K., Aizawa, K., Suzuki, M. (1989) Antimetastatic effects of N-acetyl chitohexaose on mouse-bearing Lewis lung carcinoma, in: *Chitin and Chitosan* (Skjåk-Bræk, G., Anthonsen, T., Sandford, P., Eds.), London: Elsevier Applied Science, 707–711.

Suzuki, Y., Okamoto, Y., Morimoto, M., Sashiwa, H., Saimoto, H., Tanioka, S., Shigemasa, Y., Minami, S. (2000) Influence of physico-chemical properties of chitin and chitosan on complement activation, *Carbohydr. Polym.* **42**, 307–310.

Synstad, B., Gaseidnes, S., Vriend, G., Nielsen, J.-E., Eijsink, V. G. H. (2000) On the contribution of conserved acidic residues to catalytic activity of chitinase B from *Serratia marcescens*, *Adv. Chitin Sci.* **4**, 524–529.

Takai, M., Shimizu, Y., Hayashi, J., Uraki, Y., Tokura, S. (1989) NMR and X-ray studies of chitin and chitosan in solid state, in: *Chitin and Chitosan* (Skjåk-Bræk, G., Anthonsen, T., Sandford, P., Eds.), London: Elsevier Applied Science, 431–436.

Takiguchi, Y., Sakamaki, Y., Yamaguchi, T. (1999) Depolymerization of chitosan by ultrasonic irradiation, *Kichin, Kitosan Kenkyu* **5**, 75–79 [*Chem. Abstr.* **131**, 130184].

Tanabe, T., Kawase, T., Watanabe, T., Uchida, Y., Mitsutomi, M. (2000) Purification and characterization of a 49-kDa chitinase from *Streptomyces griseus* HUT 6037, *J. Biosci. Bioeng.* **89**, 27–32.

Tanioka, S., Matsui, Y., Irie, T., Tanigawa, T., Tanaka, Y., Shibata, H., Sawa, Y., Kono, Y. (1996) Oxidative depolymerization of chitosan by hydroxyl radical, *Biosci. Biotechnol. Biochem.* **60**, 2001–2004.

Technical-Insights (1998) Chitin and chitosan: an expanding range of markets await exploitation, 3rd edn. URL: www.wiley/com/technical insights/reporttocs/chitin3.html.

Tellam, R. L., Eisemann, C. (2000) Chitin is only a minor component of the peritrophic matrix from larvae of *Lucilia cuprina*, *Insect Biochem. Mol. Biol.* **30**, 1189–1201.

Tellam, R. L., Vuocolo, T., Johnson, S. E., Jarmey, J., Pearson, R. D. (2000) Insect chitin synthase – cDNA sequence, gene organization and expression, *Eur. J. Biochem.* **267**, 6025–6042.

Terbojevich, M., Cosani, A. (1997) Molecular weight determination of chitin and chitosan, in: *Chitin Handbook* (Muzzarelli, R. A. A., Peter, M. G., Eds.), Grottammare: Atec, 87–101.

Terbojevich, M., Carraro, C., Cosani, A. (1988) Solution studies of the chitin-lithium chloride-N,N-di-methylacetamide system, *Carbohydr. Res.* **180**, 73–86.

Terbojevich, M., Cosani, A., Bianchi, E., Marsano, E. (1996) Solution behavior of chitin in dimethylacetamide/lithium chloride, *Adv. Chitin Sci.* **1**, 333–339.

Terwissscha van Scheltinga, A. C. (1997) *Structure and mechanism of the plant chitinase hevamine*, PhD Thesis, University of Groningen.

Terwissscha van Scheltinga, A. C., Armand, S., Kalk, K. H., Isogai, A., Henrissat, B., Dijkstra, B. W. (1995) Stereochemistry of chitin hydrolysis by a plant chitinase/lysozyme and x-ray structure of a complex with allosamidin. Evidence for substrate assisted catalysis, *Biochemistry* **34**, 15619–15623.

Terwissscha van Scheltinga, A. C., Hennig, M., Dijkstra, B. W. (1996) The 1.8 Å resolution structure of hevamine, a plant chitinase/lysozyme, and analysis of the conserved sequence and structure motifs of glycosyl hydrolase family 18, *J. Mol. Biol.* **262**, 243–257.

Tews, I., Perrakis, A., Oppenheim, A., Dauter, Z., Wilson, K. S., Vorgias, C. E. (1996) Bacterial chitobiase structure provides insight into catalytic mechanism and the basis of Tay-Sachs disease, *Nature Struct. Biol.* **3**, 638–648.

Tews, I., Terwissscha van Scheltinga, A. C., Perrakis, A., Wilson, K. S., Dijkstra, B. W. (1997) Substrate-assisted catalysis unifies two families of chitinolytic enzymes, *J. Am. Chem. Soc.* **119**, 7954–7959.

Thanou, M., Verhoef, J. C., Florea, B. I., Junginger, H. E. (2000) Chitosan derivatives as intestinal penetration enhancers of the peptide drug buserelin *in vivo* and *in vitro*, *Adv. Chitin Sci.* **4**, 244–249.

Thom, E., Wadstein, J. (2000) Chitosan in weight reduction: results from a large scale consumer study, *Adv. Chitin Sci.* **4**, 229–232.

Tjoelker, L. W., Gosting, L., Frey, S., Hunter, C. L., Le Trong, H., Steiner, B., Brammer, H., Gray, P. M. (2000) Structural and functional definition of the human chitinase chitin-binding domain, *J. Biol. Chem.* **275**, 514–520.

Tokura, S., Nishi, N., Nishimura, S.-I., Ikeuchi, Y., Azuma, I., Nishimura, K. (1984) Physicochemical, biochemical, and biological properties of chitin derivatives, in: *Chitin, Chitosan, and Related Enzymes* (Zikakis, J. P., Ed.), New York: Academic Press, 303–325.

Tokura, S., Ueno, K., Miyazaki, S., Nishi, N. (1997) Molecular weight-dependent antimicrobial activity by chitosan, *Macromol. Symp.* **120**, 1–9.

Tokuyasu, K., Ono, H., Hayashi, K., Mori, Y. (1999a) Reverse hydrolysis reaction of chitin deacetylase and enzymatic synthesis of β-D-GlcNAc-(1-4)-GlcN from chitobiose, *Carbohydr. Res.* **322**, 26–31.

Tokuyasu, K., Ono, H., Kitagawa, Y., Ohnishi-Kameyama, M., Hayashi, K., Mori, Y. (1999b) Selective N-deacetylation of p-nitrophenyl N,N'-diacetyl-β-chitobioside and its use to differentiate the action of two types of chitinases, *Carbohydr. Res.* **316**, 173–178.

Tokuyasu, K., Ono, H., Mitsutomi, M., Hayashi, K., Mori, Y. (2000) Synthesis of a chitosan tetramer derivative, β-D-GlcNAc-(1-4)-β-D-GlcNAc-(1-4)-β-D-GlcNAc-(1-4)-D-GlcN through a partial N-acetylation reaction by chitin deacetylase, *Carbohydr. Res.* **325**, 211–215.

Tolaimate, A., Desbrieres, J., Rhazi, M., Alagui, A., Vincendon, M., Vottero, P. (2000) On the influence of deacetylation process on the physicochemical characteristics of chitosan from squid chitin, *Polymer* **41**, 2463–2469.

Tsaih, M. L., Chen, R. H. (1997) Effect of molecular weight and urea on the conformation of chitosan molecules in dilute solutions, *Int. J. Biol. Macromol.* **20**, 233–240.

Tsaih, M. L., Chen, R. H. (1999) Effects of ionic strength and pH on the diffusion coefficients and conformation of chitosans molecule in solution, *J. Appl. Polym. Sci.* **73**, 2041–2050.

Tsigos, I., Martinou, A., Kafetzopoulos, D., Bouriotis, V. (2000) Chitin deacetylases: new, versatile tools in biotechnology, *Trends Biotechnol.* **18**, 305–312.

Tsujibo, H., Hatano, N., Endo, H., Miyamoto, K., Inamori, Y. (2000a) Purification and characterization of a thermostable chitinase from *Streptomyces thermoviolaceus* OPC-520 and cloning of the encoding gene, *Biosci. Biotechnol. Biochem.* **64**, 96–102.

Tsujibo, H., Miyamoto, J., Kondo, N., Miyamoto, K., Baba, N., Inamori, Y. (2000b) Molecular cloning of the gene encoding an outer-membrane-associated β-N-acetylglucosaminidase involved in chitin degradation system of *Alteromonas* sp. strain O-7, *Biosci. Biotechnol. Biochem.* **64**, 2512–2516.

Tsujibo, H., Okamoto, T., Hatano, N., Miyamoto, K., Watanabe, T., Mitsutomi, M., Inamori, Y. (2000c) Family 19 chitinases from *Streptomyces thermoviolaceus* OPC-520: molecular cloning and characterization, *Biosci. Biotechnol. Biochem.* **64**, 2445–2453.

Ueno, K., Yamaguchi, T., Sakairi, N., Nishi, N., Tokura, S. (1997) Antimicrobial activity by fractionated chitosan oligomers, *Adv. Chitin Sci.* **2**, 156–161.

van Aalten, D. M. F., Synstad, B., Brurberg, M. B., Hough, E., Riise, B. W., Eijsink, V. G. H., Wierenga, R. K. (2000) Structure of a two-domain chitotriosidase from *Serratia marcescens* at 1.9 Å resolution, *Proc. Natl. Acad. Sci. USA* **97**, 5842–5847.

Van Damme, E. J. M., Peumans, W. J., Barre, A., Rouge, P. (1998) Plant lectins: a composite of several distinct families of structurally and evolutionary related proteins with diverse biological roles, *Crit. Rev. Plant Sci.* **17**, 575–692.

Vander, P., Vårum, K. M., Domard, A., El Gueddari, N. E., Moerschbacher, B. M. (1998) Comparison of the ability of partially N-acetylated chitosans and chitooligosaccharides to elicit resistance reactions in wheat leaves, *Plant Physiol.* **118**, 1353–1359.

Varki, A. (1996) Does DG42 synthesize hyaluronan or chitin ?: a controversy about oligosaccharides in vertebrate development, *Proc. Natl. Acad. Sci. USA* **93**, 4523–4525.

Vårum, K. M., Smidsrød, O. (1997) Specificity in enzymic and chemical degradation of chitosans, *Adv. Chitin Sci.* **2**, 168–175.

Vårum, K. M., Anthonsen, M. W., Grasdalen, H., Smidsrød, O. (1991a) Determination of degree of N-acetylation and the distribution of N-acetyl groups in partially N-deacetylated chitins (chitosans) by high field NMR spectroscopy, *Carbohydr. Res.* **211**, 17–23.

Vårum, K. M., Anthonsen, M. W., Grasdalen, H., Smidsrød, O. (1991b) ^{13}C-NMR studies of the acetylation sequences in partially N-deacetylated chitins (chitosans), *Carbohydr. Res.* **217**, 19–27.

Vårum, K. M., Anthonsen, M. W., Nordtveit, R. J., Ottoy, M. H., Smidsrød, O. (1994) Structure–property relationships in chitosans, in: *Chitin World* (Karnicki, Z. S., Wojtasz-Pajak, A., Brzeski, M. M., Bykowski, P. J., Eds.), Bremerhaven: Wirtschaftsverlag NW, 166–174.

Vårum, K. M., Holme, H. K., Izume, M., Stokke, B. T., Smidsrød, O. (1996) Determination of enzymic hydrolysis specificity of partially N-acetylated chitosans, *Biochim. Biophys. Acta* **1291**, 5–15.

Veneroni, G., Veneroni, F., Contos, S., Tripodi, S., De Bernardi, M., Guarino, C., Marletta, M. (1996a) Effect of a new chitosan dietary integrator and hypocaloric diet on hyperlipidemia and overweight in obese patients, *Acta Toxicol. Ther.* **17**, 53–70.

Veneroni, G., Veneroni, F., Contos, S., Tripodi, S., De Bernardi, M., Guarino, C., Marletta, M. (1996b) Effect of a new chitosan on hyperlipidemia and overweight in obese patients, in: *Chitin Enzymology* (Muzzarelli, R. A. A., Ed.), Grottammare: Atec, 63–67, vol. 2.

Vincendon, M. (1997) Triphenylsilylchitin: a new chitin derivative soluble in organic solvents, *Adv. Chitin Sci.* **2**, 328–333.

Vincendon, M., Roux, J. C., Chanzy, H., Tanner, S., Belton, P. (1989) Solid-state CP/MAS ^{13}C NMR analysis of the crystalline chitin polymorphs, in: *Chitin and Chitosan* (Skjåk-Bræk, G., Anthonsen, T., Sandford, P., Eds.), London: Elsevier Applied Science, 437–438.

Vincent, J. F. V. (1990) *Structural Biomaterials*, Princeton: Princeton University Press.

Vinetz, J. M., Dave, S. K., Specht, C. A., Brameld, K. A., Xu, B., Hayward, R., Fidock, D. A. (1999) The chitinase PfCHT1 from the human malaria parasite *Plasmodium falciparum* lacks proenzyme and chitin-binding domains and displays unique substrate preferences, *Proc. Natl. Acad. Sci. USA* **96**, 14061–14066.

Vinetz, J. M., Valenzuela, J. G., Specht, C. A., Aravind, L., Langer, R. C., Ribeiro, J. M. C., Kaslow, D. C. (2000) Chitinases of the avian malaria parasite *Plasmodium gallinaceum*, a class of enzymes necessary for parasite invasion of the mosquito midgut, *J. Biol. Chem.* **275**, 10331–10341.

Vogt, W., Bachen, M., Gaumann, A., Jacob, B., Laue, C., Pommersheim, R., Schrezenmeir, J. (1996) Immobilization of enzymes and living cells by multilayer microcapsules, *DECHEMA Monogr.* **132**, 195–204.

Voigt, A., Lichtenfeld, H., Sukhorukov, G. B., Zastrow, H., Donath, E., Baeumler, H., Moehwald, H. (1999) Membrane filtration for microencapsulation and microcapsules fabrication by layer-by-layer polyelectrolyte adsorption, *Ind. Eng. Chem. Res.* **38**, 4037–4043.

Vorgias, C. E. (1997) Structural basis of chitin hydrolysis in bacteria, *Adv. Chitin Sci.* **2**, 176–187.

Wachter, R., Tesmann, H., Svenning, R., Olsen, R., Stenberg, E. (1995) Henkel KGaA, Germany, Kationische Biopolymere, EP 0737 211.

Watanabe, T., Kimura, K., Sumiya, T., Nikaidou, N., Suzuki, K., Suzuki, M., Taiyoji, M., Ferrer, S., Regue, M. (1997) Genetic analysis of the chitinase system of *Serratia marcescens* 2170, *J. Bacteriol.* **179**, 7111–7117.

Watanabe, T., Kanai, R., Kawase, T., Tanabe, T., Mitsutomi, M., Sakuda, S., Miyashita, K. (1999) Family 19 chitinases of *Streptomyces* species: characterization and distribution, *Microbiology UK* **145**, 3353–3363.

Wieczorek, A., Mucha, M. (1997) Application of chitin derivatives and their composites to biodegradable paper coatings, *Adv. Chitin Sci.* **2**, 890–896.

Win, N. N., Pengju, G., Stevens, W. F. (2000) Deacetylation of chitin by fungal enzymes, *Adv. Chitin Sci.* **4**, 55–62.

Wojtasz-Pajak, A., Brzeski, M. M., Malesa-Ciecwierz, M. (1994) Regulatory aspects of chitosan and its practical applications, in: *Chitin World* (Karnicki, Z. S., Wojtasz-Pajak, A., Brzeski, M. M., Bykowski, P. J., Eds.), Bremerhaven: Wirtschaftsverlag NW, 435–445.

Wright, J. E., Retnakaran, A. (Eds.) (1987) *Chitin and Benzoylphenyl Ureas*, Dordrecht: W. Junk.

Wu, Y., Adam, R., Williams, S. A., Bianco, A. E. (1996) Chitinase genes expressed by infective larvae of the filarial nematodes, *Acanthocheilonema viteae* and *Onchocerca volvulus*, *Mol. Biochem. Parasitol.* **75**, 207–219.

Wuolijoki, E., Hirvela, T., Ylitalo, P. (1999) Decrease in serum LDL cholesterol with microcrystalline chitosan, *Methods Find. Exp. Clin. Pharmacol.* **21**, 357–361.

Yalpani, M. (1997) Nutraceuticals. A materials-based perspective, *Front. Foods Food Ingredients* **2**, 53–70.

Yalpani, M., Pantaleone, D. (1994) An examination of the unusual susceptibilities of aminoglycans to enzymic hydrolysis, *Carbohydr. Res.* **256**, 159–175.

Yamagami, T., Ishiguro, M. (1998) Complete amino acid sequences of chitinase-1 and -2 from bulbs of genus *Tulipa*, *Biosci. Biotechnol. Biochem.* **62**, 1253–1257.

Yamagami, T., Tanigawa, M., Ishiguro, M., Funatsu, G. (1998) Complete amino acid sequence of chitinase-A from leaves of pokeweed (*Phytolacca americana*), *Biosci. Biotechnol. Biochem.* **62**, 825–828.

Yamano, N., Matsushita, Y., Kamada, Y., Fujishima, S., Arita, M. (1996) Purification and characterization of N-acetylglucosamine 6-phosphate deacetylase with activity against N-acetylglucosamine from *Vibrio cholerae* Non-O1, *Biosci. Biotechnol. Biochem.* **60**, 1320–1323.

Yoshida, M., Itano, N., Yamada, Y., Kimata, K. (2000) In vitro synthesis of hyaluronan by a single protein derived from mouse HAS1 gene and characterization of amino acid residues essential for the activity, *J. Biol. Chem.* **275**, 497–506.

Zecchi, V., Aiedeh, K., Orienti, I. (1997) Controlled drug delivery systems, in: *Chitin Handbook* (Muzzarelli, R. A. A., Peter, M. G., Eds.), Grottammare: Atec, 397–404.

Zhang, H., Du, Y. G., Yu, X. J., Mitsutomi, M., Aiba, S. (1999) Preparation of chitooligosaccharides from chitosan by a complex enzyme, *Carbohydr. Res.* **320**, 257–260.

Zhang, M., Inui, H., Hirano, S. (1997) A facile method for the preparation of 6-deoxy derivatives of chitin, *J. Carbohydr. Chem.* **16**, 673–679.

Zhang, M., Haga, A., Sekiguchi, H., Hirano, S. (2000) Structure of insect chitin isolated from beetle larva cuticle and silkworm (*Bombyx mori*) pupa exuvia, *Int. J. Biol. Macromol.* **27**, 99–105.

Zielinski, B. A., Aebischer, P. (1994) Chitosan as a matrix for mammalian cell encapsulation, *Biomaterials* **15**, 1049–1056.

5
Curdlan

Dr. In-Young Lee
Korea Research Institute of Bioscience and Biotechnology and DawMaJin Biotech Corp., P.O. Box 115, Daeduk Valley, Daejon 305-600, Korea; Tel.: +82-428620722; Fax: +82-428620702; E-mail: leeiy@dmj-biotech.com

1	**Introduction**	210
2	**Historical Outline**	210
3	**Structure**	211
3.1	Beta-1,3-glucan	211
3.2	Conformation in Solution	212
3.3	Gel Structure	213
4	**Occurrence**	213
5	**Biosynthesis**	215
6	**Molecular Genetics**	216
7	**Production**	217
7.1	Carbon Source	217
7.2	Nitrogen Effect	217
7.3	Oxygen Supply	217
7.4	Phosphate Effect	218
7.5	pH Effect	218
7.6	Batch Production	218
7.7	Continuous Production	219
7.8	Isolation Process	219
8	**Properties**	220
8.1	Gel Formation	220
8.2	Immunestimulatory Activity	221

Polysaccharides and Polyamides in the Food Industry. Properties, Production, and Patents.
Edited by A. Steinbüchel and S. K. Rhee
Copyright © 2005 WILEY-VCH Verlag GmbH & Co. KGaA, Weinheim
ISBN: 3-527-31345-1

9	Applications	221
9.1	Food Applications	221
9.2	Pharmaceutical Applications	222
9.3	Agricultural Applications	223
9.4	Other Industrial Applications	223
10	**Outlook and Perspectives**	223
11	**Patents**	226
12	**References**	228

AIDS	acquired immunodeficiency syndrome
AMP	adenosine 5'-monophosphate
ATCC	American Type Culture Collection
DMSO	dimethylsulfoxide
DP	degree of polymerization
FDA	Food and Drug Administration
HIV	human immunodeficiency virus
MNNG (also NTG)	N-methyl-N-nitro-N-nitrosoguanidine
NMR	nuclear magnetic resonance
TEM	transmission electron microscopy
UDP-glucose	uridine 5'-diphosphate glucose
UMP	uridine 5'-monophosphate

1
Introduction

Curdlan, an insoluble microbial exopolymer is composed almost exclusively of β-(1,3)-glucosidic linkages. One of the unique features of curdlan is that aqueous suspensions can be thermally induced to produce high-set gels, which will not return to the liquid state upon reheating (Harada et al., 1968), and this has attracted the attention of the food industry. In addition to this, curdlan offers many health benefits, as the beta-glucan family is well known among the scientific community to have immunestimulatory effects. Many informative reviews have been produced on this subject (Harada, 1977; Harada et al., 1993); however, apart from offering a brief review of β-(1,3)-glucan, the present chapter article provides an updated overview of the production, properties, and application of curdlan.

2
Historical Outline

Curdlan was discovered in 1966 by Professor Harada and coworkers, and given its name because of its ability to "curdle" when heated (Harada et al., 1966). At this time, Harada and his colleagues were working on the identification of organisms capable of utilizing petrochemical materials, and isolated *Alcaligenes faecalis* var. *myxogenes* 10C3 from soil. This organism was found to be capable of growing on a medium containing 10% ethylene glycol as the sole carbon source

(Harada et al., 1965), and also produced a new β-(1,3)-glucan that contained about 10% succinic acid, and which was named succinoglucan (Harada 1965; Harada and Yoshimura, 1965). They were also able to derive a spontaneous mutant that mainly produced a water-insoluble neutral polysaccharide, β-(1,3)-glucan, and which did not contain succinoglucan.

Scientists at Takeda Chemical Industries Ltd. (Osaka, Japan) have played a pioneering role in both the research and development of curdlan. Thus, as early as 1989, curdlan was approved and commercialized for food usage in Korea, Taiwan, and Japan. Upon obtaining approval in December 1996, Pureglucan™ – the tradename of curdlan – was launched in the US market as a formulation aid, processing aid, stabilizer, and thickener or texture modifier for food use (Spicer et al., 1999). No evidence of any toxicity nor carcinogenicity of Pureglucan has been observed.

3
Structure

3.1
Beta-1,3-glucan

Both chemical and enzymatic analyses have confirmed that curdlan is a homopolymer of D-glucose linked in β-(1,3) fashion (Saito et al., 1968) (Figure 1). Curdlan has an average degree of polymerization (DP) of approximately 450, and is unbranched (Naganishi et al., 1976). Nakata et al. (1998) reported that the average molecular weight of curdlan in 0.3 N NaOH is in the range of 5.3×10^4 to 2.0×10^6 daltons. Within the class of polysaccharides classified as β-(1,3)-glucans, there are a number of structural variants. The sources of glucans and their structural differences are listed in Table 1. Mycelial fungi are an abundant source of β-(1,3)-glucans; grifolan, which is produced from $Grifola\ frondosa$ and stimulates cytokine production from macrophages, is a β-(1,3)-glucan with a molecular weight of $> 4.5 \times 10^5$ daltons (Okazaki et al., 1995), while lentinan, from $Lentinus\ edodes$, has a molecular weight of 5×10^5 daltons and two glucose branches for every five β-(1,3)-glucosyl units in the backbone (Jong and Birmingham, 1993). The structure of schizophyllan is very similar to that of lentinan, but it has one glucose branch for every third glucose in the β-1,3-backbone and its molecular weight is 4.5×10^5 daltons (Misaki et al., 1993). Scleroglucan from $Sclerotium\ rolfsii$ has one glucose branch for every third glucose unit (Farina et al., 2001), whereas SSG from $Sclerotinia\ sclerotiorum$ is a highly branched β-(1,3)-glucan (Sakurai et al., 1991). Pachyman, from $Poria\ cocos$, has an average of 3.2 branch points per molecule of β-(1,3)-glucan (Okuyama et al., 1996; Zhang et al., 1997), whilst krestin is a protein-linked β-glucan with a molecular weight of c. 100,000 daltons, which can be extracted from the mycelia of $Coriolus\ versicolor$ (Azuma, 1987). β-(1,3)-Glucan is also present in the inner cell wall of the bakers' yeast $Saccha$-

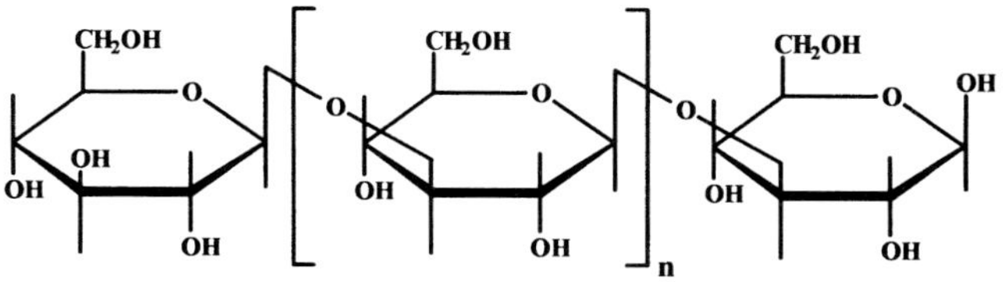

Fig. 1   Structure of curdlan.

Tab. 1 A variety of glucans having β-1,3 linkage in their backbones

Source	Branch	M_w	Reference
Bacteria			
Curdlan (*Agrobacterium sp. Alcaligens sp.*)	Exclusively β-(1,3)-glucosidic linkages	$5.3 \times 10^4 - 2.0 \times 10^6$	Nakata et al. (1998)
Fungi			
Grifolan (*Grifola frondosa*)	Branched β-1,3-gulcan	4.5×10^5	Okazaki et al. (1995)
Lentinan (*Lentinus eeodes*)	Two glucose branches for every five glucose unit	5×10^5	Jong and Birmingham (1993)
Schizophyllan (*Schizophyllum commune*)	One glucose branch for every third glucose unit	4.5×10^5	Misaki et al. (1993)
Scleroglucan (*Sclerotium glucanum*)	One glucose branch for every third glucose unit	$1.6 - 5.0 \times 10^6$	Farina et al. (2001)
SSG (*Sclerotinia sclerotiorum*)	Highly branched β-1,3-glucan	$2 \times 10^5 - 2 \times 10^6$	Okazaki et al. (1995)
Pachyman (*Poria cocos*)	Several β-1,6-linked branch points per molecule	2.06×10^4, 8.93×10^4	Zhang et al. (1997)
Krestin (*Coriolus versicolor*)	β-1,3-glucan	1.0×10^6	Azuma (1987)
Yeast (*Saccharomyces cerevisiae*)			
Soluble glucan	β(1,6) linkage to β-(1,3) backbone	$2 \times 10^5 - 2 \times 10^6$	Janusz et al. (1986)
Insoluble glucan	β-(1,6) linkage to β-(1,3) backbone	3.53×10^4, 4.57×10^6	Williams et al. (1994)
Brown algea			
Laminarin (*Laminaria digitata*)	β-1,3-glucan and β-1,6-glucan		Read et al. (1996)

romyces cerevisiae to support the structural strength of its cell wall. Unlike lentinan, schizophyllan and scleroglucan, the side branches of yeast β-(1,3)-glucans are chains of glucose molecules, and not single glucose residues. Depending on the extraction procedure used and their subsequent treatment, the yeast glucans may be either particulate water-insoluble or water-soluble macromolecules.

3.2
Conformation in Solution

Many researchers have investigated the molecular structures of curdlan in aqueous system. Three conformers of soluble curdlan have been reported, including single-helix, triple-helix, and random coil. Ogawa et al. (1972) studied the conformational behavior of curdlan in alkaline solution by measuring the optical rotatory dispersion, intrinsic viscosity and flow birefringence. At low concentrations of sodium hydroxide, curdlan has a helical (ordered) conformation, but a significant conformational change occurs at a NaOH concentration of 0.19–0.24 N NaOH. In alkaline solution >0.2 N NaOH, curdlan is completely soluble and exists as random coils, but upon neutralization the polymer adopts an 'ordered state', which is composed of a mixture of single and triple helices. A ^{13}C-NMR study supported this finding (Saito et al., 1977). Increasing the salt concentration shifts the point of conformational transition to a higher alkali concentration (Ogawa et al., 1973a), and addition of nonsolvents such as 2-chloroethanol, dioxan or water to dimethylsulfoxide (DMSO) solution also changes the conformation of curdlan to a rigid, ordered structure (Ogawa et al., 1973b). These workers also showed that the optical rotation was dependent upon the DP of the curdlan in 0.1 N sodium hydroxide (Ogawa et al., 1973c), and concluded that the content of the ordered form increases with DP until becoming constant at DP values of about 200. Electron microscopic comparison of the molecular structures of curdlan with different DPs showed that only curdlan with higher DP can form a gel when heated (Koreeda et al., 1974).

The conversion between triple-helix and single-helix conformers is mediated by different chemical or physical treatments. Treatment of the triple-helix schizophyllan with NaOH has been used to prepare single helix-rich forms (Ohno et al., 1995). Young et al. (2000) proposed a transition mechanism after an investigation using fluorescence resonance energy transfer spectroscopy, which showed that a partially opened triple-helix conformer was formed on treatment with NaOH, and that increasing degrees of strand opening were associated with increasing concentrations of NaOH. After neutralizing the NaOH, the partially opened conformers gradually reverted to the triple-helix.

3.3
Gel Structure

Some clarification of the fine structure of dispersed molecules and networks is necessary to understand the viscoelastic properties of curdlan, which forms two distinct types of gel. For both gels – described as low-set and high-set gels – transmission electron microscopy (TEM) showed them to be composed of three curdlan molecules that are associated to form a triple helix. Tada et al. (1997) proposed a mechanism of formation of the low-set gel using static light-scattering measurements. The molecular associates are formed at a NaOH concentration of 0.01–0.1 N at 25 °C, and this association progresses with as the NaOH concentration decreases. Consequently, the average molecular weight for the molecular associate in 0.01 N NaOH is higher than that in 0.1 N NaOH aqueous solution. The molecular associates consist of a dense core and hydrophilic surface at low NaOH concentrations. By contrast, Kasai and Harada (1980) proposed an annealing model to form a high-set, resilient gel upon heating to >80 °C. This annealing is associated with the irreversible loss of water, and resulted in a more tightly coiled triple helix. The structure crystallizes as a triplex of right-handed, six-fold helical chains in a hexagonal unit cell with a fiber repeating length of 18.78 Å (Chuah et al., 1983). Further removal of water from this structure, by drying under vacuum, results in further tightening of the six-fold triple helix and a decrease in the fiber period to only 5.87 Å (Deslandes et al., 1980).

4
Occurrence

As described in Section 2, β-(1-3)-D-glucans are present in a variety of living systems, including fungi, yeasts, algae, bacteria and higher plants. However, until now only bacteria belonging to the *Alcaligenes* and *Agrobacterium* species have been reported to produce the linear β-(1,3)-glucan type of homopolymer, curdlan. Figure 2 illustrates

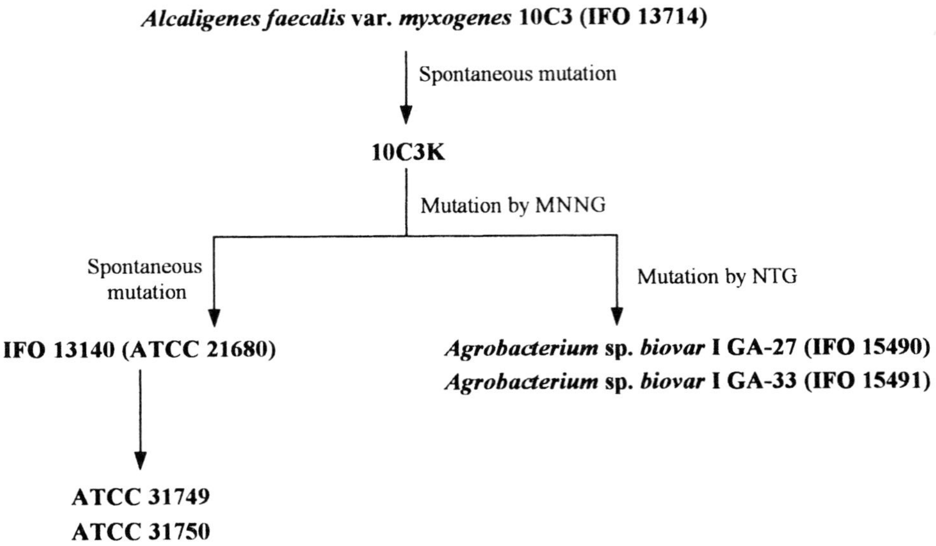

```
Alcaligenes faecalis var. myxogenes 10C3 (IFO 13714)
                    │
                    │ Spontaneous mutation
                    ▼
                  10C3K
                    │
                    │ Mutation by MNNG
        ┌───────────┴───────────┐
Spontaneous                      Mutation by NTG
mutation │                               │
         ▼                               ▼
IFO 13140 (ATCC 21680)    Agrobacterium sp. biovar I GA-27 (IFO 15490)
         │                Agrobacterium sp. biovar I GA-33 (IFO 15491)
         ▼
    ATCC 31749
    ATCC 31750
```

Fig. 2 Lineage of representative strains for curdlan production.

the lineage of the curdlan-producing strains since Harada et al. first isolated *Alcaligenes faecalis* var. *myxogenes* 10C3 during the screening of soil bacteria capable of metabolizing various petroleum fractions (Harada et al., 1965). The parent strain produced two different types of exopolysaccharide; a water-insoluble neutral homoglucan called 'curdlan', and a water-soluble acidic heteroglucan containing about 10% succinic acid, and referred to as 'succinoglucan' (Harada, 1965; Harada and Yoshimura 1965). Moreover, a mutant strain 10C3K was isolated from a stock culture of 10C3, which produced only curdlan. Strain 10C3K is a spontaneous mutant with a stable ability to produce exocellular polysaccharide. By inducing mutagenesis with *N*-methyl-*N*-nitro-*N*-nitrosoguanidine (MNNG), Takeda Chemical Industries Ltd. later isolated a uracil auxotrophic mutant from strain 10C3K, which was named *Alcaligenes faecalis* var. *myxogenes* IFO 13140 (ATCC 21680) and had improved gel-forming β-(1,3)-glucan-producing ability. Phillips and Lawford (1983) isolated a mutant strain from strain ATCC 21680 in a nitrogen-limited chemostat culture (the accession number was ATCC 31749). Unlike its auxotrophic parent strain, ATCC 31749 does not require uracil for its growth, and is not a revertant as it can be distinguished from 10C3K by its inability to hydrolyze starch and its ability to grow on citric acid as sole carbon source. ATCC 31750, which arose as a spontaneous variant of the parent ATCC 31749, produces only the water-insoluble curdlan-type glucan, while the parent strain produces both soluble and insoluble polysaccharides. All of these strains, which were formerly regarded as *Alcaligenes* species, have now been taxonomically reclassified as *Agrobacterium* species (IFO Research Communications, Vol. 15, pp. 57–75 (1991)). Takeda Chemical Industries Ltd. derived further mutant strains from strain 10C3K, which reduced the activity of the enzyme, phosphoenol pyruvic acid carboxykinase (Kanegae et al., 1996).

Naganishi et al. (1974) examined the occurrence of curdlan-type polysaccharides in microorganisms by using the water-soluble dye aniline blue, with which curdlan forms a blue complex. It was also shown that the rate of color complex formation was dependent both on the polymer concentration and DP; hence these findings provided an excellent tool for the screening of curdlan-producing bacteria. Naganishi et al. (1976) tested 687 strains of different genus of bacteria using the aniline blue staining technique. Among those examined, some strains of *Alcaligenes* and *Agrobacterium* species turned blue on agar plates containing aniline blue, and these have been used widely in the production of curdlan-type polysaccharides. Some strains of *Bacillus* formed blue complexes with aniline blue, but their polymeric constitution has not yet been studied.

5
Biosynthesis

Sutherland (1977, 1993) generalized the biosynthesis of extracellular polysaccharides into three major steps: (1) substrate uptake; (2) intracellular formation of polysaccharide; and (3) extrusion from cell. A metabolic pathway for exopolysaccharide biosynthesis is shown schematically in Figure 3. First, a carbohydrate substrate enters the cell by active transport and group translocation involving substrate phosphorylation. The substrate is then directed along either catabolic pathways, or those leading to polysaccharide synthesis. UDP-glucose, a key precursor, is synthesized by the UDP-glucose pyrophosphorylase-induced conversion of glucose-1-phosphate to UDP-glucose. Subsequently, polymer construction occurs together with the transfer of monosaccharides from UDP-glucose to a carrier lipid.

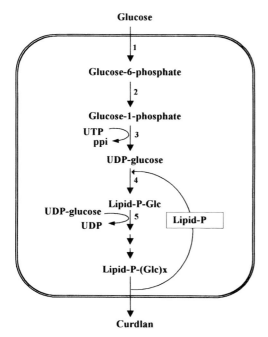

Fig. 3 Metabolic pathway for the synthesis of curdlan. 1, hexokinase; 2, phosphoglucomutase; 3, UDP-glucose pyrophosphorylase; 4, transferase; 5, polymerase. Lipid-P represents isoprenoid lipid phosphate.

After further chain elongation, the polymer is extruded from the cells.

Nitrogen-limited culture has also been employed for the production of curdlan, and has been generally explained by the roles of the carrier lipids (Sutherland, 1977; Ielpi et al., 1993). The availability of isoprenoid lipid may provide a way of regulating polysaccharide synthesis. Since curdlan biosynthesis takes place most extensively after cell growth has been stopped due to nitrogen exhaustion, isoprenoid lipids would be more available for carrying oligosaccharides instead of cellular liposaccharide and peptidoglycan.

The synthesis of precursor molecules is also of considerable importance in polysaccharide synthesis, in terms of the metabolic driving force. UDP-glucose serves as an activated precursor for glycosyl moieties in the synthesis of curdlan, a homopolysaccharide composed exclusively of β-1,3-linked glucose residues. In addition, cellular nucleotides not only play an important role in the synthesis of sugar nucleotides, but also have a widespread regulatory potential in cellular metabolism. Kim et al. (1999) examined the change of intracellular nucleotide levels and their stimulatory effects on curdlan synthesis in *Agrobacterium* species under different culture conditions. Under nitrogen-limited conditions where curdlan synthesis was stimulated, intracellular levels of UMP and AMP were at least twice as high as those occurring under nitrogen-sufficient conditions, though UDP-glucose levels were similar. The time profiles of curdlan synthesis and cellular nucleotide levels showed that curdlan synthesis is positively related with intracellular levels of UMP and AMP. *In vitro* enzyme reactions involved in the synthesis of UDP-glucose showed that a higher UMP concentration promotes the synthesis of UDP-glucose, while AMP neither inhibits nor facilitates the activity of UDP-glucose pyrophosphorylase. The addition of UMP to the medium also increased curdlan synthesis. From these results, these workers concluded that the higher intracellular UMP levels caused by nitrogen limitation enhance the metabolic flux of curdlan synthesis by promoting cellular UDP-glucose synthesis.

Kai et al. (1993) reported a study of the biosynthetic pathway of curdlan using ^{13}C-labeled glucose in *Agrobacterium* sp. ATCC 31749. By analyzing the labeled products, the biosynthesis of curdlan was interpreted as involving five routes: direct polymerization from glucose; rearrangement; isomerization of cleaved trioses; from fructose-6-phosphate; and from fructose fragments produced in various pathways of glycolysis. However, it was noted that more than 60% of curdlan is synthesized by direct polymerization, and that curdlan biosynthesis via glycolysis is comparatively low. This analysis also indicated that glycolysis occurs mainly via the pentose cycle and the Entner–Doudoroff pathway rather than the Embden–Meyerhof pathway.

6
Molecular Genetics

Little is known about the molecular genetics of bacterial curdlan biosynthesis, while there is growing information about the genes required for β-(1,3)-glucan synthesis in yeasts and filamentous fungi. Recently, Stasinopoulos et al. (1999) cloned genes that were essential for the production of curdlan and, by using comparative sequence analysis, identified them as putative curdlan synthase genes. Further genetic investigations will open up new avenues for curdlan synthesis, and these will doubtlessly be exploited to produce curdlan in higher yields.

7 Production

7.1 Carbon Source

Many crucial factors that affect curdlan production, including carbon, nitrogen, phosphate, oxygen supply, and pH have been investigated. High productivity using cheap carbon sources is important for the industrial production of curdlan. Lee, I.-Y. et al. (1997) reported that maltose and sucrose were efficient carbon sources for the production of curdlan by a strain of *Agrobacterium* species, with maximal production (60 g L^{-1}) being obtained from sucrose, with a productivity of 0.5 g L^{-1} h^{-1} when nitrogen was limited at a cell concentration of 16 g L^{-1}. Molasses, which contains large amounts of sucrose, might also be the substrate of choice, with up to 42 g L^{-1} of curdlan, at a yield of 0.35 g curdlan per gram total sugar, being obtained in 5-day cultivation. Sucrose is a less expensive substrate than glucose, and as sugar beet or sugar cane molasses are cheap byproducts widely available from the sugar industry, they are very attractive carbon sources from an economic point of view.

7.2 Nitrogen Effect

As previously described, relatively few strains of *Agrobacterium* and *Alcaligenes* species are known to produce curdlan. In such strains, curdlan production is associated with the poststationary phase of nitrogen depletion; thus, the operation involves an initial production of biomass, which is followed by curdlan production. Therefore, it is important to determine the initial concentration of the nitrogen source because it provides the limiting factor for cell growth during batch fermentation. Kim, M.-K. et al. (2000) reported that the cell growth rate decreased as the ammonium concentration increased. However, since higher cell concentrations produce more curdlan, an optimal ammonium concentration should be determined to provide an appropriate cell concentration while minimizing the inhibitory effect of the ammonium ion.

7.3 Oxygen Supply

Curdlan-producing stains are highly aerobic, and an adequate oxygen supply is therefore a key factor in production. Since curdlan is insoluble in water, the fermentation broth is of relatively low viscosity, and there is little resistance to oxygen transfer from gas to the liquid. However, a layer of insoluble exopolymer surrounding the cell mass offers resistance to oxygen transfer from the liquid into the cell, and therefore a high dissolved oxygen concentration is required for maximal productivity. Shake-flask fermentation results have shown that the specific production rate decreases as the volume of medium is increased, indicating that these cultures are limited by the relatively low oxygen transfer capacity of the system. Several investigations have been made into developing the process for curdlan production, especially with respect to reactor design (Lawford et al., 1986; Lawford and Rousseau, 1991, 1992). These workers employed two different types of impeller: a radial-flow, flat-blade impeller; and an axial-flow impeller. The radial-flow impeller was effective at providing high oxygen transfer rates to increase the production of curdlan, but the high shear characteristics of this design yielded a product of inferior quality in terms of tensile strength of the thermally induced gel. An axial-flow impeller typically produces less shear and more pumping. High volumetric oxygen transfer can be achieved by low shear designs equipped with sparging devices, which consist of microporous materials through which oxygen-enriched air is dispersed. The maximal specific production rate was 90 mg per g cells h^{-1} when 30% oxygen-enriched air was supplied in the low-shear system.

7.4
Phosphate Effect

Phosphate concentration must also be considered because it significantly influences cell growth and product formation. The production of rhamnose-containing polysaccharide by a *Klebsiella* strain was enhanced by a reduction in the phosphate content of the medium (Farres et al., 1997). In contrast, a sufficiency of phosphate resulted in good alginate yields in *Pseudomonas* strains and showed growth-associated production (Conti et al., 1994). Thus, the effect of phosphate on the production of polysaccharides is variable. Kim, M.-K. et al. (2000) investigated the influence of inorganic phosphate concentration on the production of curdlan by *Agrobacterium* species. Under nitrogen-limited conditions which allow curdlan production, the concentration of phosphate remains constant as it is not further utilized for cell growth. The optimal residual phosphate concentration for curdlan production was in the range 0.1–0.5 g L^{-1}. Relatively low concentrations appeared to be optimal for curdlan production, although without phosphate, curdlan production was extremely low. However, on increasing the cell phosphate concentration from 0.42 to 1.68 g L^{-1}, curdlan production increased from 0.44 to 2.80 g L^{-1}. Moreover, the optimal phosphate concentration range was not dependent upon cell concentration, and the specific production rate was about 70 mg curdlan per g cells h^{-1}, irrespective of cell concentration.

7.5
pH Effect

The pH of the culture is one of the most important factors because it significantly influences rates of both cell growth and product formation. High viscosity of the culture broth is often a critical problem in polysaccharide production. However, the viscosity problem can be obviated by operating the fermentation at a slightly acidic pH, since curdlan is insoluble under these conditions. Moreover, there appears to be more than one single optimal pH because fermentation of the culture for curdlan production is divided into two phases – the cell growth phase and the curdlan production phase. Lee, J.-H. et al. (1999) sought an optimal pH profile to maximize curdlan production in a batch fermentation of *Agrobacterium* species. The cell growth rate was maximal at pH 7.0, while curdlan production was maximal at pH 5.5. The pH profile provided a strategy to shift the culture pH from the optimal growth condition (pH 7.0) to the optimal production one (pH 5.5) at the time of ammonium exhaustion. By adopting the optimal pH profile in a batch process, these workers obtained a significant improvement in curdlan production (64 g L^{-1}) compared with that obtained using a constant-pH operation (36 g L^{-1}).

7.6
Batch Production

A high curdlan production was attempted by employing the optimal operation strategy with an *Agrobacterium* strain (Lee, I.-Y. et al., 1997, 1999a; Lee, J.-H. et al., 1999; Kim, M.-K. et al., 2000) that was tolerant of high concentrations of sucrose. This made batch operation possible, with an initial sucrose concentration of 140 g L^{-1}. An initial ammonium concentration 0.8 g L^{-1} was chosen, which produced 6.4 g L^{-1} of cells, the cell yield from ammonium being 8.0 g cells per g ammonium. A typical batch fermentation profile in a 300-L stirred tank reactor is shown in Figure 4. To produce a volumetric oxygen transfer coefficient of 146.5 h^{-1}, the agitation speed was set at 200 r.p.m. over the whole fermentation period, with an inner pressure of 0.2 bar. When the ammonium was exhausted in the culture broth at 20 h, the cell concentration had reached 6.8 g L^{-1}. The pH was controlled at 7.0 with 4 N NaOH/KOH at the cell growing stage. The culture pH was then shifted from 7.0 to 5.5 by adding 3 N HCl at the time of nitrogen limitation. Aeration rates were maintained at 0.5 vvm (volume volume^{-1} min^{-1}). Curdlan production began from the onset of nitrogen exhaustion, and a maximum concentration of 58 g L^{-1} was obtained in 120 h cultivation. The dissolved oxygen level fell rapidly during cell growth, and increased immediately after nitrogen exhaustion, as cell growth ceased. Subsequently, as the curdlan concentration increased, the dissolved oxygen level became limited due to the increased viscosity of the culture broth.

7.7
Continuous Production

Continuous production of curdlan was attempted in a two-stage continuous process (Phillips and Lawford, 1983). In the first stage, the curdlan-producing strain *Alcaligenes* sp. ATCC 31749 was grown aerobically in a medium containing carbon and nitrogen. However, the amount of nitrogen in the first stage was so limited that the effluent contained substantially no inorganic nitrogen. The effluent was fed into the second stage in a constant-volume fermenter, where it was mixed with a nitrogen-free medium in order to induce curdlan production without further cell growth. Curdlan production was 7 g L^{-1} at a dilution rate of 0.02 h^{-1} in a steady-state culture, providing a curdlan productivity of 0.14 g L^{-1} h^{-1}.

7.8
Isolation Process

The recovery procedure is based on the conformational transition which occurs when the concentration of alkali exceeds 0.2 N (Ogawa et al., 1972). Under alkaline conditions, the biomass can be separated from the dissolved curdlan, which remains in the supernatant. Upon neutralization of the alkaline supernatant, the polymer forms

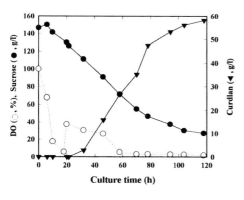
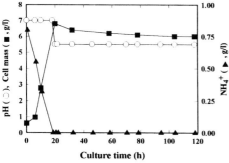

Fig. 4 Batch production of curdlan with *Agrobacterium sp.* ATCC 31750 in a 300-L jar fermenter (Lee, I.-Y. et al., 1999a).

an insoluble gel that can be recovered by centrifugation, the polymer being subsequently washed free of contaminating salt. Procedures for preserving curdlan in the dry state include both dehydration with organic solvents and spray drying.

8
Properties

8.1
Gel Formation

Curdlan is water-insoluble but soluble in alkali, and forms two-types of gels: high-set or low-set gels. The high-set gel is formed upon heating to around 80 °C, depending on its concentration in water (Maeda et al., 1967). The high-set curdlan gels are thermally irreversible and, unlike agar gel, further heating does not cause them to revert to the liquid state; they are also very elastic and resilient, whereas agar gels are brittle and relatively fragile. Curdlan forms a low-set reversible gel when its suspension is heated to 55–65 °C and then cooled. When a low-set gel is heated to 80 °C, however, it turns into a high-set gel. At the same curdlan concentration, a low-set gel has a weaker strength than a high-set gel. Changes in external factors such as pH, temperature, and ionic strength greatly affect gelation ability. Curdlan also forms a reversible gel when its alkaline solution is neutralized, and the addition of calcium or magnesium ions to a weakly alkaline solution of curdlan produces a gel with a bridged structure (Aizawa et al., 1974). Other methods to prepare curdlan gels include cooling DMSO solutions, or dialyzing its alkaline or DMSO solutions in water (Ogawa et al., 1972). The strength of the curdlan gel increases with temperature, concentration, and heating time (Maeda et al., 1967; Takeda technical report, 1997) (Figure 5). Curdlan, unlike other gelling agents, has a unique ability to form gels over a wide range of pH values (3.0–10.0).

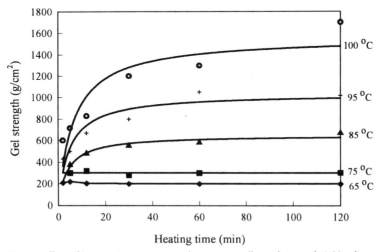

Fig. 5 Effect of heating temperature and time on curdlan gel strength (2% gel) (Takeda technical report, 1997).

8.2
Immunestimulatory Activity

β-(1,3)-Glucan has been found to be effective in stimulating macrophages (leukocytes or white blood cells). Macrophages form the immune system's first line of defense against foreign invaders, and can recognize and kill tumor cells, remove foreign debris resulting from oxidative and radiation damage, speed up the recovery of damage tissue, further activate cytokines, and initiate an immune cascade system to mobilize B and T cells (Di Luzio et al., 1979; Di Luzio, 1983; Suzuki et al., 1992; Hadden, 1993). Curdlan is a bacterial polysaccharide composed entirely of β-(1,3)-D-glucosidic linkages, and has been reported to show antitumor activity (Sasaki and Takasuka, 1976; Yoshioka et al., 1985). However, differences in biological activities seem to be dependent upon the degree of branching, molecular conformation, and molecular weight. Sasaki et al. (1978) examined the effect of chain length on antitumor activity by using different molecular weight glucans, which were obtained by the acid hydrolysis of curdlan. The results show that insoluble glucans with a number-average DP >50 have strong antitumor activity. As described in Section 2, β-glucans exist in three conformers: single-helix, triple-helix, and random coil. Among these, the ordered (helical) conformations are considered to be biologically active forms (Bohn and BeMiller, 1995). Some researchers reported that the triple-helical structure of schizophyllan, a β-glucan obtained from *Schizophyllum commune*, is essential for its anti-tumor activity (Yanaki et al., 1983; Kojima et al., 1986). However, other studies have suggested that the single-helix is more potent in this respect (Saito et al., 1991; Aketagawa et al., 1993). Ohno et al. (1995, 1996) showed that both the triple- and single-helix conformers of schizopyllan are active against tumors and leukopenia. In addition, the triple-helix conformer had a significant antagonistic effect upon zymosan-mediated hydrogen peroxide synthesis in peritoneal macrophages, whereas the single-helix conformer showed strong activity on the synthesis of tumor necrosis factor, nitric oxide, and hydrogen peroxide. It remains unclear as to which of the helical conformers is the most active, but it is likely that the structure–activity relationship of the β-glucan-mediated immunopharmacological activities vary, and are dependent upon the assay systems adopted.

9
Applications

9.1
Food Applications

Since the time of curdlan's original discovery, it has been mainly proposed for use in the food industry as a possible extender or substitute for natural plant gums in food preparations requiring a thickening or bodying additive. The food applications of curdlan are summarized in Table 2. These food additives and essential ingredients are generally divided in terms of how much curdlan is used; food additives require less than 1% curdlan, and essential ingredients more than 1%. Curdlan is useful as a gelling material to improve the textural quality, water-holding capacity and thermal stability of various foods. The polymer can be added during the production process, before heating, either as a powder, or as a suspension or slurry in either water or aqueous alcohol. The foods employing these functions include soy-bean curd (tofu), sweet bean paste jelly, boiled fish paste, noodles, sausages, jellies, and jams (Masayuki and Yukihiro, 1990; Taguchi et al., 1991; Ken and Akirou, 1992; Masanori et al., 1992; Shunsuke and Masatoshi, 1992; Hideaki et al., 1993; Hiroki and Etsuko, 1993; Masahiro and Masaru, 1993; Masatoshi et al., 1993; Yukihiro and Takahiro, 1993; Yasuhiro, 1994; Masaaki and Takaaki, 1995; Masataka et al., 1997; Akihiro and Takaaki, 1998; Hiroshi et al., 1999; Kazushi, 1999; Masao and Toshimi, 1999; Takaaki, 1999). Various tests have shown that this polymer is safe. The gel of curdlan has properties intermediate between the brittleness of agar gel and the elasticity of gelatin (Kimura et al., 1973). Furthermore, since the β-(1,3)-glucan polymer is not readily degraded by human digestive enzymes, it offers the possibility of new calorie-reduced products, such as those often referred to as "dietetic" foods. The fact that curdlan set-gels are known efficiently to absorb high concentrations of sugars from syrups and are relatively resistant to syneresis, suggests a use in sweet jellies and various other dessert-type foods. The resistance of curdlan gels to degradation by freezing and thawing also indicates potential in frozen food products. The pseudoplastic flow behavior of curdlan-containing fluids points to a possible role as thickeners and stabilizers in liquefied foods, such as salad dressings and spreads. Curdlan is also used to produce an edible casting film for foods with other water-soluble polymeric substances having heat-sealability (Akira and Atsushi, 1989).

9.2
Pharmaceutical Applications

Curdlan sulfate having a β-(1,3)-glucan backbone showed high anti-AIDS (acquired immunodeficiency syndrome) virus activity, with few adverse side effects; therefore, curdlan sulfate has a growing potential as an anti-AIDS drug (Jagodzinski et al., 1994; Takeda-Hirokawa et al., 1997). The Phase I/II trials (toxicity testing) of curdlan sulfate for human immunodeficiency virus (HIV) carriers have been carried out in the United States under the auspices of the Food and Drugs Administration (FDA) since 1992. A sulfuric acid ester of curdlan low molecular polymer, with an average DP ranging from 10 to 400 was found to be an active antiretrovirus ingredient useful for preventing and treating infection because of its high water solubility, low toxicity and relatively high inhibitory activity upon human retrovirus multiplication (Takashi and Junji, 1991). Mikio et al. (1995) disclosed that a

Tab. 2 Application of curdlan as food additives and essential ingredient (Spicer et al., 1999)

Function	Objective foods
Essential ingredient	
Gelling agent	Dessert, jelly, pudding, dry mixes
Bulking agent	Dietetic foods, diabetic foods
Low-calorie foods	Edible film, edible fiber casings
Food additive	
Improving viscoelasticity	Noodle, hamburger, sausage
Binding agent	Hamburger, starch jelly
Water-holding agent	Noodle, sausage, ham, starch jelly
Prevention of deterioration	Frozen egg products
Masking of malodors or aromas	Boiled rice
Retention of shape	Starch jelly, dry desert mixes
Thickeners and stabilizers	Salad dressing, low-calorie foods, frozen foods
Coating agent	Flavors

curdlan derivative modified by reaction with glycidol developed excellent antiviral activity whilst showing extremely low toxicity. Evans et al. (1998) subsequently reported that the low toxicity of curdlan and its marked anti-invasion activity on merozoites makes it a potential auxiliary treatment for severe malaria. Moreover, linear β-(1,3)-glucan sulfate has potential as a blood coagulation inhibitor useful for both the prevention and treatment of thrombosis, because linear β-(1,3)-glucan sulfate has been found to have a strong inhibitory action on blood coagulation (Tsuneo, 1990). Curdlan is also used as a sustained release suppository containing drug-active components, such as indomethacin, diclofenac sodium and ibuprofen (Toshiko, 1992). Kanke et al. (1992) investigated the *in vitro* release of a curdlan tablet containing theophylline; drug release from the tablet was the lowest tested, and was unaffected by pH and various other ions. Kim, B.-S. et al. (2000) proposed that hydroxyethyl derivatives of curdlan can be used as protein drug delivery vehicles. A curdlan hydrolysate with a number average molecular weight ranging from 340 to 4000 was used as an immunoactivator and did not cause any adverse side effects (Masahiro et al., 1998a). Hiroshi et al. (1997) have also reported that linear or branched β-(1,3)-glucans such as lentinan and curdlan are effective in the treatment of dementia.

9.3
Agricultural Applications

Fumio et al. (1989) reported that polysaccharides with the β-(1,3)-glucan structure, such as lentinan, schizophyllan, curdlan or laminarin are useful for the multiplication of *Bifidobacterium* bacteria and, when given to animals in the food, also inhibit the intestinal putrefying bacteria, thereby preventing the animals from aging. In addition, curdlan is incorporated into fish feed mixtures to improve immune activity (Masahiro et al., 1998b).

9.4
Other Industrial Applications

Curdlan has been studied as a support for immobilized enzymes (Murooka et al., 1977). For example, in the immobilization of an enzyme or a mold using curdlan gel, the curdlan is gelatinized by heating an aqueous suspension containing curdlan, an enzyme or a mold (Toshio, 1986). Other potential industrial applications include the use of curdlan films and fibers which are water-insoluble, biodegradable and impermeable to oxygen. Curdlan is also used to prevent nozzle clogging, and the electrostatic damage of electronic components when used as a seal in an ink-jet recoding head (Yoichi et al., 1996). Furthermore, curdlan functions as a concrete admixture to produce concrete with high fluidity but low separating properties (Toshiyuki et al., 1995; Lee, I.-Y. et al., 1999b). Applications of curdlan other than in the food industry are shown in Figure 6.

10
Outlook and Perspectives

Curdlan has been well accepted in the meat processing and noodle industries for its extremely unusual property of forming gels that are resistant to freezing or retorting. Takeda has been marketing this innovative product in Japan for several years, and sales are promising. Indeed, production has continued to grow by more than 10% each year during the past eight years such that, in recent years, Takeda Chemical Industries, Ltd. has produced 600–700 tonnes of curdlan annually.

Pharmaceutical applications

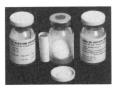
Anti-HIV agent
(www.iis.u-tokyo.ac.jp)

Immunoactivator
(Macrophage activation)

Drug delivery vehicle
(Curdlan gel)

Agricultural applications

Feed for domestic animals

Feed for fish animals

Plant fertilizer

Other industrial applications

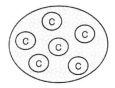
Immobilization support
(Gel entrapment)

Edible film

Concrete admixture
(Segregation reducing agent)

Fig. 6 Applications of curdlan in fields other than the food industry.

In addition, it should be noted that a variety of glucans with the β-1,3 linkage have immunestimulatory effects. Krestin, a β-glucan extracted from the mycelium of the basidiomycete *Coriolus versicolor*, has a general immunestimulatory effect in humans and has also been used as an anticancer agent (Kamisato and Nowakowski, 1988). Lentinan from *Lentinus edodes* is another glucan which inhibits tumor growth and increases resistance to infections by bacteria, viruses, and parasites (Jong and Birmingham, 1993). Schizophyllan (Numasaki et al., 1990) from *Schizophyllum commune* has the property of inducing the formation of cytotoxic macrophages and generating antitumor activity in the human body. In their preparation, these fungal glucans require extensive extraction operations to be performed in order to obtain comparatively small amounts of pure glucans, which makes the process very expensive. By contrast, curdlan as an exotype of β-(1,3)-glucan can be readily produced on a large scale by a submerged bacterial culture. Thus, if the material is correctly designed on a molecular

Tab. 3 Patent holders and publication year

Company	Year of publication													
	1988	1989	1990	1991	1992	1993	1994	1995	1996	1997	1998	1999	2000	Total
Ajinomoto	2		1							2				5
Asai							1							2
Chem Reizou	1													1
Dainippon	1	1							1	1	1			5
Daiichi		1												1
Daito										1				1
Endo Akira	1													1
Ezaki Glico														1
Fuji Oil	1				2	1						1		5
Hinoshoku												1	1	2
House foods	1								1					2
Hodaka						1								1
Ina						1					1		4	6
Japan Organo						2		2	2	7	2	2	2	19
Kanai Masako												1		1
Kanebo					2			1			1			4
Kanegafuchi				1			1				1			3
Kiteii					1									1
Kibun foods									1					1
Kyokuto				1										1
Meiji			2		1					1	1			5
Morigana												1		1
Nagai				1	1									2
Nikken									1					1
Nippon					2	1							1	4
Nitta Gelatin							1							1
Nokyo									1					1
Okada					2									2
Sanei		4		1			1		1					7
Sanyo	1													1
Shiseido							1		2					3
Snow Brand					1						2	1	2	6
Sumisho												1		1
Takeda		1	12	8	6	5	1	5	3	5	9	3	4	62
Taiyo Kagaku							1							1
Tsubuki								1						1
Tsukishima						1								1
Unie Colloid							2							2
Wako	5													5
Yamamoto											1			1
Total	13	7	15	12	18	12	8	13	9	19	19	11	15	171

11
Patents

The Japanese patent literature between 1988 and 2000 was reviewed, as most of the patents relating to the application of curdlan were first filed in Japan. The companies holding patents, together with the annual basis to enhance its immunestimulatory activity, there is a large potential market in both the nutraceutical and pharmaceutical industries. number of patents filed, are listed in Table 3. Takeda and Japan Organo each hold about one-half of total 171 patents filed, indicating their major role in the development of curdlan application. It should also be noted that 122 patents relate to the application of curdlan as food additives or ingredients. The patents described in Section 8 are listed in Table 4.

Acknowledgments

The author thanks Mi-Kyoung Kim for her help in the preparation of the manuscript.

Tab. 4 A list of the patents of curdlan application

Patent no.	Inventor(s)	Title	Year
JP-61015688	Toshio, O.	Immobilized enzyme, or the like and preparation thereof	1986
JP-01289457	Akira, K., Atsushi, I.	Edible film	1989
JP-01137990A	Fumio, I., Kimikazu, I., Satoshi, S.	Polysaccharide having activity for multiplying *Bifidobacterium*	1989
JP-02124902	Tsuneo, A., Koichi, K., Junji, K.	Blood coagulation inhibitor	1990
JP-02249466	Masayuki, T., Yukihiro, N.	Noodle made of rice powder and producing method thereof	1990
JP-03218317	Takashi, T., Junji, K.	Anti-retrovirus agent	1991
JP-03157401	Taguchi, T., Hiroshi, K., Yukihiro, N.	Production of linear gel	1991
JP-04210507	Masanori, T., Tetsuya, T., Yukihiro, N.	Noodles	1992
JP-04330256	Shunsuke, O., Masatoshi, S.	Production of uncooked Chinese noodle and boiled Chinese noodle	1992
JP-04197148	Ken, O., Akirou, M.	Soybean protein-containing solid food	1992
JP-04074115	Toshiko, S.	Sustained release suppository	1992
JP-05184310	Masatoshi, H., Masami, F., Kenichi, H., Hiromi, K.	Devil's tongue for frozen food and its production	1993
JP-05207859	Hideaki, Y., Chieko, M., Takeshi, A.	Ganmodoki and its production	1993
JP-05023143	Yukihiro, N., Takahiro, F.	Ground fish meat or fish or cattle paste product and composition therefor	1993
JP-05076290	Hiroki, O., Etsuko, M.	Noodle-like soybean protein-containing food	1993
JP-05288909	Masahiro, K., Masaru, N.	Preparation of processed edible meat	1993
JP-06335370	Yasuhiro, S.	Formed jelly-containing liquid food and its production	1994
JP-07010624	Toshiyuki, U., Yoshio, T., Minoru, Y.	Additive for concrete	1995

Tab. 4 (cont.)

Patent no.	Inventor(s)	Title	Year
JP-07274876	Masaaki, A., Takaaki, I.	Quality improver for wheat flour product and wheat flour product having improved texture	1995
JP-07228601	Mikio, K., Yoshiro, O., Hirotomo, O., Yoshikazu, K., Hiroshi, I., Hisashi, M., Takeshi, K.	Water soluble β-1,3-glucan derivative and antiviral agent containing the derivative	1995
JP-0818787	Yoichi, T., Hiroshi, S., Makiko, K.	Biodegradable component for protecting ink jet recording head	1996
USP-5508191	Kanegae, Y. Yutani, A., Nakatsui, I.	Mutant strains of *Agrobacterium* for producing beta-1,3-glucan	1996
JP-09075017	Masataka, M., Hideo, Y., Yukihiro, N.	Devil's-tongue jelly and its production	1997
Patent number	Inventors	Title	Year
JP-09255579	Hiroshi, S., Nobuyoshi, N., Yutaro, K., Toru, M.	Medicine for treating dementia	1997
JP-10194977	Masahiro, K., Masaaki, K., Shinji, M., Hiroaki, K.	Immunoactivator	1998a
JP-10313794	Masahiro, K., Mikitomo, A., Akitomo, A., Tomonori, Y.	Feed composition for fish kind cultivation	1998b
JP-10014541	Akihiro, S., Takaaki, I.	Quality improving agent for fishery paste product and production of fishery paste product	1998
JP-11187819	Hiroshi, Y., Yuichi, S., Norio, I., Nana, I.	Frozen dessert food and its production	1999
JP-11075726	Masao, K., Toshimi, T.	Jelly-like food	1999
JP-11178533	Kazushi, M.	Production of bean curd including ingredient	1999
JP-11056246	Takaaki, I.	Quality improver for bean jam product and production of bean jam product	1999

12
References

Aizawa, M., Takahashi, M., Suzuki, S. (1974) Gel formation of curdlan-type polysaccharide in DMSO-H_2O mixed solvents, *Chem. Lett.* 193–196.

Aketagawa, J., Tanaka, S., Tamura, H., Shibata, Y., Saito, H. (1993) Activation of *Limulus* coagulation factor G by several (1→3)-β-D-glucans: comparison of the potency of glucans with identical degree of polymerization but different conformations, *J. Biochem.* **113**, 683–686.

Akihiro, S., Takaaki, I. (1998) Quality improving agent for fishery paste product and production of fishery paste product. Japanese patent 10014541.

Akira, K., Atsushi, I. (1989) Edible film. Japanese patent 01289457.

Azuma, I. (1987) Development of immunostimulants in Japan, in: *Immunostimulants: Now and Tomorrow* (Azuma, I., Jolles, G., Eds.), Tokyo: Japan Sci. Soc. Press/Berlin: Springer-Verlag, 41–45.

Bohn, J. A., BeMiller, J. N. (1995) (1→3)-β-D-glucans as biological response modifiers: a review of structure–functional activity relationships, *Carbohydr. Res.* **28**, 3–14.

Chuah, C. T., Sarko, A., Deslandes, Y., Marchessault, R. H. (1983) Triple-helical crystalline structure of curdlan and paramylon hydrates, *Macromolecules* **16**, 1375–1382.

Conti, E., Flaibani, A., O'Regan, M., Sutherland, I. W. (1994) Alginate from *Pseudomonas fluorescens* and *P. putida*: production and properties, *Microbiology* **140**, 1125–1132.

Deslandes, Y., Marchessault, R. H., Sarko, A. (1980) Triple-helical structure of (1→3)-β-D-glucan, *Macromolecules* **13**, 1466–1471.

Di Luzio, N. R. (1983) Immunopharmacology of glucan: a broad spectrum enhancer of host defence mechanisms, *Trends Pharmacol. Sci.* **4**, 344–347.

Di Luzio, N. R., Williams, D. L., McNamee, R. B., Edwards, B. F., Kitahama, A. (1979) Comparative tumor-inhibitory and anti-bacterial activity of soluble and particulate glucan, *Int. J. Cancer* **24**, 773–779.

Evans, S. G., Morrison, D., Kaneko, Y., Havlik, I. (1998) The effect of curdlan sulfate on development *in vitro* of *Plasmodium falciparum*, *Trans. R. Soc. Trop. Med. Hyg.* **92**, 87–89.

Farina, J. I., Sineriz, F., Molina, O. E., Perotti, N. I. (2001) Isolation and physico-chemical characterization of soluble scleroglucan from *Sclerotium rolfsii*. Rheological properties, molecular weight and conformational characteristics, *Carbohydr. Res.* **44**, 41–50.

Farres, J., Caminal, G., Lopez-Santin, J. (1997) Influence of phosphate on rhamnose-containing exopolysaccharide rheology and production by *Klebsiella* I-714, *Appl. Microbiol. Biotechnol.* **48**, 522–527.

Fumio, I., Kimikazu, I., Satoshi, S. (1989) Polysaccharide having activity for multiplying *Bifidobacterium*. Japanese patent 01137990A.

Hadden, J. W. (1993) Immunostimulants, *Immunology Today* **14**, 275–280.

Harada, T. (1965) Succinoglucan 10C3: a new acidic polysaccharide of *Alcaligenes faecalis* var. *myxogenes*, *Arch. Biochem. Biophys.* **112**, 65–69.

Harada, T. (1977) Production, properties, and application of curdlan, in: *Extracellular Microbial Polysaccharides* (Sanford, P. A. Laskin, A., Eds.), Washington, DC: American Chemical Society, 265–283.

Harada, T., Yoshimura, T. (1965) Rheological properties of succinoglucan 10C3 from *Alcaligenes faecalis* var. *myxogenes*, *Agr. Biol. Chem.* **29**, 1027–1032.

Harada, T., Yoshimura, T., Hidaka, H., Koreeda, A. (1965) Production of a new acidic polysaccharide, succinoglucan by *Alcaligenes faecalis* var. *myxogenes*, *Agr. Biol. Chem.* **29**, 757–762.

Harada, T., Masada, M., Fujimori, K., Maeda, I. (1966) Production of a firm, resilient gel-forming polysaccharide by a mutant of *Alcaligenes faecalis* var. *myxogenes* 10C3, *Agr. Biol. Chem.* **30**, 196–198.

Harada, T., Misaki, A., Saito, H. (1968) Curdlan: a bacterial gel-forming β-1,3-glucan, *Arch. Biochem. Biophys.* **124**, 292–298.

Harada, T., Terasaki, M., Harada, A. (1993) Curdlan, in: *Industrial Gums* (Whistler, R. L., BeMiller, J. N., Eds.), San Diego, CA: Academic Press, Inc., 427–445.

Hideaki, Y., Chieko, M., Takeshi, A. (1993) Ganmodoki and its production. Japanese patent 05207859.

Hiroki, O., Etsuko, M. (1993) Noodle-like soybean protein-containing food. Japanese patent 05076290.

Hiroshi, S., Nobuyoshi, N., Yutaro, K., Toru, M. (1997) Medicine for treating dementia. Japanese patent 09255579.

Hiroshi, Y., Yuichi, S., Norio, I., Nana, I. (1999) Frozen dessert food and its production. Japanese patent 11187819.

Ielpi, L., Couso, R. O., Dankert, M. A. (1993) Sequential assembly and polymerization of the polyphenol-linked pentasaccharide repeating unit of the xanthan polysaccharide in *Xanthomonas campestris*, *J. Bacteriol.* **175**, 2490–2500.

Jagodzinski, P. P., Wiaderkiewicz, R., Kurzawski, G., Kloczewiak, M., Nakashima, H., Hyjek, E., Yamamoto, N., Uryu, T., Kaneko, Y., Posner, M. R., Kozbor, D. (1994) Mechanism of the inhibitory effect of curdlan sulfate on HIV-1 infection in vitro, *Virology* **202**, 735–745.

Janusz, M. J., Austen, K. F., Czop, J. K. (1986) Isolation of soluble yeast β-glucans that inhibit human monocyte phagocytosis mediated by β-glucan receptors, *J. Immunol.* **137**, 3270–3276.

Jong, S. C., Birmingham, J. M. (1993) Medicinal and therapeutic value of the Shiitake mushroom, *Adv. Appl. Microbiol.* **39**, 153–184.

Kai, A., Ishino, T., Arashida, T., Hatanaka, K., Akaike, Y., Matsuzaki, K., Kaneko, Y., Mimura, T. (1993) Biosynthesis of curdlan from culture media containing ^{13}C-labeled glucose as the carbon source, *Carbohydr. Res.* **240**, 153–159.

Kamisato, J. K., Nowakowski, M. (1988) Morphological and biochemical alterations of macrophages produced by a glucan, PSK, *Immunopharmacology* **16**, 88–96.

Kanegae, Y. Yutani, A., Nakatsui, I. (1996) Mutant strains of *Agrobacterium* for producing beta-1,3-glucan. US patent 5508191.

Kanke, M., Koda, K., Koda, Y., Katayama, H. (1992) Application of curdlan to controlled drug delivery. I. The preparation and evaluation of theophylline-containing curdlan tablets, *Pharmaceut. Res.* **9**, 414–418.

Kasai, N., Harada, T. (1980) Ultrastructure of curdlan. in: *Fiber Diffraction Methods*, (French, A. D., Gardner, K. H., Eds.), Washington, DC: American Chemical Society Symposium Series, 363–383.

Kazushi, M. (1999) Production of bean curd including ingredient. Japanese patent 11178533.

Ken, O., Akirou, M. (1992) Soybean protein-containing solid food. Japanese patent 04197148.

Kim, B.-S., Jung, I.-D., Kim, J.-S., Lee, J.-H., Lee, I.-Y., Lee, K.-B. (2000) Curdlan gels as protein drug delivery vehicles, *Biotechnol. Lett.* **22**, 1127–1130.

Kim, M.-K., Lee, I.-Y., Ko, J.-H., Rhee, Y.-H., Park, Y.-H., (1999) Higher intracellular levels of uridine monophosphate under nitrogen-limited conditions enhance metabolic flux of curdlan synthesis in *Agrobacterium* species, *Biotechnol. Bioeng.* **62**, 317–323.

Kim, M.-K., Lee, I.-Y., Lee, J.-H, Kim, K.-T., Rhee, Y.-H., Park, Y.-H. (2000) Residual phosphate concentration under nitrogen-limiting conditions regulates curdlan production in *Agrobacterium* species, *J. Ind. Microbiol. Biotechnol.* **25**, 180–183.

Kimura, H., Moritaka, S., Misaki, M. (1973) Polysaccharide 13140: a new thermo-gelable polysaccharide, *J. Food Sci.* **38**, 668–670.

Kojima, T., Tabata, K., Itoh, W., Yanaki, T. (1986) Molecular weight dependence of the antitumor activity of schizophyllan, *Agr. Biol. Chem.* **50**, 231–232.

Koreeda, A., Harada, T., Ogawa, K., Sato, S. Kasai, N. (1974) Study of the ultrastructure of gel-forming (1→3)-β-D-glucan (curdlan-type polysaccharide) by electron microscopy, *Carbohydr. Res.* **33**, 396–399.

Lawford, H. G., Rousseau, J. D. (1991) Bioreactor design considerations in the production of high-quality microbial exopolysaccharide, *Appl. Biochem. Biotechnol.* **28/29**, 667–684.

Lawford, H. G., Rousseau, J. D. (1992) Production of β-1,3-glucan exopolysaccharide in low shear systems, *Appl. Biochem. Biotechnol.* **34/35**, 597–612.

Lawford, H., Keenan, J., Phillips, K., Orts, W. (1986) Influence of bioreactor design on the rate and amount of curdlan-type exopolysaccharide production by *Alcaligenes faecalis*, *Biotechnol. Lett.* **8**, 145–150.

Lee, J.-H., Lee, I.-Y., Kim, M.-K., Park, Y.-H. (1999) Optimal pH control of batch processes for production of curdlan by *Agrobacterium* species, *J. Ind. Microbiol. Biotechnol.* **23**, 143–148.

Lee, I.-Y., Seo, W.-T., Kim, K.-G., Kim, M.-K., Park, C.-S., Park, Y.-H. (1997) Production of curdlan using sucrose or sugar cane molasses by two-step fed-batch cultivation of *Agrobacterium* species, *J. Ind. Microbiol. Biotechnol.* **18**, 255–259.

Lee, I.-Y, Kim, M.-K., Lee, J.-H., Seo, W.-T., Jung, J.-K., Lee, H.-W., Park, Y.-H., (1999a) Influence of agitation speed on production of curdlan by *Agrobacterium* species, *Bioprocess Eng.* **20**, 283–287.

Lee, I.-Y., Kim, S.-W., Lee, J.-H., Kim, M.-K., Cho, I.-S., Park, Y.-H. (1999b) A high viscosity of curdlan at alkaline pH increases segregational resistance of concrete, *Korean J. Biotechnol. Bioeng.* **14**, 1–5.

Maeda, I., Saito, H., Masda, M., Misaki, A., Harada, T. (1967) Properties of gels formed by heat treatment of curdlan, a bacterial β-1,3 glucan, *Agr. Biol. Chem.* **31**, 1184–1188.

Masaaki, A., Takaaki, I. (1995) Quality improver for wheat flour product and wheat flour product having improved texture. Japanese patent 07274876.

Masahiro, K., Masaru, N. (1993) Preparation of processed edible meat. Japanese patent 05288909.

Masahiro, K., Masaaki, K., Shinji, M., Hiroaki, K. (1998a) Immunoactivator. Japanese patent 10194977.

Masahiro, K., Mikitomo, A., Akitomo, A., Tomonori, Y. (1998b) Feed composition for fish kind cultivation. Japanese patent 10313794.

Masanori, T., Tetsuya, T., Yukihiro, N. (1992) Noodles. Japanese patent 04210507.

Masao, K., Toshimi, T. (1999) Jelly-like food. Japanese patent 11075726.

Masataka, M., Hideo, Y., Yukihiro, N. (1997) Devil's-tongue jelly and its production. Japanese patent 09075017.

Masatoshi, H., Masami, F., Kenichi, H., Hiromi, K. (1993) Devil's tongue for frozen food and its production. Japanese patent 05184310.

Masayuki, T., Yukihiro, N. (1990) Noodle made of rice powder and producing method thereof. Japanese patent 02249466.

Mikio, K., Yoshiro, O., Hirotomo, O., Yoshikazu, K., Hiroshi, I., Hisashi, M., Takeshi, K. (1995) Water soluble β-1,3-glucan derivative and antiviral agent containing the derivative. Japanese patent 07228601.

Misaki, A., Kishida, E., Kakuta, M., Tabata, K. (1993) Antitumor fungal (1→3)-β-D-glucans: structural diversity and effects of chemical modification. In: *Carbohydrates and Carbohydrate Polymers* (Yalpani, M., Ed.), Mount Prospect, IL: ATL Press, 116–129.

Murooka, Y., Yamada, T., Harada, T. (1977) Affinity chromatography of *Klebsiella* arylsulfatase on tyrosyl-hexamethylenediamine-β-1,3-glucan and immunoadsorbent, *Biochim. Biophys. Acta* **485**, 134–140.

Naganishi, I., Kimura, K., Kusui, S., Yamazaki, E. (1974) Complex formation of gel-forming bacterial (1→3)-β-D-glucan (curdlan type polysaccharide) with dyes in aqueous solution, *Carbohydr. Res.* **32**, 47–52.

Naganishi, I., Kimura, K., Suzuki, T., Ishikawa, M., Banno, I., Sakene, T., Harada, T. (1976) Demonstration of curdlan-type polysaccharide and some other β-1,3-glucan in microorganisms with aniline blue, *J. Gen. Appl. Microbiol.* **22**, 1–11.

Nakata, M., Kawaguchi, T., Kodama, Y., Konno, A. (1998) Characterization of curdlan in aqueous sodium hydroxide, *Polymer* **39**, 1475–1481.

Numasaki, Y., Kikuchi, M., Sugiyama, Y., Ohba, Y. (1990) A glucan Sizofiran: T cell adjuvant property and antitumor and cytotoxic macrophage inducing activities, *Oyo Yakuri Pharmacometrics* **39**, 39–48.

Ogawa, K., Watanabe, T., Tsurugi, J., Ono, S. (1972) Conformational behavior of a gel-forming (1→3)-β-D-glucan in alkaline solution, *Carbohydr. Res.* **23**, 399–405.

Ogawa, K., Tsurugi, J., Watanabe, T. (1973a) Effect of salt on the conformation of gel-forming β-1,3-D-glucan in alkaline solution, *Chem. Lett.* 95–98.

Ogawa, K., Miyagi, M., Fukumoto, T., Watanabe, T. (1973b) Effect of 2-chloroethanol, dioxane, or water on the conformation of a gel-forming β-1,3-glucan in DMSO, *Chem. Lett.* 943–946.

Ogawa, K., Tsurugi, J., Watanabe, T. (1973c) The dependence of the conformation of a (1→3)-β-D-glucan on chain length in alkaline solution, *Carbohydr. Res.* **29**, 397–403.

Ohno, N., Miura, N. N., Chiba, N., Adachi, Y., Yadomae, T. (1995) Comparison of the immunopharmacological activities of triple and single-helical schizophyllan in mice, *Biol. Pharm. Bull.* **18**, 1242–1247.

Ohno, N., Hashimoto, T., Adachi, Y., Yadomae, T. (1996) Conformation dependency of nitric oxide synthesis of murine peritoneal macrophages by β-glucans *in vitro*, *Immunol. Lett.* **52**, 1–7.

Okazaki, M., Adachi, Y., Ohno, N., Yadomae, T. (1995) Structure–activity relationship of (1→3)-β-D-glucans in the induction of cytokine production from macrophages, *in vitro*, *Biol. Pharm. Bull.* **18**, 1320–1327.

Okuyama, K., Obata, Y., Noguchi, K., Kusaba, T., Ito, Y., Ohno, S. (1996) Single structure of curdlan triacetate, *Biopolymers* **38**, 557–566.

Phillips, K. R., Lawford, H. G. (1983) Curdlan: Its properties and production in batch and continuous fermentations, in: *Progress Industrial Microbiology* (Bushell, D. E., Ed.), Amsterdam: Elsevier Scientific Publishing Co., 201–229.

Read, S. M., Currie, G., Bacic, A. (1996) Analysis of the structural heterogeneity of laminarin by electrospray-ionisation-mass spectrometry, *Carbohydr. Res.* **281**, 187–201.

Saito, H., Misaki, A., Harada, T. (1968) Comparison of structure of curdlan and pachyman, *Agr. Biol. Chem.* **32**, 1261–1269.

Saito, H., Ohki, T., Sasaki, T. (1977) A ^{13}C Nuclear Magnetic Resonance study of gel-forming $(1 \rightarrow 3)$-β-D-glucans. Evidence of the presence of single-helical conformation in a resilient gel of a curdlan-type polysaccharide 13140 from *Alcaligenes faecalis* var. *myxogenes* IFO 13140, *Biochemistry* **16**, 908–914.

Saito, H., Yoshioka, Y., Uehara, N., Aketagawa, J., Tanaka, S., Shibata, Y. (1991) Relationship between conformation and biological response for $(1 \rightarrow 3)$-β-D-glucans in the activation of coagulation factor G from limulus amebocyte lysate and host-mediated antitumor activity. Demonstration of single-helix conformation as a stimulant, *Carbohydr. Res.* **217**, 181–190.

Sakurai, T., Suzuki, I., Kinoshita, A., Oikawa, S., Masuda, A., Ohsawa, M., Tadomae, T. (1991) Effect of intraperitoneally administered β-1,3-glucan, SSG, obtained from *Sclerotinia sclerotiorum* IFO 9395 on the functions of murine alveolar macrophages, *Chem. Pharm. Bull.* **39**, 214–217.

Sasaki, T., Takasuka, N. (1976) Further study of the structure of lentinan, an anti-tumor polysaccharide from *Lentinus edodes*, *Carbohydr. Res.* **47**, 99–104.

Sasaki, T., Abiko, N., Sugino, Y., Nitta, K. (1978) Dependence on chain length of antitumor activity of $(1 \rightarrow 3)$-β-D-glucan from *Alcaligenes faecalis* var. *myxogenes*, IFO 13140, and its acid-degraded products, *Cancer Res.* **38**, 379–383.

Shunsuke, O., Masatoshi, S. (1992) Production of uncooked Chinese noodle and boiled Chinese noodle. Japanese patent 04330256.

Spicer, E. J. F., Goldenthal, E. I., Ikeda, T. (1999) A toxicological assessment of curdlan. *Fd. Chem. Toxicol.* **37**, 455–479.

Stasinopoulos, S. J., Fisher, P. R., Stone, B. A., Stanisich, V. A. (1999) Detection of two loci involved in $(1 \rightarrow 3)$-β-glucan (curdlan) biosynthesis by *Agrobacterium* sp. ATCC 31749, and comparative sequence analysis of the putative curdlan synthase gene, *Glycobiology* **9**, 21–31.

Sutherland, I. W. (1977) Microbial exopolysaccharide synthesis, in: *Extracellular Microbial Polysaccharides* (Sanford, P. A., Laskin, A., Eds.), Washington, DC: American Chemical Society, 40–57.

Sutherland, I. W. (1993) Biosynthesis of extracellular polysaccharides, in: *Industrial Gums* (Whistler, R. L., BeMiller, J. N., Eds.), San Diego, CA: Academic Press, Inc., 69–85.

Suzuki, T., Ohno, N., Saito, K., Yamdomae, T. (1992) Activation of the complement system by (1,3)-β-D-glucan having different degrees of branching and different ultrastructures, *J. Pharmacobiodyn.* **15**, 277–285.

Tada, T., Matsumoto, T., Masuda, T. (1997) Influence of alkaline concentration on molecular association structure and viscoelastic properties of curdlan aqueous systems, *Biopolymers* **42**, 479–487.

Taguchi, T., Hiroshi, K., Yukihiro, N. (1991) Production of linear gel. Japanese patent 03157401.

Takaaki, I. (1999) Quality improver for bean jam product and production of bean jam product. Japanese patent 11056246.

Takashi, T., Junji, K. (1991) Anti-retrovirus agent. Japanese patent 03218317.

Takeda technical report. (1997) Pureglucan: basic properties and food applications. Takeda Chemical Industries, Ltd. Japan.

Takeda-Hirokawa, N., Neoh, L. P., Akimoto, H., Kaneko, H., Hishikawa, T., Sekigawa, I., Hashimoto, H., Hirose, S.-I., Murakami, T., Yamamoto, N., Mimura, T., Kaneko, Y. (1997) Role of curdlan sulfate in the binding of HIV-1 gp120 to CD4 molecules and the production of gp120-mediated THF-α, *Microbiol. Immunol.* **41**, 741–745.

Toshio, O. (1986) Immobilized enzyme, or the like and preparation thereof. Japanese patent 61015688.

Toshiko, S. (1992) Sustained release suppository. Japanese patent 04074115.

Toshiyuki, U., Yoshio, T., Minoru, Y. (1995) Additive for concrete. Japanese patent 07010624.

Tsuneo, A., Koichi, K., Junji, K. (1990) Blood coagulation inhibitor. Japanese patent 02124902.

Williams, D. L., Pretus, H. A., Ensley, H. E., Browder, I. W. (1994) Molecular weight analysis of a water-insoluble, yeast-derived $(1 \rightarrow 3)$-β-D-glucan by organic-phase size-exclusion chromatography, *Carbohydr. Res.* **253**, 293–298.

Yanaki, T., Ito, W., Tabata, K., Kojima, T., Norysuye, T., Takano, N., Fujita, H. (1983) Correlation between the antitumor activity of a polysaccharide schizophyllan and its triple-helical conformation in dilute aqueous solution, *Biophys. Chem.* **17**, 337–342.

Yasuhiro, S. (1994) Formed jelly-containing liquid food and its production. Japanese patent 06335370.

Yoichi, T., Hiroshi, S., Makiko, K. (1996) Biodegradable component for protecting ink jet recording head. Japanese patent 0818787.

Yoshioka, Y., Tabeta, R., Saito, R., Uehara, N., Fukuoka, F. (1985) Antitumor polysaccharides from *P. ostreatus* (Fr.) Quel: isolation and structure of a beta-glucan, *Carbohydr. Res.* **140**, 93–100.

Young, S.-H., Dong, W.-J., Jacobs, R. R. (2000) Observation of a partially opened triple-helix conformation in $(1 \rightarrow 3)$-β-glucan by fluorescence resonance energy transfer spectroscopy, *J. Biol. Chem.* **275**, 11874–11879.

Yukihiro, N., Takahiro, F. (1993) Ground fish meat or fish or cattle paste product and composition therefor. Japanese patent 05023143.

Zhang, L., Ding, Q., Zhang, P., Zhu, R., Zhou, Y. (1997) Molecular weight and aggregation behaviour in solution of β-D-glucan from *Poria cocos* sclerotium, *Carbohydr. Res.* **303**, 193–197.

6
Dextran

Dr. Timothy D. Leathers
Fermentation Biochemistry Research Unit, National Center for Agricultural Utilization Research, Agricultural Research Service, United States Department of Agriculture;1815 N. University St.; Peoria, IL, 61604, USA; Tel. +1-309-681-6377; Fax: +1-309-681-6427; E-mail: leathetd@ncaur.usda.gov

1	Introduction	234
2	Historical Outline	234
3	Chemical Structure	235
4	Physiological Function	237
5	Chemical Analyses	237
6	Occurrence	237
7	Biosynthesis	238
8	Genetics and Molecular Biology	239
9	Biodegradation	240
10	Production	241
11	Properties and Applications	242
12	Patents	244
13	Outlook and Perspectives	246
14	References	247

Polysaccharides and Polyamides in the Food Industry. Properties, Production, and Patents.
Edited by A. Steinbüchel and S. K. Rhee
Copyright © 2005 WILEY-VCH Verlag GmbH & Co. KGaA, Weinheim
ISBN: 3-527-31345-1

ATP	adenosine 5′-triphosphate
CM	carboxymethyl
DEAE	diethylaminoethyl
EC	enzyme commission
FTIR	fourier-transform infrared
GRAS	Generally Regarded as Safe
HIV	human immunodeficiency virus
IU	international units
K_m	Michaelis-Menten constant
MRI	magnetic resonance imaging
NMR	nuclear magnetic resonance
QAE	diethyl(2-hydroxypropyl)aminoethyl
SP	sulfopropyl

Names are necessary to report factually on available data; however, the USDA neither guarantees nor warrants the standard of the product, and the use of the name by USDA implies no approval of the product to the exclusion of others that also may be suitable.

1
Introduction

Dextrans are defined as homopolysaccharides of glucose that feature a substantial number of consecutive α-(1→6) linkages in their major chains, usually more than 50% of total linkages. These α-D-glucans also possess side chains stemming from α-(1→2), α-(1→3), or α-(1→4) branch linkages. The exact structure of each type of dextran depends on its specific microbial strain of origin. Dextrans are produced by certain lactic acid bacteria, particularly strains of *Streptococcus* species and *Leuconostoc mesenteroides*. Dextrans from oral *Streptococcus* species are of clinical interest as components of dental plaque. Dextrans from *L. mesenteroides* are of commercial interest, primarily as specialty chemicals for clinical, pharmaceutical, research, and industrial uses. Numerous reviews have appeared on dextrans, including those by Evans and Hibbert (1946), Neely (1960), Jeanes (1966, 1978), Murphy and Whistler (1973), Sidebotham (1974), Walker (1978), Alsop (1983), Robyt (1986, 1992, 1995), de Belder (1990, 1993), and Cote and Ahlgren (1995).

2
Historical Outline

Because dextrans are formed from sucrose, they have long been known as troublesome contaminants of food products and sugar refineries. Pasteur made an early applied study of dextran formation in wine, proving that this phenomenon was caused by microbial activity (Pasteur, 1861). Scheibler (1874) determined that dextran was a carbohydrate of the empirical formula $(C_6H_{10}O_6)_n$ having a positive optical rotation and thus coined the term "dextran". Van Tieghem (1878) identified a dextran-forming bacterium and named it *Leuconostoc mesenteroides*. Beijerinck (1912) investigated the phenomenon, and Hehre (1941) demonstrated dextran synthesis by a cell-free culture filtrate. Scientific interest in dextrans was stimulated by studies suggesting its value as a blood-plasma volume expander (Gronwall and Ingelman, 1945, 1948). In the late 1940s, an extensive research program on dextrans was initiated at the Northern Regional Research Laboratory (now the National Center for Agricultural Utilization Research) of the Agricultural Research Service, U.S. Department of Agriculture, in Peoria, Illinois. Numerous dextran-producing strains were characterized, including *L. mesenteroides* strain NRRL B-512F used today for commercial dextran production in North America and Western Europe.

3
Chemical Structure

Within the general definition of dextran as a glucan in which α-(1→6) linkages predominate, chemical structures vary considerably as a function of the specific microbial strain of origin. The article of commerce is the product of a single strain of *L. mesenteroides*, NRRL B-512F. As shown in Figure 1, dextran from this strain features α-(1→6) linkages in the main chains with a relatively low level (about 5%) of α-(1→3) branch linkages (Van Cleve et al., 1956; Jeanes et al., 1954; Slodki et al., 1986). Larm et al. (1971) estimated that 40% of these side chains are one subunit long and 45% are two subunits long. The remaining side chains are probably greater than 30 subunits long, and branches appear to be distributed randomly (Bovery, 1959; Covacevich and Richards, 1977; Taylor et al., 1985; Kuge et al., 1987).

Dextrans are produced by numerous additional strains of bacteria, and the structures of these dextrans are diverse. In the classic study of Jeanes et al. (1954), 96 strains of *Leuconostoc* and *Streptococcus* were surveyed for the formation of polysaccharides from sucrose. In this study it was reported that α-(1→6) linkages in dextran varied from 50% to 97% of total linkages. The balance represented α-(1→2), α-(1→3), or α-(1→4) linkages, usually at branch points. Several isolates produced more than one type of polysaccharide, which were named on the basis of their greater (fraction S, for soluble) or lesser (fraction L) degree of solubility in water–ethanol mixtures.

Leuconostoc citreum strain NRRL B-742 (formerly *L. mesenteroides*; Takahashi et al., 1992) produces a fraction L dextran that contains approximately 15% α-(1→4) branch linkages and a fraction S dextran that contains a variable (30% to 45%) percentage of α-(1→3) branches (Jeanes et al., 1954; Seymour et al., 1979a; Côté and Robyt, 1983; Slodki et al., 1986). Dextran from *L. mesenteroides* strain NRRL B-1299 has a high percentage (27% to 35%) of α-(1→2) linked single-glucose branches (Jeanes et al., 1954; Kobayashi and Matsuda, 1977; Slodki et al., 1986). Soluble dextran produced by *Streptococcus sobrinus* strain 6715 (formerly *Streptococcus mutans* strain 6715) appears to be structurally similar to the S dextran from *L. citreum* strain NRRL B-742 in that it is an α-(1→6) glucan with a relatively high percentage of α-(1→3) branch linkages (Shimamura et al., 1982). Many strains of oral *Streptococcus* species, and some strains of *L. mesenteroides*, also make α-D-glucans containing linear sequences of consecutive α-(1→3) linkages. These low solubility glucans were once proposed to be "Class 3" dextrans (Seymour and Knapp, 1980) but now are considered to be forms of mutan, an important component of dental plaque.

L. mesenteroides strains NRRL B-1355, NRLL B-1498, and NRRL B-1501 produce an S fraction glucan with a unique backbone structure of regularly alternating α-(1→3) and α-(1→6) linkages (Côté and Robyt, 1982; Misaki et al., 1980; Seymour and

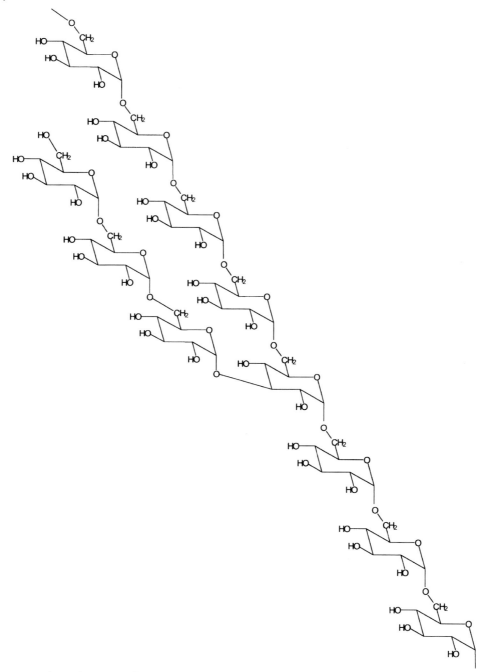

Fig. 1 Chemical structure of a representative portion of dextran from *Leuconostoc mesenteroides* strain NRRL B-512F. Figure courtesy of Dr. Gregory L. Côté.

Knapp, 1980). Because this polysaccharide does not contain significant regions of contiguous α-(1→6) linkages, it is not considered to be a true dextran and has been named alternan (Côté and Robyt, 1982). Alternan has distinctive properties of significant basic and applied interest, and it is the subject of Chapter 13, this volume. A mutant derivative of strain NRRL B-1355 recently was reported to produce a third polysaccharide, an insoluble α-D-glucan containing linear (1→3) and (1→6) linkages with (1→2) and (1→3) branch points (Smith et al., 1998; Côté et al., 1999).

4
Physiological Function

Oral *Streptococcus* species produce both dextran and mutan, an insoluble glucan in which α-(1→3) linkages predominate. These glucans comprise the matrix of dental plaque, which provides an environment for the proliferation of these bacteria (Hamada and Slade, 1980; Shimamura et al., 1982; Loesche, 1986). The physiological function of dextrans produced by *L. mesenteroides* is unknown. Since dextran-producing bacteria do not break down the polymers, dextrans presumably do not serve as storage materials. It is possible that these polysaccharides serve to protect cells from dessication or help them adhere to environmental substrates.

5
Chemical Analyses

General methods applicable to polysaccharides and polyglucans may be used to detect and quantitate dextrans. More specifically, dextrans show high positive specific optical rotation and exhibit characteristic infrared absorption bands at about 917, 840, and 768 cm^{-1} (Barker et al., 1956; Jeanes, 1966). Since dextrans are homopolysaccharides of glucose subunits, many studies have focused on the analysis of specific linkage patterns. Methods have included periodate analysis (Rankin and Jeanes, 1954; Dimler et al., 1955), methylation analysis (Van Cleve et al., 1956; Lindberg and Svensson, 1968; Seymour et al., 1977; Jeanes and Seymour, 1979; Seymour et al., 1979a; Slodki et al., 1986), nuclear magnetic resonance (NMR) spectroscopy (Seymour et al., 1976; Seymour et al., 1979b; Cheetham et al., 1991), and Fourier-transform infrared (FTIR) spectroscopy (Seymour and Julian, 1979). Enzymes that attack dextran in a specific fashion also have been exploited to reveal useful structural information (Covacevich and Richards, 1977; Sawai et al., 1978; Taylor et al., 1985; Pearce et al., 1990).

6
Occurrence

Several lactic acid bacteria have been reported to produce dextrans, principally including *Streptococcus* species and *L. mesenteroides* (Jeanes, 1966; Sidebotham, 1974; Cerning, 1990). *Leuconostoc* and *Streptococcus* are related genera, both composed of Gram-positive, facultatively anaerobic cocci. *L. mesenteroides* (incorporating as subspecies the former species *L. dextranicum* and *L. cremoris*) generally is found on plant materials, particularly on mature or harvested crops, and often plays a role in spoilage (Holzapfel and Schillinger, 1992; Stiles and Holzapfel, 1997). Because sucrose is the natural substrate for dextran synthesis, contamination problems are most evident in food products containing this sugar. *L. mesenteroides* can cause significant problems in sugar (particularly cane) refineries, where dextrans can clog filters and inhibit sugar crystallization (Jeanes, 1977). *Leuconostoc* species are considered Generally Regarded as Safe (GRAS) organisms because of their common appearance in natural fermented foods. In fact, certain strains are valued as starter cultures for buttermilk, cheese, and other dairy products (Holzapfel and Schillinger, 1992). *L. mesenteroides* also plays a role in the production of sauerkraut and other fermented vegetables. Recently, a strain of *L. mesenteroides* that produces both dextran and the related glucan alternan was isolated from the traditional fermented food Kim-Chi (Jung et al., 1999). Oral *Streptococcus* species produce dextran and the insoluble glucan mutan as components of dental plaque (Hamada and Slade, 1980; Shimamura et al., 1982). In addition, certain strains of the ubiquitous Gram-negative bacterium *Gluconobacter oxydans* (formerly *Acetobacter capsulatus*) produce dextran from starch-derived dextrins (Hehre and Hamilton, 1951; Kooi, 1958; Yamamoto et al., 1993a,b; Mountzouris et al., 1999).

7
Biosynthesis

Dextrans are produced extracellularly by secreted enzymes commonly referred to as glucansucrases or, more specifically, dextransucrases. These enzymes are glycosyltransferases (EC 2.4.1.5) that catalyze the transfer of D-glucopyranosyl subunits from sucrose to dextrans. Fructose is released and consumed by growing cells, if they are present. No adenosine 5′-triphosphate (ATP) or cofactors are required for these reactions, as the enzymes utilize energy available in the glycosidic bond between glucose and fructose. Glucansucrase synthesis in wild-type strains of *L. mesenteroides* is induced by growth on sucrose, while *Streptococcus* species produce this enzyme constitutively. Dextransucrase production by *L. mesenteroides* strain NRRL B-512F appears to be regulated at the transcriptional level (Quirasco et al., 1999).

Characterizations of dextransucrases have been complicated by the appearance of multiple enzyme species in culture fluids. Many of these species appear to be proteolytic-processing or degradation products, although some retain activity (Sanchez-Gonzalez et al., 1999). However, strains that produce more than one type of glucan appear to produce a separate glucansucrase for each. Furthermore, enzymes may exist as aggregates and typically are associated with their polysaccharide products, making purifications difficult. Despite these problems, a number of glucansucrases have been studied. Dextransucrase from *L. mesenteroides* strain NRRL B-512F has been purified and characterized (Robyt and Walseth, 1979; Kobayashi and Matsuda, 1980; Paul et al., 1984; Miller et al., 1986; Fu and Robyt, 1990; Kitaoka and Robyt, 1998a; Kim and Kim, 1999). The enzyme appears to have an initial molecular mass of 170 kDa, a pI value of 4.1, and a Michaelis-Menten constant (K_m) for sucrose of approximately 12 to 16 mM. Optimal reaction conditions are pH 5.0 to 5.5 and 30 °C. Kim and Kim (1999) reported a specific activity of up to 250 IU mg^{-1} protein for highly purified enzyme. Low levels of calcium are necessary for optimal enzyme production and activity. Various forms of dextransucrase have been described from *L. mesenteroides* strain NRRL B-1299, with molecular masses of 48 to 79 kDa, temperature optima of 35 °C to 45 °C, pH optima of 5.0 to 6.5, and K_m values for sucrose of 13 to 30 mM (Kobayashi and Matsuda, 1975, 1976; Dols et al., 1997). Multiple forms of dextransucrase have been described from *Streptococcus* species, having molecular masses of 94 to 170 kDa, pI values of approximately 4.0, temperature optima of

34 °C to 42 °C, pH optima of 5.0 to 5.7, and K_m values for sucrose of 2 to 9 mM (Chludzinski et al., 1974; Fukui et al., 1974; Shimamura et al., 1982; Furuta et al., 1985; McCabe, 1985). In addition, dextran-dextrinase (also called dextrin dextranase) has been purified and characterized from *G. oxydans* strain ATTC 11894 (Yamamoto et al., 1992; Suzuki et al., 1999).

Robyt and colleagues have developed a model for the reaction mechanism of dextransucrase from *L. mesenteroides* strain NRRL B-512F (Robyt et al., 1974; Robyt, 1992, 1995; Su and Robyt, 1994). According to this model, two nucleophilic reaction sites exist in the catalytic domain of the enzyme. Sucrose is hydrolyzed at one or both sites, and the glucosyl residues are bound covalently to the enzyme in high-energy bonds conserved from sucrose. A dextran chain grows by successive glucosyl insertions between the enzyme and the reducing end of the chain, which remains bound to the enzyme. Branches are formed when glucosyl units or dextran chains are transferred to secondary hydroxyl positions on the dextran chains (Robyt and Taniguchi, 1976). Termination of chain extension occurs by transfer to an acceptor molecule.

Sucrose is strongly preferred as the glucosyl donor, although other natural and synthetic donors have been identified (Hehre and Suzuki, 1966; Binder and Robyt, 1983). However, a number of sugars and derivatives may function as alternative acceptors, including maltose, isomaltose, nigerose, α-methyl glucoside, and others (Robyt and Taniguchi, 1976; Robyt and Walseth, 1978; Robyt and Eklund, 1983; Fu et al., 1990). These acceptor reactions can be utilized to produce dextrans of lower average molecular weights, including clinical dextrans (Koepsell et al., 1955; Tsuchiya et al., 1955; Remaud et al., 1991; Robyt, 1992) and oligosaccharides of interest (Pelenc et al., 1991; Remaud et al., 1992; Remaud-Simeon et al., 1994; Dols et al., 1999). Maltose, the most effective alternative acceptor, accepts a glucosyl residue to form the trisaccharide panose (Killey et al., 1955; Heincke et al., 1999). The acceptor reaction with fructose, the natural co-product of dextran synthesis, has been studied for production of the disaccharide leucrose, a potential alternative sweetener and substrate for industrial conversions (Stodola et al., 1956; Swengers, 1991; Reh et al., 1996; Heincke et al., 1999). A minor product of this reaction is isomaltulose, also known as palatinose, likewise of interest as an alternative sweetener (Sharpe et al., 1960; Takazoe, 1989).

Robyt and Martin (1983) found evidence that a similar reaction mechanism exists for glucansucrases from *S. sobrinus* strain 6715. Alternative models for the glucansucrase reaction mechanism have been reviewed (Monchois et al., 1999). Because the glucansucrases from *Leuconostoc* and *Streptococcus* species appear to be closely related on a molecular level, it seems likely that they share a common reaction mechanism. If so, differences among the polymer structures might be determined by subtle differences in the stereochemistry of the reaction sites.

8
Genetics and Molecular Biology

L. mesenteroides strain NRRL B-512F, used for commercial production of dextran, has been described as a laboratory "substrain" that supplanted natural isolate NRRL B-512 in 1950 (Van Cleve et al., 1956). A dextransucrase hyperproducer mutant of NRRL B-512F was isolated as NRRL B-512FM (Miller and Robyt, 1984). Wild-type strains of *L. mesenteroides* form glucansucrases only when cultured on sucrose, and further mutations were obtained that allowed strain

B-512FMC to produce dextransucrase constitutively (Kim and Robyt, 1994). Further improvements in enzyme productivity were obtained through additional rounds of mutagenesis (Kim et al., 1997; Kitaoka and Robyt, 1998b). Although commercial dextran currently is produced by a fermentative process, such strains would be particularly valuable for dextran production by an enzymatic process. Similar glucansucrase mutants have been obtained for other strains of *Leuconostoc*, including NRRL B-742, NRRL B-1142, NRRL B-1299, and NRRL B-1355 (Kim and Robyt, 1994, 1995a,b, 1996; Kitaoka and Robyt, 1998b). Dextransucrase-deficient mutants of strain NRRL B-1355 also have been isolated for improved production of alternan (Smith et al., 1994; Leathers et al., 1995, 1997, 1998). Alternan is the subject of Chapter 13, this volume.

Because of clinical interest in developing anti-caries vaccines, a number of glucansucrase genes have been cloned from oral *Streptococcus* species (Shiroza et al., 1987; Ueda et al., 1988; Honda et al., 1990). Glucansucrase genes have been cloned and sequenced from *L. mesenteroides* NRRL B-512F (Wilke-Douglas et al., 1989; Bhatnagar and Singh, 1999; Arguello-Morales et al., 2000a; Funane et al., 2000; Ryu et al., 2000), *L. mesenteroides* strain NRRL B-1299 (Monchois et al., 1996, 1998), *L. citreum* strain NRRL B-742 (Kim et al., 2000), and *L. mesenteroides* strain NRRL B-1355 (Arguello-Morales et al., 2000b; Kossman et al., 2000). Interestingly, some of these genes apparently do not specify enzymes normally secreted *in vivo*. Glucansucrase genes appear to be closely related and exhibit a common organizational structure, with a conserved N-terminal catalytic domain and a C-terminal glucan-binding domain that contains a series of direct tandem repeat sequences (Monchois et al., 1999; Remaud-Simeon et al., 2000). Based on site-directed mutageneses and consensus sequences, potentially important catalytic sites have been proposed (Monchois et al., 1997; Arguello-Morales, 2000b; Monchois et al., 2000; Remaud-Simeon et al., 2000). On a broader scale, glucansucrases resemble enzymes in glycosyl hydrolase family 13, which includes α-amylases (Fujiwara et al., 1998; Janecek et al., 2000; Remaud-Simeon et al., 2000).

9
Biodegradation

A variety of fungi produce dextranases, including *Aspergillus* species (Carlson and Carlson, 1955b; Hiraoka et al., 1972), *Chaetomium gracile* (Hattori et al., 1981), *Fusarium* species (Simonson and Liberta, 1975; Shimizu et al., 1998), *Lipomyces starkeyi* (Webb and Spencer-Martins, 1983; Koenig and Day, 1988), *Paecilomyces lilacinus* (Lee and Fox, 1985; Sun et al., 1988; Galvez-Mariscal and Lopez-Munguia, 1991), and *Penicillium* species (Tsuchiya et al., 1956; Chaiet et al., 1970). These enzymes are endodextranases with specificity for internal α-(1→6) linkages, and they produce mainly isomaltose or isomaltotriose from dextran. Dextranases from *C. gracile* and *Penicillium* sp. are produced commercially and used for treatment of dextran contamination problems in sugar processing (Godfrey, 1983). Endodextranases have shown potential for the enzymatic production of specific molecular weight fractions of dextran (Carlson and Carlson, 1955a; Corman and Tsuchiya, 1957; Novak and Stoycos, 1958; Day and Kim, 1992; Kim and Day, 1995; Kim and Robyt, 1996). These enzymes also have been tested for the treatment of dental plaque (Fitzgerald et al., 1968; Caldwell et al., 1971), although the more highly branched dextrans are far less susceptible to endodextranase digestion. Limit endodextranase digestion of the branched dextran from *L. citreum* strain NRRL B-742 produces an interesting branched fraction with rheological characteristics similar to those of polydextrose (Cote et al., 1997).

Dextranases also have been reported from a number of bacteria. *Arthrobacter globiformis* produces isomaltodextranase, an exodextranase that successively releases isomaltose from the non-reducing ends of dextrans and oligosaccharides (Torii et al., 1976; Okada et al., 1988). This enzyme recognizes not only α-(1 → 6) linkages but also α-(1 → 2), α-(1 → 3), and α-(1 → 4) linkages. Unlike endodextranases, isomaltodextranase is able to partially hydrolyze alternan, producing an interesting limit alternan (Sawai et al., 1978; Cote, 1992). An isomaltodextranase from *Actinomadura* sp. exhibits slightly different specificities (Sawai et al., 1981). A dextran α-(1 → 2) debranching enzyme also has been described from a *Flavobacterium* sp. (Mitsuishi et al., 1979). Dextran from *L. mesenteroides* strain NRRL B-512F is degraded by intestinal bacteria and enzymes in mammalian tissues other than blood (Sery and Hehre, 1956; Fischer and Stein, 1960). Intravenously administered clinical dextrans are metabolized slowly and completely in the body.

10
Production

To date, commercial production of dextran has employed primarily simple batch fermentation methods, using live cultures grown on sucrose. Methods and conditions for dextran fermentation have been detailed (Tarr and Hibbert, 1931; Hehre et al., 1959; Jeanes, 1965b, 1966; Alsop, 1983; de Belder, 1993). *L. mesenteroides* is a fastidious organism, and its special nutritional requirements include glutamic acid, valine, biotin, nicotinic acid, thiamine, and pantothenic acid (Holzapfel and Schillinger, 1992). In dextran production, these needs are met by combinations of complex medium components, such as yeast extract, corn steep liquor, casamino acids, malt extract, peptone, and tryptone. Sucrose serves as a carbon source, inducer of dextransucrase, and substrate for dextran production. Low levels of calcium (e.g., 0.005%) are necessary for optimal enzyme and dextran yields, and other basal salts, including a source of phosphate, complete the medium. Operative production factors include initial pH (typically pH 6.7 to 7.2), temperature (about 25 °C), initial sucrose concentration (usually 2%), and time (usually 24 to 48 h). Dextran branching appears to increase at elevated temperatures (Sabatie et al., 1988). High levels of sucrose (10% to 50%) reduce the yield of high-molecular-weight dextran, and this observation has been exploited for the production of intermediate sized dextran (Tsuchiya et al., 1955; Alsop, 1983). The organism is facultatively anaerobic or microaerophilic, and fermentations are not aerated. During the first 20 h of fermentation, culture pH falls to approximately 5.0 because of the formation of organic acids, favorably near the optimal pH of dextran sucrase. Dextran may be recovered by precipitation with solvents, particularly alcohols (Hehre et al., 1959; Jeanes, 1965b).

It has long been recognized that dextran also can be produced enzymatically, using cell-free culture supernatants that contain dextransucrase (Hehre, 1941; Tsuchiya and Koepsell, 1954; Hellman et al., 1955; Behrens and Ringpfeil, 1962; Jeanes, 1965a). Accordingly, improved dextransucrase production and purification methods have been developed (Lawford et al., 1979; Paul et al., 1984; Miller et al., 1986; Fu and Robyt, 1990; Kim and Kim, 1999). Glucansucrases also have been immobilized, although this approach may be most useful for production of oligosaccharides (Kaboli and Reilly, 1980; Monsan et al., 1987; Cote and Ahlgren, 1994; Reh et al., 1996; Alcalde et al., 1999). Enzymatic synthesis offers advantages of product molecular weight and quality control, as well as the benefit of obtaining fructose as a valuable co-product. However, this approach has been largely ignored for commercial production, presumably for economic reasons. Dextran production from maltodextrins, using dextran-dextrinase from *Gluconobacter oxydans*, also has attracted interest (Hehre and Hamilton, 1951; Kooi, 1958; Yamamoto et al., 1993a,b; Mountzouris et al., 1999).

Clinical dextran fractions are produced primarily by simple methods of partial acid hydrolysis followed by differential fractionation in solvents (Wolff et al., 1955; Gronwall, 1957; de Belder, 1990). Attractive alternative methods to produce these fractions include the use of dextranases (Carlson and Carlson, 1955a; Corman and Tsuchiya, 1957; Novak and Stoycos, 1958; Day and Kim, 1992; Kim and Day, 1995; Kim and Robyt, 1996) and chain-terminating acceptor reactions (Koepsell et al., 1955; Tsuchiya et al., 1955; Remaud et al., 1991; Robyt, 1992).

Dextran has been produced commercially for many years and by a number of companies, including Dextran Products, Ltd., Toronto, Canada; Pfeifer und Langen, Dormagen, Germany; Pharmachem Corp., Bethlehem, Pennsylvania, USA; and Pharmacia, Uppsala, Sweden. Annual world production was recently estimated at 2000 tons per year (Vandamme et al., 1996). The wholesale price of dextran varies, but recently it has been near 3 USD per pound.

11
Properties and Applications

Purified dextrans are white, tasteless solids. Other physical and chemical properties vary depending on the specific chemical structure, which is determined by the microbial strain of origin and method of production. Dextrans with the highest percentages of α-$(1 \rightarrow 6)$ linkages are generally the most soluble in water. Dextran from *L. mesenteroides* strain NRRL B-512F is freely soluble in water and other solvents, including 6 M urea, 2 M glycine, formamide, glycerol, etc. (Jeanes, 1966; de Belder, 1990). Dextran solutions behave as Newtonian fluids, and their viscosity is a function of concentration, temperature, and average molecular weight (Granath, 1958; Gekko and Noguchi, 1971; Carrasco et al., 1989). Native dextran is polydisperse and typically of high average molecular weight (generally between 10^6 and 10^9 daltons). However, many of the current applications for dextran depend on the convenience with which it can be broken down to fractions of specific weight ranges. The relative linearity of dextran from strain NRRL B-512F is crucial for the production of such fractions. Dextrans exhibit characteristic serological reactions, apparently related to their molecular weight and degree of branching (Gronwall, 1957; Kabat and Bezer, 1958; Jeanes, 1986). However, intravenously administered clinical dextrans are of relatively low antigenicity, although individuals can exhibit hypersensitivity. The pharmacological properties of clinical dextrans have been reviewed recently (de Belder, 1996). Free hydroxyl groups in dextran are potential targets for chemical derivatizations, and dextran from strain NRRL B-512F is particularly suitable for these reactions because of its low level of branch linkages.

A number of bulk chemical applications have been demonstrated for dextran, including uses in oil-drilling operations, agriculture, food products, and the manufacture of photographic films and other products (Murphy and Whistler, 1973; Alsop, 1983; Glicksman, 1983). It should be noted that dextrans are not explicitly approved as food additives in the United States or Europe, although *L. mesenteroides* is a GRAS organism commonly found in fermented foods. Currently, dextran and dextran derivatives are used primarily as specialty chemicals in clinical, pharmaceutical, research, and industrial applications (Yalpani, 1986; de Belder, 1996; Vandamme et al., 1996). Early work by Gronwall and Ingelman (1945, 1948) established the potential of using a hydrolyzed dextran fraction as a blood-plasma volume expander (Figure 2). Clinical dextrans used today are Dextran 40 and Dextran 70, which are 40,000 and 70,000 dalton average molecular weight fractions, respectively. Dextrans are less expensive than the albumins and starch derivatives also used in plasma therapies (Lilley and Aucker, 1999). These colloids essentially replace normal blood proteins in providing osmotic pressure to pull fluid from the interstitial space into the plasma. This treatment is useful to prevent shock from hemorrhage, burns, surgery, or trauma and to reduce the risk of thrombosis and embolisms. Dextran 40 also improves blood flow and inhibits the aggregation of erythrocytes (de Belder, 1996).

Iron dextran is a colloidal preparation used especially in veterinary medicine for the treatment of anemia, particularly in newborn piglets. Special iron dextran preparations also have been developed to enhance magnetic resonance imaging (MRI) techniques (de Belder, 1996). Dextran sulfate has been used as a substitute for heparin in anticoagulant therapy, and, more recently, it is being studied as an antiviral agent, particularly in the treatment of human immunodeficiency virus (HIV) (Mitsuya et al., 1988; Piret et al., 2000). Dextran can be crosslinked by epichlorohydrin (Flodin and Porath, 1961) to form beads (Sephadex®) that have become widely used in

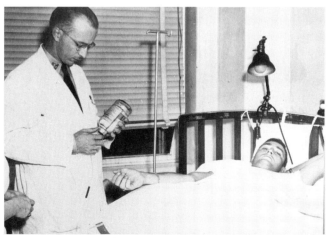

Fig. 2 Administration of dextran to a soldier at Walter Reed General Hospital, 1952. U.S. Dept. of Agriculture photograph.

research and industry for separations based on gel filtration. Anion and cation exchange resins based on Sephadex derivatives are widely used, including carboxymethyl (CM) Sephadex, diethylaminoethyl (DEAE) Sephadex, diethyl(2-hydroxypropyl)aminoethyl (QAE) Sephadex, and sulfopropyl (SP) Sephadex (Figure 3). Dextran is also an important component of many aqueous two-phase extraction systems, usually used in conjunction with polyethylene glycol (Tjerneld, 1992; Sinha et al., 2000).

Recently, oligosaccharides have received a great deal of attention as potential prebiotic compounds in food products, animal feeds, and cosmetics (Hidaka and Hirayama, 1991; Kohmoto et al., 1991; Monsan and Paul, 1995; Lamothe et al., 1996; Monsan et al., 2000). In contrast to clinical dextran preparations, prebiotic oligosaccharides must be resistant to digestion and preferentially utilized by beneficial bifidobacteria and lactic acid bacteria in the intestinal or skin microflora. Accordingly, dextran oligosaccharides of interest as prebiotics include the more branched varieties, containing α-(1 → 3) linkages from *L. citreum* strain NRRL B-742 (Remaud et al., 1992), α-(1 → 2) linkages from *L. mesenteroides* strain NRRL B-1299 (Remaud-Simeon et al., 1994; Dols et al., 1999), or the alternating α-(1 → 3) and α-(1 → 6) linkages from *L. mesenteroides* strain NRRL B-1355 (Pelenc et al., 1991).

12
Patents

Numerous patents claim methods for the production of dextran and dextran derivatives. The following examples, also summarized in Table 1, are illustrative. An early patent by Gronwall and Ingelman (1948) suggested that dextran might be useful as a blood-plasma volume expander. Hehre et al. (1959) described methods for dextran production by fermentation. Enzymatic production of dextran using dextransucrase also has been claimed (Tsuchiya and Koepsell, 1954; Hellman et al., 1955; Behrens and Ringpfeil, 1962). Wolff et al. (1955) described partial acid hydrolysis and fractionation of dextran for clinical applications. Alternative methods for the production of clinical dextrans include the use of dextranases (Carlson and Carlson, 1955a; Corman and Tsuchiya, 1957; Novak and Stoycos, 1958; Day and Kim, 1992) and chain-terminating acceptor reactions (Koepsell et al., 1955). Flodin and Porath (1961) described the cross-linking of dextran to form beads (Sephadex®) useful for gel filtration. Methods have been patented for the production of therapeutic iron dextrans (London and Twigg, 1958; Herb, 1979) and dextran sulfate (Morii et al., 1964; Usher, 1989). Recently, dextran oligosaccharides have garnered interest as potential prebiotic compounds (Lamothe et al., 1996).

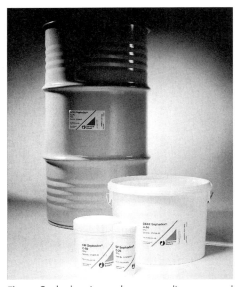

Fig. 3 Sephadex ion-exchange media are used widely for process scale applications. Photograph courtesy of Amersham Pharmacia Biotech, Inc.

Tab. 1 Selected patents related to dextran

Patent number	Holder	Inventors	Title	Date
U.S. Patent 2,437,518	Pharmacia AB, Sweden	A. Gronwall, B. Ingelman	Manufacture of infusion and injection fluids	1948
U.S. Patent 2,686,147	U.S. Dept. Agriculture	H. M. Tsuchiya, H. J. Koepsell	Production of dextransucrase	1954
U.S. Patent 2,709,150	Enzmatic Chemicals, Inc., Delaware, USA	V. W. Carlson, W. W. Carlson	Method of producing dextran material by bacteriological and enzymatic action	1955
U.S. Patent 2,712,007	U.S. Dept. Agriculture	I. A. Wolff, R. L. Mellies, C. E. Rist	Fractionation of dextran products	1955
U.S. Patent 2,726,190	U.S. Dept. Agriculture	H. J. Koepsell, N. N. Hellman, H. M. Tsuchiya	Modification of dextran synthesis by means of alternate glucosyl acceptors	1955
U.S. Patent 2,726,985	U.S. Dept. Agriculture	N. N Hellman, H. M. Tsuchiya, S. P. Rogovin, R. W. Jackson, F. R. Senti	Controlled enzymatic synthesis of dextran	1955
U.S. Patent 2,841,578	The Commonwealth Engineering Co. of Ohio	L. J. Novak, G. S. Stoycos	Method for producing clinical dextran	1958
U.S. Patent 2,776,925	U.S. Dept. Agriculture	J. Corman, H. M. Tsuchiya.	Enzymic production of dextran of intermediate molecular weights	1957
U.S. Patent 2,820,740	Benger Laboratories Ltd., England	E. London, G. D. Twigg	Therapeutic preparation of iron	1958
U.S. Patent 2,906,669	U.S. Dept. Agriculture	E. J. Hehre, H. M. Tsuchiya, N. N. Hellman, F. R. Senti	Production of dextran	1959
U.S. Patent 3,002,823	Pharmacia AB, Sweden	P. G. M. Flodin, J. O. Porath	Process of separating materials having different molecular weights and dimensions	1961
U.S. Patent 3,044,940	VEB Serum-Werk Bernburg, Germany	U. Behrens, M. Ringpfeil	Process for enzymatic synthesis of dextran	1962
U.S. Patent 3,141,014	Meito Sangyo Kabushiki Kaishu, Japan	E. Morii, K. Iwata, H. Kokkoku	Sodium and potassium salts of the dextran sulfate acid ester having substantially no anticoagulant activity but having lipolytic activity and the method of preparation thereof	1964
U.S. Patent 4,180,567	Pharmachem Corp., USA	J. R. Herb	Iron preparations and methods of making and administering the same	1979

Tab. 1 (cont.)

Patent number	Holder	Inventors	Title	Date
U.S. Patent 4,855,416	Polydex Pharmaceuticals, Ltd., The Bahamas	T. C. Usher	Method for the manufacture of dextran sulfate and salts thereof	1989
U.S. Patent 5,229,277	Louisiana State Univ.	D. F. Day, D. Kim	Process for the production of dextran polymers of controlled molecular size and molecular size distributions	1992
U.S. Patent 5,518,733.	Bioeurope, France	J.-P. Lamothe, Y. G. Marchenay, P. F. Monsan, F. M. B. Paul, V. Pelenc	Cosmetic compositions containing oligosaccharides	1996

13 Outlook and Perspectives

Advances in the molecular biology of glucansucrases promise not only to resolve fundamental questions concerning enzyme structure, function, and regulation but also to open new avenues for dextran applications. Recombinant organisms that overproduce dextransucrases may reduce the cost of dextran production by enzymatic synthesis, making dextrans more competitive for bulk chemical applications. Alternatively, dextrans might be produced in transgenic crops, as has been demonstrated recently for fructans (Caimi et al., 1996; Pilon-Smits et al., 1996; Sevenier et al., 1998) and *Streptococcus* glucans (Nichols, 2000a,b,c). Novel dextransucrases might be created by site-directed mutagenesis, chimeric recombination, or shuffling of dextransucrase genes. At the same time, dextran oligosaccharides appear to have considerable potential to find new and expanded markets as prebiotic supplements in foods, cosmetics, and animal feeds.

14
References

Alcalde, M., Plou, F. J., Gomez de Segura, A., Remaud-Simeon, M. Willemot, R. M., Monsan, P., Ballesteros, A. (1999) Immobilization of native and dextran-free dextransucrases from *Leuconostoc mesenteroides* NRRL B-512F for the synthesis of glucooligosaccharides, *Biotechnol. Tech.* **13**, 749–755.

Alsop, R. M. (1983) Industrial production of dextrans, in: *Progress in Industrial Microbiology*, (Bushell, M. E., Ed.), London: Elsevier, 1–44, Vol. 18.

Arguello-Morales, M. A., Remaud-Simeon, M., Pizzut, S., Sarcabal, P., Willemot, R.-M., Monsan, P. (2000a) *Leuconostoc mesenteroides* NRRL B-1355 dsrC gene for dextransucrase. GenBank Accession No. AJ250172.

Arguello-Morales, M. A., Remaud-Simeon, M., Pizzut, S., Sarcabal, P., Willemot, R.-M., Monsan, P. (2000b) Sequence analysis of the gene encoding alternansucrase, a sucrose glucosyltransferase from *Leuconostoc mesenteroides* NRRL B-1355. *FEMS Microbiol. Lett.* **182**, 81–85.

Barker, S. A., Bourne, E. J., Whiffen, D. H. (1956) Use of infrared analysis in the determination of carbohydrate structure, in: *Methods of Biochemical Analysis*, (Glick, D., Ed.), New York: Interscience Publishers, Inc., 213–245, Vol. 3.

Behrens, U., Ringpfeil, M. (1962) Process for enzymatic synthesis of dextran. U.S. Patent 3,044,940.

Beijerinck, M. W. K. (1912) Mucilaginous substances of the cell wall produced from cane sugar by bacteria. *Folia Microbiol.* **1**, 377.

Bhatnagar, R., Singh, D. K. S. (1999) Cloning and characterization of dextransucrase gene from *Leuconostoc mesenteroides* NRRL B-512F. GenBank Accession No. U81374.

Binder, T. P., Robyt, J. F. (1983) p-nitrophenyl α-D-glucopyranoside, a new substrate for glucansucrases, *Carbohydr. Res.* **124**, 287–299.

Bovery, F. A. (1959) Enzymatic polymerization. I. Molecular weight and branching during the formation of dextran, *J. Polymer Sci.* **35**, 167–182.

Caimi, P. G., McCole, L. M., Klein, T. M., Kerr, P. S. (1996) Fructan accumulation and sucrose metabolism in transgenic maize endosperm expressing a *Bacillus amyloliquefaciens* SacB gene, *Plant Physiol.* **110**, 355–363.

Caldwell, R. C., Sandham, H. J., Mann, W. V., Finn, S. B., Formicola, A. J. (1971) The effect of a dextranase mouthwash on dental plaque in young adults and children, *J. Amer. Dent. Assoc.* **82**, 124–131.

Carlson, V. W., Carlson, W. W. (1955a) Method of producing dextran material by bacteriological and enzymatic action. U.S. Patent 2,709,150.

Carlson, V. W., Carlson, W. W. (1955b) Production of endodextranase by *Aspergillus wentii*. U.S. Patent 2,716,084.

Carrasco, F., Chornet, E., Overend, R. P., Costa, J. (1989) A generalized correlation for the viscosity of dextrans in aqueous solutions as a function of temperature, concentration, and molecular weight at low shear rates, *J. Appl. Polymer Sci.* **37**, 2087–2098.

Cerning, J. (1990) Exocellular polysaccharides produced by lactic acid bacteria, *FEMS Microbiol. Rev.* **87**,113–130.

Chaiet, L., Kempf, A. J., Harman, R., Kaczka, E., Weston, R., Nollstadt, K., Wolf, F. J. (1970) Isolation of a pure dextranase from *Penicillium funiculosum*, *Appl. Microbiol.* **20**, 421–426.

Cheetham, N. W. H., Fiala-Beer, E., Walker, G. J. (1991) Dextran structural details from high-field proton NMR spectroscopy, *Carbohydr. Polymers* **14**, 149–158.

Chludzinski, A. M., Germaine, G. R., Schachtele, C. F. (1974) Purification and properties of dextransucrase from *Streptococcus mutans*, *J. Bacteriol.* **118**, 1–7.

Corman, J., Tsuchiya, H. M. (1957) Enzymic production of dextran of intermediate molecular weights. U.S. Patent 2,776,925.

Côté, G. L. (1992) Low-viscosity α-D-glucan fractions derived from sucrose which are resistant to enzymatic digestion, *Carbohydr. Polym.* **19**, 249–252.

Côté, G. L., Ahlgren, J. A. (1994) Production, isolation, and immobilization of alternansucrase. Amer. Chem. Soc. 207 Natl. Meeting, Abstract CARB#12.

Côté, G. L., Ahlgren, J. A. (1995) Microbial polysaccharides, in: *Kirk-Othmer Encyclopedia of Chemical Technology* (Kroschvitz, J. I., Howe-Grant, M., Eds), New York: John Wiley & Sons, Inc., 578–612, 4th Ed., Vol. 16.

Côté, G. L., Robyt, J. F. (1982) Isolation and partial characterization of an extracellular glucansucrase from *L. mesenteroides* NRRL B-1355 that synthesizes an alternating $(1 \rightarrow 6)$, $(1 \rightarrow 3)$-α-D-glucan, *Carbohyd. Res.* **101**, 57–74.

Côté, G. L., Robyt, J. F. (1983) The formation of α-D-$(1 \rightarrow 3)$ branch linkages by an exocellular glucansucrase from *Leuconostoc mesenteroides* NRRL B-742, *Carbohydr. Res.* **119**, 141–156.

Côté, G. L., Leathers, T. D., Ahlgren, J. A., Wyckoff, H. A., Hayman, G. T., Biely, P. (1997) Alternan and highly branched limit dextrans: Low-viscosity polysaccharides as potential new food ingredients, in: *Chemistry of Novel Foods* (Spanier, A. M., Tamura, M., Okai, H., Mills, O., Eds), Carol Stream, IL: Allured Publishing Corp., 95–110.

Côté, G. L., Ahlgren, J. A., Smith, M. R. (1999) Some structural features of an insolube α-D-glucan from a mutant strain of *Leuconostoc mesenteroides* NRRL B-1355, *J. Ind. Microbiol. Biotechnol.* **23**, 656–660.

Covacevich, M. T., Richards, G. N. (1977) Frequency and distribution of branching in a dextran: an enzymic method, *Carbohydr. Res.* **54**, 311–315.

Day, D. F., Kim, D. (1992) Process for the production of dextran polymers of controlled molecular size and molecular size distributions. U.S. Patent 5,229,277.

de Belder, A. N. (1990) *Dextran*. Uppsala, Sweden: Pharmacia.

de Belder, A. N. (1993) Dextran, in: *Industrial Gums. Polysaccharides and Their Derivatives*, Third Edition (Whistler, R. L., BeMiller, J. N., Eds.), San Diego, CA: Academic Press, 399–425.

de Belder, A. N. (1996) Medical applications of dextran and its derivatives, in: *Polysaccharides. Medical Applications* (Dumitriu, S., Ed.), New York: Marcel Dekker, 505–523.

Dimler, R. J., Wolff, I. A., Sloan, J. W., Rist, C. E. (1955) Interpretation of periodate oxidation data on degraded dextran, *J. Am. Chem. Soc.* **77**, 6568–6573.

Dols, M., Remaud-Simeon, M., Willemot, R.-M., Vignon, M., Monsan, P. F. (1997) Characterization of dextransucrases from *Leuconostoc mesenteroides* NRRL B-1299, *Appl. Biochem. Biotechnol.* **62**, 47–59.

Dols, M., Remaud-Simeon, M., Willemot, R.-M., Demuth, B., Joerdening, H.-J., Buchholz, K., Monsan, P. (1999) Kinetic modeling of oligosaccharide synthesis catalyzed by *Leuconostoc mesenteroides* NRRL B-1299 dextransucrase, *Biotechnol. Bioeng.* **63**, 308–315.

Evans, T. H., Hibbert, H. (1946) Bacterial polysaccharides, in: *Adv. Carbohydr. Chem.* (Pigman, W. W., Wolfrom, M. L., Eds.). New York: Academic Press, 203–233, Vol. 2.

Fischer, E. H., Stein, E. A. (1960) Cleavage of O- and S-glycosidic bonds (survey), in: *The Enzymes* (Boyer, P. D., Lardy, H., Myrback, K., Eds.), New York: Academic Press, 301–312, Vol. 4.

Fitzgerald, R. J., Spinell, D. M., Stoudt, T. H. (1968) Enzymatic removal of artificial plaques, *Arch. Oral Biol.* **13**, 125–128.

Flodin, P. G. M., Porath, J. O. (1961) Process of separating materials having different molecular weights and dimensions. U.S. Patent 3,002,823.

Fu, D., Robyt, J. F. (1990) A facile purification of *Leuconostoc mesenteroides* B-512FM dextransucrase, *Prep. Biochem.* **20**, 93–106.

Fu, D., Slodki, M. E., Robyt, J. F. (1990) Specificity of acceptor binding to *Leuconostoc mesenteroides* B512F dextransucrase: binding and acceptor-product structure of α-methyl-D-glucopyranoside analogs modified at C-2, C-3, and C-4 by inversion of the hydroxyl and by replacement of the hydroxyl with hydrogen, *Arch. Biochem. Biophys.* **276**, 460–465.

Fujiwara, T. Terao, Y., Hoshino, T., Kawabata, S., Ooshima, T., Sobue, S., Kimura, S., Hamada, S. (1998) Molecular analyses of glucosyltransferase genes among strains of *Streptococcus mutans*, *FEMS Microbiol. Lett.* **161**, 331–336.

Fukui, K., Fukui, Y., Moriyama, T. (1974) Purification and properties of dextransucrase and invertase from *Streptococcus mutans*, *J. Bacteriol.* **118**, 796–804.

Funane, K., Mizuno, K., Takahara, H., Kobayashi, M. (2000) Gene encoding a dextransucrase-like protein in *Leuconostoc mesenteroides* NRRL B-512F, *Biosci. Biotechnol. Biochem.* **64**, 29–38.

Furuta, T., Koga, T., Nisizawa, T., Okahashi, N., Hamada, S. (1985) Purification and characterization of glucosyltransferases from *Streptococcus mutans* 6715, *J. Gen. Microbiol.* **131**, 285–293.

Galvez-Mariscal, A., Lopez-Munguia, A. (1991) Production and characterization of a dextranase from an isolated *Paecilomyces lilacinus* strain, *Appl. Microbiol. Biotechnol.* **36**, 327–331.

Gekko, K., Noguchi, H. (1971) Physicochemical studies of oligodextran. I. Molecular weight dependence of intrinsic viscosity, partial specific compressibility and hydrated water, *Biopolymers* **10**, 1513–1524.

Glicksman, M. (1983) Dextran, in: *Food Hydrocolloids* (Glicksman, M., Ed.), Boca Raton: CRC Press 157–166.

Godfrey, T. (1983) Dextranase and sugar processing, in: *Industrial Enzymology. The Application of Enzymes in Industry* (Godfrey, T., Reichelt, T., Eds.), New York: Nature Press, 422–424.

Granath, K. A. (1958) Solution properties of branched dextrans, *J. Colloid Sci.* **13**, 308–328.

Gronwall, A. (1957) *Dextran and Its Use in Colloidal Infusion Solutions.* Stockholm: Almqvist & Wiksell.

Gronwall, A., Ingelman, B. (1945) Dextran as a substitute for plasma, *Nature* **155**, 45.

Gronwall, A., Ingelman, B. (1948) Manufacture of infusion and injection fluids. U.S. Patent 2,437,518.

Hamada, S., Slade, H. D. (1980) Biology, immunology, and cariogenicity of *Streptococcus mutans*, *Microbiol. Rev.* **44**, 331–384.

Hattori, A., Ishibashi, K., Minato, S. (1981) The purification and characterization of the dextranase from *Chaetomium gracile*, *Agric. Biol. Chem.* **45**, 2409–2416.

Hehre, E. J. (1941) Production from sucrose of a serologically reactive polysaccharide by a sterile bacterial extract, *Science* **93**, 237–238.

Hehre, E. J., Hamilton, D. M. (1951) The biological synthesis of dextran from dextrins, *J. Biol. Chem.* **192**, 161–174.

Hehre, E. J., Tsuchiya, H. M., Hellman, N. N., Senti, F. R. (1959) Production of dextran. U.S. Patent 2,906,669.

Hehre, E. J., Suzuki, H. (1966) New reactions of dextransucrase: α-D-glucosyl transfers to and from the anomeric sites of lactulose and fructose, *Arch. Biochem. Biophys.* **113**, 675–683.

Heincke, K., Demuth, B., Jordening, H.-J., Buchholz, K. (1999) Kinetics of the dextransucrase acceptor reaction with maltose - experimental results and modeling, *Enzyme Microbial Technol.* **24**, 523–534.

Herb, J. R. (1979) Iron preparations and methods of making and administering the same. U.S. Patent 4,180,567.

Hellman, N. N., Tsuchiya, H. M., Rogovin, S. P., Jackson, R. W., Senti, F. R. (1955) Controlled enzymatic synthesis of dextran. U.S. Patent 2,726,985.

Hidaka, H., Hirayama, M. (1991) Useful characteristics and commercial applications of fructo-oligosaccharides, *Biochem. Soc. Trans.* **19**, 561–565.

Hiraoka, N., Fukumoto, J., Tsuru, D. (1972) Studies on mold dextranases: III. Purification and some enzymatic properties of *Aspergillus carneus* dextranase, *J. Biochem.* **71**, 57–64.

Holzapfel, W. H., Schillinger, U. (1992) The genus *Leuconostoc*, in: *The Procaryotes*, 2nd Ed., (Ballows, A., Truper, H. G., Dworkin, M., Harder, W., Schleifer, K.-H., Eds.), New York: Springer-Verlag, 1508–1534, Vol. 2.

Honda, O., Kato, C., Kuramitsu, H. K. (1990) Nucleotide sequence of the *Streptococcus mutans gtfD* gene encoding the glucosyltransferase-S enzyme, *J. Gen. Microbiol.* **136**, 2099–2105.

Janecek, S., Svensson, B, Russell, R. R. B. (2000) Location of repeat elements in glucansucrases of *Leuconostoc* and *Streptococcus* species, *FEMS Microbiol. Lett.* **192**, 53–57.

Jeanes, A. (1965a) Dextrans. Preparation of a water soluble dextran by enzymic synthesis, in: *Methods in Carbohydrate Chemistry* (Whistler, R. L., BeMiller, J. N., Eds.), New York: Academic Press, 127–132, Vol. 5.

Jeanes, A. (1965b) Dextrans. Preparation of dextrans from growing *Leuconostoc* cultures, in: *Methods in Carbohydrate Chemistry* (Whistler, R. L., BeMiller, J. N., Eds.), New York: Academic Press, 118–126, Vol. 5.

Jeanes, A. (1966) Dextran, in: *Encyclopedia of Polymer Science and Engineering,* (Mark, H. F.; Bikales, N. M.; Overberger, C. G.; Menges, G.; Kroschwitz, J. I., Eds.), New York: John Wiley & Sons, 752–767, Vol. 4.

Jeanes, A. (1977) Dextrans and pullulans: industrially significant α-D-glucans, in: *ACS Symp. Series No. 45, Extracellular Microbial Polysaccharides* (Sandford, P. A., Laskin, A., Eds.), Washington, D. C.: American Chemical Society, 284–298.

Jeanes, A. (1978) *Dextran Bibliography.* Washington, D. C.: U. S. Dept. Agriculture.

Jeanes, A. (1986) Immunochemical and related interactions with dextrans reviewed in terms of improved structural information. *Mol. Immun.* **23**, 999–1028.

Jeanes, A., Seymour, F. R. (1979) The α-D-glucopyranosidic linkages of dextrans: comparison of percentages from structural analysis by periodate oxidation and by methylation, *Carbohydr. Res.* **74**, 31–40.

Jeanes, A., Haynes, W. C., Wilham, C. A., Rankin, J. C., Melvin, E. H., Austin, M. J., Cluskey, J. E., Fisher, B. E., Tsuchiya, H. M., Rist, C. E. (1954) Characterization and classification of dextrans from ninety-six strains of bacteria, *J. Amer. Chem. Soc.* **76**, 5041–5052.

Jung, H-K., Kim, K-N., Lee, H-S., Jung, S-H. (1999) Production of alternan by *Leuconostoc mesenteroides* CBI-110, *Kor. J. Appl. Microbiol. Biotechnol.* **27**, 35–40.

Kabat, E. A., Bezer, A. E. (1958) The effect of variation in molecular weight on the antigenicity of dextran in man, *Arch. Biochem. Biophys.* **78**, 306–318.

Kaboli, H., Reilly, P. J. (1980) Immobilization and properties of *Leuconostoc mesenteroides* dextransucrase, *Biotechnol. Bioeng.* **22**, 1055–1069.

Killey, M., Dimler, R. J., Cluskey, J. E. (1955) Preparation of panose by the action of NRRL B-512 dextransucrase on a sucrose-maltose mixture, *J. Amer. Chem. Soc.* **77**, 3315–3318.

Kim, D., Day, D. F. (1995) Isolation of a dextranase constitutive mutant of *Lipomyces starkeyi* and its use for the production of clinical size dextran, *Lett. Appl. Microbiol.* **20**, 268–270.

Kim, D., Kim, D-W. (1999) Facile purification and characterization of dextransucrase from *Leuconostoc mesenteroides* B-512FMCM, *J. Microbiol. Biotechnol.* **9**, 219–222.

Kim, D., Robyt, J. F. (1994) Production and selection of mutants of *Leuconostoc mesenteroides* constitutive for glucansucrases, *Enzyme Microb. Technol.* **16**, 659–664.

Kim, D., Robyt, J. F. (1995a) Dextransucrase constitutive mutants of *Leuconostoc mesenteroides* B-1299, *Enzyme Microb. Technol.* **17**, 1050–1056.

Kim, D., Robyt, J. F. (1995b) Production, selection, and characteristics of mutants of *Leuconostoc mesenteroides* B-742 constitutive for dextransucrases, *Enzyme Microb. Technol.* **17**, 689–695.

Kim, D., Robyt, J. F. (1996) Properties and uses of dextransucrases elaborated by a new class of *Leuconostoc mesenteroides* mutants, *Prog. Biotechnol.* **12**, 125–144.

Kim, D., Kim, D-W., Lee, J-H., Park, K-H., Day, L. M., Day, D. F. (1997) Development of constitutive dextransucrase hyper-producing mutants of *Leuconostoc mesenteroides* using the synchrotron radiation in the 70–1000 eV region, *Biotechnol. Tech.* **11**, 319–321.

Kim, H., Kim, D., Ryu, W-H., Robyt, J. F. (2000) Cloning and sequencing of the α-1 → 6 dextransucrase gene from *Leuconostoc mesenteroides* B-742CB, *J. Microbiol. Biotechnol.* **10**, 559–563.

Kitaoka, M., Robyt, J. F. (1998a) Large-scale preparation of highly purified dextransucrase from a high-producing constitutive mutant of *Leuconostoc mesenteroides* B-512FMC, *Enzyme Microb. Technol.* **23**, 386–391.

Kitaoka, M., Robyt, J. F. (1998b) Use of a microtiter plate screening method for obtaining *Leuconostoc mesenteroides* mutants constitutive for glucansucrase, *Enzyme Microb. Technol.* **22**, 527–531.

Kobayashi, M., Matsuda, K. (1975) Purification and characterization of two activities of the intracellular dextransucrase from *Leuconostoc mesenteroides* NRRL B-1299, *Biochim. Biophys. Acta* **397**, 69–79.

Kobayashi, M., Matsuda, K. (1976) Purification and properties of the extracellular dextransucrase from *Leuconostoc mesenteroides* NRRL B-1299, *J. Biochem.* **79**, 1301–1308.

Kobayashi, M., Matsuda, K. (1977) Structural characteristics of dextrans synthesized by dextransucrases from *Leuconostoc mesenteroides* NRRL B-1299, *Agric. Biol. Chem.* **41**, 1931–1937.

Kobayashi, M., Matsuda, K. (1980) Characterization of the multiple forms and main component of dextransucrase from *Leuconostoc mesenteroides* NRRL B-512F, *Biochim. Biophys. Acta* **614**, 46–62.

Koenig, D. W., Day, D. F. (1988) Production of dextranase by *Lipomyces starkeyi*, *Biotechnol. Lett.* **10**, 117–122.

Koepsell, H. J., Hellman, N. N., Tsuchiya, H. M. (1955) Modification of dextran synthesis by means of alternate glucosyl acceptors. U.S. Patent 2,726,190.

Kohmoto, T., Fukui, F., Takaku, H., Mitsuoka, T. (1991) Dose-response test of isomaltooligosaccharides for increasing fecal bifidobacteria, *Agric. Biol. Chem.* **55**, 2157–2159.

Kooi, E. R. (1958) Production of dextran-dextrinase. U.S. Patent 2,833,695.

Kossman, J., Welsh, T., Quanz, M., Knuth, K. (2000) Nucleic acid molecules encoding alternansucrase. PCT Patent WO00/47727.

Kuge, T., Kobayashi, K., Kitamura, S., Tanahashi, H. (1987) Degrees of long-chain branching in dextrans, *Carbohydr. Res.* **160**, 205–214.

Lamothe, J.-P., Marchenay, Y. G., Monsan, P. F., Paul, F. M. B., Pelenc, V. (1996) Cosmetic compositions containing oligosaccharides. U.S. Patent 5,518,733.

Larm, O., Lindberg, B., Svensson, S. (1971) Studies on the length of the side chains of the dextran elaborated by *Leuconostoc mesenteroides* NRRL B-512, *Carbohydr. Res.* **20**, 39–48.

Lawford, G. R., Kligerman, A., Williams, T. (1979) Dextran biosynthesis and dextransucrase production by continuous culture of *Leuconostoc mesenteroides*, *Biotechnol. Bioeng.* **21**, 1121–1131.

Leathers, T. D., Hayman, G. T., Cote, G. L. (1995) Rapid screening of *Leuconostoc mesenteroides* mutants for elevated proportions of alternan to dextran. *Curr. Microbiol.* **31**, 19–22.

Leathers, T. D., Hayman, G. T., Cote, G. L. (1997) Microorganism strains that produce a high proportion of alternan to dextran. U.S. Patent 5,702,942.

Leathers, T. D., Hayman, G. T., Cote, G. L. (1998) Rapid screening method to select microorganism strains that produce a high proportion of alternan to dextran. U.S. Patent 5,789,209.

Lee, J. M., Fox, P. F. (1985) Purification and characterization of *Paecilomyces lilacinus* dextranase, *Enzyme Microb. Technol.* **7**, 573–577.

Lilley, L. L., Aucker, R. S. (1999) Fluids and electrolytes, in: *Pharmacology and the Nursing Process*. St. Louis, MO: Mosby, Inc., 335–348.

Lindberg, B., Svensson, S. (1968) Structural studies on dextran from *Leuconostoc mesenteroides* NRRL B-512, *Acta Chem. Scand.* **22**, 1907–1912.

Loesche, W. J. (1986) Role of *Streptococcus mutans* in human dental decay, *Microbiol. Rev.* **50**, 353–380.

London, E., Twigg, G. D. (1958) Therapeutic preparation of iron. U.S. Patent 2,820, 740.

McCabe, M. M. (1985) Purification and characterization of a primer-independent glucosyltransferase from *Streptococcus mutans* 6715-13 mutant 27, *Infect. Immun.* **50**, 771–777.

Miller, A. W., Robyt, J. F. (1984) Stabilization of dextransucrase from *Leuconostoc mesenteroides* NRRL B-512F by nonionic detergents, poly(ethylene glycol) and high-molecular-weight dextran, *Biochim. Biophys. Acta* **785**, 89–96.

Miller, A. W., Eklund, S. H., Robyt, J. F. (1986) Milligram to gram scale purification and characterization of dextransucrase from *Leuconostoc mesenteroides* NRRL B-512F, *Carbohydr. Res.* **147**, 119–133.

Misaki, A., Torii, M., Sawai, T., Goldstein, I. J. (1980) Structure of the dextran of *Leuconostoc mesenteroides* B-1355, *Carbohydr. Res.* **84**, 273–285.

Mitsuishi, Y., Kobayashi, M., Matsuda, K. (1979) Dextran α-1, 2 debranching enzyme from *Flavobacterium* sp. M-73: its production and purification, *Agric. Biol. Chem.* **43**, 2283–2290.

Mitsuya, H., Looney, D. J., Kuno, S., Ueno, R., Wong-Staal, F., Broder, S. (1988) Dextran sulfate suppression of viruses in the HIV family: inhibition of virion binding to CD4$^+$ cells, *Science* **240**, 646–649.

Monchois, V., Willemot, R.-M., Remaud-Simeon, M., Croux, C., Monsan, P. (1996) Cloning and sequencing of a gene coding for a novel dextransucrase from *Leuconostoc mesenteroides* NRRL B-1299 synthesizing only a α(1-6) and α(1-3) linkages, *Gene* **182**, 23–32.

Monchois, V., Remaud-Simeon, M., Russell, R. R. B., Monsan, P., Willemot, R.-M. (1997) Characterization of *Leuconostoc mesenteroides* NRRL B512F dextransucrase (DSRS) and identification of amino-acid residues playing a key role in enzyme activity, *Appl. Microbiol. Biotechnol.* **48**, 465–472.

Monchois, V., Remaud-Simeon, M., Monsan, P., Willemot, R.-M. (1998) Cloning and sequencing of a gene coding for an extracellular dextransucrase (DSRB) from *Leuconostoc mesenteroides* NRRL B-1299 synthesizing only α(1-6) glucan, *FEMS Microbiol. Lett.* **159**, 307–315.

Monchois, V., Willemot, R.-M., Monsan, P. (1999) Glucansucrases: mechanism of action and structure-function relationships, *FEMS Microbiol. Rev.* **23**, 131–151.

Monchois, V., Vignon, M., Russell, R. R. B. (2000) Mutagenesis of asp-569 of glucosyltransferase I glucansucrase modulates glucan and oligosaccharide synthesis, *Appl. Environ. Microbiol.* **66**, 1923–1927.

Monsan, P., Paul, F. (1995) Enzymatic synthesis of oligosaccharides, *FEMS Microbiol. Rev.* **16**, 187–192.

Monsan, P., Paul, F., Auriol, D., Lopez, A. (1987) Dextran synthesis using immobilized *Leuconostoc mesenteroides* dextransucrase, in: *Methods in Enzymology: Immobilized Enzymes and Cells* (Mosbach, K., Ed.), Orlando, FL: Academic Press, Inc, 239–254, Vol. 136.

Monsan, P., Potocki de Montalk, G., Sarcabal, P., Remaud-Simeon, M., Willemont, R.-M. (2000) Glucansucrases: efficient tools for the synthesis of oligosaccharides of nutritional interest, in: *Food Biotechnology* (Bielecki, S., Tramper, J., Polak, J., Eds.), Amsterdam: Elsevier Science B. V., 115–122.

Morii, E., Iwata, K., Kokkoku, H. (1964) Sodium and potassium salts of the dextran sulphate acid ester having substantially no anticoagulant activity but having lipolytic activity and the method of preparation thereof. U.S. Patent 3,141,014.

Mountzouris, K. C., Gilmour, S. G., Jay, A. J., Rastall, R. A. (1999) A study of dextran production from maltodextrin by cell suspensions of *Gluconobacter oxydans* NCIB 4943, *J. Appl. Microbiol.* **87**, 546–556.

Murphy, P. T., Whistler, R. L. (1973) Dextrans, in *Industrial Gums*, Second Ed. (Whistler, R. L., BeMiller, J. N., Eds.), New York: Academic Press, 513–542.

Neely, W. B. (1960) Dextran: structure and synthesis, in: *Adv. Carbohydr. Chem.* (Wolfrom, M. L., Tipson, R. S., Eds.), New York: Academic Press, 341–369, Vol. 15.

Nichols, S. E. (2000a) Plant cells and plants transformed with *Streptococcus mutans* gene encoding glucosyltransferase C enzyme. U.S. Patent 6,127,603.

Nichols, S. E. (2000b) Plant cells and plants transformed with *Streptococcus mutans* genes encoding wild-type or mutant glucosyltransferase B enzymes. U.S. Patent 6,087,559.

Nichols, S. E. (2000c) Plant cells and plants transformed with *Streptococcus mutans* genes encoding wild-type or mutant glucosyltransferase D enzymes. U.S. Patent 6,127,602.

Novak, L. J., Stoycos, G. S. (1958) Method for producing clinical dextran. U.S. Patent 2,841,578.

Okada, G., Takayanagi, T., Sawai, T. (1988) Improved purification and further characterization of an isomaltodextranase from *Arthrobacter globiformis* T6, *Agric. Biol. Chem.* **52**, 495–501.

Pasteur, L. (1861) On the viscous fermentation and the butyrous fermentation, *Bull. Soc. Chim. Paris*, 30–31.

Paul, F., Auriol, D., Oriol, E. Monsan, P. (1984) Production and purification of dextransucrase from *Leuconostoc mesenteroides* NRRL B-512(F). *Ann. N. Y. Acad. Sci.* **434**, 267–270.

Pearce, B. J., Walker, G. J., Slodki, M. E., Schuerch, C. (1990) Enzymic and methylation analysis of dextrans and (1-3)-α-D-glucans, *Carbohydr. Res.* **203**, 229–246.

Pelenc, V., Lopez-Munguia, A., Remaud, M., Biton, J., Michel, J. M., Paul, F., Monsan, P. (1991) Enzymatic synthesis of oligoalternans, *Sci. Aliments* **11**, 465–476.

Pilon-Smits, E. A. H., Ebskamp, M. J. M., Jeuken, M. J. W., van der Meer, I. M., Visser, R. G. F., Weisbeek, P. J., Smeekens, S. C. M. (1996) Microbial fructan production in transgenic potato plants and tubers, *Ind. Crops Products* **5**, 35–46.

Piret, J., Lamontagne, J., Bestman-Smith, J., Roy, S., Gourde, P., Desormeaux, A., Omar, R. F., Juhasz, J., Bergeron, M. G. (2000) In vitro and in vivo evaluations of sodium lauryl sulfate and dextran sulfate as microbicides against herpes simplex and human immunodeficiency viruses, *J. Clin. Microbiol.* **38**, 110–119.

Quirasco, M., Lopez-Munguia, A., Remaud-Simeon, M., Monsan, P., Farres, A. (1999) Induction and transcriptional studies of the dextransucrase gene in *Leuconostoc mesenteroides* NRRL B-512F, *Appl. Environ. Microbiol.* **65**, 5504–5509.

Rankin, J. C., Jeanes, A. (1954) Evaluation of periodate oxidation method for structural analysis of dextrans, *J. Am. Chem. Soc.* **76**, 4435–4441.

Reh, K.-D., Noll-Borchers, M., Buchholz, K. (1996) Productivity of immobilized dextransucrase for leucrose formation, *Enzyme Microb. Technol.* **19**, 518–524.

Remaud, M., Paul, F., Monsan, P., Heyraud, A., Rinaudo, M. (1991) Molecular weight characterization and structural properties of controlled molecular weight dextrans synthesized by acceptor reaction using highly purified dextransucrase, *J. Carbohydr. Chem.* **10**, 861–876.

Remaud, M., Paul, F., Monsan, P. (1992) Characterization of α-(1 → 3) branched oligosaccharides synthesized by acceptor reaction with the extracellular glucosyltransferases from *L. mesenteroides* NRRL B-742, *J. Carbohydr. Chem.* **11**, 359–378.

Remaud-Simeon, M., Lopez-Munguia, A., Pelenc, V., Paul, F., Monsan, P. (1994) Production and use of glucosyltransferases from *Leuconostoc mesenteroides* NRRL B-1299 for the synthesis of oligosaccharides containing α-(1 → 2) linkages, *Appl. Biochem. Biotechnol.* **44**, 101–117.

Remaud-Simeon, M., Willemot, R.-M., Sarcabal, P., Potocki de Montalk, G., Monsan, P. (2000) Glucansucrases: molecular engineering and oligosaccharide synthesis, *J. Mol. Catalysis B: Enzymatic* **10**, 117–128.

Robyt, J. F. (1986) Dextran, in: *Encyclopedia of Polymer Science and Engineering*, (Mark, H. F., Gaylord, N. G., Bikales, N. M., Eds.), New York: John Wiley & Sons, 752–767, Vol. 4.

Robyt, J. F. (1992) Structure, biosynthesis, and uses of nonstarch polysaccharides: dextran, alternan, pullulan, and algin, in: *Developments in Biochemistry and Biophysics* (Alexander, R. J., Zobel, H. F., Eds.), St. Paul: Amer. Assoc. Cereal Chemists, 261–292.

Robyt, J. F. (1995) Mechanisms in the glucansucrase synthesis of polysaccharides and oligosaccharides from sucrose, in: *Adv. Carbohydr. Chem. Biochem.* (Horton, D., Ed.), San Diego: Academic Press, 133–168, Vol. 51.

Robyt, J. F., Eklund, S. H. (1983) Relative, quantitative effects of acceptors in the reaction of *Leuconostoc mesenteroides* B-512F dextransucrase, *Carbohydr. Res.* **121**, 279–286.

Robyt, J. F., Martin, P. J. (1983) Mechanism of synthesis of D-glucans by D-glucosyltransferases from *Streptococcus mutans* 6715, *Carbohydr. Res.* **113**, 301–315.

Robyt, J. F., Taniguchi, H. (1976) The mechanism of dextransucrase action. Biosynthesis of branch linkages by acceptor reactions with dextran, *Arch. Biochem. Biophys.* **174**, 129–135.

Robyt, J. F., Walseth, T. F. (1978) The mechanism of acceptor reactions of *Leuconostoc mesenteroides* B-512F dextransucrase, *Carbohydr. Res.* **61**, 433–445.

Robyt, J. F., Walseth, T. F. (1979) Production, purification and properties of dextransucrase from *Leuconostoc mesenteroides* NRRL B-512F, *Carbohydr. Res.* **68**, 95–111.

Robyt, J. F., Kimble, B. K., Walseth, T. F. (1974) The mechanism of dextransucrase action. Direction of dextran biosynthesis, *Arch. Biochem. Biophys.* **165**, 634–640.

Ryu, H-J., Kim, D., Kim, D-W., Moon, Y-Y., Robyt, J. F. (2000) Cloning of a dextransucrase gene (*fmcmds*) from a constitutive dextransucrase hyper-producing *Leuconostoc mesenteroides* B-512FMCM developed using VUV, *Biotechnol. Lett.* **22**, 421–425.

Sabatie, J., Choplin, L., Moan, M., Doublier, J. L., Paul, F., Monsan, P. (1988) The effect of synthesis temperature on the structure of dextran NRRL B 512F, *Carbohydr. Polymers* **9**, 87–101.

Sanchez-Gonzalez, M., Alagon, A., Rodriguez-Sotres, R., Lopez-Munguia, A. (1999) Proteolytic processing of dextransucrase of *Leuconostoc mesenteroides*, *FEMS Microbiol. Lett.* **181**, 25–30.

Sawai, T., Tohyama, T., Natsume, T. (1978) Hydrolysis of fourteen native dextrans by *Arthrobacter* isomaltodextranase and correlation with dextran structure, *Carbohydr. Res.* **66**, 195–205.

Sawai, T., Ohara, S., Ichimi, Y., Okaji, S., Hisada, K., Fukaya, N. (1981) Purification and some properties of the isomaltodextranase of *Actinomadura* strain R10 and comparison with that of *Arthrobacter globiformis* T6, *Carbohydr. Res.* **89**, 289–299.

Scheibler, C. (1874) Investigation on the nature of the gelatinous excretion (so-called frog's spawn) which is observed in production of beet-sugar juices, *Z. Ver. Dtsch. Zucker-Ind.* **24**, 309–335.

Sery, T. W., Hehre, E. J. (1956) Degradation of dextrans by enzymes of intestinal bacteria, *J. Bacteriol.* **71**, 373–380.

Sevenier, R., Hall, R. D., van der Meer, I. M., Hakkert, H. J. C., van Tunen, A. J., Koops, A. J. (1998) High level fructan accumulation in a transgenic sugar beet, *Nature Biotechnol.* **16**, 843–846.

Seymour, F. R., Julian, R. L. (1979) Fourier-transform, infrared difference-spectrometry for structural analysis of dextrans, *Carbohydr. Res.* **74**, 63–75.

Seymour, F. R., Knapp, R. D. (1980) Unusual dextrans: 13. Structural analysis of dextrans from strains of *Leuconostoc* and related genera, that contain 3-O-α-D-glucosylated α-D-glucopyranosyl residues at the branch points, or in consecutive linear position, *Carbohyd. Res.* **81**, 105–129.

Seymour, F. R., Knapp, R. D., Bishop, S. H. (1976) Determination of the structure of dextran by ^{13}C-nuclear magnetic resonance spectroscopy, *Carbohydr. Res.* **51**, 179–194.

Seymour, F. R., Slodki, M. E., Plattner, R. D., Jeanes, A. (1977) Six unusual dextrans: methylation structural analysis by combined G. L. C.-M. S. of per-O-acetylaldononitriles, *Carbohydr. Res.* **53**, 153–166.

Seymour, F. R., Chen, E. C. M., Bishop, S. H. (1979a) Methylation structural analysis of unusual dextrans by combined gas-liquid chromatography-mass spectrometry, *Carbohydr. Res.* **68**, 113–121.

Seymour, F. R., Knapp, R. D., Bishop, S. H. (1979b) Correlation of the structure of dextrans to their ^{1}H-NMR spectra, *Carbohydr. Res.* **74**, 77–92.

Sharpe, E. S., Stodola, F. H., Koepsell, H. J. (1960) Formation of isomaltulose in enzymatic dextran synthesis, *J. Org. Chem.* **25**, 1062–1063.

Shimamura, A., Tsumori, H., Mukasa, H. (1982) Purification and properties of *Streptococcus mutans* extracellular glucosyltransferase, *Biochim. Biophys. Acta* **702**, 72–80.

Shimizu, E., Unno, T., Ohba, M., Okada, G. (1998) Purification and characterization of an isomaltotriose-producing *endo*-dextranase from a *Fusarium* sp., *Biosci. Biotechnol. Biochem.* **62**, 117–122.

Shiroza, T., Ueda, S., Kuramitsu, H. K. (1987) Sequence analysis of the *gftB* gene from *Streptococcus mutans*, *J. Bacteriol.* **169**, 4263–4270.

Sidebotham, R. L. (1974) Dextrans, in: *Adv. Carbohydr. Chem. Biochem.* (Tipson, R. S., Horton, D., Eds.), New York: Academic Press, 371–444, Vol. 30.

Simonson, L. G., Liberta, A. E. (1975) New sources of fungal dextranase. *Mycologia* **4**, 845–851.

Sinha, J., Dey, P. K., Panda, T. (2000) Aqueous two-phase: the system of choice for extractive fermentations, *Appl. Microbiol. Biotechnol.* **54**, 476–486.

Slodki, M. E., England, R. E., Plattner, R. D., Dick Jr., W. E. (1986) Methylation analyses of NRRL dextrans by capillary gas-liquid chromatography, *Carbohyd. Res.* **156**, 199–206.

Smith, M. R., Zahnley, J., Goodman, N. (1994) Glucosyltransferase mutants of *Leuconostoc mesenteroides* NRRL B-1355, *Appl. Environ. Microbiol.* **60**, 2723–2731.

Smith, M. R., Zahnley, J. C., Wong, R. Y., Lundin, R. E., Ahlgren, J. A. (1998) A mutant strain of *Leuconostoc mesenteroides* B-1355 producing a glucosyltransferase synthesizing $\alpha(1 \to 2)$ glucosidic linkages, *J. Ind. Microbiol. Biotechnol.* **21**, 37–45.

Stiles, M. E., Holzapfel, W. H. (1997) Lactic acid bacteria of foods and their current taxonomy, *Int. J. Food Microbiol.* **36**, 1–29.

Stodola, F. H., Sharpe, E. S., Koepsell, H. J. (1956) The preparation, properties and structure of the disaccharide leucrose, *J. Amer. Chem. Soc.* **78**, 2514–2518.

Su, D., Robyt, J. F. (1994) Determination of the number of sucrose and acceptor binding sites for *Leuconostoc mesenteroides* B-512FM dextransucrase, and the confirmation of the two-site mechanism for dextran synthesis, *Arch. Biochem. Biophys.* **308**, 471–476.

Sun, J., Cheng, X., Zhang, Y., Yan, Z., Zhang, S. (1988) A strain of *Paecilomyces lilacinus* producing high quality dextranase, *Ann. N. Y. Acad. Sci.* **542**, 192–194.

Suzuki, M., Unno, T., Okada, G. (1999) Simple purification and characterization of an extracellular dextrin dextranase from *Acetobacter capsulatum* ATTC 11894, *J. Appl. Glycosci.* **46**, 469–473.

Swengers, D. (1991) Leucrose, a ketodisaccharide of industrial design, in: *Carbohydrates as Organic Raw Materials* (Lichtenthaler, F. W., Ed.), Weinheim, Germany: VCH, 183–195.

Takahashi, M., Okada, S., Uchimura, T., Kozaki, M. (1992) *Leuconostoc amelibiosum* Schillinger, Holzapfel, and Kandler 1989 is a later subjective synonym of *Leuconostoc citreum* Farrow, Facklam, and Collins 1989, *Int. J. Syst. Bacteriol.* **42**, 649–651.

Takazoe, I. (1989) Palatinose - an isomeric alternative to sucrose, in: *Progress in Sweeteners* (Grenby, T. H., Ed.), New York: Elsevier Science, 143–167.

Tarr, H. L. A., Hibbert, H. (1931) Studies on reactions relating to carbohydrates and polysaccharides. XXXVII. The formation of dextran by *Leuconostoc mesenteroides*, *Can. J. Res.* **5**, 414–427.

Taylor, C., Cheetham, N. W. H., Walker, G. J. (1985) Application of high-performance liquid chromatography to a study of branching dextrans, *Carbohydr. Res.* **137**, 1–12.

Tjerneld, F. (1992) Aqueous two-phase partitioning on an industrial scale, in: *Poly(Ethylene Glycol) Chemistry: Biotechnical and Biomedical Applications* (Harris, J. M., Ed.), New York: Plenum Press, 85–102.

Torii, M., Sakakibara, K., Misaki, A., Sawai, T. (1976) Degradation of alpha-linked D-gluco-oligosaccharides and dextrans by an isomaltodextranase preparation from *Arthrobacter globiformis* T6, *Biochem. Biophys. Res. Comm.* **70**, 459–464.

Tsuchiya, H. M., Koepsell, H. J. (1954) Production of dextransucrase. U.S. Patent 2,686,147.

Tsuchiya, H. M., Hellman, N. N., Koepsell, H. J., Corman, J., Stringer, C. S., Rogovin, S. P., Bogard, M. O., Bryant, G., Feger, V. H., Hoffman, C. A., Senti, F. R., Jackson, R. W. (1955) Factors affecting molecular weight of enzymatically synthesized dextran, *J. Amer. Chem. Soc.* **77**, 2412–2419.

Tsuchiya, H. M., Jeanes, A., Bricker, H. M., Wilham, C. A. (1956) Production of dextranase. U.S. Patent 2,742,399.

Ueda, S., Shiroza, R., Kuramitsu, H. K. (1988) Sequence analysis of the *gtfC* gene from *Streptococcus mutans* GS-5, *Gene* **69**, 101–109.

Usher, T. C. (1989) Method for the manufacture of dextran sulfate and salts thereof. U.S. Patent 4,855,416.

Van Cleve, J. W., Schaefer, W. C., Rist, C. E. (1956) The structure of NRRL B-512 dextran. Methylation studies, *J. Amer. Chem. Soc.* **78**, 4435–4438.

Vandamme, E. J., Bruggeman, G., De Baets, S., Vanhooren, P. T. (1996) Useful polymers of microbial origin, *Agro-Food-Industry Hi-Tech* **Sept./Oct.**, 21–25.

Van Tieghem, P. (1878) On sugar-mill gum, *Ann. Sci. Nat. Bot. Biol. Veg.* **7**, 180–203.

Walker, G. J. (1978) Dextrans, in: *International Review of Biochemistry. Biochemistry of Carbohydrates II*, (Manners, D. J., Ed.), Baltimore, MD: University Park Press, 75–125, Vol. 16.

Webb, E., Spencer-Martins, I. (1983) Extracellular endodextranase from the yeast *Lipomyces starkeyi*, *Can. J. Microbiol.* **29**, 1092–1095.

Wilke-Douglas, M., Perchorowicz, J. T., Houck C. M., Thomas, B. R. (1989) Methods and compositions for altering physical characteristics of fruit and fruit products. PCT Patent WO89/12386.

Wolff, I. A., Mellies, R. L., Rist, C. E. (1955) Fractionation of dextran products. U.S. Patent 2,712,007.

Yalpani, M. (1986) Preparation and applications of dextran-derived products in biotechnology and related areas, *CRC Crit. Rev. Biotechnol.* **3**, 375–421.

Yamamoto, K., Yoshikawa, K., Kitahata, S., Okada, S. (1992) Purification and some properties of dextrin dextranase from *Acetobacter capsulatus* ATTC 11894, *Biosci. Biotechnol. Biochem.* **56**, 169–173.

Yamamoto, K., Yoshikawa, K., Okada, S. (1993a) Dextran synthesis from reduced maltooligosaccharides by dextrin dextranase from *Acetobacter capsulatus* ATTC 11894, *Biosci. Biotechnol. Biochem.* **57**, 136–137.

Yamamoto, K., Yoshikawa, K., Okada, S. (1993b) Effective dextran production from starch by dextrin dextranase with debranching enzyme, *J. Ferment. Bioeng.* **76**, 411–413.

7
Exopolysaccharides of Lactic Acid Bacteria

Ir. Isabel Hallemeersch[1], Ir. Sophie De Baets[2], Prof. Dr. Ir. Erick J. Vandamme[3]

[1] Laboratory of Industrial Microbiology and Biocatalysis, Department of Biochemical and Microbial Technology, Faculty of Agricultural and Applied Biological Sciences, Ghent University, Coupure links 653, B-9000 Gent, Belgium; Tel.: +32-9-264-6029; Fax: +32-9-264-6231; E-mail: Isabel.Hallemeersch@rug.ac.be

[2] Laboratory of Industrial Microbiology and Biocatalysis, Department of Biochemical and Microbial Technology, Faculty of Agricultural and Applied Biological Sciences, Ghent University, Coupure links 653, B-9000 Gent, Belgium; Tel.: +32-9-264-6028; Fax: +32-9-264-6231; E-mail: Sophie.DeBaets@rug.ac.be

[3] Laboratory of Industrial Microbiology and Biocatalysis, Department of Biochemical and Microbial Technology, Faculty of Agricultural and Applied Biological Sciences, Ghent University, Coupure links 653, B-9000 Gent, Belgium; Tel.: +32-9-264-6027; Fax: +32-9-264-6231; E-mail: Erick.Vandamme@rug.ac.be

1	Introduction	258
2	Historical Outline	258
3	Chemical Structure	259
4	Occurrence	263
5	Physiological Function	263
6	Chemical Analysis and Detection	264
6.1	Separation of EPS and Microbial Cells	264
6.2	Isolation and Purification	264
6.3	Structural Analysis	265
7	Biosynthesis	265
8	Genetics and Regulation	266

Polysaccharides and Polyamides in the Food Industry. Properties, Production, and Patents.
Edited by A. Steinbüchel and S. K. Rhee
Copyright © 2005 WILEY-VCH Verlag GmbH & Co. KGaA, Weinheim
ISBN: 3-527-31345-1

9	**Factors Influencing Growth and EPS Production**	270
9.1	Effect of Physico-chemical Parameters	270
9.2	Effect of Nutritional Parameters	271
9.3	Kinetics of EPS Biosynthesis	273
10	**Applications**	273
11	**References**	276

EPS	exopolysaccharides
GRAS	generally recognized as safe
LAB	lactic acid bacteria
NMR	nuclear magnetic resonance

1
Introduction

The exopolysaccharides (EPS) produced by lactic acid bacteria (LAB) can be divided into three major groups, based on their composition: (1) glucans, namely dextrans, alternans, and mutans; (2) fructans such as levan; and (3) heteropolysaccharides produced by mesophilic and thermophilic LAB (Cerning, 1990). As details of the glucans and fructans are described in Chapters 12–14 of this Volume, only the latter group will be discussed here. These EPS are composed of linear and branched repeating units, and vary in size from disaccharides to heptasaccharides (Petry et al., 2000).

Over the past few decades, there has been a growing interest in ropy LAB, such as *Lactoccocus lactis* subsp. *cremoris*, *L. lactis* subsp. *lactis*, *Lactobacillus delbrueckii* subsp. *bulgaricus*, *Lb. helveticus*, *Streptococcus salivarius* subsp. *thermophilus*, etc. Based on the "generally recognized as safe" (GRAS) status of these bacteria, their EPS preparations are widely applied as thickening, gelling and stabilizing agents in the food industry (Sutherland, 1994). Their most important application area in this respect is undoubtedly in the dairy industry, where they are used in the production of various fermented milk products and contribute to both texture and mouthfeel (Cerning, 1995). Moreover, it was suggested that they also possess advantageous biological functions and are beneficial for human health (Nakajima et al., 1992; Kitazawa et al., 1993; Gibson and Roberfroid, 1995; Hosono et al., 1997).

2
Historical Outline

Although ropy LAB such as *S. thermophilus* and *Lb. delbreuckii* subsp. *bulgaricus* have been available commercially as dairy starter cultures since the early 1900s, the chemical composition and structures of their EPS were reported in detail only as recently as the 1980s (Marshall and Rawson, 1999). Little information existed regarding the level of polysaccharide produced, the culture conditions and the rheological properties, in contrast to the situation with glucans such as dextran, on which most attention was focused and which had already undergone intensive study (Forsén and Häivä, 1981).

Nonetheless, during the 1950s and 1970s, a number of articles were published on the composition of heteropolysaccharides produced by LAB. These early investigations did not refer to any precise culture conditions, nor analytical details, and should therefore be treated with some caution (Sundman, 1953; Nilsson and Nilsson, 1958; Groux, 1973; Tamime, 1978).

It was not until the 1980s that EPS from LAB received the attention of numerous investigators, particularly in France and the Netherlands, where the use of stabilizing agents from plant or animal origin was prohibited. Among the early publications on the chemical composition of EPS no clear picture emerged (Cerning et al., 1986), and EPS were seen either as proteinaceous material, a carbohydrate–protein complex, or simply as a complex carbohydrate. These differences were due to different isolation and purification methods, each with their variable efficiencies. This was especially so because EPS purification was hampered by the use of complex culture media, such as milk or whey; moreover, some bacterial strains were able to synthesize more than one type of EPS (Marshall et al., 1995).

Subsequently, a great deal of research was carried out on various aspects of EPS synthesis, such as the influence of nutritional and physico-chemical fermentation parameters, the isolation and characterization of the EPS, and the rheological properties. Detailed structural and rheological studies were performed in order to provide a better insight into the mechanisms by which the LAB and their EPS influence the consistency of dairy products (Staaf et al., 2000; Faber et al., 2001).

During the late 1980s, investigations into the instability of the mucoid character of LAB revealed the involvement of plasmids in EPS synthesis (Vedamuthu and Neville, 1986; Neve et al., 1988). This was the beginning of a series of studies on the molecular biology and genetics of EPS produced by various LAB.

During the past decade, a large number of genes involved in the synthesis of repeating units, polymerization, chain length determination and export have been characterized and functionally analyzed. In addition, the entire genome of *L. lactis* has been reported recently (Bolotin et al., 1998; Kleerebezem et al., 2000; Ricciardi and Clementi, 2000). Models were developed in order to predict the behavior and the effect of EPS addition to food products, and this has led to the identification of several important properties, though these models require further validation and fine-tuning (Kleerebezem et al., 1999).

Nowadays, the challenge is to increase production levels and to modify the structure and hence the properties of EPS, by using genetic approaches and enzyme technology. The purpose of this type of polysaccharide engineering is the synthesis of tailor-made oligo- and polysaccharides, for a variety of specific applications (De Vuyst and Degeest, 1999a; Kleerebezem et al., 2000).

3 Chemical Structure

Heteropolysaccharides produced by LAB have been investigated to a lesser extent when compared with other polymers. The overall level of EPS production is low and is often characteristically variable (Cerning, 1990). Most strains produce a limited quantity of polymer, perhaps up to 200 mg L^{-1}, although some strains have been found to produce up to 4 g L^{-1}.

Many LAB synthesize heteropolysaccharides with a molecular weight in excess of 10^6 daltons. In addition, their composition, structure and physico-chemical properties are highly variable, these are being influenced by the composition of the culture medium (Ricciardi and Clementi, 2000).

Today, it is generally recognized that EPS from LAB are composed of linear and branched repeating units that vary in size from disaccharides to heptasaccharides, and which contain α- and β- linkages. Most EPS contain D-glucose, D-galactose and/or L-rhamnose in different ratios, but other hexoses such as D-mannose, D-fructose and pentoses such as L-fucose, D-xylose and D-arabinose also appear. Hexosamines and uronic acids are also found in minor quantities (Andaloussi et al., 1995; Marshall et al., 1995; Petry et al., 2000). The chemical composition of a few EPS is summarized in Table 1, while Table 2 shows the structure of typical repeating units.

The EPS from mesophilic LAB have a more varied composition than those from thermophilic LAB, and contain sometimes acetylated and phosphorylated residues (Ricciardi and Clementi, 2000). Different authors have described the chemical composition of EPS produced by *Lactobacillus* strains (Toba et al., 1991; Kojic et al., 1992; Cerning, 1994; Van den Berg et al., 1995; Robijn et al., 1996). Several lactococci have also been described that produce EPS; these usually contain glucose and galactose, but rhamnose and charged residues are also found quite commonly (Cerning, 1994). Most *Pediococcus* strains, which occur most often as spoilers in beer and wine, produce β-glucans of high molecular weight (Ricciardi and Clementi, 2000).

Galactose is seen as the major monosaccharide in the EPS produced by thermophilic

Tab. 1 Chemical composition of several exopolysaccharides (EPS) synthesized by lactic acid bacteria (LAB)

Strain	Gal	Glc	Rha	Other	Reference
Lb. delbreukii subsp bulgaricus NCFB 2772	+	+	+	–	Grobben et al. (1995, 1996)
Lb. delbreukii subsp. *bulgaricus* CNRZ 1187	+	+	–	–	Petry et al. (2000)
Lb. delbreukii subsp. *bulgaricus* CNRZ 416	+	+	+	–	Petry et al. (2000)
Lb. casei CG11	–	+	+	–	Kojic et al. (1992)
Lb. casei CG11	+	+	+	–	Cerning (1994)
Lb. helveticus var. *jugurti*	+	+	–	–	Oda et al. (1983)
Lb. helveticus LB161	+	+	–	Ac, P	Staaf et al. (2000)
Lb. kefiranofaciens	+	+	–	–	Toba et al. (1991)
Lb. paracasei	+	+	+	–	Van Calsteren (2001)
Lb. paracasei 34-1	+	–	–	GP	Robijn et al. (1996)
Lb. rhamnosus	+	+	+	–	Van Calsteren (2001)
Lb. sake 0-1	–	+	+	Ac, GP	Robijn et al. (1996)
Lb. sake 0-1	–	+	+	–	Van den Berg et al. (1995)
L. lactis subsp. *cremoris* LC330	+	+	–	GlcNAc	Marshall et al. (1995)
L. lactis subsp. *cremoris* LC330	+	+	+	GlcNAc P	Marshall et al. (1995)
L. lactis subsp. *cremoris* SBT 0495	+	+	+	P	Nakajima and Toyoda (1990)
Pediococcus	–	+	–	–	Cerning (1994)
S. thermophilus EU20	+	+	+	–	Marshall et al. (2001)
S. thermophilus CNCMI 733	+	+	–	GalNAc	Doco et al. (1991)
S. thermophilus OR 901	+	–	+	–	Ariga et al. (1992)
S. thermophilus S3	+	–	+	Ac	Faber et al. (2001)

Gal, galactose ; Glc, glucose ; Rha, rhamnose ; Ac, acetate ; P, phosphate ; GP, glycerol-3-phosphate ; Nac, N-acetyl.

3 Chemical Structure

Tab. 2 Primary structure of several EPS synthesized by LAB

Strain	Structure	Reference
Lb. delbrueckii subsp. *bulgaricus* NCFB 2772	β-D-Gal*p* β-D-Gal*p* α-L-Rha*p* ↓ ↓ ↓ →2)-α-D-Gal*p*-(1→3)-β-D-Glc*p*-(1→3)-β-D-Gal*p*-(1→4)-α-D-Gal*p*-(1→	Grobben et al. (1995, 1996)
Lb. helveticus LB161	β-D-Glc*p* β-D-Glc*p* ↓ ↓ →4)-α-D-Glc*p*-(1→4)-β-D-Gal*p*-(1→3)-α-D-Gal*p*-(1→2)-α-D-Glc*p*-(1→3)-β-D-Glc*p*-(1→	Staaf et al. (2000)
Lb. paracasei 34-1	*sn*-Glycerol-3-phosphate-3 ↓ →3)-β-D-Gal*p*NAc-(1→4)-β-D-Gal*p*-(1→6)-β-D-Gal*p*-(1→6)-β-D-Gal*p*-(1→	Robijn et al. (1996)

Lb. sake 0-1	β-D-Glc*p*-(1→6) (Ac)₀.₈₅
	             ↓ ↓
	→4)-β-D-Glc*p*-(1→4)-α-D-Glc*p*-(1→3)-β-L-Rha*p*-(1→
	                                     ↑
	-*sn*-Glycerol-3-phosphate-3→4)-α-L-Rha*p*-(1→3)
		Robijn et al. (1996)

L. lactis subsp. *cremoris* SBT 0495	α-L-Rha*p*-(1→2)
	               ↓
	→4)-β-D-Glc*p*-(1→4)-β-D-Gal*p*-(1→4)-β-D-Glc*p*-(1→
	               ↑
	α-D-Gal*p*-1-phosphate
		Nakajima and Toyoda, 1990

Tab. 2 (cont.)

Strain	Structure	Reference
S. thermophilus CNCMI 733	→3)-β-D-Gal*p*-(1→3)-β-D-Glc*p*-(1→3)-α-D-Gal*p*NAc-(1→ β-D-Gal*p*-(1→6)-β-D-Gal*p*-(1→4)↑ α-D-Gal*p*-(1→6)↑	Doco et al., 1991
S. thermophilus OR 901	→2)-α-D-Gal*p*-((1→3)-α-D-Gal*p*-(1→3)-α-L-Rha*p*-(1→2)-α-L-Rha*p*-(1→ β-D-Gal*p*2Ac↑	Ariga et al., 1992
S. thermophilus S3	→3)-β-D-Gal*p*-((1→3)-α-D-Gal*p*-(1→3)-α-L-Rha*p*-(1→2)-α-L-Rha*p*-(1→2)- α-D-Gal*p*-(1→	Faber et al., 2001

Gal = Galactose, Glc = Glucose, Rha = Rhamnose, Ac = Acetate

LAB, most likely because it is metabolized less quickly than glucose and hence is available for polymer synthesis (Cerning, 1990).

Many EPS types from *S. thermophilus* strains have been investigated as to their composition and characteristics. Most such heteropolysaccharides contain D-glucose and D-galactose and, on occasion, also L-rhamnose.

Heteropolysaccharide production was also observed in other thermophilic LAB such as *Lactobacillus helveticus* and *Lactobacillus delbreuckii* subsp. *bulgaricus*. Grobben et al. (1995, 1996) demonstrated with a *Lb. delbreuckii* subsp. *bulgaricus* strain, that the EPS composition varied with different sugars present in the growth medium. When using glucose or lactose as a carbon source, the EPS consisted of glucose, galactose, and rhamnose in a ratio of 1:6.8:0.7, but in the presence of fructose, the EPS formed contained only glucose and galactose in a ratio of 1:2.4. However, when glucose and fructose are used together, the composition of the EPS is similar to that obtained when either glucose or lactose was the carbon source. Marshall et al. (1995) isolated two EPS types from the culture broth of the same strain–a high and a low molecular fraction. Finally, De Vuyst and Degeest (1999b) showed that both fractions were similar in their monomeric composition and that the growth medium was in fact influencing EPS composition.

Petry et al. (2000) also investigated the EPS of two *Lb. delbreuckii* subsp. *bulgaricus* strains and concluded that the monomeric sugar composition remained the same, but that the relative proportions of the individual monosaccharides varied under different fermentation conditions.

In order to obtain insight into the structure–function relationship, it is necessary to study the conformation of the polymer in solution as well as its dynamic behavior. The three-dimensional structure of a polysaccharide is dependent on the time-averaged ring conformation of the monosaccharides and the relative orientations of adjacent monosaccharides. Besides the conformations around the glycosidic linkages, the intramolecular hydrogen bonds also contribute to the conformation, and this can be important for a complete understanding of the solution behavior. NMR spectroscopy does not provide sufficient information to elucidate the complete conformation (Vliegenthart et al., 2001). As the secondary and tertiary conformation of a polysaccharide is mainly dependent on its primary (chemical) structure, relatively small changes in the chemical structure might have a major effect on the conformation and the physical and chemical properties of the polysaccharide. However, despite all information already currently available, prediction of the properties on basis of polymer structure is not yet possible.

Cheese and fermented meat and vegetables are also an important source of ropy LAB (Kojic et al., 1992; Van den Berg et al., 1993). Recently, Smitinont et al. (1999) isolated two EPS producing *Pediococcus* strains from traditional Thai fermented foods; although grown on sucrose, the EPS differed from dextran.

EPS-producing LAB can also be disadvantageous however, for example when causing spoilage of beer, wine, and vacuum-packed cooked meat products (Korkeala et al., 1988; Morin, 1998).

More severe problems arise when biofilm formation occurs on heat exchanger plates in cheese and milk factories, resulting in either excessive openness in cheese texture, or taste changes in milk. In particular, thermophilic LAB can cause significant problems by contaminating heat exchanger plates on the downstream side of the pasteurizers, whereby already pasteurized milk may become re-contaminated (Bouwman et al., 1982; Neu and Marshall, 1990).

4
Occurrence

The LAB form a diverse group of bacteria, including the genera *Lactobacillus*, *Lactococcus*, *Streptococcus*, *Enterococcus*, *Leuconostoc*, and *Pediococcus*. They are Gram positive, have a low DNA GC-content (32–53 mol.%), are catalase-, reductase-, and oxidase-negative, and are also nonmotile and nonsporulating. *Bifidobacteria* are often associated with these genera as they are also added to traditional yogurt cultures (Dellaglio, 1994).

EPS-producing LAB can be isolated from dairy products such as the well-known Scandinavian fermented milk products "viili" and "longfil", yogurts, fermented milk drinks, and kefir (Kandler and Kunath, 1983; Toba et al., 1991; Ariga et al., 1992).

5
Physiological Function

Although EPS are not essential to ensure the viability of bacterial cells, their exact function in nature has not yet been clearly defined and is most likely highly complex (Cerning, 1994).

In the past, it has generally been accepted that EPS do not serve as storage materials, since most EPS-producing strains are not able to catabolize EPS, nor use them as a carbon source. In the case of LAB this remains doubtful however, as the amount of EPS produced often decreases upon prolonged fermentation. One possible explanation is that the synthesized EPS are degraded enzymatically by the LAB themselves. Recently, Pham et al. (2000) reported on the synthesis of several glycohydrolases by a *Lb. rhamnosus* strain which was able to hydrolyze the EPS to a limited extent.

The role of EPS in pathogenesis has been investigated, since their synthesis–which results mainly in capsule formation–is widespread among pathogenic bacteria. The presence of EPS alone does not appear to be sufficient to turn bacterial cells virulent. It has been suggested that microorganisms producing EPS are more hydrophilic, and thus are less susceptible to phagocytosis (Jann and Jann, 1977). EPS also seem to play a role in the protection against phage attack, antibiotics and other toxic compounds.

Another proposed function is the protection against desiccation and other extreme physical conditions; this is due to the fact that EPS are present as a highly hydrated layer surrounding the bacterial cells. EPS also play an important role in the adhesion of microorganisms onto solid surfaces, although the adsorption process is complex and the exact impact of EPS in the process has not yet been completely elucidated (Cerning, 1994).

6
Chemical Analysis and Detection

6.1
Separation of EPS and Microbial Cells

The first step in the purification of EPS is the separation of microbial cells from the EPS-containing culture broth. This is normally done by centrifugation, and when working at laboratory scale, ultracentrifugation can be used. It is difficult to define an optimal g-value as this depends on the specific viscosity of the culture broth (Sutherland, 1972). Sometimes a too-high viscosity may prevent easy sedimentation of bacterial cells. The addition of electrolytes (NaCl) may facilitate the separation by neutralizing the charges on the polysaccharides (Cerning, 1994). Because EPS are thermostable, heat treatment can also be used to reduce the viscosity.

When present as a capsule, the EPS must first be dissociated from the cells and, depending on the nature of the association between the cells and the capsule, extreme conditions such as alkaline treatment, sonication or heating may be required (Morin, 1998).

6.2
Isolation and Purification

EPS are mainly recovered by precipitation with organic solvents, such as ethanol, acetone, or isopropanol. Organic solvents permit separation by lowering the solubility of the EPS. Although the EPS:solvent ratio is variable, about one up to three volumes of solvents are normally used. The precipitate is finally recovered by centrifugation, filtration, and pressing or settling, after which dialysis and freeze-drying of the final product is carried out (Garcia-Garibay and Marshall, 1991; Marshall et al., 1995).

EPS from LAB are often synthesized in complex media such as milk, milk ultrafiltrate and whey; the proteins present in these media, such as casein, seem to co-precipitate with the EPS and should be removed prior to the isolation step. This can be achieved enzymatically by the addition of proteases such as pronase, trypsin, or proteinase K (Cerning, 1994). Residual peptides can also be removed by repeated trichloroacetic acid precipitation, followed by centrifugation and precipitation of the EPS. Gel filtration is another possibility, although it is much slower and problems often arise due to the high viscosity of the media. Further purification of the EPS can be achieved by anion-exchange chromatography (Nakajima and Toyoda, 1990; Doco et al., 1991; Andaloussi et al., 1995).

Although EPS yields are higher in milk or whey-based media, the EPS isolation from such complex media is often tedious and time-consuming. More recently, several semi-defined and chemically defined media have been developed; these are more suitable to investigate the influence of nutrients on the growth and on the EPS production, the metabolic pathways involved, the composition of the EPS, and the rheological properties (Petry et al., 2000).

6.3
Structural Analysis

Several structures of repeating units of branched EPS from LAB have been elucidated. General methods used to study the composition and specific linkages present in polysaccharides are normally applied, including acid hydrolysis, periodate oxidation, methylation analysis, enzymatic degradation, Smith-degradation and NMR-spectroscopy. With these methods it is possible to determine not only the neutral sugars but also the anomeric configuration, the specific linkages present, and the sequence of the repeating units.

The molecular weights of polysaccharides are determined by gel filtration and high-performance liquid chromatography.

On occasion, a strain produces more than one type of EPS, and these can usually be separated using chromatography based on the differences in either molecular weight or charge (Toba et al., 1991; Marshall et al., 1995; Robijn et al., 1996; Grobben et al., 1997).

7
Biosynthesis

Unlike the biochemical pathways involved in the biosynthesis of exocellular homopolysaccharides, those of exocellular heteropolysaccharides are much more complex (Cerning, 1994).

Heteropolysaccharides are synthesized at the cytoplasmic membrane through polymerization of precursors, and are formed in the cytoplasm. These precursors are mainly UDP-nucleotide diphosphate sugars. As with the large number of enzymes involved, the sugar nucleotide precursors are not all unique to EPS synthesis; some are also involved in the synthesis of cell-wall polymers such as peptidoglycans, lipopolysaccharides, and teichoic acids. However, as they are freely soluble in the cytoplasm, they can be readily channeled to the appropriate biosynthetic process (Sutherland, 1990).

The sugar nucleotides serve different functions. First, they play an important role in sugar activation, supplying energy for the assembly of glycosyl units on appropriate carrier molecules, with the release of a diphosphate nucleotide. Second, they play a role in sugar interconversions, which involve several mechanisms such as epimerization (UDP-D-glucose → UDP-D-galactose), oxidation (UDP-D-glucose → UDP-D-glucuronic acid), decarboxylation (UDP-D-glucose → UDP-D-xylose), reduction and rearrangement (GDP-D-mannose → GDP-L-fucose) (Sutherland, 1972, 1990).

LAB can metabolize a variety of mono- and disaccharides, but as EPS are mainly produced in milk or whey-based media, the main carbon source present is lactose. Based on studies investigating the influence of sugars on EPS synthesis and the enzymes involved in their anabolism, it might be concluded that glucose or the glucose moiety of lactose is the most important sugar for

EPS production in LAB (De Vos and Vaughan, 1994; Grobben et al., 1996). Hence, glucose-1-phosphate serves as an important precursor for EPS synthesis, and phosphoglucomutase might be seen as a key enzyme linking energy generation and sugar nucleotide formation (Kleerebezem et al., 2000) (Figure 1).

De Vos (1996) suggested that if the flux via phosphoglucomutase was high enough, then EPS overproduction could be achieved. As such, the galactose moiety of lactose would be catabolized via glycolysis, while the glucose moiety would serve for EPS production. A major problem is the inability of *S. thermophilus* and *Lb. delbreuckii* subsp. *bulgaricus* to catabolize galactose, this being due to the absence of the enzyme galactokinase; hence galactose is excreted via the lactose/galactose antiport transport system. However, the gene coding for this enzyme is present but is not transcribed. Galactose-fermenting mutants were constructed and isolated by Kleerebezem et al. (1999) through repair of the promotor mutations, and as a result galactose can be fully metabolized.

The composition and amount of EPS produced are not only dependent on the type of carbon source and on the sugar nucleotide level, but also on the assembly process of the repeating units.

The involvement of an isoprenoid glycosyl lipid carrier was reported in 1971 by Troy et al. This lipid is a C55-isoprenyl phosphate (bactoprenyl phosphate, undecaprenyl phosphate), and is the same acceptor lipid that functions in the formation of several cell-wall polymers (Figure 2). As such, the availability of the isoprenoid carrier is one of the most important factors affecting EPS production. Because of competition for the same acceptor lipid, it was postulated that EPS production is stimulated under conditions which lead to a reduction in growth (e.g., a lower temperature), and so cell-wall polymer biosynthesis is reduced (Sutherland, 1972).

Few authors have reported the exact role of the lipid carrier, and details of the mechanism remain unknown. However, possibilities include facilitation of the formation of the repeating units, solubilization of hydrophilic oligosaccharides in a hydrophobic membrane environment, and transport across the membrane (Troy, 1979; Cerning, 1990).

Once transported through the cell membrane, the polysaccharide can be excreted into the environment or will remain attached to the cell as a capsule. The transport of polysaccharides is an energy-demanding process and has not been fully clarified (Van den Berg et al., 1995).

Recently, *in vitro* experiments have been conducted to elucidate the biosynthesis of the polysaccharide backbone of *L. lactis* NIZO B40. As shown in Figure 3, EPS synthesis occurs through sequential addition of sugar residues by specific glycosyl transferases from sugar nucleotides to a growing repeating unit anchored to the lipid acceptor, thereby yielding the EPS (Kleerebezem et al., 1999, 2000).

8
Genetics and Regulation

A well-known problem of LAB in the dairy industry is the instability of EPS production at the genetic level, as well as the instability of the ropy texture itself (Cerning 1990, 1994). Loss of the ropy character may occur after numerous transfers and prolonged incubation periods, even at optimum growth temperatures. Therefore, strains have to be reselected regularly from the master culture to conserve the ropy character in industrially applied strains.

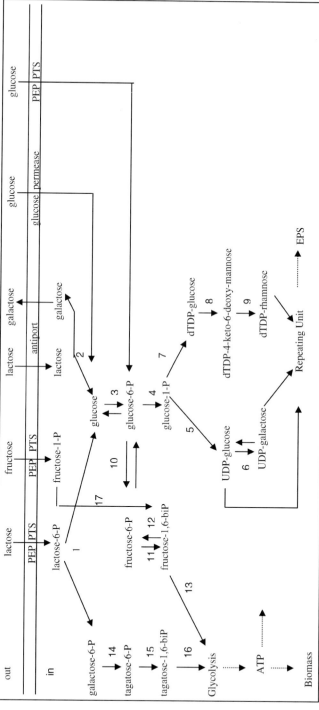

Fig. 1 Schematic representation of the lactose-, glucose-, and fructose-utilizing pathways and exopolysaccharide biosynthesis. *Lactococcus lactis* transports lactose via the lactose phosphoenolpyruvate (PEP) -dependent phosphotransferase system (PTS), while the galactose-negative *Streptococcus thermophilus* and *Lactobacillus delbrueckii* subsp. *bulgaricus* transport lactose via a lactose/galactose antiport transport system. Enzymes involved are: 1, phospho-β-galactosidase; 2, β-galactosidase; 3, glucokinase; 4, phosphoglucomutase; 5, UDP-glucose pyrophosphorylase; 6, UDP-galactose-4-epimerase; 7, dTDP-glucose pyrophosphorylase; 8, dehydratase; 9, epimerase reductase; 10, phosphoglucoisomerase; 11, 6-phosphofructokinase; 12, fructose-1,6-biphosphatase; 13, fructose-1,6-biphosphate aldolase; 14, galactose-6-phosphate isomerase; 15, tagatose-6-phosphate kinase; 16, tagatose-1,6-biphosphate aldolase; 17, fructokinase (Kleerebezem et al., 1999).

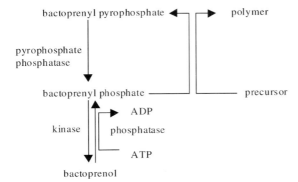

Fig. 2 Structure of the active form of the isoprenoid lipid carrier. (b) Activation and deactivation of the isoprenoid lipid; the polymer can be EPS, lipopolysaccharide, or peptidoglycan (Cerning, 1994).

Numerous plasmids encoding for EPS production have already been identified in mesophilic LAB. Moreover, it has been proven that the loss of the EPS-producing capacity in these bacteria is associated with the loss of plasmids (Vedamuthu and Neville, 1986; Neve et al., 1988). In thermophilic LAB, the EPS production is not encoded by plasmids, but the required genes are located on the chromosome. It is likely that the genetic instability is due to mobile genetic elements or to a generalized genomic instability, including deletions and rearrangements. Both phenomena were observed for *Lb. delbreuckii* subsp. *bulgaricus* and *S. thermophilus* (Stingele et al., 1996).

During the past few years, specific gene clusters have been characterized for some EPS-producing lactococci and streptococci; in addition, the entire genome sequence of *L. lactis* has recently been completed (Kleerebezem et al., 2000). These gene clusters contain *eps* genes, which are involved in the synthesis of the repeating units, export, polymerization and chain length determination. In addition to these *eps* genes, a number of housekeeping genes are also required for EPS synthesis. These genes are involved in the metabolic pathways leading to the EPS building blocks, namely the sugar nucleotides. Elucidation of the function of the glycosyltransferase genes might create opportunities for EPS engineering, while the identification and characterization of the housekeeping genes allows the design of metabolic engineering strategies, leading to increased EPS production levels (Kleerebezem et al., 1999).

The chromosomally located gene cluster of *S. thermophilus* Sfi6 consists of a 15.25 kb region coding for 15 genes; only 13 genes, *eps*A to *eps*M, were involved in EPS synthesis (Stingele et al., 1996). Low et al. (1998) partially identified and characterized the *eps* gene cluster of *S. thermophilus* MR-1C and found an organization similar to that of *S. thermophilus* Sfi6.

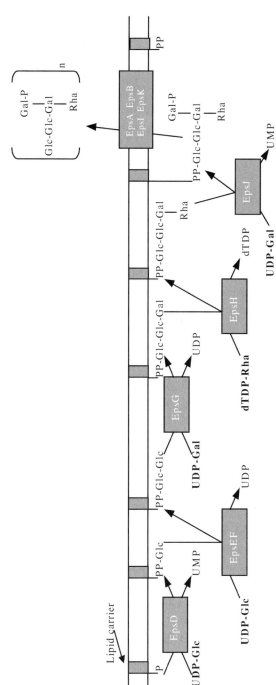

Fig. 3 Schematic representation of EPS biosynthesis by *L. lactis* NIZO B-40. Nucleotide sugars are indicated in bold; enzymes are indicated in the gray boxes by their encoding genes using established genetic nomenclature. Gal, galactose; Glc, glucose; Rha, rhamnose (Kleerebezem et al., 1999).

For *L. lactis*, the *eps* genes are encoded on large plasmids (>20 kb) and can be transferred from one strain to another by conjugation. The best-characterized lactococcal strain is *L. lactis* NIZO B40, which produces a phosphopolysaccharide. A 12-kb gene cluster specific for EPS biosynthesis, contains 14 coordinately expressed genes and is inserted between IS sequences on a 40-kb plasmid.

Major similarities exist between the organization of the genes in the different gene clusters involved in EPS biosynthesis; four functional regions can be distinguished: a region involved in regulation is located at the beginning of the gene cluster; a central region which is involved in the synthesis of the EPS repeating units; and two flanking regions that are involved in polymerization and export (Stingele et al., 1996).

Further detailed studies can offer perspectives for EPS engineering and evaluation of the possibilities and restrictions towards the production of tailor-made EPS. This concerns not only changing the chemical structure of the polymers but also other relevant features, such as its chain length.

It was shown recently that a *L. lactis* NIZO B40 EpsD deletion mutant could functionally be complemented by the homologous *Eps*D gene itself, but more important also by heterologous glucosyltransferase encoding genes, yielding an EPS with a similar molecular weight but with a different structure (Van Kranenburg et al., 1999). Other examples, which indicate possibilities for the synthesis of heterologous or new EPS structures, have also been reported (Gilbert et al., 1998; Stingele, 1998).

Another approach would be to engineer the metabolic pathways in order to increase EPS production levels, perhaps by increasing the flux towards sugar nucleotides. The so-called "housekeeping" genes are involved in the synthesis of sugar nucleotides; in this respect, the physiology of sugar nucleotide synthesis should be investigated in relation to EPS production levels (Kleerebezem et al., 1999).

At present, all genes of *L. lactis* MG1363 encoding for enzymes involved in the biosynthesis of sugar nucleotides from glucose-1-phosphate have been cloned, and specific enzyme activities, which form bottlenecks in EPS synthesis, can now be investigated. The potential of these metabolic engineering strategies to improve production levels can also be evaluated (Boels et al., 1998). The role of the genes which encode for important enzymes in these metabolic pathways is now also being investigated in other LAB.

It is expected that engineering of EPS and their biosynthetic pathways in LAB will be an interesting challenge in future metabolic research. However, the use of genetically modified microorganisms or their products will require not only legal approval, but also acceptance by the consumer (De Vuyst and Degeest, 1999b; Kleerebezem et al., 2000).

9
Factors Influencing Growth and EPS Production

9.1
Effect of Physico-chemical Parameters

Generally, EPS are produced at temperatures below the optimal growth temperature. For mesophilic LAB, EPS production increased by 50–60% if the strains are cultured at 25°C instead of at 30°C (Cerning et al., 1992). *L. lactis* subsp. *lactis* exhibited the highest EPS production at 22°C; at temperatures higher than 37°C, no further EPS production occurred. The highest EPS production for *L. lactis* subsp. *cremoris* was noted at 20°C (Marshall et al., 1995), while for *Lb. casei* EPS are produced at 25°C as well as at 30°C (Cerning, 1994; Mozzi et al., 1996).

Mozzi et al. (1995) investigated the influence of the temperature on EPS production for the thermophilic species *Lb. delbreuckii* subsp. *bulgaricus*, *Lb. acidophilus* and *S. salivarius* subsp. *thermophilus*. In milk-based media without pH control, the optimum temperature for EPS production was 42°C for *Lb. acidophilus* and 37°C for *Lb. delbreuckii* subsp. *bulgaricus*, which coincides with optimal growth temperatures. Garcia-Garibay and Marshall (1991) found a maximal EPS production for *Lb. delbreuckii* subsp. *bulgaricus* at 45°C. Grobben et al. (1995) noted an increased EPS production with an increased temperature during batch fermentation, using a defined medium with controlled pH; the EPS production was maximal at 40°C.

For *S. salivarius* subsp. *thermophilus*, the highest EPS production was noted at 30°C. These results were confirmed by Gancel and Novel (1994) and Gassem et al. (1995).

As mentioned previously, EPS synthesis is expected to be stimulated by reduced growth temperatures; as such, cell growth is reduced as well as the synthesis of cell-wall polymers. This would result in a better availability of the lipid isoprenoid carrier which is necessary for export of EPS (Sutherland, 1982).

For some LAB, the effect of the initial pH of the growth medium on EPS biosynthesis was investigated; during this type of experiment the pH was no longer controlled during fermentations and was allowed to vary freely. On occasion, no clear effect was noted, but for *S. thermophilus* S22 an optimal initial pH of 7.0 for lactose as a carbon source and an optimal initial pH of 5.5 for sucrose was determined (Gancel and Novel, 1994).

A maximal EPS yield was obtained for *Lb. casei* CRL 87 when the pH was controlled at 6.0 during the fermentation; moreover, no decrease of EPS level occurred upon prolonged incubation (Mozzi et al., 1994).

In a continuous fermentation with *Lb. delbreuckii* subsp. *bulgaricus*, and using a whey-based medium, maximal EPS production occurred at pH 6.5, while the optimal pH for EPS production in batch fermentations was 5.8 (Van den Berg et al., 1995).

With *Lb. delbreuckii* subsp. *bulgaricus* CNRZ 1187 and CNRZ 416, about three-fold and four-fold respectively more EPS was produced when the pH was controlled at 6.0 (Petry et al., 2000).

In general, higher EPS yields are obtained when fermentations are carried out under controlled pH conditions. The optimal pH control range for EPS production by LAB is often close to pH 6–6.5 (Mozzi et al., 1994), and this may be a limiting factor when considering industrial use of these bacteria in fermented milk drinks, as the pH is allowed to vary freely during such fermentations (De Vuyst and Degeest, 1999a).

With regard to temperature, EPS production appears to increase under less than optimal pH conditions for growth, though this might also be explained by a better availability of the lipid carriers (Ricciardi and Clementi, 2000).

According to Cerning (1990), EPS biosynthesis by LAB is highest under aerobic growth conditions, and this explains why more EPS is excreted during growth on solid media rather than in liquid cultures. However, another reason might be that adhesion of the bacterial cells onto a solid surface stimulates EPS formation.

Mäkelä and Korkeala (1992) investigated the effect of aeration, temperature and medium composition for different LAB. Aerobic or anaerobic conditions resulted in better EPS production, depending on the strain.

9.2
Effect of Nutritional Parameters

In general, milk-based media or other complex media containing peptones or yeast extract are used for culturing EPS-producing

LAB. Higher EPS yields are indeed obtained, but isolation and purification of the polymers are hampered. Recently, some chemically defined media were developed which greatly facilitate the study of the influence of nutrients on EPS production, growth, chemical composition, and rheological properties.

Cerning et al. (1986) found that the addition of casein enhanced growth, but not EPS synthesis, although contrasting results were subsequently reported which indicated a positive effect on EPS production by *Lb. delbrueckii* subsp. *bulgaricus* following the addition of hydrolyzed casein (Garcia-Garibay and Marshall, 1991; Cerning, 1994).

Another point identified was that the carbon source influences not only the composition of the EPS produced by LAB but also the total amount produced. *Lb. casei* CG11 produced more EPS when glucose was added to milk or milk ultrafiltrate; moreover, the monomeric composition changed when glucose was used as the major monosaccharide rather than galactose (Cerning et al., 1992). In a later study, Mozzi et al. (1995) obtained the highest EPS yield for *Lb. casei* CRL 87 when using galactose as the carbon source.

Considerably more EPS was synthesized in a chemically defined medium by *Lb. delbreuckii* subsp. *bulgaricus* NCFB 2772 when grown on glucose, lactose, or glucose and fructose rather than on mannose or fructose. A continuous culture was used to study the effect of glucose and/or fructose on EPS production and composition. When grown on glucose or glucose and fructose, the EPS contained rhamnose, while in the presence of fructose alone, no rhamnose was present in the EPS and the overall EPS yield remained low. It was found that in this latter case, the enzymes involved in the synthesis of rhamnose were absent; moreover, when glucose was present in the culture medium, higher activities were determined for various enzymes involved in EPS synthesis (Grobben et al., 1995, 1996).

Later, it was shown that this strain produces two types of EPS: a high-molecular fraction, which appears to be influenced by the carbon source; and a low-molecular fraction, which is produced more consistently (Grobben et al., 1997).

For another *Lb. delbreuckii* subsp. *bulgaricus* strain, the composition of the EPS was found to be independent of the nature of the carbon source (Manca de Nadra et al., 1985). No influence of the carbon source on the amount of EPS produced by *Lb. sake* 01 was noted by Van den Berg et al. (1995).

Petry et al. (2000) designed a chemically defined medium that allowed good growth and EPS production for several *Lb. delbreuckii* subsp. *bulgaricus* strains. The presence of excess sugar in the growth medium (10–20 g L^{-1}) did not stimulate EPS production, as was proven for *Lb. casei* and *Lb. rhamnosus*. Independently of the carbon source, the monomeric composition remained similar.

In the case of *S. thermophilus* LY03, a clear influence of the type of carbon source on the amount of EPS produced was noted, while the composition remained constant. This strain produces two polysaccharides with identical composition, but different molecular weights. The proportion in which these polysaccharides are synthesized is strongly dependent on the C/N ratio of the culture medium (De Vuyst and Degeest, 1999b).

Other nutrients such as vitamins, amino acids and minerals might also affect the composition and quantity of EPS produced. *Lb. delbreuckii* subsp. *bulgaricus* NCFB 2722 required only riboflavin, nicotinic acid and calcium pantothenate for EPS synthesis. The absence of glutamine, asparagine and threonine reduced growth considerably (Grobben et al., 1997).

Clearly, contrasting results have often been obtained, and so it is advisable to evaluate separately the influence of physicochemical and nutritional parameters on EPS synthesis for each strain. Moreover, additional studies on the effect of medium composition and of culture conditions on the amount and composition of EPS might result in a better understanding of the biosynthesis regulatory mechanisms.

9.3
Kinetics of EPS Biosynthesis

In general, mesophilic LAB produce greater amounts of EPS under conditions not optimal for growth, and so EPS production appears to be nongrowth-associated. In contrast, in thermophilic LAB, EPS production appears to be growth-associated (Andaloussi et al., 1995; Grobben et al., 1995; Manca de Nadra et al., 1985).

Growth-associated EPS production was noted for *S. thermophilus* LY03 (De Vuyst and Degeest, 1999a) and for *Lb. delbreuckii* subsp. *bulgaricus* NCFB 2772 (Grobben et al., 1995). However, Gancel and Novel (1994) reported that the EPS production of a *S. thermophilus* strain was not growth-associated, and Petry et al. (2000) noted that EPS production occurred mainly during the stationary growth phase for a *Lb. delbreuckii* subsp. *bulgaricus* strain.

For some mesophilic LAB, a growth-associated EPS production was observed, this being the case for *Lb. sake* 01 and a *Lb. rhamnosus* strain (Van den Berg et al., 1995). Other investigations observed a continued EPS synthesis during the stationary growth phase (Manca de Nadra et al., 1985; Gancel and Novel 1994).

Often, the amount of EPS produced declines upon prolonged fermentation. Several explanations were proposed, such as an enzymatic degradation, or a change in the physical parameters of the culture. Moreover, this trend is not rare with regard to some other microbial EPS types such as gellan (Kennedy and Sutherland, 1994) or sphingan (Hashimoto and Murata, 1998).

Pham et al. (2000) investigated this phenomenon for a *Lb. rhamnosus* strain and identified the presence of a large spectrum of glycohydrolases, three of them extracellular and the others either cell-wall bound or intracellular. When incubated with EPS, the enzymes were capable of liberating reducing sugars and of lowering the viscosity; indeed, a viscosity loss of 33% was noted after 27 h.

Gassem et al. (1995) observed a reduction in viscosity of the culture liquid after 18 h of fermentation with *S. salivarius* subsp. *thermophilus* ST3. Cerning (1994) noted similar results for *S. salivarius* subsp. *thermophilus*, although EPS degradation was also influenced by pH and temperature.

Various authors have reported this phenomenon, but few have investigated it further; as such, the complete elucidation of this degradation mechanism requires further study.

10
Applications

One way to improve the rheological properties and functional characteristics of dairy products, such as yogurt, is the addition of stabilizers. Normally, hydrocolloids of animal or plant origin are applied, but their use is prohibited in several countries, including France and the Netherlands. EPS from LAB, which are GRAS bacteria, may serve as alternative biothickeners (Grobben et al., 1995). In order to be suitable, the polymers should be compatible with other food components and with applied processing conditions, and should exhibit thixotropic or pseudoplastic behavior, notably when considering aspects such as processing costs, mouthfeel and texture (Van den Berg et al., 1995).

Since the amount of EPS produced by LAB is rather limited, these bacteria are mostly used to produce EPS *in situ*, to improve consistency and viscosity as well as to avoid syneresis (whey separation) in yogurts. Also, for the production of other fermented milk products and cheeses, EPS-producing LAB are used in starter cultures (Cerning, 1995; Marshall and Rawson, 1999; Ricciardi and Clementi, 2000).

Over the past 40 years, yogurt production has evolved substantially. One variation, which is now dominating the market, is stirred yogurt in various forms, obtained by further processing of set yogurt. Due to different mechanical processing steps, stirred yogurt tends to lose some of its initial viscosity. This might be solved by using ropy LAB as starter cultures, because yogurt containing viscosifying EPS is less susceptible to damage by mechanical pumping, blending and filling (Cerning, 1990; Rawson and Marshall, 1997).

Based on rheological studies, it was found that no clear correlation existed between the EPS concentration and apparent viscosity. The effect of EPS appeared to be much more complex, however. The contribution of EPS to specific food properties depends not only on the properties of the EPS itself, such as molar mass and structural composition, but also on the interaction of the EPS with various milk constituents, and in particular caseins (Kleerebezem et al., 1999; Marshall and Rawson, 1999).

Yogurt consists of a network of casein micelles, associated to form a gel-like structure within which small spaces are filled with liquid whey phase, while larger spaces contain the EPS and the LAB. The typical viscoelastic property of yogurt is obtained by retention of the whey. By using scanning electron microscopy, EPS strands were observed between the cells and the protein network; hence, besides the protein strands and protein–protein bonds, protein–polysaccharide bonds also occur (Wacher-Rodarte et al., 1993; Ricciardi and Clementi, 2000).

The use of ropy strains results in a yogurt which is more stable towards shear forces. Due to EPS synthesis, protein strand formation and protein–protein bond formation is partly prevented, resulting in less rigid gels with an increased viscosity, an ability to recover lost viscosity, and a certain adhesiveness. Although these characteristics are strain-dependent, it appears likely that EPS synthesis contributes to adhesiveness while the protein network has a greater influence on firmness and elasticity (Rawson and Marshall, 1997).

The thermophilic *S. thermophilus* and *Lb. delbrueckii* subsp. *bulgaricus* are traditionally used in yogurt and are both able to form EPS. Ropy, mesophilic LAB are also widely used in starter cultures, but they produce less EPS and their ropiness is often an unstable characteristic (Wacher-Rodarte et al., 1993).

In Scandinavian countries, slime-producing mesophilic LAB are used in the manufacture of different fermented milks. In Finnish "viili" and Nordic "longfil", a ropy strain of *L. lactis* subsp. *cremoris* is essential for correct consistency (Forsén and Häivä, 1981; Kontusaari and Forsén, 1988; Toba et al., 1990).

Kefir is an acidic, alcoholic beverage that is popular in Eastern Europe. The EPS produced by *Lb. kefiranofaciens* comprises the mass of the kefir grain, and was named kefiran. Other microorganisms are also associated with kefir, such as *Lb. acidophilus*, *Lb. kefirgranum*, *Lb. kefir*, *Lb. parkefir* and the yeast *Candida kefir* (Yokoi et al., 1990).

The capacity of EPS to increase water retention was used to improve the functional properties of cheese. Perry et al. (1988) produced low-fat mozzarella with a significantly higher moisture content and a better stretching ability, using capsulated nonropy cultures of *S. thermophilus* and *Lb. delbreuckii* subsp. *bulgaricus*. Eventually, Low et al. (1998) showed that *S. thermophilus* was mainly responsible for the increased moisture content. Scanning electron microscopy was used to elucidate the positive effect on the microstructure of the cheese, and revealed that bridges were formed between the EPS filaments and the casein fibers, thus preventing collapse of the whey channels.

EPS produced by LAB may also contribute to human health because of their prebiotic (Gibson and Robertfroid, 1995), cholesterol-lowering (Nakajima et al., 1992), immuno-modulating (Kitazawa et al., 1993; Hosono et al., 1997) or anti-tumor (Oda et al., 1983; Toba et al., 1991) activities.

Other health-promoting activities can be attributed to the LAB themselves, for example the removal of lactose in lactose-intolerant persons. While the pH permits metabolic activity of the LAB, lactose is utilized; however, by re-routing the sugar metabolism in *L. lactis* towards L-alanine rather than towards lactic acid, complete conversion of lactose was obtained. Moreover, the sweetness and general taste of the yogurt were enhanced (Kleerebezem et al., 2000).

Although LAB are known to be vitamin-auxotrophic bacteria, *S. thermophilus* is able to produce folic acid, an essential compound in human nutrition, and which is further metabolized by another yogurt bacterium, *Lb. delbreuckii* subsp. *bulgaricus*. Applying high folic acid-producing strains or relatively high numbers of *S. thermophilus* should lead to the production of yogurt with an increased folic acid content (Kleerebezem et al., 2000).

During the fermentation of dairy products, LAB are responsible for the synthesis of typical flavor compounds such as acetaldehyde in yogurt and diacetyl in buttermilk (Kleerebezem et al., 2000). Moreover, LAB also produce antimicrobial peptides ("bacteriocins") which have an antagonistic effect against genetically closely related strains; they are mostly thermostable, digestible and active in low concentrations. The antimicrobial spectrum of most bacteriocins is restricted to Gram-positive bacteria, but bacteria responsible for food spoilage, such as *Bacillus cereus*, *Clostridium perfringens*, *Listeria monocytogenes* and *Staphylococcus aureus* are often included. Consequently, bacteriocins may serve as a natural food preservative (De Vuyst and Vandamme, 1994).

It is clear that the metabolic activity of LAB results in "totally natural" products with a clear added value, and which have not only the desired nutritional properties but also contribute to human health in general. However, most bacterial activities may still be improved as they do not reach maximal functionality during *in situ* milk fermentation.

11
References

Andaloussi, S. A., Talbaoui, H., Marczak, R., Bonaly, R. (1995) Isolation and characterization of exocellular polysaccharides produced by *Bifidobacterium longum*, *Appl. Microbiol. Biotechnol.* **43**, 995.

Ariga, H., Urashima, T., Michihata, E., Ito, M., Morizono, N., Kimura, T., Takahashi, S. (1992) Extracellular polysaccharide from encapsulated *Streptococcus salivarius* subsp. *thermophilus* OR 901 isolated from commercial yogurt, *J. Food Sci.* **57**, 625–628.

Boels, I. C., Kleerebezem, M., Hugenholtz, J., de Vos, W. M. (1998) Metabolic engineering of exopolysaccharide production in *Lactococcus lactis*, in: Proceedings, 5th ASM on the genetics and molecular biology of streptococci, enterococci and lactococci, 66.

Bolotin, A., Mauger, S., Malarme, K., Sorokin, A., Ehrlich, D. S. (1998) *Lactococcus lactis* IL1403 diagnostic genomics, in: Proceedings, 5th ASM on the genetics and molecular biology of streptococci, enterococci and lactococci, 10–11.

Bouwman, S., Lund, D., Driessen, F. M., Schmidt, D. G. (1982) Growth of thermoresistant streptococci and deposition of milk constituents on plates of heat exchangers during long operating times, *Food Prot.* **45**, 806–812.

Cerning, J. (1990) Exocellular polysaccharides produced by lactic acid bacteria, *FEMS Microbiol. Rev.* **87**, 113–130.

Cerning, J. (1994) Polysaccharides exocellulaires produits par les bactéries lactiques, in: *Bactéries Lactiques: Aspects fondamentaux et technologiques* (de Roissart, H., Luquet, F. M., Eds), Uriage: Lorica, 309–329.

Cerning, J. (1995) Production of exopolysaccharides by lactic acid bacteria and dairy propionibacteria, *Le Lait* **75**, 463–472.

Cerning, J., Bouillanne, C., Desmazeaud, M. J. (1986) Isolation and characterization of exocellular polysaccharide produced by *Lactobacillus bulgaricus*, *Biotechnol. Lett.* **8**, 625–628.

Cerning, J., Bouillanne, C., Landon, M., Desmazeaud, M. J. (1990) Comparison of exocellular polysaccharide production by thermophilic lactic acid bacteria, *Sciences des Aliments* **10**, 443–451.

Dellaglio, F. (1994) Caractéristiques générales des bactéries lactiques, in: *Bactéries Lactiques: Aspects fondamentaux et technologiques* (de Roissart, H., Luquet, F. M., Eds), Uriage: Lorica, 10–58.

De Vos, W. M. (1996) Metabolic engineering of sugar catabolism in lactic acid bacteria, *Antonie van Leeuwenhoek* **70**, 223–242.

De Vos, W. M., Vaughan, E. E. (1994) Genetics of lactose utilization in lactic acid bacteria, *FEMS Microbiol. Rev.* **15**, 216–237.

De Vuyst, L., Degeest, B. (1999a) Exopolysaccharides from lactic acid bacteria: technological bottlenecks and practical solutions, *Macromol. Symp.* **140**, 31–41.

De Vuyst, L., Degeest, B. (1999b) Heteropolysaccharides from lactic acid bacteria, *FEMS Microbiol. Rev.* **23**, 153–177.

De Vuyst, L., Vandamme, E. J. (1994) Lactic acid bacteria and bacteriocins: their practical importance, in: *Bacteriocins of lactic acid bacteria, microbiology, genetics and applications* (De Vuyst, L., Vandamme, E. J., Eds), London: Blackie Academic and Professional, 1–11.

Doco, T., Carcano, D., Ramos, P., Loones, A., Fournet, B. (1991) Rapid isolation and estimation of polysaccharide from fermented skim milk with *Streptococcus salivarius* subsp. *thermophilus* by coupled anion exchange and gel permeation high-performance liquid chromatography, *J. Dairy Res.* **58**, 147–150.

Faber, E. J., van den Haak, M. J., Kamerling, J. P., Vliegenthart, J. F. G. (2001) Structure of the exopolysaccharide produced by *Streptococcus thermophilus* S3, *Carbohydr. Res.* **331**, 173–182.

Forsén, R., Häivä, V. (1981) Induction of stable slime-forming and mucoid states by p-fluorophenylalanine in lactic streptococci, *FEMS Microbiol. Lett.* **12**, 409–413.

Gancel, F., Novel, G. (1994) Exopolysaccharide production by *Streptococcus salivarius* subsp. *thermophilus* cultures: two distinct modes of polymer production and degradation among clonal variants, *J. Appl. Bacteriol.* **77**, 689–695.

Garcia-Garibay, M., Marshall, V. M. (1991) Polymer production by *Lactobacillus delbrueckii* ssp. *bulgaricus*, *J. Appl. Bacteriol.* **70**, 325–328.

Gassem, M. A., Schmidt, K. A., Frank, J. F. (1995) Exopolysaccharide production in different media by lactic acid bacteria, *Cult. Dairy Prod. J.* **30**, 18–21.

Gibson, G. R., Roberfroid, M. D. (1995) Dietary modulation of the human colonic microbiota: introducing the concept of prebiotics, *J. Nutr.* **124**, 1401–1412.

Gilbert, C., Robinson, K., LePage, R. W. F., Wells, J. M. (1998) Heterologous biosynthesis of pneumococcal type 3 capsule in *Lactococcus lactis* and immunogenicity studies, in: Proceedings, 5th ASM on the genetics and molecular biology of streptococci, enterococci and lactococci, 67.

Grobben, G. J., Sikkema, J., Smith, M. R., De Bont, J. A. M. (1995) Production of extracellular polysaccharides by *Lactobacillus delbreuckii* ssp. *bulgaricus* NCFB 2772 grown in a chemically defined medium, *J. Appl. Bacteriol.* **79**, 103–107.

Grobben, G. J., Smith, M. R., Sikkema, J., De Bont, J. A. M. (1996) Influence of fructose and glucose on the production of exopolysaccharides and the activities of enzymes involved in the sugar metabolism and the synthesis of sugar nucleotides in *Lactobacillus delbreuckii* subsp. *bulgaricus* NCFB 2772, *Appl. Microbiol. Biotechnol.* **46**, 279–284.

Grobben, G. J., Van Casteren, W. H. M., Schols, H. A., Oosterveld, A., Sala, G., Smith, M. R., Sikkema, J., De Bont, J. A. M. (1997) Analysis of the exopolysaccharides produced by *Lactobacillus delbreuckii* subsp. *bulgaricus* NCFB 2772 grown in continuous culture on glucose and fructose, *Appl. Microbiol. Biotechnol.* **48**, 516–521.

Groux, M. (1973) Kritische betrachtungen der heutigen joghurt-herstellung mit berücksichtigung des proteinabbaues, *Schweiz. Milchz.* **4**, 2–8.

Hashimoto, W., Murata, K. (1998) α-L-Rhamnosidase of *Sphingomonas* sp. R1 producing an unusual exopolysaccharide of sphingan, *Biosci. Biotechnol. Biochem.* **62**, 1068–1074.

Hosono, A., Lee, J., Ametani, A., Natsume, M., Adachi, T., Kaminogawa, S. (1997) Characterization of a water-soluble polysaccharide fraction with immunopotentiating activity from *Bifidobacterium adolescentis* M101-4, *Biosci. Biotechnol. Biochem.* **61**, 312–316.

Jann, K., Jann, B. (1977) Bacterial polysaccharide antigens, in: *Surface Carbohydrates of the Prokaryotic Cell* (Sutherland, I. W., Ed.), New York: Academic Press, 247–287.

Kandler, O., Kunath, P. (1983) *Lactobacillus kefir* sp. nov., a component of the microflora of kefir, *Syst. Appl. Microbiol.* **4**, 286–294.

Kennedy, L., Sutherland, I. W. (1994) Gellan lyases – novel polysaccharide lyase, *Microbiology* **140**, 3007–3013.

Kitazawa, H., Yamaguchi, T., Miura, M., Saito, T., Itoh, H. (1993) B-cell mitogen produced by slime-forming, encapsulated *Lactococcus lactis* subspecies *cremoris* isolated from ropy sour milk, viili, *J. Dairy Sci.* **76**, 1514–1519.

Kleerebezem, M., van Kranenburg, R., Tuinier, R., Boels, I. C., Zoon, P., Looijesteijn, E., Hugenholtz, J., de Vos, W. M. (1999) Exopolysaccharides produced by *Lactococcus lactis*: from genetic engineering to improved rheological properties, *Antonie van Leeuwenhoek* **76**, 657–665.

Kleerebezem, M., Hols, P., Hugenholtz, J. (2000) Lactic acid bacteria as a cell factory: rerouting of carbon metabolism in *Lactococcus lactis* by metabolic engineering, *Enzyme Microb. Technol.* **26**, 840–848.

Kojic, M., Vujcic, M., Banina, A., Cocconcelli, P., Cerning, J., Topisirovic, L. (1992) Analysis of exopolysaccharide production by *Lactobacillus casei* CG11, isolated from cheese, *Appl. Environ. Microbiol.* **58**, 4086–4088.

Kontusaari, S., Forsèn, R. (1988) Finnish fermented milk "Viili": involvement of two cell surface proteins in production of slime by *Streptococcus lactis* spp. *cremoris*, *J. Dairy Sci.* **71**, 3197–3202.

Korkeala, H., Suortti, T., Mäkelä, P. (1988) Ropy slime formation in vacuum-packed cooked meat products caused by homofermentative *Lactobacilli* and a *Leuconostoc* species. *Int. J. Food Microbiol.* **7**, 339–347.

Low, D., Ahlgren, J. A., Horne, D., McMahon, D. J., Oberg, C., Broadbent, J. R. (1998) Role of *Streptococcus thermophilus* MR-1C capsular exopolysaccharide in cheese moisture retention, *Appl. Environ. Microbiol.* **64**, 2140–2147.

Mäkelä, P. M., Korkeala, H. J. (1992) The ability of the ropy slime-producing lactic acid bacteria to form ropy colonies on different culture media and at different incubation temperatures and atmospheres, *Int. J. Food Microbiol.* **16**, 161–166.

Manca de Nadra, M. C., Strasser de Saad, A. M., Pesce de Ruiz Holgado, A. A., Oliver, G. (1985) Extracellular polysaccharide production by *Lb. bulgaricus* CRL 420, *Milchwissenschaft* **40**, 409–411.

Marshall, V. M., Rawson, H. L. (1999) Effects of exopolysaccharide-producing strains of thermophilic lactic acid bacteria on the texture of stirred yoghurt, *Int. J. Food Sci. Technol.* **34**, 137–143.

Marshall, V. M., Cowie, E. N., Moreton, R. S. (1995) Analysis and production of two exopolysaccharides from *Lactobacillus casei, J. Dairy Res.* **62**, 621–628.

Marshall, V. M., Dunn, H., Elvin, M., McLay, N., Gu, Y., Laws, A. P. (2001) Structural characterisation of the exopolysaccharide produced by *Streptococcus thermophilus* EU20, in: Proceedings, 1st International Symposium on Exopolysaccharides from Lactic acid bacteria. Brussels, Belgium: VUB, 12.

Morin, A. (1998) Screening for polysaccharide-producing microorganisms, factors influencing the production, and recovery of microbial polysaccharides, in: *Polysaccharides: Structural Diversity and Functional Versatility* (Dumitriu, S., Ed.), New York: Marcel Dekker, Inc., vol. 8, 275–296.

Mozzi, F., de Giori, G. S., Oliver, G., de Valdez, G. F. (1994) Effect of culture pH on the growth characteristics and polysaccharide production by *Lactobacillus casei, Milchwissenschaft* **49**, 667–670.

Mozzi, F., Oliver, G., de Giori, G. S., de Valdez, G. F. (1995) Influence of temperature on the production of exopolysaccharides by thermophilic lactic acid bacteria, *Milchwissenschaft* **50**, 80–82.

Mozzi, F., de Giori, G. S., Oliver, G., de Valdez, G. F. (1996) Exopolysaccharide production by *Lactobacillus casei* under controlled pH, *Biotechnol. Lett.* **18**, 435–439.

Nakajima, H., Toyoda, S. (1990) A novel phosphopolysaccharide from slime-forming *Lactococcus lactis* subspecies *cremoris* SBT 0495, *J. Dairy Sci.* **73**, 1472–1477.

Nakajima, H., Suzuki, Y., Kaizu, H., Hirota, T. (1992) Cholesterol-lowering activity of ropy fermented milk, *J. Food Sci.* **57**, 1327–1329.

Neu, T. R., Marshall, K. C. (1990) Bacterial polymers: physicochemical aspects of their interactions at interfaces, *J. Biomater. Appl.* **5**, 107–133.

Neve, H., Geis, A., Teuber, M. (1988) Plasmid-encoded functions of ropy lactic acid streptococcal strains from Scandinavian fermented milk, *Biochimie* **70**, 437–442.

Nilsson, R., Nilsson, G. (1958) Studies concerning Swedish ropy milk, the antibiotic qualities of ropy milk, *Arch. Microbiol.* **31**, 191–197.

Oda, M., Hasegawa, H., Komatsu, S., Kambe, M., Tsuchiya, F. (1983) Anti-tumor polysaccharide from *Lactobacillus* sp., *Agr. Biol. Chem.* **47**, 1623–1625.

Perry, D. B., McMahon, D. J., Oberg, C. J. (1998) Manufacture of low fat mozzarella cheese using exopolysaccharide-producing starter cultures, *J. Dairy Sci.* **81**, 561–563.

Petry, S., Furlan, S., Crepeau, M. J., Cerning, J., Desmazeaud, M. (2000) Factors affecting exocellular polysaccharide production by *Lactobacillus delbrueckii* subsp. *bulgaricus* grown in a chemically defined medium, *Appl. Environ. Microbiol.* **66**, 3427–3431.

Pham, P. L., Dupont, I., Roy, D., Lapointe, G., Cerning, J. (2000) Production of exopolysaccharide by *Lactobacillus rhamnosus* R and analysis of its enzymatic degradation during prolonged fermentation, *Appl. Environ. Microbiol.* **66**, 2302–2310.

Rawson, H. L., Marshall, V. M. (1997) Effect of ropy strains of *Lactobacillus delbrueckii* ssp. *bulgaricus* and *Streptococcus thermophilus* on rheology of stirred yogurt, *Int. J. Food Sci. Technol.* **32**, 213–220.

Ricciardi, A., Clementi, F. (2000) Exopolysaccharides from lactic acid bacteria: structure, production and technological applications, *Ital. J. Food Sci.* **12**, 23–45.

Robijn, G. W., Wienk, H. L. J., Van den Berg, D. J. C., Haas, H., Kamerling, J. P., Vliegenthart, J. F. G. (1996) Structural studies of the exopolysaccharide produced by *Lactobacillus paracasei* 34-1, *Carbohydr. Res.* **285**, 129–139.

Smitinont, T., Tansakul, C., Tanasupawat, S., Keeratipibul, S., Navarini, I., Bosco, M., Cescutti, P. (1999) Exopolysaccharide-producing lactic acid bacteria strains from traditional Thai fermented foods: isolation, identification and exopolysaccharide characterization, *Int. J. Food Microbiol.* **51**, 105–111.

Staaf, M., Yang, Z., Huttunen, E., Widmalm, G. (2000) Structural elucidation of the viscous exopolysaccharide produced by *Lactobacillus helveticus* Lb161, *Carbohydr. Res.* **326**, 113–119

Stingele, F. (1998) Exopolysaccharide production and engineering in dairy *Streptococcus* and *Lactococcus*, in: Proceedings, 5th ASM on the genetics and molecular biology of streptococci, enterococci and lactococci, 31–32.

Stingele, F., Neeser, J. R., Mollet, B. (1996) Identification and characterization of the eps gene cluster from *Streptococcus thermophilus* Sfi6, *J. Bacteriol.* **178**, 1680–1690.

Sundman, V. (1953) On the protein character of a slime produced by *Streptococcus cremoris* in Finnish ropy sour milk, *Acta Chem. Scand.* **7**, 558–560.

Sutherland, I. W. (1972) Bacterial exopolysaccharides, *Adv. Microbiol. Physiol.* **8**, 143.

Sutherland, I. W. (1982) Biosynthesis of microbial exopolysaccharides, in: *Advances in Microbial Physiology* (Rose, A.H., Ed.), London: Academic Press, Vol. 33, 78–150.

Sutherland, I. W. (1990) *Biotechnology of Microbial Exopolysaccharides*. Cambridge, Sydney: Cambridge University Press, 163.

Sutherland, I. W. (1994) Structure–function relationships in microbial exopolysaccharides, *Biotech. Adv.* **12**, 393–448.

Tamime, A. Y. (1978) Some aspects of the production of a concentrated yoghurt (labneh) popular in the Middle East, *Milchwissenschaft* **33**, 209–212.

Toba, T., Nakajima, H., Tobitani, A., Adachi, S. (1990) Scanning electron microscopic and texture studies on characteristic consistency of Nordic ropy sour milk, *Int. J. Food Microbiol.* **11**, 313–320.

Toba, T., Kotani, T., Adachi, S. (1991) Capsular polysaccharide of a slime-forming *Lactococcus lactis* ssp. *cremoris* LAPT 3001 isolated from Swedish fermented milk longfil, *Int. J. Food Microbiol.* **12**, 167–172.

Troy, F. A. (1979) The chemistry and biosynthesis of selected bacterial polymers, *Annu. Rev. Microbiol.* **33**, 519–560.

Troy, F. A., Frerman, F. A., Heath, E. C. (1971) Biosynthesis of capsular polysaccharides in *Aerobacter aerogenes*, *J. Biol. Chem.* **243**, 118–133.

Van Calsteren, M. (2001) Structural characterisation of the exopolysaccharide produced by different strains of *Lactobacillus rhamnosus* and *Lactobacillus paracasei*, in: Proceedings, 1st International Symposium on Exopolysaccharides from Lactic acid bacteria, 13.

Van den Berg, D. J. C., Smits, A., Pot, B., Ledeboer, A. M., Kersters, K., Verbrakel, J. M. A., Verrips, C. T. (1993) Isolation, screening and identification of lactic acid bacteria from traditional food fermentation processes and culture collections, *Food Biotechnol.* **7**, 189–205.

Van den Berg, D. J. C., Robijn, G. W., Janssen, A. C., Giusepin, M. L. F., Vreeker, R., Kamerling, J. P., Vliegenthart, J. F. G., Ledeboer, A. M., Verrips, C. T. (1995) Production of a novel extracellular polysaccharide by *Lactobacillus sake* 0-1 and characterization of the polysaccharide, *Appl. Environ. Microbiol.* **61**, 2840–2844.

Van Kranenburg, R., Vos, H. R., Van Swam, I., Kleerebezem, M., De Vos, W. M. (1999) Functional analysis of glycosyltransferase genes from *Lactococcus* and other Gram-positive cocci: complementation, expression and diversity, *J. Bacteriol.* **181**, 6347–6353.

Vedamuthu, E. R., Neville, J. M. (1986) Involvement of a plasmid in production of ropiness in milk cultures by *Streptococcus cremoris*, *Appl. Environ. Microbiol.* **61**, 2840–2844.

Vliegenthart, J. F. G., Faber, E. J., Kamerling, J. P. (2001) Studies on structure of exopolysaccharides from lactic acid bacteria, in: Proceedings, 1st International Symposium on Exopolysaccharides from Lactic acid bacteria. Brussels, Belgium: VUB, 11.

Wacher-Rodarte, C., Galvan, M., Farres, A., Gallardo, F., Marshall, V. M. E., Garcia-Garibay, M. (1993) Yogurt production from reconstituted skim milk powders using different polymer and non-polymer forming starter cultures, *J. Dairy Res.* **60**, 247–254.

Yokoi, H., Watanabe, T., Fuji, Y., Toba, T., Adachi, S. (1990) Isolation and characterization of polysaccharide-producing bacteria from kefir grains, *J. Dairy Sci.* **73**, 1684–1689.

8
Inulin

Dr. Anne Franck, Ir. Leen De Leenheer
ORAFTI, Aandorenstraat 1, 3300 Tienen, Belgium; Tel.: +32-16-801-218;
Fax: +32-16-801-359; E-mail: anne.franck@orafti.com
ORAFTI, Aandorenstraat 1, 3300 Tienen, Belgium; Tel.: +-32-16-801-351;
Fax: +32-16-801-496; E-mail: leen.de.leenheer@orafti.com

1	**Introduction**	283
2	**Historical Outline**	283
3	**Chemical Structure**	284
4	**Natural Occurrence**	286
5	**Physiological Function**	288
6	**Chemical Analysis and Detection**	288
6.1	High-performance Liquid Chromatography (HPLC)	288
6.2	Gas Chromatography	288
6.3	HPAEC Analysis (Dionex)	289
6.4	Permethylation	291
6.5	Quantitative Determination of Inulin and Oligofructose in Food	291
6.6	Quantitative Determination of Inulin in Food	291
7	**Biosynthesis**	292
7.1	Synthesis of Microbial Fructan	293
7.2	*In vitro* Synthesis of FOS	293
7.3	Synthesis of Inulin from Plant Origin (Asteraceae)	293
7.3.1	Biochemistry	293
7.3.2	Molecular Genetics	294
8	**Biodegradation**	295
8.1	Plant Endogenous Degradation	295

Polysaccharides and Polyamides in the Food Industry. Properties, Production, and Patents.
Edited by A. Steinbüchel and S. K. Rhee
Copyright © 2005 WILEY-VCH Verlag GmbH & Co. KGaA, Weinheim
ISBN: 3-527-31345-1

8.1.1	Biochemistry	295
8.1.2	Molecular Genetics	296
8.2	*In vitro* Hydrolysis by Yeast and Mold Enzymes	296
9	**Production**	**297**
9.1	FOS Production Starting from Sucrose	297
9.2	Commercial Inulin of Plant Origin	297
9.2.1	Agricultural Aspects	297
9.2.2	Processing	298
9.3	Commercial Production of Inulin and FOS	300
9.4	Scale of Production	300
10	**Properties**	**300**
10.1	Physical and Chemical Properties	301
10.2	Material Properties	303
10.3	Biological Properties	303
10.3.1	Nondigestibility	303
10.3.2	Caloric Value	304
10.3.3	Improvement of Lipid Metabolism	304
10.3.4	Effects on Gut Function	305
10.3.5	Modulation of Gut Microflora	305
10.3.6	Suitability for Diabetics	306
10.3.7	Reduction of Cancer Risk	306
10.3.8	Increase in Mineral Absorption	306
10.3.9	Intestinal Acceptability	307
11	**Food Applications**	**307**
12	**Non-food Developments and Applications**	**309**
13	**Outlook and Perspectives**	**310**
14	**Patents**	**311**
15	**References**	**315**

DFAI	di-D-fructofuranose 2′,1; 2,1′ -dianhydride
DFAII	di-D-fructofuranose 2′,1; 2,3′-dianhydride
DP	degree of polymerization
DP5+ or DP ≥ 5	fructan molecules with a DP of 5 or more
DP_n	a fructan with a degree of polymerization of n
EFA	European Fructan Association
F	fructose (only in reactions)
F2 to F9	fructan molecule consisting of only fructofuranosyl units (2 to 9 indicates the number of units present)

FEH	fructan exohydrolase
FFT	fructan:fructan fructosyltransferase
F_m	fructofuranosyl-only fructan molecule with a DP of m
F-G-F	neo-kestose (only in reactions)
FOS	fructo-oligosaccharide
Fru	fructose
G	glucose (only in reactions)
GF	sucrose
$GF2, GF8...GF_n$	fructan molecule consisting of 2, 8...n fructofuranosyl units and containing one terminal glucose
G-F	sucrose (only in reactions)
$G-F-(F)_n$	fructan molecule with a DP of $n+2$ and containing one terminal glucose
G-F-F	1-kestose (only in reactions)
Glc	glucose
6G-FFT	fructan:fructan 6G-fructosyltransferase
PAD	pulsed amperometric detector
PED	pulsed electrochemical detector
RI	refractive index
SST	sucrose:sucrose fructosyltransferase

1
Introduction

Inulin, a nondigestible carbohydrate, is a fructan that is not only found in many plants as a storage carbohydrate, but has also been part of man's daily diet for several centuries. It is present in many regularly consumed vegetables, fruits and cereals, including leek, onion, garlic, wheat, chicory, artichoke, and banana. Industrially, inulin is obtained from chicory roots, and is used as a functional food ingredient that offers a unique combination of interesting nutritional properties and important technological benefits. In food formulations, inulin significantly improves the organoleptic characteristics, allowing an upgrading of both taste and mouthfeel in a wide range of applications. In particular, this taste-free fructan increases the stability of foams and emulsions, as well as showing an exceptional fat-like behavior when used in the form of a gel in water. By contrast, as an ever-increasing amount of information becomes available on inulin, its nutritional attributes continue to amaze both researchers and nutritionists alike. Consequently, fat and carbohydrate replacement with inulin offers the advantage of not having to compromise on taste and texture, while delivering further nutritional benefits. Hence, inulin represents a key ingredient that offers new opportunities to a food industry which is constantly seeking well-balanced, yet better tasting, products of the future.

2
Historical Outline

Rose, a German scientist, first isolated a "peculiar substance of plant origin" from a boiling water extract of *Inula helenium* in 1804, and the substance was later called inulin by Thomson (1818). The German plant physiologist Julius Sachs (1864) was a pioneer in fructan research and, by using only a microscope, was able to detect the spherocrystals of inulin in the tubers of *Dahlia*, *Helianthus tuberosus* and *Inula helenium* after ethanol precipitation.

Although today, chicory is the major crop used for the industrial production of inulin, the first reference to chicory being consumed by humans was made during the first century by Pedanios Dioscoride (Leroux, 1996) who, as a physician in the Roman army, praised the plant for its beneficial effects on the stomach, liver, and kidneys. Much later, Baillargé (1942) stated that in about 1850, Jerusalem artichoke (*Helianthus tuberosus*) pulp, when prepared by cooking and drying the tubers, was added in a 50:50 ratio to flour when baking bread to provide cheap food for laborers.

On a more physiological basis, Külz reported in 1874 that no sugar appeared in the urine of diabetics who ate 50–120 g of inulin per day, and by the end of the nineteenth century the feeding of diabetic patients with pure inulin in doses of 40–100 g daily was reported to be "with much benefit" (Von Mehring, 1876). The first studies on the effects of inulin in healthy humans appeared during the early twentieth century (Lewis, 1912), whilst the nontoxicity of inulin was demonstrated dramatically some years later (Shannon and Smith, 1935) when one of the authors injected himself intravenously with 160 g inulin. In particular, during the past 10 years there has been a spectacular increase in the number of publications relating to the functional and nutritional benefits of inulin.

Subsequently, as inulin changed from a subject of mere scientific interest into an industrial product with many applications, there was a major stimulation of research related to its production and use.

3
Chemical Structure

Inulin has been defined as a polydisperse carbohydrate material consisting mainly, if not exclusively, of $\beta(2 \rightarrow 1)$ fructosyl-fructose links (Waterhouse and Chatterton, 1993). A starting glucose moiety can be present, but is not necessary. In contrast, levan – which is formed by certain bacteria – consists mainly or exclusively of $\beta(2 \rightarrow 6)$ fructosyl-fructose links. As is the case for inulin, glucose can be present, but again it is not necessary. Fructan is a more general name which is used for any compound in which one or more fructosyl-fructose links constitute the majority. The term "fructan" therefore covers both inulin and levan.

When referring to the definition of inulin, both GF_n and F_m compounds are considered to be included under this same nomenclature. In chicory inulin, n (the number of fructose units linked to a terminal glucose) can vary from two to 70 (De Leenheer and Hoebregs, 1994). This also means that inulin is a mixture of oligomers and polymers. The molecular structure of inulin compounds is shown in Figure 1.

The degree of polymerization (DP) of inulin, as well as the presence of branches, are important properties since they influence the functionality of most inulin to a striking extent. Thus, a strict distinction must be made between inulin of plant origin and that of bacterial origin. The DP of plant inulin is rather low (maximally < 200) and varies according to the plant species, weather conditions and the physiological age of the plant (see Section 9).

Native inulin always contains glucose, fructose, sucrose, and small oligosaccharides. The term "native" refers to inulin that, before its analysis, is extracted from fresh roots, taking precautions to inhibit the plant's own inulinase activity as well as acid

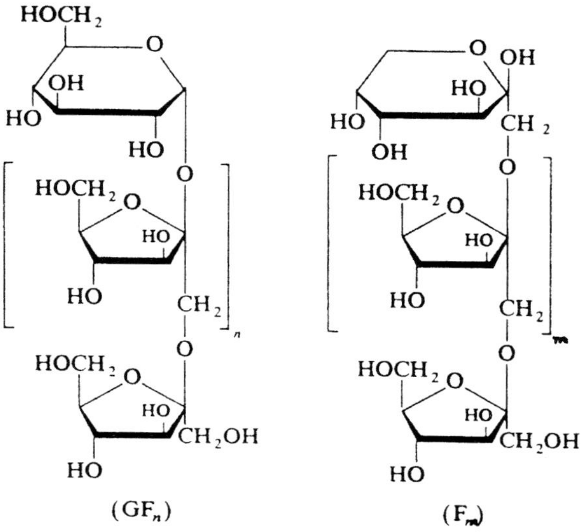

Fig. 1 Chemical structure of inulin.

hydrolysis. Moreover, no fractionation procedure is applied to eliminate the smaller oligosaccharides and monomers that are naturally present. In this respect, the commercially available inulin (Sigma Chemical Co.) that is derived from *Dahlia*, Jerusalem artichoke or chicory is not considered to be "native" as these products barely represent the inulin typical of the plants from which it is extracted, the average DP being 27–29 (for all three products). This DP value is not only very high, but chains of < 10 units are also absent (De Leenheer, 1996); this difference is shown clearly in Figure 2.

Until recently, (plant) inulin was considered to be a linear molecule, but by using optimized permethylation analysis it was possible to show that even native chicory inulin (DP 12) has a very small degree of branching (1–2%), and this was also the case for inulin from *Dahlia* (De Leenheer and Hoebregs, 1994).

In contrast to plant inulin, bacterial inulin has a very high DP, ranging from 10,000 to over 100,000; moreover, this inulin is highly branched (≥15%). Although Harada et al. (1993) reported that the inulin derived from the spores of *Aspergillus sydowi* was linear, this could not be confirmed by permethylation analysis. The fact that inulin has a small intrinsic viscosity in spite of its high molecular weight (as do levans), and that it appears to adopt a compact, globular shape rather than a coil is another indication of its nonlinearity.

From a structural/polymeric viewpoint, (linear) inulin can be considered as a polyoxyethylene backbone to which fructose moieties are attached, as are the steps of a spiral staircase. Inulin crystallizes along a pseudohexagonal, six-fold symmetry with an advance of 0.24 nm per monomer. Moreover, two inulin crystalline allomorphs exist: semi-hydrated and hydrated. The difference between the unit cells seems not to correlate with any change in the conformation of the six-fold helix, but rather to a variation in water content (Andre et al., 1996).

Oligomers with a DP up to 5 can adopt structures resembling the conformation of cyclo-inulohexaose. Oligomers with DP between 7 and 8 most likely adopt a conformational change because they form helical structures that become more rigid as the

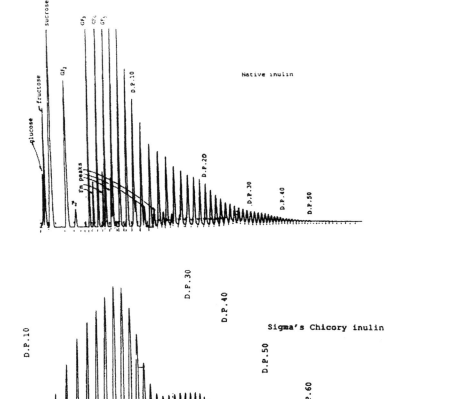

Fig. 2 Dionex chromatogram of native inulin and commercially available inulin (Sigma) obtained from chicory.

DP increases (Timmermans et al., 1997). This hypothesis of changing conformation also provides a very reasonable explanation for the observation that, at DP values between 6 and 9, the elution sequence of the oligomers on a reversed-phase C_{18} Nucleosil column is completely reversed (De Leenheer and Hoebregs, 1994).

4
Natural Occurrence

After starch, fructans are the most abundant nonstructural polysaccharides found naturally, being present in a wide variety of plants, and in some bacteria. Most reports on the natural occurrence of fructans do not differentiate between levan and inulin, however.

Fructan-producing plants are commonly present among the grasses (1200 species), whereas 15% of flowering plants produce fructans in significant amounts. They are widely spread within the Liliaceae (3500 species), and most frequently among the Compositae (25,000 species) (Hendry and Wallace, 1993). Strictly speaking, β(2→1)-defined inulin is typical of the Compositae.

Inulin-containing plants that are commonly used for human nutrition belong mainly to either the Liliacea, e.g., leek, onion, garlic and asparagus, or the Compositae, e.g., Jerusalem artichoke, dahlia, chicory and yacon (Table 1).

The occurrence of endogenous inulin in fungi appears doubtful: inulin-type molecules can be produced by incubating spores in sucrose (Harada et al., 1993), although as sucrose is not a true fungal carbohydrate it is unlikely that sufficient starting material is present for the synthesis of inulin via the sucrose:sucrose fructosyltransferase route (Lewis, 1991)

In bacteria, the presence of fructan-producing genera is common. Bacterial fructans are almost by definition of the levan type, and are found among the Pseudomonaceae, Enterobacteriaceae, Streptococcaceae, Actinomycetes and Bacillaceae. Recently, fructans were also detected within the Lactobacilli, more specific in *Lactobacillus reuteri* (van Geel-Schutten, 2000). Two fructosyltransferase genes have been described: one gene product is excreted by the bacterium during growth and is of the levan-type, while the second fructosyltransferase gene could only be brought to expression in *Escherichia coli* and produces an inulin-type fructan. Full characterization of these fructans is ongoing.

Until 1999, *Streptococcus mutans* was the only bacterium known to produce inulin-type molecules of which the linearity is, to date, unclear (Ponstein and Van Leeuwen, 1993).

One plant of special status is the *Agave Azul Tequila Weber* (Liliaceae). Although tequila is an alcoholic drink that is known world-wide, few people realize that it is made

Tab. 1 Inulin content (% of fresh weight) of plants that are commonly used in human nutrition (Van Loo et al., 1995)

Source	Edible parts	Dry solids content	Inulin content
Onion	Bulb	6–12	2–6
Jerusalem artichoke	Tuber	19–25	14–19
Chicory	Root	20–25	15–20
Leek	Bulb	15–20*	3–10
Garlic	Bulb	40–45*	9–16
Artichoke	Leaves-heart	14–16	3–10
Banana	Fruit	24–26	0.3–0.7
Rye	Cereal	88–90	0.5–1*
Barley	Cereal	NA	0.5–1.5*
Dandelion	Leaves	50–55*	12–15
Burdock	Root	21–25	3.5–4.0
Camas	Bulb	31–50	12–22
Murnong	Root	25–28	8–13
Yacon	Root	13–31	3–19
Salsify	Root	20–22	4–11

NA, data not available. *Estimated value.

by the fermentation of a type of "inulin". In fact, the fructan molecule is highly branched (24%), containing $\beta(2 \rightarrow 1)$ linkages as well as $\beta(2 \rightarrow 6)$ linkages (L. De Leenheer, unpublished results).

5
Physiological Function

Despite major advances having been made in the elucidation of the metabolism of fructans, their precise physiological function remains a subject of debate. The most documented role is that of a long-term reserve carbohydrate stored in underground, over-wintering organs. Two other functions are often quoted: first, as a cryoprotectant, and second as an osmotic regulator. Together, these roles allow not only survival but also growth under conditions of water shortage, whether induced by drought or low temperatures (Hendry and Wallace, 1993).

De Roover et al. (2000) have reported on the effects of drought on inulin metabolism in chicory. Glucose, fructose and sucrose contents were increased in the roots and leaves of stressed plants, whereas the inulin concentration was found to be ten-fold higher than in control plants, with inulin content being normal in roots but absent in leaves. In a cold environment (3 weeks at 4°C), chicory inulin is clearly degraded, and this results in lower-DP inulin and mainly fructose which, osmotically, is more active than inulin.

The role of fructans as true cryoprotectors is under discussion, as the increase in hexoses and sucrose upon depolymerization of the fructan would only account for a freezing point decrease of 0.2–0.5°C (Van Den Ende, 1996). In contrast, inulin was seen to interact directly with membrane lipids upon freeze-drying, thereby preserving the membranes in a liquid-crystalline phase at room temperature and preventing phase transition and solute leakage during rehydration (Hincha et al., 2000).

6
Chemical Analysis and Detection

Although several techniques are available for the analysis of inulin, no single method provides a complete and quantitative analysis of all the compounds present; hence, a combination of different methods is often necessary.

6.1
High-performance Liquid Chromatography (HPLC)

For HPLC analysis, two columns in series in the K+ form are used (Aminex HPX-87 K+) for optimal separation. The separations into fructose, glucose, difructose-dianhydride (DFA), sucrose (GF), F2 and F3 are optimal, but further separation into DP 3, DP 4 and DP ≥ 5 is not very precise. DP 3 and DP 4 fractions are not pure, but include GF2 plus F4 and GF3 plus F5, respectively. The DP ≥ 5 fraction is the integrated sum of DP 5 and higher-DP molecules. As this fraction might include small oligosaccharides as well as high-DP inulins, a fixed response factor cannot be determined, which makes this analysis unsuited for quantitative inulin determination. The method is well adapted to evaluate the relative amounts of the different compounds present, especially the amounts of the non-inulin compounds glucose, fructose and sucrose (Table 2).

6.2
Gas Chromatography

A high-temperature capillary gas chromatographic method was developed for the

Tab. 2 HPLC and gas chromatographic (GC) analysis of RAFTILOSE®L95

GC analysis	[%]	HPLC analysis	[%]
Fructose	2.0	Fructose	2.1
Glucose	<0.1	Glucose	<0.1
DFA	<0.1	Saccharose	0.6
Saccharose	0.3	DP2	2.4
F2	2.4	F3	30.0
GF2	0.4	DP3	26.7
F3	29.6	DP4	14.4
GF3	3.6	DP5+	23.8
F4	26.8		
GF4	7.5		
F5	10.9		
GF5	5.9		
F6	7.1		
GF6	2.0		
F7	1.1		
GF7	0.1		
F8	0.1		
GF8	0.1		
F9	<0.1		
DP10	0.1		

quantitative determination of fructo-oligosaccharides (FOS) with DP <10 (Joye and Hoebregs, 2000). Sample preparation involves oxymation and silylation of the extracted sugars. The oximetrimethylsilyl derivatives are analyzed on an apolar capillary aluminum-clad column with temperature programming up to 440°C and detection by flame ionization. The method is accurate and specific, as malto-, isomalto- and galacto-oligosaccharides, all of which are commonly present in foods, do not interfere. Moreover, $\beta(2 \rightarrow 6)$ oligosaccharides (levan) can be clearly distinguished from $\beta(2 \rightarrow 1)$ compounds, and GF_n compounds from F_n compounds (Figure 3; see also Table 2).

6.3
HPAEC Analysis (Dionex)

High-pressure anion exchange chromatography (HPAEC) is another technique which can be used to differentiate between GF_n and F_n compounds; moreover, the method also provides a "fingerprint" of the molecular weight distribution of inulin (see Figure 2). This analytical technique uses a Dionex series 4000 ion chromatograph (Carbo-Pac PA-1 column) coupled with a pulsed amperometric detector (PAD). During the analysis, the carbohydrates are eluted with a NaOH/NaAc gradient; the high pH (13–14) of the NaOH converts the hydroxyl groups into oxy-anions. The degree of oxyanion interaction with the anion-exchange resin determines the carbohydrate retention times. To reduce the retention times, a competing ion such as acetate is added to the eluant. The PAD system oxidizes and detects the now separated carbohydrates as they pass through the detector.

The major drawback of HPAEC-PAD is that it is very difficult to quantify the high-DP oligomers, due on the one hand to the lack of appropriate standards and on the other hand to the reduced sensitivity of the PAD detector for high-DP polymers. The detector measures in fact the electrons released during oxidation of the carbohydrates at the gold electrode. Chatterton et al. (1993) have suggested that, as carbohydrates become larger, then proportionally fewer electrons are released per fructosyl unit, and so the PAD output per µg sugar decreases as the DP is increased

Timmermans et al. (1993) also used HPAE-chromatography, but coupled with a pulsed electrochemical detector (PED). The sensitivity of the PED detector decreased clearly from DP 2 to DP 5, but these authors observed only a slow decrease for DP 10–17. From this, they calculated the PED respons-

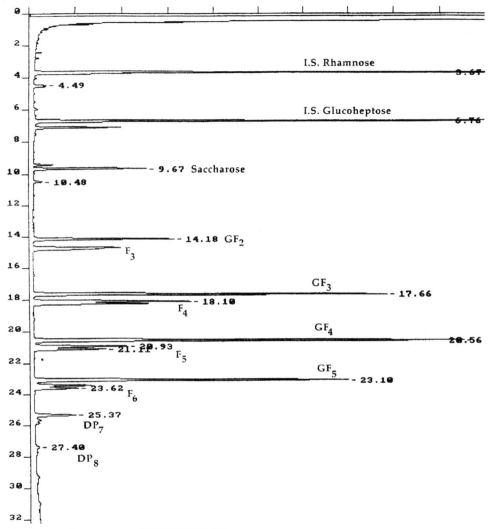

Fig. 3 Gas chromatogram of RAFTILOSE® L95.

es for inulin oligomers with different DPs relative to sucrose, and this enabled quantification of oligomers up to DP 17. Based on these relative responses, it was then possible to calculate the weight fraction of each compound present.

In a further study (Timmermans et al., 1997), the same group developed a HPAEC method with modified gradient elution in combination with a refractive index (RI) detector. In order to allow RI and PAD detection, the gradient was adapted to give a constant RI: the sodium acetate concentration was increased in order to obtain the desired fractionation, and the sodium hydroxide concentration decreased to keep the

RI constant. Application of this gradient enables comparison of RI (mass concentration-related) and PAD responses, and thus the quantitative determination of mixtures of oligomers.

In most cases however, to determine the percentage distribution of different inulin compounds measured by HPAEC-PAD, this is simply calculated by integration of the peaks of interest versus the total surface under the curve, and not really quantified.

The average DP is determined by the fructose to glucose ratio after complete hydrolysis with the NOVO inulinase Fructozyme® (previously called SP230), at pH 4.5 and 60°C during 30 min:

$$\overline{DPn} = F/G + 1.$$

6.4
Permethylation

The type of linkage and the occurrence of branching are checked by permethylation, followed by reductive cleavage and *in situ* acetylation. The method is based on previously described procedures (Ciucanu and Kerek, 1984; Jun and Gray, 1987; Mischnick-Lübbecke and König, 1989). Capillary gas chromatography (CGC) analysis is carried out on a gas chromatograph (Carlo-Erba) with on-column injector, equipped with an OV1701 column, temperature programming up to 250°C, and with flame ionization detection.

6.5
Quantitative Determination of Inulin and Oligofructose in Food

The Association of Official Analytical Chemists (AOAC) method no. 997.08 (Hoebregs, 1997) was developed because inulin and FOS are classified as dietary fiber but cannot be measured using the classical AOAC fiber method. An overview of the method is given in Figure 4.

If inulin is the only compound present in the sample, the method consists only of steps 1 and 3. The inulin is extracted from a substrate at 85°C for 10 min; one part of the extract is removed for determination of free fructose, glucose and sucrose using any reliable chromatographic method available (HPLC, gas chromatography or HPAEC-PAD), while the other part is submitted to enzymatic hydrolysis, identical to that described for the DP determination. Following this hydrolysis step, the resultant fructose and glucose are determined again by chromatography. By subtracting the initial glucose, fructose and sucrose content from the final contents, the following formula can be applied:

$$Inu = k \, (G_{inu} + F_{inu})$$

where k (<1) depends on the DP of the inulin analyzed and corrects for the water gain after hydrolysis. G_{inu} and F_{inu} are respectively glucose and fructose strictly originating from inulin.

If a complex sample needs to be analyzed – as is often the case when dealing with food products – an amyloglucosidase treatment must be included before step 3, and an extra sugar analysis performed to avoid overestimation of the glucose originating from any starch or maltodextrin present.

6.6
Quantitative Determination of Inulin in Food

Although the AOAC method 997.08 is very reliable, it is very labor-intensive, and requires the use of chromatographic apparatus. The AOAC method 999.03 (McCleary et al. 2000), which is completely based on the use of enzymes for both the hydrolysis and sugar determinations, simply requires a spectrophotometer and other standard labo-

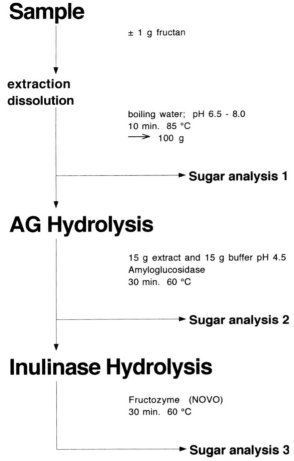

Fig. 4 Flow diagram of the enzymatic method of fructan analysis in food. AG, amyloglucosidase.

ratory equipment. This method is reliable for the determination of inulin, but cannot be used to determine oligosaccharides obtained by the hydrolysis of inulin, or any form of oligofructose that contains F_n-type compounds, as these are significantly underestimated (H. Hoebregs, unpublished results).

7
Biosynthesis

A clear distinction should be made between inulin of bacterial or fungal origin, which vary between high molecular-weight fructans and oligosaccharides, and inulin of plant origin.

7.1
Synthesis of Microbial Fructan

In bacteria and fungal spores, fructan synthesis is the result of the action of only one sucrase enzyme – inulosucrase or levansucrase – and is dependent upon the linkage formed between adjacent fructose molecules, whether $\beta(2 \rightarrow 1)$ or $\beta(2 \rightarrow 6)$ (Uchiyama, 1993).

$$G\text{-}F + G\text{-}F\text{-}(F)_m \rightarrow G + G\text{-}F\text{-}(F)_{m+1}$$

Fructose is directly transferred from the donor sucrose to an acceptor sucrose or fructan, with high-DP fructans being the most efficient acceptors. Low-DP FOS are often barely detectable during the reaction. As mentioned previously, the high-DP inulins ($> 10^7$) are mainly of the levan type, and branched.

7.2
In vitro Synthesis of FOS

Special attention must be given to the industrial production of FOS by the *in vitro* reaction of some fructosyltransferases on sucrose. The basic reaction is the same as mentioned previously, with low-DP oligosaccharides being formed exclusively.

Most FOS-producing fructosyltransferases are of fungal origin, e.g., *Aspergillus*, *Aureobasidium*, *Penicillium*, and produce a FOS composed primarily of GF2 + GF3 + GF4 for some 55%, while the remaining 45% is composed of nonreacted sucrose and glucose, plus fructose as byproducts (Hirayama and Hidaka, 1993). More recently, Kim et al. (1998) reported the production of mainly GF4 by the bacterium *Bacillus macerans*, with GF2 and GF3 not being detectable during the reaction.

7.3
Synthesis of Inulin from Plant Origin (Asteraceae)

7.3.1
Biochemistry

In recent years, significant progress has been made in the elucidation of inulin metabolism in Asteraceae.

The process began when Edelman and Jefford (1968) suggested that in Jerusalem artichoke, synthesis proceeds via the concerted action of two fructosyltransferases, namely SST (sucrose:sucrose fructosyltransferase; EC 2.4.1.99) and FFT (fructan:fructan fructosyltransferase; EC 2.4.1.100).

1-SST catalyzes the first step by the formation of 1-kestose:

$$G\text{-}F + G\text{-}F \rightarrow G\text{-}F\text{-}F + G$$

1-FFT then utilizes the formed 1-kestose as substrate for further chain elongation into inulin-type fructans:

$$G\text{-}F\text{-}(F)_n + G\text{-}F\text{-}(F)_m$$
$$\rightleftarrows G\text{-}F\text{-}(F)_{n+1} + G\text{-}F\text{-}(F)_{m-1}$$

In this reaction, sucrose can act only as an acceptor, and not as a donor. The reaction is reversible, and sucrose as well as high-DP inulins are recognized as being the best acceptors.

This hypothesis was heavily criticized as it was formulated without *in vitro* evidence for the synthesis of inulin beyond a DP >3, with no published evidence for the existence of the key enzyme SST, without use of completely purified enzymes, and without using physiologically acceptable substrate conditions (Cairns, 1993). However, this criticism inspired others to demonstrate the de-novo synthesis of inulin with DP >3, using pure enzymes under physiological conditions, and this resulted in products that were identical to those found in the plants from which the enzymes or genes originated.

1-SST and 1-FFT (both enzymes and genes) have been isolated from, e.g., *Helianthus tuberosus* (Jerusalem artichoke) (Koops and Jonker, 1996), chicory (Van den Ende et al., 1996), and *Cynara scolymus* (globe artichoke) (Hellwege et al., 1998). Indeed, at the enzyme level, it was shown that 1-SST, by using a long incubation time (96 h) could also produce oligosaccharides with a DP >3.

The concerted action of SST and FFT, in yielding plant-similar inulin, was proven *in vitro* using *Helianthus* (Koops and Jonker, 1996), chicory (Van den Ende, 1996) and *Cynara* enzymes (Heyer et al., 1999).

In addition to inulin from the 1-kestose series, inulin can also be formed from the neo-kestose series – the type of fructan found in onions. In this case a third enzyme is needed, 6G-FFT (fructan:fructan 6G-fructosyltransferase), which catalyzes the transfer of a fructose residue of 1-kestose to C6 of the glucose moiety of sucrose, forming neo-kestose (F-G-F) (Vijn et al., 1998).

7.3.2
Molecular Genetics

In vivo inulin production was demonstrated by isolating 1-SST and 1-FFT genes of *Helianthus* and expressing them in sugar beet (Van Tunen et al., 1996; Koops et al., 1999), and also by expressing both genes of *Cynara* in potato (Heyer et al., 1999). The inulin profile from extracts from globe artichoke and the corresponding extracts from transformed potato were identical (Heyer et al., 1999). 1-SST from *Helianthus*, when expressed in sugar beet, produced an oligofructose of DP 3 to 6 (Sevenier et al., 1998).

The homology between the Asteraceae fructosyltransferases is very high: within the 1-SST genes, homologies over 80% are observed between *Helianthus tuberosus*, *Cynara scolymus*, *Taraxacum officinale* and *Cichorium intybus*. The homology with 1-SST of *Allium cepa* (Leliaceae) varies between 46 and 49%; for acid invertases of dicotyledon plants 54–61% is reported, whereas homology to bacterial fructosyltransferases is low (17–22%) (Hellwege et al., 1997; Van den Ende et al., 2000a).

In previously quoted reports on cloning, the level of regulation by comparing mRNA and enzyme concentrations and enzyme activities has rarely been identified. Van den Ende et al. (2000a) clearly demonstrated a correlation between mRNA level and protein concentration in *Taraxacum officinale*, which confirmed the transcriptional regulation of 1-SST.

Most Asteraceae fructosyltransferases are heterodimers that originate from the cleavage of a single polypeptide. Both forms may be present in the plant, with the monomeric form dominating in young seedlings and plants, and the heterodimers present during later stages. This indicates a developmental regulation and extends the possibility of post-transcriptional processes being present.

It is likely that, in the near future, more extensive reports will be made detailing the relationship between gene induction/expression and protein level/activity. A "DNA chip" is a microarray of thousands of small DNA spots printed onto a microscope slide, allowing identification of target DNA sequences and analysis of gene expression level (Koops et al., 2000). DNA chips are currently used to analyze inline metabolism during the development of chicory tap roots. Using this technique, it will be possible to study both up- and down-regulation of metabolic enzymes of interest upon any natural or imposed growth condition, thereby generating information that might be valuable in the optimization of inulin production.

8
Biodegradation

The degradation of inulin, as is the case with its synthesis, is completely different when the endogenous plant processes are compared with *in vitro* hydrolyzing procedures.

8.1
Plant Endogenous Degradation

8.1.1
Biochemistry

The third important enzyme postulated in the model of Edelman and Jefford was the fructan exohydrolase (FEH; EC.2.4.1.80):

$$G\text{-}F\text{-}(F)_n \rightarrow G\text{-}F\text{-}(F)_{n-1} + F$$

This enzyme catalyzes inulin breakdown through cleavage between the terminal fructosyl group and its adjacent fructose residue, under the assumption that it does not catalyze sucrose hydrolysis. This hypothesis has now been confirmed, and two FEHs have been isolated from chicory (Claessens et al., 1990; De Roover et al., 1999). Both enzymes showed no activity towards sucrose, indicating that these FEHs are neither invertases nor β-fructofuranosidases. Both enzymes are induced by cold temperature; however, FEH II is expressed more in roots which are forced (3 weeks at 16°C) after cold induction (3 weeks at 1°C), whereas FEH I is present in larger amounts in cold-induced roots but less so in forced roots.

In order to observe the evolution of inulin in chicory, extracts were made during a period from late August until mid-December. Although the data listed in Table 3 refer to the analysis of 1999, a similar trend is seen each year. When the first frost occurs, fructose and sucrose levels increase significantly, while glucose levels remain unchanged. This would imply that, through the action of fructan hydrolases a surplus of fructose residues (which are not necessary for metabolic or osmoregulatory purposes) are released and reconverted to glucose, from which more sucrose can then be produced. While SST and FFT remain active, new inulin molecules can be created, albeit of a smaller molecular size than the originals. The fact that the DP changes as a function of harvest time, and not of the inulin concentration, confirms this theory.

The presence of inulo-n-oligosaccharides in native chicory extract was first reported by De Leenheer and Hoebregs in 1994. It was noted that the concentration of these fructose oligomers increased slightly from late September onwards, and were doubled by the end of November. Van den Ende et al. (1996) identified a biochemical explanation for this occurrence, in that under high

Tab. 3 Evaluation of the composition of inulin in chicory as a function of harvesting time

Date (1999)	$\overline{DP}$ inulin	Inulin [% roots]	Fructose [% carbohydrates]	Sucrose [% carbohydrates]
End August	14.5	17.4	0.6	1.9
Mid September	15	17.7	0.4	2.2
End September	14.6	18.2	0.4	2.4
Mid October	14	17.9	0.6	2.7
End October	12.4	17.7	1.1	3.4
Mid November	11.7	16.9	1.5	4
End November	10.9	16.8	1.6	5.2
Mid December	9.5	16.7	2.2	5.7

fructose concentration (formed by the action of FEH), the following reaction takes place in the presence of FFT:

$$G\text{-}F\text{-}(F)_n + F \rightarrow G\text{-}F\text{-}(F)_{n-1} + F_2$$

Inulobiose can be further used as acceptor to form inulotriose, etc. The reaction was proven *in vitro*, and corresponds very well with the observations made *in vivo*.

8.1.2
Molecular Genetics

Fructan 1-exohydrolase I (1-FEH I) is the first plant FEH for which cloning was reported (Van den Ende et al., 2000b). Functionality of DNA was demonstrated by heterologous expression in potato tubers. Surprisingly, this FEH was found to be more homologous to cell-wall invertases (44–53%) than to vacuolar invertases (38–41%) and fructosyltransferases (33–38%). As the FEH I enzyme could not be purified from the apoplastic fluid at levels higher than could be explained by cellular leakage, a vacuolar localization was suggested. Unlike plant fructosyltransferases, which are believed to have evolved from a vacuolar invertase, 1-FEH I may have evolved from a cell-wall invertase-like ancestor that later obtained a vacuolar-targeting signal. The gene expression pattern correlates well with the FEH enzyme activity observed upon cold storage.

8.2
***In vitro* Hydrolysis by Yeast and Mold Enzymes**

Hydrolysis is obtained either by exo-inulinases (EC 3.2.1.80), a combined action of exo- and endo- inulinases, or solely by endo-inulinases (EC 3.2.1.7) (Vandamme and Derijcke, 1983). While the best known yeast in this context, *Kluyveromyces fragilis*, produces only an exo-inulinase, most inulin-hydrolyzing fungi form both exo- and endo-inulinases (Uchiyama, 1993). The most well-described are the enzymes from *Aspergillus niger* (Nakamura et al., 1978), *Aspergillus ficuum* (Zittan, 1981), *Chrysosporium pannorum* (Xiao et al., 1989) and *Penicillium purpurogenum* (Onodera and Shiomi, 1988). All these enzymes form a mixture of F3, F4 and F5 compounds. For industrial fructose and oligofructose production, the enzymes of *A. niger* or *A. ficuum* are used (see Section 9). Although the "endo-inulinase" action of *A. ficuum* results in the formation of oligosaccharides upon inulin hydrolysis, it cannot be considered as a truly endo-acting enzyme. Upon degradation of high molecular-weight microbial inulin, only F3 and F4 compounds were formed, and no internal cleavage was observed.

Other products from inulin are made by intramolecular (depolymerizing) fructose transferases to form difructose-dianhydrides (DFAs), which are cyclic forms of difructose. Two types of DFA can be formed when inulin is the substrate: DFAI is di-D-fructofuranose 2′,1;2′,1′-dianhydride; and DFAIII is di-D-fructofuranose 2′,1;2,3′-dianhydride. The corresponding fructotransferases were purified from *Arthobacter globiformis* and *Arthobacter urefaciens* or *Pseudomonas* (Tamura et al., 1988; Norinsho et al., 1989; Uchiyama, 1993). DFAs have half the sweetness of sucrose and are difficult to digest; they are therefore well suited for use in low-calorie foods, though a bitter after-taste is often experienced. They have only minimal industrial application.

Finally, cyclofructan is produced (by the Mitsubishi Kasei company) using an extracellular enzyme of *Bacillus circulans*. This enzyme forms mainly cycloinulo-hexaose (CFR-6) in addition to small amounts of cyclinulo-heptaose and -octaose from inulin by an intramolecular transfructosylation reaction (Uchiyama, 1993). However, the cavity in the hexamer is not large enough to harbor a guest molecule, as can cyclodextrins

(French, 1993). At present, there is no evidence of their industrial application.

9
Production

Two main types of industrial product are available commercially: (1) the FOS and (2) inulin products derived from chicory. Of the two types, the latter currently dominates the world market

9.1
FOS Production Starting from Sucrose

The best known FOS are those made by the fructosyltransferase of *Aspergillus niger* (Hidaka et al., 1991). The product is sold commercially under the trade-name of Actilight® or Neosugar®. The primary reaction product contains only 55% of FOS, of which 25–30% is 1-kestose (GF2), 10–15% nystose (GF3) and 5–10% 1F-fructofuranosylnystose (GF4). It is interesting to note that not only $\beta(2 \rightarrow 1)$ fructosyl compounds are present, but also some $\beta(2 \rightarrow 6)$ compounds, branches and even molecules of the neokestose type (Joye and Hoebregs, 2000; unpublished results obtained by permethylation).

9.2
Commercial Inulin of Plant Origin

9.2.1
Agricultural Aspects

Given their high inulin content (>10%) *Dahlia*, Jerusalem artichoke (*Helianthus tuberosus*) and chicory (*Cichorium intybus*) might each be considered as good candidates for the industrial production of inulin in temperate regions.

Dahlia

Many *Dahlia* cultivars (cultivated varieties) are available, but they have all been selected for their flowers, rather than for their inulin production. The tuberous roots have no buds, and can be propagated only if attached to a piece of stem tissue, which hampers crop establishment of tubers. When propagated from seed, sowing must be delayed until late spring, given the extreme sensitivity of *Dahlia* to frost (Meyer et al., 1993). Mechanical harvesting of the tubers is feasible only on sandy grounds. Although the mean DP of *Dahlia* inulin is higher than that of chicory inulin, its yield is only half that of chicory (Table 4). For these reasons, *Dahlia* has become less important as an inulin-producing crop.

Tab. 4 Yields and composition of *Dahlia*, Jerusalem artichoke and chicory

		Dahlia	Jerusalem artichoke	Chicory
Roots/tubers	Tonnes per ha	25	42	43.5
	Range		35–60	25–75
DM %	Average	18	22	22.4
	Range	15–22	19–25	20–25
Inulin [%]	Average	11	16	16.6
	Range	10–12	14–18	14.9–18.3
Inulin [tonnes per ha]		2.5–3	4.5–8.5	5–11
Mean DPn		13–20	6–10	10–14

DM, dry matter; DP, degree of polymerization.

Jerusalem artichoke

Jerusalem artichoke is another candidate with a rather high inulin content (14–19%), although a relatively large fraction of the total dry matter produced is present as structural stem dry matter (ca. 4–9 tonnes ha^{-1}). Only a small part of the plant is devoted to inulin storage (4–8 tonnes ha^{-1}) (Zubr et al., 1993). The production of Jerusalem artichoke tubers is comparable with potato production. It is not recommended that Jerusalem artichoke be cultivated in clay soil as the tubers are small and irregular, and large amounts of soil remain attached to the tubers. They are frost tolerant, and so (in theory) could be harvested during winter and early spring, such that a very long processing period would be possible. In practice however, the inulin metabolism is even more sensitive to cold than that for chicory. As the inulin has only 20% of its chains longer than DP 10, the product is unsuited for many functional food and non-food applications (see Section 12).

Chicory

Chicory is the third candidate for industrial inulin production, and indeed the favored one! Chicory is a biennial plant, and during the first season the chicory plants remain in vegetative phase, producing only leaves, taproots and fibrous roots. This makes them very efficient plants as they store their assimilates in the tap roots, which have the appearance of small, oblong sugar beet. The inulin content is high and fairly constant from year to year for any given region. During the period 1989–2000, the mean inulin content of the industrially harvested chicory crops varied slightly, from 16.0 to 17.6%. This variation is not dependent on the type or quantity of fertilizer used, as high-N dosages (>80 U) improve neither inulin concentration, DP, nor global dry matter yield.

The yields (tonnes of roots per hectare) show a much wider variation, however. The data listed in Table 4 refer to industrial yields, i.e., tonnes of chicory roots received at the processing plant per hectare harvested. These yields are 15–20% less than theoretical yields, because during harvesting the small ends of the roots and tiny side-roots are broken off and remain in the ground. Harvesting is carried out using modified sugar-beet harvesters and specially designed harvesters, as chicory roots are thinner and more brittle. The chicory used for inulin production (*Cichorium intybus*) is of the same type as that used for the production of coffee substitute.

Originally, it was thought that growing chicory would be similar to growing sugar beet, but in practice each step of cultivation was found to be more delicate and demanding, from seed bed preparation to harvest; hence, the culture appeared more as a vegetable crop than a root crop. Fortunately, the culture is more drought-resistant; indeed, some drought is even beneficial for inulin synthesis.

There are no major problems related to bacterial or viral diseases with chicory, but in order to keep the chicory culture healthy a strict crop rotation is imposed (once every 5 years). Moreover, an efficient herbicide control program was developed based on mini-doses (ORAFTI, 2001); these are harmless to the plant and almost eliminate the need for manual weeding. Prices of 43 Euro tonne^{-1} roots and yields of 45 tonnes ha^{-1} make this crop an attractive proposition, and consequently each year more farmers offer to cultivate chicory plants.

9.2.2
Processing

The production of inulin involves two phases: (1) the extraction and primary purification that results in a raw impure syrup; and

(2) the refining phase, where the end product is >99.5% pure.

Phase 1
The first phase of the ORAFTI chicory process is very similar to the sugar beet process. Roots are harvested and stored in piles on the field. To minimize losses, no more than 7 days must elapse between harvest and delivery date. The roots are transported to the factory, weighed and carefully stored. Chicory's metabolic losses during storage are twice as high as those in sugar beet. During a storage trial under a controlled atmosphere at 10°C, CO_2 escape was ~250 L per day per tonne for chicory, compared with 130 L per tonne for sugar beet; by comparison, bruised chicory roots produced 300–320 L CO_2 per day per tonne. The roots are then transported on a stream of water to the factory, where they are washed and sliced. Raw inulin is extracted from the resulting "chips" with hot water in a countercurrent diffuser. The leached chips are dried and sold as feed. A first purification step is applied to the extraction juice by liming and carbonation at high pH. The resulting $CaCO_3$ sludge precipitates easily, with peptides, some anions, degraded proteins and colloids being trapped in the flocs. This generates a foam-type product which is used by the farmers to improve their soil structure, as it is rich in calcium and organic matter.

Phase 2
In the second step, the raw juice is further refined using cationic and anionic ion-exchange resins for demineralization, and active carbon for decolorization. The technology used is comparable with that used in starch processing. Major points of concern are strict control of pH and temperature. After demineralization and decolorization, the juice is passed over a 0.2-µm filter to be sterilized, evaporated and spray-dried. Spray-drying has been, until now, the most convenient technology for converting the refined inulin into a storable, microbiologically stable and commercial end product.

Chicory processing is clearly more difficult than sugar beet processing. The chicory process can be considered as one large compromise, balancing maximal removal of impurities against degradation of the inulin chain, color formation, Maillard reaction, contamination and incorrect removal of taste and odor.

The ORAFTI process
The use of ion-exchange resins for demineralization purposes has often been criticized as being environmentally unacceptable, but the ORAFTI process proves that the opposite can be true, provided that the necessary investments are made. The chemicals used for regeneration of the ion exchangers are not NaOH and HCl (as normal), but NH_3 and H_2SO_4. The price of regeneration is higher with the latter products, but the advantage is that the effluents can be converted into reusable byproducts, and for this reason all effluents are carefully collected and evaporated. At high concentrations, easily crystallized salts such as $(NH_4)_2SO_4$ and K_2SO_4 precipitate; they are separated from the mother liquor and sold commercially as fertilizer (as they contain N and K). The mother liquor is further evaporated into a stable and storable end product and sold as feed, based on its high organic matter content. In this way the cycle between the processing industry and agriculture is fully closed.

The resulting inulin (RAFTILINE®) has a DP which reflects the original DP present in chicory, and varies between 10 and 14. A special grade inulin, RAFTILINE HP®, with DP >23 is also available; this is made by physical elimination of the small DP fraction (Smits et al., 1997).

The process described above is used not only for the production of RAFTILINE®, but also for the production of fructose syrups (RAFTISWEET®) and oligofructose syrups (RAFTILOSE®). The only additional process required is an enzymatic hydrolysis.

The fructose syrup is produced using a mold inulinase from NOVO (Fructozyme®) which has both endo- and exo-activity. The operating conditions are pH 4.5 and 60°C, and the resulting fructose syrup has a high fructose content (average $\geq 85\%$), low oligosaccharide content ($\leq 2\%$), and no aftertaste. In contrast, with an acid hydrolysis process, these criteria cannot be met without degrading fructose further into hydroxymethylfurfural (HMF), and also creating a "burned" after-taste.

In the production of RAFTILOSE®, a purified endo-inulinase is used since the aim is to produce inulin oligomers with the least possible formation of monomers. However, the presence of small quantities of glucose, fructose and sucrose cannot be avoided, as they are contained in the initial crude chicory extract.

9.3
Commercial Production of Inulin and FOS

Today, there are three industrial inulin processors, all of which use chicory as the raw material. ORAFTI (Belgium) is the name of the activity group of the RAFFINERIE TIRLEMONTOISE, dealing with the production and commercialization of all three types of previously described products. WARCOING is Belgium's oldest inulin producer, and is producing fructose syrups and inulin. The same applies to SENSUS, which is a subsidiary of COSUN and is located in The Netherlands.

As mentioned previously, synthesized FOS are marketed by Beghin-Say (France) and Meiji Seika (Japan).

9.4
Scale of Production

The fact that chicory processing has evolved from a laboratory or pilot-plant scale to a full industrial scale is proven amply, given the data listed in Table 5.

While the production of inulin and oligofructose is not regulated by quota, that of fructose is regulated. As these quota are officially attributed by the European Union, they provide an idea of the scale of production, but do not reveal, for each processor, the secrets of actual or future production capacity and sales.

The fact that the land area assigned to chicory for inulin/fructose production has evolved from a few hundred hectares in 1990 to 20,000 ha in 2000 is clearly also a reference to the growth of inulin-related business.

10
Properties

It is not only the functional properties, but more importantly the nutritional properties, that suggest inulin may become "the" food ingredient of the 21st century.

Tab. 5 Quota, in fructose equivalents, attributed to the three chicory processors

	Quota A in fructose equivalents [tonnes DS]
ORAFTI	54.500
WARCOING	34.800
SENSUS	33.650
Total	122.950

10.1
Physical and Chemical Properties

In standard chicory inulin, the DP ranges from 2 to > 60, with an average of about 10–12. Long-chain inulin, from which the lower-DP fraction has been physically removed and which has an average DP of ~25 is also available for high-performance fat replacement and texture improvement (Table 6).

Chicory inulin is available as white, odorless powders with a high purity and well-known chemical composition. The taste is neutral, without any off-flavor or after-taste. Standard chicory inulin is slightly sweet (10% sweetness compared with sugar), whereas long-chain inulin is not at all sweet (Franck, 1993). Inulin combines easily with other ingredients, it does not modify delicate dairy (milk) flavors, and can even suitably enhance fruit aromas. Inulin behaves as a bulk ingredient, contributing to body and mouthfeel. It is often used in recipes, sweetened with high-potency sweeteners, such as aspartame and acesulfame K, to provide a rounder mouthfeel and a better-sustained flavor with reduced after-taste, as well as an improved stability. Combinations of acesulfame-aspartame blends with inulin (from 2 to 10% in solution) also exhibit a significant quantitative synergy (Wiedmann and Jager, 1997).

Inulin is moderately soluble in water (maximum 10% at room temperature) which allows its incorporation in watery systems where the precipitation of other fibers often leads to problems. To prepare a solution of inulin, the use of warm water (50–100°C) is suggested. The viscosity of inulin solutions is rather low, e.g., 1.65 mPa·s at 10°C for a 5% dry matter (d.m.) solution, and 100 mPa·s for a 30% d.m. solution (De Leenheer, 1996). Inulin

Tab. 6 Physico-chemical properties of inulin

	Standard inulin	*Long-chain inulin*
Chemical structure	GF_n ($2 \leq n \leq 60$)	GF_n ($10 \leq n \leq 60$)
Average degree of polymerization	12	25
Dry matter (d.m.) [%]	> 95	> 95
Inulin content [% d.m.]	92	99.5
Sugar content [% d.m.]	8	0.5
pH [10% w/w]	5–7	5–7
Sulfated ash [% d.m.]	< 0.2	< 0.2
Heavy metals [ppm d.m.]	< 0.2	< 0.2
Appearance	White powder	White powder
Taste	Neutral	Neutral
Sweetness [versus sucrose = 100%]	10	None
Solubility in water at 25°C [g L^{-1}]	120	10
Viscosity in water (5%) at 10°C [mPa.s]	1.6	2.4
Heat stability	Good	Good
Acid stability	Fair	Good
Functionality in foods	Fat replacement	Fat replacement
	Body and mouthfeel	Body and mouthfeel
	Texture improvement	Texture improvement
	Foam stabilization	Foam stabilization
	Emulsion stabilization	Emulsion stabilization
	Synergy with gelling agents	Synergy with gelling agents

F, fructosyl unit; G, glucosyl unit.

exerts a small effect on the freezing and boiling point of water (e.g., 15% inulin decreases the freezing point by 0.5°C).

In very acid conditions, the β(2 → 1) bonds between the fructose units of inulin can be (partially) hydrolyzed. Fructose is formed in this process, which is more pronounced at low pH, high temperature and low dry substance conditions. Inulin is stable in applications with a pH higher than 4. Even at lower pH values, the hydrolysis of inulin can be limited to less than 10% if the products have a high dry substance content ($>70\%$), are stored at a low temperature ($<10°C$), or have a short shelf-life.

At high concentration ($>25\%$ in water for standard inulin and $>15\%$ for long-chain inulin), inulin has gelling properties and forms a particle gel network after shearing. When the fructan is thoroughly mixed with water or another aqueous liquid, using a shearing device such as a rotor-stator mixer or a homogenizer, a white creamy structure results which can easily be incorporated in foods to replace fat (up to 100%) (Franck, 1993). This process has been patented by ORAFTI. In several cases, the creamy opaque structure can even be produced directly *in situ* by submitting a mixture of inulin, water and the other ingredients of the food product to an appropriate mixing device included in the processing line. The particle gel of inulin imitates fat extremely well; it provides a short spreadable texture, a smooth fatty mouthfeel, as well as a glossy aspect and a well-balanced flavor release. As far as fat replacement is concerned, long-chain inulin shows about twice the functionality compared with standard chicory inulin, thus allowing for lower dosage levels and ingredient costs. Special instant qualities, which do not require shearing to give stable homogeneous gels also have been developed (and patented) using specific spray-drying technology.

The gel strength obtained depends on the inulin concentration and total dry substance content, on the shearing parameters such as temperature, time, speed or pressure and also on the type of shearing device used, but is not influenced by pH (between pH 4 and 9). The gel strength also increases during the first 24 h after shearing. Electron cryomicroscopy has shown that such an inulin gel is composed of a three-dimensional network of insoluble submicron inulin particles in water. These particles of about 100 nm in size aggregate to form larger clusters having a diameter of 1–5 µm. Large amounts of water are immobilized in that network, as determined by NMR experiments, which assures the physical stability of the gel in function of the time. The X-ray diffraction pattern confirms the crystalline nature of the gel particles, whereas the starting inulin powder is essentially amorphous. The thermal stability of inulin gels is limited, but it is significantly improved in the presence of other ingredients, allowing therefore heat treatments as well as freeze–thaw cycles.

An inulin gel exhibits a viscoelastic rheological behavior that is characterized by a certain elasticity as well as a viscosity. The gel also shows shear-thinning and thixotropic properties, and is characterized by a relatively low yield stress (e.g., 1540 Pa for a gel of 30% standard inulin in water at 25°C). When submitted to an increasing deformation (oscillatory rheological experiments), the gel gradually loses its solid properties: the elasticity modulus decreases, whereas the fluid properties – and thus the viscosity modulus – which are initially lower, increase more and more. The dynamic behavior of an inulin gel is quite similar to that of a margarine-type product, and rather different from that of typical polymer gels such as starch. Furthermore inulin displays synergy with most gelling agents (e.g., gelatin, alginate, k- and i-carrageenan, gellan gum, maltodextrin), that is, the gel strength obtained by combining them is higher than the sum of the gel strengths obtained for the two gels made separately (Franck and Coussement, 1997).

Inulin also improves the stability of foams and emulsions, such as aerated dairy desserts, ice creams, table spreads and sauces. It can therefore replace other stabilizers in different food products.

10.2 Material Properties

Commercial inulin products are available as white powders, with different particle size distribution and density. The content of the different chain length components also differs depending on the target application. Indeed, longer chains behave more like polysaccharides, and exhibit fat-replacing and stabilizing properties, whereas shorter chains show technological properties closer to those of sucrose and glucose syrups.

10.3 Biological Properties

Inulin has been shown to provide several interesting nutritional properties to animal and human volunteers (Van Loo et al., 1999) (Table 7).

10.3.1 Nondigestibility

Due to its specific chemical structure ($\beta(2 \to 1)$ bonds), which cannot be hydrolyzed by human digestive enzymes (the small intestinal sucrase-maltase complex), inulin passes through the mouth, stomach and small intestine without undergoing any significant change, and without being metabolized. This has been confirmed in human studies with ileostomized volunteers (Knudsen and Hessov, 1995; Ellegård et al., 1997). Inulin thus enters the colon almost quantitatively ($>90\%$), where it is completely metabolized by the intestinal bacteria (Roberfroid, 1993). Even at high-intake doses, no significant amounts have been detected in the feces.

Tab. 7 Biological and nutritional properties of inulin

Strong evidence:

Non-digestibility and low caloric value (1.5 kcal g^{-1})
Suitable for diabetics
Soluble dietary fiber
Stool bulking effect: increase in stool weight and stool frequency, relief of constipation
Modulation of the gut flora composition, stimulating beneficial bacteria (*Bifidobacterium*) and repressing harmful bacteria (*Clostridium*): prebiotic/bifidogenic effect
Improvement of calcium (and magnesium) bioavailability

Promising evidence:
Reduction of serum triglycerides and insulin levels
Reduction of colon cancer risk (in animal models)
Modulation of immune response
Protection against intestinal disorders and infections

10.3.2
Caloric Value

The nondigestibility of inulin is at the basis of its reduced caloric value compared with its component monosaccharide moieties. Inulin indeed is completely converted mainly to short-chain fatty acids (SCFA: acetate, propionate, butyrate), lactate, bacterial biomass, and gases. Only the SCFA and lactate contribute to the host's energy metabolism. This corresponds to a limited fraction of the original energy content of the component monosaccharides. Moreover, the SCFA and lactate are less effective energy substrates than sugars. Taken together, these findings explain the reduced caloric value of inulin.

Based on studies in humans, the caloric value of shorter-chain ^{14}C-labeled fructo-oligosaccharides was calculated by Hosoya et al. (1988) to be 1.5 kcal g^{-1}. Experimental *in vitro* data (fermentation) and *in vivo* data (in rats) allowed Roberfroid et al. (1993) to calculate the caloric value of inulin and oligofructose according to basic biochemical principles. When 1 mole of fructose is completely metabolized into CO_2, 38 moles of ATP are produced. When the fructose is bound (as in inulin), 40% of it is converted into bacterial biomass, 5% into CO_2, 40% into SCFA, and 15% into lactate. Only 90% of these metabolites are absorbed into the blood; their subsequent metabolism in the liver produces 14 moles of ATP, which is 35% of the caloric content of free fructose. Hence, the caloric value is 1.4 kcal g^{-1} inulin or oligofructose. Others have even suggested lower caloric values (Delzenne et al., 1995; Ellegård et al., 1997; Castiglia-Delavaud et al., 1998), and consequently a caloric value of 1–1.5 kcal g^{-1} of inulin is used currently for food labeling.

10.3.3
Improvement of Lipid Metabolism

In rats, distinct effects on lipid metabolism have been observed, with serum triglycerides mainly affected. The observed metabolic changes originate at the level of the liver (Bhattathiry, 1971; Delzenne et al., 1993; Levrat et al., 1994; Fiordaliso et al., 1995; Delzenne and Kok, 1998).

Biochemical studies with isolated hepatocytes have shown that inulin or oligofructose consumption reduces the activity of key hepatic enzymes which are related to lipogenesis (de-novo synthesis of fatty acids or assembly of triglycerides from acyl groups and glycerol) (Kok et al., 1996a, b, 1998a). Further research revealed that altered gene expression is at the basis of the downregulation. It is postulated that inulin and oligofructose, or their bacterial metabolites, affect the hormone status (insulin and/or glucose-dependent insulinotropic polypeptide) of the rats (Kok et al., 1998a). More recently, a dose–effect study in golden Syrian hamsters also showed a significant decrease in serum triglycerides and cholesterol (VLDL-cholesterol) (Trautwein et al., 1998).

Human volunteer studies have indicated that inulin (and oligofructose) has a modulating effect on lipid metabolism. Indeed, it has been reported that the consumption of fructans reduced serum triglycerides and sometimes also cholesterol (mostly LDL-cholesterol) in healthy volunteers who were (slightly) hyperlipidemic. In contrast, the optimal lipid parameters of healthy normolipidemic young adults were usually not affected. Recent studies have indicated that the triglyceride-lowering effect might take time to establish (about 8 weeks) (Canzi et al., 1995; Davidson et al., 1998; Brighenti et al., 1999; Jackson et al., 1999; Causey et al., 2000).

An alternate line of thought is to monitor the impact of inulin or oligofructose on an unbalanced diet (a chronic problem of Western society) in normolipidemic subjects. Initially, rat experiments showed that the addition of 10% inulin or oligofructose to a fat-rich diet reduced the postprandial serum triglyceride levels, as well as serum cholesterol levels, by more than 50% compared with control groups. Enhanced triglyceride-rich lipoprotein catabolism appears to be at the base of this observation (Kok et al., 1998b).

10.3.4
Effects on Gut Function

Inulin is a soluble dietary fiber (Roberfroid, 1993; Prosky, 1999) that is not hydrolyzed by human digestive enzymes, and induces typical fiber effects on gut function, such as a reduction of intestinal pH, relief of constipation, and increased stool weight and frequency (bulking effect). This fecal bulking effect is similar to that caused by other soluble fibers such as pectin and guar gum (Roberfroid, 1997). Each gram of ingested inulin increased the fecal wet weight by 1.5–2 g, and this was also reflected by an increased fecal dry weight excretion. The latter was mainly caused by an increased excretion of bacterial biomass (Gibson et al., 1995; Castiglia-Delavaud et al., 1998).

The stimulation of gut function also resulted in increased stool frequency, the increase being higher in volunteers with a low initial frequency. Relief of constipation has been reported in different studies with inulin (Gibson et al., 1995; Kleessen et al., 1997; Den Hond et al., 2000). The effects on intestinal transit time are more contradictory, as some researchers have observed an increased and others a decreased, transit time.

10.3.5
Modulation of Gut Microflora

As it is nondigestible, inulin reaches the colon almost quantitatively. The colon is an energy-deficient ecosystem, and the import of easily fermentable carbohydrates has a major impact on the composition and metabolic activity of its complex microflora. The latter is composed of over 400 different types of bacteria, and represents over 50% of the dry solids content in the colon.

Inulin selectively promotes the growth and metabolic activity of beneficial bacteria, mainly bifidobacteria, while repressing harmful ones (e.g., clostridia) – this is called the "prebiotic" or "bifidogenic" effect. The bifidogenicity of inulin has been demonstrated in *in vitro* models (Wang and Gibson, 1993; Gibson and Wang, 1994) and in several *in vivo* human volunteer studies (Gibson et al., 1995; Kleessen et al., 1997; Roberfroid et al., 1998; Kruse et al., 1999). The increase in the number of bifidobacteria appears to be inversely correlated with their counts at the start of the studies. A daily intake of 5–10 g of inulin is sufficient to enhance significantly the bifidobacterial population (Roberfroid et al., 1998).

By using *in vitro* experiments, the bifidogenicity of the high-DP fractions of inulin (DP > 10) has been confirmed. This fraction is fermented twice as slowly as the low-DP fraction (DP < 10), and hence the long-chain inulin fraction has interesting potential for the stimulation of metabolic activity in the distal colon (Roberfroid et al., 1998).

During the past decade, many possible beneficial effects have been attributed to bifidobacteria, and recent objective observations have begun to support their potential health-promoting activity. As bifidobacteria counts were increased in several studies with inulin, a concomitant decrease in potentially pathogenic populations was often observed (*Clostridium* sp., *E. coli*, in human studies;

Salmonella in poultry; *Veillonella*, *Shigella*, and *Listeria* in *in vitro* models). In experiments with gnotobiotic quails (a model for preterm babies), animals administered a necrotizing enterocolitis-inducing flora (comprising *Clostridium butyricum*) died, but those which received bifidobacteria made a complete recovery. This was the first objective demonstration of a health effect of bifidobacteria in living beings (Butel et al., 1998). It has also recently been observed that patients suffering from Crohn's disease systematically have low counts of bifidobacteria (Favier et al., 1997).

Inulin, in being able to stimulate probiotic strains, is considered a prebiotic. It appears that the combination of the two (synbiotic) has synergistic effects, and recent observations have indicated that synbiotics can contribute to the health-food concept (Gibson and Roberfroid, 1995).

10.3.6
Suitability for Diabetics

As inulin is only hydrolyzed negligibly during passage through the mouth, stomach and small intestine in humans, it has no influence on blood glucose and insulin levels when ingested orally (Beringer and Wenger, 1955).

The potential use of inulin as a food for diabetics has been recognized since the early twentieth century; in 1912, Lewis referred to Persia (1905), who recommended inulin to diabetics and stated that the product was well digested and assimilated by those people in large doses, and for long periods of time. Strauss (1911) also reported the feeding of 40–100 g per day of pure inulin to be very beneficial for these patients, and this was confirmed by Root and Baker (1925), McCance and Lawrence (1929) and Wise and Heyl (1931). Since then, many more applications for diabetics have been described, including inulin-based diabetic bread and pastry (Beringer and Wenger, 1955; Kuppers-Sonnenberg, 1971) and inulin-based diabetic jam (Birch and Soon, 1973).

10.3.7
Reduction of Cancer Risk

Very recent research has indicated that inulin has a significant chemopreventive potential (Roland et al., 1994a, b, 1995, 1996; Delzenne et al., 1995; Taper et al., 1995, 1997, 1998; Reddy et al., 1997; Rowland et al., 1998). Indeed, it can prevent to a significant extent the formation of chemically induced colonic precancerous lesions in rats. It clearly delays the initiation phase of carcinogenesis, and long-chain inulin (average DP ~25) may be more effective than the shorter-chain fructans in preventing lesion formation in the colon, where cancer incidence is highest in humans. This effect may be due to slower fermentation of the long-chain molecules, which in turn stimulates bacterial activity in the distal colon (Reddy et al., 1997).

Other experiments indicate that long-chain inulin is also able to delay the propagation phase of colon carcinogenesis in rats. Synergistic effects were observed when inulin and bifidobacteria were co-administered, this being the first experimental observation of a synbiotic effect *in vivo* (a synergistic physiological effect obtained by combining a probiotic with a prebiotic) (Rowland et al., 1998). It was also shown that inulin inhibits the development of cancer cells transplanted in the thigh and peritoneum of mice (Taper et al., 1997, 1998).

10.3.8
Increase in Mineral Absorption

Inulin and oligofructose significantly increase the intestinal absorption of calcium, iron, and magnesium in rats (Delzenne et al., 1995; Scholz-Ahrens et al., 1998). Experiments in rats showed that the increased calcium absorption resulted in increased bone mineral density (Lemort and Roberfroid, 1997; Scholz-Ahrens et al., 1998), which means that the increased absorption effectively results in a build-up of the body's calcium reserve. It was also shown in ovariectomized rats (a model for postmenopausal women) that calcium uptake from food was improved, and bone-mineral density increased, by feeding inulin-type fructans (Scholz-Ahrens et al., 1998).

In an initial study in healthy adult volunteers given 40 g per day inulin, a significant increase in calcium absorption was seen using the balance technique (Coudray et al., 1997). Recently, a method using dual stable isotopes of calcium has been validated for use in human volunteers. In analogy with rat experiments, where data on increased mineral absorption were obtained mainly with growing animals, the experiments in humans were performed with adolescents. This stage of life is characterized by increased mineral absorption. It was found in 11 teenagers that an intake of 15 g per day of a chicory fructan significantly increased the absorption of calcium (Van den Heuvel et al., 1999). A significant increase in calcium absorption has now been confirmed after ingestion of only 8 g per day in 30 adolescent girls (S. Abrams, personal communication). This is an important finding, as it is during this particular period of life that humans build up their calcium reserves. The higher the peak bone mass becomes at this period of life, the less the probability of osteoporosis occurring in later life.

Taken together, these results provide promising evidence that inulin increases calcium absorption in humans, and might therefore contribute actively to reducing the later risk of osteoporosis (Franck, 1998).

10.3.9
Intestinal Acceptability

Intestinal acceptability of nondigestible components is mainly determined by two phenomena. First, the osmotic effect, leading to an increased presence of water in the colon. Smaller molecules exert a higher osmotic pressure and bring more water into the colon, and this is probably why sorbitol, for example, has a significantly higher laxative potential than inulin (Hata and Nakajima, 1984). The second effect is caused by the fermentation products, mainly SCFA and gases. Slowly fermenting compounds appear easier to tolerate than their fast-fermenting analogues; this may explain why inulin is easier to tolerate than polyols and shorter-chain oligofructose.

Flatulence is a well-known and often accepted side effect of the intake of vegetables. Dietary fibers, in general, are known and rewarded for their properties of "stool softening". Research has shown that portions of 5–10 g inulin are well tolerated by most people, and that daily doses of 10–15 g cause no significant discomfort. At higher doses, flatulence may cause some discomfort to sensitive people. Inulin rarely causes diarrhea (Absolonne et al., 1995).

11
Food Applications

Inulin can be used for either its nutritional advantages or its technological properties, but it is often used to offer a double benefit, namely an improved organoleptic quality and a better-balanced nutritional composition. An overview of the applications of inulin in foods and drinks is provided in Table 8.

The use of inulin as a fiber ingredient is straightforward, and often leads to an improved taste and texture (Franck and Cousse-

Tab. 8 Food applications of inulin

Application	Functionality	Dosage level [% w/w]
Dairy products (yogurts, cheeses, desserts, drinks)	Fat replacement Body and mouthfeel Foam stabilization Fiber and prebiotic	2–10
Frozen desserts	Fat replacement Texture improvement Melting behavior Low caloric value	2–10
Table spreads and butter-like products	Fat replacement Texture and spreadability Emulsion stabilization Replacement of gelatin Fiber and prebiotic	2–10
Baked goods and breads	Fiber and prebiotic Moisture retention	2–15
Breakfast cereals and extruded snacks	Fiber and prebiotic Crispness and expansion Low caloric value	2–20
Fillings	Fat replacement Texture improvement	2–30
Salad-dressings and sauces	Fat replacement Mouthfeel and body Emulsion stabilization	2–10
Meat products	Fat replacement Texture and stability Fiber	2–10
Dietetic products and meal replacers	Fat replacement Synergy with intense sweeteners Body and mouthfeel Fiber and low calorie	2–15
Chocolate	Sugar replacement Fiber Heat resistance	5–30
Tablets	Sugar replacement Fiber and prebiotic	5–75

ment, 1997). When used in bakery products and breakfast cereals, this presents a major progress in comparison with other fibers. Inulin provides more crispness and expansion to extruded snacks and cereals, and also increases the bowl-life. It also keeps breads and cakes moist and fresh for longer. Its solubility allows fiber incorporation in watery systems such as drinks, dairy products, and table spreads. Increasingly, inulin is being used in functional foods, and especially in many dairy products, as a prebiotic ingredient to stimulate the growth of the beneficial intestinal bacteria.

Due to its specific gelling characteristics, inulin allows the development of low-fat foods without compromising on taste and texture. This is especially true in spreadable products. In table spreads (both fat and water-continuous), inulin allows the replacement of significant amounts of fat and the stabilization of the emulsion, while providing a short spreadable texture. Long-chain inulin (from 2 to 10% depending on the recipe) provides excellent results in water-in-oil spreads, with a fat content ranging from 20 to 60%, as well as in water-continuous formulations containing ≤10% fat. It can also be applied in fat-reduced spreads containing dairy proteins, as well as in butter-like products and other dairy-based spreads. In low-fat dairy products, such as fresh cheese, cream cheese or processed cheese, the addition of a few percent of inulin increases the body, gives a creamier mouthfeel, and also imparts a better-balanced round flavor. Inulin also is destined to be used as a fat replacer in frozen desserts, providing an easy processing, a fatty mouthfeel, excellent melting properties, as well as freeze–thaw stability, without any unwanted off-flavor. Fat replacement can further be applied in meal replacers, meat products, sauces and soups. Fat-reduced meat products, such as sausages, pâtés and other meat-based spreads, with a creamier and juicier mouthfeel and an improved stability due to water immobilization can be obtained.

The synergistic effect of inulin with other gelling agents constitutes an additional advantage in all these applications. In several products, inulin – and especially its high-performance (or long-chain) version – can even (partially) replace gelatin, starch, maltodextrin, alginate, caseinate, and other stabilizers. The replacement of gelatin by inulin, an all-vegetable ingredient, is particularly interesting in dairy desserts, yogurts, cheese products and table spreads.

In the yogurt market, low-fat products are showing the strongest growth, and in particular diet yogurts with fruit. The incorporation of (instant) inulin (1–3%) in the recipe, possibly through the fruit preparation, improves the mouthfeel, reduces syneresis and offers a synergistic taste effect in combination with aspartame and acesulfame K, without increasing significantly the caloric content. Inulin also increases the stability of foams and mousses: its incorporation at 1–5% into dairy-based aerated desserts (chocolate, fruit, yogurt or fresh cheese-based mousses) improves their processability and upgrades the quality. The resulting products retain their typical structure for a longer time and give a fat-like feeling, even in the case of low-fat or fat-free formulations.

Inulin has found an interesting application as a low-calorie bulk ingredient in chocolate without added sugar, not to reduce fat but to replace sugar, often in combination with a polyol. It is also used as a dietary fiber or sugar replacer in tablets. Hence, inulin has become a key ingredient offering new opportunities to the food industry looking for the well-balanced and yet better tasting products of the future.

12
Non-food Developments and Applications

As an answer to the growing demand for products made from "renewable resources" and with (modified) starch as a reference in mind, much research has been devoted to the chemical modification of inulin (Meriggi et al., 2001). Although during the past few years a substantial effort has been made, the production and application of these inulin derivatives on industrial scale remain at present limited to two examples.

The first example is the introduction of carboxyl groups into the inulin polymer by carboxymethylation. This reaction is carried out by heating an aqueous solution of inulin and monochloroacetic acid in an alkaline medium (Verraest et al., 1995a). The resulting product, carboxymethylinulin (CMI) is an inhibitor of calcium carbonate formation. A clear advantage of CMI over other saccharide-based polycarboxylates [for example carboxymethylcellulose (CMC) or carboxymethylsucrose (CMSU)] arises from the fact that CMC has a relatively high viscosity, and CMSU has no influence at all on the $CaCO_3$ crystallization. It was also noted that carboxymethylation of inulin with a high degree of polymerization results in better inhibitory properties (Verraest et al., 1996). In addition to these scale-inhibiting properties, CMI can also be applied as dispersing agent or as a metal ion carrier.

The second example is the polysaccharide-based polycarboxylate, dicarboxy-inulin (DCI). This is produced by the oxidation of inulin, and has excellent calcium-binding properties. It is therefore a potential candidate to replace other currently used (and nonbiodegradable) builders or co-builders in detergent formulations, e.g., polyacrylates. Although many oxidation methods exist (Meriggi et al., 2001), the one-step method using NaOCl as the oxidant and bromide as the catalyst is preferred (Besemer, 1993; Besemer and van Bekkum, 1994). Due to the fructose ring-opening and direct formation of the calcium-complexing oxydiacetate (ODA) structure, DCI has a higher calcium-complexing capacity than dicarboxystarch. Moreover, due to the formation and presence of this ODA structure, it is much easier to compromise between the oxidation degree (and calcium-sequestering capacity) and the biodegradability.

Besides the chemically modified inulin, the native fructose polymer (without derivatization) might also have a promising future in the non-food sector. Possible applications may be found in the field of ceramics processing (Ingelbert, 1999). In this type of industry, inulin can function as a temporary binder and may be an alternative for the widely used polyvinylalcohol (PVA). Using inulin in a ceramics-slurry provides a "green form" which, after pressing, has an acceptable "green strength"; this allows the "green form" to be processed easily, and to be completely removed by burning out during the sintering stage.

13
Outlook and Perspectives

A major increase in the number of scientific reports related to inulin research has been noted during the past decade. During this time there has been a shift from information regarding "crops" or "products" to communications dealing with the nutritional aspects of inulin, and the elucidation of genetic/biochemical mechanisms of inulin synthesis in plants. Undoubtedly, this trend will continue in the near future.

In the field of crop improvement, it is clear that the use of genetic engineering as a research tool will also increase. The desire to produce high-DP fructans is closely linked with ongoing developments in the chemical modification of inulin on the one hand, and the improvement of its functional properties on the other hand.

Fundamental mechanisms governing the nutritional benefits of inulin also need to be further investigated. The impact of inulin on the composition of the colonic microflora and, more importantly, the implication of such an altered bacterial ecosystem on the host's health status, will be better elucidated. Modulation of the immune response and protection against infectious diseases will be assessed as potential benefits of regular inulin consumption. Mineral absorption and bone health is another topic worthy of continued research, together with investigations into inulin's reduction of the incidence of cancer, notably of the colon. Specifically, more attention will be paid to the roles of inulin components, each with their different chain length.

With regard to the derivatization reactions of inulin, further developments are expected that are focused not only on new reactions, but also on new applications. Chemically modified inulins will most likely become available commercially as specialty chemicals with specific functionalities. It must be borne in mind however, that functionality must be linked to an industrially affordable and environmentally friendly process.

New developments will in time generate new products, and older products will be redesigned to fulfil changing market demands. In the current era of functional foods, inulin has already become a "reference" compound, and it is clear that its uses and applications will increase in number. The extent of this increase is more difficult to predict however.

14
Patents

For a complete overview of existing relevant patents, the reader is referred to the excellent database, which is maintained by the European Fructan Association (EFA), and is available at:
http:/www.mpimp-golm.mpg.de/efadbase.
Over 200 patents are stored in this database, which takes into account the past 10 years.

A summary of the most significant patents, dealing with mainly composition or production aspects, is listed in Table 9. The majority of patents based on genetic modification are also included. Food applications and chemical modification patents are listed in Table 10.

Acknowledgements

The authors wish to thank B. Levecke for his contribution to Section 12.

Tab. 9 Composition, production and GMO patents (EFA database)

Maruzen, Kazei (1991)	New cyclic inulo-oligosaccharide preparation from inulin or fructose using enzyme or microorganism, e.g. *Bacillus circulans*. Patent Application: J02252701.
Mitsubishi Chemicals (1991)	Difructose dianhydride-I production from inulin using inulinase from *Arthrobacter* sp. Patent Application: J03247295.
Mitsui Toatsu Chemicals (1991)	Production of inulo-oligosaccharide, inulin decomposition using *Penicillium purpurogenum* var. *rubri-sclerotium* and *Aspergillus ficuum* inulinase. Patent Application: EP429077.
Ajinomoto (1992)	Liquid composition containing polyfructan, polyfructan production by *Aspergillus sydowii* fermentation or by fructose-transferase action on sugar or inulin. Patent Application: JP4311378.
Ajinomoto (1992)	Solid composition containing polyfructan, polyfructan production by *Aspergillus sydowii* fermentation or by fructose-transferase action on sugar of inulin. Patent Application: JP4311371.

Tab. 9 (cont.)

Institut Genbiologische Forschung (1994) Berlin, Germany	Polyfructan-sucrase DNA sequence from *Erwinia amylovora*, expression in transgenic plant for polyfructan production. Patent Application: DE4227061.
Netherlands Science Research Foundation (1994) The Netherlands	Method for transgenic plant construction with modified fructan content potato transformation with plant, *Bacillus subtilis* or *Streptococcus mutans* tolerance. Patent Application: WO 94 14970.
Gleiss, A., Schneider, H. (1994) Südzucker, Germany	Preparation of long chain inulin and fructose and glucose by enzymatic treatment of crude inulin with a hydrolase. Patent Application: DE 4316425.
Caimi, P. G., Hershey, H. P., Kerr, P. S. (1995) Du Pont	Fructose polymers are produced by transgenic crop plants fructan, dextran, alternan production in maize, potato and tobacco; DNA construct containing tissue-specific promoter, vacuole targeting sequence and fructosyltransferase gene. Patent Application: WO 95 13389.
Van Tunen, A. J., Van der Meer, I. M., Koops, A. J. (1996) Wageningen, The Netherlands	DNA sequences encoding fructan-producing enzymes, Jerusalem artichoke-derived fructosyltransferase expression in transgenic plant for low calorie sweetener production. Patent Application: WO 96 21023.
Smeekens, J. C. M., Ebskamp, M. J., Geerts, H. A. M., Weisbeek, P. J. (1996) The Hague, The Netherlands	Production of oligosaccharides by transgenic plants by *Streptococcus mutans*, *Bacillus subtilis*, onion, barley or sunflower fructosyltransferase gene transfer. Patent Application: WO 96 01904.
Partida, V. Z., Lopez, A. C., Gomez, A. J. M. (1997) Aspen, CO, United States	Fructose syrup production from agave plant pulp by extraction and inulinase treatment. Patent Application: WO 97 34017.
Turk, S., Gerrits, N., Smeekens, J. C. M., Weisbeek, P. J. (1997) Kapelle, The Netherlands	Production of modified polysaccharides by treatment with sugar donor and transfer enzyme, starch production by treatment with enzyme or by expression in transgenic plant expressing fructosyltransferase or in a suspension cell culture prepared from the transgenic plant. Patent Application: WO 97 29186.
Yanai, K., Nakane, A., Nakumara, H., Baba, Y., Watabe, A., Hirayama, M. (1997) Tokyo, Japan	Recombinant beta-fructofuranosidase from gene derived from *Aspergillus niger* and enzyme engineering for use in fructooligosaccharide production from sucrose. Patent Application: WO 97 34004.
Heyer, A. G., Wendenburg, R. (1997) Mannheim, Germany	Modified fructosyltransferase gene encoding protein with inulin-sucrase activity, *Streptococcus mutans* chimeric enzyme expression in a maize, rice, wheat, barley, sugarbeet, sugarcane or potato transgenic plant, for inulin production. Patent Application: DE 19617687.
De Leenheer, L., Booten, K. (1998) Brussels, Belgium	Poly-disperse saccharide composition, low in glucose, fructose, sucrose, fructo-oligosaccharide production using endo-inulinase for use as a sweetener. Patent Application: WO 98 05793.

Tab. 10 Food application and non-food patents (authors' database)

Frippiat, A., Smits, G. (1993) Brussels, Belgium Brussels, Belgium Raffinerie Tirlemontoise	Compositions having a creamy structure and containing fructan, preparation method therefor and uses thereof. Patent Application: WO 93 06744
Smits, G., Daenekindt, L., Booten, K. (1994) Brussels, Belgium Raffinerie Tirlemontoise	Pelletized composition, preparation method therefor and food products containing same. Patent Application: WO 94 19973
Frippiat, A. (1994) Brussels, Belgium Raffinerie Tirlemontoise	Stabilized food composition. Patent Application: WO 94 22327
Van Loo, J., Booten, K., Smits, G. (1994) Raffinerie Tirlemontoise	Method for separating a polydispersed saccharide composition, resulting products and use thereof in food compositions. Patent Application: WO 94 12541
Kunz, M., Munir, M., Vogel, M. (1994) Mannheim, Germany Südzucker	Verfahren zur Herstellung von langkettigem Inulin. Patent Application: EP 627 490
Smits, G., Daenekindt, L., Booten, K. (1996) Brussels, Belgium Raffinerie Tirlemontoise	Fractionated polydisperse compositions. Patent Application: WO 96 01849
Roberfroid, M., Van Loo, J., Delzenne, N., Coussement, P. (1996) Brussels, Belgium Raffinerie Tirlemontoise	Use of a composition containing inulin or oligofructose in cancer treatment. Patent Application: EP 692 252
Booten, K., De Soete, J., Frippiat, A. (1998) Brussels, Belgium Raffinerie Tirlemontoise	Fructan-and/or polydextrose-containing dairy powders, process for preparing same and their use. Patent Application: EP 821 885
De Soete J., Booten, K., Daenekindt, L. (1998) Brussels, Belgium Raffinerie Tirlemontoise	Low density fructan composition. Patent Application: WO 98 38223
Frippiat, A. (1998) Brussels, Belgium Raffinerie Tirlemontoise	Inulin based hydrocolloid compositions. Patent Application: EP 867 470
Sherianne, M. (1998) Northfield, Illinois Kraft Foods	Method for producing fat-free and low-fat viscous dressings using inulin. Patent Application: US 5, 721, 004
Van Lengerich, B., Larson, M. (1999) Minneapolis, Minnesota General Mills	Cereal products with inulin and methods of preparation. Patent Application: WO 99 346 88
Kunz, M., Haji Begli, A. (1995) Mannheim, Germany Südzucker	Inulinderivate, Verfahren zu deren Herstellung sowie ihre Verwendung. Patent Application: EP 638 589

Tab. 10 (cont.)

Verraest, D., Batelaan, J., Peters, J., Van Bekkum, H. (1995) Arnhem, The Netherlands Akzo Nobel	Carboxymethyl inulin. Patent Application: WO 95 15984
Stevens, C., Booten, K., Laquiere, I., Daenekindt, L. (1999) Brussels, Belgium Raffinerie Tirlemontoise	Surface-Active Alkylurethanes of fructans. Patent Application: WO 99 64549

15
References

Absolonne, J., Jossart, M., Coussement, P., Roberfroid, M. (1995) Digestive acceptability of oligofructose, *Proceedings, First ORAFTI Research Conference*, Brussels, 151–160.

Ajinomoto (1992) Liquid composition containing polyfructan, polyfructan production by *Aspergillus sydowii* fermentation or by fructose-transferase action on sugar or inulin, Patent Application JP4311378.

Ajinomoto (1992) Solid composition containing polyfructan, polyfructan production by *Aspergillus sydowii* fermentation or by fructose-transferase action on sugar or inulin, Patent Application JP4311371.

Andre, I., Putaux, J. L., Chanzy, H., Taravel, F. R., Timmermans, J. W., de Wit, D. (1996) Single crystals of inulin, *Int. J. Biol. Macromol.* **18**, 195–204.

Baillargé, E. (1942) Le topinambour, ses usages, sa culture, in: *La Terre Encyclopédie Paysanne* (Flammarion, Ed.).

Beringer, A., Wenger, R. (1955) Inulin in der Ernährung des Diabetikers, *Dtsch. Z. Verdauungs- u. Stoffwechselkrankh.* **15**, 268–272.

Besemer, A. C. (1993) *The bromide-catalyzed hypochlorite oxidation of starch and inulin. Calcium complexation of oxidized fructans*, PhD Thesis, Delft University of Technology, The Netherlands.

Besemer, A. C., van Bekkum, H. (1994) The hypochlorite oxidation of inulin, *Recl. Trav. Chim.* **113**, 398–402.

Bhattathiry, E. P. M. (1971) Effects of polysaccharides on the biosynthesis of lipids in adult rats, *Far East Med. J.*, **7**, 187–190.

Birch, G. G., Soon, E. B. T. (1973) Composition and properties of diabetic jams, *Confect. Prod.* **39**, 73–76.

Booten, K., De Soete, J., Frippiat, A. (1998) Fructan- and/or polydextrose-containing dairy powders, process for preparing same and their use, Patent Application EP 821 885.

Brighenti, F., Casiraghi, M. C., Canzi, E., Ferrari, A. (1999) Effect of consumption of a ready-to-eat breakfast cereal containing inulin on the intestinal milieu and blood lipids in healthy male volunteers, *Eur. J. Clin. Nutr.* **53**, 726–733.

Butel, M., Roland, N., Hibert, A., Papot, F., Favre, A., Tessedre, A., Bensaada, M., Rimbault, A., Szylit, O. (1998) Clostridial pathogenicity in experimental necrotising enterocolitis in gnotobiotic quails and protective role of bifidobacteria, *J. Med. Microbiol.* **47**, 391–399.

Caimi, P. G., Hershey, H. P., Kerr, P. S. (1995) Fructose polymers are produced by transgenic crop plants fructan, dextran, alternan production in maize, potato and tobacco; DNA construct containing tissue-specific promoter, vacuole targeting sequence and fructosyltransferase gene, Patent Application WO 95 13389.

Cairns, A. J. (1993) Evidence for the de novo synthesis of fructan by enzymes from higher plants: a reappraisal of the SSF/FFT model, *New Phytologist* **123**, 15–24.

Canzi, E., Brighenti, F., Casiraghi, M., Del Puppo, E., Ferrari, A. (1995) Prolonged consumption of inulin in ready-to-eat breakfast: effect on intestinal ecosystem, bowel habits and lipid metabolism, *COST '92 Workshop on Dietary Fibre and Fermentation*, Helsinki, Finland.

Castiglia-Delavaud, C., Verdier, E., Besle, J. M., Vernet, J., Boirie, Y., Beaufrere, B., De Baynast, R., Vermorel, M. (1998) Net energy value of non-starch polysaccharide isolates (sugarbeet fibre and commercial inulin) and their impact on nutrient digestive utilization in healthy human subjects, *Br. J. Nutr.* **80**, 343–352.

Causey, J. L., Feirtag, J. M., Gallaher, D. D., Tungland, B. C., Slavin, J. L. (2000) Effects of dietary inulin on serum lipids, blood glucose and the gastrointestinal environment in hypercholesterolemic men, *Nutr. Res.* **2**, 191–201.

Chatteron, N. J., Harrison, P. A., Thornley, W. R., Bennet, J. H. (1993) Separation and quantification of fructan (inulin) oligomers by anion exchange chromatography, in: *Inulin and Inulin-containing Crops* (Fuchs, A., Ed.), Amsterdam: Elsevier Science Publishers, vol. 3, 93–99.

Ciucanu, I., Kerek, F. (1984) A simple and rapid method for the permethylation of carbohydrates, *Carbohydr. Res.* **131**, 209–217.

Claessens, G., Van Laere, A., De Proft, M. (1990) Purification and properties of an inulinase from chicory roots, *J. Plant Physiol.* **136**, 35–39.

Coudray, C., Bellanger, J., Castiglia-Delavaud, C., Rémésy, C., Vermorel, M., Rayssignuier, Y. (1997) Effects of soluble or partly soluble dietary fibres supplementation on absorption and balance of calcium, magnesium, iron and zinc in healthy young men, *Eur. J. Clin. Nutr.* **51**, 375–380.

Davidson, M., Maki, K., Synecki, C., Torri, S. A., Drennan, K. B. (1998) Effects of dietary inulin on serum lipids in men and women with hypercholesterolemia, *Nutr. Res.* **18**, 503–517.

De Leenheer, L. (1996) Production and use of inulin: industrial reality with a promising future, in: *Carbohydrates as Organic Raw Materials III* (Van Bekkum, H., Röper, H., Voragen, F., Eds.), New York: VCH, 67–92.

De Leenheer, L., Hoebregs, H. (1994) Progress in the elucidation of the composition of chicory inulin, *Starch* **46**, 193–196.

De Roover, J., Van Laere, A., De Winter, M., Timmermans, J., Van den Ende, W. (1999) Purification and properties of a second fructan exohydrolase from the roots of *Cichorium intybus*, *Physiol. Planta* **106**, 28–34.

De Roover, J., Vandenbranden, K., Van Laere, A., Van den Ende, W. (2000) Drought induces fructan synthesis and 1 SST in roots and leaves of chicory seedlings, *Planta* **210**, 808–814.

De Soete J., Booten, K., Daenekindt, L. (1998) Low density fructan composition, Patent Application WO 98 38223.

Delzenne, N., Kok, N. (1998) Effect of non-digestible fermentable carbohydrates on hepatic fatty acid metabolism, *Biochem. Soc. Trans.* **26**, 228–230.

Delzenne, N., Kok, N., Fiordaliso, M. F., Deboyser, D. M., Goethals, F. M., Roberfroid, M. R. (1993) Dietary fructo-oligosaccharides modify lipid metabolism in rats, *Am. J. Clin. Nutr.* **57**, 820S.

Delzenne, N., Aertssens, J., Verplaetse, N., Roccaro, M., Roberfroid, M. (1995) Effect of fermentable fructo-oligosaccharides on energy and nutrients absorption in the rat, *Life Sci.* **57**, 1579–1587.

Den Hond, E., Geypens, B., Ghoos, Y. (2000) Effect of high performance chicory inulin on constipation, *Nutr. Res.* **20**, 731–736.

Edelman, J., Jefford, T. G. (1968) The mechanism of fructan metabolism in higher plants as exemplified in *Helianthus tuberosus* L, *New Phytologist* **123**, 15–24.

Ellegård, L., Andersson, H., Bosaeus, I. (1997) Inulin and oligofructose do not influence the absorption of cholesterol, Fe, Ca, Mg, and bile acids but increase energy excretion in man. A blinded, controlled cross-over study in ileostomy subjects, *Eur. J. Clin. Nutr.* **51**, 1–5.

Favier, C., Neut, C., Mizon, A., Cortot, A., Colombel, J.-F., Mizon, J. (1997) Fecal β-D-galactosidase production and Bifidobacteria are decreased in Crohn's disease, *Dig. Dis. Sci.* **42**, 817–822.

Fiordaliso, M., Kok, N., Desager, J. P., Goethals, F., Deboyser, D., Roberfroid, M., Delzenne, N. (1995) Dietary oligofructose lowers triglycerides, phospholipids and cholesterol in serum and very low density lipoproteins of rats, *Lipids* **30**, 163–167.

Franck, A. (1993) Rafticreming: the new process allowing to turn fat into dietary fibre in: *FIE 1992 Conference Proceedings*. Maarssen: Expoconsult Publishers, 193–197.

Franck, A. (1998) Prebiotics stimulate calcium absorption: a review, *Milchwissenschaft* **53**, 427–429.

Franck, A., Coussement, P. (1997) Multi-functional inulin, *Food Ingredients and Analysis International*, October, 8–10.

French, A. D. (1993) Recent Advances in the structural chemistry of inulin, in: *Inulin and Inulin-containing Crops, Studies in Plant Science* (Fuchs, A, Ed.), Amsterdam: Elsevier Science Publishers, Vol. 3, 121–127.

Frippiat, A. (1994) Stabilized food composition, Patent Application WO 94 22327.

Frippiat, A. (1998) Inulin based hydrocolloid compositions, Patent Application EP 867 470.

Frippiat, A., Smits, G. (1993) Compositions having a creamy structure and containing fructan, preparation method therefor and uses thereof, Patent Application WO 93 06744.

Gibson, G. R., Roberfroid, M. B. (1995) Dietary modulation of the human colonic microbiota – Introducing the concept of prebiotics, *J. Nutr.* **125**, 1401–1412.

Gibson, G. R., Wang, X. (1994) Bifidogenic properties of different types of fructo-oligosaccharides, *Food Microbiol.* **11**, 491–498.

Gibson, G. R., Beatty, E. R., Wang, X., Cummings, J. H. (1995) Selective stimulation of bifidobacteria in the human colon by oligofructose and inulin, *Gastroenterology* **108**, 975–982.

Gleiss, A., Schneider, H. (1994) Preparation of long chain inulin and fructose and glucose by enzymatic treatment of crude inulin with a hydrolase, Patent Application DE 4316425.

Harada, T., Suzuki, S., Toniguchi, H., Sasaki, T. (1993) Characteristics and applications of polyfructan synthesized from sucrose by *Aspergillus sydowi* conidia, *Food Hydrocoll.* **7**, 23–38.

Hata, H., Nakajima, K. (1984) Fructo-oligosaccharides intake and effect on the digestive tract, *Proceedings of the 2nd Neosugar Research Conference*, Tokyo, Japan.

Hellwege, E., Gitscher, D., Wilmitzer, L., Heyer, A.G. (1997) Transgenic potato tubers accumulate high levels of 1-kestox and nystose, *Plant J.* **12**, 1057–1065.

Hellwege, E., Gritscher, D., Willmitzer, L., Heyer, A. (1998) Functional characterization of fructosyltransferase genes from Artichoke and Jerusalem Artichoke, in: *Proceedings of the Seventh Seminar of Inulin* (Fuchs, A., Van Laere, A., Eds.), European Fructan Association, 121–125.

Hendry, G. A., Wallace, R. K. (1993) The origin, distribution and evolutionary significance of fructans, in: *Science and Technology of Fructans* (Suzuki, M., Chatteron, N. J., Eds.), Boca Raton, FL: CRC Press, 119–139.

Heyer, A. G., Wendenburg, R. (1997) Modified fructosyltransferase gene encoding protein with inulin-sucrase activity, *Streptococcus mutans* chimeric enzyme expression in a maize, rice, wheat, barley, sugarbeet, sugarcane or potato transgenic plant, for inulin production, Patent Application DE 19617687.

Heyer, A. G., Hellwege, E. M., Willmitzer, L., Czapla, S., Sittlinger, B. (1999) Confirmation of the SST/FFT model of inulin synthesis by transgene expression in potato, in: *Proceedings of the Eight Seminar on Inulin* (Fuchs, A., Ed.), European Fructan Association, 94–101.

Hidaka, H., Hirayama, M., Yamada, K. (1991) Fructo-oligosaccharides, enzymatic preparation and biofunctions, *J. Carbohydr. Chem.* **10**, 509–516.

Hincha, D. K., Hellwege, E. M., Heyer, A. G., Crowe, J. H. (2000) Plant fructans stabilize phosphatidylcholine liposomes during freeze-drying, *Eur. J. Biochem.* **267**, 535–540.

Hirayama, M., Hidaka, H. (1993) Production and utilization of microbial fructans, in: *Science and Technology of Fructans* (Suzuki, M., Chatteron, N. J., Eds.), Boca Raton, FL: CRC Press, 273–302.

Hoebregs, H. (1997) Fructans in foods and food products, ion-exchange chromatographic method: collaborative study, *J. AOAC* **5**, 80, 1029–1037.

Hosoya, N., Dhorranintra, B., Hidaka, H. (1988) Utilisation of U-14C fructo-oligosaccharides in man as energy resources, *J. Clin. Biochem. Nutr.* **5**, 67–74.

Ingelbert, G. (1999) Inutec, ein alternatives Bindmittel in sprügetrockneten Granulaten für die keramische Industrie, *Keramische Zeitschrift* **51**, 3.

Institut Genbiologische Forschung (1994) Polyfructan-sucrase DNA sequence from *Erwinia amylovora* expression in transgenic plant for polyfructan production, Patent Application DE4227061.

Jackson, K. G., Taylor, G. R. J., Clohessy, A. M., Williams, C. M. (1999) The effect of the daily intake of inulin on fasting lipid, insulin and glucose concentrations in middle-aged men and women, *Br. J. Nutr.* **82**, 23–30.

Joye, D., Hoebregs, H. (2000) Determination of oligofructose, a soluble dietary fiber, by high-temperature capillary gas chromatography, *J. AOAC International* **83**, 1020–1025.

Jun, J. G., Gray, G. R. (1987) A new catalyst for reductive cleavage of methylated glycans, *Carbohydr. Res.* **163**, 247–261.

Kim, W., Won Choi, J., Won Yun, J. (1998) Selective production of GF_4-fructooligosaccharide from sucrose by a new transfructosylating enzyme, *Biotechnol. Lett.* **20**, 1031–1034.

Kleessen, B., Sykura, B., Zunft, H. J. (1997) Effect of inulin and lactose on fecal microflora, microbial activity, and bowel habit in elderly constipated persons, *Am. J. Clin. Nutr.* **65**, 1397–1402.

Knudsen, K. E. B., Hessov, I. (1995) Recovery of inulin from Jerusalem artichoke (*Helianthus tubersosus* L.) in the small intestine of man, *Br. J. Nutr.* **74**, 101–113.

Kok, N., Roberfroid, M., Delzenne, N. (1996a) Dietary oligofructose modifies the impact of fructose on hepatic triacylglycerol metabolism, *Metabolism* **45**, 1547–1550.

Kok, N., Roberfroid, M., Robert, A., Delzenne, N. (1996b) Involvement of lipogenesis in the lower VLDL secretion induced by oligofructose in rats, *Br. J. Nutr.* **76**, 881–890.

Kok, N., Morgan, L., Williams, C., Roberfroid, M., Thissen, J. P., Delzenne, N. (1998a) Insulin, glucagon-like peptide 1, glucose-dependent insulinotropic polypeptide and insulin-like growth factor I as putative mediators of the hypolipidemic effect of oligofructose in rats, *J. Nutr.* **128**, 1099–1103.

Kok, N., Taper, H. S., Delzenne, N. M. (1998b) Oligofructose modulates lipid metabolism alterations induced by a fat-rich diet in rats, *J. Appl. Toxicol.* **18**, 47–53.

Koops, A. J., Jonker, H. H. (1996) Purification and characterisation of the enzymes of fructan biosynthesis in tubers of *Helianthus tuberosus*, *Plant Physiol.* **110**, 1167–1175.

Koops, A., Sevenier, R., Van Tunen, A., De Leenheer, L. (1999) Patent WO 99/54480 Applicant: "Stichting Dienst Landbouwkundig Onderzoek" and Tiense Suikerraffinaderij.

Koops, A., Van Arkel, J. Hakkert, H., Vorst, O. (2000) DNA chips: useful tool to analyze fructan metabolism in chicory tap roots, *Fourth International Fructan Symposium*, Abstract **3.2**

Kruse, H. P., Kleessen, B., Blaut, M. (1999) Effects of inulin on faecal bifidobacteria in human subjects, *Br. J. Nutr.* **82**, 375–382.

Kulz, E. (1874) Beitrage zur Path. Therap. der Diabetes, *Jahrb. Tierchem.* **4**, 448.

Kunz, M., Haji Begli, A. (1995) Inulinderivate, Verfahren zu deren Herstellung sowie ihre Verwendung, Patent Application EP 638 589.

Kunz, M., Munir, M., Vogel, M. (1994) Verfahren zur Herstellung von langkettigern Inulin, Patent Application EP 627 490.

Kuppers-Sonnenberg, G. A. (1971) Topinambur in der Diabetiker-Diät, *Selecta* **37**, 2836–2840.

Lemort, C., Roberfroid, M. (1997) Effect of chicory fructo-oligosaccharides on Ca balance, Abstracts of NDO symposium, Wageningen, 163.

Leroux, X. (1996) La vie c'est bon comme tout, *Commercial Publication*, 5.

Levrat, M.-A., Favier, M.-L., Moundras, C., Rémésy, C., Demigné, C., Morand, C. (1994) Role of dietary propionic acid and bile acid excretion in the hypocholesterolemic effects of oligosaccharides in rats, *J. Nutr.* **124**, 531–538.

Lewis, D. H. (1991) Fungi and sugars – a suite of interactions, *Mycol. Res.* **95**, 897.

Lewis, H. B. (1912) The value of inulin as a foodstuff, *J. Am. Med. Assoc.* **58**, 176–177.

Maruzen, Kazei (1991) New cyclic inulo-oligosaccharide preparation form inulin or fructose using enzyme or microorganism, e.g. *Bacillus circulans*, Patent Application J02252701.

McCance, R. A., Lawrence, R. D. (1929) The carbohydrate content of foods – inulin and the fructosans, *Medical Research Council, Special Report Series* 135, 58.

McCleary, B., Murphy, A., Mugford, D. (2000) Measurement of total fructan in foods by enzymatic/spectrophotometric method: collaborative study, *J. AOAC* **83**, 356–364.

Merrigi, A., Stevens, C. V., Booten, K. (2001) Chemical modification of inulin, a valuable remarkable resource, and its industrial applications, accepted for publication.

Meyer, W. J. M., Matthysen, E. W. J. M., Borm, G. E. L. (1993) Crop characteristics and inulin production of Jerusalem Artichoke and Chicory, in: *Inulin and Inulin-containing Crops, Studies in Plant Science* (Fuchs, A, Ed.), Amsterdam: Elsevier Science Publishers, Vol. 3, 29–44.

Mischnick-Lübbecke, P., König, W. A. (1989) Determination of the substitution pattern of modified polysaccharides. Part I. Benzyl starches, *Carbohydr. Res.* **185**, 113–118.

Mitsubishi Chemicals (1991) Difructose dianhydride-I production from inulin using inulinase from *Arthrobacter* sp., Patent Application J03247295.

Mitsui Toatsu Chemicals (1991) Production of inulo-oligosaccharide, inulin decomposition using *Penicillium purpurogenum* var. *rubri-sclerotium* and *Aspergillus ficuum* inulinase, Patent Application EP429077.

Nakamura, T., Maruki, S., Nakatsu, S., Veda, S. (1978) General properties of an extracellular inulinase from *Aspergillus* sp., *Nippon Nogei Kagaku Kaishi* **52**, 581–587.

Netherlands Science Research Foundation (1994) Method for transgenic plant construction with modified fructan content potato transformation with plant, *Bacillus subtilis* or *Streptococcus mutans* tolerance, Patent Application WO 94 14970.

Norinsho, Nihon Denpun Kogyokk (1989) Difructose di-anhydride, obtained by heating inulin (plant extract) with inulin fructo-transferase fixed on solid support of chitosan and silica, Japanese Patent JO1285195.

Onodera, S., Shiomi, N. C. (1988) Purification and substrate specificity of endo-type inulinase from *Penicillium purpurogenum*, *Agric. Biol. Chem.* **52**, 2569–2575.

ORAFTI (2001) Golden rules for success in chicory, *Vademecum for the farmer* (ORAFTI, Ed.).

Partida, V. Z., Lopez, A. C., Gomez, A. J. M. (1997) Fructose syrup production from agave plant pulp by extraction and inulinase treatment, Patent Application WO 97 34017.

Persia (1905) Reference of Lewis (1912), *Nuova Revista Clin. Therapeut.* 8.

Ponstein, A. S., Van Leeuwen, M. B. (1993) In vitro synthesis of inulin by the inulosucrase from *Streptococcus mutans*, in: *Inulin and Inulin containing crops, Studies in Plant Science* (Fuchs, A., Ed.), Amsterdam: Elsevier Science Publishers, Vol. 3, 281–287.

Prosky, L. (1999) Inulin and oligofructose are part of the dietary fiber complex, *J. AOAC Int.* **82**, 223–226.

Reddy, D. S., Hamid, R., Rao, C. V. (1997) Effect of dietary oligofructose and inulin on colonic preneoplastic aberrant crypt foci inhibition, *Carcinogenesis* **18**, 1371–1374.

Roberfroid, M. (1993) Dietary fiber, inulin and oligofructose: a review comparing their physiological effects, *Crit. Rev. Food Sci. Nutr.* **33**, 103–148.

Roberfroid, M. B. (1997) Health benefits of non-digestible oligosaccharides, in: *Dietary Fiber in Health and Disease* (Kritchevsky, D., Bonefield, C., Eds.), New York: Plenum Press, 211–219.

Roberfroid, M., Gibson, G. R., Delzenne, N. (1993) Biochemistry of oligofructose, a non-digestible fructooligosaccharide: an approach to estimate its caloric value, *Nutr. Rev.* **51**, 137–146.

Roberfroid, M., Van Loo, J., Delzenne, N., Coussement, P. (1996) Use of a composition containing inulin or oligofructose in cancer treatment, Patent Application EP 692 252.

Roberfroid, M., Van Loo, J., Gibson, G. (1998) The bifidogenic nature of chicory inulin and its hydrolysis products, *J. Nutr.* **128**, 11–19.

Roland, N., Nugon-Baudon, L., Szylit, O. (1994a) Influence of dietary fibres on two intestinal transferases in rats inoculated with a whole human faecal flora, *R. Soc. Chem.* **123**, 369–373.

Roland, N., Nugon-Baudon, L, Flinois, J. P., Beaune, Ph. (1994b) Hepatic and intestinal cytochrome P450, glutathione-S-transferase and UDP-glucuronyltransferase are affected by six types of dietary fibre in rats inoculated with a human whole faecal flora, *J. Nutr.* **124**, 1581–1587.

Roland, N., Nugon-Baudon, L., Andrieux, C., Szylit, O. (1995) Comparative study of the fermentative characteristics of inulin and different types of fiber in rats inoculated with a human whole fecal flora, *Br. J. Nutr.* **74**, 239–249.

Roland, N., Rabot, S., Nugon-Baudon, L. (1996) Modulation of the biogenical effects of glucosinolates by inulin and oat fibre in gnotobiotic rats inoculated with a whole human flora, *Food Chem. Toxicol.* **34**, 671–677.

Root, H., Baker, M. (1925) Inulin and artichokes in the treatment of diabetes, *Arch. Intern. Med.* **36**, 126–145.

Rose, V. (1804) Über eine eigenthumliche vegetablische Substanz, *Gehlens Neues Algem. Jahrb. Chem.* **3**, 217.

Rowland, I. R., Rummey, C. J., Coutts, J. T., Lievense, L. (1998) Effects of *Bifidobacterium longum* and inulin on gut bacterial metabolism and carcinogen-induced aberrant crypt foci in rats, *Carcinogenesis* **19**, 281–285.

Sachs, J. (1864) Über die Sphärokristalle des Inulins und dessen mikroskopische Nachweisung in den Zellen, *Bot. Z.* **22**, 77.

Scholz-Ahrens, K., Van Loo, J., Schrezenmeir, J. (1998) Oligofructose stimuliert die Femurmineralisation in Abhängigkeit von der Calciumzufuhr bei der ovariektomisierten Ratte, Symposium Deutsche Gesellschaft für Ernährungsforschung, Karlsruhe.

Sevenier, R., Hall, R. D., Van der Meer, I. M., Hakkert, H. J., Van Tunen, A. J., Koops, A. J. (1998) High level fructan accumulation in transgenic sugar beet, *Nature Biotechnol.* **16**, 843–846.

Shannon, J. A., Smith, H. W. (1935) The excretion of inulin, xylose and urea by normal and man, *J. Clin. Invest.* **14**, 393–401.

Sherianne, M. (1998) Method for producing fat-free and low-fat viscous dressings using inulin, Patent Application US 5, 721, 004.

Smeekens, J. C. M., Ebskamp, M. J., Geerts, H. A. M., Weisbeek, P. J. (1996) Production of oligosaccharides by transgenic plants by *Streptococcus mutans*, *Bacillus subtilis*, onion, barley or sunflower fructosyltransferase gene transfer, Patent Application WO 96 01904.

Smits, G., Daenekindt, L., Booten, K. (1994) Fractionated polydisperse compositions, Patent Application WO 96 01849.

Smits, G., Daenekindt, L., Booten, K. (1994) Pelletized composition, preparation method therefor and food products containing same, Patent Application WO 94 19973.

Smits, G., Daenekindt, L., Booten, K. (1997) Fractionated polydisperse composition, Patent Application EPO 679026B1.

Stevens, C., Booten, K., Laquiere, I., Daenekindt, L. (1999) Surface-active alkylurethanes of fructans, Patent Application WO 99 64549.

Strauss, H. (1911) Zur verwendung inulinreicher Gemüse bei Diabetikern, *Therapie der Gegenwart III*, 347–351.

Tamura, K., Kuramoto, T., Kitahata, S. (1988) Enzymatic manufacture of di-D-fructosylfuranose 1,2; 2,3-dianhydride, Japanese Patent 63219389.

Taper, H. S., Delzenne, N., Tshilombo, A., Roberfroid, M. (1995), Dietary fructo-oligosaccharides (FOS) protect young rats against the atrophy of exocrine pancreas induced by high fructose and partial copper deficiency, *Food Chem. Toxicol.* **33**, 631–639.

Taper, H., Delzenne, N., Roberfroid, M. B. (1997) Growth inhibition of transplantable mouse tumours by non digestible carbohydrates, *Int. J. Cancer* **71**, 1109–1112.

Taper, H., Lemort, C., Roberfroid, M. (1998) Inhibition effect of dietary inulin and oligofructose on the growth of transplantable mouse tumor, *Anticancer Res.* **18**, 4123–4126.

Taper, H., Frippiat, A, Van Loo, J., Roberfroid, M. (1999) Synergistic composition for use in the treatment of cancer, Patent Application WO 99 59600.

Thomson, T. (1818) *A System of Chemistry*, 5th London edition, Abraham Small, Philadelphia, **4**, 65.

Timmermans, J. W., van Leeuwen, M. B., Tournois, H., de Wit, D., Vliegenthart, J. F. G. (1993) Proceedings of the fourth seminar on inulin, Wageningen, 12–15.

Timmermans, J. W., Bitter, M. G., de Wit, D., Vliegenthart, J. F. G. (1997) The interaction of inulin oligosaccharides with Ba^{2+} studied by ^{1}H-NMR spectroscopy, *J. Carbohydr. Chem.* **2**, 213–230.

Trautwein, E. A., Rieckhoff, D., Erbersdobler, H. F. (1998) Dietary inulin lowers plasma cholesterol and triacylglycerol and alters biliary bile acid profile in hamsters, *J. Nutr.* **128**, 1937–1943.

Turk, S., Gerrits, N., Smeekens, J. C. M., Weisbeek, P. J. (1997) Production of modified polysaccharides by treatment with sugar donor and transfer enzyme, starch production by treatment with enzyme or by expression in transgenic plant expressing fructosyltransferase or in a suspension cell culture prepared from the transgenic plant, Patent Application WO 97 29186.

Uchiyama, T. (1993) Metabolism in microorganism, Part II: Biosynthesis and degradation of fructans by microbial enzymes other than levansucrase, in: *Science and Technology of Fructans* (Suzuki, M., Chatteron, N. J., Eds.), Boca Raton, FL: CRC Press, 169–190.

Vandamme, E. J., De Rijcke, D. (1983) Microbial inulinases: fermentation processes, properties and applications, *Adv. Appl. Microbiol.* **29**, 139–176.

Van den Ende, W. (1996) *Fructan metabolism in chicory roots (Cichoriumintybus L.)*, PhD Thesis, University of Leuven (B), 11.

Van den Ende, W., De Roover, J., Van Laere, A. (1996) In vitro synthesis of fructo-oligosaccharide from inulin and fructose by purified chicory root FFT, *Physiol. Plant* **97**, 346–352.

Van den Ende, W., Michiels, A., Van Wonterghem, D., Vergauwen, R., Van Laere, A. (2000a) Cloning developmental and tissue-specific expression of sucrose:sucrose 1-Fructosyltransferase from *Taraxacum officinale*. Fructan localisation in roots, *Plant Physiol.* **123**, 71–79.

Van den Ende, W., Michiels, A., De Roover, J., Verhaert, P., Van Laere, A. (2000b) Cloning and functional analysis of chicory root fructan 1-exohydrolase I: a vacuolar enzyme derived from a cell-wall invertase ancestor? *Plant J.* **24**, 447–456.

Van den Heuvel, E., Muys, T., Van Dokkum, W., Schaafsma, G. (1999) Oligofructose stimulates calcium absorption in adolescents, *Am. J. Clin. Nutr.* **69**, 544–548.

Van Geel-Schutten, I. (2000) Expolysaccharide synthesis by *Lactobacillus reuteri*, molecular characterisation of a fructosyltransferase and a glucansucrase, PhD Thesis, University of Groningen, The Netherlands, 53–73.

Van Lengerich, B., Larson, M. (1999) Cereal products with inulin and methods of preparation, Patent Application WO 99 346 88.

Van Loo, J., Booten, K., Smits, G. (1994) Method for separating a polydispersed saccharide composition, resulting products and use thereof in food compositions, Patent Application WO 94 12541.

Van Loo, J., Coussement, P., De Leenheer, L., Hoebregs, H., Smits, G. (1995) On the presence of inulin and oligofructose as natural ingredients in the Western diet, *Crit. Rev. Food Sci. Nutr.* **35**, 525–552.

Van Loo, J., Cummings, J., Delzenne, N., Englyst, H., Franck, A., Hopkins, M., Kok, N., Macfarlane, G., Newton, D., Quigley, M., Roberfroid, M., van Vliet, T., van den Heuvel, E. (1999) Functional food properties of non-digestible oligosaccharides: a consensus report from the ENDO project (DGXII AIRII-CT94-1095), *Br. J. Nutr.* **81**, 121–132.

Van Tunen, A. J., Van der Meer, I. M., Koops, A. J. (1996) DNA sequences encoding fructan-producing enzymes, Jerusalem artichoke-derived fructosyltransferase expression in transgenic plant for low calorie sweetener production, Patent Application WO 96 21023.

Verraest, D., Batelaan, J., Peters, J., Van Bekkum, H. (1995a) Carboxymethyl inulin, Patent Application WO 95 15984.

Verraest, D. L., Peters, J. A., Batelaan, J. G., van Bekkum, H. (1995b) Carboxymethylation of inulin, *Carbohydr. Res.* **271**, 101–112.

Verraest, D. L., Peters, J. A., van Bekkum, H., van Rosmalen, G. M. (1996) Carboxymethyl inulin: a new inhibitor for calcium carbonate precipitation, *J. Am. Oil Chem. Soc.* **73**, 55–62.

Vijn, I., Van Dijken, A., Lüscher, M., Bos, A., Smeets, E., Weisbeek, P., Wiemken, A., Smeerkens, S. (1998) Cloning of sucrose: sucrose 1-fructosyltransferase from onion and synthesis of structurally defined fructan molecules from sucrose, *Plant Physiol.* **117**, 1507–1513.

Von Mehring, P. (1876) Referenced by Lewis (1912).

Wang, X., Gibson, G. R. (1993) Effects of the *in vitro* fermentation of oligofructose and inulin by bacteria growing in the human large intestine, *J. Appl. Bacteriol.* **75**, 373–380.

Watherhouse, A. L., Chatterton, N. J. (1993) Glossary of fructan terms, in: *Science and Technology of Fructans* (Suzuki, M., Chatteron, N. J., Eds.), Boca Raton, FL: CRC Press, 2–7.

Wiedmann, M., Jager, M. (1997) Synergistic sweeteners, *Food Ingredients and Analysis International*, November–December, 51–56.

Wise, E., Heyl, F. (1931) Failure of a diabetic to utilize inulin, *J. Am. Pharm. Soc.* **20**, 26–29.

Xiao, R., Tanida, M., Takao, S. (1989) Purification and some properties of endo-inulinase from *Chrysosporium pannorum*, *J. Ferment. Bioeng.* **67**, 244–251.

Yanai, K., Nakane, A., Nakumara, H., Baba, Y., Watabe, A., Hirayama, M. (1997) Recombinant beta-fructofuranosidase from gene derived from *Aspergillus niger* and enzyme engineering for use in fructooligosaccharide production from sucrose, Patent Application WO 97 34004.

Zittan, L. (1981) Enzymatic hydrolysis of inulin, an alternative way to fructose production, 32° Starch convention, Detmold, Germany.

Zubr, J., Pedersen, H.S. (1993) Characteristics of growth and development of different Jerusalem Artichoke cultivars, in: *Inulin and Inulin-containing Crops, Studies in Plant Science* (Fuchs, A., Ed.), Amsterdam: Elsevier Science Publishers, Vol. 3, 11–19.

9
Levan

Dr. Sang-Ki Rhee[1], Dr. Ki-Bang Song[2], Dr. Chul-Ho Kim[3], Dr. Buem-Seek Park[4], Ms. Eun-Kyung Jang[5], Dr. Ki-Hyo Jang[6]

[1] Biomolecular Engineering Laboratory, Korea Research Institute of Bioscience and Biotechnology (KRIBB), 52 Eoeun-dong, Yuseong, Daejeon 305-333, Korea; Tel.: +82-42-860-4450; Fax: +82-42-860-4594; E-mail: rheesk@mail.kribb.re.kr

[2] Biomolecular Engineering Laboratory, Korea Research Institute of Bioscience and Biotechnology (KRIBB), 52 Eoeun-dong, Yuseong, Daejeon 305-333, Korea; Tel.: +82-42-860-4457; Fax: +82-42-860-4594; E-mail: songkb@mail.kribb.re.kr

[3] Biomolecular Engineering Laboratory, Korea Research Institute of Bioscience and Biotechnology (KRIBB), RealBioTech Co., Ltd., #202 Bioventure Center, KRIBB, 52 Eoeun-dong, Yuseong, Daejeon 305-333, Korea; Tel.: +82-42-860-4452; Fax: +82-42-860-4594; E-mail: kim3641@mail.kribb.re.kr

[4] Biomolecular Engineering Laboratory, Korea Research Institute of Bioscience and Biotechnology (KRIBB), 52 Eoeun-dong, Yuseong, Daejeon 305-333, Korea; Tel.: +82-42-860-4454; Fax: +82-42-860-4594; E-mail: buemseekpk@mail.kribb.re.kr

[5] RealBioTech Co., Ltd., #202 Bioventure Center, KRIBB, 52 Eoeun-dong, Yuseong, Daejeon 305-333, Korea; Tel.: +82-42-863-4381; Fax: +82-42-863-4382; E-mail: levanis@realbio.com

[6] Department of Food and Nutrition, Samcheok National University, Samcheok, Gangwon 245-711, Korea; Tel.: +82-33-570-6882; Fax: +82-33-570-6881; E-mail: kihyojang@samcheok.ac.kr

1	Introduction	325
2	Historical Outline	326
3	Chemical Structures of Levan	326
4	Occurrence	328
5	Physiological Functions of Levan	329

Polysaccharides and Polyamides in the Food Industry. Properties, Production, and Patents.
Edited by A. Steinbüchel and S. K. Rhee
Copyright © 2005 WILEY-VCH Verlag GmbH & Co. KGaA, Weinheim
ISBN: 3-527-31345-1

6	**Chemical Analysis and Detection**	329
6.1	Spectrophotometry	329
6.2	High-Performance Liquid Chromatography (HPLC)	329
6.3	Other Methods	330
7	**Biosynthesis of Levan**	330
7.1	Enzymology of Levan Synthesis	330
7.2	Genetic Basis of Levan Synthesis	331
7.3	Regulation of Levan Synthesis	333
7.3.1	Regulation at the Protein Level	333
7.3.2	Regulation at Transcriptional and Translational Levels	333
8	**Biodegradation of Levan**	334
8.1	Enzymology of Levan Degradation	334
8.1.1	Levanase	334
8.1.2	Levansucrase	335
8.1.3	Levan Fructotransferase	335
8.2	Genetic Basis of Levan Degradation	335
8.3	Regulation of Levan Degradation	336
9	**Biotechnological Production of Levan**	337
9.1	Isolation and Screening for Levan-producing Strains	337
9.2	Fermentative Production of Levan	337
9.3	*In vitro* Biosynthesis of Levan	337
9.4	Recovery and Purification of Levan	338
9.5	Commercial Production of Levan	338
9.6	Market Analysis and Cost of Levan Production	339
9.7	Levan Competitors	339
10	**Properties of Levan**	339
11	**Applications of Levan**	340
11.1	Medical Applications	340
11.2	Pharmaceutical Applications	340
11.3	Agricultural Applications	340
11.4	Food Applications	341
11.5	Other Applications	341
12	**Patents**	342
13	**Current Problems and Limits**	343
14	**Outlook and Perspectives**	344
15	**References**	345

1-kestose O-β-D-fructofuranosyl-(2 → 1)-β-D-fructofuranosyl-(2 → 1)-β-D-glucopyranoside
1-SST   sucrose:sucrose 1-fructosyltransferase
6-SFT   sucrose:fructan 6-fructosyltransferase
DFA   di-β-D-fructofuranose dianhydride
DP   degree of polymerization
EPS   exopolysaccharides
FFT   fructan:fructan fructosyltransferase
FOS   fructo-oligosaccharides
HPr   histidine-containing phosphocarrier protein
LBT   levanbiosyl transfer
LFT   levanfructosyl transfer
LFTase   levan fructotransferase
PEG   polyethylene glycol
PTS   phosphoenolpyruvate-dependent carbohydrate
RBT   RealBioTech Co., Ltd.
TLC   thin-layer chromatography

1
Introduction

Fructan, one of the most highly distributed biopolymers in nature, is a homopolysaccharide composed of D-fructofuranosyl residues joined by β-(2,6) and β-(2,1) linkages. Two types of fructan, distinguishable by the type of linkage present, are inulin and levan. The term levan is used to describe the microbial polyfructan which consists of D-fructofuranosyl residues linked predominantly by β-(2,6) linkage as a main chain, but with some β-(2,1) branching points. The other polyfructan, inulin, is mainly isolated from natural vegetable sources and serves as a reserve carbohydrate in the Compositate and Gramineae (Vandamme and Derycke, 1983), although inulins from the microbial origin have also been reported in *Streptococcus mutans* and *Streptococcus sanguis*, the human pathogens involved in dental caries (Birkhed et al., 1979). The fructose homopolymer, levan, is found in plants and especially in bioproducts of microorganisms. Plant levans, graminans, and phleins have shorter residues (varying from 10 to ~200 fructose residues) than microbial levans, of which molecular weights are up to several million daltons, with multiple branches. Microbial levans are produced extracellularly from sucrose- and raffinose-based substrates by levansucrase (sucrose 6-fructosyltransferase, EC 2.4.1.10) from a wide range of taxa such as bacteria, yeasts, and fungi (Han, 1990; Hendry and Wallace, 1993). Microbial levans are produced mainly by bacteria such as *Bacillus subtilis*, *Zymomonas mobilis*, *Bacillus polymyxa*, *Aerobacter levanicum*, *Erwinia amylovora*, *Rhanella aquatilis* and *Pseudomonas*. The production and utilization of levan in the industrial field have been strictly limited until very recently, and very few reports have been made on the production of levan using fermentation techniques (Elisashvili, 1984; Beker et al., 1990; Han, 1990; Keith et al., 1991; Ohtsuka et al., 1992; Uchiyama, 1993). Recently, great interest in this fructan has been renewed to discover novel applications for levan as a new industrial gum in the fields of cosmetics, foods (e.g., as dietary fiber), and pharmaceuticals. In this chapter, we describe the production and degradation of levan by use of enzymatic reactions, the genetic regulation and control of such reactions, and outline the properties of levan and the current status of its industrial applications.

2
Historical Outline

Historically, levan was generally considered to be an undesirable byproduct of sugar and juice processing because it increases the viscosity of the processing liquor (Fuchs, 1959; Avigad, 1965). Levan was first described by Lippmann in Germany in 1881, when the name "laevulan" was proposed. Greig-Smith (1901) later showed that a strain of *Bacillus*, when grown on sucrose, produced fructans, and the name "levan" was then introduced as being analogous to dextran. The term laevulan now denotes partially degraded levan fractions. However, early reports on levan were confusing because the microbial nomenclature was unsystematic and the materials were inadequately described.

The biosynthesis of levan was elucidated some years later. The mechanism was shown to involve two enzymes, sucrose fructosyltransferase and fructan fructosyltransferase, and was proposed by Edelman and Jefford in 1968. The enzyme kinetics of the transfructosylation reaction was revealed by Chambert and Gonzy-Treboul in 1976. The enzyme which is now generally recognized as levansucrase was named by Hestrin et al. in 1943, and is responsible for the synthesis of levan from sucrose. The most extensive studies of levansucrase were performed in *B. subtilis*, and focused on the localization of the enzyme as well as its properties, expression regulation, genetic organization, and kinetics (Suzuki and Chatterton, 1993). As levan began to receive more attention based on its potential applications, many levan-producing microorganisms were identified (Han, 1990; Hendry and Wallace, 1993). The mass production of levan from *Z. mobilis*, and the secretion of levansucrase were reported relatively recently in detail by Song and coworkers (1996) and Ananthalakshmy and Gunasekaran (1999).

Although extensive research and searches for industrial applications have been conducted with dextran (which is also known as a bacterial biopolymer), much less attention has been focused on levan, mainly because of the very poor yields obtained in its industrial production. Nonetheless, great interest has been expressed in the diverse aspects of this fructose homopolymer over the past decade, despite the applications of levan having remained relatively few in number because of the limited supplies. Levans are now available commercially in reagent grade from microbial sources (e.g., from Sigma Chemical Co., IGI Biotechnology), but these are used only for research purposes. Since the time when levan was first produced on a large scale by using levansucrase from genetically engineered *Escherichia coli* (Song et al., 1996), attention has been renewed on the potential industrial application of levan and its derivatives in the fields of agriculture, cosmetics, food ingredients, animal feed and pharmaceuticals (Clarke et al., 1997; Kim et al., 1998; Vijn and Smeekens, 1999; Rhee et al., 2000d), as well as being a good source of pure fructose and di-β-D-fructofuranose dianhydride (DFA) (Saito and Tomita, 2000).

3
Chemical Structures of Levan

Fructans are chemically versatile molecules, and consist of a single glucose unit attached to two or more fructose units. Three fructan trisaccharides are known, each being produced through a glycosidic linkage of fructose to one of the three primary hydroxyl groups of sucrose. Fructose linked to the primary carbon of the fructose moiety of sucrose forms 1-kestose (also called isokestose), while fructose linked to the sixth carbon of the fructose moiety of sucrose forms 6-kestose (also called kestose). Both of these trisaccharides have a terminal glucose and a terminal fructose. Linkage of a fructose moiety to the sixth carbon of glucose moiety of sucrose forms neokestose, with both end groups being fructose (Nelson and Spollen, 1987).

Chemically, levan consists of β-D-fructofuranosyl residues linked predominantly through β-(2,6) as 6-kestose of the basic trisaccharide, with extensive branching through β-(2,1) linkages (Figure 1). In contrast, inulin is composed of β-D-fructofuranose attached by β-(2,1) linkages. The first monomer of the chain is either a β-D-glucopyranosyl or a β-D-fructofuranosyl residue. Although they are similar fructose homopolymers, it is evident that levan is different from inulin-type fructan since microbial inulin contains 5–7% of β-(2,6)-linked branches (Wolff et al., 2000).

The molecular shape of levan, as visualized by electron microscopy (Newbrun et al., 1971), is spheroidal, indicating that the constituent chains are extended radially at the same synthetic rate. The molecular

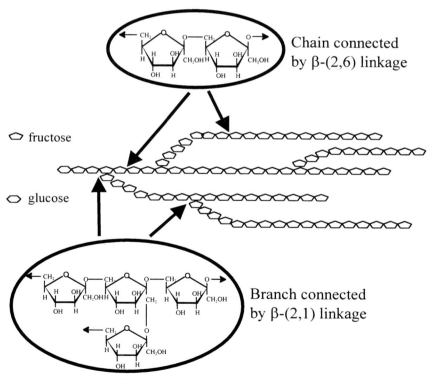

Fig. 1 Structure of levan. The main chain is connected by β-(2,6) linkages and the branch is connected to the main chain by a β-(2,1) linkage; the branch then continues with β-(2,6) linkages.

weight of bacterial levans is typically in the range of 2×10^6 to 10^8 (Keith et al., 1991), with the final molecular size being influenced by the synthesizing conditions such as ionic strength, temperature, and co-solutes. Although microbial levans have the similar structure, several types of (IX) levan are produced by different microorganisms, and this may be attributed to a varying degree of polymerization (DP) and branching of the repeating unit.

A cell-free enzyme system could be used to synthesize levans which have both β-(2,6) and β-(2,1) linked fructosyl units, and with similar structure to those of a whole-cell enzyme system. However, the structure of levans synthesized in the cell-free enzyme system is also known to differ in length compared with that synthesized by whole-cell systems (Han, 1990).

4
Occurrence

Various polysaccharides are produced as structural components in living organisms, and levan is one of the most diversely distributed components in plants, yeasts, fungi, and bacteria in particular. Levan is produced by grass (*Dactylis glomerata*, *Poa secunda* and *Agropyron cristatum*), wheat and barley (*Hordeum vulgare*), fungi (*Aspergillus sydawi* and *Aspergillus versicolor*) and yeasts (Han, 1990). Levans produced by microorganisms have been reported by Han (1990), and Hendry and Wallace (1993), and are listed in Table 1.

Previously, oral bacteria such as *Streptococcus*, *Rothis* and *Odontomyces* had received much attention due to their presence in human dental caries, together with soil microorganisms, especially *Bacillus*. Subsequently, focus was centered on the biological and functional aspects of levan rather than on its oral accumulation. The most extensive studies of levan were performed using *B. subtilis* (Suzuki and Chatterton, 1993). Furthermore, levans from *Bacillus polymyxa* (Aymerich, 1990) and *Pseudomonas* sp. (Hettwer et al., 1995, 1998) were identified as playing a role in the plant defense response. Levan from *B. subtilis* was shown to be tolerant against salt stress (Kunst and Rapoport, 1995), while that obtained from *Z. mobilis* exhibited antitumor activity (Calazans et al., 2000). The synthesis of levan using the genus *Lactobacillus* was also recently reported (Van Geel-Schutten et al., 1999).

Tab. 1 Levan-producing microorganisms

Microorganism	Reference
Acetobacter xylinum	Tajima et al. (1998)
Actinomyces naeslundii	Bergeron et al. (2000)
Bacillus circulans	Perez Oseguera et al. (1996)
Bacillus stearothermophilus	Li et al. (1997)
Gluconacetobacter (formerly *Acetobacter*) *diazotrophicus*	Arrieta et al. (1996)
Lactobacillus reuteri	Van Geel-Schutten et al. (1999)
Pseudomonas syringae pv. phaseolicola	Hettwer et al. (1995)
Pseudomonas syringae pv. glycinea	Hettwer et al. (1998)
Rahnella aquatilis	Ohtsuka et al. (1992); Song et al. (1998)
Serratia levanicum	Kojima et al. (1993)
Zymomonas mobilis	Song et al. (1993)

5
Physiological Functions of Levan

Bacterial polysaccharides are found either as a dense layer of more or less regularly arranged polymer structures attached to the bacterial cell walls (capsules) or as loosely associated exopolysaccharides (EPS) (Beveridge and Graham, 1991). Levans produced microbiologically have a number of interesting features. The levan which is synthesized extracellularly by bacteria may be visualized in a sucrose-containing medium, giving rise to a typical mucoid morphology. This type of mucoid feature provides a role in the symbiosis, phytopathogenesis, or participation in the defense mechanism against cold and dry conditions (Kunst and Rapoport, 1995). Extracellular levan produced by bacterial plant pathogens increases bacterial fitness and also acts as a detoxifying barrier against plant defence compounds (Hettwer et al., 1998). Among natural polysaccharides, glucans and fructans possess antitumor activity, and levan is included in this list. The antitumor activity of levan against sarcoma 180 depends on the molecular weight of the polysaccharides (Calazans et al., 2000), and this may indicate the polydiversity of levan.

Levans produced in plants are present as storage carbohydrates in the stem and leaf sheaths, and are degraded in a later stage of the growing season to provide plants with carbohydrates for grain filling (Pollock and Cairns, 1991). The biological role of polysaccharides in protection is less clearly understood, but a hypothesis has recently emerged. Levan penetrates into lipid membranes composed of monomolecular lipid layers, after which interactions occur which are orders of magnitude greater than the interaction between disaccharides and lipids. An extended layer of levan adheres to the lipids and partially protrudes into the aqueous phase. It is also possible that the membranes present are coated with levan; this coating of membranes imparts a reduction in accessibility of the membrane surface to proteins. In this way, in a biological system levan is able to protect membranes by interacting with the membrane lipid fraction (Vereyken et al., 2001).

6
Chemical Analysis and Detection

Several methods can be used in the qualitative and quantitative analysis of levan and the estimation of its concentration in solutions, with spectrophotometry and chromatography being the major techniques.

6.1
Spectrophotometry

Low concentrations of levan are measured by monitoring the optical density at 450–550 nm, as the presence of levan creates turbidity within the enzyme reaction mixture.

6.2
High-Performance Liquid Chromatography (HPLC)

The HPLC method is employed for both qualitative and quantitative determination of levan and other components (oligosaccharides, sucrose, fructose, and glucose). Details of the method are described here. The enzyme reaction mixture or levan solution is filtered using a 0.45 μm pore size membrane filter, and the filtrate is analyzed by HPLC equipped with a gel filtration column and refractive index detector (Shodex Ionpack KS-802, 300×8 mm; Showa Denko Co., Japan) (Jang et al., 2000). Deionized water is used as a mobile phase at a flow rate of

0.4 mL min^{-1}. The DP of levan is also determined by HPLC equipped with successive columns, GPC 4000–GPC 1000 (Polymer Laboratories, USA), and a refractive index detector (Jang et al., 2001). The analyses of sugar components and linkage type of levan are determined by using acid hydrolysis, methylation and nuclear magnetic resonance (NMR) shift experiments (Suzuki and Chatterton, 1993). In ^{13}C-NMR, signals of carbons of levan obtained from *Z. mobilis* and *Aerobacter levanicum* are identical, and show six main resonances at 104.9, 81.0, 76.9, 64.1, and 60.6 p.p.m. (Song and Rhee, 1994), these signals differing from those obtained with inulin.

6.3
Other Methods

Levan (nonmobiles) can be distinguished from oligosaccharides, sucrose, and other byproducts, either qualitatively or quantitatively, by the use of thin-layer chromatography (TLC). The sucrose-hydrolyzing activity of bacterial levansucrase is also used in the determination of levan concentration. The methods established are based on the fact that glucose is formed stoichiometrically in relation to the amount of fructose incorporated into levan (major product) and oligosaccharides (minor products). The amounts of glucose generated by the enzymatic reaction can be determined quantitatively by commercially available kits from the suppliers (Song et al., 1993).

7
Biosynthesis of Levan

The biosynthesis of levan requires the involvement of an extracellular enzyme levansucrase, which shows specificity for sucrose. Genetic characterization of the enzyme and the regulation of levan synthesis have been extensively studied, mostly using levansucrase genes from *B. subtilis* and *Z. mobilis*.

7.1
Enzymology of Levan Synthesis

Levansucrase (sucrose:2,6-β-D fructan:6-D-fructosyltransferase, sucrose 6-fructosyltransferase, EC 2.4.1.10.) was first named by Hestrin et al. (1943), and is responsible for the synthesis of levan from sucrose. Levansucrase exists as constituent intracellular and inducible extracellular forms in microorganisms (Han, 1990). The function of the levan-producing enzyme located intracellularly in some bacteria is not yet understood. The most abundant substrate for levansucrase in nature is sucrose, but raffinose also serves as a substrate.

Levansucrase is a type of transferase which catalyzes a fructosyl transfer from sucrose to various acceptor molecules. The enzyme catalyzes the following reactions:

1. Polymerization:

(Sucrose)$_n$ → (Glucose)$_n$ + Levan + Oligosaccharides

2. Hydrolysis:

Sucrose + H$_2$O → Fructose + Glucose

(Levan)$_n$ + H$_2$O → (Levan)$_{n-1}$ + Fructose

3. Acceptor:

Sucrose + Acceptor molecules
  → Fructosyl-acceptor + Glucose

4. Exchange:

Sucrose + [^{14}C]Glucose
  → Fructose-[^{14}C]Glucose + Glucose

5. Disproportionation:

[Levan]$_m$ + [Levan]$_n$
$\rightarrow$ [Levan]$_{m-1}$ + [Levan]$_{n+1}$

The enzyme catalyzes hydrolysis and polymerization reactions concomitantly (Reaction 1), resulting in a fructose homopolymer (levan) and free glucose. This reaction occurs when sucrose exists as the sole fructosyl donor and acceptor, and involves three steps: initiation, propagation, and termination (Chambert et al., 1974). The chains of levan grow step-wise by repeated transfer of a hexosyl group from the donor to growing acceptor molecules. The enzyme primarily catalyzes a coupled reaction by a ping-pong mechanism, i.e., sucrose hydrolysis followed by transfructosylation involving a fructosyl-enzyme intermediate (Chambert et al., 1976).

When water acts as an acceptor, a free fructose is generated from both sucrose and levan (Reaction 2). This reaction occurs in all the levansucrase-catalyzed reactions mentioned above, but the rate is much slower when compared with a sugar acceptor. Reaction 3 occurs in the presence of an acceptor in the environment. The enzyme transfers the fructosyl residue of sucrose specifically to the C-1 hydroxyl group of aldose in the acceptor. Compounds containing hydroxyl groups, such as methanol, glycerol and oligosaccharides, can act as fructosyl acceptors.

The reaction mechanism yields a non-reducing sugar compound and a series of oligosaccharides, in which the sugar molecule with one more fructose moiety remains as a major reaction product. The reaction occurs predominantly in the presence of a high concentration of fructosyl donors, such as sucrose or raffinose. Reaction 4 might be considered analogous to Reactions 2 and 3, but differs in the regeneration of sucrose, which has a high-energy bond. The enzyme also catalyzes Reaction 5, a disproportionation reaction, in which the degree of polydispersity of levan or oligomers is modified. The above five reactions compete with one another, yielding a specific major product with some minor products but they are predominantly controlled by environmental factors.

At present, little is known of plant levans, and their biosynthesis is not fully understood (Heyer et al., 1999). One plant levan, known as "graminan", is synthesized by sucrose:fructan 6-fructosyltransferase (6-SFT) which catalyzes the formation and extension of β-(2,6)-linked fructans. The 6-SFT is closely related to vacuolar invertase and transfers the fructosyl residues from sucrose preferentially to 1-kestose or larger fructans (Sprenger et al., 1995). However, most fructan synthesis in plants occurs in two steps (Edelman and Jefford, 1968). Initially, sucrose:sucrose 1-fructosyltransferase (1-SST; EC 2.4.1.99.) catalyzes the formation of the trisaccharide 1-kestose and glucose from two molecules of sucrose. Later, fructan:fructan 1-fructosyltransferase (1-FFT; EC 2.4.1.100.) reversibly transfers fructosyl residues from one fructan with a DP of ≥ 3 to another DP of ≥ 2, producing a mixture of fructans with different chain lengths.

7.2
Genetic Basis of Levan Synthesis

As yet, levansucrase genes have been cloned and biochemically characterized in seven Gram-negative strains; namely, *Acetobacter diazotrophicus* (Arrieta et al., 1996), *Acetobacter xylinum* (Tajima et al., 2000), *Erwinia amylovora* (Geier and Geider, 1993), *P. syringae* pv. glycinea, *P. syringae* pv. phaseolicola (Hettwer et al., 1998), *Rahnella aquatilis* (Song et al., 1998) and *Z. mobilis* (Song et al., 1993). Several levansucrase genes have also been cloned in Gram-positive strains, such as *Bacillus* (Gay et al., 1983; Li et al., 1997) and *Streptococcus* species (Sato et al., 1984). All levansucrase genes share several conserved regions, which are thought to be important for the enzyme activity. Although conservation is observed, dissimilarity exists depending on the source of the enzyme. Levansucrase genes from a Gram-negative origin show relatively high similarity (>50%) when compared with the genes from Gram-positive bacterial enzymes. However, very little similarity (<30%) exists among the genes from two different sources (Song and Rhee, 1994). The deduced amino acid sequences are aligned in Figure 2.

Although the amino acid sequences of levansucrases do not show any considerable homology to those of sucrose-related enzymes, the third (-EWS/AGT/SP/A-) and the

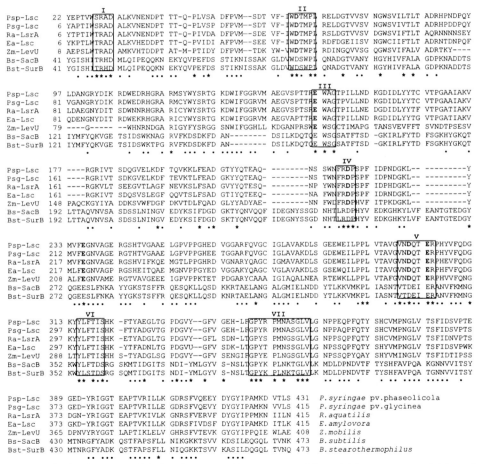

Fig. 2 Multiple alignment of deduced amino acid sequences of bacterial levansucrases. Origins of levansucrase are indicated in ends of the sequences. Asterisks indicate identical- and similar-residues in all levansucrases. Regions considered as important for activity are boxed (I–VII). Amino acid residues that are different between Gram-negative and Gram-positive origin are indicated by dots.

fourth (-FRDP-) conserved regions are found in all fructosyl- and glucosyltransferases, sucrase, sucrose-phosphate hydrolase and even in fructan-hydrolyzing enzymes. The fact that the regions are preserved in all of the sucrose-related enzymes implies that they may be catalytically important regions for the hydrolysis of sucrose. The serine residue in the sixth region (-YLFTI/DS-) has been proposed as the putative residue of the catalytic site (Chambert and Petit-Glatron, 1991).

7.3
Regulation of Levan Synthesis

The genes encoding sucrose-hydrolyzing enzymes may not be linked to each other on the chromosome, but linked only with accessory genes coding for proteins belonging to the phosphoenolpyruvate-dependent carbohydrate phosphotransferase (PTS) system. The expression of these genes is regulated by many regulatory protein systems such as the *glk* operon of *Z. mobilis* and the pleiotropic system of *B. subtilis*, etc.

7.3.1
Regulation at the Protein Level

In microorganisms, two types of sucrose utilization were found: intra- and extracellular. Commonly, these sucrose-uptake utilization systems exist within the cell; sucrose is transported by the PTS system. The sucrase system of this type is well known in *B. subtilis* (Klier and Rapoport, 1988). In contrast, some bacteria such as *Z. mobilis* that lack the PTS system first hydrolyze sucrose extracellularly to monomeric sugars, after which these sugars are transported inside the cell (Di Marco and Romano, 1985).

The co-contribution of both saccharolytic enzymes for the sucrose utilization of *Z. mobilis* has been well characterized. The glucose uptake and utilization system (*glk* operon), which is located very close to the *levU* operon and is also linked metabolically with the sucrose utilization system of *Z. mobilis*, is also regulated by the mechanism of tightly linked gene expression (Liu et al., 1992). In the intervening sequence of the *levU* and *glf* operon, two putative ORFs, encoding Lrp-like regulatory protein and aspartate racemase respectively, were found (Song et al., 1999).

At the molecular level, the genes encoding sucrose-hydrolyzing enzymes reported to date are not linked to each other on the chromosome, but are linked only with accessory genes coding for proteins belonging to the PTS system (Bruckner et al., 1993). The expression of these genes is modulated by regulatory mechanisms, such as anti-termination or repression, which is controlled by the complex regulatory network system including many regulatory proteins (Klier and Rapoport, 1988).

7.3.2
Regulation at Transcriptional and Translational Levels

The genes encoding the extracellular levansucrase and sucrase have been isolated and characterized. The nucleotide sequences of the DNA segment containing the genes encoding extracellular levansucrase and sucrase of *Z. mobilis* and *B. subtilis* were reported recently (Kyono et al., 1995). The two genes are located together in an operon on the chromosome, whereas almost all other genes coding for saccharolytic enzymes in other bacteria and yeasts are dispersed on the chromosomes (Carson and Botstein, 1983). The levansucrase gene of *B. subtilis* is activated in the presence of an inducer (sucrose or fructose), and is under a pleiotropic regulatory system controlling the expression of the sucrose operon (Lepesant et al., 1976; Shimotsu and Henner, 1986).

The pleiotropic system involving the *degS/degU, degQ* (formerly *sacU* and *sacQ*) and *degR* genes affects the expression of *sacB* (Débarbouillé et al., 1991). Levansucrase is encoded by the *sacB* gene and expressed from a constitutive promoter in the closely linked *sacR* locus. The *sacR* locus contains a palindromic structure acting as a transcription terminator. In the presence of sucrose, an anti-terminator, the *sacY* gene product that belongs to the *sacS* operon allows transcription of the *sacB* gene. The expression of this gene is also controlled by other regulatory genes such as two-component system *degS/degU* and also by *degQ* (Rapoport and Klier, 1990) (Figure 3).

8
Biodegradation of Levan

Although the biodegradation of levan involves several enzymes including levanase, levansucrase, and levan fructotransferase, the genetic characterization of these is limited to levanase.

8.1
Enzymology of Levan Degradation

Levan is degraded to D-fructose, levanbiose, sucrose, levan oligomers or low molecular-weight levan by the hydrolytic activity of levanase, levansucrase or levan fructotransferase from some plants and microorganisms. The mode and degree of hydrolysis depend on the enzyme sources and the reaction conditions.

8.1.1
Levanase

Many levan-forming microorganisms also produce hydrolytic enzymes–levanases–that degrade levan (Hestrin and Goldblum, 1953; Avigad, 1965). Certain strains of *Bacillus, Pseudomonas, Actinomyces, Aerobacter, Clostridium* and *Streptococcus* produce exocellular levanase (2,6-β-fructan 6-levanbiohydrolase, EC 3.2.1.64.) (Fuchs, 1959; Uchiyama, 1993). The enzyme hydrolyzes only levan, and the resulting product is usually levanbiose, indicating that a terminal fructosyl unit is removed. An exo-hydrolytic enzyme

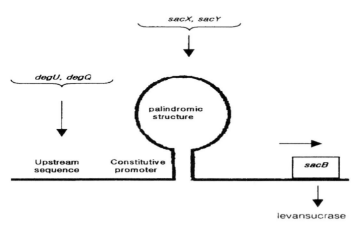

Fig. 3 Specific and pleiotropic control mechanisms affecting the *sacB* gene in *Bacillus subtilis*. *degU*: transcriptional regulator of degradation enzymes; *degQ*: pleiotropic regulatory gene; *sacX*: negative regulatory protein of *sacY*; *sacY*: positive levansucrase synthesis regulatory protein; *sacB*: gene encoding levansucrase. (Reprinted from Débarbouillé et al., 1991, p. 758, with permission from Elsevier Science.)

which has a 2,6-β-linkage-specific fructan-β-fructosidase activity (Marx et al., 1997) was reported from the grass *Lolium perenne*, and a 2,1-β-linkage-specific exohydrolase from Jerusalem artichoke. The other exo-levanase (fructan-β-fructosidase, EC 3.2.1.80; beta-D-fructofuranosidase, EC 3.2.1.26.) hydrolyzes levan to produce D-fructose. Endo-levanases (2,6-β-D-fructan fructanohydrolase, EC 3.2.1.65.) hydrolyze levan and levan oligomers consisting of more than three fructosyl units.

8.1.2
Levansucrase

Levan may be degraded not only by levanases, but also by levansucrase itself, which may catalyze the hydrolysis under certain conditions (Rapoport and Dedonder, 1963). The degree of levan hydrolysis depends on the enzyme sources and reaction conditions. For example, levansucrase from *R. aquatilis* showed a higher degradation activity than did that from *Z. mobilis* (Song et al., 1998), though for both enzymes a higher degradation activity of levan was seen as the reaction temperature was increased from 4 °C to 30 °C.

Although indirect evidence of the reversal of enzymatic synthesis of levan has been observed, little is known regarding the nature of such enzymatic degradation. Smith (1976) showed that beta-fructofuranosidase present in tall fescue degraded levan by removing one fructose residue at a time until a molecule of sucrose remained. Levansucrase of *B. subtilis* has a hydrolytic effect on small levans (Dedonder, 1966), the hydrolytic action stopping at branch points. Neither inulin, inulobiose, inulintriose, nor methyl D-fructofuranoside is hydrolyzed, despite these substrates being hydrolyzed by inulinase and yeast invertase. This hydrolytic activity may be responsible for the appearance of heterogeneous short-chain polysaccharides, rather than uniform high molecular-weight polymers, in the final product of many levan preparations.

8.1.3
Levan Fructotransferase

Microbial levan is an interesting starting material for the production of valuable oligosaccharides such as DFA IV (Yun, 1996; Saito and Tomita, 2000). DFA IV (di-D-fructose-2,6′:6,2′-dianhydride) is an oligosaccharide which is produced from levan by microbial enzymes, i.e., levan fructotransferase (LFTase) and a type of levanase (Tanaka et al., 1981, 1983; Saito et al., 1997). Currently, two LFTases have been isolated and cloned from *Arthrobacter nicotinovorans* GS-9 (Saito et al., 1997) and *A. ureafaciens* (Tanaka et al., 1981; Song et al., 2000). The enzymes have also been shown to degrade levan molecules from the nonreducing fructose end of the outer chains, and to catalyze intermolecular levanbiosyl and levanfructosyl transfer (LBT and LFT, respectively) reactions (Tanaka et al., 1983).

8.2
Genetic Basis of Levan Degradation

In *B. subtilis*, the expression of the levanase operon is inducible by fructose and is subjected to catabolite repression. A fructose-inducible promoter has been characterized 2.7 kb upstream from the gene *sacC*, which encodes levanase. The *sacC* gene is the distal gene of an operon containing five genes: *levD*, *levE*, *levF*, *levG*, and *sacC* (Martin et al., 1989; Débarbouillé et al., 1991) and is expressed under the regulated control of *sacR*, the inducible levansucrase leader region. The first four gene products are involved in a fructose-PTS system. In *Pseudomonas*, levanase is an exohydrolase of levan and produces levanbiose as a sole product; the limits of hydrolysis of levan from Z. mobilis and Serratia sp. were 65% and 80%, respectively (Jung et al., 1999).

8.3
Regulation of Levan Degradation

There are two levels on which the expression of the levanase operon in B. subtilis is controlled: (1) an induction by fructose, which involves a positive regulator, LevR, and the fructose phosphotransferase system encoded by this operon (lev-PTS); and (2) a global regulation of catabolite repression (Débarbouillé et al., 1991) (Figure 4).

The LevR protein is an activator for the expression of the levanase operon from B. subtilis. RNA polymerase containing the sigma 54-like factor sigma L recognizes the promoter of this operon. One domain of the LevR protein is homologous to activators of the NtrC family, and another resembles anti-terminator proteins of the BglG family (Débarbouillé et al., 1991). It has been proposed that the domain, which is similar to anti-terminators, is a target of phosphoenolpyruvate:sugar phosphotransferase system (PTS)-dependent regulation of LevR activity. The LevR protein is not only negatively regulated by the fructose-specific enzyme IIA/B of the phosphotransferase system encoded by the levanase operon (lev-PTS), but is also positively controlled by the histidine-containing phosphocarrier protein (HPr) of PTS (Martin et al., 1990; Stülke et al., 1995). This second type of control of LevR activity depends on phosphoenolpyruvate-dependent phosphorylation of HPr histidine 15, as demonstrated with point mutations in the ptsH gene encoding HPr. In vitro phosphorylation of partially purified LevR was obtained in the presence of phosphoenolpyruvate, enzyme I, and HPr. The dependence of truncated LevR polypeptides on stimulation by HPr indicates that the domain homologous to anti-terminators is the target of HPr-dependent regulation of LevR activity. This domain appears to be duplicated in the LevR protein. The first anti-terminator-like domain seems to be the target of enzyme I and HPr-dependent phosphorylation and the site of LevR activation, whereas the carboxy-terminal anti-terminator-like domain could be the target for negative regulation by lev-PTS (Débarbouillé et al., 1990).

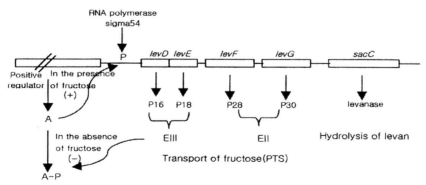

Fig. 4 Regulation model of levanase operon from Bacillus subtilis. P represents the fructose-inducible promoter. The levD, levE, levF, and levG gene products correspond to a fructose-specific phosphoenolpyruvate-dependent carbohydrate (PTS). LevR encodes a positive regulator. The activator may exist in two forms: (A-P), an inactive phosphorylated form; or (A), an active non-phosphorylated form. (Reprinted from Débarbouillé et al., 1991, p. 759, with permission from Elsevier Science.)

9
Biotechnological Production of Levan

Levan can be produced by either microbial fermentation or enzymatic synthesis, but the conversion yield of sucrose to levan is higher in the latter process than in the former.

9.1
Isolation and Screening for Levan-producing Strains

In the screening of levan-producing microbial strains, the levan formation activity can be determined using solid agar plates containing sucrose; the strains producing levan are then isolated by following the analytical procedures. The presence of levansucrase is positively selected by inducing mucoid morphology to the microorganisms. Subsequently, levan is collected from the agar plates by precipitation with alcohol (methanol, ethanol, or isopropanol). The identity of levan is then determined after acid hydrolysis of the polymers, followed by TLC analysis; fructose is identified as a single spot on the TLC plates.

9.2
Fermentative Production of Levan

The microbial production of levan requires fermentation and handling of highly viscous solutions. The conditions for producing levan by growing cultures of bacteria vary according to the microorganisms used, but yields of levan production are fairly low (Table 2); this is due to the utilization of sucrose as energy source, the formation of byproducts, the low level of levansucrase production, and the presence of levanase activity in bacteria. In addition, the recovery process of levan from the fermentation broth is often very difficult due to the high viscosity of levan. In theory, the yield of levan production by levansucrase is 50% (w/w) when sucrose is used as a substrate. Routinely, the yields of levan production based on the amount of sucrose consumed are no higher than 58% of the theoretical yield by fermentation (Table 2).

9.3
In vitro Biosynthesis of Levan

In the *in vitro* formation of levan by bacterial levansucrase, sucrose serves as fructosyl donor while the released glucose inhibits levan formation. The inhibitory action is influenced by competition with the glucose moiety of sucrose for the enzyme activity. The glucose moiety of sucrose can be replaced by D-xylose, L-arabinose, lactose, etc. Although the catalytic properties of levansucrase vary, the substrate specificity for acceptors is relatively broad where alco-

Tab. 2 Levan production by fermentation processes.

Strains	Type of production[a]	Substrate conc. [%]	Levan Yield [%,w/w][b]	Reference
Bacillus spp.	BF	12	23.5	Elisashvili (1984)
Bacillus polymyxa	BF	15	26.6	Han (1990)
Erwinia herbicola	CF	5	19.2	Keith et al. (1991)
Gluconobacter oxydans	BF	6.2	23.3	Uchiyama (1993)
Rahnella aquatilis	BF	10	29	Ohtsuka et al. (1992)
Zymomonas mobilis	CF	12	23	Beker et al. (1990)

[a] BF, batch fermentation; CF, continuous fermentation. [b] Based on sucrose consumed.

hol, monosaccharides, disaccharides, sugar alcohols and levan are available, but not for levanbiose, levantriose, and levantetraose.

The optimal temperature range for the *in vitro* synthesis of levan is from 0 °C to 40 °C. Levansucrase prepared from *Z. mobilis* displayed an optimum levan formation activity at 0 °C (Song et al., 1996), from *B. subtilis* at > 10 °C (Elisashvili, 1984), from *Pseudomonas* at 18 °C (Hettwer et al., 1995), and from *Rahnella* at 40 °C (Ohtsuka et al., 1992). Most levansucrase activities are inactivated at temperatures higher than 45 °C, with the exception of *R. aquatilis* ATCC 33071, which shows the maximum velocity of levan formation at 50 °C within 3 h, after which the rate declines slightly. Interestingly, at lower temperatures, transfructosylation rather than hydrolysis of sucrose is preferentially catalyzed; however, at higher temperatures hydrolysis is preferentially catalyzed, and this thermolabile feature may have advantages for large-scale levan production. In particular, levan production by *Z. mobilis* levansucrase was most active at the lowest temperature (0 °C), so that it could provide a stable operating condition with minimized contamination opportunities (Song et al., 1996). Plant fructosyltransferases lose 50% of their activity at 5 °C compared with that obtained at the optimum temperature of 20–25 °C (Koops and Jonker, 1996). The *Z. mobilis* levansucrase is stable at pH 4–7, and no activity is observed below pH 3 and above pH 9, similar to the enzymes from *Bacillus*, *Pseudomonas*, and *Rahnella*, the optimum pH of which was 6.0. However, the enzyme from *B. licheniformis* NRRL B-18962 retains 50% of its maximal activity at 55 °C and pH 4.

When *Z. mobilis* levansucrase was immobilized onto hydroxyapatite, the enzymatic and biochemical properties were similar to those of native enzyme towards salt and detergent effects (Jang et al., 2000). However, immobilization of the enzyme on the surface of a matrix shifts the optimum pH to acidic conditions (pH 4.0). The cell-free system synthesized two types of levan which differ in molecular weight. Levans produced by the immobilized system consisted of a higher proportion of low molecular-weight levan to total levan generated than those obtained by the native enzyme. Toluene-permeabilized whole-cell systems produced levan similarly to immobilized systems (Jang et al., 2001).

9.4
Recovery and Purification of Levan

In the microbial production of levan, the yield is low and, as a consequence, costly processes are required to extract the levan from the fermentation broth. The separation process of levan with high purity from the reaction mixture containing sucrose, glucose, fructose, and fructo-oligosaccharides is both laborious and inefficient. Likewise, the separation of levan by using solvents requires huge amounts of ethanol, methanol, isopropanol, or acetone to be used (Rhee et al., 1998). Subsequently, this solvent is lost as waste or recovered by distillation. Recently, membrane processes have been developed to separate polysaccharides from the fermentation broth or enzymatic reaction mixture, without organic waste. However, the resulting solution contains a low concentration (< 5%) of levan, and this must be recovered using various types of drier.

9.5
Commercial Production of Levan

Currently, a Korean start-up company, Real-BioTech Co., Ltd. (RBT), is the first and only company worldwide to produce levan on a commercial basis. RBT produces levan in large-scale quantities for supply to companies as a moisturizing ingredient in cosmetics, as a dietary fiber, as food and feed additives, and as a fertilizer (http://www.realbio.com). High-purity levan (>99%) is used in cosmetics and functional foods, while low-grade levan (<15%) containing glucose and oligosaccharides is used as a supplement for feeds and fertilizers.

9.6
Market Analysis and Cost of Levan Production

As levan became available commercially only recently, it is difficult to estimate the size of its current market. It is clear, however, that the expected world market is huge, since levan has a variety of functions and applications within the bioindustries. The production cost of levan depends mainly on the cost of the raw materials, including sucrose and levansucrase, for which the depreciation costs account for 30% and 18%, respectively (RealBioTech Co., unpublished data).

9.7
Levan Competitors

Many types of oligosaccharides and polysaccharides may represent potential competitors of levan in industrial applications. The strongest competitors in the food industry are the fructo-oligosaccharides (FOS), and especially inulin, which belongs to the same fructan category as levan. The use of inulin is limited as a dietary fiber by its insolubility in water at room temperature, whereas levan is a water-soluble and viscous fructan. Other potential competitors are dextran in the food and pharmaceutical industries, β-glucans in the feed industry, and hyaluronic acid in the cosmetic industry. In addition, levan could replace other potential competitors such as xanthan gum, pullulan, and mannan in the food industry. While most commercially available polysaccharides are produced either by microbial fermentation or direct extraction from natural sources, it is possible to produce levan from sucrose by a simple one-step enzyme reaction, which in turn makes it more competitive in terms of production costs.

10
Properties of Levan

While levan is highly soluble in water at room temperature, inulin is almost insoluble (<0.5%), this difference being most likely due to the presence of β-(2,6) linkages in levan. Despite their highly branched molecular structures, microbial levans have several common interesting features in soluble form (Kasapis et al., 1994), including an exceptionally low intrinsic viscosity for a polymer of high molecular weight, an unusual sensitivity of viscosity to increasing concentration between the beginning and end of the intermediate zone, and an extreme concentration-dependence of viscosity at the intermediate zone (Figure 5). These properties may be derived from intermolecular interaction by physical entanglement rather than from any form of specific noncovalent association.

The viscosity of a solution of bacterial levan, originating from *P. syringae* pv. phaseolicola, exhibited Newtonian characteristics up to a concentration of 20%. The concentration-dependence of the 'zero-shear' specific viscosity for the levan solution was unusually high, as would be expected from the molecule's branched structure (Kasapis et al., 1994). In Figure 5, there are three linear regions with changes of slope at levan concentrations of about 4% and 20%. The linear region, including the intermediate region, was also observed in xanthan gum (Milas et al., 1990). However, the concentration-dependence of viscosity may

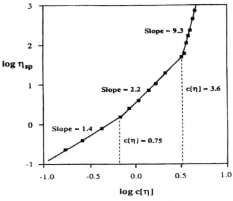

Fig. 5 Concentration dependence of 'zero-sheara-pos; specific viscosity for levan at 20 °C. The parameter η is used for solution viscosity and η_{sp} for specific viscosity which defines the fractional increase in viscosity due to the presence of the polymer ($\eta_{sp} = (\eta-\eta_s)/\eta_s = \eta_{rel}-1$). η_{rel} indicates relative viscosity, which is the ratio of η to η_s (η_s for solvent viscosity). (Reprinted from Kasapis et al., 1994, p. 59, with permission from Elsevier Science.)

vary with DP, pH, temperature, and salt concentrations (Vina et al., 1998).

11
Applications of Levan

Levan has a wide range of industrial applications, for example in medicine, pharmacy, agriculture, and food. However, it is likely that the low production cost of levan will permit a much increased use of levan in the near future.

11.1
Medical Applications

Microbially produced levan has a direct effect on tumor cells that is related to a modification in the cell membrane, including changes in cell permeability (Leibovici and Stark, 1985; Calazans et al., 2000), as well as radioprotective and antibacterial activities (Vina et al., 1998). Levan derivatives have also been suggested as inhibitors of smooth muscle cell proliferation, as excipients in making tablets, and as agents to transit water into gels. Sulfated, phosphated and acetylated levans have also been suggested as anti-AIDS agents (Clarke et al., 1997).

11.2
Pharmaceutical Applications

Water-soluble polymers, including levan, can be used in a wide variety of applications in the pharmaceutical industry. Water-soluble polymers such as cellulose derivatives, pectin and carrageenan play key roles in the formulation of solid, liquid, semisolid and even controlled release dosage forms (Guo et al., 1998). The viscosity of levan varies with its DP and degree of branching, which relates to the number of side fructose chains attached to one fructose unit in the main fructose chain, and in this respect levan can be used in pharmaceutical formulations in various ways. Low molecular-weight, less branched levan usually provides a low viscosity, and can be used as a tablet binder in immediate-release dosage forms, while levans of medium- and high-viscosity grade are used in controlled-release matrix formulations. Levan has also been suggested as a possible substitute for blood expanders (Imam and Abd-Allah, 1974).

11.3
Agricultural Applications

Microbial levans were introduced in plants to promote their agronomic performance in temperature zones, as well as their natural storage capacities (Vijn and Smeekens, 1999). Transgenic tobacco plants expressing levansucrase genes from *B. subtilis* (Pilon-

Smits et al., 1995) or *Z. mobilis* (Park et al., 1999) showed an increased tolerance to drought and cold stresses. Transgenic plants accumulating fructan have been suggested as novel nutritional feed for ruminants (Biggs and Hancock, 1998, 2001). Recently, microbial levan produced enzymatically was developed as an animal feed (Rhee et al., 2000d) and also as a soil conditioner to improve the germination of various seeds (Imam and Abd-Allah, 1974).

11.4
Food Applications

A number of novel applications of levan have been suggested, particularly in food (Han, 1990; Suzuki and Chatterton, 1993). Levan may act as a prebiotic to change the intestinal microflora, thereby offering beneficial effects when present in the human diet. Levan and its partially hydrolyzed products are fermented by intestinal bacteria including bifidobacteria and *Lactobacillus* sp. (Müller and Seyfarth, 1997; Yamamoto et al., 1999; Marx et al., 2000). Levanheptaose was also suggested as a carbon source for selective intestinal microflora, including *Bifidobacterium adolescentis*, *Lactobacillus acidophilus*, and *Eubacterium limosum*, whereas *Clostridium perfringens*, *E. coli* and *Staphylococcus aureus* did not utilize levan (Kang et al., 2000). Cholesterol- and triacylglycerol-lowering effects of levan have also been reported (Yamamoto et al., 1999) and may be applied to develop levans as health foods or nutraceuticals.

Today, it also seems possible that levan might be used in the dairy industry, as *Lactobacillus reuteri* produces levan-type EPS. EPS-producing lactic acid bacteria, including the genera of *Streptococcus*, *Lactobacillus*, and *Lactococcus*, are used *in situ* to improve the texture of fermented dairy products such as yogurt and cheese. This group of food-grade bacteria produces a wide variety of structurally different polymers, including levan with potential use for new applications. A number of Japanese companies use microbial levans as additives in their milk products which contain *Lactobacillus* species. In addition, the replacement with levan of thickeners or stabilizers that are produced by nonfood-grade bacteria has recently emerged (Van Kranenburg et al., 1999).

11.5
Other Applications

One of the striking consequences of the densely branched structure of levan is its effectiveness in resisting interpenetration by other polymers, leading to macroscopic phase-separation (Kasapis et al., 1994; Chung et al., 1997). Dextran is often used to create two-phase liquid systems (e.g., polyethylene glycol (PEG)/dextran), and to purify biological materials of interest by selectively partitioning them into one phase (Albertsson and Tjerneld, 1994). Microbial levans also display phase-separation phenomena with pectin, locust bean gum, and PEG. Solutions of levan and locust bean gum showed a substantial reduction in viscosity, similar to the mixture levan/pectin. Levan/locust bean gum phases can be separated into discrete phases in mixed solutions in which the lower one consisted predominantly of the denser polymer, the levan phase (Kasapis et al., 1994). The PEG/levan two-phase system was prepared by combining PEG (60%, w/w) and levan (6.77%, w/w) (Chung et al., 1997). This aqueous two-phase system showed a good partitioning with six model proteins including horse heart cytochrome c, horse hemoglobin, horse heart myoglobin, hen egg albumin, bovine serum albumin, and hen egg lysozyme.

12
Patents

Many patents have been filed for the application of levan as functional food additives (Table 3). It was claimed that levan from *S. salivarius* can be used as a food additive with a hypocholesterolemic effect (Kazuoki, 1996). New applications of levan as food and feed additives (Rhee et al., 2000d) as well as a raw material for the production of difructose dianhydride IV (Rhee et al., 2000c) were also developed. Levan derivatives such as sulfated, phosphated or acetylated levans, were claimed to be anti-AIDS agents (Robert and Garegg, 1998), while a fructan N-alkylurethane which has excellent surface-active properties in combination with good biodegradability was patented as a surfactant for household use and industrial applications by means of replacing a hydroxyl group of fructose with an alkylaminocarbonyloxy group (Stevens et al., 1999). A glycol/levan aqueous two-phase system which can substitute the glycerol/dextran system was developed for the partitioning of proteins (Rhee et al., 2000b). Besides levan, the fructosyl transferase activity of levansucrase has been used in the production of alkyl β-D-fructoside (Rhee et al., 2000a).

In spite of many studies on the production and application of levan, few of the levansucrase genes isolated from the microorganisms *Z. mobilis* (Rhee and Song, 1998), *R. aquatilis* (Rhee et al., 1999) and *A. diazotrophicus* (Juan et al., 1998) have been patented. A method for the production of levan using a recombinant levansucrase from *Z. mobilis* was claimed (Rhee et al., 1998), and a creative patent preparing transgenic plants harboring levansucrase gene was applied for by German researchers (Roeber et al., 1994). In the case of levansucrase, a process was claimed for the production of acid-stable

Tab. 3 Relevant patents for levan production and applications

Publication number	Applicants	Inventors	Title of invention	Date
US 4,769,254	IBI	Mays, T.D., Dally, E.L.	Microbial production of polyfructose.	September 6, 1988
US 4,927,757	IRFI	Hatcher, H.J. et al.	Production of substantially pure fructose.	May 22, 1990
WO 94/04692	IGF; Roeber, M.; Geier, G.; Geider, K.; Willmitzer, L.	Roeber, M. et al.	DNA sequences which lead to the formation of polyfructans (levans), plasmids containing these sequences as well as a process for preparing transgenic plants.	March 3, 1994
US 5,334,524	SOLVAY ENZYMES INC.	Robert, L.C.	Process for producing levansucrase using *Bacillus*.	August 2, 1994
US 5,527,784	–	Kazuoki, I.	Antihyperlipidemic and antiobesity agent comprising levan or hydrolysis products thereof obtained from *Streptococcus salivarius*.	June 17, 1996
US 5,547,863	USASA	Han, Y.W., Clarke, M.A.	Production of fructan (levan) polyfructose polymers using *Bacillus polymyxa*.	August 20, 1996

Tab. 3 (cont.)

Publication number	Applicants	Inventors	Title of invention	Date
WO 98/03184	Clarke, G; Margaret, A; SPRI -	Robert, E.J. et al.	Levan derivatives, their preparation, composition and applications including medical and food applications.	January 29, 1998
US 5,731,173		Juan, G.A.S. et al.	Fructosyltransferase enzyme, method for its production and DNA encoding the enzyme.	March 24, 1998
Korean patent 145946	KRIBB	Rhee, S. K. et al.	Method for production of levan using levansucrase.	May 6, 1998
Korean patent 176410	KRIBB	Rhee, S. K. et al.	A novel levansucrase.	November 13, 1998
Korean patent 0207960	KRIBB	Rhee, S. K. et al.	Base and amino acid sequence of levansucrase derived from *Rahnella aquatilis*.	April 14, 1999
EP 0964054 A1	TS N.V.	Stevens, C.V. et al.	Surface-active alkylurethanes of fructans.	December 15, 1999
Korean patent 0257118	KRIBB	Rhee, S. K. et al.	A process for preparation of alkyl β-D-fructoside using levansucrase.	Febraury 28, 2000
Korean patent 262769	KRIBB	Rhee, S. K. et al.	Novel polyethylene glycol/levan aqueous two-phase system and protein partitioning method using thereof.	May 6, 2000
WO 01/29185	KRIBB and RBT	Rhee, S. K. et al.	Enzymatic production of difructose dianhydride IV from sucrose and elevant enzymes and genes coding for them.	October 19, 2000
WO 01/49127	KRIBB and RBT	Rhee, S. K. et al.	Animal feed containing simple polysaccharides.	December 29, 2000

EP: European Patent; IBI: Igene Biotechnology Institute; IGF: INST GENBIOLOGISCHE FORSCHUNG (DE); IRFI: Idaho Research Foundation Institute; KRIBB: Korea Research Institute of Bioscience and Biotechnology; PCT: World Intellectual Property Organization; RBT: RealBioTech Co., Ltd.; SPRI: Sugar Processing Research Institute; TS N.V: Tiense Suikerraffinaderij N.V.; US: United States Patent; USASA: The United States of America as represented by the Secretary of the Agriculture; WO: World IPO(PCT).

levansucrase from *Bacillus* which is not induced by sucrose (Robert, 1994).

13
Current Problems and Limits

Although levan is a water-soluble and low-viscosity polysaccharide, its applications might be limited due to its turbidity and low fluidity. In order to expand the application areas of levan, the development of biosynthetic methods, including the involvement of novel enzymes, will be essential in order to control the DP (molecular weight) and the degree of branching. More important factors from a commercial aspect include process development for the large-scale production which is comparable with that used for other competitive polysaccharides, especially inulin from chicory. In order to produce levan more economically, a cost-effective purification process of levan from the reaction mixture containing sucrose, glucose, fructose and fructo-oligosaccharides must be developed. The membrane processes will likely serve as one of these solutions.

14
Outlook and Perspectives

Levan has a great potential as a functional biomaterial in the food, cosmetic, pharmaceutical and other industries. However, use of this biopolymer has yet not been practicable due to a paucity of information on its polymeric properties highlighting its industrial applicability, as well as a lack of feasible processes for large-scale production.

For technical applications, fructans with a high molecular mass and a low degree of branching would be desirable. However, microbial levans and their oligomers have been less well characterized with regard to carbohydrate structure analysis. In order to utilize the versatility of water-soluble levans to a maximum, a broader understanding of the behavior of levan is required. The characterization of polysaccharides has advanced considerably during the past two decades due to the introduction of powerful methods such as mass spectrometry, nuclear magnetic resonance, atomic force microscopy, scanning probe microscopy, small-angle X-ray scattering, small-angle neutron scattering, and molecular-mechanic-based carbohydrate modeling (Brant, 1999). As a consequence, the complete characterization of levan should be attained in the near future. Furthermore, the fundamental rheological properties of levan in solution, for example viscosity, thixotropy, dilatancy, elasticity, pseudoelasticity, and viscoelasticity will become increasingly important for new applications of this compound.

15
References

Albertsson, P. A., Tjerneld, F. (1994) Phase diagrams, *Methods Enzymol.* **228**, 3–13.

Ananthalakshmy, V. K., Gunasekaran, P. (1999) Overproduction of levan in *Zymomonas mobilis* by using cloned *sacB* gene, *Enzyme Microb. Technol.* **25**, 109–115.

Arrieta, J., Hernández, L., Coego, A., Suárez, V., Balmori, E., Menéndez, C., Petit-Glatron, M. F., Chambert, R., Selman-Housein, G. (1996) Molecular characterization of the levansucrase gene from the endophytic sugarcane bacterium *Acetobacter diazotrophicus* SRT4, *Microbiology* **142**, 1077–1085.

Avigad, G. (1965) In *Methods in Carbohydrate Chemistry*, New York, London: Academic Press, vol. V, 161–165.

Aymerich, S. (1990) What is the role of levansucrase in *Bacillus subtilis*? *Symbiosis* **9**, 179–184.

Beker, M. J., Shvinka, J. E., Pankova, L. M., Laivenieks, M. G., Mezhbarde, I. N. (1990) A simultaneous sucrose bioconversion into ethanol and levan by *Zymomonas mobilis*, *Appl. Biochem. Biotechnol.* **24/25**, 265–274.

Bergeron, L. J., Morou-Bermudez, E., Burne, R. A. (2000) Characterization of the fructosyltransferase gene of *Actinomyces naeslundii* WVU45, *J. Bacteriol.* **182**, 3649–3654.

Beveridge, T. J., Graham, L. L. (1991) Surface layers of bacteria, *Microbiol. Rev.* **55**, 684–705.

Biggs, D. R., Hancock, K. R. (1998) *In vitro* digestion of bacterial and plant fructans and effects on ammonia accumulation in cow and sheep rumen fluids, *J. Gen. Appl. Microbiol.* **44**, 167–171.

Biggs, D. R., Hancock, K. R. (2001) Fructan 2000, *Trends Plant Sci.* **6**, 8–9.

Birkhed, D., Rosell, K.-G, Granath, K. (1979) Structure of extracellular water-soluble polysaccharides synthesized from sucrose by oral strains of *Streptococcus mutans*, *Streptococcus salivarius*, *Streptococcus sanguis*, and *Actinomyces viscosus*, *Arch. Oral Biol.* **24**, 53–61.

Brant, D. A. (1999) Novel approaches to the analysis of polysaccharide structures, *Curr. Opin. Struct. Biol.* **9**, 556–562.

Bruckner, R., Wagner, E., Gotz, F. (1993) Cloning and characterization of the scrA gene encoding the sucrose-specific Enzyme II of the phosphotransferase system from *Staphylococcus xylosus*, *Mol. Gen. Genet.* **241**, 33–41.

Calazans, G. M. T., Lima, R. C., de França, F. P., Lopes, C. E. (2000) Molecular weight and antitumour activity of *Zymomonas mobilis* levans, *Int. J. Biol. Macromol.* **27**, 245–247.

Carson, M., Botstein, D. (1983) Organization of the SUC gene family in *Saccharomyces*, *Mol. Cell. Biol.* **3**, 351–359.

Chambert, R. G., Gonzy-Treboul, G. (1976) Levansucrase of *Bacillus subtilis*. Characterization of a stabilized fructosyl-enzyme complex and identification of an aspartyl residue as the binding site of the fructosyl group, *Eur. J. Biochem.* **71**, 493–508.

Chambert, R., Petit-Glatron, M. F. (1991) Polymerase and hydrolase activities of *Bacillus subtilis* levansucrase can be separately modulated by site-directed mutagenesis, *Biochem. J.* **279**, 35–41.

Chambert, R. G., Gonzy-Treboul, G., Dedonder, R. (1974) Kinetic studies of levansucrase of *Bacillus subtilis*, *Eur. J. Biochem.* **41**, 285

Chung, B. H., Kim, W. K., Song, K. B., Kim, C. H., Rhee, S. K. (1997) Novel polyethylene glycol/levan aqueous two-phase system for protein partitioning, *Biotech. Techn.* **11**, 327–329.

Clarke, M. A., Roberts, E. J., Garegg, P. J. (1997) New compounds from microbiological products of sucrose, *Carbohydr. Polym.* **34**, 425.

Débarbouillé, M., Arnaud, M., Fouet, A., Klier, A., Rapoport, G. (1990) The *sacT* genes regulating the *sacPA* operon in *Bacillus subtilis* shares strong homology with transcriptional antiterminators, *J. Bacteriol.* **172**, 3966–3973.

Débarbouillé, M., Martin. V, Arnaud, M., Klier, A., Rapoport, G. (1991) Positive and negative regulation controlling expression of the sac genes in *Bacillus subtilis*, *Res. Microbiol.* **142**, 757–764.

Dedonder, R. (1966) Levansucrase from *Bacillus subtilis*, *Methods Enzymol.* **8**, 500–505.

Di Marco, A. A., Romano, A. H. (1985) D-Glucose transport system of *Zymomonas mobilis*, *Appl. Environ. Microbiol.* **49**, 151–157.

Edelman, J., Jefford, T. G. (1968) The mechanism of fructosan metabolism in higher plants as exemplified in *Helianthus tuberosus*, *New Phytol.* **67**, 517–531.

Elisashvili, V. I. (1984) Levan synthesis by *Bacillus* sp., *Appl. Biochem. Microbiol.* **20**, 82–87.

Fuchs, A. (1959) *On the synthesis and breakdown of levan by bacteria*, Thesis, Uitgeverij Waltman, Delft.

Gay, P., Le Coq, D., Steinmetz, M., Ferrari, E., Hoch, J. A. (1983) Cloning structural gene *sacB*, which codes for exoenzyme levansucrase of *Bacillus subtilis*: expression of the gene in *Escherichia coli*, *J. Bacteriol.* **153**, 1424–1431.

Geier, G., Geider, K. (1993) Characterization and influence on virulence of levansucrase gene from the fireblight pathogen *Erwinia amylovora*, *Physiol. Mol. Plant Pathol.* **42**, 387–404.

Greig-Smith, R. (1901) The gum fermentation of sugar cane juice, *Proc. Linn. Soc. N.S.W.* **26**, 589.

Guo, J.-H, Skinner, G. W., Harcum, W. W., Barnum, P. E. (1998) Pharmaceutical applications of naturally occurring water-soluble polymers, *Pharmaceutical Science & Technology Today* **1**, 254–261.

Han, Y. W. (1989) Levan production by *Bacillus polymyxa*, *J. Ind. Microbiol.* **4**, 447–452.

Han, Y. W. (1990) Microbial levan, *Adv. Appl. Microbiol.* **35**, 171–194.

Han, Y. W., Clarke, M. A. (1996) Production of fructan(levan) polyfructose polymers using *Bacillus polymyxa*, U. S. Patent 5,547,863.

Hatcher, H. J., Gallian, J. J., Leeper, S. A. (1990) Production of substantially pure fructose, U. S. Patent 4,927,757.

Hendry, G. A. F., Wallace, R. K. (1993) The origin, distribution, and evolutionary significance of fructans, in: *Science and Technology of Fructans* (Suzuki, M., Chatterton, N. J., Eds.), Boca Raton: CRC Press, 119–139.

Hestrin, S., Goldblum, J. (1953) Levanpolyase, *Nature* **172**, 1047–1064.

Hestrin, S., Avineri-Shapiro, S., Aschner, M. (1943) The enzymatic production of levan, *Biochem. J.* **37**, 450.

Hettwer, U., Gross, M., Rudolph, K. (1995) Purification and characterization of an extracellular levansucrase from *Pseudomonas syringae* pv. phaseolicola, *J. Bacteriol.* **177**, 2834–2839.

Hettwer, U., Jaeckel, F. R., Boch, J., Meyer, M., Rudolph, K., Ullrich, M. S. (1998) Cloning, nucleotide sequence, and expression in *Escherichia coli* of levansucrase genes from the plant pathogens *Pseudomonas syringae* pv. glycinea and *P. syringae* pv. phaseolicola, *Appl. Environ. Microbiol.* **64**, 3180–3187.

Heyer, A. G., Lloyd, J. R., Kossmann, L. (1999) Production of polymeric carbohydrates, *Curr. Opin. Biotechnol.* **10**, 169–174.

Imam, G. M., Abd-Allah, N. M. (1974) Fructosan, a new soil conditioning polysaccharide isolated from the metabolites of *Bacillus polymyxa* AS-1 and its clinical applications, *Egypt. J. Bot.* **17**, 19–26.

Jang, K. H., Kim, J. S., Song, K. B., Kim, C. H., Chung, B. H., Rhee, S. K. (2000) Production of levan using recombinant levansucrase immobilized on hydroxyapatite, *Bioprocess Eng.* **23**, 89–93.

Jang, K. H, Song, K. B., Kim, C. H., Chung, B. H., Kang, S. A., Chun, U. H., Choue, R. W., Rhee, S. K. (2001) Comparison of characteristics of levan produced by different preparations of levansucrase from *Zymomonas mobilis*, *Biotechnol. Lett.* **23**, 339–344.

Juan, G. A. S., Lazaro, H. G., Alberto, C. G., Guillermo, S. H. S. (1998) Fructosyltransferase enzyme, method for its production and DNA encoding the enzyme, U. S. Patent 5,731,173.

Jung, K. E., Lee, S. O., Lee, J. D., Lee, T. H. (1999) Purification and characterization of a levanbiose-producing levanase from *Pseudomonas* sp. No. 43, *Biotechnol. Appl. Biochem.* **28**, 263–268.

Kang, S. K., Park, S. J., Lee, J. D., Lee, T. H. (2000) Physiological effects of levanoligosaccharide on growth of intestinal microflora, *J. Korean Soc. Food Sci. Nutr.* **29**, 35–40.

Kasapis, S., Morris, E. R., Gross, M., Rudolph, K. (1994) Solution properties of levan polysaccharide from *Pseudomonas syringae* pv. phaseolicola, and its possible primary role as a blocker of recognition during pathogenesis, *Carbohydr. Polym.* **23**, 55–64.

Kazuoki, I. (1996) Antihyperlipidemic and anti-obesity agent comprising levan or hydrolysis products thereof obtained from *Streptococcus salivarius*, U. S. Patent 5,527,784.

Keith, K., Wiley, B., Ball, D., Arcidiacono, S., Zorfass, D., Mayer, J., Kaplan, D. (1991) Continuous culture system for production of biopolymer levan using *Erwinia herbicola*, *Biotechnol. Bioeng.* **38**, 557–560.

Kim, M. G., Seo, J. W., Song, K.-B., Kim, C. H., Chung, B. H, Rhee, S. K. (1998) Levan and fructosyl derivatives formation by a recombinant levansucrase from *Rahnella aquatilis*, *Biotechnol. Lett.* **20**, 333–336.

Klier, A. F., Rapoport, G. (1988) Genetics and regulation of carbohydrate catabolism in *Bacillus*, *Annu. Rev. Microbiol.* **42**, 65–95.

Kojima, I., Saito, T., Iizuka, M., Minamiura, N., Ono, S. (1993) Characterization of levan produced by *Serratia* sp., *J. Ferment. Bioeng.* **75**, 9–12.

Koops, A. J., Jonker, H. H. (1996) Purification and characterization of the enzymes of fructan biosynthesis in tubers of *Helianthus tuberosus* Colombia. II. Purification of sucrose:sucrose 1-fructosyltransferase and reconstitution of fructan synthesis *in vitro* with purified sucrose:sucrose 1-fructosyltransferase and fructan:fructan 1-fructosyltransferase, *Plant Physiol.* **110**, 1167–1175.

Kunst, F., Rapoport, G. (1995) Salt stress is an environmental signal affecting degradative enzyme synthesis in *Bacillus subtilis*, *J. Bacteriol.* **177**, 2403–2407.

Kyono, K., Yanase, H., Tonomura, K., Kawasaki, H., Sakai, T. (1995) Cloning and characterization of *Zymomonas mobilis* genes encoding extracellular levansucrase and invertase, *Biosci. Biotechnol. Biochem.* **59**, 289–293.

Leibovici, J., Stark, Y. (1985) Increase in cell permeability as a cytotoxic agent by the polysaccharide levan, *Cell. Mol. Biol.* **31**, 337–341.

Lepesant, J. A., Kunst, F., Pascal, M., Steinmetz, M., Dedonder, R. (1976) Specific and pleiotropic regulatory mechanisms in the sucrose system of *Bacillus subtilis*, *Microbiology* **168**, 58–69.

Li, Y., Triccas, J. A., Ferenci, T. (1997) A novel levansucrase-levanase gene cluster in *Bacillus stearothermophilus* ATCC 12980, *Biochim. Biophys. Acta* **1353**, 203–208.

Liu, J., Barnell, W. O., Conway, T. (1992) The polycistronic mRNA of the *Zymomonas mobilis glf-zwf-edd-glk* operon is subject to complex transcript processing, *J. Bacteriol.* **174**, 2824–2833.

Martin, I., Débarbouillé, M., Klier, A., Rapoport, G. (1989) Induction and metabolic regulation of levanase synthesis in *Bacillus subtilis*, *J. Bacteriol.* **171**, 1885–1892.

Martin, I., Débarbouillé, M., Klier, A., Rapoport, G. (1990) Levanase operon of *Bacillus subtilis* includes a fructose-specific phosphotransferase system regulating the expression of the operon, *J. Mol. Biol.* **214**, 657–671.

Marx, S. P., Nosberger, J., Frehner, M. (1997) Hydrolysis of fructan in grasses: 2,6-β-linkage-specific fructan-β-fructosidase from stubble of *Lolium perenne*, *New Phytol.* **135**, 279–290.

Marx, S. P., Winkler, S., Hartmeier, W. (2000) Metabolization of β-(2,6)-linked fructose-oligosaccharides by different bifidobacteria, *FEMS Microbiol. Lett.* **182**, 163–169.

Mays, T. D., Dally, E. L. (1988) Microbial production of polyfructose, U.S. Patent 4,769,254.

Milas, M., Rinaudo, M., Knipper, M., Schuppise, J. L. (1990) Flow and viscoelastic properties of xanthan gum solutions, *Macromolecules* **23**, 2506–2511.

Müller, M., Seyfarth, W. (1997) Purification and substrate specificity of an extracellular fructan-hydrolase from *Lactobacillus paracasei* ssp. *paracasei* P 4134, *New Phytol.* **136**, 89–96.

Nelson, C. J., Spollen, W. G. (1987) Fructans, *Physiol. Plant.* **71**, 512–516.

Newbrun, E., Lacy, R., Christie, T. M. (1971) The morphology and size of the extracellular polysaccharide from oral streptococci, *Arch. Oral Biol.* **16**, 863–872.

Ohtsuka, K., Hino, S., Fukushima, T., Ozawa, O., Kanematsu, T., Uchida, T. (1992) Characterization of levansucrase from *Rahnella aquatilis* JCM-1638, *Biosci. Biotechnol. Biochem.* **56**, 1373–1377.

Park, J. M., Kwon, S. Y., Song, K. B., Kwak, J. W., Lee, S. B., Nam, Y. W., Shin, J. S., Park, Y. I., Rhee, S. K., Paek, K. H. (1999) Transgenic tobacco plants expressing the bacterial levansucrase gene show enhanced tolerance to osmotic stress, *J. Microbiol. Biotechnol.* **9**, 213–218.

Perez Oseguera, M. A., Guereca, L., Lopez-Munguia, A. (1996) Properties of levansucrase from *Bacillus circulans*, *Appl. Microbiol. Biotechnol.* **45**, 465–471.

Pilon-Smits, E. A. H., Ebskamp, M. J. M., Paul, M. J., Jeuken, M. J. W., Weisbeek, P. J., Smeekens, J. C. M. (1995) Improved performance of transgenic fructan-accumulating tobacco under drought stress, *Plant Physiol.* **107**, 125–130.

Pollock, C. J., Cairns, A. J. (1991) Fructan metabolism in grasses and cereals, *Annu. Rev. Plant Physiol. Plant Mol. Biol.* **42**, 77–101.

Rapoport, G., Dedonder, R. (1963) Le lévane-sucrase de *B. subtilis* III. Reaction d'hydrolyse, de transfert et d'échange avec des analogues du saccharose, *Bull. Soc. Chim. Biol.* (France) **45**, 515–535.

Rapoport, G., Klier, A. (1990) Gene expression using *Bacillus*, *Curr. Opin. Biotechnol.* **1**, 21–27.

Rhee, S. K., Song, K. B. (1998) A novel levansucrase, Korean Patent 176410.

Rhee, S. K., Song, K. B., Kim, C. H. (1998) Method for production of levan using levansucrase, Korean Patent 145946.

Rhee, S. K., Song, K. B., Seo, J. W., Kim, C. H., Chung, B. H. (1999) Base and amino acid sequence of levansucrase derived from *Rahnella aquatilis*, Korean Patent 0207960.

Rhee, S. K., Kim, C. H., Song, K. B., Kim, M. G., Seo, J. W., Chung, B. H. (2000a) A process for preparation of alkyl β-D-fructoside using levansucrase, Korean Patent 0257118.

Rhee, S. K, Chung, B. H., Kim, W. K., Song, K. B., Kim, C. H. (2000b) Novel polyethylene glycol/levan aqueous two-phase system and protein partitioning method using thereof, Korean Patent 262769.

Rhee, S. K., Song, K. B., Kim, C. H., Ryu, E. J., Lee, Y. B. (2000c) Enzymatic production of difructose dianhydride IV from sucrose and elevant enzymes and genes coding for them, PCT-KR00-01183.

Rhee, S. K., Song, K. B., Yoon, B. D., Kim, C. H. (2000d) Animal feed containing simple polysaccharides, PCT-KR00-01556.

Robert, E. J., Garegg, P. J. (1998) Levan derivatives, their preparation, composition and applications including medical and food applications, WO 98/03184.

Robert, L. C. (1994) Process for producing levan sucrase using *Bacillus*, U. S. Patent 5,334,524.

Roeber, M., Geier, G., Geider, K., Willmitzer, L. (1994) DNA sequences which lead to the formation of polyfructans (levans), plasmids containing these sequences as well as a process for preparing transgenic plants, WO-94/004692.

Saito, K., Tomita, F. (2000) Difructose anhydrides: their mass-production and physiological functions, *Biosci. Biotechnol. Biochem.* **64**, 1321–1327.

Saito, K., Goto, H., Yokoda, A., Tomita, F. (1997) Purification of levan fructotransferase from *Arthrobacter nicotinovorans* GS-9 and production of DFA IV from levan by the enzyme, *Biosci. Biotechnol. Biochem.* **61**, 1705–1709.

Sato, S., Koga, T., Inoue, M. (1984) Isolation and some properties of extracellular D-glucosyltransferases and D-fructosyltransferase from *Streptococcus mutans* serotypes c, e, and f, *Carbohydr. Res.* **134**, 293–304.

Shimotsu, H., Henner, D. J. (1986) Modulation of *Bacillus subtilis* levansucrase gene expression by sucrose and regulation of the steady-state mRNA level by *sacU* and *sacQ* genes, *J. Bacteriol.* **168**, 380–388.

Smith, A. E. (1976) Beta-fructofuranosidase and invertase activity in tall fescue culm bases, *J. Agric. Food Chem.* **24**, 476–478.

Song, K. B., Rhee, S. K. (1994) Enzymatic synthesis of levan by *Zymomonas mobilis* levansucrase overexpressed in *Escherichia coli*, *Biotechnol. Lett.* **16**, 1305–1310.

Song, K. B., Joo, H. K., Rhee, S. K. (1993) Nucleotide sequence of levansucrase gene (*levU*) of *Zymomonas mobilis* ZM1 (ATCC10988), *Biochim. Biophys. Acta* **1173**, 320–324.

Song, K. B., Belghith, H., Rhee, S. K. (1996) Production of levan, a fructose polymer, using an overexpressed recombinant levansucrase, *Ann. N. Y. Acad. Sci.* **799**, 601–607.

Song, K. B., Seo, J. W., Kim, M. K., Rhee, S. K. (1998) Levansucrase from *Rahnella aquatilis*: gene cloning, expression and levan formation, *Ann. N. Y. Acad. Sci.* **864**, 506–511.

Song, K. B., Seo, J. W., Rhee, S. K. (1999) Transcriptional analysis of *levU* operon encoding saccharolytic enzymes and two apparent genes involved in amino acid biosynthesis in *Zymomonas mobilis*, *Gene* **232**, 107–114.

Song, K. B., Bae, K. S., Lee, Y. B., Lee, K. Y., Rhee, S. K. (2000) Characteristics of levan fructotransferase from *Arthrobacter ureafaciens* K2032 and difructose anhydride IV formation from levan, *Enzyme Microb. Technol.* **27**, 212–218.

Sprenger, N., Bortlik, K., Brandt, A., Boller, T., Wiemken, A. (1995) Purification, cloning, and functional expression of sucrose:fructan 6-fructosyltransferase, a key enzyme of fructan synthesis in barley, *Proc. Natl. Acad. Sci. USA* **92**, 11652–11656.

Stevens, C. V., Karl, B., Isabelle, M.-A., Lucien, D. (1999) Surface-active alkylurethanes of fructans, EP 0964054 A1.

Stülke, J., Martin, V. I., Charrier, V., Klier, A., Deutscher, J., Rapoport, G. (1995) The HPr protein of the phosphotransferase system links induction and catabolite repression of the *Bacillus subtilis* levanase operon, *J. Bacteriol.* **177**, 6928–6936.

Suzuki, M., Chatterton, N. J. (1993) *Science and Technology of Fructans*. Boca Raton: CRC Press.

Tajima, K., Uenishi, N., Fujiwara, M., Erata, T., Munekata, M., Takai, M. (1998) The production of a new water-soluble polysaccharide by *Acetobacter xylinum* NCI 1005 and its structural analysis by NMR spectroscopy, *Carbohydr. Res.* **305**, 117–122.

Tajima, K., Tanio, T., Kobayashi, Y., Kohno, H., Fujiwara, M., Shiba, T., Erata, T., Munekata, M., Takai, M. (2000) Cloning and sequencing of the levansucrase gene from *Acetobacter xylinum* NCI 1005, *DNA Res.* **7**, 237–242.

Tanaka, K., Kawaguchi, H., Ohno, K., Shohji, K. (1981) Enzymatic formation of difructose IV from bacterial levan, *J. Biochem.* **90**, 1545–1548.

Tanaka, K., Karigane, T., Yamaguchi, F., Nishikawa, S., Yoshida, N. (1983) Action of levan fructotransferase of *Arthrobacter ureafaciens* on levanoligosaccharides, *J. Biochem.* **94**, 1569–1578.

Uchiyama, T. (1993) Metabolism in microorganisms. Part II. Biosynthesis and degradation of fructans by microbial enzymes other than levansucrase. in: *Science and Technology of Fructans* (Suzuki, M., Chatterton, N. J., Eds.), Boca Raton: CRC Press, 169–190.

Vandamme, E. J., Derycke, D. G. (1983) Microbial inulinase: fermentation process, properties, and applications, *Adv. Appl. Microbiol.* **29**, 139–176.

Van Geel-Schutten, G. H., Faber, E. J., Smit, E., Bonting, K., Smith, M. R., Ten Brink, B., Kamerling, J. P., Vliegenthart, J. F. G., Dijkhuizen, L. (1999) Biochemical and structural characterization of the glucan and fructan exopolysaccharides synthesized by the *Lactobacillus reuteri* wild-type strain and mutant strains, *Appl. Environ. Microbiol.* **65**, 3008–3014.

Van Kranenburg, R., Boels, I. C., Kleerebezem, M., de Vos, W. M. (1999) Genetics and engineering of microbial exopolysaccharides for food: approaches for the production of existing and novel polysaccharides, *Curr. Opin. Biotechnol.* **10**, 498–504.

Vereyken, I. J., Chupin, V., Demel, R. A., Smeekens, S. C. M., Kruijff, B. (2001) Fructans insert between the headgroups of phospholipids, *Biochim. Biophys. Acta* **1510**, 307–320.

Vijn, I., Smeekens, S. (1999) Fructan: more than reserve carbohydrate? *Plant Physiol.* **120**, 351–359.

Vina, I., Karsakevich, A., Gonta, S., Linde, R., Bekers, M. (1998) Influence of some physicochemical factors on the viscosity of aqueous levan solutions of *Zymomonas mobilis*, *Acta Biotechnol.* **18**, 167–174.

Wolff, D., Czapla, S., Heyer, A. G., Radosta, S., Mischnick, P., Springer, J. (2000) Globular shape of high molar mass inulin revealed by static light scattering and viscometry, *Polymer* **41**, 8009–8016.

Yamamoto, Y., Takahashi, Y., Kawano, M., Iizuka, M., Matsumoto, T., Saeki, S., Yamaguchi, H. (1999) *In vitro* digestibility and fermentability of levan and its hypocholesterolemic effects in rats, *J. Nutr. Biochem.* **10**, 13–18.

Yun, J. W. (1996) Fructooligosaccharides – occurrence, preparation, and application, *Enzyme Microb. Technol.* **19**, 107–117.

10
Pectins

Dr. Marie-Christine Ralet, Dr. Estelle Bonnin, Dr. Jean-François Thibault
Unité de Recherche sur les Polysaccharides, leurs Organisations et Interactions,
Institut National de la Recherche Agronomique B.P. 71627,
F-44316 Nantes Cedex 03, France; Tel.: + 33-2-40-67-50-64; Fax: + 33-2-40-67-50-66;
E-mail: ralet@nantes.inra.fr
Unité de Recherche sur les Polysaccharides, leurs Organisations et Interactions,
Institut National de la Recherche Agronomique B.P. 71627,
F-44316 Nantes Cedex 03, France; Tel.: + 33-2-40-67-50-58; Fax: + 33-2-40-67-50-66;
E-mail: bonnin@nantes.inra.fr
Unité de Recherche sur les Polysaccharides, leurs Organisations et Interactions,
Institut National de la Recherche Agronomique B.P. 71627,
F-44316 Nantes Cedex 03, France; Tel.: + 33-2-40-67-50-61; Fax: + 33-2-40-67-50-66;
E-mail: thibault@nantes.inra.fr

1	Introduction	354
2	Historical Outline	354
3	Chemical Structure of Pectin	355
3.1	Primary Structure	355
3.1.1	Backbone	355
3.1.2	Side Chains	357
3.1.3	Distribution of Structural Elements	357
3.1.4	Nonsugar Substituents	357
3.2	Occurrence and Distribution of Pectins in Cell Walls	358
3.3	Macromolecular Features	359
3.3.1	Molar Mass	359
3.3.2	Conformation	360
4	Pectin Characterization	361
4.1	Extraction	361
4.2	Analysis of Pectin Constituents	361
4.2.1	Galacturonic Acid	361

Polysaccharides and Polyamides in the Food Industry. Properties, Production, and Patents.
Edited by A. Steinbüchel and S. K. Rhee
Copyright © 2005 WILEY-VCH Verlag GmbH & Co. KGaA, Weinheim
ISBN: 3-527-31345-1

| 4.2.2 | Neutral Sugars | 361 |
| 4.2.3 | Substituents | 362 |

5 Pectin Biosynthesis ... 362

6 Molecular Genetics ... 363

7 Pectin Degradation ... 363
7.1	Chemical Degradation	364
7.1.1	Acidic Medium	364
7.1.2	Neutral or Alkaline Medium	364
7.2	Biodegradation by Pectolytic Enzymes	364
7.2.1	HG-degrading Enzymes	364
7.2.2	"Hairy" Region-degrading Enzymes	366
7.3	Pectins and Plant Product Transformation	368
7.3.1	Texture	368
7.3.2	Industrial Applications of Pectolytic Enzymes in Fruit and Vegetable Processing	369

8 Industrial Pectin ... 373
8.1	Current Raw Materials	373
8.2	Extraction of Pectin	373
8.3	Regulation and World Market	374

9 Pectin Gelling Properties and Applications ... 375
| 9.1 | HM Pectin | 375 |
| 9.2 | LM Pectin | 376 |

10 Stabilizing Properties ... 378

11 Outlook and Perspectives ... 378

12 References ... 380

ADI	admissible daily intake
AF	arabinofuranosidase
Ara	arabinose
Ara$f$	arabinofuranose
β-Gal	β-galactosidase
DAc	degree of acetylation
DAm	degree of amidation
DM	degree of methylation
EC	enzyme commission
endo-A	*endo*-arabinanase
endo-Gal	*endo*-galactanase
exo-PAL	*exo*-pectate lyase
exo-PG	*exo*-polygalacturonase

FAE	feruloyl-esterase
FerA	ferulic acid
Gal	galactose
GalA	galacturonic acid
Gal*p*	galactopyranose
Gal*p*A	galacturonic acid (pyranose form)
GDP-Glc	guanosine diphosphoglucose
GRAS	generally regarded as safe
[η]	intrinsic viscosity
HG	homogalacturonan
HM	high-methoxyl
^{1}H-NMR	proton nuclear magnetic resonance
HPAEC	high performance anion-exchange chromatography
HPSEC	high performance size-exclusion chromatography
IR	infra-red
kDa	kilodalton
LALLS	low-angle laser light scattering
LM	low-methoxyl
MALDI-TOF MS	matrix-assisted laser desorption/ionization time-of-flight mass spectrometry
MALLS	multi-angle laser light scattering
M_η	viscosity-average molar mass
M_w	weight-average molar mass
mRNA	messenger ribonucleic acid
PAE	pectin acetyl-esterase
PAL	*endo*-pectate lyase
PG	*endo*-polygalacturonase
PL	*endo*-pectin lyase
PME	pectin methyl-esterase
PMT	pectin methyl transferase
RG	rhamnogalacturonan
RGAE	rhamnogalacturonan acetyl-esterase
RG-galacturonohydrolase	rhamnogalacturonan-galacturonohydrolase
RG-hydrolase	rhamnogalacturonan-hydrolase
RG-lyase	rhamnogalacturonan-lyase
RG-rhamnohydrolase	rhamnogalacturonan-rhamnohydrolase
Rg_w	radius of gyration
Rha	rhamnose
Rha*p*	rhamnopyranose
rin	ripening inhibited
SS	slow set
UDP-Glc	uridine diphosphoglucose
URS	ultra-rapid set
Xyl	xylose
Xyl*p*	xylopyranose

1 Introduction

Pectins are a complex family of heterogeneous branched polysaccharides that arise from the primary cell walls and intercellular regions of higher plants. The term pectins or pectic substances describes a group of polysaccharides in which the presence of partly methyl-esterified galacturonic acid and rhamnose is a distinctive feature (Voragen et al., 1995). Pectins are classified as high- methoxyl (HM) and low-methoxyl (LM), depending on their degree of methyl-esterification.

Several aspects of plant physiology (Dey and Brinson, 1984), plant pathology (Darvill et al., 1980) and fruits and vegetables texture (Van Buren, 1991) are related to pectic substances, the remarkable gelling properties of which make them a key additive within the food industries. The origins of pectin production date back to the practice by preserve manufacturers of making a pectin-rich extract from apple residues–from the juice and cider industries–in order to supplement the gelling power of "difficult" fruits (May, 1990). The process begins with an acidic extraction of pectins from citrus or apple wastes. This first step leads to HM pectins that can be de-esterified by a further regulated acid, alkaline or enzymatic treatment to produce LM pectins. About 80% of the world production of HM pectins is used in the manufacture of jams and jellies, but numerous other applications (confectionery, milk products, etc.) exist for pectins in food manufacture.

2 Historical Outline

The word "pectin" comes from the Greek word "pectos", meaning "coagulum", and was first given by the French chemist Braconnot in 1824 to a substance he found present in many fruits. This substance was obtained as an extract, and produced a gel when mixed with alcohol and sugar. Since this initial report, considerable effort has been devoted by scientists to different aspects of the pectins, including molecular structure, location within the cell walls, and the rheological properties of the extracted material. The production of commercial pectins from selected raw materials began in the early twentieth century, and consequently an abundant literature exists. The work of Kertesz (1951) provided an extensive review of the knowledge of all aspects relating to pectins. At that time, the presence of methylated galacturonic acid residues linked in $\alpha(1,4)$ was seen as evidence for the existence of pectins. The presence of other components (methylpentose, xylose, acetic acid, etc.) and of accompanying arabinans and galactans was shown, but their incorporation into the pectic backbone was not proven.

The structure of the rhamnogalacturonans and of the side chains was elucidated during the early 1970s, at about the time of the first hypothesis on the structure of the cell wall. By 1965, pectins had been used in industry essentially as gelling agents in jams or jellies, though thereafter other application areas have been identified such that nowadays, pectins are used as fat substitutes, dietary fiber, and as stabilizers in acidified milk systems.

Today, reviews on all aspects of the pectins are published on a regular basis (see Voragen et al., 1995 for some references).

3
Chemical Structure of Pectin

Pectins contain a high proportion of galacturonic acid together with various neutral sugars, mainly rhamnose, arabinose and galactose. In addition, nonsugar components such as methanol, acetic acid and ferulic acid are often associated with pectic polysaccharides (Table 1).

3.1
Primary Structure

Pectin structure involves a succession of long "smooth" homogalacturonan (HG) regions and short "hairy" rhamnogalacturonan (RG) regions. Complex neutral sugars side chains are attached to the RG region (Figure 1). Galacturonic acid units present in the homogalacturonic and rhamnogalacturonic regions can be methyl-esterified at position 6, and acetylated at position 2 and/or 3.

3.1.1
Backbone

The dominant feature of pectins consists of a linear chain of α-(1→4)-linked D-galacturonic acid units (Aspinall, 1980; Voragen et al., 1995) (Figure 1). The HGs, which can also be referred to as polygalacturonic acid or pectic acid, have been isolated from various plant tissues (Voragen et al., 1995). Homogalacturonans can also be substituted by neutral sugars units or short chains. Xylogalacturonan is a galacturonan in which galacturonic acid units are substituted on O-2 and/or O-3 by xylose units or short 2-linked xylose side chains (Aspinall and Baillie, 1963; Bouveng, 1965; Siddiqui and Wood, 1976; Matsuura and Hatanaka, 1988; Ralet et al., 1993; Weightman et al., 1994; Schols et al., 1995; Renard et al., 1997; Le Goff et al., 2001) (see Figure 1).

The second fundamental feature of the pectin structure is the recurrent presence of small amounts of rhamnose (typically 1–4%) within the backbone. α-(1→4)-linked D-galacturonic acid units are interrupted by the insertion of (1→2)-linked rhamnopyr-

Tab. 1 Galacturonic acid (GalA), rhamnose (Rha), arabinose (Ara), xylose (Xyl), galactose (Gal) and ferulic acid (FerA) contents (mg g^{-1}), degree of methylation (DM) and degree of acetylation (DAc) of some acid-extracted pectins

Origin	GalA	Rha	Ara	Xyl	Gal	FerA	DM	DAc
Appel[a]	731	23	44	17	42	ND	74	<1
Citrus[a]	792	14	11	2	24	ND	72	<1
Beet[b]	624	54	51	2	92	ND	54	16
Beet[b]	558	28	124	3	47	7	64	23
Carrot[c]	613	43	35	3	79	ND	63	14
Grape[d]	533	58	77	6	44	ND	68	5
Pea[e]	469	102	144	155	75	ND	ND	ND
Potato[f]	350	ND	ND	ND	ND	ND	31	14
Pear[f]	380	ND	ND	ND	ND	ND	13	14
Glasswort[g]	289	22	462	3	43	17	65	45
Quinoa[h]	625	38	33	18	51	3	63	20

[a] Axelos and Thibault (1991a); [b] Guillon and Thibault (1988); [c] Massiot et al. (1988); [d] Saulnier and Thibault (1987); [e] Renard et al. (1997); [f] Voragen et al. (1986); [g] Renard et al. (1993a); [h] Renard et al. (1999); ND, Not determined.

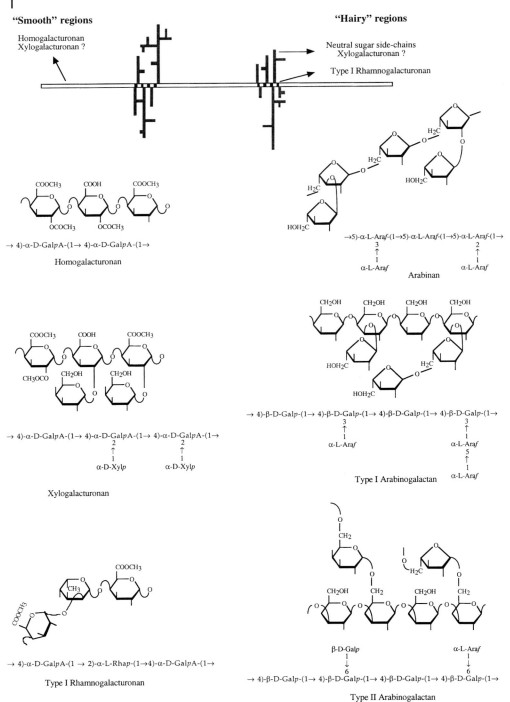

Fig. 1 Schematic representation and chemical structure of the main structural elements of pectins.

anosyl residues giving a type I RG (Aspinall et al., 1967a, b, 1968; Schols et al., 1990a, b) (see Figure 1). It has been established that the anomeric configuration of the rhamnosyl residues in these regions is α, which implies a chain deviation of 62° (Lau et al., 1985, 1987). Another type of RG (rhamnogalacturonan II), which exhibits an extremely complex structure with at least 12 different monomers, is a very minor component of plant cell walls (O'Neill et al., 1990).

3.1.2
Side Chains

Neutral sugar side chains are generally associated with type I RGs. The most commonly found sugars are galactose and arabinose, which form complex chains of various degree of polymerization. Side chains are usually linked on O-4, and sometimes on O-3, of the rhamnosyl units (Aspinall, 1980; Selvandran, 1985). The proportion of branched rhamnosyl residues is variable: 10–50% in carrot, 30% in pea hull, 60% in beet, and 25–100% in apple (Voragen et al., 1995). Among neutral sugars, side chain arabinans and arabinogalactans (type I and II) can usually be distinguished (Selvendran and O'Neill, 1987; Van Buren, 1991).

Arabinans are branched polysaccharides with a backbone of α-(1 → 5)-linked L-arabinofuranosyl units carrying α-L-arabinofuranosyl single units or short α-(1 → 3)-linked L-arabinofuranosyl side chains on O-3 and/or O-2 (see Figure 1). Pectins with arabinans attached have been isolated from several sources (Voragen et al., 1995).

Type I arabinogalactans exhibit a linear β-(1 → 4)-linked D-galactopyranosyl backbone. α-L-Arabinofuranosyl single units or short α-(1 → 5)-linked L-arabinofuranosyl side chains, which can be themselves substituted on O-3 by α-L-arabinofuranosyl units, can be attached on O-3, or sometimes on O-6, of β-D-galactopyranosyl residues (Figure 1).

Type II arabinogalactans are highly branched polymers with a backbone of β-(1 → 3)-linked D-galactopyranosyl units carrying short β-(1 → 6)-linked D-galactopyranosyl side chains on O-6. Galactosyl residues from main and secondary chains can be substituted by α-L-arabinofuranosyl single units, β-D-galactopyranosyl single units or short α-(1 → 5)-linked L-arabinofuranosyl side chains (Figure 1).

3.1.3
Distribution of Structural Elements

Numerous studies about the exact location of rhamnosyl residues have been carried out. It has been shown, mainly by using pectolytic enzymes (see Section 7.2) (De Vries et al., 1982), that rhamnosyl units are not randomly distributed along the pectic molecule but in fact are concentrated in blocks of rhamnogalacturonic "hairy" regions separated by long (~72–100 residues) (Thibault et al., 1993) unsubstituted "smooth" regions containing almost exclusively galacturonic acid units. Pectic molecules are therefore constituted by an alternation of "smooth" HG regions and of "hairy" RG regions carrying neutral sugar side chains (Voragen et al., 1995) (see Figure 1).

3.1.4
Nonsugar Substituents

Methyl-esterification of carboxyl groups and acetyl-esterification of alcoholic functions are the two main substitutions naturally present in pectic substances. Commercial pectins can also be amidated through chemical modifications (May, 1990; Pilnik and Voragen, 1992).

Carboxyl groups are partly methyl-esterified, and the degree of methylation (DM; the percentage of carboxyl groups esterified with methanol) permits the pectins to be classified. If >50% of the carboxyl groups are methylated, the pectins are called high-methoxyl (HM) pectins; if <50% are methylated, they are called low-methoxyl (LM) pectins. Native pectins are usually highly methylated (see Table 1). LM pectins are generally obtained by controlled acidic de-esterification but other means, including alkali, enzymes and ammonia can also be used. Treatment of pectin with acid, alkali, or acidic microbial (*Aspergillus japonicus, Aspergillus niger, Aspergillus foetidus*) pectin methylesterases (PME) leads to pectins with a random-like distribution of free carboxyl groups (Kohn, 1975; Kohn et al., 1983; Thibault and Rinaudo, 1985), whereas the action of alkaline PMEs from higher plants (tomato, orange, alfalfa, apple) and from fungi (*Trichoderma reesei*) results in a blockwise arrangement of free carboxyl groups (Kohn et al., 1983; Thibault and Rinaudo, 1985; Denès et al., 2000). The distribution pattern of free and esterified carboxyl groups has a profound effect on gelling properties (Kohn et al., 1983; Thibault and Rinaudo, 1985; Limberg et al., 2000a; Ralet et al., 2001). Quantitative information about the distribution of free carboxyl groups have recently been obtained (Daas et al., 1998, 1999, 2000, 2001a, b; Limberg et al., 2000a, b) (see also Section 3.2.3). The method using ammonia produces a different type of LM pectins in which some carboxylic groups have been amidated. The degree of amidation (DAm) is defined as the number of amidated galacturonic acid units for each 100 galacturonic acid units. The resulting amidated LM pectins exhibit peculiar gelling properties (see Section 9). A blockwise distribution of the amide groups was suggested (Racapé et al., 1989).

Secondary alcoholic functions of galacturonic acid residues may also be acetyl-esterified, though the precise location (on O-2 and/or O-3) of the acetyl groups is still unknown (Voragen et al., 1995). The degree of acetylation (DAc) is defined as the number of acetyl groups for each 100 galacturonic acid units. DAc is generally low in native pectins, but pectins with high DAc such as sugar beet, pear, carrot, glasswort and potato pectins, have been reported (see Table 1).

The presence of ferulic acid-esterifying arabinose and galactose units in the pectin side chains has been observed in the Chenopodiacea family (sugar-beet, spinach, glasswort, quinoa) (see Table 1) (Fry, 1983; Rombouts and Thibault, 1986; Renard et al., 1993a, 1999; Ralet et al., 1994a). This phenolic acid was shown to be linked on the O-2 of the arabinosyl and/or on the O-6 of the galactosyl residues (Ishii and Tobita, 1993; Colquhoun et al., 1994; Micard et al., 1997).

3.2
Occurrence and Distribution of Pectins in Cell Walls

Cell walls are major structural and functional components surrounding plant cells. Plant cell walls are composed of structurally complex macromolecular assemblies of predominantly carbohydrates and phenolics in a diverse array of linkages. Pectic substances are essentially present, together with cellulose and hemicellulosic xyloglucans, in dicots in which the parenchymatous tissues form the main elements of fruits and vegetables. Within these tissues the cells are not (or are only minimally) differentiated, and the cell walls display the composition and properties of primary cell walls (thinness, extensibility, and high hydration).

The plant cell wall is a dynamic compartment composed of various zones that differ in their structure and their setting schedule: the middle lamella forms the interface between the primary walls of neighboring cells; the new primary cell wall, which is formed during cell division; and finally, when cells have achieved their final size and shape, the secondary cell wall. Primary and secondary walls allow the cells to withstand external and internal stresses (Jarvis and McCann, 2000). Pectins are essentially present in the middle lamella and the primary cell walls. The middle lamella is the first compartment excreted by plant cells; in fact, contact between cells is maintained through the middle lamella, and the cell corners are often filled with pectin-rich polysaccharides. Pectins present in the middle lamella are thought to be LM in nature and highly interconnected through calcium ions, thereby helping to regulate cell–cell adhesion (Carpita and McCann, 2000). Primary cell walls are composed of a framework of cellulose microfibrils and xyloglucans (typically 5 to 10 strata) embedded in a matrix of pectic polysaccharides (McCann and Roberts, 1991; Carpita and Gibeaut, 1993; Carpita and McCann, 2000). Some primary cell walls also contain large amounts of various structural proteins which can form intermolecular bridges (Carpita and Gibeaut, 1993; Carpita and McCann, 2000). Our current appreciation of the complexity and dynamic of the primary cell wall has been essentially formed during the past decade by direct microscopic visualization of cell walls using monoclonal antibody probes (Liners et al., 1989; Knox, 1997; Mogami et al., 1999; Willats et al., 2000). Current knowledge of the molecular mechanisms underlying the coordination of polysaccharides biosynthesis and cell wall assembly remains very limited, however. In secondary cell walls, the cellulose microfibrils are interlocked with glucuronoarabinoxylans instead of xyloglucans. These cell walls are pectin-poor and have very little structural protein, but they can accumulate extensive interconnecting networks of phenylpropanoids, particularly as the cells stop expanding (Carpita and McCann, 2000). As with the primary cell walls, current understanding of the mechanisms of the formation of secondary cell walls is similarly restricted.

3.3
Macromolecular Features

The macromolecular characteristics of pectins mainly include their molar mass and conformation, and these are determinants of the industrial applications of pectins.

3.3.1
Molar Mass

Molar mass measurement is difficult because of the heterogeneity, the water (in)-solubility, and the aggregation of pectin; the polyelectrolyte character of pectin must also be taken into account (Thibault et al., 1991). Molar mass values, which range from a few tens to several hundreds of kilodaltons (kDa), depend not only on the method used but also on the origin of the pectin and the means of extraction (Table 2). Owens et al. (1946) established the first Mark-Houwink-Sakurada equation for pectins in 0.155 M NaCl ($[\eta] = 1.4 \times 10^{-6} M_\eta^{1.34}$, where $[\eta]$ is the intrinsic viscosity in $dL\,g^{-1}$, and M_η is the viscosity average molar mass). The particularly high power index value suggests a considerable chain stiffness. Recent developments in high-performance size-exclusion chromatography coupled with laser light scattering or viscosimetric detection, led to an improvement in pectin characterization, although aggregation phenomena disturb light-scattering data. Recent results permit the establishment of Mark-Houwink-Sakudara equations with power index values of around 0.8–0.9, which are in better agreement with the semi-flexible character of pectins (Rinaudo, 1996).

Tab. 2 Weight-average molar masses (M_w) and macromolecular characteristics of some pectic samples

Origin	DM	$[\eta]$ [mL g^{-1}]	M_w [kDa]	Method used for M_w	Rg_w [nm]	Reference
Citrus	68	604	178	HPSEC-viscosimetry univeral calibration		Deckers et al. (1986)
Citrus	38		380		32–37	Acelos et al. (1987)
Apple	65		5330	HPSEC-MALLS		Kontamins and Kokini (1990)
Citrus	70		97	HPSEC-sedimentation		Harding et al. (1991)
Citrus	70		184	HPSEC-MALLS	53	Corredig et al. (2000)
	71		188	HPSEC-MALLS	54	Corredig et al. (2000)
	27–33		110	HPSEC-MALLS	57	Corredig et al. (2000)
	27–33	(DAm 20–25)	185	HPSEC-MALLS	50	Corredig et al. (2000)
Citrus	78	345	176	Seidmentation epuilibrium		Morris et al. (2000)
			186	HPSEC-MALLS		Morris et al. (2000)
	54	402	217	Sedimentation equilibrium		Morris et al. (2000)
			211	HPSEC-MALLS		Morris et al. (2000)
	28	228	165	Sedimentation equilibrium		Morris et al. (2000)
			187	HPSEC-MALLS		Morris et al. (2000)

DAm, degree of amidation.

3.3.2
Conformation

The linkage between galacturonic acid residues is axial–axial, and this imposes a severe conformational constraint; consequently, HG molecules tend to adopt a pseudohelical conformation. The overall conformation of a polygalacturonic chain is still uncertain. According to the authors, the investigation methods, and the counterion nature (Na$^+$, H$^+$, Ca^{2+}), right-handed 3_1 or 2_1 helices or left-handed 3_2 helices were proposed. Molecular modeling indicated that right or left-handed 3-fold helices, 2-fold helices and even 4-fold helices can be generated with low energy values and easy interconversion (Cros et al., 1992). These four helices have a maximum extension, as indicated by an almost constant magnitude of the pitch per monomer, of 0.435 nm. The calculated difference in energy between the 2-fold and 3-fold conformation for Ca^{2+}-homogalacturonates indicates that both types of helical conformation are almost equally favorable. Thus, depending on the environment, one or the other form may exist (Braccini et al., 1999; Pérez et al., 2000). Walkinshaw and Arnott (1981a, b) have shown by X-ray diffraction that the Na$^+$- and Ca^{2+}- salts of pectate (DM < 5) adopt a 3_1 conformation in the dry state, and that the same was probably true for the acid and methyl forms. Recently, solid-state NMR has claimed to provide evidence for the presence of 2_1 and 3_1 helices in Na$^+$-homogalacturonate, 3_1 helices for H$^+$-homogalacturonate, and exclusively 3_1 helices for Ca^{2+}-homogalacturonate (Renard and Jarvis, 1999). The evolution of the spectra of the Ca salts of partially methylated HGs show the disappearance of the 2_1 helices as the DM increases.

Experimental results from viscosimetry, laser light-scattering, small-angle X-ray or neutron scattering have shown that pectins are semi-flexible polymers with a persistence length (2–30 nm) that is intermediate between very flexible polymers such as amylose (persistence length <2 nm) and very stiff polymers such as xanthan (persistence length >40 nm). Axelos and Thibault (1991a), Cros et al. (1996) and Catoire et al. (1997) clearly showed that rhamnose insertion does not induce any change in the global orientation of the pectin chain.

4
Pectin Characterization

Different methods can be used to quantify specific constituent of pectins – typically galacturonic acid – in entire cell walls. However, the fine structural characterization of pectins has, until now, only been achieved on extracted and purified molecules.

4.1
Extraction

Chemical sequential extractions have been widely used and reviewed in detail (Selvendran and O'Neill, 1987; Thibault et al., 1994; Voragen et al., 1995). These sequential extractions aim at the release of pectins bound to the cell wall matrix by different means, without extensive degradation. Pectin fractions are usually obtained by sequential extraction of the purified cell walls with: (1) cold and/or hot water or buffer solutions; (2) cold and/or hot solutions of chelating agents; (3) hot diluted acids; and (4) cold dilute alkali. Enzymatic extractions have also been used, mainly for structural studies, and to obtain a better understanding of the way that pectins are bound to the cell wall matrix (Renard et al., 1991a, b, 1993b; Schols and Voragen, 1996).

4.2
Analysis of Pectin Constituents

The most important chemical and physicochemical characteristics of pectins for manufacturers and users are their galacturonic and neutral sugar contents, their degree of substitution (methylation, acetylation, amidation, and eventually feruloylation), and their molar mass (see Section 3.3.1).

4.2.1
Galacturonic Acid

Colorimetric procedures are the most widely used methods for galacturonic acid content determination. In concentrated acidic medium, furfuric derivatives are generated and conjugated with aromatic compounds (*m*-hydroxybiphenyl, 3,5-dimethylphenol) (Blumenkrantz and Asboe-Hansen, 1973; Scott, 1979) to give colored products that are quantified by colorimetry. The *m*-hydroxybiphenyl method was automated by Thibault (1979). Titrimetry is also widely used, and is the reference method for commercial pectins. Pectins, in their acidic form, are titrated by a dilute alkali, saponified, and titrated again. This "double titration" gives access to both DM and galacturonic acid content, but it suffers from a lack of specificity, and cannot be used for acetylated pectins. The method also requires large quantities of pectins (from 100 mg to 1 g) for its operation.

4.2.2
Neutral Sugars

The most widely used method for neutral sugars determination is gas-liquid chromatography after hydrolysis of the pectin or cell wall sample in their constitutive neutral sugars and derivatization in volatile compounds. Strong acids (usually trifluoroacetic acid or sulfuric acid) can break the glycosidic linkages, releasing the individual monomers. Precise hydrolysis conditions have to be carefully determined, as the various glycosidic linkages present in pectins show differing degrees of resistance to acid hydrolysis (Micard et al., 1996). When the monosaccharides are produced they can be chemically reduced with borohydride, and then acetylated to produce alditol acetates (Blakeney et al., 1983; Englyst and Cummings, 1984). The alditol acetates derived from different sugars are volatile at different temperatures and can be separated on that basis by gas-liquid chromatography. Other methods of breaking the glycosidic bonds include methanolysis (Quemener and Thibault, 1990).

Complex polysaccharides cannot be fully identified from their derivative content alone, though their linkage structure can be investigated using methylation analysis (Hakomori, 1964). With this approach, the polysaccharides are chemically methylated on free hydroxyl groups before acid hydrolysis and acetylation; the partially methylated alditol acetates are then separated by gas-liquid chromatography. Unequivocal determination of the structure of the partially methylated alditol acetates is then possible using mass spectrometry.

4.2.3
Substituents

Many methods have been developed to quantify methoxyl and acetyl substituents in pectins. Some direct determinations of the DM have been proposed using ^{1}H-NMR (Grasladen et al., 1988) or by infra-red (IR) (Gnanasambandam and Proctor, 2000). However, most of the methods require a saponification to be performed before the quantification of methanol and/or acetic acid by various methods, including gas-liquid chromatography, high-performance liquid chromatography, enzymatic methods, titrimetry or colorimetry after derivation (Thibault et al., 1994; Voragen et al., 1995). The DM is also commonly determined, together with the galacturonic acid content by titration (see Section 4.2.1).

The amide content can be measured by quantifying ammonia released after alkaline distillation, or by one of the methods used for total nitrogen determination (Voragen et al., 1995).

The ferulic acid content can be determined spectrophotometrically, by high-performance liquid chromatography after saponification and extraction, or by gas-liquid chromatography after derivatization (Micard et al., 1994; Ralet et al., 1994b; Ralph et al., 1994; Saulnier et al., 1995).

Recently, the distribution of methyl esters along the HG backbone has been extensively studied. Pure *endo*-polygalacturonase, *exo*-polygalacturonase or *endo*-pectin lyase were used to produce specific oligosaccharides. High-performance anion-exchange chromatography (HPAEC) at slightly acidic pH (5 or 6) was used to separate nonesterified and partially-methylesterified galacturonic acid oligomers of various degrees of polymerization arising from enzymatic digests. Matrix-assisted laser desorption/ionization time-of-flight mass spectrometry (MALDI-TOF MS) was then used to: (1) identify the different oligosaccharides produced; (2) discriminate between pectins of similar DM but which differed by the repartition of methoxyl groups along the homogalacturonic backbone; and (3) introduce the concept of "degree of blockiness" in pectin (Daas et al., 1998, 1999, 2000; 2001a, b; Limberg et al., 2000a, b; Voragen et al., 2001)

5
Pectin Biosynthesis

The plant Golgi apparatus is the site of synthesis of the noncellulosic polysaccharides. Two types of enzymes are required for the synthesis of plant cell-wall polysaccharides. The first type catalyzes the production of the energetically activated glycosyl residues for polysaccharide synthesis, while the second group transfers the glycosyl residues from the activated donors onto a growing polysaccharide chain (Mohnen, 1999). From synthesized UDP-Glc and GDP-Glc, pathways for nucleotide sugar interconversion produce various nucleotide sugars through enzyme-catalyzed reactions (Carpita and McCann, 2000; Gibeaut, 2000). Many of these interconversion enzymes (epimerases, reductases, decarboxylases) are membrane-bound and localized to the endoplasmic reticulum–Golgi apparatus. The accurate location of the polysaccharide synthases, on or within the Golgi membranes, has not yet been clearly established, and at least three possible means exist for the delivery and utilization of nucleotide sugars by the plant Golgi apparatus (Gibeaut, 2000). The synthesis of complex polysaccharides must clearly be coordinated with the transport of certain nucleotide sugars into the Golgi apparatus.

Current knowledge of pectic polysaccharide biosynthesis is rather limited. There are at least 46 glycosyltransferases required for the synthesis of pectin, based on the one linkage–one enzyme assumption and on the structure of pectin (see Section 3) (Mohnen, 1999). An α-1,4-galacturonosyl transferase (EC 2.4.1.43) involved in HG biosynthesis has been identified and partially characterized (Doong et al., 1995; Doong and Mohnen, 1998; Scheller et al., 1999). No studies have been reported on other galacturonosyltransferases that synthetize the backbone of type I RG, nor on rhamnosyltransferases. The synthesis of HGs appears to be coordinated with its methyl-esterification. Methyl-ester groups (from S-adenosyl-methionine) are added to galacturonans by the action of homogalacturonan- (pectin-) methyltransferase (PMT, EC 2.1.1.18) during pectin synthesis (Bruyant-Vannier et al., 1996; Goubet et al., 1998; Ishikawa et al., 2000). A rhamnogalacturonan-acetyltransferase (Pauly and Scheller, 2000) and a galactosyltransferase (Goubet and Morvan, 1993, 1994) were also identified. After the biosynthesis steps, polysaccharides destined for the cell wall are probably packed into secretory vesicles, transported to the cell surface, and integrated within the cell wall.

6
Molecular Genetics

Genetic engineering may concern not only endogenous enzymes for biosynthesis and biodegradation, but also (more directly) pectins. As shown in Section 5, the biosynthetic pathways of pectins are not clear. However, gene technology and protein sequencing form part of the studies of biosynthesis and may help to assign one function to one gene. In the same way, molecular biology is a powerful tool which has enabled the role of endogenous degradative enzymes in fruit ripening and texture to be identified (see Section 7.3.1). In Section 9 it is shown that the industrial applications of pectins have their origins in the compositions and macromolecular characteristics, though at present these techniques are not capable of producing modified pectins useful for industrial applications.

7
Pectin Degradation

Pectic substances can undergo two main types of degradation, depolymerization and de-esterification, both of which can be achieved either chemically or enzymatically.

7.1
Chemical Degradation

One peculiarity of pectins compared with other polysaccharides (with the exception of alginates) is their relative stability in acidic medium (Pilnik and Zwiker, 1970). Pectins show their greatest stability at slightly acidic pH (3–5) and low temperature, but at higher pH and/or temperature, they can undergo degradation either by hydrolysis or β-elimination.

7.1.1
Acidic Medium

An acidic medium (pH 1–3) favors the hydrolysis of glycosidic and ester linkages. At low temperature, de-esterification reactions are predominant; an increase in temperature promotes the hydrolysis of glycosidic bonds. The nature of the linkages and the sugars involved also plays a major role in the susceptibility of pectin to acid hydrolysis. For example, bonds between arabinofuranosyl rings are more acid-labile than bonds between other sugar moieties. Furthermore, linkages between galacturonic acid residues are very acid-resistant; they are notably more stable than linkages between uronic acid and a neutral sugar. Neutral sugar side chains are therefore degraded more easily than rhamnogalacturonic regions, with homogalacturonic regions being the most acid-resistant zones of pectic molecules (Thibault et al., 1993).

7.1.2
Neutral or Alkaline Medium

In a neutral or alkaline medium, two degradation reactions can occur with pectin: (1) saponification of esters (methyl, acetyl and feruloyl); and (2) β-elimination. The latter reaction induces the water-free breaking of glycosidic bonds between galacturonic acid residues and the appearance of a double bond between galacturonic acid C_4 and C_5, conjugated with that carried by the carbonyl group. As one of the conditions for β-elimination is the presence of methyl esters, the saponification and β-elimination reactions compete one with another. At low temperature ($< 4\,°C$) and in alkaline medium, pectins are de-esterified without any major incidence on their degree of polymerization. A temperature increase favors β-elimination reactions, and these become predominant at high temperature and neutral to slightly acidic medium (Albersheim et al., 1960). HM pectins are particularly prone to β-elimination.

7.2
Biodegradation by Pectolytic Enzymes

Due to the complexity of pectins, many different enzymes are required to achieve their degradation (Figure 2). These enzymes occur in many higher plants but are also produced by several microorganisms. They are especially involved in the colonization phenomena developed by the phytopathogenic fungi, and are classified into two groups according to their specificity towards the pectic substrate: (1) HG-degrading enzymes; and (2) "hairy" region-degrading enzymes.

Some authors (Sakai, 1992) also referred to "protopectinases" as enzymes capable of solubilizing a fraction of pectins that remains in the cell wall after extraction with water and a calcium-chelating agent. Several "protopectinases" have been studied in detail and have been shown to exhibit the same specificity as other pectinases towards the different subunits of pectins.

7.2.1
HG-degrading Enzymes

Three different reactions are catalyzed by enzymes towards the "smooth" regions of

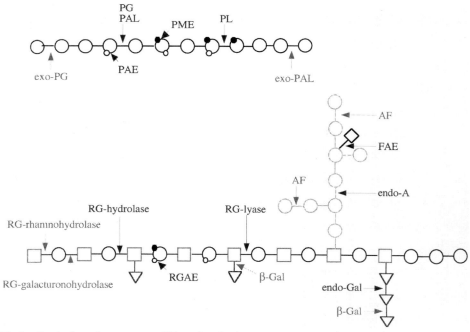

Fig. 2 Pectin-degrading enzymes. PG: *endo*-polygalacturonase; PAL: *endo*-pectate lyase; PME: pectin methyl-esterase; PL: *endo*-pectin lyase; *exo*-PG: *exo*-polygalacturonase; *exo*-PAL: *exo*-pectate lyase; PAE: pectin acetyl-esterase; GR-hydrolyse: rhamnogalacturonan-rhamnohydrolase; RG-galacturonohydrolase: rhamnogalacturonan-galacturonohydrolase; GR-lyase: rhamnogalacturonan-lyase; RGEA: rhamnogalacturonan acetyl-esterase; endo-A: *endo*-arabinase; AF: arabinofuranosidase*; FAE: feruloyl-esterase*; endo-Gal: *endo*-galactanase; β-Gal: β-galactosidase*; ○: galacturonic acid; ▫: rhamnose; ●: methoxyl group; ○: arabinose; ○: acetyl group; ▽: galactose; ◇: ferulic acid; *: enzymes acting on oligosaccharides.

pectins: saponification is ensured by esterases, while depolymerization can be achieved either by hydrolases or by lyases.

Pectin Esterases

Pectin methylesterases (or pectin methylhydrolases, EC 3.1.1.11) catalyze the release of methoxyl groups from the "smooth" regions of pectins. They have been described in several higher plants such as tomato, citrus or apple, but are also produced by various bacteria and fungi.

Plant pectin methylesterases have a pH optimum around 7. They release a methoxyl group adjacent to a free galacturonic acid and slide along the HG backbone to produce pectins with blocks of free carboxyl groups. As outlined in Section 3.1.4, this blockwise distribution of methoxyl groups confers on pectins an extreme sensitivity towards calcium ions.

In contrast, fungal pectin methylesterases are more active at more acidic pH (~4–5), and saponify methyl esters in a more or less random fashion. Therefore, pectins produced by fungal methylesterase treatment do not exhibit similar sensibility to calcium as do those produced by plant methylesterase treatment (Ralet et al., 2001).

Acetyl groups from HGs can be released by pectin acetylesterases. In contrast to methylesterases, very few acetylesterases have yet been described. Such activities have been found in plants such as orange (Williamson, 1991) and mung bean (Bordenave et al., 1995), as well as in microorganisms such as *Aspergillus niger* (Searle-van Leeuwen et al., 1996).

Polygalacturonases
Polygalacturonases hydrolyze the linkage between two unsubstituted galacturonic acids in the "smooth" regions of pectins. They are widely distributed among microorganisms and plants, and have been the subject of many publications due to their extensive technological applications (see Section 7.3.2). They can be separated into two groups according to their mode of action:

1) The *endo*-polygalacturonases poly (α-1,4-D-galacturonide) glycanohydrolases (EC 3.2.1.15) act in the chain of homogalacturonan to release mainly monomer, dimer and trimer of galacturonic acid as end products. When acting according to a single chain attack, oligomers are produced from the beginning of the reaction; when acting according to a multi-chain attack, products with a high degree of polymerization are first released and are then further degraded in oligomers.
2) The *exo*-polygalacturonases generally attack the nonreducing end of the substrate and slide along the chain up to a methoxylated residue, a rhamnose residue or a side chain that block their action. Two types of *exo*-polygalacturonases can be distinguished according to the reaction products: the poly (1,4-α-D-galacturonide) galacturonohydrolases (EC 3.2.1.67) release only the monomer, while the poly (1,4-α-D-galacturonide) digalacturonohydrolases (EC 3.2.1.82) release the dimer.

Polygalacturonases belong to the family 28 of glycosyl hydrolases (Henrissat, 1991) and act according to an inverting mechanism (Biely et al., 1996); this means that the product released from the cleavage of an α linkage is in a β anomery.

Lyases
Lyases catalyze the breakdown of HG by β-elimination cleavage, producing Δ4,5 unsaturated oligosaccharides at the nonreducing end. Most often, they originate from bacteria, but lyases of fungal origin are also known.

Pectate lyases (poly-α-1,4- D-galacturonide lyase) are specific for polygalacturonic acid or LM-pectins as they can accommodate methyl-esterified galacturonic acid residues in their active site. They have a high pH optimum (8–9) and an absolute requirement for calcium ions. In this group, both *endo*- (EC 4.2.2.2) and *exo*-enzymes (EC 4.2.2.9) can be found, the latter enzymes releasing unsaturated dimer.

Pectin lyases (poly-α-1,4- D-methoxylgalacturonide lyase; EC 4.2.2.10), split the glycosidic linkage between two methyl-esterified galacturonic acid residues, and therefore are specific for HM-pectins. They function optimally at pH ~6, and their affinity for the substrate decreases with decreasing DM. This last group occurs predominantly in fungi.

7.2.2
"Hairy" Region-degrading Enzymes
The degradation of the "hairy" regions requires enzymes that are active on the rhamnogalacturonic backbone and on the neutral sugar side chains. Feruloyl esterases capable of releasing ferulic acid from the side chains have also been described.

RG I-degrading Enzymes

During the past decade, these enzymes have been studied in a commercial preparation from *Aspergillus aculeatus* (Schols et al., 1990a). The first enzyme found was a hydrolase which cuts the $1 \rightarrow 2$ linkage between the galacturonic acid and the rhamnose residues, releasing oligomers with a strict alternation of galacturonic acid and rhamnose and a rhamnose residue at the nonreducing end.

Since then, other RG I-degrading enzymes from the same microorganism have been described:

- A RG lyase which splits in an *endo* fashion the $1 \rightarrow 4$ linkage between the rhamnose and the galacturonic acid units and releases oligomers with a reducing rhamnose and an unsaturated galacturonic acid at the nonreducing end (Mutter et al., 1996).
- A RG rhamnohydrolase which is an *exo*-enzyme releasing rhamnose from the nonreducing end of the polymer (Mutter et al., 1994).
- A RG galacturonohydrolase which is an *exo*-enzyme releasing galacturonic acid from the nonreducing end of the polymer (Mutter et al., 1998).
- A RG acetylesterase which releases acetyl groups from the rhamnogalacturonic region (Searle-van Leeuwen et al., 1992). It acts synergistically with the rhamnogalacturonan hydrolase or the rhamnogalacturonan lyase for the degradation of apple "hairy" region.

Until now, none of these enzymes has been referred by the Enzyme Commission.

Arabinan-degrading Enzymes

These enzymes have been reviewed by Beldman et al. (1997):

- The *endo*-α-L-arabinanases (EC 3.2.1.99) hydrolyze randomly the α-$(1 \rightarrow 5)$ linkage in the arabinan chains of pectin. Monomer, dimer or trimer of arabinose are released depending on the bacterial of fungal origin of the enzyme.
- The *exo*-arabinanases release an arabinosyl residue from the nonreducing end of arabinan or arabinogalactan, whatever the linkage, $1 \rightarrow 3$ or $1 \rightarrow 5$. They are not active towards low molar mass substrates. They have been isolated from numerous strains of bacteria (optimal pH 6–7) or from fungi (optimal pH 2–3).
- The α-L-arabinofuranosidases (EC 3.2.1.55) hydrolyze $1 \rightarrow 3$ or $1 \rightarrow 5$ linkages only in arabino-oligosaccharides or *p*-nitrophenyl-α-L-arabinofuranoside. When acting on oligomers, they release the nonreducing arabinosyl unit.

Galactan-degrading Enzymes

This group also includes *endo*-enzymes, *exo*-enzymes and glycosidases:

- The *endo*-(1,4)-β-D-galactanases (EC 3.2.1.89) hydrolyze randomly the $1 \rightarrow 4$ linkage between two galactose units in the galactan backbone, releasing monomers or galacto-oligosaccharides with a DP lower than 4.
- Some *endo*-(1,6)-galactanases were also described, as for example in *Aspergillus niger* (Brillouet et al., 1991).
- During the past decade, some *exo*-galactanases specific for $1 \rightarrow 4$ or $1 \rightarrow 3$ linkages were found in some bacteria (*Bacillus subtilis*; Nakano et al., 1990), fungi (*Aspergillus niger*; Pellerin and Brillouet, 1994; Bonnin et al., 1995) or plants (*Lupinus angustifolius*; Buckeridge and Reid, 1994).
- The β-D-galactosidases (EC 3.2.1.23) release galactosyl units from the nonreducing end of galactan or arabinogalactan of low molar mass. They were isolated from

several sources and studied in detail because of their industrial application in lactose hydrolysis.

Feruloyl Esterases

Ferulic acid esterifies either galactose or arabinose residues of pectic side chains (see Section 3.1.4). Some enzymes able to release these substituents (feruloyl esterase) have recently been purified and characterized. As reviewed by Williamson et al. (1998), they exhibit various preferences with regard to the nature of the sugar moiety linked to ferulic acid (arabinose or galactose) and to the size of the feruloylated substrate. These enzymes originate mainly from *Aspergillus niger* strains, and can be classified according to their preferred substrate.

7.3
Pectins and Plant Product Transformation

In the cell wall, the pectins form a 3-D network which does not hinder the cellular extension or differentiation. This is produced by molecular events, leading to textural changes.

7.3.1
Texture

Plant product texture is strongly influenced by cell-wall structure, cell-to-cell adhesion, and turgor pressure generated by the cell. The mechanical strength and the texture of cell walls change dramatically during the fruit-softening process, including ripening and storage. Generally, fruit and vegetable cell walls contain a large amount of pectins (see Section 3.2), and the relationship between pectin degradation and softening has been studied extensively (Van Buren, 1991; Wakabayashi, 2000). During processing, texture modifications are primarily due to β-elimination reactions (Voragen et al., 1995).

Changes During Ripening and Storage

An extensive solubilization and depolymerization of pectins, correlated with a decrease in fruit tissue firmness, has been observed during the ripening of many fruits such as tomato, kiwi, blackberry, plum and strawberry (Redgwell et al., 1997). Enzymes participating in pectin degradation include polygalacturonases – mainly *endo*-polygalacturonases – and pectin methylesterases. Both enzyme activities were shown to increase drastically during the ripening of several fruits (Huber, 1983; Campbell et al., 1990; Huber and O'Donoghue, 1993; Blumer et al., 2000; Wakabayashi et al., 2000). However, molecular genetic approaches have revealed that the reduction of *endo*-polygalacturonase mRNA accumulation in tomato did not prevent either pectin solubilization or fruit softening (Smith et al., 1990). Moreover, increasing the *endo*-polygalacturonase activity in the tomato *rin* (ripening inhibited) mutant led to extensive depolymerization and solubilization of pectins, without promoting fruit softening (Giovannoni et al., 1989). Various factors, such as the cell-wall physico-chemical environment, are probably involved in the regulation of endogenous enzymatic activities (Almeida and Huber, 1999; Wakabayashi, 2000). The enzymatic degradation of other cell-wall polysaccharides such as xyloglucans could be involved in the early stage of fruit softening, while in the late ripening stages, *endo*-polygalacturonase-mediated pectin degradation might become predominant (Wakabayashi, 2000). The loss of galactose and/or arabinose during ripening is often very important (Voragen et al., 1995). The removal of neutral-sugar pectic side chains may facilitate HG degradation in ripening fruits (Wakabayashi, 2000).

Changes with Processing

The quality of processed fruits and vegetables depends greatly on the cell-wall structure of the tissues involved. Pectins will play a role in canned fruits and vegetables texture, pulps and purée consistencies, fruits and oleaginous seeds pressability and extractability, fruit juice and nectar filterability, clarity or cloud stability (see also Section 7.3.2), fibrous byproducts water-holding capacity, digestibility and nutritional value (Pilnik and Voragen, 1991; Voragen et al., 1995; Thibault et al., 2001; Thibault and Ralet, 2001). Fruit and vegetable processing often implies some cooking, leading to tissue softening and cell separation.

Pectin involvement in the texture of processed fruits and vegetables results from the interaction of three factors: (1) the degradation of pectins leading to tissue softening (see also Section 7.3.1); (2) the presence of endogenous pectin methylesterases; and (3) the presence of calcium ions. The main reaction occurring during cooking is β-elimination of intercellular pectins (Van Buren, 1991); the pectins are then partly solubilized and their molar mass decreases (Greve et al., 1994; Ng and Waldron, 1997). Ions can also be involved in softening either in a positive or a negative manner (Voragen et al., 1995). Calcium in particular, by favoring pectin intermolecular association, can limit texture losses. The calcium-binding strength depends on the distribution of the methoxyl groups along the pectic "smooth" regions, with a blockwise distribution leading to higher calcium binding properties (see also Section 3.1.4). Hence, a pre-cooking phase at 60–70 °C permits an activation of endogenous pectin methylesterases, leading to a blockwise de-esterification of homogalacturonic regions. This pre-cooking phase, together with calcium addition, limits excessive texture losses (Gierschner et al., 1994; Ng and Waldron, 1997). Similarly, vegetable blanching before freezing leads to pectin methylesterases activation and contributes to a well-preserved shape and texture after thawing (Fuchigami et al., 1995). This method of obtaining fairly rigid networks through calcium ions is also used to favor the pressability of pectin-rich products such as citrus peels or sugar-beet pulps (Voragen et al., 1995).

7.3.2
Industrial Applications of Pectolytic Enzymes in Fruit and Vegetable Processing

Commercial pectolytic preparations contain essentially the enzymes described above (Section 7.2), albeit at different ratios, and are used as processing aids in the industrial treatment of fruit and vegetable pulps or juices (Figure 3). The enzymes have long been used to increase juice yield and to clarify juices in the production of nonalcoholic juices, wine or cider. Pectic enzymes are also used in the maceration of fruit tissues to produce nectars, and also in liquefaction processes. The nature of enzymatic activities involved changes according to the chosen process (Table 3). It is important that the exogenous enzymes used in such productions should not hide the presence of endogenous enzymes in fruits or vegetables. Their effect can be negligible compared with that of added enzymes (as in clarification), or even inhibited (as in citrus juice preparations). However, the enzymes can also be exploited or reinforced, as for example in the preparation of preserved food or in sugar preserving.

Pressing

The successive steps of juice making are to crush the fruit, to press it, to centrifuge or filter and, according to the final product, to clarify, concentrate and pasteurize. The crushing step induces a partial solubilization of pectins. The pulp consists of a heterogeneous system in which the liquid phase is very viscous and the solid phase exhibits a very high water-holding capacity, which

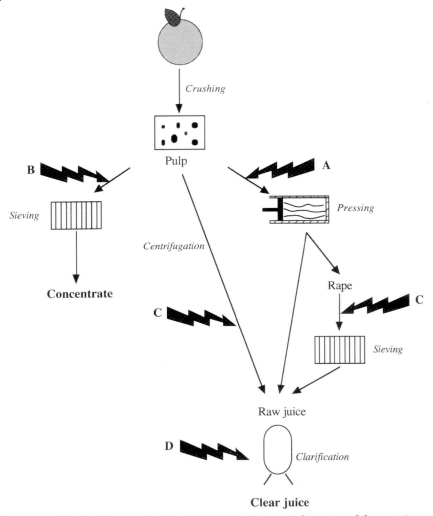

Fig. 3 Use of polysaccharide-degrading enzymes in the transformation of fruits and vegetables. A: pectinases for pressing; B: pectinases for maceration; C: pectinases + cellulases for liquefaction; D: pectinases for clarification. A: Pectinases for pressing; B: Pectinases for maceration; C: Pectinases + Cellulases for liquefaction.

limits the pressing yield. Therefore, the addition of enzymes at this step allows a partial degradation of the cell walls, with a subsequent increase in juice yield, as well as a better extraction of pigments and aroma. This process was initially developed for the production of berry juices (blackcurrant, strawberry, raspberry, grape). These fruits are very rich in pectins and give a highly viscous pulp that is impossible to filter or centrifuge on an industrial scale without the addition of enzymes.

More recently, the use of enzymes was also developed for apple juice production, and an

Tab. 3 Pectins clasification and gelation conditions

Pectin type	DM	DAm	Gelation conditions			Commercial name	Principal uses
			General conditions	pH	Setting time [min]		
HM*	74-77	0	Brix >55	3.1–3.4	1–3	Ultra-rapid set	Jams with whole fruits, „Dundee" jelllies
HM	71–84	0	Brix >55	3.0–3.3	3.0–7	Rapid set	„Classical" jams
HM	66–69	0	Brix >55	2.8–3.1	15–25	Medium-rapid set	Acid jams and jellies (raspberry)
HM	58-65	0	Brix >55	2.6–2.9	30–120	Slow set	Acid to very acid jams and jellies (black currant)
LM**	50	0	Calcium ions (100 mg g^{-1} of pectin at 30% soluble solids)			Slow set	Low-sugar products (hypocaloric jams and jellies, baby food)
LM	40	0	Calcium ions (50 mg g^{-1} of pectin at 30% soluble solids)			Medium set	Calcium-rich gelled products
LM	30	0	Calcium Ions (25 mg g^{-1} of pectin at 30% soluble solids)			Rapid set	
amidated LM	35	15	Calcium ions			Slow set	Yoghurt fruit bases
amidated LM	30	20	Calcium ions			Rapid set	Heat-reversible bakery toppings

HM, high-methoxyl; LM, low-methoxyl.

increasing number of apples is now stored for several months before being transformed. Despite the low storage temperature, the fruit continues to evolve, and the juice yield decreases in line with the storage time. Although the addition of enzymes reduces the costs and increases the yield of apple juice, the raw juice obtained from enzymatically treated apples contains more pectins, and hence requires more drastic treatment at the clarification step.

Clarification

Raw press juice is a viscous and highly turbid liquid that is currently clarified by filtration; however, the cloud particles cause the filter rapidly to clog. The use of enzymes as a clarification aid has been investigated for over seven decades (Kertesz, 1930). According to Yamasaki et al. (1964), cloud particles consist mainly of proteins and pectins. At the juice's natural pH (~2 for cranberry juice, 4 for citrus, 4.5 for banana, up to 5.9 for papaya; Grassin and Fauquembergue, 1996), proteins are positively charged and coated by negatively charged pectins. Pectin degradation reveals the positive charges from one particle to the negative charges of another particle, allowing them to aggregate and flocculate. When this degradation is ensured by a combination of polygalacturonase and pectin methylesterase, some methanol is released in the juice. The replacement of these enzymes by a pectin lyase – which de-polymerizes without methanol release – was shown to be efficient in the case of orange or apple juice, but clarification of grapefruit juice was incomplete due to the low DM of those pectins.

From a technological point of view, clarification can be performed using either a discontinuous or a continuous process. The latter has the advantage in that it allows a combination of enzymatic reaction and simultaneous ultrafiltration of the treated juice. Moreover, this process allows enzyme recycling. Although ultrafiltration of enzyme-treated juices has been introduced increasingly in the juice industry (Moslang, 1984), it is not applicable to very viscous juices such as peach juice.

Liquefaction

Liquefaction induces a complete degradation of the cell wall, leading to the release of protoplasts that are disrupted under an osmotic pressure effect. This technology is applied to fruits that cannot be easily pressed, especially tropical fruits (mango, guava, banana) (Schreier et al., 1985). Liquefaction is also intensely studied for stored apples, which yield less juice than fresh apples (Pilnik, 1988) (see Section 7.3.2.1), and also improves the extraction of pulp-associated components such as flavor and color (e.g., carotene, anthocyans). The complete breakdown of the cell wall requires the simultaneous and synergistic action of pectin- and cellulose-degrading enzymes. However, the ratio between all these activities influences the efficiency and the yield of liquefaction, and must be adapted to the cell wall composition. At the same time as the cell walls disrupt, complete liquefaction induces a reduction in viscosity of the product, which can be further clarified using more common techniques.

Maceration

In the maceration process, the fruit tissue is transformed in a cell suspension, resulting in pulpy products that can be used as a base for nectars, baby foods and ingredients for dairy products. Since the aim of this process is to obtain intact cells, the enzymes should affect exclusively the middle lamella pectins and should therefore either contain polygalacturonase (Zetelaki-Horvath and Gatai, 1977a, b) or pectin lyase (Ishii and Yokotsuka, 1971). The endogenous pectin methylesterase is very important in this process because it can have two different effects:

- It can de-esterify the solubilized pectins, which will react with calcium to form a gel carrying the cloud particles.
- It can also act synergistically with the added polygalacturonase and turn the maceration into a "pulp enzymatic process".

The inactivation of endogenous pectin methylesterase is thus important for the maceration of many products.

The mechanical aspect is also a key point of the maceration process, as it can disrupt many cells, with negative effect on juice quality in terms of flavor or loss of nutritive elements. Compared with clarification or liquefaction, only a limited degradation of pectin occurs in the maceration process. Macerated products contain therefore larger amounts of desirable dietary fiber and may be more attractive for the consumer.

Whichever process is used, the enzymatic tool should be adapted to the fruit to be treated. The sugar composition of the cell wall to be degraded is directly related to the enzyme(s) to be used, and the level of each activity must be carefully adjusted. The physiological status of the fruit (unripe, ripe, and overripe) also influences the efficiency of the enzymatic treatment. Enzyme level is dependent not only on the fruit but also on the technological conditions (temperature, treatment duration, and type of press), but are most often in the range of 100–

150 g ton^{-1} of pulp from pressing. Moreover, the need for enzymes which are specific to each process or product has resulted in the major developments in the industrial production of enzymes.

Enzymatic Valorization of Plant Byproducts

The first transformation of plant material produces huge amounts of byproducts such as fruit pulps and peels. For example, sugar beet pulp is the main byproduct of the sugar-refining industry, and cannot be easily stored (because of its high water content) or dried (due to the expensive cost of drying). The development of new and profitable channels of uses might valorize this biomass. Sugar beet pulp can be used as a source of pectin which may be converted to its monomers by enzymatic saccharification combined with adapted fractionation and purification processes to produce galacturonic acid, rhamnose, arabinose, and ferulic acid (Micard et al., 1996). These monomers can be transformed into various products and may therefore be considered as high added-value molecules. Galacturonic acid can be transformed into a tensioactive agent by esterification with various fatty acids (Petit et al., 1994), while rhamnose can be used as a precursor of aroma and applied as caramel, roasted or fruit flavor (Wong et al., 1983). Arabinose has anti-Parkinsonian properties (Vogel, 1991), while ferulic acid can be converted into vanillin by filamentous fungi (Thibault et al., 1998).

8
Industrial Pectin

Pectin is commercially extracted from citrus peels and apple pomace under mildly acidic conditions, and may be further chemically de-esterified. In this way a whole range of pectins is produced which vary essentially by the degree of substitution (methyl or amide) of carboxyl groups (see Table 3).

8.1
Current Raw Materials

Several raw materials are currently used widely in industrial processes. For example, the dried press cake of apple juice manufacture (known as apple pomace) contains ~10–15% of extractable pectins; apple pomace must be rapidly dried to avoid the development of various fungi, which produce pectolytic enzymes. A second raw material is the wet or dried peels and rags obtained after the extraction of citrus juice (wet or dry citrus pomace contains ~20–30% of extractable pectins). Citrus (lime, lemon, orange or grapefruit) peels must be blanched in order to avoid de-esterification reactions due to particularly active endogenous pectin methylesterases.

Other interesting sources considered for pectin production are sugar beet pulp, sunflower heads, potato fiber, onion skins, tobacco leaves, wastes from the processing of tropical fruits and more particularly papaya, mango, coffee and cocoa (Voragen et al., 1995). These sources are not only interesting with respect to the quality and quantity of pectin that can be extracted, but also to the availability of this material in logistically favored locations. For this reason, sugar beet pulp has always attracted the attention of pectin manufacturer (Thibault et al., 2001).

8.2
Extraction of Pectin

Pectins are industrially extracted from citrus peels and apple pomace by hot, acidified water. Mineral acids, mainly nitric acid, are used. The extraction conditions (pH, temperature, time) must be optimized in order to provide good yields of material that also has the desired gelling capacity and DM. The extraction conditions are in the range of pH 1.5 to 3.0, 60 to 100 °C and 0.5 to 6 h. The solid:liquid ratio has to be well defined for an efficient liquid/solid separation. Ratios of 1:17 for apple and 1:35 for citrus are often used. Separation of the viscous extract from the strongly swollen and partly disintegrated plant material can be achieved by a combination of centrifugation and filtration, and this remains a key problem in pectin manufacture. Currently, rotary drum vacuum filtration is commonly used in the industry (Joye and Luzio, 2000), but extraction technologies are under continuous development (Ralet and Thibault, 1994; Ralet et al., 1994c; Hwang et al., 1998; Fishman et al., 2000; Joye and Luzio, 2000).

Extracts are rapidly brought to pH 3–4, whereupon the temperature is lowered to avoid pectin demethylation and depolymerization. Weak bases such as sodium carbonate or ammonia are used to minimize β-elimination reactions. When pectins are extracted from apple pomace, amylases must be added to degrade co-extracted starch. Pectins are recovered from the clarified extract by alcoholic (ethanol or isopropanol) precipitation. The precipitate is separated by filtration, washed to remove contaminants (heavy metals, acid, sugars, polyphenols, pigments), and finally dried and milled to the desired particle size. During processing, the pectin may undergo an ion-exchange step to create the sodium form for ease of use in food applications (Joye and Luzio, 2000).

Pectins suspended in alcohol are in a very suitable form for further modification. By using acid treatment in isopropanol, they can be saponified to the desired DM under conditions of low pH and temperatures not exceeding 50 °C to avoid depolymerization reactions. This treatment can yield HM pectins with DM values in the range of 55–75% or LM pectins in the DM range of 20–45%. LM pectins can also be obtained by ammonia treatment of the alcoholic suspensions. Under these conditions, methoxyl groups are partly saponified and partly replaced by amide groups, giving amidated pectins (May, 1990). Regulations require that the DAm does not exceed 25%.

Because of variations in the raw materials, there may be large differences in the gelling capacity of pectins. HM pectins are therefore standardized with sucrose, glucose or lactose to a given gelling power defined as degree sag. The standard of 150 ° sag means that 1 g of pectin is able to gel 150 g of sucrose under defined conditions of pH and temperature. Due to their large application range, LM pectins are not necessarily standardized.

8.3
Regulation and World Market

The annual world production of pectins can be estimated as 20,000–25,000 tons, with a current growth rate of about 5% per annum. The HM pectins represent about two-thirds of the total pectins, though the LM pectin market (including amidated pectins) is growing more rapidly than the HM market. Citrus pectins are produced in greater quantities than apple pectins. Pectins were originally produced in the apple- or citrus-growing areas from fresh raw materials, but today an increasing amount of dried raw materials is used, and as a result factories are often located in other areas. The main producing areas include North and Central Europe for apple pomace, and USA (California, Florida), South America (Brazil, Argentina, Mexico) and southern Europe for citrus peels. Average prices for the HM and LM pectins are 10 $ kg^{-1} and 12 $ kg^{-1}, respectively.

The main companies producing pectins (those which have factories in Europe as well as in America) are CP-Kelco (US, amalgamating Copenhagen Pektin Fabrik in Denmark, Pomosin Pectin in Germany and Citrus Colloids in UK), Danisco-Cultor (Denmark), Degussa Texturant Systems (France), Herbstreith and Fox KG (Germany) and Obipektin (Switzerland).

Pectins have been approved by the Food and Drug Administration in US, and by a European Directive. They have "GRAS" (generally regarded as safe) status, and have no "ADI" (admissible day intake), although some national regulations may limit the amount of pectin added in some applications. In Europe, all pectins (HM pectins, conventional and amidated LM pectins) are referred to as additive number E440.

9
Pectin Gelling Properties and Applications

A good solubilization of pectins is the first step of the gelation process. Pectin solubility in aqueous media depends on many parameters, the main ones being the counterion nature, ionic strength, and pH. Solubility is indeed favored by the dissociation of carboxyl groups, which leads to molecule individualization through electrostatic repulsion. Once pectins are solubilized, they may gel through inter-chain associations. The numerous hydroxyl groups and free, methylated or amidated carboxyl functions, can lead to the setting up of hydrogen bonds, hydrophobic interactions or ionic bonds between molecules. The DM value is an important criterion because it governs the gelation conditions: HM pectins (DM $>55\%$) will gel with at least 55% of soluble solids and at pH ≤ 3.5; conventional and amidated LM pectins (DM $<45\%$) will gel in the presence of calcium ions (see Table 3).

9.1
HM Pectin

Jam manufacture is the main use of industrially extracted pectins, taking benefit of the ability of HM pectins to form gels with high amounts of sugar ($>55\%$) and acid. Junction zone formation is made possible through the "smooth" regions of pectins. The high sugar concentration creates conditions of low water activity, and this promotes chain–chain interactions rather than chain–solvent interactions (Rees, 1972). The acid reduces the dissociation of carboxyl groups, thus diminishing electrostatic chain repulsion. To avoid turbidity, synadirectional synresis and precipitation, there must be junction zone-terminating structural elements present in the chain, and rhamnose and "hairy" regions are thought to play such a role. The junction zones are stabilized by hydrogen bonds between nondissociated carboxyl and secondary alcohol groups (Morris et al., 1982) and by hydrophobic interactions between methoxyl groups (Oakenfull and Scott, 1984) (Figure 4).

The analytical parameter that allows prediction of gelling behavior, and particularly of gel setting time, is the DM. This fact has led to a further subdivision of HM pectins based on the setting time and temperature with DM ranging from 77 (for ultra-rapid-set pectins; URS) to 58 (for slow-set pectins; SS) (see Table 3). Setting times vary from 1–3 min to more than 1 h, and the pH necessary to achieve gelation decreases from 3.4 to 2.6, for URS and SS pectins, respectively.

HM pectins require 55–85% sugar and pH 2.5–3.8 in order to gel, and these conditions limit the possible uses of HM pectins as a gelling agent in sweetened fruit products. In fact, about 80% of the world production of HM pectins is used in the manufacture of jams and jellies (Figure 5),

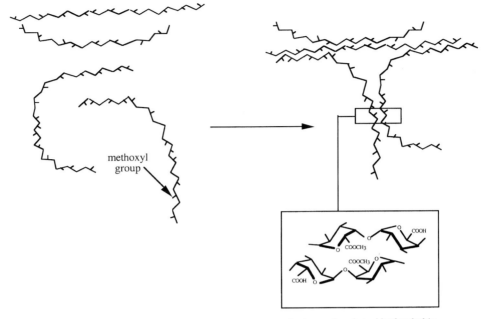

Fig. 4 Gelation of high-methoxyl (HM) pectins.

the pectin being added to supplement the gelling power of fruits (May, 1990). It is the art of the jam manufacturer to choose the pH conditions and the correct type of pectin to achieve the desired setting time or setting temperature. For example, the production of jams with whole fruits requires URS pectin in order to ensure a uniform distribution of fruit particles in the continuous jelly phase (see Table 3). In contrast, the production of jams from very acid fruits such as blackcurrant requires the use of SS pectins to avoid pre-gel formation (Table 3).

HM pectins are also used in the confectionery industry for making fruit jellies and jelly centers, in fruit juices and fruit drink concentrates as a stabilizer and/or to provide a "natural mouthfeel", and also in fermented and directly acidified dairy products. The "protective colloid" effect of HM pectin allows the stabilization of sour milk products, whereby pectins react with caseins and prevent coagulation at acidic pH.

9.2
LM Pectin

LM pectins form gels in the presence of divalent ions (calcium for food purposes; Axelos and Thibault, 1991b; Garnier et al., 1994). To achieve gelation, the required pH range is much larger than for HM pectins (2.0–6.0), and sugar addition is not necessary. The formation of junction zones between the "smooth" regions of different chains leads to gelation (Morris et al., 1978) (Figure 6). Because of the electrostatic nature of the bonds, LM pectin gels are very

Fig. 5 Application of citrus high-methoxyl (HM) pectins in jellies.

sensitive to structural parameters which can modify the environment of the carboxyl groups, such as the nature and the amounts of substituents along the homogalacturonic backbone (Kohn et al., 1983; Thibault and Rinaudo, 1985; Renard and Jarvis, 1999; Ralet et al., 2001). The gel-forming ability increases with decreasing DM; furthermore, LM pectins with a blockwise distribution of free carboxyl groups are very sensitive to low calcium levels (Kohn et al., 1983; Thibault and Rinaudo, 1985; Ralet et al., 2001). The pH influences the texture of the gel; when the pH is below 3.5 (pK_a of galacturonic acid), there is a predominance of nondissociated acid groups, which leads to more hydrogen bonding in the gel network. This gives rise to a more rigid, nonshear-reversible gel network. When the pH is above 3.5, there is a predominance of ionized acid groups, which favors calcium cross-linking (Figure 6), and this leads to the formation of a more spreadable, shear-reversible gel network. The soluble solids content also has some effect on the gelling performances of LM pectin; with a given LM pectin, as the soluble solids increases, the calcium requirement decreases and the calcium "bandwidth" becomes more narrow.

Amidation increases or improves the gelling ability of LM pectins; amidated pectins need less calcium to gel and are less prone to precipitation by high calcium levels (May, 1990). The amide groups are thought to be distributed in a blockwise pattern along the pectic chain (Racapé et al., 1989), and it has been suggested (Voragen et al., 1995) that these blocks can promote association through hydrogen bonding in addition to the calcium binding (Figure 7).

The applications of LM pectins are directly based on their peculiar gelling properties. The traditional application of LM pectin is in jams with soluble solids below 55% (low/reduced-calorie jams, jellies preserves and conserves) which is the limit for the use of HM pectin (Table 3). The type of LM pectin must be carefully selected according to the soluble solids/pH conditions in the application medium. The heat reversibility of LM pectin gels may be utilized in bakery jams and jellies for glazing purposes. LM pectins also find applications in the production of fruit preparations for yogurt. Finally, because of their calcium reactivity, LM pectins, and notably those that are amidated, are well adapted for gelled milk products and in fruit/milk desserts.

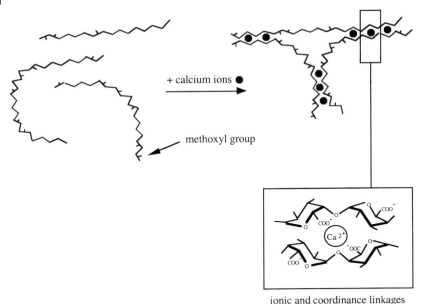

Fig. 6 Gelation of low-methoxyl (LM) pectins.

10
Stabilizing Properties

Pectins are increasingly used as stabilizers in acidified milk systems, including sour milk products and mixtures of fruit juice and milk. Pectins react with caseins and prevent coagulation at acidic pH. At the natural pH of milk (6.8), the casein particles exhibit a net negative charge and repel each other; however, when the pH is lowered, electrostatic repulsion decreases and casein particles tend to aggregate. Thus, in the absence of stabilizer, high viscosity, whey exsudation and sandy mouthfeel are likely to occur in low-pH milk drinks. HM pectins can minimize protein–protein interactions, thereby reducing protein coagulation. This property is explained by the remaining carboxylic functions that are sufficiently dissociated at the pH of acidified milk products (4.0). Pectins will then electrostatically stick to the positive areas of casein particles, producing a highly hydrated layer next to the surface of the casein, and this prevents aggregation by "steric stabilization".

11
Outlook and Perspectives

Pectins are complex heteropolysaccharides that arise from plant cell walls. Although they are still widely used as gelling agents in jam manufacture, their field of application has extended widely in recent years. Furthermore, the implication of pectin in the quality of fresh and preserved fruits and vegetables, as well as the rational use of pectolytic enzymes in fruit and vegetable transformation is largely sustained by an increasing knowledge regarding their fine chemical structure. In the future, there will be a need for pectic molecules for which molar mass and chemical structure can be controlled, including their content in neutral

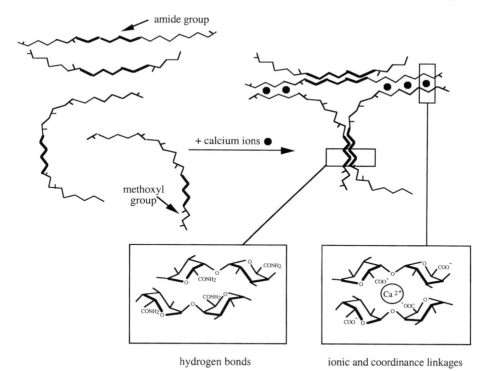

Fig. 7 Gelation of amidated low-methoxyl (LM) pectins.

sugars and in substituents. It is likely that these materials would be obtained either by chemical modifications of raw pectins, or perhaps in time by virtue of gene technology. Irrespective of these advances, knowledge of the biosynthesis of pectins still needs to be improved. A better understanding of such biosynthesis and the interactions between polymers within the cell-wall network might also lead to the identification of new extraction methods.

12
References

Albersheim, P., Neukom, H., Deuel, H. (1960) Splitting of pectin chain molecules in neutral solutions, *Arch. Biochem. Biophys.* **90**, 46–51.

Almeida, D. P. F., Huber, D. J. (1999) Apoplastic pH and inorganic ion levels in tomato fruit: a potential mean for regulation of cell wall metabolism during ripening, *Physiol. Plant.* **105**, 506–512.

Aspinall, G. O. (1980) Chemistry of cell wall polysaccharides, in: *The Biochemistry of Plants* (Preiss, J., Ed.), New York: Academic Press, 473–500.

Aspinall, G. O., Baillie, J. (1963) Gum tragacanth. Part I. Fractionation of the gum and structure of tragacanthic acid, *J. Chem. Soc.* 1702–1714.

Aspinall, G. O., Begbie, R., Hamilton, A., Whyte, J. N. C. (1967a) Polysaccharides of soya beans. Part III. Extraction and fractionation of polysaccharides from cotyledon-meal, *J. Chem. Soc.* (**C**), 1065–1070.

Aspinall, G. O., Cottrell, I. W., Egan, S. V., Morrison, I. M., Whyte, J. N. C. (1967b) Polysaccharides of soya beans. Part IV. Partial hydrolysis of the acidic polysaccharide complex from cotyledon-meal, *J. Chem. Soc.* (**C**), 1071–1080.

Aspinall, G. O., Craig, J. W. T., Whyte, J. L. (1968) Lemon peel pectin. Part I. Fractionation and partial hydrolysis of water soluble pectins, *Carbohydr. Res.* **7**, 442–452.

Axelos, M. A. V., Thibault, J.-F. (1991a) Influence of the substituents of the carboxyl groups and of the rhamnose content on the solution properties and flexibility of pectins, *Int. J. Biol. Macromol.* **13**, 77–82.

Axelos, M. A. V., Thibault, J.-F. (1991b) The chemistry of low-methoxyl pectin gelation, in: *The Chemistry and Technology of Pectin* (Walter, R.H., Ed.), New York: Academic Press, 109–117.

Axelos, M. A. V., Lefebvre, J., Thibault, J.-F. (1987) Structural and mechanical characterization of citrus and apple sodium pectin sols, *Food Hydrocolloids* **5/6**, 569–570.

Beldman, G., Schols, H. A., Pitson, S. M., Searle-van Leeuwen, M. J. F., Voragen, A. G. J. (1997) Arabinans and arabinan-degrading enzymes, in: *Advances in Macromolecular Carbohydrate Research* (Sturgeon, R. J., Ed.), London: JAI Press Inc., Vol.1, 1–64.

Biely, P., Benen, J. A. E., Heinrichova, K., Kester, H. C. M., Visser, J. (1996) Inversion of configuration during hydrolysis of α-1,4-galacturonidic linkage by three *Aspergillus* polygalacturonases, *FEBS Lett.* **382**, 249–255.

Blakeney, A. B., Harris, P. J., Henry, R. J., Stone, B. A. (1983) A simple and rapid preparation of alditol acetates for monosaccharide analysis, *Carbohydr. Res.* **113**, 291–299.

Blumenkrantz, N., Asboe-Hansen, G. (1973) New method for quantitative determination of uronic acids, *Anal. Biochem.* **54**, 484–489.

Blumer, J. M., Clay, R. P., Bergmann, C. W., Albersheim, P., Darvill, A. (2000) Characterization of changes in pectin methyl-esterase expression and pectin esterification during tomato fruit ripening, *Can. J. Bot.* **78**, 607–618.

Bonnin, E., Lahaye, M., Vigouroux, J., Thibault, J.-F. (1995) Preliminary characterization of a new exo(1,4)β-D-galactanase with transferase activity, *Int. J. Biol. Macromol.* **17**, 345–351.

Bordenave, M., Goldberg, R., Huet, J.-C., Pernollet, J.-C. (1995) A novel protein from mung bean hypocotyl cell walls with acetyl esterase activity, *Phytochemistry* **38**, 315–319.

Bouveng, H. O. (1965) Polysaccharides in pollen. II. The xylogalacturonan from mountain pine (*Pinus mugo* Turra) pollen. *Acta Chem. Scand.* **19**, 953–963.

Braccini, I., Grasso, R. P., Pérez, S. (1999) Conformational and configurational features of acidic polysaccharides and their interactions with calcium ions: a molecular modeling investigation, *Carbohydr. Res.* **317**, 119–130.

Brillouet, J.-M., Williams, P., Moutounet, M. (1991) Purification and some properties of an exo-β (1-6) D galactanase from *Aspergillus niger*, *Agric. Biol. Chem.* **55**, 1565–1571.

Bruyant-Vannier, M.-P., Gaudinet-Schaumann, A., Bourlard, T., Morvan, C. (1996) Solubilization and partial characterization of pectin methyltransferase from flax cells, *Plant Physiol. Biochem.* **34**, 489–499.

Buckeridge, M. S., Reid, J. S. G. (1994) Purification and properties of a novel β-galactosidase or exo(1,4)β-D-galactanase from the cotyledons of germinated *Lupinus angustifolius* L. seeds, *Planta* **192**, 502–511.

Campbell, A. D., Huysamer, M., Stotz, H. U., Greve, L. C., Labavitch, J. M. (1990) Comparison of ripening processes in intact tomato fruit and excised pericarp discs, *Plant Physiol.* **94**, 1582–1589.

Carpita, N. C., Gibeaut, D. M. (1993) Structural models of primary cell walls in flowering plants: consistency of molecular structure with the physical properties of the walls during growth, *Plant J.* **3**, 1–30.

Carpita, N. C., McCann, M. C. (2000) The cell wall, in: *Biochemistry and Molecular Biology of Plants*, (Buchanan, B., Gruissem, W., Jones, R., Eds.), Rockville, MD: American Society of Plant Physiologists, 52–108.

Catoire, L., Derouet, C., Redon, A.-M., Goldberg, R., Hervé du Penhoat, C. (1997) An NMR study of the dynamic single-strand conformation of sodium pectate, *Carbohydr. Res.* **300**, 19–29.

Colquhoun, I. J., Ralet, M.-C., Thibault, J.-F., Faulds, C. B., Williamson, G. (1994) Feruloylated oligosaccharides from cell-wall polysaccharides. Part II. Structure identification of feruloylated oligosaccharides from sugar-beet pulp, *Carbohydr. Res.* **263**, 243–256.

Corredig, M., Kerr, W., Wicker, L. (2000) Molecular characterization of commercial pectins by separation with linear mix gel permeation columns in-line with multi-angle light scattering detection, *Food Hydrocolloids* **14**, 41–47.

Cros, S., Hervé du Penhoat, C., Bouchemal, N., Ohassan, H., Imberty, A., Pérez, S. (1992) Solution conformation of a pectin fragment disaccharide using molecular modelling and nuclear magnetic resonance, *Int. J. Biol. Mol.* **14**, 313–320.

Cros, S., Garnier, C., Axelos, M. A. V., Imberty, A., Pérez, S. (1996) Solution conformations of pectin polysaccharides: determination of chain characteristics by small angle neutron scattering, viscometry, and molecular modelling, *Biopolymers* **39**, 339–352.

Daas, P. J. H., Arisz, P. W., Schols, H. A., De Ruiter, G. A., Voragen, A. G. J. (1998) Analysis of partially methyl-esterified galacturonic acid oligomers by high-performance anion-exchange chromatography and matrix-assisted laser desorption/ionization time-of-flight mass spectrometry, *Anal. Biochem.* **257**, 195–202.

Daas, P. J. H., Meyer-Hansen, K., Schols, H. A., De Ruiter, G. A., Voragen, A. G. J. (1999) Investigation of the non-esterified galacturonic acid distribution in pectin with endopolygalacturonase, *Carbohydr. Res.* **318**, 135–145.

Daas, P. J. H., Voragen, A. G. J., Schols, H. A. (2000) Characterization of non-esterified galacturonic acid sequences in pectin with endopolygalacturonase, *Carbohydr. Res.* **326**, 120–129.

Daas, P. J. H., Boxma, B., Hopman, A. M. C. P., Voragen, A. G. J., Schols, H. A. (2001a) Non-esterified galacturonic acid sequence homology of pectins, *Biopolymers* **58**, 1–8.

Daas, P. J. H., Voragen, A. G. J., Schols, H. A. (2001b) Study of the methyl ester distribution in pectin with *endo*-polygalacturonase and high-performance size-exclusion chromatography, *Biopolymers* **58**, 195–203.

Darvill, A. G., McNeil, M., Albersheim, P., Delmer, D. P. (1980) The primary cell walls of flowering plants, in: *The Biochemistry of Plants* (Tolbert, N. E., Ed.), New York: Academic Press, Vol. 1, 91–116.

Deckers, H. A., Olieman, C., Rombouts, F. M., Pilnik, W. (1986) Calibration and application of high-performance size exclusion columns for molecular weight distribution of pectins, *Carbohydr. Polym.* **6**, 361–378.

Denes, J.-M., Baron, A., Renard, C. M. G. C., Pean, C., Drilleau, J.-F. (2000) Different action patterns for apple pectin methyl-esterase at pH 7.0 and 4.5, *Carbohydr. Res.* **327**, 385–393.

De Vries, J. A., Rombouts, F. M., Voragen, A. G. J., Pilnik, W. (1982) Enzymic degradation of apple pectins, *Carbohydr. Polym.* **2**, 25–33.

Dey, P. M., Brinson, K. (1984) Plant cell-walls, in: *Advances in Carbohydrate Chemistry and Biochemistry* (Tipson, R. S., Horton, D., Eds.), Orlando, FL: Academic Press, Vol. 42, 266–382.

Doong, R. L., Mohnen, D. (1998) Solubilization and characterization of a galacturonosyltransferase that synthetizes the pectic polysaccharide homogalacturonan, *Plant J.* **13**, 363–374.

Doong, R. L., Liljebjelke, K., Fralish, G., Kumar, A., Mohnen, D. (1995) Cell-free synthesis of pectin, *Plant Physiol.* **109**, 141–152.

Englyst, H. N., Cummings, J. H. (1984) Simplified method for the measurement of total non-starch polysaccharides by gas-liquid chromatography of constituent sugars as alditol acetates, *Analyst* **109**, 937–942.

Fishman, M. L., Chau, H. K., Hoagland, P., Ayyad, K. (2000) Characterization of pectin, flash-extracted from orange albedo by microwave heating, under pressure, *Carbohydr. Res.* **323**, 126–138.

Fry, S. C. (1983) Feruloylated pectins from the primary cell wall: their structure and possible functions, *Planta* **157**, 111–123.

Fuchigami, M., Miyazaki, K., Hyakumoto, N. (1995) Frozen carrot texture and pectic component as affected by low temperature blanching and quick freezing, *J. Food Sci.* **60**, 132–136.

Garnier, C., Axelos, M. A. V., Thibault, J.-F. (1994) Selectivity and cooperativity in the binding of calcium ions by pectins, *Carbohydr. Res.* **256**, 71–81.

Gibeaut, D. M. (2000) Nucleotide sugars and glycosyltransferases for synthesis of cell wall matrix polysaccharides, *Plant Physiol. Biochem.* **38**, 69–80.

Gierschner, K., Jahn, W., Philippos, S. (1994) Specific modification of cell wall hydrocolloids in a new technique for processing high quality canned vegetables, *Dtsch. Lebensm.-Rundschau* **91**, 103–109.

Giovannoni, J. J., DellaPenna, D., Bennett, A. B., Fischer, R. L. (1989) Expression of a chimeric polygalacturonase gene in transgenic rin (ripening inhibitor) tomato fruit results in polyuronide degradation but not fruit softening, *Plant Cell* **1**, 53–63.

Gnanasambandam, R., Proctor, A. (2000) Determination of pectin degree of esterification by diffuse reflectance Fourier transform infrared, *Food Chem.* **68**, 327–332.

Goubet, F., Morvan, C. (1993) Evidence for several galactan synthases in flax (*Linum usitassimum* L.) suspension-cultured cells, *Plant Cell Physiol.* **34**, 1297–1303.

Goubet, F., Morvan, C. (1994) Synthesis of cell wall galactans from flax (*Linum usitassimum* L.) Suspension-cultured cells, *Plant Cell Physiol.* **35**, 719–727.

Goubet, F., Council, L. N., Mohnen, D. (1998) Identification and partial characterization of the pectin methyltransferase "homogalacturonan-methyltransferase" from membranes of tobacco cell suspensions, *Plant Physiol.* **116**, 337–347.

Grasladen, H., Bakoy, O. E., Larsen, B. (1988) Determination of the degree of esterification and the distribution of methylated and free carboxyl groups in pectins by ^{1}H-N.M.R. spectroscopy, *Carbohydr. Res.* **184**, 183–191.

Grassin, C., Fauquembergue, P. (1996) Fruit juices, in: *Industrial Enzymology* (Godfrey, T., West, S., Eds.), London: MacMillan Press, 227–264.

Greve, L. C., McArdle, R. N., Gohlke, J. R., Labavitch, J. M. (1994) Impact of heating on carrot firmness: changes in cell wall components, *J. Agric. Food Chem.* **42**, 2900–2906.

Guillon, F., Thibault, J.-F. (1988) Further characterization of acid- and alkali-soluble pectins from sugar beet pulp, *Lebensmittelwiss. Technol.* **21**, 198–205.

Hakomori, S. (1964) A rapid permethylation of a glycolipid and polysaccharide catalyzed by methyl-sulfinyl carbanion in dimethylsulfoxide, *J. Biochem.* **55**, 205–208.

Harding, S. E., Berth, G., Ball, A., Mitchell, J. R., de la Torre, J. C. (1991) The molecular weight distribution and conformation of citrus pectins in solution studied by hydrodynamics, *Carbohydr. Polym.* **16**, 1–15.

Henrissat, B. (1991) A classification of glycosyl hydrolases based on amino acid sequence similarities, *Biochem. J.* **280**, 309–316.

Huber, D. J. (1983) The role of cell wall hydrolases in fruit softening, *Hortic. Rev.* **5**, 169–219.

Huber, D. J., O'Donoghue, E. M. (1993) Polyuronides in avocado (*Persea americana*) and tomato (*Lycopersicon esculentum*) fruits exhibit markedly different patterns of molecular weight downshifts during ripening, *Plant Physiol.* **102**, 473–480.

Hwang, J.-K., Kim, C.-J., Kim, C.-T. (1998) Extrusion of apple pomace facilitates pectin extraction, *J. Food Sci.* **63**, 841–844.

Ishii, S., Yokotsuka, T. (1971) Maceration of plant tissues by pectin transeliminase, *Agric. Biol. Chem.* **35**, 1157–1159.

Ishii, T., Tobita, T. (1993) Structural characterization of feruloyl oligosaccharides from spinach-leaf cell walls, *Carbohydr. Res.* **248**, 179–190.

Ishikawa, M., Kuroyama, H., Takeuchi, Y., Tsumuraya, Y. (2000) Characterization of pectin methyltransferase from soybean hypocotyls, *Planta* **210**, 782–791.

Jarvis, M. C., McCann, M. C. (2000) Macromolecular biophysics of the plant cell wall: concepts and methodology, *Plant Physiol. Biochem.* **38**, 1–13.

Joye, D. D., Luzio, G. A. (2000) Process for selective extraction of pectins from plant material by different pH, *Carbohydr. Polym.* **43**, 337–342.

Kertesz, Z. I. (1930) A new method for enzymic clarification of unfermented apple juice, US Patent no. 1.932.833.

Kertesz, Z. I. (1951) *The Pectic Substances*. New York: Wiley Interscience.

Knox, J. P. (1997) The use of antibodies to study the architecture and developmental regulation of plant cell walls, *Int. Rev. Cytol.* **171**, 79–120.

Kohn, R. (1975) Ion binding on polyuronates-alginate and pectin, *Pure Appl. Chem.* **42**, 371–397.

Kohn, R., Markovic, O., Machova, E. (1983) Deesterification mode of pectin by pectin esterases of *Aspergillus foetidus*, tomatoes and alfalfa, *Collec. Czech. Chem. Commun.* **48**, 790–797.

Kontaminas, M. G., Kokini, J. L. (1990) Measurement of molecular parameters of water soluble apple pectin using low angle laser light scattering, *Lebensmittelwiss. Technol.* **23**, 174–177.

Lau, J. M., McNeil, M., Darvill, A. G., Albersheim, P. (1985) Structure of the backbone of rhamnogalacturonan I, a pectic polysaccharide in the primary cell walls of plants, *Carbohydr. Res.* **137**, 111–125.

Lau, J. M., McNeil, M., Darvill, A. G., Albersheim, P. (1987) Selective degradation of the glycosyluronic acid residues of complex carbohydrates by lithium dissolved in ethylenediamine, *Carbohydr. Res.* **168**, 219–243.

Le Goff, A., Renard, C. M. G. C., Bonnin, E., Thibault, J.-F. (2001) Extraction, purification and chemical characterisation of xylogalacturonans from pea hulls, *Carbohydr. Polym.* **45**, 325–334.

Limberg, G., Körner, R., Buchholt, H. C., Christensen, T. M. I. E., Roepstorff, P., Mikkelsen, J. D. (2000a) Analysis of different de-esterification mechanisms for pectin by enzymatic fingerprinting using endopectin lyase and endopolygalacturonase II from *A. niger*, *Carbohydr. Res.* **327**, 293–307.

Limberg, G., Körner, R., Buchholt, H. C., Christensen, T. M. I. E., Roepstorff, P., Mikkelsen, J. D. (2000b) Quantification of the amount of galacturonic acid residues in block sequences in pectin homogalacturonan by enzymatic fingerprinting with exo- and endo-polygalacturonase II from *A. niger*, *Carbohydr. Res.* **327**, 321–332.

Liners, F., Letesson, J.-J., Didembourg, C., Van Cutsem, P. (1989) Monoclonal antibodies against pectins. Recognition of a conformation induced by calcium, *Plant Physiol.* **91**, 1419–1424.

Massiot, P., Rouau, X., Thibault, J.-F. (1988) Characterisation of the extractable pectins and hemicelluloses of the cell wall of carrot, *Carbohydr. Res.* **172**, 229–242.

Matsuura, Y., Hatanaka, C. (1988) Isolation and characterisation of a xylose-rich pectic polysaccharide from Japanese radish, *Agric. Biol. Chem.* **52**, 3215–3216.

May, C. (1990) Industrial pectins: sources, production and applications, *Carbohydr. Polym.* **12**, 79–99.

McCann, M. C., Roberts, K. (1991) Architecture of the primary cell wall, in: *The Cytoskeletal Basis of Plant Growth and Form* (Lloyd, C. W., Ed.), London: Academic Press, 109–129.

Micard, V., Renard, C. M. G. C., Thibault, J.-F. (1994) Studies on enzymic release of ferulic acid from sugar-beet pulp, *Lebensmittelwiss. Technol.* **27**, 59–66.

Micard, V., Renard, C. M. G. C., Thibault, J.-F. (1996) Enzymic saccharification of sugar beet pulp, *Enzyme Microbial Technol.* **19**, 162–170.

Micard, V., Renard, C. M. G. C., Colquhoun, I. J., Thibault, J.-F. (1997) End-products of enzymic saccharification of beet pulp, with a special attention to feruloylated oligosaccharides, *Carbohydr. Polym.* **32**, 283–292.

Mogami, N., Nakamura, S., Nakamura, N. (1999) Immunolocalization of the cell wall components in *Pinus densiflora* pollen, *Protoplasma* **206**, 1–10.

Mohnen, D. (1999) Biosynthesis of pectins and galactomannans, in: *Comprehensive Natural Products Chemistry, Vol. 3, Carbohydrates and Their Derivatives Including Tannins, Cellulose and Related Lignins* (Pinto, B. M., Ed.), Amsterdam: Elsevier Science, 497–527.

Morris, E. R., Rees, D. A., Thom, D., Boyd, J. (1978) Chiroptical and stoichiometric evidence of a specific primary dimerisation process in alginate gelation, *Carbohydr. Res.* **66**, 145–154.

Morris, E. R., Powell, D. A., Gidley, M., Rees, D. A. (1982) Conformations and interactions of pectins I. Polymorphism between gel and solid states of calcium polygalacturonate, *J. Mol. Biol.* **155**, 507–516.

Morris, G. A., Foster, T. J., Harding, S. E. (2000) The effect of the degree of esterification on the hydrodynamic properties of citrus pectin, *Food Hydrocolloids* **14**, 227–235.

Moslang, H. (1984) Ultrafiltration in the fruit juice industry, *Confructa* **28**, 219–224.

Mutter, M., Beldman, G., Schols, H. A., Voragen, A. G. J. (1994) Rhamnogalacturonan α-L-rhamnopyranosylhydrolase. A novel enzyme specific for the terminal non-reducing rhamnosyl unit in rhamnogalacturonan regions of pectin, *Plant Physiol.* **106**, 241–250.

Mutter, M., Colquhoun, I. J., Schols, H. A., Voragen, A. G. J. (1996) Rhamnogalacturonase B from *Aspergillus aculeatus* is a rhamnogalacturonan α-L-rhamnopyranosyl-(1-4)-α-D-galactopyranosyluronide lyase, *Plant Physiol.* **110**, 73–77.

Mutter, M., Beldman, G., Pitson, S. M., Schols, H. A., Voragen, A. G. J. (1998) Rhamnogalacturonan α-D-galactopyranosyluronohydrolase. An enzyme that specifically removes the terminal non reducing galacturonosyl residue in rhamnogalacturonan regions of pectin, *Plant Physiol.* **117**, 153–163.

Nakano, H., Takenishi, S., Kitahata, S., Kinugasa, H., Watanabe, Y. (1990) Purification and characterisation of an exo-1,4-β-galactanase from a strain of *Bacillus subtilis*, *Eur. J. Biochem.* **193**, 61–67.

Ng, A., Waldron, K. W. (1997) Effect of cooking and precooking on cell wall chemistry in relation to firmness of carrot tissues, *J. Sci. Food Agric.* **73**, 503–512.

Oakenfull, D., Scott, A. (1984) Hydrophobic interaction in the gelation of high methoxyl pectins, *J. Food Sci.* **49**, 1093–1098.

O'Neill, M., Albersheim, P., Darvill, A. G. (1990) The pectic polysaccharides of primary cell walls, *Methods Plant Biochem.* **2**, 415–441.

Owens, H. S., Lotzkar, H., Schultz, T. H., McClay, W. D. (1946) Shape and size of pectinic acid molecules deduced from viscometric measurements, *J. Am. Chem. Soc.* **68**, 1628–1632.

Pauly, M., Scheller, H. V. (2000) O-acetylation of plant cell wall polysaccharides: identification and partial characterization of a rhamnogalacturonan O-acetyl-transferase from potato suspension-cultured cells, *Planta* **210**, 659–667.

Pellerin, P., Brillouet, J.-M. (1994) Purification and properties of an exo-(1, 3)-β-D-galactanase from *Aspergillus niger*, *Carbohydr. Res.* **264**, 281–291.

Petit, S., Ralainirina, R., Favre, S., De Baynast, R. (1994) Galacturonic acid derivatives, process for their preparation and applications thereof. Patent WO 93/02092.

Pérez, S., Mazeau, K., Hervé du Penhoat, C. (2000) The three-dimensional structures of the pectic polysaccharides, *Plant Physiol. Biochem.* **38**, 37–55.

Pilnik, W. (1988) From traditional ideas to modern fruit juice winning process, *X Intern. Congr. Fruit Juices*, vol. X, 159–179.

Pilnik, W., Voragen, A. G. J. (1991) The significance of endogenous and exogenous pectic enzymes in fruit and vegetable processing, in: *Food Enzymology* (Fox, J.P., Ed.), London: Elsevier Applied Science, Vol. 1, 303–336.

Pilnik, W., Voragen, A. G. J. (1992) Gelling agents (pectins) from plants for the food industry, *Adv. Plant Cell Biochem. Biotechnol.* **1**, 219–270.

Pilnik, W., Zwiker, P. (1970) Pektine, *Gordian* **70**, 202–204, 252–257, 302–305, 343–346.

Quemener, B., Thibault, J.-F. (1990) Assessment of methanolysis for the determination of sugars in pectins, *Carbohydr. Res.* **206**, 277–287.

Racapé, E., Thibault, J.-F., Reitsma, J. C. E., Pilnik, W. (1989) Properties of amidated pectins. II. Polyelectrolyte behavior and calcium binding of amidated pectins and amidated pectic acids, *Biopolymers* **28**, 1435–1448.

Ralet, M.-C., Thibault, J.-F. (1994) Extraction and characterisation of very highly methylated pectins from lemon cell walls, *Carbohydr. Res.* **260**, 283–296.

Ralet, M.-C., Saulnier, L., Thibault, J.-F. (1993) Raw and extruded fibre from pea hulls. II. Structural study of the water-soluble polysaccharides, *Carbohydr. Polym.* **20**, 25–34.

Ralet, M.-C., Faulds, C. B., Williamson, G., Thibault, J.-F. (1994a) Feruloylated oligosaccharides from cell-wall polysaccharides. Part I. Isolation and purification of feruloylated oligosaccharides from cell walls of sugar-beet pulp, *Carbohydr. Res.* **263**, 227–241.

Ralet, M.-C., Faulds, C. B., Williamson, G., Thibault, J.-F. (1994b) Feruloylated oligosaccharides from cell-wall polysaccharides. Part III. Degradation of feruloylated oligosaccharides from wheat bran and sugar-beet pulp by ferulic acid esterases from *Aspergillus niger*, *Carbohydr. Res.* **263**, 257–269.

Ralet, M.-C., Axelos, M. A. V., Thibault, J.-F. (1994c) Gelation properties of extruded lemon cell walls and their water-soluble pectins, *Carbohydr. Res.* **260**, 271–282.

Ralet, M.-C., Dronnet, V., Buchholt, H. C., Thibault, J.-F. (2001) Enzymatically and chemically de-esterified lime pectins: characterisation, polyelectrolyte behaviour and calcium binding properties, *Carbohydr. Res.* **336**, 117–125.

Ralph, J., Quideau, S., Grabber, J. H., Hatfield, R. D. (1994) Identification and synthesis of new ferulic acid dehydrodimers present in grass cell walls, *J. Chem. Soc. Perkin Trans.* **1**, 3485–3498.

Redgwell, R. J., McRae, E., Hallett, I., Fischer, M., Perry, J., Harker, R. (1997) *In vivo* and *in vitro* swelling of cell walls during fruit ripening, *Planta* **203**, 162–173.

Rees, D. A. (1972) Polysaccharide gel. A molecular view, *Chem. Ind.* **16**, 630–636.

Renard, C. M. G. C., Jarvis, M. C. (1999) Acetylation and methylation of homogalacturonans. Part II. Effect on ion-binding properties and conformations, *Carbohydr. Polym.* **39**, 209–216.

Renard, C. M. G. C., Searle-Van Leeuwen, M., Voragen, A. G. J., Thibault, J.-F., Pilnik, W. (1991a) Studies on apple protopectin II: apple cell walls degradation by pure polysaccharidases and their combinations, *Carbohydr. Polym.* **14**, 295–314.

Renard, C. M. G. C., Schols, H. A., Voragen, A. G. J., Thibault, J.-F., Pilnik, W. (1991b) Studies on apple protopectin III: Characterization of the material extracted by pure polysaccharidases from apple cell walls, *Carbohydr. Polym.* **15**, 13–32.

Renard, C. M. G. C., Champenois, Y., Thibault, J.-F. (1993a) Characterization of the extractable pectins and hemicelluloses of the cell wall of glasswort, *Salicornia ramosissima*, *Carbohydr. Polym.* **22**, 239–245.

Renard, C. M. G. C., Thibault, J.-F., Voragen, A. G. J., Van den Broek, L. A. M., Pilnik, W. (1993b) Studies on apple protopectin VI: extraction of pectins from apple cell walls with rhamnogalacturonase, *Carbohydr. Polym.* **22**, 203–210.

Renard, C. M. G. C., Weightman, R. M., Thibault, J.-F. (1997) Structure of the xylose-rich pectins from pea hulls, *Int. J. Biol. Macromol.* **21**, 155–162.

Renard, C. M. G. C., Wende, G., Booth, E. J. (1999) Cell wall phenolics and polysaccharides in different tissues of quinoa (*Chenopodium quinoa* Willd), *J. Sci. Food Agric.* **79**, 2029–2034.

Rinaudo, M. (1996) Physicochemical properties of pectins in solution and gel states, in: *Pectins and pectinases, Progress in Biotechnology* (Visser, J., Voragen, A. G. J., Eds.), Amsterdam: Elsevier Sciences, Vol. 14, 21–34.

Rombouts, F. M., Thibault, J.-F. (1986) Feruloylated pectic substances from sugar-beet pulp, *Carbohydr. Res.* **154**, 177–188.

Sakai, T. (1992) Degradation of pectins, in: *Microbial Degradation of Natural Products* (Vinkelmann, G., Ed.), Weinheim: VCH, 57–81.

Saulnier, L., Thibault, J.-F. (1987) Extraction and characterization of pectic substances from pulp of grape berries, *Carbohydr. Polym.* **7**, 329–343.

Saulnier, L., Vigouroux, J., Thibault, J.-F. (1995) Isolation and partial characterization of feruloylated oligosaccharides from maize bran, *Carbohydr. Res.* **272**, 241–253.

Scheller, H. V., Doong, R. L., Ridley, B. L., Mohnen, D. (1999) Pectin biosynthesis: a solubilized $\alpha 1,4$-galacturonosyltransferase from tobacco catalyses the transfer of galacturonic acid from UDP-galacturonic acid onto the non-reducing end of homogalacturonan, *Planta* **207**, 512–517.

Schols, H. A., Voragen, A. G. J. (1996) Complex pectins: structure elucidation using enzymes, in: *Pectins and Pectinases, Progress in Biotechnology* (Visser, J., Voragen, A. G. J., Eds.), Amsterdam: Elsevier Sciences, Vol. 14, 3–20.

Schols, H. A., Geraeds, C. C. J. M., Searle-Van Leeuwen, M. F. S., Kormelink, F. J. M., Voragen, A. G. J. (1990a) Rhamnogalacturonase: a novel enzyme that degrades the hairy regions of pectins, *Carbohydr. Res.* **206**, 105–115.

Schols, H. A., Posthumus, M. A., Voragen, A. G. J. (1990b) Structural features of hairy regions of pectins isolated from apple juice produced by the liquefaction process, *Carbohydr. Res.* **206**, 117–129.

Schols, H. A., Bakx, E. J., Schipper, D., Voragen, A. G. J. (1995) A xylogalacturonan subunit present in the modified hairy regions of apple pectin, *Carbohydr. Res.* **279**, 265–279.

Schreier, R., Kiltsteiner-Eberle, R., Idstein, H. (1985) Untersuchungen zur enzymatischen Verflüssigung tropischer Fruchtpülpen: guava, papaya, und mango, *Flüss. Ost.* **52**, 365–370.

Scott, R. W. (1979) Colorimetric determination of hexuronic acids in plant materials, *Anal. Chem.* **51**, 936–941.

Searle-Van Leeuwen, M. J. F., Van Den Broek, L. A. M., Schols, H. A., Beldman, G., Voragen, A. G. J. (1992) Rhamnogalacturonan acetylesterase: a novel enzyme from *Aspergillus aculeatus*, specific for the deacetylation of hairy (ramified) regions of pectins, *Appl. Microbiol. Biotechnol.* **38**, 347–349.

Searle-van Leeuwen, M. J. F., Vincken, J.-P., Schipper, D., Voragen, A. G. J., Beldman, G. (1996) Acetyl esterases of *Aspergillus niger*: purification and mode of action on pectins, in: *Pectins and Pectinases, Progress in Biotechnology* (Visser, J., Voragen, A. G. J., Eds.), Amsterdam: Elsevier Sciences, Vol. 14, 793–798.

Selvendran, R. R. (1985) Developments in the chemistry and biochemistry of pectic and hemicellulosic polymers, *J. Cell Sci. Suppl.* **2**, 51–88.

Selvendran, R. R., O'Neill, M. A. (1987) Isolation and analysis of cell walls from plant material, in: *Methods of Biochemical Analysis* (Glick, D., Ed.), New York: John Wiley & Sons, Vol. 32, 25–153.

Siddiqui, I. R., Wood, P. J. (1976) Structural investigation of oxalate-soluble rapeseed (*Brassica campestris*) polysaccharides. Part IV. Pectic polysaccharides, *Carbohydr. Res.* **50**, 97–107.

Smith, C. J. S., Watson, C. F., Morris, P. C., et al. (1990) Inheritance and effect on ripening of antisense polygalacturonase genes in transgenic tomatoes, *Plant Mol. Biol.* **14**, 369–379.

Thibault, J.-F. (1979) Automatisation du dosage des substances pectiques par la méthode au méta-hydroxydiphényle, *Lebensmittelwiss. Technol.* **12**, 247–251.

Thibault, J.-F., Ralet, M.-C. (2001) Pectins: their origin, structure and function, in: *Advanced Dietary Fibres* (McCleary, B. V., Prosky, L., Eds.), Oxford: Blackwell Science, 369–378.

Thibault, J.-F., Rinaudo, M. (1985) Interactions of mono- and divalent counterions with alkali- and enzyme-deesterified pectins in salt-free solutions, *Biopolymers* **24**, 2131–2143.

Thibault, J.-F., Saulnier, L., Axelos, M. A. V., Renard, C. M. G. C. (1991) Difficultés expérimentales de l'étude des macromolécules pectiques, *Bull. Soc. Bot. Fr.* **138**, 319–337.

Thibault, J.-F., Renard, C. M. G. C., Axelos, M. A. V., Roger, P., Crépeau, M.-J. (1993) Studies of the length of homogalacturonic regions in pectins by acid hydrolysis, *Carbohydr. Res.* **238**, 271–286.

Thibault, J.-F., Renard, C. M. G. C., Guillon, F. (1994) Physical and chemical analysis of dietary fibres in sugar beet and vegetables, in: *Modern Methods of Plant Analysis, Vegetables and Vegetable Products* (Jackson, J. F., Linskens, H. F., Eds.), Berlin-Heidelberg: Springer-Verlag. Vol.16, 23–55.

Thibault, J.-F., Asther, M., Colonna Ceccaldi, B. (1998) Fungal bioconversion of agricultural by-products to vanillin, *Lebensmittelwiss. Technol.* **31**, 530–536.

Thibault, J.-F., Renard, C. M. G. C., Guillon, F. (2001) Sugar beet fiber: production, composition, physicochemical properties, physiological effects, safety, and food applications, in: *Handbook of Dietary Fiber in Functional Foods Development* (Cho, S. S., Dreher, M., Eds.), New-York: Marcel Dekker, 553–582.

Van Buren, J. P. (1991) Function of pectin in plant tissue structure and firmness, in: *The Chemistry and Technology of Pectin* (Walter, R. H., Ed.), San Diego, CA: Academic Press, 1–22.

Vogel, M. (1991) Alternative utilization of sugar beet pulp, *Zuckerindustrie* **116**, 265–270.

Voragen, A. G. J., Pilnik, W., Thibault, J.-F., Axelos, M. A. V., Renard, C. M. G. C. (1995) Pectins, in: *Food Polysaccharides* (Stephen, A. M., Dea, Y., Eds.), London: Marcel Dekker, 287–339.

Voragen, A. G. J., Beldman, G., Schols, H. A. (2001) Chemistry and enzymology of pectins, in: *Advanced Dietary Fibre Technology* (McCleary, B. V., Prosky, L., Eds.), Oxford: Blackwell Science, 379–398.

Wakabayashi, K. (2000) Changes in cell wall polysaccharides during fruit ripening, *J. Plant Res.* **113**, 231–237.

Wakabayashi, K., Chun, J.-P., Huber, D. J. (2000) Extensive solubilization and depolymerisation of cell wall polysaccharides during avocado (*Persea americana*) ripening involves concerted action of polygalacturonase and pectinmethyl-esterase, *Physiol. Plant.* **108**, 345–352.

Walkinshaw, M. D., Arnott, S. (1981a) Conformations and interactions of pectins. Part I. X-Ray diffraction analyses of sodium pectate in neutral and acidified forms, *J. Mol. Biol.* **153**, 1055–1073.

Walkinshaw, M. D., Arnott, S. (1981b) Conformations and interactions of pectins. Part II. Models for junction zones in pectinic acid and calcium pectate gels, *J. Mol. Biol.* **153**, 1075–1085.

Weightman, R. M., Renard, C. M. G. C., Thibault, J.-F. (1994) Structure and properties of the polysaccharides from pea hulls. Part I. Chemical extraction and fractionation of the polysaccharides, *Carbohydr. Res.* **24**, 139–148.

Willats, W. G. T., Limberg, G., Buchholt, H. C., Van Alebeek, G.-J., Benen, J., Christensen, T. M. I. E., Visser, J., Voragen, A. G. J., Mikkelsen, J. D., Knox, J. P. (2000) Analysis of pectic epitopes recognised by hybridoma and phage display monoclonal antibodies using defined oligosaccharides, polysaccharides, and enzymic degradation, *Carbohydr. Res.* **327**, 309–320.

Williamson, G. (1991) Purification and characterization of pectin acetylesterase from orange peel, *Phytochemistry* **30**, 445–449.

Williamson, G., Kroon, P. A., Faulds, C. B. (1998) Hairy plant polysaccharides: a close shave with microbial esterases, *Microbiology* **144**, 2011–2023.

Wong, C.-H., Mazenod, F. P., Whitesides, G. M. (1983) Chemical and enzymatic syntheses of 6-deoxyhexoses. Conversion of 2,5-dimethyl-4-hydroxy-2,3-dihydrofuran-3-one (furaneol) and analogues, *J. Org. Chem.* **48**, 3493–3497.

Yamasaki, M., Yasui, T., Amarina, K. (1964) Pectic enzymes in the clarification of apple juice. I. Study on the clarification reaction in a simplified mode, *Agric. Biol. Chem.* **28**, 779–787.

Zetelaki-Horvath, K., Gatai, K. (1977a) Disintegration of vegetable tissues by endo-polygalacturonase, *Acta Aliment.* **6**, 227–240.

Zetelaki-Horvath, K., Gatai, K. (1977b) Application of endopolygalacturonase in vegetable and fruit, *Acta Aliment.* **6**, 335–376.

11
Pullulan

Dr. Timothy D. Leathers
Fermentation Biotechnology Research Unit, National Center for Agricultural Utilization Research, Agricultural Research Service, United States Department of Agriculture; 1815 N. University St.; Peoria, IL 61604, USA; Tel.: +1-309-681-6377; Fax: +1-309-681-6427; E-mail: leathetd@ncaur.usda.gov

1	Introduction	388
2	Historical Outline	388
3	Chemical Structure	389
4	Physiological Function	391
5	Chemical Analyses	391
6	Occurrence	392
7	Biosynthesis	393
7.1	Culture Conditions and Cell Morphology	393
7.2	Biosynthetic Mechanism	396
8	Genetics and Molecular Biology	397
9	Biodegradation	399
10	Production	401
11	Properties and Applications	402
12	Patents	405
13	Outlook and Perspectives	408
14	References	409

Polysaccharides and Polyamides in the Food Industry. Properties, Production, and Patents.
Edited by A. Steinbüchel and S. K. Rhee
Copyright © 2005 WILEY-VCH Verlag GmbH & Co. KGaA, Weinheim
ISBN: 3-527-31345-1

ADPG   adenosine 5′-diphosphate-glucose
ATP   adenosine 5′-triphosphate
EC   enzyme commission
IR   infrared
UDPG   uridine 5′-diphosphate-glucose

Names are necessary to report factually on available data; however, the USDA neither guarantees nor warrants the standard of the product, and the use of the name by USDA implies no approval of the product to the exclusion of others that also may be suitable.

1
Introduction

Pullulan is a linear homopolysaccharide of glucose that often is described as an α-(1 → 6) linked polymer of maltotriose subunits. This unique linkage pattern endows pullulan with distinctive physical traits. Pullulan has adhesive properties and can be used to form fibers, compression moldings, and strong, oxygen-impermeable films. Pullulan is derivatized easily to control its solubility or provide reactive groups. Consequently, pullulan and its derivatives have numerous potential food, pharmaceutical, and industrial applications. Pullulan is produced as a water-soluble, extracellular polysaccharide by certain strains of the polymorphic fungus *Aureobasidium pullulans* (De Bary) Arnaud (formerly known as *Pullularia pullulans* (De Bary) Berkhout or *Dematium pullulans* De Bary). A number of reviews on pullulan have appeared, including those by Yuen (1974), Jeanes (1977), Zajic and LeDuy (1977), Slodki and Cadmus (1978), Catley (1979), Kondrat'eva (1981), Sandford (1982), LeDuy et al. (1988), Deshpande et al. (1992), Pollock (1992), Robyt (1992), Seviour et al. (1992), Tsujisaka and Mitsuhashi (1993), Dillon and Martin (1994), Côté and Ahlgren (1995), Lachke and Rale (1995), De Baets and Vandamme (1999), and Israilides et al. (1999).

2
Historical Outline

Aureobasidium pullulans was first described (as *Dematium pullulans*) by De Bary in 1866 (Cooke, 1959). Bauer (1938) made early observations on polysaccharide formation by *A. pullulans*, and Bernier (1958) isolated and began to characterize the polymer. Bender et al. (1959) studied the novel polysaccharide and named it "pullulan". During the 1960s, the basic structure of pullulan was resolved (Bender and Wallenfels, 1961; Wallenfels et al., 1961, 1965; Bouveng et al., 1962, 1963a; Sowa et al., 1963; Ueda et al., 1963). Bender and Wallenfels (1961) discovered the enzyme pullulanase, which specifically hydrolyzes the α-(1 → 6) linkages in pullulan and converts the polysaccharide almost quantitatively to maltotriose. Thus, pullulan is commonly viewed as an α-(1 → 6) linked polymer of maltotriose subunits. Catley and coworkers subsequently established the occurrence of a minor percentage of randomly distributed maltotetraose subunits in pullulan (Catley et al., 1966; Catley, 1970; Catley and Whelan, 1971; Carolan et al., 1983). Commercial production of pullulan began in 1976 by the Hayashibara Company, Ltd., in Okayama, Japan (Tsujisaka and Mitsuhashi, 1993).

3 Chemical Structure

Bernier (1958) isolated water-soluble polysaccharides from cultures of A. pullulans and reported that glucose is the major product of acid hydrolysis. Based on the positive optical rotation and infrared (IR) spectrum of pullulan, Bender et al. (1959) concluded that the polymer is an α-glucan in which α-(1 → 4) linkages predominate. Subsequent studies using IR, periodate oxidation, and methylation analysis established that pullulan is essentially a linear glucan containing α-(1 → 4) and α-(1 → 6) linkages in a ratio of 2:1 (Wallenfels et al., 1961, 1965; Bouveng et al., 1962; Sowa et al., 1963). Partial acid hydrolysates of pullulan include isomaltose, maltose, panose, and isopanose (Bender et al., 1959; Bouveng et al., 1963a; Sowa et al., 1963). The discovery of the enzyme pullulanase provided a critical tool for the analysis of the structure of pullulan (Bender and Wallenfels, 1961). Pullulanase specifically hydrolyzes the α-(1 → 6) linkages of pullulan and converts the polymer almost quantitatively to maltotriose (Bender and Wallenfels, 1961; Wallenfels et al., 1961, 1965). Based on this result, pullulan frequently is described as a polymer of α-(1 → 6) linked maltotriose subunits (Figure 1). However, pullulan also can be viewed as a polymer of panose or

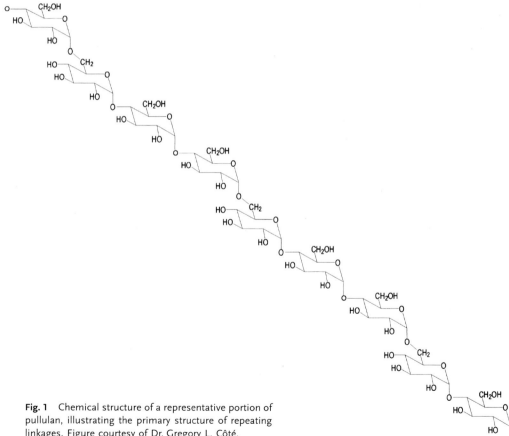

Fig. 1 Chemical structure of a representative portion of pullulan, illustrating the primary structure of repeating linkages. Figure courtesy of Dr. Gregory L. Côté.

isopanose subunits, which may reflect more accurately the biosynthetic origins of the molecule. Indeed, a number of enzymes have been described since that produce panose or isopanose from pullulan (see Section 9).

Catley and coworkers established that pullulan contains maltotetraose subunits (Figure 2) in addition to the predominant maltotriose subunits (Catley et al., 1966; Catley, 1970; Catley and Whelan, 1971). The frequency of maltotetraose subunits appears to vary on a strain-specific basis, from about 1% to 7% of total residues (Taguchi et al., 1973a; Catley et al., 1986). Evidence suggests that maltotetraose subunits are distributed randomly throughout the molecule (Carolan et al., 1983). Unlike the maltotriose subunits in pullulan, maltotetraose residues are substrates for many α-amylases, and it has been proposed that hydrolysis of pullulan at these sites accounts for the decrease in molecular weight commonly observed in late cultures (see Section 9).

It is possible that pullulan contains additional minor structural features, perhaps depending on culture conditions and strain differences. Sowa et al. (1963) reported that pullulan fro one source contains 6% α-(1 → 3) linkages, and Fujii et al. (1984) indicated that pullulan from *A. pullulans* strain FERM-P4257 includes some α-(1 → 3) linkages. Fungal glucoamylases generally attack pullulan in a sequential, exo-fashion from the non-reducing end, hydrolyzing both α-(1 → 4) and α-(1 → 6) linkages. Marshall (1975) found that a glucoamylase from *Cladosporium resinae* digests pullulan rapidly but incompletely and concluded that the molecule may contain anomalous linkages. Similarly, Catley et al. (1986) reported that pullulan from *A. pullulans* strain P-50 is partially resistant to digestion by glucoamylase from *Aspergillus niger* and suggested that the molecule may contain randomly distributed branches or alternative subunit residues. Zemek et al. (1980) proposed that pullulan from six strains of *A. pullulans* contains glycogen-like stretches of 52–135 glucosyl residues at the non-reducing termini.

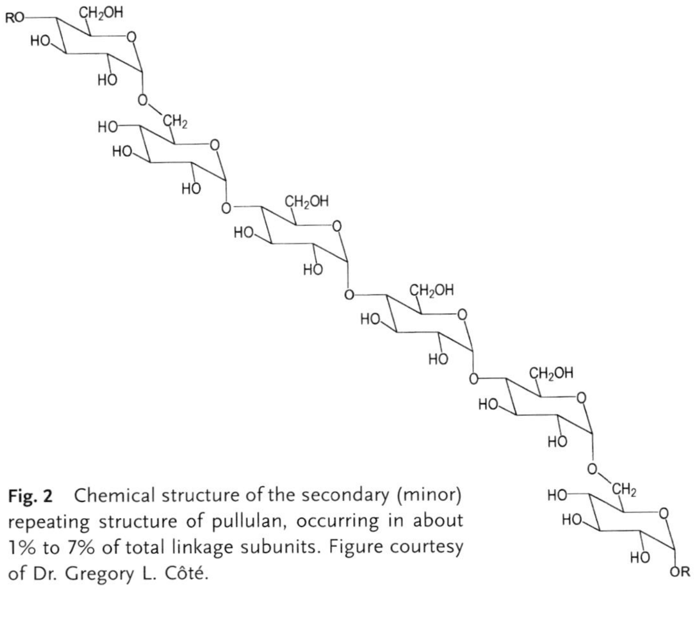

Fig. 2 Chemical structure of the secondary (minor) repeating structure of pullulan, occurring in about 1% to 7% of total linkage subunits. Figure courtesy of Dr. Gregory L. Côté.

Many reports describe additional (non-pullulan) polysaccharides from cultures of *A. pullulans*. Bernier (1958) noted the production of a distinct insoluble "jelly" associated with mycelial mats, and Wallenfels et al. (1965) developed a method to better isolate pullulan (termed "restpullulan") from this material. Bouveng et al. (1962) reported that cultures grown on sucrose produce both pullulan and an acidic glucan that contains uronic acid, while cultures grown on maltose produce a heteropolysaccharide that contains glucose, galactose, and mannose. On xylose, cultures made three polysaccharides: (1) pullulan; (2) an acidic heteropolysaccharide that contained glucose, galactose, mannose, and glucuronic acid; and (3) a β-linked glucan containing $(1 \rightarrow 3)$ and $(1 \rightarrow 6)$ linkages (Bouveng et al., 1963b). Taguchi et al. (1973a) reported that the production of acidic extracellular polysaccharide varies among three strains of *A. pullulans*. Cell walls of *A. pullulans* contain both heteropolysaccharides and β-glucan with $(1 \rightarrow 3)$ and $(1 \rightarrow 6)$ linkages (Brown and Lindberg, 1967a,b; Brown et al., 1973). Kikuchi et al. (1973) described an insoluble heteropolysaccharide from *A. pullulans* strain S-1 that contains glucose, mannose, and galactose in ratios similar to polysaccharides from cell-wall extracts, and they suggested that this material is released by autolysis, possibly during cell morphogenesis. Elinov and Matveeva (1972) and Elinov et al. (1974, 1975) reported that *A. pullulans* strain VKPM F-448 produces a homoglucan with a $\beta\text{-}(1 \rightarrow 3)$ linked backbone and $\alpha\text{-}(1 \rightarrow 4)$ linked side chains attached by $\beta\text{-}(1 \rightarrow 6)$ linkages. This potentially useful polysaccharide was later termed "aubasidan" (Elinov et al., 1987). Yurlova and de Hoog (1997) subsequently identified a distinct taxonomic group of *A. pullulans* strains that make both pullulan and aubasidan-like polysaccharides. Fujii et al. (1984) reported that strain FERM-P4257 secretes a $\beta\text{-}(1 \rightarrow 3)$-linked glucan as an alternative to pullulan, depending on the medium used. Leal-Serrano et al. (1980) reported that *A. pullulans* strain CBS 105.22 produces an acid β-glucan with 68% $(1 \rightarrow 3)$ and 32% $(1 \rightarrow 6)$ linkages and containing malic acid. Hamada and Tsujisaka (1983) found that *Aureobasidium* sp. strain K-1 forms an acidic glucan containing sulfoacetic acid, with a $\beta\text{-}(1 \rightarrow 3)$ linked backbone and single $\beta\text{-}(1 \rightarrow 6)$ branch linkages on three of four backbone residues. Promma et al. (1997) described a marine isolate of *A. pullulans* that makes an acidic α-glucan containing malic acid, with alternating $(1 \rightarrow 6)$ and $(1 \rightarrow 4)$ backbone linkages and regular $(1 \rightarrow 3)$ branch linkages.

4
Physiological Function

The physiological function of pullulan is uncertain. Since *Aureobasidium* strains cannot break down pullulan significantly to metabolizable sugars, pullulan presumably does not serve as a storage material. It is generally supposed that pullulan and similar polysaccharides serve to protect cells from dessication or help them adhere to environmental substrates. Andrews et al. (1994) reported that pullulan helps *A. pullulans* adhere to leaf surfaces. Bardage and Bjurman (1998) concluded that pullulan is important in the adhesion of *A. pullulans* blastospores to painted wood.

5
Chemical Analyses

General methods applicable to polysaccharides can be used to detect and quantitate pullulan. As described above, the basic structure of pullulan was deduced in part from its specific optical rotation and IR absorption characteristics. Periodate analysis, methylation analysis, and partial acid hydrolysis were also important. However, studies using pullulanase, which specifically hydrolyzes the α-(1→6) linkages in pullulan, were definitive in resolving the fundamental structure of the polymer (Bender and Wallenfels, 1961). Assays based on pullulanase remain the principal method for the specific detection and measurement of pullulan. Catley (1971a) developed a sensitive radioassay for pullulan. *A. pullulans* cultures were fed ^{14}C-glucose to produce labeled pullulan. After pullulanase digestion, maltotriose and maltotetraose were separated by paper chromatography and measured by liquid scintillation counting. Finkelman and Vardanis (1982b) described a simplified radioassay based on the solubilization by pullulanase of radioactive counts into ethanol. Leathers et al. (1988) described a quantitative pullulan assay based on reducing sugars (maltotriose equivalents) released by pullulanase digestion. Israilides et al. (1994) suggested the complete hydrolysis of pullulan by means of pullulanase and glucoamylase, followed by the specific measurement of glucose using glucose oxidase. Imshenetskii and Kondrat'eva (1978) devised a rapid assay based on the turbidity of pullulan suspensions in water–ethanol mixtures. An enzyme sensor for measuring pullulan or pullulanase also has been reported, utilizing immobilized glucoamylase and glucose oxidase (Renneberg et al., 1985). The molecular weight of pullulan has been estimated by various methods, including light scattering (Ueda et al., 1963), sedimentation behavior (Wallenfels et al., 1965), viscosity (Wallenfels et al., 1965), and chromatography (Catley, 1970).

6
Occurrence

A. pullulans is a ubiquitous fungus, commonly isolated from environmental samples (Cooke, 1959; Hermanides-Nijhof, 1977). It is found in soil and water, particularly as an early-colonizing saprophyte on decaying leaf litter, wood, and many other plant materials. *A. pullulans* often is reported as a (usually mild) plant pathogen, notably of grapes. It is important in the post-harvest decay of fruits and other crops. *A. pullulans* has been studied for its role in the deterioration of house paints (Brand and Kemp, 1973) and in the rotting of military equipment in the tropics (Marsh and Bollenbacher, 1949). It also has been reported as a troublesome slime-producing contaminant of paper mills. It can colonize optical lenses and electronic equipment and has been implicated in the biodeterioration of plasticized polyvinyl chloride (Webb et al., 2000). Clinical isolates are known, and insect associations have been reported. The fungus has been described as omnivorous, and isolates produce an impressive array of degradative enzymes, including amylases (Federici, 1982; Leathers, 1987, 1993; Linardi and Machado, 1990; Saha et al., 1993) proteases, (Ahearn et al., 1968; Federici, 1982), esterases (Federici, 1982), pectinases (Dennis and Buhagiar, 1973; Finkelman and Zajic, 1978; Federici, 1982), and hemicellulases, including xylanase (Flannigan, 1970; Biely et al., 1978, 1979; Saha and Bothast, 1998) and mannanase (Kremnicky et al., 1996; Kremnicky and Biely, 1997). So-called "color-variant" strains of *Aureobasidium*, some of which produce pullulan, are natural overproducers of endoxylanase (Leathers et al., 1984; Leathers, 1986; 1989). *A. pullulans* utilizes cellobiose, but not cellulose (White et al., 1948; Marsh et al., 1949; Flannigan, 1970; Dennis, 1972; de Hoog and Yurlova,

1994; Saha et al., 1994). Surveys suggest that many, but not all, strains are capable of producing pullulan (Ueda et al., 1963; Leathers et al., 1988; Augustin et al., 1997). However, optimal culture conditions for pullulan production appear to vary considerably on a strain-specific basis (Silman et al., 1990; Yurlova et al., 1995).

Pullulan also has been isolated from the saprophytic (and sometimes mycoparasitic) fungus *Tremella mesenterica* (Fraser and Jennings, 1971), from the obligate tree parasitic fungi *Cyttaria harioti* and *C. darwinii* (Waksman et al., 1977; Oliva et al., 1986), and from the fungal agent of chestnut blight, *Cryphonectria parasitica* (Corsaro et al., 1998). Interestingly, Corsaro et al. (1998) reported that pullulan from *C. parasitica* exhibits mild phytotoxic effects against both host and non-host (tomato) tissues.

7
Biosynthesis

7.1
Culture Conditions and Cell Morphology

Much of the published research on pullulan concerns the control of pullulan synthesis by culture conditions and the relationship between pullulan production and cell morphology. Overall, this literature is confusing and seemingly contradictory. In part, this may be because multiple factors interact in the regulation of pullulan biosynthesis. Perhaps more important may be the role of strain variability. In comparative studies, *A. pullulans* strains have been observed to differ considerably with respect to growth, pullulan yields, and cell morphology (Ueda et al., 1963; Cernakova et al., 1980; Kockova-Kratochvilova et al., 1980; Park, 1982b; Gadd and Cooper, 1984; Leathers et al., 1988; Silman et al., 1990; Augustin et al., 1997).

Optimal conditions for pullulan production may depend on the particular isolate of *A. pullulans* employed.

A. pullulans strains typically produce pullulan when cultured on sucrose, glucose, fructose, maltose, starch, or maltooligosaccharides (Bender et al., 1959; Ueda et al., 1963; Catley, 1971b; Behrens and Lohse, 1977; Imshenetskii et al., 1981a, 1985b; Leathers et al., 1988; West and Reed-Hamer, 1991; Badr-Eldin et al., 1994). Sucrose often has been described as the optimal substrate. Less frequently, sugars such as xylose, arabinose, mannose, galactose, rhamnose, and lactose have been reported to support pullulan production, usually in reduced yields (Bouveng et al., 1962, 1963b; Ueda et al., 1963; Imshenetskii et al., 1981a; LeDuy et al., 1983). Perhaps the best-characterized pullulan-producing strain is *A. pullulans* QM 3090 (also known as NRRL Y-2567 or ATCC 9348; in some publications, this strain is referred to as QM 3092). Using this strain, Catley (1972) found that pullulan synthesis is inducible by glucose, fructose, or saccharides that can be broken down to these sugars. Induction of pullulan synthesis was inhibited by cycloheximide, indicating that new protein synthesis is required (Catley, 1972). Catley (1971b) reported that glycerol can contribute to pullulan synthesis, but only in the presence of the inducer glucose. This result also suggests that potential contamination by glucose is a concern in substrate-utilization studies.

A. pullulans has been exploited for some time as a model organism for studies of fungal morphogenesis. Although the organism is actually polymorphic, studies usually have focused on dimorphic transitions between yeast-like cells and mycelia, or between yeast-like cells and chlamydospores (large resting cells). Again, much of this literature appears to be inconsistent, possibly owing to differences among strains of *A.*

pullulans. Brown et al. (1973) described a strain of *A. pullulans* that is mainly yeast-like during growth on nitrate as a nitrogen source and is more filamentous during growth on ammonium. On the other hand, Park (1982b) found under different conditions that ammonium is not required for morphogenesis from yeast-like to filamentous growth, although it does favor maintenance of mycelial growth. Sevilla et al. (1977) described a yeast-like strain of *A. pullulans* that grows in a filamentous form only with cysteine as a nitrogen source. However, Park (1982a) characterized a different strain in which mycelial germination is supported by a number of amino acids. Vinroot and Torzilli (1988) found that high pH (7.5) inhibits yeast-like growth. In contrast, Park (1982a) concluded that pH is not correlated with cell morphology. Low inoculum densities appear to favor mycelial development in some strains (Ramos and Garcia Acha, 1975; Park, 1984), while high inoculum densities seem to favor mycelial growth in other strains (Catley, 1980; Vinroot and Torzilli, 1988). Similarly, high agitation rates may favor either yeast-like (Leal-Serrano et al., 1980) or mycelial (Vinroot and Torzilli, 1988) growth. A number of other factors have been found to affect the morphology of *A. pullulans*. Moragues et al. (1988) observed that n-alkanols promote hyphal development of strain CECT 2660 (ATCC 48433). Copper and yeast extract were reported to promote filamentous growth of strain IMI 45533 (Gadd and Griffiths, 1980; Cooper and Gadd, 1984). Conditions of limiting nitrogen and low buffering capacity (allowing a drop in culture pH) promoted a shift in *A. pullulans* strain CECT 2660 from yeast-like cells to chlamydospores (Dominguez et al., 1978; Bermejo et al., 1981a,b).

Correlations between *A. pullulans* morphology and pullulan production often have been reported, although it is not clear that these are cause-and-effect relationships. Catley (1980) concluded that the yeast-like form of strain QM 3090 is the primary producer of pullulan. During growth on a suitable carbon source, batch cultures exhibited a drop in pH and a shift from mycelial to yeast-like morphology upon limitation of ammonium nitrogen (Catley, 1971a, 1973). Pullulan was produced in late exponential and early stationary-phase cultures, and thus primarily by yeast-like cells. Ono et al. (1977b) made similar observations using *A. pullulans* strain S-1 and concluded likewise that yeast-like cells are the source of pullulan. Kelly and Catley (1977) isolated mutants of QM 3090 that exhibit an increased proportion of yeast-like cells and increased pullulan yields. Consistent with the notion that ammonium inhibits pullulan synthesis, Imshenetskii et al. (1981a) found that derivatives of *A. pullulans* strain 1125_4 produce more pullulan on nitrate than on ammonium nitrogen sources. Seviour and Kristiansen (1983) also reported that free ammonium ions inhibit pullulan formation by strain QM 3092 (ATCC 201428). Heald and Kristiansen (1985) decided that the yeast-like form of QM 3092 is the primary source of pullulan and found that the rate of pullulan synthesis is independent of pH. Using this strain, McNeil and Kristiansen (1987) reported that pullulan yield and molecular weight, as well as the percentage of yeast-like cells, increase as a function of fermentor impeller speed. Using strain QM 3090, Bulmer et al. (1987) concluded that pullulan is a secondary metabolite produced by yeast-like cells during ammonium limitation.

In contrast, McNeil et al. (1989) found that pullulan production by strain QM 3092 under chemostat conditions is optimal at pH 4.5, which favors a 50% mixture of yeast-like and hyphal cells. Auer and Seviour (1990) confirmed that ammonium ions repress pullulan formation by strain QM 3092. Nevertheless, they found no clear correlations among morphology, pH, and pullulan yields. Reeslev et al. (1991, 1993) and Reeslev and Jensen (1995) reported that pullulan production by batch cultures of QM 3092 is correlated with the formation of yeast-like cells, promoted by low levels of Zn^{2+}, Fe^{3+}, or yeast extract. However, under chemostat conditions that favored yeast-like growth from pH 3 to 7, pullulan production was found to be optimal only at pH 4.0 (Reeslev et al., 1997). Gibbs and Seviour (1992) concluded that bioreactor design and nitrogen source affect pullulan formation by strain QM 3092, but not its morphology. LeDuy and Boa (1983) reported that strains 2552 and 140B grow with different morphologies on peat hydrolysate medium but make equivalent amounts of pullulan. Studies with *A. pullulans* strain ATCC 42023 (ppKM-3) found that ammonium and complex nitrogen sources are superior to nitrate for pullulan production (West and Reed-Hamer, 1991; Reed-Hamer and West, 1994). Lacroix et al. (1985) developed a two-stage fermentation process for pullulan, based on the finding that strains 2552 and 140B grow best at pH 2.0 but make pullulan only at pH 5.5. Lee and Yoo (1993) reported that *A. pullulans* strain IFO 4464 provides optimal yields of pullulan at an initial pH of 6.0, although pullulan molecular weights are highest at an initial pH of 3.0. Imshenetskii et al. (1981b) determined that an initial pH of 8.0 is optimal for pullulan synthesis by strain 1125_4 and its derivatives. Using *A. pullulans* strain IMI 145194, Madi et al. (1996) found that pH 4.5 is optimal for pullulan production, while yeast-like growth is favored at pH 6.5. Yamasaki et al. (1993a,b) observed that strain S-2 is primarily yeast-like during active growth, associated with pullulan accumulation, and becomes partially mycelial in late cultures. On the other hand, Simon et al. (1993, 1995) concluded that pullulan formation by *A. pullulans* strain CBS 70076 is associated mainly with swollen cells and chlamydospores. Similarly, Andrews et al. (1994) reported that strains Y-117 (ATCC 90393) and FS 43–262 (ATCC 90394) make pullulan on conversion to swollen cells and chlamydospores, mediated by conditions of limiting nitrogen or medium acidification.

A. pullulans is a strict aerobe, and oxygen is required for both its growth and pullulan production (Ono et al., 1977b; Rho et al., 1988; Dufresne et al., 1990; Madi et al., 1996). However, Wecker and Onken (1991) reported that optimal pullulan yields from strain QM 3090 are obtained at decreased dissolved oxygen conditions, and Gibbs and Seviour (1996) observed that high oxygen levels dramatically reduce yields. Audet et al. (1996) concluded that intermediate dissolved oxygen concentrations are optimal for pullulan synthesis by strain 2552, as high concentrations enhance yields at the expense of molecular weight. It also has been suggested that shear stress associated with fermentor agitation may influence both cell morphology and pullulan formation (McNeil and Kristiansen, 1987).

Optimal temperatures for pullulan production appear to vary slightly on a strain-specific basis, usually in the range of 24 °C to 30 °C (Zajic, 1967; Imshenetskii et al., 1981b; McNeil and Kristiansen, 1990; Tsujisaka and Mitsuhashi, 1993; West and Reed-Hamer, 1993a). Vitamins and minerals also may influence pullulan synthesis. Bender et al. (1959) found that thiamine increases pullulan yields. Biotin, ferric chloride, manganese chloride, and zinc chloride were reported to enhance pullulan formation by *A. pullulans* strain ATCC 42023 (West and Reed-Hamer, 1992; West and Strophus, 1997a). However, while using strain QM 3092, Reeslev et al. (1993) and Reeslev and

Jensen (1995) determined that Fe^{3+} and Zn^{2+} inhibit the development of both yeast-like cells and polysaccharide.

Although the growth of *A. pullulans* is not always easy to measure reliably, many studies support the idea that pullulan production occurs primarily during the stationary-growth phase (Ueda et al., 1963; Ono et al., 1977b; McNeil and Kristiansen, 1987, 1990). In a survey of 18 strains of *A. pullulans*, Leathers et al. (1988) reported an inverse relationship between biomass and pullulan yields, consistent with the notion that pullulan is a secondary metabolite generated during growth limitation in excess carbon. On the contrary, a survey of 31 different strains under different conditions found that biomass and pullulan yields vary independently (Augustin et al., 1997). Kristiansen et al. (1982) determined that pullulan production follows intermediate kinetics. Other studies show that pullulan formation is associated principally with active cell growth. A kinetic analysis by Klimek and Ollis (1980) concluded that pullulan synthesis is wholly growth-associated.

Some studies have suggested an inverse relationship between pullulan yields and apparent molecular weights (Leathers, 1987; Pollock et al., 1992; Audet et al., 1996). The molecular weight of pullulan may depend not only on the synthetic characteristics of the culture but also on endogenous glucanases that partially degrade pullulan. Pullulan molecular weight generally falls with culture time, presumably because of the action of these enzymes (see Section 9). Interestingly, Madi et al. (1997) reported that strain IMI 145195 produces two distinct pullulan fractions that differ only in molecular weight (2×10^6 and 4.3×10^5). Similarly, McNeil and Kristiansen (1987) observed two molecular weight species in some pullulan preparations. Strains have been isolated that favor production of high-molecular-weight pullulan (Pollock et al., 1992; Thorne et al., 1993, 2000), and molecular weight can be influenced by control of culture conditions (Yuen, 1974; Kato and Shiosaka, 1975b; Kaplan et al., 1987; Wiley et al., 1987, 1993). Low initial phosphate and pH levels have been reported to favor the formation of high-molecular-weight pullulan (Sugimoto, 1978; Tsujisaka and Mitsuhashi, 1993). Apparent pullulan molecular weights as high as approximately 10^6 to 10^7 have been reported.

A. pullulans is considered to be one of the "black yeasts", and pullulan contamination by fungal melanin is a common problem. Melanin production appears to vary considerably, depending on strain characteristics and culture conditions (Kockova-Kratochvilova et al., 1980; Leathers et al., 1988; Silman et al., 1990). Pigment accumulation typically occurs late in culture growth, possibly associated with the formation of chlamydospores. Naturally occurring and mutant strains have been isolated that produce reduced levels of melanin (Leathers et al., 1988; Silman et al., 1990; Bock et al., 1991; Pollock et al., 1992; Tarabasz-Szymanska and Galas, 1993; West and Reed-Hamer, 1993b).

7.2
Biosynthetic Mechanism

Relatively little is understood about the mechanism of pullulan biosynthesis. Unlike bacterial dextrans, which are synthesized extracellularly by secreted glucansucrases, pullulan is synthesized intracellularly and secreted by *A. pullulans*. Ueda and Kono (1965) found that acetone-dried cells of *A. pullulans* transform maltose to form panose, maltotriose, and maltotetraose. Taguchi et al. (1973b) reported pullulan synthesis by cell-free extracts of *A. pullulans* from uridine 5′-diphosphate-glucose (UDPG) in the presence of adenosine 5′-triphosphate

(ATP). Extracts did not form pullulan from sucrose or adenosine 5'-diphosphate-glucose (ADPG), and ATP absolutely was required. Acetone-dried cells produced pullulan from sucrose and incorporated ^{14}C-labeled sucrose into organic extracts that included lipid fractions, predicting the involvement of a lipid intermediate (Taguchi et al., 1973b). Further analysis indicated that the lipid is a discrete species that contains a glucose moiety and pyrophosphate. Cold sucrose quickly chased the label from this fraction, indicating a rapid turnover of this intermediate. Ono et al. (1977a) confirmed that glucolipids are formed in mycelia believed to be accumulating pullulan. Catley and McDowell (1982) found that *A. pullulans* incorporates ^{14}C-labeled glucose into lipid-linked glucose, isomaltose, panose, and isopanose. Based on these results, a reaction mechanism was proposed in which pullulan is formed by the polymerization of either panosyl or isopanosyl moieties (Catley and McDowell, 1982). An occasional direct linkage of panosyl and isopanosyl moieties was postulated to form the minor maltotetraosyl elements in pullulan. Mechanisms involving polyprenyl-linked saccharides are well established in the synthesis of oligosaccharides and polysaccharides of diverse organisms, including *Klebsiella* (Sutherland and Norval, 1970) and *Saccharomyces* (Lehle and Tanner, 1978). This model contrasts with the transglucosylation mechanism invoked in the synthesis of dextran (Robyt, 1995). Nevertheless, a glucosyltransferase that catalyzes the synthesis of panose and isomaltose from maltose has been characterized from *A. pullulans* (Hayashi et al., 1994a,b).

Catley and Hutchison (1981) prepared spheroplasts of *A. pullulans* that were impaired in pullulan elaboration and concluded that the assembly or secretion of the polysaccharide may be associated with the cell wall, plasma membrane, or periplasmic space. However, Finkelman and Vardanis (1982a) prepared protoplasts of *A. pullulans* that were able to produce pullulan and concluded to the contrary. No enzymes involved in pullulan biosynthesis have been identified at this time, and attempts to repeat the cell-free synthesis of pullulan as described by Taguchi et al. (1973b) have proven unsuccessful (Catley and McDowell, 1982; Israilides et al., 1999).

8 Genetics and Molecular Biology

A. pullulans is genetically imperfect and traditionally has been considered to be among the Fungi Imperfecti or Deuteromycetes (Wynne and Gott, 1956; Cooke, 1962; Hermanides-Nijhof, 1977). More recently, *Aureobasidium* has been described as a filamentous ascomycete (Euascomycetes, order Dothideales, family Dothideaceae) capable of growing yeast-like in culture (de Hoog and Yurlova, 1994; de Hoog, 1998). *Aureobasidium* commonly is referred to as a yeast-like fungus and often is considered to be one of the "black yeasts" because many cultures produce fungal melanin. However, the fungus is more properly polymorphic and has a complex life cycle featuring a range of morphological forms (Cooke and Matsuura, 1963; Ramos and Garcia Acha, 1975; Pechak and Crang, 1977; Kockova-Kratochvilova et al., 1980). Cultures commonly contain a mixture of cell morphologies, ranging from yeast-like budding cells to highly mycelial forms and including "swollen cells" and large chlamydospores. The distribution of cell morphologies varies considerably as a function of culture conditions and strain differences.

As previously discussed, strain variability appears to be an important factor in pullulan production and other characteristics of *A.*

pullulans. However, it is uncertain how much genetic diversity is represented by such strain differences. Cernakova et al. (1980) organized 43 strains of *A. pullulans* into three major groups based on biochemical characteristics. Based on DNA polymorphisms, Mokrousov (1995a,b) proposed that 40 strains classified as *A. pullulans* be divided into three separate species. Using similar techniques, Yurlova et al. (1995) sorted 42 strains of *A. pullulans* into four groups that also differ in optimal culture conditions for polysaccharide production. Subsequently, Yurlova and de Hoog (1997) proposed that strains capable of producing both pullulan and aubasidan-like polysaccharides comprise a new taxonomic variety, *A. pullulans* var. *aubasidani* Yurlova. Leathers et al. (1988) reported that one ecologically and morphologically distinct group of *Aureobasidium* isolates, the so-called "color-variant" strains, exhibit only 37% to 44% DNA relatedness to typically pigmented strains and therefore may represent a recently diverged sibling species. Although morphologically similar to typically pigmented strains of *A. pullulans*, which are off-white to black, naturally occurring color variants display brilliant pigments of red, yellow, pink, or purple (Wickerham and Kurtzman, 1975). Unlike typically pigmented strains, color variants thus far have been found only in tropical latitudes, although they may be isolated side-by-side with typical strains. Notably, some color-variant strains produce authentic pullulan with little or no melanin contamination (Leathers et al., 1988; Silman et al., 1990). Pollock et al. (1992) isolated new strains of *A. pullulans* in San Diego County, California, some of which produce pullulan with little pigmentation. Although the new isolates fell into two groups based on DNA polymorphisms, it is unclear whether any of these strains are genetically similar to the color variants. Certain of these strains produce particularly high molecular weight pullulan (Pollock et al., 1992; Thorne et al., 1993, 2000).

Mutation strategies have been employed in attempts to improve pullulan production. Kelly and Catley (1977) isolated mutants of *A. pullulans* strain QM 3090 that are more yeast-like than the parent strain and produce more pullulan. Pollock et al. (1992) isolated *A. pullulans* mutants that are more yeast-like and less pigmented than parental strains. Mutants of strain ATCC 42023 have been isolated that show reduced pigmentation (West and Reed-Hamer, 1993b) or enhanced pullulan yields (West and Reed-Hamer, 1994b). Other *A. pullulans* mutants have reduced melanin as well as enhanced pullulan yields (Bock et al., 1991; Tarabasz-Szymanska and Galas, 1993). Imshenetskii et al. (1982, 1983b, 1985a) described methods for the isolation of *A. pullulans* mutants, and polyploid derivatives were generated that exhibit enhanced yields of pullulan (Imshenetskii and Kondrat'eva, 1979; Imshenetskii et al., 1983a, 1985b). However, the authors concluded that improved yields are not directly related to ploidy levels but rather to concurrent mutations. Although the parental *A. pullulans* strain 1125_4 used in these studies is believed to be haploid, the ploidy of *A. pullulans* in general is uncertain and may vary among strains (Pollock, 1992). Although it would be difficult to screen mutants directly for improvements in pullulan molecular weight, it has been suggested that amylase-deficient mutants may produce higher molecular weight pullulan (Leathers, 1991, 1993). Strains of *A. pullulans* used in commercial production reportedly have been mutated for more rapid production of pullulan and improved yields (Sugimoto, 1978).

Molecular methods have been developed for *A. pullulans*, including the production of protoplasts (Finkelman et al., 1980; White and Gadd, 1984) and the transformation of cells (Cullen et al. 1991; Thornewell et al., 1995b). Genes have been cloned from *A. pullulans* and expressed in other systems (Thornewell et al., 1995a; Peery and Skatrud, 1996). However, molecular genetic studies of pullulan biosynthesis are currently limited by a lack of information concerning the enzymes involved.

9
Biodegradation

In early studies it was recognized that pullulan is resistant to digestion by both dextranases, which recognize consecutive internal α-(1→6) linkages, and β-amylases, which successively cleave maltose subunits from the non-reducing ends of starch (Ueda et al., 1963; Wallenfels et al., 1965). The predominant maltotriose subunit of pullulan is also resistant to most α-amylases (such as human salivary and porcine pancreatic α-amylases), which recognize consecutive internal α-(1→4) linkages in starch (Ueda et al., 1963; Wallenfels et al., 1965). However, Catley and coworkers identified minor maltotetraose subunits in pullulan and demonstrated that these residues are substrates for hydrolysis by α-amylase (Catley et al., 1966; Catley, 1970; Catley and Whelan, 1971). Furthermore, Catley (1970) observed amylase activity in *A. pullulans* culture supernatants and proposed that this activity is responsible for the reduction in pullulan molecular weight and viscosity observed in late cultures. In support of this idea, Miura et al. (1978) found that the addition of amylase inhibitor to growing cultures increases the molecular weight of pullulan produced. Since maltotetraose residues are distributed randomly throughout pullulan (Carolan et al., 1983), hydrolysis at these points should effect a rapid decrease in polymer viscosity. It is possible that this system allows *A. pullulans* to respond to growth or environmental conditions. Pullulan-productive strains have been reported to secrete low, constitutive levels of amylase when grown on a variety of non-starch substrates, and only slightly higher levels on starch (Leathers, 1987, 1993; Saha et al., 1993). Amylases from *Aureobasidium* are less active against pullulan than against starch. However, the enzymes appear to be more active than human salivary amylase against pullulan (Leathers, 1987). A comparative study of eight strains concluded that low amylase levels may be necessary but not sufficient for production of high-molecular-weight pullulan (Leathers, 1987).

Pullulan is attacked by glucoamylases (amyloglucosidase, EC 3.2.1.3) from numerous fungi, including species of *Aspergillus*, *Candida*, *Rhizopus*, and *Sclerotium* (Ueda et al., 1963; Wallenfels et al., 1965; Marshall, 1975; Saha et al., 1979; McCleary and Anderson, 1980; De Mot et al., 1985; Kelkar and Deshpande, 1993). Glucoamylases are progressive exo-acting enzymes that attack from the non-reducing end to produce glucose. *Aureobasidium* also has been reported to produce glucoamylases, and it has been postulated that these enzymes may be involved in pullulan degradation in late cultures (Saha et al., 1993; West and Strohfus, 1996a).

Bender and Wallenfels (1961) first described a pullulanase (pullulan 6-glucanohydrolase, EC 3.2.1.41) from *Klebsiella planticola* (also *K. pneumoniae*, and formerly *Aerobacter aerogenes* or *Enterobacter aerogenes*) with specificity for the internal α-(1→6) linkages of pullulan. This enzyme converts pullulan almost quantitatively to maltotriose, which is readily broken down by amylases (Catley, 1978). Because pullulanases also recognize α-(1→6) branch linkages in amylopectin, they sometimes have been referred to as "bacterial isoamylases". However, isoamylases (EC 3.2.1.68) lacking pullulanase activity are widely produced by many organisms, including bacteria, fungi, plants, and animals (Manners, 1971). Many plants produce both isoamylases and pullulanases as separate activities.

Pullulanase is of commercial interest as a debranching enzyme in the enzymatic saccharification of starches (Marshall, 1975; Norman, 1982; Woods and Swinton, 1995). In combination with glucoamylase, α-amylase, or β-amylase, pullulanase improves saccharification rates and yields. Diverse bacterial species produce pullulanase, including *Escherichia intermedia* (Ueda and Nanri, 1967), *Streptococcus mitis* (Walker, 1968), *Streptomyces flavochromogenes* (Yagisawa et al., 1972), *Bacillus cereus* (Takasaki, 1976), *Micrococcus* sp. (Kimura and Horikoshi, 1990), and *Ruminobacter amylophilus* (Anderson, 1995).

Unlike the enzyme from *Klebsiella* sp., pullulanases from some bacteria also attack α-(1 → 4) linkages in starch or starch-derived oligosaccharides (Saha et al., 1991). Sources include *Bacillus circulans* (Sata et al., 1989), *B. subtilis* (Takasaki, 1987), *Thermoanaerobacter thermohydrosulfuricus* (formerly *Clostrium thermohydrosulfuricum*) (Melasniemi, 1988; Mathupala et al., 1990), *Thermoanaerobacter brockii* (formerly *Thermoanaerobium brockii*) (Coleman et al., 1987; Plant et al., 1987), and *Thermococcus hydrothermalis* (Gantelet and Duchiron, 1998). These enzymes have been called type II pullulanases, amylopullulanases, or amylase-pullulanases.

Pullulanase from *Klebsiella* sp. is optimally active near 50 °C and at pH 5.0 to 6.5 (Bender and Wallenfels, 1961; Ueda and Ohba, 1972). Pullulanases from other organisms, particularly *Bacillus* species, have been identified that are more suitable for alkaline (Nakamura et al., 1975; Kimura and Horikoshi, 1990; Ara et al., 1992; Lee et al., 1997) or acidic conditions (Kusano et al., 1988; Tomimura, 1991; Castro et al., 1993). Some of these enzymes also have higher optimal temperatures or thermostabilities than pullulanase from *Klebsiella*, typically up to 60 °C. More highly thermophilic pullulanases (70 °C to 85 °C) have been identified from *Bacillus flavocaldarius* (Suzuki et al., 1991), *Thermoactinomyces sacchari* (formerly *T. thalpophilus*) (Odibo and Obi, 1988), *Thermoanaerobacter brockii* (Coleman et al., 1987; Plant et al., 1987), *Thermoanaerobacter thermohydrosulfuricus* (Melasniemi, 1988; Mathupala et al., 1990; Zeikus and Hyun, 1988), *Thermus aquaticus* (Plant et al., 1986), and other organisms (Koch et al., 1987; Saha and Zeikus, 1989). Recently, pullulanases from deep-sea hydrothermal vent archaea, such as *Thermococcus hydrothermalis*, have been isolated with optimal activities near 100 °C (Gantelet and Duchiron, 1998; Leveque et al., 2000).

Sakano et al. (1971, 1972, 1973) discovered an isopullulanase (pullulan 4-glucanohydrolase, EC 3.2.1.57) from *Aspergillus niger* that cleaves α-(1 → 4) linkages adjacent to the α-(1 → 6) linkages of pullulan to produce isopanose. Sakano et al. (1978) subsequently reported that a purified α-amylase from a strain of *Themoactinomyces vulgaris* hydrolyzes pullulan to produce panose. Kukuri et al. (1988) described a neopullulanase from *Bacillus stearothermophilus* that produces mainly panose from pullulan. Neopullulanases since have been identified from strains of *Bacteroides thetaiotaomicron* (Smith and Salyers, 1991) and *Bacillus polymyxa* (Yebra et al., 1997). An amylase from *Bacillus licheniformis* also has been characterized with neopullulanase activity (Kim et al., 1992). In addition, Bender (1994) described a decycling maltodextrinase from *Flavobacterium* sp. that produces panose from pullulan. A number of pullulanase genes have been cloned, including those from *A. niger* (Aoki and Sakano, 1997), alkalophilic *Bacillus* sp. (Lee et al., 1994), *B. licheniformis* (Kim et al., 1992), *B. polymyxa* (Yebra et al., 1999), *Fervidobacterium pennavorans* (Bertoldo et al., 1999), *K. pneumoniae* (Janse and Pretorius, 1993), *Thermoanaerobacter brockii* (Coleman and McAlister, 1986; Coleman et al., 1987), and *Thermoanaerobacterium saccharolyticum* (Ramesh et al., 1994).

10
Production

Methods and conditions for pullulan production have been detailed (Wallenfels and Bender, 1961; Zajic, 1967; Yuen, 1974; Kato and Shiosaka, 1974, 1975b; Kato and Nomura, 1976, 1977; Sugimoto, 1978; Catley, 1979; Kondrat'eva and Lobacheva, 1990; Tsujisaka and Mitsuhashi, 1993; McNeil and Harvey, 1993; Thorne et al., 1993, 2000; Ozaki et al., 1996; Murofushi et al., 1998). In commercial production, *A. pullulans* is cultivated batch-wise on medium containing starch hydrolysates of dextrose equivalent 40–50 at 10% to 15% concentration (Tsujisaka and Mitsuhashi, 1993). The medium includes peptone, phosphate, and basal salts. Culture pH is initially adjusted to pH 6.5, and falls, especially during the first 24 h, to a final pH of approximately 3.5. Maximal culture growth occurs within 75 h, and optimal pullulan yields are obtained within about 100 h (Tsujisaka and Mitsuhashi, 1993). Cultures are stirred and aerated, and temperature is held at 30 °C. Yields of greater than 70% of initial substrate are claimed. Culture conditions and strain selection are important in obtaining high-molecular-weight pullulan that is relatively free of melanin. *A. pullulans* cells are removed by filtration of diluted culture broth. Melanin is removed by treatment with activated charcoal, and pullulan is recovered and purified by precipitation with organic solvents, particularly alcohols. Pullulan may be further purified through the use of ultrafiltration and ion-exchange resins.

Research has suggested potential alternative methods for pullulan production. Numerous carbon sources have been tested successfully, including beet molasses (Roukas, 1998; Roukas and Liakopoulou-Kyriakides, 1999), carob pod (Roukas and Biliaderis, 1995), cornmeal hydrolysates (Imshenetskii et al., 1985b), corn syrup (West and Reed-Hamer, 1991, 1993a), fuel ethanol fermentation stillage (Leathers and Gupta, 1994), grape skin pulp (Israilides et al., 1998), olive oil and sucrose (Youssef et al., 1998), peat hydrolysate (Boa and LeDuy, 1984, 1987), hydrolyzed potato starch (Barnett et al., 1999), soy bean oil (Shabtai and Mukmenev, 1995), spent grain liquor (Roukas, 1999a), spent sulfite liquor (Zajic et al., 1979), and deproteinized, enzyme-hydrolyzed whey (Roukas, 1999b). Alternative nitrogen sources include corn steep solids, hydrolyzed soy protein, and fuel ethanol stillage (West and Reed-Hamer, 1994a; West and Strophus, 1996d, 1997c, 1999). In addition, mixed-culture techniques have been tested for the utilization of lactose (LeDuy et al., 1983) and inulin (Shin et al., 1989).

Fed-batch (Shin et al., 1987; Moscovici et al., 1996) and continuous (McNeil et al., 1989; Schuster et al., 1993; Reeslev et al., 1997) fermentations for pullulan have been demonstrated. Immobilization of *A. pullulans* cells for pullulan production has been studied (Mulchandani et al., 1989; West and Strohfus, 1996b, 1996c, 1996e, 1997b, 1998; West, 2000). Audet et al. (1996) demonstrated the use of a reciprocating plate bioreactor for pullulan fermentations, and Yamasaki et al. (1993a, 1993b) reported on cross-flow membrane technology for pullulan recovery by filtration.

Hayashibara Co., Ltd. remains the principal commercial source of pullulan. Production is currently at approximately 300 metric tons per year, and growth is anticipated during the next two years (personal communication, S. Endo, Hayashibara Co., Ltd.). Food-grade pullulan (PF-20) wholesales in Japan for approximately 20 USD per kg, and pharmaceutical-grade (deionized) pullulan (PI-20) sells for approximately 25 USD per kg (personal communication, S. Endo, Hayashibara Co., Ltd.).

11
Properties and Applications

Pullulan has numerous demonstrated uses in foods, pharmaceuticals, manufacturing, and electronics (Yuen, 1974; Sugimoto, 1978; Cottrell, 1980; Paul et al., 1986; Tsujisaka and Mitsuhashi, 1993; Vandamme et al., 1996). The regular introduction of α-$(1 \rightarrow 6)$ linkages in pullulan interrupts what would otherwise be a linear amylose chain. This difference is thought to impart structural flexibility and enhanced solubility, resulting in distinct film- and fiber-forming characteristics that allow pullulan to mimic synthetic polymers derived from petroleum.

Dry pullulan powders are white and non-hygroscopic and dissolve readily in hot or cold water. Pullulan is non-toxic, non-mutagenic, odorless, tasteless, and edible (Fujii and Shinohara, 1986; Kimoto et al., 1997). Presumably because of its resistance to mammalian amylases, it provides few calories and appears to be treated as dietary fiber in rats and humans (Oku et al., 1979; Yoneyama et al., 1990). Studies suggest that dietary pullulan functions as a prebiotic, promoting the growth of beneficial bifidobacteria (Mitsuhashi et al., 1990; Yoneyama et al., 1990; Sugawa-Katayama et al., 1994).

Pullulan can be used as a partial replacement for starch in pastas or baked goods (Yuen, 1974; Hijiya and Shiosaka, 1975b; Kato and Shiosaka, 1975a; Hiji, 1986). It is claimed that pullulan inhibits fungal growth in foods (Yuen, 1974).

The solution properties of pullulan have been studied (Buliga and Brant, 1987; Kato et al.1982; Kawahara et al., 1984; Nishinari et al., 1991). Pullulan appears to behave as a random expanded flexible coil in aqueous solutions. Modeling studies suggest that polymer flexibility is imparted by α-$(1 \rightarrow 6)$ linkages (Brant and Burton, 1981). Pullulan solutions are of relatively low viscosity, resembling gum arabic (Tsujisaka and Mitsuhashi, 1993). Pullulan can be used as a low-viscosity filler in beverages and sauces and to stabilize the quality and texture of mayonnaise. The viscosity of pullulan solutions is stable to heating, changes in pH, and most metal ions, including sodium chloride. Pullulan has demonstrated uses in cosmetics, lotions, and shampoos (Nakashio et al., 1976b). The rheological properties of whole fermentation broths also have been characterized (LeDuy et al., 1974; Miura et al., 1976).

Pullulan and its derivatives exhibit adhesive properties (Hijiya and Shiosaka, 1975a). Pullulan can be used as a denture adhesive and a binder and stabilizer in food pastes; it also can be used to adhere nuts to cookies. Pullulan solutions impart strength to paper or wood and can be used as an effective adhesive for glass, metal, and concrete (Tsujisaka and Mitsuhashi, 1993). Pullulan has been used in the manufacture of plywood (Ichikawa and Yamada, 1982) and as a glue for postage stamps and envelopes (Yuen, 1974). Pullulan can be used as a binder for tobacco (Miyaka, 1979), seed coatings and plant fertilizers (Matsunaga et al., 1977a, 1978), and foundry sand (Mori et al., 1978).

Pullulan films (Figure 3) are formed by drying a pullulan solution (usually 5% to 10%) onto an appropriate smooth surface. Films can be as thin as 5 to 60 μm. Pullulan films are clear and highly oxygen-impermeable and have excellent mechanical properties (Yuen, 1974). Underivatized films readily dissolve in water and thus melt in the mouth as edible food coatings (Conca and Yang, 1993). The oxygen resistance of pullulan films is particularly well suited for the protection of readily oxidized fats and vitamins in foods. Pullulan films can be used for coating or packaging dried foods, including nuts, noodles, confectionaries, sardines, vegetables, and meats (Krochta and De Mulder-Johnston, 1997). Films are resistant to oil and grease and can be printed on with edible inks. They also can be heat-sealed to form single-serving packets that dissolve in water. Properties can be modified by blending with other agents such as polyvinyl alcohol, amylose, or gelatin. Pullulan films can be complexed with polyethylene glycol (Hijiya and Miyake, 1990) or enriched with rice protein (Shih, 1996). Specialty films may include colors or flavors, and decorative pullulan chips are produced for food uses. Alternatively, pullulan can be applied directly to foods as a protective glaze. Some applications of pullulan films are illustrated in Figure 4.

Pullulan also can be used in pharmaceutical coatings, including sustained-release formulations (Miyamoto et al., 1986; Izutsu et al., 1987; Childers et al., 1991). Oral care products based on pullulan films recently have been commercialized (Anonymous, 2001).

Fig. 3 Rolls of pullulan film. Photograph courtesy of Hayashibara Co., Ltd.

Fig. 4 Processed foods utilizing pullulan films. Photograph courtesy of Hayashibara Co., Ltd.

Pullulan can be formed by wet or dry spinning into fibers that resemble nylon or rayon. It also can be molded by compression or extrusion into goods that resemble polystyrene or polyvinyl alcohol (Hijiya and Shiosaka, 1974, 1975c; Nakashio et al., 1975b; Matsunaga et al., 1976, 1977b; Tsuji et al., 1976). Copolymers are not necessary, so products are completely biodegradable. Contact lenses made of pullulan have been demonstrated for the delivery of ophthalmic medicines. Pullulan is also useful as an industrial flocculating agent (Zajic, 1967; Zajic and LeDuy, 1973). It can be used in the production of paper (Nakashio et al., 1976a; Nomura, 1976) and can improve the characteristics of paint (Nakashio et al., 1975a). In addition, pullulan and its derivatives have demonstrated photographic, lithographic, and electronic applications (Sano et al., 1976; Tsukada et al., 1978; Shimizu et al., 1983; Sasago et al., 1988; Vermeersch et al., 1995).

Pullulan resembles commercial dextran in that it is an essentially linear, neutral, soluble α-D-glucan (dextran is slightly branched). Sized pullulan fractions having molecular weights of 30,000 to 90,000 can be used as a blood-plasma volume expander, in place of dextran (Igarashi et al., 1983). Also like dextran, pullulan is readily derivatized. Numerous derivatizations of pullulan have been described, most intended to reduce its water solubility or to introduce charged or reactive groups for functionality. The water solubility of pullulan can be progressively reduced by esterification (Hijiya and Shiosaka, 1975a, 1975c) or etherification (Fujita et al., 1979a). Hydrogenation reportedly increases the heat stability of pullulan (Kato and Shiosaka, 1976), and carboxylation enhances its solubility in cold water (Tsuji et al., 1978). Cross-linked pullulan beads (analogous to Sephadex®) have been demonstrated to be useful in gel permeation chromatography (Nagase et al., 1979; Motozato et al., 1986). Polyanionic and cationic derivatives have been prepared (Fujita et al., 1979b; Oonishi, 1985), and pullulan gels have been used for enzyme immobilization (Hirohara et al., 1981). Cyanoethylated pullulan has potential uses in electronic devices (Onda et al., 1982). Pullulan has been sulfated (Mocanu et al., 1985), chlorinated (Mayer et al., 1990), and sulfinethylated (Imai et al., 1991). Azidopullulan (Mayer et al., 1990) and siloxane derivatives have been prepared (Uchida et al., 1996). Pullulan substituted with cholesterol or fatty acids can be used to stabilize fatty emulsions (Yamaguchi and Sunamoto, 1991). Pullulan derivatives are promising as nontoxic conjugates for vaccines (Yamaguchi et al., 1985; Mitsuhashi and Koyama, 1987) and can facilitate liposome delivery (Takada et al., 1984; Sunamoto et al., 1987). Although pullulan derivatives generally are considered to be biodegradable, Ball et al. (1993) reported that certain derivatives are resistant to glucoamylase and pullulanases.

Pullulan is valuable as a specific substrate for the assay of pullulanases, which are commercially employed in the saccharification of starches. In theory, pullulan could be used as a substrate for the enzymatic production of panose or isopanose, which have value as prebiotic nutritional supplements (Kohmoto et al., 1991). However, these oligosaccharides also can be derived from amylopectin or synthesized using transglycosylation reactions (Kuriki et al., 1993; Hayashi et al., 1994b; Yun et al., 1994). Accurately sized pullulan molecular weight species are commercially produced for their value as chromatography standards (Kawahara et al., 1984).

12
Patents

Literally hundreds of patents describe the production and use of pullulan and pullulan derivatives. The following examples, also summarized in Table 1, are illustrative. Methods for pullulan production have been described (Wallenfels and Bender, 1961; Zajic, 1967; Kato and Shiosaka, 1974; Ozaki et al., 1996). Pullulan has demonstrated applications in foods (Hijiya and Shiosaka, 1975b; Kato and Shiosaka, 1975a; Hiji, 1986) and personal care products (Nakashio et al., 1976b). Pullulan and its derivatives exhibit useful adhesive properties (Hijiya and Shiosaka, 1975a). Pullulan films are widely used in food preservation, and pullulan can be used in pharmaceutical coatings (Miyamoto et al., 1986; Izutsu et al., 1987; Childers et al., 1991). Pullulan can be formed into fibers or molded products (Hijiya and Shiosaka, 1974; Nakashio et al., 1975b; Matsunaga et al., 1977b; Tsuji et al., 1976). In addition, pullulan and its derivatives have demonstrated photographic, lithographic, and electronic applications (Sano et al., 1976; Tsukada et al., 1978; Shimizu et al., 1983; Sasago et al., 1988; Vermeersch et al., 1995). Numerous derivatizations of pullulan have been patented, including esterification (Hijiya and Shiosaka, 1975a, 1975c), etherification (Fujita et al., 1979a), hydrogenation (Kato and Shiosaka, 1976), and carboxylation (Tsuji et al., 1978). Pullulan can be crosslinked for use in gel permeation chromatography (Nagase et al., 1979). Cyanoethylated pullulan has been prepared for use in electronic devices (Onda et al., 1982). Pullulan has been sulfated (Mocanu et al., 1985) and sulfinethylated (Imai et al., 1991), and siloxane derivatives have been prepared (Uchida et al., 1996).

Tab. 1 Selected patents related to pullulan

Patent number	Holder	Inventors	Title	Date
German Patent 1,096,850	Farbwerke Hoechst Aktiengesellschaft, Germany	K. Wallenfels, H. Bender	Procedure for the production of a dextran-like polysaccharide from *Pullularia pullulans*	1961
German Patent 2,512,110	Hayashibara Biochemical Laboratories, Inc., Japan; Sumitomo Chemical Co. Ltd., Japan	S. Nakashio, K. Tsuji, N. Toyota, F. Fujita, T. Nomura	Process for the production of pullulan-containing fibers	1975
Japanese Patent 3,021,602	Shin-Etsu Chemical Co., Ltd., Japan	K. Imai, T. Shiomi, Y. Tesuka	Sulfinyl ethyl pullulan and production thereof	1991
Japanese Patent 58,098,909	Matsushita Electrical Ind. Co. Ltd., Japan	T. Shimizu, M. Moriwaki, W. Shimoma	Condenser	1983
Romanian Patent 88,034	Inst. Chimii Macromolecular, Romania	G. C. Mocanu, G. Stanciulescu, A. Carpov, D. Mihai, R. P. Ghiocel, M. Moscovici	Procedure of obtaining sulfur esters of pullulan	1985
U.S. Patent 3,320,136	Kerr-McGee Oil Industries, Inc., USA	J.E. Zajic	Process for preparing a polysaccharide flocculating agent	1967

Tab. 1 (cont.)

Patent number	Holder	Inventors	Title	Date
U.S. Patent 3,784,390	Hayashibara Biochemical Laboratories, Inc., Japan	H. Hijiya, M. Shiosaka	Shaped bodies of pullulan and their use	1974
U.S. Patent 3,827,937	Hayashibara Biochemical Laboratories, Inc., Japan	K. Kato, M. Shiosaka	Method of producing pullulan	1974
U.S. Patent 3,871,892	Hayashibara Biochemical Laboratories, Inc., Japan	H. Hijiya, M. Shiosaka	Shaped bodies of pullulan esters and their use	1975
U.S. Patent 3,872,228	Hayashibara Biochemical Laboratories, Inc., Japan	H. Hijiya, M. Shiosaka	Process for the preparation of food containing pullulan and amylose	1975
U.S. Patent 3,873,333	Hayashibara Biochemical Laboratories, Inc., Japan	H. Hijiya, M. Shiosaka	Adhesives and pastes	1975
U.S. Patent 3,875,308	Hayashibara Biochemical Laboratories, Inc., Japan	K. Kato, M. Shiosaka	Food compositions containing pullulan	1975
U.S. Patent 3,931,146	Hayashibara Biochemical Laboratories, Inc., Japan	K. Kato, M. Shiosaka	Hydrogenated pullulan.	1976
U.S. Patent 3,960,685	Sumitomo Chemical Co. Ltd., Japan	T. Sano, Y. Uemura, A. Furuta	Photosensitive resin composition containing pullulan or esters thereof	1976
U.S. Patent 3,972,997	Sumitomo Chemical Co. Ltd., Japan	S. Nakashio, K. Tsuji, N. Toyota, F. Fujita	Novel cosmetics containing pullulan	1976
U.S. Patent 3,993,840	Sumitomo Chemical Co. Ltd., Japan	K. Tsuji, N. Toyota, F. Fujita	Molded pullulan type resins coated with thermosetting films	1976
U.S. Patent 4,045,388	Sumitomo Chemical Co. Ltd., Japan	H. Matsunaga, K. Tsuji, T. Saito	Resin composition of hydrophilic pullulan, hydrophobic thermoplastic resin, and plasticizer	1977
U.S. Patent 4,090,016	Sumitomo Chemical Co. Ltd., Japan	K. Tsuji, M. Fujimoto, F. Masuko, T. Nagase	Carboxylated pullulan and method for producing same	1978
U.S. Patent 4,095,525	Sumitomo Chemical Co. Ltd., Japan	N. Tsukada, K. Hagihara, K. Tsuji, M. Fujimoto, T. Nagase	Protective coating material for lithographic printing plate	1978

Tab. 1 (cont.)

Patent number	Holder	Inventors	Title	Date
U.S. Patent 4,152,170	Hayashibara Biochemical Laboratories, Inc., Japan; Sumitomo Chemical Co. Ltd., Japan	T. Nagase, K. Tsuji, M. Fujimoto, F. Masuko	Cross-linked pullulan	1979
U.S. Patent 4,167,623	Sumitomo Chemical Co. Ltd., Japan; Hayashibara Biochemical Laboratories, Inc., Japan	F. Fujita, K. Fukami, M. Fujimoto	Pullulan aminoalkyl ether	1979
U.S. Patent 4,322,524	Shin-Etsu Chemical Co., Ltd., Japan	Y. Onda, H. Muto, H. Suzuki	Cyanoethylpullulan	1982
U.S. Patent 4,610,891	Zeria Shinyaku Kogyo Kabushiki Kaisha, Japan	Y. Miyamoto, H. Goto, H. Sato, H. Okano, M. Iijima	Process for sugar-coating solid preparation	1986
U.S. Patent 4,629,725	Y. Hiji	Y. Hiji	Method for inhibiting increase in blood sugar content	1986
U.S. Patent 4,650,666	Dainippon Pharmaceutical Co., Japan	Y. Izutsu, K. Sogo, S. Okamoto, T. Tanaka	Pullulan and sugar-coated pharmaceutical composition	1987
U.S. Patent 4,745,042	Matsushita Electrical Ind. Co. Ltd., Japan	M. Sasago, M. Endo, K. Takeyama, N. Nomura	Water-soluble photopolymer and method of forming pattern by use of same	1988
U.S. Patent 5,015,480	Eli Lilly and Co, USA	R. F. Childers, P. L. Oren, W. M. K. Seidler	Film-coating formulations	1991
U.S. Patent 5,402,725	AGFA-Gevaert, N.V., Belgium	J. T. Vermeersch, P. J. Coppens, G. I. Hauquier, E. H. Schacht	Lithographic base with a modified dextran or pullulan hydrophobic layer	1995
U.S. Patent 5,518,902	Hayashibara Biochemical Laboratories, Inc., Japan	Y. Ozaki, T. Nomura, T. Miyake	High pullulan content product, and its preparation and uses	1996
U.S. Patent 5,583,244	Shin-Etsu Chemical Co., Ltd., Japan	S. Uchida, A. Yamamoto, I. Fukui, M. Endo, H. Umezawa, S. Nagura, T. Kubota	Siloxane-containing pullulan and method for the preparation thereof	1996

13
Outlook and Perspectives

Pullulan is a unique polysaccharide with a multitude of demonstrated practical applications. However, it may be considered too costly for many of these potential uses, including those in which it serves as biodegradable replacement for petroleum-derived polymers. Bulk applications, for example as an industrial cement or flocculating agent, are not economical. Although pullulan is produced efficiently by fermentation of relatively inexpensive substrates, further reductions in cost might be expected from an expanded production scale. However, markets for pullulan have been fairly stable for a number of years. Currently, a major outlet for pullulan is in specialty food applications, particularly in edible films. Nevertheless, emerging clinical and pharmaceutical uses for pullulan hold promise for expanded markets. The mechanism of pullulan biosynthesis is not well understood, and molecular approaches appear to hold little promise at this time for improvements in pullulan production. However, traditional strain improvements, particularly for reduction of contaminating melanin and for control of pullulan molecular weight, could improve the economics of production and open new avenues for pullulan utilization.

14
References

Ahearn, D. G., Meyers, S. P., Nichols, R. A. (1968) Extracellular proteinases of yeasts and yeast like fungi, *Appl. Microbiol.* **16**, 1370–1374.

Anderson, K. L. (1995) Biochemical analysis of starch degradation by *Ruminobacter amylophilus* 70, *Appl. Environ. Microbiol.* **61**, 1488–1491.

Andrews, J. H., Harris, R. F., Spear, R. N., Lau, G. W., Nordheim, E. V. (1994) Morphogenesis and adhesion of *Aureobasidium pullulans, Can. J. Microbiol.* **40**, 6–17.

Anonymous (2001) Press Release. Hayashibara in worldwide license and supply agreement with Pfizer using pullulan for oral use. http://www.hayashibara.co.jp/eng/contents_hn.html.

Aoki, H., Sakano, Y. (1997) Molecular cloning and heterologous expression of the isopullulanase gene from *Aspergillus niger* A.T.C.C. 9642, *Biochem. J.* **323**, 757–764.

Ara, K., Igarashi, K., Saeki, K., Kawai, S., Ito, S. (1992) Purification and some properties of an alkaline pullulanase from alkalophilic *Bacillus* sp. KSM–1876, *Biosci. Biotechnol. Biochem.* **56**, 62–65.

Audet, J., Lounes, M., Thibault, J. (1996) Pullulan fermentation in a reciprocating plate bioreactor, *Bioprocess Eng.* **15**, 209–214.

Auer, D. P. F., Seviour, R. J. (1990) Influence of varying nitrogen sources on polysaccharide production by *Aureobasidium pullulans* in batch culture, *Appl. Microbiol. Biotechnol.* **32**, 637–644.

Augustin, J., Kuniak, L., Hudecova, D. (1997) Screening of yeasts and yeast-like organisms *Aureobasidium pullulans* for pullulan production, *Biologia, Bratislava* **52**, 399–404.

Badr-Eldin, S. M., El-Tayeb, O. M., El-Masry, H. G., Mohamad, F. H. A., Abd El-Rahman, O. A. (1994) Polysaccharide production by *Aureobasidium pullulans*: factors affecting polysaccharide formation, *World J. Microbiol. Bacteriol.* **10**, 423–426.

Ball, D. H., Reese, E. T., Stenhouse, P. J. (1993) Synthesis and susceptibility to enzymes of pullulan derivatives, in: *Biodegradable Polymers and Packaging* (Ching, C., Kaplan, D. L., Thomas, E. L., Eds.), Lancaster: Technomic, 111–118.

Bardage, S. L., Bjurman, J. (1998) Isoation of an *Aureobasidium pullulans* polysaccharide that promotes adhesion of blastospores to water-borne paints, *Can. J. Microbiol.* **44**, 954–958.

Barnett, C., Smith, A., Scanlon, B., Israilides, C. J. (1999) Pullulan production by *Aureobasidium pullulans* growing on hydrolysed potato starch waste, *Carbohydr. Polymers* **38**, 203–209.

Bauer, R. (1938) Physiology of *Dematium pullulans* de Bary, Zentralbl. Bacteriol., Parasitenkd., Infektionskr. Hyg., Abt. 2 **98**, 133–167.

Behrens, U., Lohse, R. (1977) Investigations for the production of pullulan, *Die Nahrung* **21**, 199–204.

Bender, H., Lehmann, J., Wallenfels, K. (1959) Pullulan, an extracellular glucan from *Pullularia pullulans*. *Biochim. Biophys. Acta* **36**, 309–316.

Bender, H., Wallenfels, K. (1961) Investigations on pullulan. II. Specific degradation by means of a bacterial enzyme, *Biochem. Zeitschrift* **334**, 79–95.

Bender, H. (1994) Studies of the degradation of pullulan by the decycling maltodextrinase of *Flavobacterium* sp., *Carbohydr. Res.* **260**, 119–130.

Bermejo, J. M., Dominguez, J. B., Goni, F. M., Uruburu, F. (1981a) Influence of carbon and nitrogen sources on the transition from yeast-like cells to chlamydospores in *Aureobasidium pullulans, Antonie van Leeuwenhoek J. Microbiol.* **47**, 107–119.

Bermejo, J. M., Dominguez, J. B., Goni, F. M., Uruburu, F. (1981b) Influence of pH on the transition from yeast-like cells to chlamydospores in *Aureobasidium pullulans, Antonie van Leeuwenhoek J. Microbiol.* **47**, 385–392.

Bernier, B. (1958) The production of polysaccharides by fungi active in the decomposition of wood and forest litter, *Can. J. Microbiol.* **4**, 195–204.

Bertoldo, C., Duffner, F., Jorgensen, P. L., Antranikian, G. (1999) Pullulanase type I from *Fervidobacterium pennavorans* Ven5: cloning, sequencing, and expression of the gene and biochemical characterization of the recombinant enzyme, *Appl. Environ. Microbiol.* **65**, 2084–2091.

Biely, P., Kratky, Z., Kockova-Kratochilova, A., Bauer, S. (1978) Xylan-degrading activity in yeasts: growth on xylose, xylan and hemicelluloses, *Folia Microbiol.* **23**, 366–371.

Biely, P., Kratky, Z., Petrakova, E., Bauer, S. (1979) Growth of *Aureobasidium pullulans* on waste water hemicelluloses, *Folia Microbiol.* **24**, 328–333.

Boa, J. M., LeDuy, A. (1984) Peat hydrolysate medium optimization for pullulan production, *Appl. Environ. Microbiol.* **48**, 26–30.

Boa, J. M., LeDuy, A. (1987) Pullulan from peat hydrolyzate. Fermentation kinetics, *Biotechnol. Bioeng.* **30**, 463–470.

Bock, A., Lechner, K., Huber, O. (1991) *Aureobasidium pullulans* strain, process for its preparation, and use thereof. U.S. Patent 5,019,514.

Bouveng, H. O., Kiessling, H., Lindberg, B., McKay, J. (1962) Polysaccharides elaborated by *Pullularia pullulans*. I. The neutral glucan synthesised from sucrose solutions, *Acta Chem. Scand.* **16**, 615–622.

Bouveng, H. O., Kiessling H., Lindberg, B., McKay, J. (1963a) Polysaccharides elaborated by *Pullularia pullulans*. II. The partial acid hydrolysis of the neutral glucan synthesised from sucrose solutions, *Acta Chem. Scand.* **17**, 797–800.

Bouveng, H. O., Kiessling, H., Lindberg, B., McKay, J. (1963b) Polysaccharides elaborated by *Pullularia pullulans*. III. Polysaccharides synthesized from xylose solutions, *Acta Chem. Scand.* **17**, 1351–1356.

Brand, B. G., Kemp, H. T. (1973) *Mildew Defacement of Organic Coatings. A Review of Literature Describing the Relationship Between Aureobasidium pullulans and Paint Films*. Philadelphia, PA: Paint Research Institute.

Brant, D. A., Burton, B. A. (1981) The configurational statistics of pullulan and some related glucans, in: *Solution Properties of Polysaccharides. ACS Symp. Series 150* (Brant, D. A., Ed.), Washington, D.C.: American Chemical Society, 81–99.

Brown, R. G., Lindberg, B. (1967a) Polysaccharides from cell walls of *Aureobasidium (Pullularia) pullulans*. Part I. Glucans, *Acta Chem. Scand.* **21**, 2379–2382.

Brown, R. G., Lindberg, B. (1967b) Polysaccharides from cell walls of *Aureobasidium (Pullularia) pullulans*. Part II. Heteropolysaccharide, *Acta Chem. Scand.* **21**, 2383–2389.

Brown, R. G., Hanic, L. A., Hsiao, M. (1973) Structure and chemical composition of yeast chlamydospores of *Aureobasidium pullulans*, *Can. J. Microbiol.* **19**, 163–168.

Buliga, G. S., Brant, D. A. (1987) Temperature and molecular weight dependence of the unperturbed dimensions of aqueous pullulan, *Int. J. Biol. Macromol.* **9**, 71–76.

Bulmer, M. A., Catley, B. J., Kelly, P. J. (1987) The effect of ammonium ions and pH on the elaboration of the fungal extracellular polysaccharide, pullulan, by *Aureobasidium pullulans*, *Appl. Microbiol. Biotechnol.* **25**, 362–365.

Carolan, G., Catley, B. J., McDougal, F. J. (1983) The location of tetrasaccharide units in pullulan, *Carbohydr. Res.* **114**, 237–243.

Castro, G. R., Ducrey Santopietro, L. M., Sineriz, F. (1993) Acid pullulanase from *Bacillus polymyxa* MIR-23, *Appl. Biochem. Biotechnol.* **37**, 227–233.

Catley, B. J., Robyt, J. F., Whelan, W. J. (1966) A minor structural feature of pullulan, *Biochem. J.*, 5P–8P.

Catley, B. J. (1970) Pullulan, a relationship between molecular weight and fine structure, *FEBS Lett.* **10**, 190–193.

Catley, B. J. (1971a) Role of pH and nitrogen limitation in the elaboration of the extracellular polysaccharide pullulan by *Pullularia pullulans*, *Appl. Microbiol.* **22**, 650–654.

Catley, B. J. (1971b) Utilization of carbon sources by *Pullularia pullulans* for the elaboration of extracellular polysaccharides, *Appl. Microbiol.* **22**, 641–649.

Catley, B. J., Whelan, W. J. (1971) Observations on the structure of pullulan, *Arch. Biochem. Biophys.* **143**, 138–142.

Catley, B. J. (1972) Pullulan elaboration, an inducible system of *Pullularia pullulans*, *FEBS Lett.* **20**, 174–176.

Catley, B. J. (1973) The rate of elaboration of the extracellular polysaccharide, pullulan, during growth of *Pullularia pullulans*, *J. Gen. Microbiol.* **78**, 33–38.

Catley, B. J. (1978) An assay for pullulanase in the presence of other carbohydrases. *Carbohydr. Res.* **61**, 419–424.

Catley, B. J. (1979) Pullulan synthesis by *Aureobasidium pullulans*, in: *Microbial Polysaccharides and Polysaccharases* (Berkeley, R. C. W., Gooday,

G. W., Ellwood, D. C., Eds.), London: Academic Press, 69–84.

Catley, B. J. (1980) The extracellular polysaccharide, pullulan, produced by *Aureobasidium pullulans*: a relationship between elaboration rate and morphology, *J. Gen. Microbiol.* **120**, 265–268.

Catley, B. J., Hutchison, A. (1981) Elaboration of pullulan by spheroplasts of *Aureobasidium pullulans*, *Trans. Br. Mycol. Soc.* **76**, 451–456.

Catley, B. J., McDowell, W. (1982) Lipid-linked saccharides formed during pullulan biosynthesis in *Aureobasidium pullulans*, *Carbohydr. Res.* **103**, 65–75.

Catley, B. J., Ramsay, A., Servis, C. (1986) Observations on the structure of the fungal extracellular polysaccharide, pullulan, *Carbohydr. Res.* **153**, 79–86.

Cernakova, M., Kockova-Kratochvilova, A., Suty, L., Zemek, J., Kuniak, E. (1980) Biochemical similarities among strains of *Aureobasidium pullulans* (de Bary) Arnaud, *Folia Microbiol.* **25**, 68–73.

Childers, R. F., Oren, P. L., Seidler, W. M. K. (1991) Film coating formulations. U.S. Patent 5,015,480.

Coleman, R. D., McAlister, M. P. (1986) Plasmids containing a gene coding for a thermostable pullulanase and pullulanase-producing strains of *Escherichia coli* and *Bacillus subtilis* containing the plasmids. U.S. Patent 4,612,287.

Coleman, R. D., Yang, S-S., McAlister, M. P. (1987) Cloning of the debranching-enzyme gene from *Thermoanaerobium brockii* into *Escherichia coli* and *Bacillus subtilis*, *J. Bacteriol.* **169**, 4302–4307.

Conca, K. R., Yang, T. C. S. (1993) Edible food barrier coatings, in: *Biodegradable Polymers and Packaging* (Ching, C., Kaplan, D. L., Thomas, E. L., Eds.), 357–369. Lancaster: Technomic.

Cooke, W. B. (1959) An ecological life history of *Aureobasidium pullulans* (de Bary) Arnaud, *Mycopathol. Mycol. Appl.* **12**, 1–45.

Cooke W. B. (1962) A taxonomic study in the "black yeasts", *Mycopathol. Mycol. Appl.* **17**, 1–43.

Cooke, W. B., Matsuura, G. (1963) Physiological studies in the black yeasts, *Mycopathol. Mycol. Appl.* **21**, 15–271.

Cooper, L. A., Gadd, G. M. (1984) The induction of mycelial development in *Aureobasidium pullulans* (IMI 45533) by yeast extract, *Antonie van Leeuwenhoek J. Microbiol.* **50**, 249–260.

Corsaro, M. M., De Castro, C., Evidente, A., Lanzetta, R., Molinaro, A., Parrilli, M., Sparapano, L. (1998) Phytotoxic extracellular polysaccharide fractions from *Cryphonectria parasitica* (Murr.) Barr strains, *Carbohydr. Polymers* **37**, 167–172.

Côté, G. L., Ahlgren, J. A. (1995) Microbial polysaccharides, in: *Kirk-Othmer Encyclopedia of Chemical Technology*, 4th Ed. (Kroschwitz, J. I., Howe-Grant, M., Eds.), New York: John Wiley & Sons, 578–612, Vol. 16.

Cottrell, I. W. (1980) Industrial potential of fungal and bacterial polysaccharides. *Fungal Polysaccharides. ACS Symp. Ser. 126* (Sandford, P. A., Matsuda, K., Eds.), Washington, D.C.: American Chemical Society, 251–270.

Cullen, D., Yang, V., Jeffries, T., Bolduc, J., Andrews, J. H. (1991) Genetic transformation of *Aureobasidium pullulans*, *J. Biotechnol.* **21**, 283–288.

De Baets, S., Vandamme, E. J. (1999) Yeasts as producers of polysaccharides with novel application potential, *SIM News* **49**, 321–328.

de Hoog, G. S., Yurlova, N. A. (1994) Conidiogenesis, nutritional physiology and taxonomy of *Aureobasidium* and *Hormonema*, *Antonie van Leeuwenhoek Int. J. Gen. Mol. Microbiol.* **65**, 41–54.

de Hoog, G. S. (1998) A key to the anamorph genera of yeastlike archi- and euascomycetes, in: *The Yeasts, A Taxonomic Study* (Kurtzman, C. P., Fell, J. W., Eds.), Amsterdam: Elsevier, 123–125.

De Mot, R., Van Oudendijck, E., Verachtert, H. (1985) Purification and characterization of an extracellular glucoamylase from the yeast *Candida tsukubaensis* CBS 6389, *Antonie van Leeuwenhoek J. Microbiol.* **51**, 275–287.

Dennis, C. (1972) Breakdown of cellulose by yeast species, *J. Gen. Microbiol.* **71**, 409–411.

Dennis C., Buhagiar, R. W. M. (1973) Comparative study of *Aureobasidium pullulans*, *A. prunorum* sp. nov. and *Trichosporon pullulans*, *Trans. Br. Mycol. Soc.* **60**, 567–575.

Deshpande, M. S., Rale, V. B., Lynch, J. M. (1992) *Aureobasidium pullulans* in applied microbiology: a status report, *Enzyme Microb. Technol.* **14**, 514–527.

Dillon, R., Martin, A. M. (1994) Production of polymers by microorganisms. A review, *Agro-Food-Industry Hi-Tech* July-August, 27–30.

Dominguez, J. B., Goni, F. M., Uruburu, F. (1978) The transition from yeast-like to chlamydospore cells in *Pullularia pullulans*, *J. Gen. Microbiol.* **108**, 111–117.

Dufresne, R., Thibault, J., LeDuy, A., Lencki, R. (1990) The effects of pressure on the growth of *Aureobasidium pullulans* and the synthesis of pullulan, *Appl. Microbiol. Biotechnol.* **32**, 526–532.

Elinov, N. P., Matveeva, A. K. (1972) Extracellular glucan produced by *Aureobasidium pullulans*, *Biokhimiya* **37**, 255–257.

Elinov, N. P., Neshataeva, E. V., Dranishnikov, A. N., Matveeva, A. K. (1974) Effect of certain inorganic salts in a synthetic nutrient medium on the structure and properties of glucan formed by *Aureobasidium pullulans*, *Priklad. Biokhim. Mikrobiol.* **10**, 557–562.

Elinov, N. P., Marikhin, V. A., Dranishnikov, A. N., Myasnikova, L. P., Maryukhta, Y. B. (1975) Peculiarities of the glucan produced by a culture of *Aureobasidium (Pullularia) pullulans*, *Dokl. Akad. Nauk S. S. S. R.* **221**, 213–216.

Elinov., N. P., Glazova, N. V., Kravchenko, S. B., Potekhina, T. S., Siluyanova, N. A. (1987) Method of producing aubasidan. U.S.S.R. Patent 1,339,129.

Federici F. (1982) Extracellular enzymatic activities in *Aureobasidium pullulans*, *Mycologia* **74**, 738–743.

Finkelman, M. J., Zajic, J. E. (1978) Pectinase from *Aureobasidium pullulans*, in: *Developments in Industrial Microbiology* (Underkofler, L. A., Ed.), Arlington, VA: Society for Industrial Microbiology, 459–464, Vol. 19.

Finkelman, M. A. J., Zajic, J. E., Vardanis, A. (1980) New method of producing protoplasts of *Aureobasidium pullulans*, *Appl. Environ. Microbiol.* **39**, 923–925.

Finkelman, M. A. J., Vardanis, A. (1982a) Pullulan elaboration by *Aureobasidium pullulans* protoplasts, *Appl. Environ. Microbiol.* **44**, 121–127.

Finkelman, M. A. J., Vardanis, A. (1982b) Simplified microassay for pullulan synthesis, *Appl. Environ. Microbiol.* **43**, 483–485.

Flannigan, B. (1970) Degradation of arabinoxylan and carboxymethyl cellulose by fungi isolated from barley kernels, *Trans. Br. Mycol. Soc.* **55**, 277–281.

Fraser, C. G., Jennings, H. J. (1971) A glucan from *Tremella mesenterica* NRRL-Y6158, *Can. J. Chem.* **49**, 1804–1807.

Fujii, N., Shinohara, S., Ueno, H., Imada, K. (1984) Polysaccharide produced by *Aureobasidium* sp. (black yeast), *Kenkyu Hokuku - Miyazaki Daigaku Nogakubu* **31**, 253–262.

Fujii, N., Shinohara, S. (1986) Polysaccharide produced by *Aureobasidium pullulans* FERM-P4257. II. Toxicity test and antitumor effect, *Kenkyu Hokuku - Miyazaki Daigaku Nogakubu* **33**, 243–248.

Fujita, F., Fukami, K., Fujimoto, M. (1979a) Pullulan aminoalkyl ether. U S. Patent 4,167,623.

Fujita, F., Fukami, K., Fujimoto, M., Nagase, T. (1979b) Ionic pullulan gels and production thereof. U.S. Patent 4,174,440.

Gadd, G. M., Griffiths, A. J. (1980) Effect of copper on morphology of *Aureobasidium pullulans*, *Trans. Br. Mycol. Soc.* **74**, 387–392.

Gadd, G. M., Cooper, L. A. (1984) Strain and medium-related variability in the yeast-mycelial transition of *Aureobasidium pullulans*, *FEMS Microbiol. Lett.* **23**, 47–49.

Gantelet, H., Duchiron, F. (1998) Purification and properties of a thermoactive and thermostable pullulanase from *Thermococcus hydrothermalis*, a hyperthermophilic archaeon isolated from a deep-sea hydrothermal vent, *Appl. Microbiol. Biotechnol.* **49**, 770–777.

Gibbs, P. A., Seviour, R. J. (1992) Influence of bioreactor design on exopolysaccharide production by *Aureobasidium pullulans*, *Biotechnol. Lett.* **14**, 491–494.

Gibbs, P. A., Seviour, R. J. (1996) Does the agitation rate and/or oxygen saturation influence exopolysaccharide production by *Aureobasidium pullulans* in batch culture? *Appl. Microbiol. Biotechnol.* **46**, 503–510.

Hamada, N., Tsujisaka, Y. (1983) The structure of the carbohydrate moiety of an acidic polysaccharide produced by *Aureobasidium* sp. K-1, *Agric. Biol. Chem.* **47**, 1167–1172.

Hayashi, S., Hayashi, T., Takasaki, Y., Imada, K. (1994a) Purification and properties of glucosyltransferase from *Aureobasidium*, *J. Ind. Microbiol.* **13**, 5–9.

Hayashi, S., Hinotani, T., Takasaki, Y., Imada, K. (1994b) The enzymatic reaction for the production of panose and isomaltose by glucosyltransferase from *Aureobasidium*, *Lett. Appl. Microbiol.* **19**, 247–248.

Heald, P. J., Kristiansen, B. (1985) Synthesis of polysaccharide by yeast-like forms of *Aureobasidium pullulans*, *Biotechnol. Bioeng.* **27**, 1516–1519.

Hermanides-Nijhof, E. J. (1977) *Aureobasidium* and allied genera, *Studies Mycol.* **15**, 141–177.

Hiji, Y. (1986) Method for inhibiting increase in blood sugar content. U.S. Patent 4,629,725.

Hijiya, H., Shiosaka, M. (1974) Shaped bodies of pullulan and their use. U.S. Patent 3,784,390.

Hijiya, H., Shiosaka, M. (1975a) Adhesives and pastes. U.S. Patent 3,873,333.

Hijiya, H., Shiosaka, M. (1975b) Process for the preparation of food containing pullulan and amylose. U.S. Patent 3,872,228.

Hijiya, H., Shiosaka, M. (1975c) Shaped bodies of pullulan esters and their use. U.S. Patent 3,871,892.

Hijiya, H., Miyake, T. (1990) Association complex comprising pullulan and polyethylene glycol, and preparation and uses of same. U.S. Patent 4,927,636.

Hirohara, H., Nabeshima, S., Fujimoto, M., Nagase, T. (1981) Enzyme immobilization with pullulan gel. U.S. Patent 4,247,642.

Ichikawa, T., Yamada, H. (1982) Manufacture of plywood. Japanese Patent 57,165,201.

Igarashi, T., Nomura, K., Naito, K., Yoshida, M. (1983) Plasma extender. U.S. Patent 4,370,472.

Imai, K., Shiomi, T., Tesuka, Y. (1991) Sulfinyl ethyl pullulan and production thereof. Japanese Patent 3,021,602.

Imshenetskii, A. A., Kondrat'eva, T. F. (1978) Rapid method of quantitative determination of pullulan in culture fluid of *Pullularia pullulans*, *Mikrobiol.* **47**, 566–568.

Imshenetskii, A. A., Kondrat'eva, T. F. (1979) More active synthesis of the polysaccharide pullulan by polyploid cultures of *Pullularia pullulans*, *Mikrobiol.* **48**, 319–323.

Imshenetskii, A. A., Kondrat'eva, T. F., Smut'ko, A. N. (1981a) Influence of carbon and nitrogen sources on pullulan biosynthesis by polyploid strains of *Pullularia pullulans*, *Mikrobiol.* **50**, 102–105.

Imshenetskii, A. A., Kondrat'eva, T. F., Smut'ko, A. N. (1981b) Influence of the acidity of the medium, conditions of aeration, and temperature on pullulan biosynthesis by polyploid strains of *Pullularia (Aureobasidium) pullulans*, *Mikrobiol.* **50**, 471–475.

Imshenetskii, A. A., Kondrat'eva, T. F., Smut'ko, A. N. (1982) Spontaneous and ultraviolet-induced variability of the pullulan-synthesizing activity of *Pullularia (Aureobasidium) pullulans* strains with different levels of ploidy, *Mikrobiol.* **51**, 964–967.

Imshenetskii, A. A., Kondrat'eva, T. F., Kudryashev, L. I., Yarovaya, S. M., Smut'ko, A. N., Alekseeva, G. S. (1983a) A comparative study of pullulans synthesized by strains of *Pullularia (Aureobasidium) pullulans* of differing levels of ploidy, *Mikrobiol.* **52**, 816–820.

Imshenetskii, A. A., Kondrat'eva, T. F., Smut'ko, A. N. (1983b) Colchicine-induced synthesis of pullulan in *Pullularia (Aureobasidium) pullulans*, *Mikrobiol.* **52**, 404–407.

Imshenetskii, A. A., Kondrat'eva, T. F., Dul'tseva, N. M. (1985a) A rapid method for selecting mutants of *Pullularia pullulans* with high pullulan-synthesizing activity, *Mikrobiol.* **54**, 647–650.

Imshenetskii, A. A., Kondrat'eva, T. F., Dvadtsamova, E. A. and Vorontsova, N. N. (1985b) Activity of pullulan synthesis by diploid culture of *Pullularia pullulans* on media with different carbon sources, *Mikrobiol.* **54**, 927–929.

Israilides, C., Bocking, M., Smith, A., Scanlon, B. (1994) A novel rapid coupled enzyme assay for the estimation of pullulan, *Biotechol. Appl. Biochem.* **19**, 285–291.

Israilides, C. J., Smith, A., Harthill, J. E., Barnett, C., Bambalov, G., Scanlon, B. (1998) Pullulan content of the ethanol precipitate from fermented agro-industrial wastes, *Appl. Microbiol. Biotechnol.* **49**, 613–617.

Israilides, C., Smith, A., Scanlon, B., Barnett, C. (1999) Pullulan from agro-industrial wastes, *Biotechnol. Genet. Eng. Rev.* **16**, 309–324.

Izutsu, Y., Sogo, K., Okamoto, S., Tanaka, T. (1987) Pullulan and sugar coated pharmaceutical composition. U.S. Patent 4,650,666.

Janse, B. J. H., Pretorius, I. S. (1993) Expression of *Klebsiella pneumoniae* pullulanase-encoding gene in *Saccharomyces cerevisiae*, *Curr. Genet.* **24**, 32–37.

Jeanes, A. (1977) Dextrans and pullulans: industrially significant α-D-glucans, in: *Extracellular Microbial Polysaccharides*. ACS Symposium Series No. 45. (Sandford, P. A., Laskin, A., Eds.), Washington, D.C.: American Chemical Society, 284–298.

Kaplan, D. L., Wiley, B. J., Arcidiacono, S., Mayer, J., Sousa, S. (1987) Controlling biopolymer molecular weight distribution – pullulan and chitosan, in: *First Materials Biotechnology Symposium*. Technical Report NATICK/TR-88/033 (Kaplan, D. L., Ed.), Natick, MA: U.S. Army, 149–173.

Kato, K., Shiosaka, M. (1974) Method of producing pullulan. U.S. Patent 3,827,937.

Kato, K., Shiosaka, M. (1975a) Food compositions containing pullulan. U.S. Patent 3,875,308.

Kato, K., Shiosaka, M. (1975b) Process for the production of pullulan. U.S. Patent 3,912,591.

Kato, K., Nomura, T. (1976) Method for continuously purifying pullulan. U.S. Patent 3,959,009.

Kato, K., Shiosaka, M. (1976) Hydrogenated pullulan. U.S. Patent 3,931,146.

Kato, K., Nomura, T. (1977) Method for purifying pullulan. U.S. Patent 4,004,977.

Kato, T., Okamoto, T., Tokuya, T., Takahashi, A. (1982) Solution properties and chain flexibility of pullulan in aqueous solution, *Biopolymers* **21**, 1623–1633.

Kawahara, K., Ohta, K., Miyamoto, H., Nakamura, S. (1984) Preparation and solution properties of pullulan fractions as standard samples for water-soluble polymers, *Carbohydr. Polymers* **4**, 335–356.

Kelkar, H. S., Deshpande, M. V. (1993) Purification and characterization of a pullulan-hydrolyzing glucoamylase from *Sclerotium rolfsii*, *Starch/Stärke* **45**, 361–368.

Kelly, P. J., Catley, B. J. (1977) The effect of ethidium bromide mutagenesis on dimorphism, extracellular metabolism and cytochrome levels in *Aureobasidium pullulans*, *J. Gen. Microbiol.* **102**, 249–254.

Kikuchi, Y., Taguchi, R., Sakano, Y., Kobayashi, T. (1973) Comparison of extracellular polysaccharide produced by *Pullularia pullulans* with polysaccharides in the cells and cell wall, *Agric. Biol. Chem.* **37**, 1751–1753.

Kim, I.-C., Cha, J.-H., Kim, J.-R., Jang, S.-Y., Seo, B.-C., Cheong, T.-K., Lee, D. S., Choi, Y. D., Park, K.-H. (1992) Catalytic properties of the cloned amylase from *Bacillus licheniformis*, *J. Biol. Chem.* **267**, 22108–22114.

Kimoto, T., Shibuya, T., Shiobara, S. (1997) Safety studies of a novel starch, pullulan: chronic toxicity in rats and bacterial mutagenicity, *Food Chem. Toxicol.* **35**, 323–329.

Kimura, T., Horikoshi, K. (1990) Characterization of pullulan-hydrolysing enzyme from an alkalopsychrotrophic *Micrococcus* sp., *Appl. Microbiol. Biotechnol.* **34**, 52–56.

Klimek, J., Ollis, D. F. (1980) Extracellular microbial polysaccharides: kinetics of *Pseudomonas* sp., *Azotobacter vinelandii*, and *Aureobasidium pullulans* batch fermentations, *Biotechnol. Bioeng.* **22**, 2321–2342.

Koch, R., Zablowski, P., Antranikian, G. (1987) Highly active and thermostable amylases and pullulanases from various anaerobic thermophiles, *Appl. Microbiol. Biotechnol.* **27**, 192–198.

Kockova-Kratochvilova, A., Cernakova, M. and Slavikova, E. (1980) Morphological changes during the life cycle of *Aureobasidium pullulans* (de Bary) Arnaud, *Folia Microbiol.* **25**, 56–67.

Kohmoto, T., Fukui, F., Takaku, H., Mitsuoka, T. (1991) Dose-response test of isomaltooligosaccharides for increasing fecal bifidobacteria, *Agric. Biol. Chem.* **55**, 2157–2159.

Kondrat'eva, T. F. (1981) Production of the polysaccharide pullulan by *Aureobasidium* (*Pullularia*) *pullulans*, *Usp. Mikrobiol.* **16**, 175–193.

Kondrat'eva, T. F., Lobacheva, N. A. (1990) Use of a mathematical planning method for optimizing the growth medium composition to increase the quantity of pullulan synthesized by *Pullularia pullulans*, *Mikrobiol.* **59**, 1004–1009.

Kremnicky, L., Slavikova, E., Mislovicova, D., Biely, P. (1996) Production of extracellular β-mannanases by yeasts and yeast-like microorganisms, *Folia. Microbiol.* **41**, 43–47.

Kremnicky, L., Biely, P. (1997) β-mannanolytic system of *Aureobasidium pullulans*, *Arch. Microbiol.* **167**, 350–355.

Kristiansen, B., Charley, R. C., Seviour, B., Harvey, L., Habeeb, S., Smith J. E. (1982) Over-production of metabolites by filamentous fungi, in: (Krumphanzl, V., Sikyta, B., Vanek, Z., Eds.), *Overproduction of Microbial Products*, London: Academic Press, 196–210.

Krochta, J. M., De Mulder-Johnston, C. (1997) Edible and biodegradable polymer films: challenges and opportunities, *Food Technol.* **51**, 61–74.

Kukuri, T., Okada, S., Imanaka, T. (1988) New type of pullulanase from *Bacillus stearothermophilus* and molecular cloning and expression of the gene in *Bacillus subtilis*, *J. Bacteriol.* **170**, 1554–1559.

Kuriki, T., Yanase, M., Takata, H., Takesada, Y., Imanaka, T., Okada, S. (1993) A new way of producing isomalto-oligosaccharide syrup by using the transglycosylation reaction of neopullulanase, *Appl. Environ. Microbiol.* **59**, 953–959.

Kusano, S., Nagahata, N., Takahashi, S-I., Fujimoto, D., Sakano, Y. (1988) Purification and properties of *Bacillus acidopullulyticus* pullulanase, *Agric. Biol. Chem.* **52**, 2293–2298.

Lachke, A. H., Rale, V. B. (1995) Trends in microbial production of pullulan and its novel applications in the food industry, in: *Food Biotechnology: Microorganisms* (Hui, Y. H., Khachatourians, G. G., Eds.), New York: VCH, 589–604.

Lacroix, C., LeDuy, A., Noel, G., Choplin, L. (1985) Effect of pH on the batch fermentation of pullulan from sucrose medium, *Biotechnol. Bioeng.* **27**, 202–207.

Leal-Serano, G., Ruperez, P., Leal, J. A. (1980) Acidic polysaccharide from *Aureobasidium pullulans*, *Trans. Br. Mycol. Soc.* **75**, 57–62.

Leathers, T. D., Kurtzman, C. P., Detroy, R. W. (1984) Overproduction and regulation of xylanase in *Aureobasidium pullulans* and *Cryptococcus albidus*, *Biotech. Bioeng. Symp.* **14**, 225–240.

Leathers, T. D. (1986) Color variants of *Aureobasidium pullulans* overproduce xylanase with extremely high specific activity, *Appl. Environ. Microbiol.* **52**, 1026–1030.

Leathers, T. D. (1987) Host amylases and pullulan production, in: *First Materials Biotechnology Symposium*. Technical Report NATICK/TR-88/033 (Kaplan, D. L., Ed.), Natick, MA: U.S. Army, 175–185.

Leathers, T. D., Nofsinger, G. W., Kurtzman, C. P. and Bothast, R. J. (1988) Pullulan production by color variant strains of *Aureobasidium pullulans*, *J. Ind. Microbiol.* **3**, 231–239.

Leathers, T. D. (1989) Purification and properties of xylanase from *Aureobasidium*, *J. Ind. Microbiol.* **4**, 341–348.

Leathers, T. D. (1991) Genetic strategies to enhance pullulan production. Proceedings, U.S.-Japan Natural Resources Protein Panel 20th Annual Meeting, Tsukuba, Japan, October 21–24, UJNR Protein Resources Panel, pp. 18–31.

Leathers, T. D. (1993) Substrate regulation and specificity of amylases from *Aureobasidium* strain NRRL Y-12,974, *FEMS Microbiol. Lett.* **110**, 217–222.

Leathers, T. D., Gupta, S. C. (1994) Production of pullulan from fuel ethanol byproducts by *Aureobasidium* sp. strain NRRL Y-12,974, *Biotechnol. Lett.* **16**, 1163–1166.

LeDuy, A., Marsan, A. A., Coupai, B. (1974) A study of the rheological properties of a non-Newtonian fermentation broth, *Biotechnol. Bioeng.* **16**, 61–76.

LeDuy, A., Boa, J. M. (1983) Pullulan production from peat hydrolyzate, *Can. J. Microbiol.* **29**, 143–146.

LeDuy, A., Varmoff, J-J., Chagraoui, A. (1983) Enhanced production of pullulan from lactose by adaptation and by mixed culture techniques, *Biotechnol. Lett.* **5**, 49–54.

LeDuy, A., Choplin, L., Zajic, J. E., Luong, J. H. T. (1988) Pullulan, in: *Encyclopedia of Polymer Science and Engineering* (Mark, H. F., Bikales, N. M., Overberger, C. G., Menges, G., Kroschwitz, J. I., Eds.), New York: John Wiley & Sons, 650–660, Vol. 13.

Lee, K. Y., Yoo, Y. J. (1993) Optimization of pH for high molecular weight pullulan, *Biotechnol. Lett.* **15**, 1021–1024.

Lee, M-J., Lee, Y-C., Kim, C-H. (1997) Intracellular and extracellular forms of alkaline pullulanase from an alkaliphilic *Bacillus* sp. S-1, *Arch. Biochem. Biophys.* **337**, 308–316.

Lee, S-P., Morikawa, M., Takagi, M., Imanaka, T. (1994) Cloning of the *aapT* gene and characterization of its product, α-amylase-pullulanase (AapT), from thermophilic and alkaliphilic *Bacillus* sp. strain XAL601, *Appl. Environ. Microbiol.* **60**, 3764–3773.

Lehle, L., Tanner, W. (1978) Biosynthesis and characterization of large dolichyl diphosphate-linked oligosaccharides in *Saccharomyces cerevisiae*, *Biochim. Biophys. Acta* **539**, 218–229.

Leveque, E., Janecek, S., Haye, B., Belarbi, A. (2000) Thermophilic archaeal amylolytic enzymes, *Enzyme Microbial Technol.* **26**, 3–14.

Linardi, V. R, Machado, K. M. G. (1990) Production of amylases by yeasts, *Can. J. Microbiol.* **36**, 751–753.

Madi, N. S., McNeil, B., Harvey, L. M. (1996) Influence of culture pH and aeration on ethanol production and pullulan molecular weight by *Aureobasidium pullulans*, *J. Chem. Technol. Biotechnol.* **66**, 343–350.

Madi, N. S., Harvey, L. M., Mehlert, A., McNeil, B. (1997) Synthesis of two distinct exopolysaccharide fractions by cultures of the polymorphic fungus *Aureobasidium pullulans*, *Carbohydr. Polymers* **32**, 307–314.

Manners, D. J. (1971) Specificity of debranching enzymes, *Nature New Biol.* **234**, 150–151.

Marsh, P. B., Bollenbacher, K. (1949) The fungi concerned in fiber deterioration. I. Their occurrence, *Text. Res. J.* **19**, 313–324.

Marsh, P. B., Bollenbacher, K., Butler, M. L., Raper, K. B. (1949) The fungi concerned in fiber deterioration. II. Their ability to decompose cellulose, *Text. Res. J.* **19**, 455–484.

Marshall, J. J. (1975) Starch-degrading enzymes, old and new, *Die Starke* **27**, 377–383.

Mathupala, S., Saha, B. C., Zeikus, J. G. (1990) Substrate competition and specificity at the active site of amylopullulanase from *Clostridium thermohydrosulfuricum*, *Biochem. Biophys. Res. Comm.* **166**, 126–132.

Matsunaga, H., Tsuji, K., Saito, T. (1976) Foamed plastics of resin compositions comprising pullulan type resins and thermoplastic resins and process for producing the same. U.S. Patent 3,932,192.

Matsunaga, H., Fujimura, S., Namioka, H., Tsuji, K., Watanabe, M. (1977a) Fertilizer composition. U.S. Patent 4,045,204.

Matsunaga, H., Tsuji, K., Saito, T. (1977b) Resin composition of hydrophilic pullulan, hydrophobic thermoplastic resin, and plasticizer. U.S. Patent 4,045,388.

Matsunaga, H., Tsuji, K., Watanabe, M. (1978) Coated seed containing pullulan-based resin used as binder. U.S. Patent 4,067,141.

Mayer, J. M., Greenberger, M., Ball, D. H., Kaplan, D. L. (1990) Polysaccharides, modified polysaccharides and polysaccharide blends for biodegradable materials, *Polym. Mater. Sci. Eng.* **63**, 732–735.

McCleary, B. V., Anderson, M. A. (1980) Hydrolysis of α-D-glucans and α-D-gluco-oligosaccharides by *Cladosporium resinae* glucoamylases, *Carbohydr. Res.* **86**, 77–96.

McNeil, B., Kristiansen, B. (1987) Influence of impeller speed upon the pullulan fermentation, *Biotechnol. Lett.* **9**, 101–104.

McNeil, B., Kristiansen, B., Seviour, R. J. (1989) Polysaccharide production and morphology of *Aureobasidium pullulans* in continuous culture, *Biotechnol. Bioeng.* **33**, 1210–1212.

McNeil, B., Kristiansen, B. (1990) Temperature effects on polysaccharide formation by *Aureobasidium pullulans* in stirred tanks, *Enzyme Microb. Technol.* **12**, 521–526.

McNeil, B., Harvey, L. M. (1993) Viscous fermentation products, *CRC Crit. Rev. Biotechnol.* **13**, 275–304.

Melasniemi, H. (1988) Purification and some properties of the extracellular α-amylase-pullulanase produced by *Clostridium thermohydrosulfuricum*, *Biochem. J.* **250**, 813–818.

Mitsuhashi, M., Koyama, S. (1987) Process for the production of virus vaccine. U.S. Patent 4,659,569.

Mitsuhashi, M., Yoneyama, M., Sakai, S. (1990) Growth promoting agent for bacteria containing pullulan with or without dextran. Canadian Patent 2,007,270.

Miura, Y., Fukushima, S., Sambuichi, M., Ueda, S. (1976) Relation between rheological properties of the broth and time course in the pullulan production, *J. Ferment. Technol.* **54**, 166–170.

Miura, Y., Irie, T., Fujio, Y., Ueda, S. (1978) Effect of amylase inhibitor on pullulan production, *Hakkokogaku* **56**, 136–138.

Miyaka, T. (1979) Shaped matters of tobaccos and process for producing the same. Canadian Patent 1,049,245.

Miyamoto, Y., Goto, H., Sato, H., Okano, H., Iijima, M. (1986) Process for sugar-coating solid preparation. U.S. Patent 4,610,891.

Mocanu, G. C., Stanciulescu, G., Carpov, A., Mihai, D., Ghiocel, R. P., Moscovici, M. (1985) Procedure of obtaining sulfur esters of pullulan. Romanian Patent 88,034.

Mokrousov, I. V. (1995a) Intraspecific differentiation of *Aureobasidium* spp. using universally primed polymerase chain reaction, *Genetika* **31**, 908–914.

Mokrousov, I. V. (1995b) Phylogenetic analysis and differentiation of species of *Aureobasidium pullulans* (De Bary) fungus, *Genetika* **31**, 780–785.

Moragues, M. D., Asturias, J. A., Sevilla, M-J. (1988) Morphogenetic potency of short-chain n-alkanols related to *Aureobasidium pullulans* dimorphism, *Curr. Microbiol.* **16**, 243–245.

Mori, A., Tusji, K., Nakae, K. (1978) Sand mold composition for metal casting. U.S. Patent 4,080,213.

Moscovici, M., Ionescu, C., Oniscu, C., Fotea, O., Protopopescu, P., Hanganu, L. D., Vitan, C. (1996) Improved exopolysaccharide production in fed-batch fermentation of *Aureobasidium pullulans*, with increased impeller speed, *Biotechnol. Lett.* **18**, 787–790.

Motozato, Y., Ihara, H., Tomoda, T., Hirayama, C. (1986) Preparation and gel permeation chromatographic properties of pullulan spheres, *J. Chromatography* **355**, 434–437.

Mulchandani, A., Luong, J. H. T., LeDuy, A. (1989) Biosynthesis of pullulan using immobilized *Aureobasidium pullulans* cells, *Biotechnol. Bioeng.* **33**, 306–312.

Murofushi, K., Nagura, S., Moriya, J. (1998) Removal of liquid from pullulan. U.S. Patent 5,709,801.

Nagase, T., Tsuji, K., Fujimoto, M., Masuko, F. (1979) Cross-linked pullulan. U.S. Patent 4,152,170.

Nakamura, N., Watanabe, K., Horikoshi, K. (1975) Purification and some properties of alkaline pullulanase from a strain of *Bacillus* no. 202–1, an alkalophilic microorganism, *Biochim. Biophys. Acta* **397**, 188–193.

Nakashio, S., Sekine, N., Toyota, N., Fujita, F. (1975a) Paint containing pullulan. U.S. Patent 3,888,809.

Nakashio, S., Tsuji, K., Toyota, N., Fujita, F., Nomura, T. (1975b) Process for the production of pullulan containing fibers. German Patent 2,512,110.

Nakashio, S., Sekine, N., Toyota, N., Fujita, F., Domoto, M. (1976a) Paper coating material containing pullulan. U.S. Patent 3,932,192.

Nakashio, S., Tsuji, K., Toyota, N., Fujita, F. (1976b) Novel cosmetics containing pullulan. U.S. Patent 3,972,997.

Nishinari, K., Kohyama, K., Williams, P. A., Phillips, G. O., Burchard, W., Ogino, K. (1991) Solution properties of pullulan, *Macromolecules* **24**, 5590–5593.

Nomura, T. (1976) Paper composed mainly of pullulan fibers and method for producing the same. U.S. Patent 3,936,347.

Norman, B. E. (1982) The use of debranching enzymes in dextrose syrup production, in: *Maize: Recent Progress in Chemistry and Technology* (Inglett, G. E., Ed.), New York: Academic Press, 157–179.

Odibo, F. J. C., Obi, S. K. C. (1988) Purification and characterization of a thermostable pullulanase from *Thermoactinomyces thalpophilus*, *J. Ind. Microbiol.* **3**, 343–350.

Oku, T., Yamada, K., Hosoya, N. (1979) Effects of pullulan and cellulose on the gastrointestinal tract of rats, *Nutr. Food Sci.* **32**, 235–241.

Oliva, E. M., Cirelli, A. F., De Lederkremer, R. M. (1986) Characterization of a pullulan in *Cyttaria darwinii*, *Carbohydr. Res.* **158**, 262–267.

Onda, Y., Muto, H., Suzuki, H. (1982) Cyanoethylpullulan. U.S. Patent 4,322,524.

Ono, K., Kawahara, Y., Ueda, S. (1977a) Effect of pH on the content of glycolipids in *Aureobasidium pullulans* S-1, *Agric. Biol. Chem.* **41**, 2313–2317.

Ono, K., Yasuda, N., Ueda, S. (1977b) Effect of pH on pullulan elaboration by *Aureobasidium pullulans* S-1, *Agric. Biol. Chem.* **41**, 2113–2118.

Oonishi, Y. (1985) Pullulan polyelectrolyte complex. Japanese Patent 60,156,702.

Ozaki, Y., Nomura, T., Miyake, T. (1996) High pullulan content product, and its preparation and uses. U.S. Patent 5,518,902.

Paul, F., Morin, A., Monsan, P. (1986) Microbial polysaccharides with actual potential industrial applications, *Biotechnol. Adv.* **4**, 245–259.

Park, D. (1982a) Amino acid nutrition and yeast-mycelial dimorphism in *Aureobasidium pullulans*, *Trans. Br. Mycol. Soc.* **79**, 170–172.

Park, D. (1982b) Inorganic nitrogen nutrition and yeast-mycelial dimorphism in *Aureobasidium pullulans*, *Trans. Br. Mycol. Soc.* **78**, 385–388.

Park, D. (1984) Population density and yeast mycelial dimorphism in *Aureobasidium pullulans*, *Trans. Br. Mycol. Soc.* **82**, 39–44.

Pechak, D. G., Crang, R. E. (1977) An analysis of *Aureobasidium pullulans* developmental stages by means of scanning electron microscopy, *Mycologia* **69**, 783–792.

Peery, R. B., Skatrud, P. L. (1996) Multiple drug resistance gene of *Aureobasidium pullulans*. U.S. Patent 5,516,655.

Plant, A. R., Morgan, H. W., Daniel, R. M. (1986) A highly stable pullulanase from *Thermus aquaticus* YT-1, *Enzyme Microb. Technol.* **8**, 668–672.

Plant, A. R., Clemens, R. M., Daniel, R. M., Morgan, H. W. (1987) Purification and preliminary characterization of an extracellular pullulanase from *Thermoanaerobium* Tok6-B1, *Appl. Microbiol. Biotechnol.* **26**, 427–433.

Pollock, T. J. (1992) Pullulan from polymorphic *Aureobasidium pullulans*, *SIM News* **42**, 147–156.

Pollock, T. J., Thorne, L., Armentrout, R. W. (1992) Isolation of new *Aureobasidium* strains that produce high-molecular-weight pullulan with reduced pigmentation, *Appl. Environ. Microbiol.* **58**, 877–883.

Promma, K., Matsuda, M., Okutani, K. (1997) Structural analysis of a malic acid-containing polysaccharide from an *Aureobasidium pullulans* of marine origin, *J. Mar. Biotechnol.* **5**, 219–224.

Ramesh, M. V., Podkovyrov, S. M., Lowe, S. E., Zeikus, J. G. (1994) Cloning and sequencing of the *Thermoanaerobacterium saccharolyticum* B6A-RI *apu* gene and purification and characterization of the amylopullulanase from *Escherichia coli*, *Appl. Environ. Microbiol.* **60**, 94–101.

Ramos S., Garcia Acha, I. (1975) A vegetative cycle of *Pullularia pullulans*, *Trans Br. Mycol. Soc.* **64**, 129–135.

Reed-Hamer, B., West, T. P. (1994) Effect of complex nitrogen sources on pullulan production relative to carbon source, *Microbios* **80**, 83–90.

Reeslev, M., Nielsen, J. C., Olsen, J., Jensen, B., Jacobsen, T. (1991) Effect of pH and the initial concentration of yeast extract on regulation of dimorphism and exopolysaccharide formation of *Aureobasidium pullulans* in batch culture, *Mycol. Res.* **95**, 220–226.

Reeslev, M., Jorgensen, B. B., Jorgensen, B. B. (1993) Influence of Zn^{2+} on yeast-mycelium dimorphism and exopolysaccharide production by the fungus *Aureobasidium pullulans* grown in a defined medium in continuous culture, *J. Gen. Microbiol.* **139**, 3065–3070.

Reeslev, M., Jensen, B. (1995) Influence of Zn^{2+} and Fe^{3+} on polysaccharide production and mycelium/yeast dimorphism of *Aureobasidium pullulans* in batch cultivations, *Appl. Microbiol. Biotechnol.* **42**, 910–915.

Reeslev, M., Strom, T., Jensen, B., Olsen, J. (1997) The ability of the yeast form of *Aureobasidium pullulans* to elaborate exopolysaccharide in chemostat culture at various pH values, *Mycol. Res.* **101**, 650–652.

Renneberg, R., Kaiser, G., Scheller, F., Tsujisaka, Y. (1985) Enzyme sensor for pullulan and pullulanase activity, *Biotechnol. Lett.* **7**, 809–812.

Rho, D., Mulchandani, A., Luong, J. H. T., LeDuy, A. (1988) Oxygen requirement in pullulan fermentation, *Appl. Microbiol. Biotechnol.* **28**, 361–366.

Robyt, J. F. (1992) Structure, biosynthesis, and uses of nonstarch polysaccharides: dextran, alternan, pullulan, and algin, in: *Developments in Biochemistry and Biophysics* (Alexander, R. J., Zobel, H. F., Eds.), St. Paul, MN: American Association of Cereal Chemists, 261–292.

Robyt, J. F. (1995) Mechanisms in the glucansucrase synthesis of polysaccharides and oligosaccharides from sucrose, in: *Adv. Carbohydr. Chem. Biochem.* (Horton, D., Ed.), San Diego, CA: Academic Press, 133–168, Vol. 51.

Roukas, T., Biliaderis, C. G. (1995) Evaluation of carob pod as a substrate for pullulan production by *Aureobasidium pullulans*, *Appl. Biochem. Biotechnol.* **55**, 27–44.

Roukas, T. (1998) Pretreatment of beet molasses to increase pullulan production, *Process Biochem.* **33**, 805–810.

Roukas, T. (1999a) Pullulan production from brewery wastes by *Aureobasidium pullulans*, *World J. Microbiol. Biotechnol.* **15**, 447–450.

Roukas, T. (1999b) Pullulan production from deproteinized whey by *Aureobasidium pullulans*, *J. Ind. Microbiol. Biotechnol.* **22**, 617–621.

Roukas, T., Liakopoulou-Kyriakides, M. (1999) Production of pullulan from beet molasses by *Aureobasidium pullulans* in a stirred tank fermentor, *J. Food Eng.* **40**, 89–94.

Saha, B. C., Mitsue, T., Ueda, S. (1979) Glucoamylase produced by submerged culture of *Aspergillus oryzae*, *Starch/Stärke* **31**, 307–314.

Saha, B. C., Zeikus, J. G. (1989) Novel highly thermostable pullulanase from thermophiles, *Trends Biotechnol.* **7**, 234–239.

Saha, B. C., Mathupala, S. P., Zeikus, J. G. (1991) Comparison of amylopullulanase to α-amylase and pullulanase, in: *Enzymes in Biomass Conversion*. ACS Symposium Ser. 460 (Leatham, G. F., Himmel, M. E., Eds.), Washington, D.C.: American Chemical Society, 362–371.

Saha, B. C., Silman, R. W., Bothast, R. J. (1993) Amylolytic enzymes produced by a color variant strain of *Aureobasidium pullulans*, *Curr. Microbiol.* **26**, 267–273.

Saha, B. C., Freer, S. N., Bothast, R. J. (1994) Production, purification, and properties of a thermostable β-glucosidase from a color variant strain of *Aureobasidium pullulans*, *Appl. Environ. Microbiol.* **60**, 3774–3780.

Saha, B. C., Bothast, R. J. (1998) Purification and characterization of a novel thermostable alpha-L-arabinofuranosidase from a color-variant strain of *Aureobasidium pullulans*, *Appl. Environ. Microbiol.* **64**, 216–220.

Sakano, Y., Masuda, N., Kobayashi, T. (1971) Hydrolysis of pullulan by a novel enzyme from *Aspergillus niger*, *Agric. Biol. Chem.* **35**, 971–973.

Sakano, Y., Higuchi, M., Kobayashi, T. (1972) Pullulan 4-glucanohydrolase from *Aspergillus niger*, *Arch. Biochem. Biophys.* **153**, 180–187.

Sakano, Y., Higuchi, M., Masuda, N., Kobayashi, T. (1973) Production of pullulan 4-glucanohydrolase by *Aspergillus niger*, *J. Ferment. Technol.* **51**, 726–733.

Sakano, Y., Kogure, M., Kobayashi, T., Tamura, M., Suekane, M. (1978) Enzymic preparation of panose and isopanose from pullulan, *Carbohydr. Res.* **61**, 175–179.

Sandford, P. A. (1982) Potentially important microbial gums, in: *Food Hydrocolloids* (Glicksman, M., Ed.), Boca Raton, FL: CRC Press, 167–202.

Sano, T., Uemura, Y., Furuta, A. (1976) Photosensitive resin composition containing pullulan or esters thereof. U.S. Patent 3,960,685.

Sasago, M., Endo, M., Takeyama, K., Nomura, N. (1988) Water-soluble photopolymer and method of forming pattern by use of the same. U.S. Patent 4,745,042.

Sata, H., Umeda, M., Kim, C-H., Taniguchi, H., Maruyama, Y. (1989) Amylase-pullulanase enzyme produced by *B. circulans* F-2, *Biochim. Biophys. Acta* **991**, 388–394.

Schuster, R., Wenzig, E., Mersmann, A. (1993) Production of the fungal exopolysaccharide pullulan by batch-wise and continuous fermentation, *Appl. Microbiol. Biotechnol.* **39**, 155–158.

Sevilla, M. J., Isusi, P., Gutierrez, R., Egea, L., Uruburu, F. (1977) Influence of carbon and nitrogen sources on the morphology of *Pullularia pullulans*, *Trans. Br. Mycol. Soc.* **68**, 300–303.

Seviour, R. J., Kristiansen, B. (1983) Effect of ammonium ion concentration on polysaccharide production by *Aureobasidium pullulans* in batch culture, *Eur. J. Appl. Microbiol. Biotechnol.* **17**, 178–181.

Seviour, R. J., Stasinopoulos, S. J., Auer, D. P. F., Gibbs, P. A. (1992) Production of pullulan and other exopolysaccharides by filamentous fungi, *Crit. Rev. Biotechnol.* **12**, 279–298.

Shabtai, Y., Mukmenev, I. (1995) Enhanced production of pigment-free pullulan by a morphogenetically arrested *Aureobasidium pullulans* (ATTC 42023) in a two-stage fermentation with shift from soy bean oil to sucrose, *Appl. Microbiol. Biotechnol.* **43**, 595–603.

Shih, F. F. (1996) Edible films from rice protein concentrate and pullulan, *Cereal Chem.* **73**, 406–409.

Shimizu, T., Moriwaki, M., Shimoma, W. (1983) Condenser. Japanese Patent 58,098,909.

Shin, Y. C., Kim, Y. H., Lee, H. S., Kim, Y. N., Byun, S. M. (1987) Production of pullulan by a fed-batch fermentation, *Biotechnol. Lett.* **9**, 621–624.

Shin, Y. C., Kim, Y. H., Lee, H. S., Cho, S. J., Byun, S. M. (1989) Production of exopolysaccharide pullulan from inulin by a mixed culture of *Aureobasidium pullulans* and *Kluyveromyces fragilis*, *Biotechnol. Bioeng.* **33**, 129–133.

Silman, R. W., Bryan, W. L., Leathers, T. D. (1990) A comparison of polysaccharides from strains of *Aureobasidium pullulans*, *FEMS Microbiol. Lett.* **71**, 65–70.

Simon, I., Caye-Vaugien, C., Bouchonneau, M. (1993) Relation between pullulan production, morphological state and growth conditions in

Aureobasidium pullulans: new observations, *J. Gen. Microbiol.* **139**, 979–985.

Simon, I., Bouchet, B., Caye-Vaugien, C., Gallant, D. J. (1995) Pullulan elaboration and differentiation of the resting forms in *Aureobasidium pullulans*, *Can. J. Microbiol.* **41**, 35–45.

Slodki, M. E., Cadmus, M. C. (1978) Production of microbial polysaccharides, in: *Advances in Applied Microbiology* (Perlman, D., Ed.), New York: Academic Press, 19–54, Vol. 23.

Smith, K. A., Salyers, A. A. (1991) Characterization of a neopullulanase and an α-glucosidase from *Bacteriodes thetaiotaomicron* 95-1, *J. Bacteriol.* **173**, 2962–2968.

Sowa, W., Blackwood, A. C., Adams, G. A. (1963) Neutral extracellular glucan of *Pullularia pullulans* (de Bary) Berkhout, *Can. J. Chem.* **41**, 2314–2319.

Sugawa-Katayama, Y., Kondou, F., Mandai, T., Yoneyama, M. (1994) Effects of pullulan, polydextrose and pectin on cecal microflora, *Oyo Toshitsu Kagaku* **41**, 413–418.

Sugimoto, K. (1978) Pullulan. Production and applications, *Ferment. Ind.* **36**, 98–108.

Sunamoto, J., Sato, T., Hirota, M., Fukushima, K., Hiratani, K., Hara, K. (1987) A newly developed immunoliposome - an egg phosphatidylcholine liposome coated with pullulan bearing both a cholesterol moiety and an IgMs fragment, *Biochim. Biophys. Acta* **898**, 323–330.

Sutherland, I. W., Norval, M. (1970) The synthesis of exopolysaccharide by *Klebsiella aerogenes* membrane preparations and the involvement of lipid intermediates, *Biochem. J.* **120**, 567–576.

Suzuki, Y., Hatagaki, K., Oda, H. (1991) A hyperthermostable pullulanase produced by an extreme thermophile, *Bacillus flavocaldarius* KP 1228, and evidence for the proline theory of increasing protein thermostability, *Appl. Microbiol. Biotechnol.* **34**, 707–714.

Taguchi, R., Kikuchi, Y., Sakano, Y., Kobayashi, T. (1973a) Structural uniformity of pullulan produced by several strains of *Pullularia pullulans*, *Agric. Biol. Chem.* **37**, 1583–1588.

Taguchi, R., Sakano, Y., Kikuchi, Y., Sakuma, M., Kobayashi, T. (1973b) Synthesis of pullulan by acetone-dried cells and cell-free enzyme from *Pullularia pullulans*, and the participation of lipid intermediate, *Agric. Biol. Chem.* **37**, 1635–1641.

Takada, M., Yuzuriha, T., Katayama, K., Iwamoto, K., Sunamoto, J. (1984) Increased lung uptake of liposomes coated with polysaccharides, *Biochem. Biophys. Acta* **802**, 237–244.

Takasaki, Y. (1976) Productions and utilizations of β-amylase and pullulanase from *Bacillus cereus* var. *mycoides*, *Agric. Biol. Chem.* **40**, 1515–1522.

Takasaki, Y. (1987) Pullulanase-amylase complex enzyme from *Bacillus subtilis*, *Agric. Biol. Chem.* **51**, 9–16.

Tarabasz-Szymanska, L., Galas, E. (1993) Two-step mutagenesis of *Pullularia pullulans* leading to clones producing pure pullulan with high yield, *Enzyme Microb. Technol.* **15**, 317–320.

Thorne, L. P., Pollock, T. J., Armentrout, R. W. (1993) High molecular weight pullulan. U.S. Patent 5,268,460.

Thorne, L. P., Pollock, T. J., Armentrout, R. W. (2000) High molecular weight pullulan and method for its production. U.S. Patent 6,010,899.

Thornewell, S. J., Peery, R. B., Skatrud, P. L. (1995a) Cloning and characterization of the gene encoding translation elongation factor 1α from *Aureobasidium pullulans*, *Gene* **162**, 105–110.

Thornewell, S. J., Peery, R. B., Skatrud, P. L. (1995b) Integrative and replicative genetic transformation of *Aureobasidium pullulans*, *Curr. Genet.* **29**, 66–72.

Tomimura, E. (1991) Thermoduric and aciduric pullulanase enzyme and method for its production. U.S. Patent 5,055,403.

Tsuji, K., Toyota, N., Fujita, F. (1976) Molded pullulan type resins coated with thermosetting films. U.S. Patent 3,993,840.

Tsuji, K., Fujimoto, M., Masuko, F., Nagase, T. (1978) Carboxylated pullulan and method for producing same. U.S. Patent 4,090,016.

Tsujisaka, Y., Mitsuhashi, M. (1993) Pullulan, in: *Industrial Gums. Polysaccharides and Their Derivatives*. Third Edition (Whistler, R. L., BeMiller, J. N., Eds.), San Diego, CA: Academic Press, 447–460.

Tsukada, N., Hagihara, K., Tsuji, K., Fujimoto, M., Nagase, T. (1978) Protective coating material for lithographic printing plate. U.S. Patent 4,095,525.

Uchida, S., Yamamoto, A., Fukui, I., Endo, M., Umezawa, H., Nagura, S., Kubota, T. (1996) Siloxane-containing pullulan and method for the preparation thereof. U.S. Patent 5,583,244.

Ueda, S., Fujita, K., Komatsu, K., Nakashima, Z. (1963) Polysaccharide produced by the genus *Pullularia*. I. Production of polysaccharide by growing cells, *Appl. Microbiol.* **11**, 211–215.

Ueda, S., Kono, H. (1965) Polysaccharide produced by the genus *Pullularia* II. Trans-α-glucosidation by acetone cells of *Pullularia*, *Appl. Microbiol.* **13**, 882–885.

Ueda, S., Nanri, N. (1967) Production of isoamylase by *Escherichia intermedia*, *Appl. Microbiol.* **15**, 492–496.

Ueda, S., Ohba, R. (1972) Purification, crystallization and some properties of extracellular pullulanase from *Aerobacter aerogenes*, *Agric. Biol. Chem.* **36**, 2381–2391.

Vandamme, E. J., Bruggeman, G., De Baets, S., Vanhooren, P. T. (1996) Useful polymers of microbial origin, *Agro-Food-Industry Hi-Tech* **5**, 21–25.

Vermeersch, J. T., Coppens, P. J., Hauquier, G. I., Schacht, E. H. (1995) Lithographic base with a modified dextran or pullulan hydrophobic layer. U.S. Patent 5,402,725.

Vinroot, S., Torzilli, A. P. (1988) Interactive effects of inoculum density, agitation, and pH on dimorphism in a salt marsh isolate of *Aureobasidium pullulans*, *Mycologia* **80**, 376–381.

Waksman, N., De Lederkremer, R. M., Cerezo, A. S. (1977) The structure of an α-D-glucan from *Cyttaria harioti* Fischer, *Carbohydr. Res.* **59**, 505–515.

Walker, G. J. (1968) Metabolism of the reserve polysaccharide of *Streptococcus mitis*. Some properties of a pullulanase, *Biochem. J.* **108**, 33–40.

Wallenfels, K., Bender, H. (1961) Procedure for the production of a dextran-like polysaccharide from *Pullularia pullulans*. German Patent 1,096,850.

Wallenfels, K., Bender, H., Keilich, G., Bechtler, G. (1961) On pullulan, the glucan of the slime coat of *Pullularia pullulans*, *Angew. Chem.* **73**, 245–246.

Wallenfels, K., Keilich, G., Bechtler, G., Freudenberger, D. (1965) Investigations on pullulan. IV. Resolution of structural problems using physical, chemical and enzymatic methods, *Biochem. Z.* **341**, 433–450.

Webb, J. S., Nixon, M., Eastwood, I. M., Greenhalgh, M., Robson, G. D., Handley, P. S. (2000) Fungal colonization and biodeterioration of plasticized polyvinyl chloride, *Appl. Environ. Microbiol.* **66**, 3194–3200.

Wecker, A., Onken, U. (1991) Influence of dissolved oxygen concentration and shear rate on the production of pullulan by *Aureobasidium pullulans*, *Biotechnol. Lett.* **13**, 155–160.

West, T. P., Reed-Hamer, B. (1991) Ability of *Aureobasidium pullulans* to synthesize pullulan upon selected sources of carbon and nitrogen, *Microbios* **67**, 117–124.

West, T. P., Reed-Hamer, B. (1992) Influence of vitamins and mineral salts upon pullulan synthesis of *Aureobasidium pullulans*, *Microbios* **71**, 115–123.

West, T. P., Reed-Hamer, B. (1993a) Effect of temperature on pullulan production in relation to carbon source, *Microbios* **75**, 261–268.

West, T. P., Reed-Hamer, B. (1993b) Polysaccharide production by a reduced pigmentation mutant of the fungus *Aureobasidium pullulans*, *FEMS Microbiol. Lett.* **113**, 345–349.

West, T. P., Reed-Hamer, B. (1994a) Effect of complex nitrogen sources on pullulan production relative to carbon source, *Microbios* **80**, 83–90.

West, T. P., Reed-Hamer, B. (1994b) Elevated polysaccharide production by mutants of the fungus *Aureobasidium pullulans*, *FEMS Microbiol. Lett.* **124**, 167–172.

West, T. P., Strohfus, B. (1996a) A pullulan-degrading enzyme activity of *Aureobasidium pullulans*, *J. Basic Microbiol.* **36**, 377–380.

West, T. P., Strohfus, B. (1996b) Fungal cell immobilization on ion exchange resins for pullulan production, *Microbios* **88**, 177–187.

West, T. P., Strohfus, B. R-H. (1996c) Polysaccharide production by sponge-immobilized cells of the fungus *Aureobasidium pullulans*, *Lett. Appl. Microbiol.* **22**, 162–164.

West, T. P., Strohfus, B. (1996d) Pullulan production by *Aureobasidium pullulans* grown on ethanol stillage as a nitrogen source, *Microbios* **88**, 7–18.

West, T. P., Strohfus, B. R-H. (1996e) Use of adsorption in immobilizing fungal cells for pullulan production, *Microbios* **85**, 117–125.

West, T. P., Strohfus, B. (1997a) Effect of manganese on polysaccharide production and cellular pigmentation in the fungus *Aureobasidium pullulans*, *World J. Microbiol. Biotechnol.* **13**, 233–235.

West, T. P., Strohfus, B. (1997b) Fungal cell immobilization on zeolite for pullulan production, *Microbios* **91**, 121–130.

West, T. P., Strohfus, B. (1997c) Pullulan production by *Aureobasidium pullulans* grown on corn steep solids as a nitrogen source, *Microbios* **92**, 171–181.

West, T. P., Strohfus, B. (1998) Polysaccharide production by *Aureobasidium pullulans* cells immobilized by entrapment, *Microbiol. Res.* **153**, 253–256.

West, T. P., Strohfus, B. (1999) Effect of nitrogen source on pullulan production by *Aureobasidium pullulans* grown in a batch bioreactor, *Microbios* **99**, 147–159.

West, T. P. (2000) Exopolysaccharide production by entrapped cells of the fungus *Aureobasidium pullulans* ATCC 201253, *J. Basic Microbiol.* **40**, 5–6.

White, C., Gadd, G. M. (1984) Isolation and purification of protoplasts from yeast-like cells, hyphae and chlamydospores of *Aureobasidium pullulans*, *J. Gen. Microbiol.* **130**, 1031–1034.

White, W. L., Darby, R. T., Stechert, G. M., Sanderson, K. (1948) Assay of cellulolytic activity of molds isolated from fabrics and related items exposed in the tropics, *Mycologia* **40**, 34–84.

Wickerham L. J., Kurtzman, C. P. (1975) Synergistic color variants of *Aureobasidium pullulans*, *Mycologia* **67**, 342–361.

Wiley, B. J., Arcidiacono, S., Sousa, S., Mayer, J. M., Kaplan, D. L. (1987) Control of molecular weight distribution of the biopolymer pullulan produced by the fungus *Aureobasidium pullulans*. Technical Report NATICK/TR-88/012. Natick, MA: U.S. Army.

Wiley, B. J., Ball, D. H., Arcidiacono, S. M., Sousa, S., Mayer, J. M., Kaplan, D. L. (1993) Control of molecular weight distribution of the biopolymer pullulan produced by *Aureobasidium pullulans*, *J. Environ. Polymer Degrad.* **1**, 3–9.

Woods, L. F. J., Swinton, S. J. (1995) Enzymes in the starch and sugar industries, in: *Enzymes in Food Processing*, Second Edition (Tucker, G. A., Woods, L. F. J., Eds.), London: Blackie Academic & Professional, 250–267.

Wynne, E. S., Gott, C. L. (1956) A proposed revision of the genus *Pullularia*, *J. Gen. Microbiol.* **14**, 512–519.

Yagisawa, M., Kato, K., Koba, Y., Ueda, S. (1972) Pullulanase of *Streptomyces* no. 280, *J. Ferment. Technol.* **50**, 572–579.

Yamaguchi, R., Iwai, H., Otsuka, Y., Yamamoto, S., Ueda, K., Usui, M., Taniguchi, Y., Matuhasi, T. (1985) Conjugation of Sendai virus with pullulan and immunopotency of the conjugated virus, *Microbiol. Immunol.* **29**, 163–168.

Yamaguchi, S., Sunamoto, J. (1991) Fatty emulsion stabilized by a polysaccharide derivative. U.S. Patent 4,997,819.

Yamasaki, H., Lee, M-S., Tanaka, T., Nakanishi, K. (1993a) Characteristics of cross-flow filtration of pullulan broth, *Appl. Microbiol. Biotechnol.* **39**, 26–30.

Yamasaki, H., Lee, M-S., Tanaka, T., Nakanishi, K. (1993b) Improvement of performance for cross-flow membrane filtration of pullulan broth, *Appl. Microbiol. Biotechnol.* **39**, 21–25.

Yebra, M. J., Arroyo, J., Sanz, P., Prieto, J. A. (1997) Characterization of novel neopullulanase from *Bacillus polymyxa*, *Appl. Biochem. Biotechnol.* **68**, 113–120.

Yebra, M. J., Blasco, A., Sanz, P. (1999) Expression and secretion of Bacillus polymyxa neopullulanase in *Saccharomyces cerevisiae*, *FEMS Microbiol. Lett.* **170**, 41–49.

Yoneyama, M., Okada, K., Mandai, T., Aga, H., Sakai, S., Ichikawa, T. (1990) Effects of pullulan intake in humans, *Denpun Kagaku* **37**, 123–127.

Youssef, F., Biliaderis, C. G., Roukas, T. (1998) Enhancement of pullulan production by *Aureobasidium pullulans* in batch culture using olive oil and sucrose as carbon sources, *Appl. Biochem. Biotechnol.* **74**, 13–30.

Yuen, S. (1974) Pullulan and its applications, *Process Biochem.* **9**, 7–9.

Yun, J. W., Lee, M. G., Song, S. K. (1994) Production of panose from maltose by intact cells of *Aureobasidium pullulans*, *Biotechnol. Lett.* **16**, 359–362.

Yurlova, N. A., Mokrousov, I. V., de Hoog, G. S. (1995) Intraspecific variability and exopolysaccharide production in *Aureobasidium pullulans*, *Antonie van Leeuwenhoek Int. J. Gen. Mol. Microbiol.* **68**, 57–63.

Yurlova, N. A., de Hoog, G. S. (1997) A new variety of *Aureobasidium pullulans* characterized by exopolysaccharide structure, nutritional physiology and molecular features, *Antonie van Leeuwenhoek Int. J. Gen. Mol. Microbiol.* **72**, 141–147.

Zajic, J. E. (1967) Process for preparing a polysaccharide flocculating agent. U.S. Patent 3,320,136.

Zajic, J. E., LeDuy, A. (1973) Flocculant and chemical properties of a polysaccharide from *Pullularia pullulans*, *Appl. Microbiol.* **25**, 628–635.

Zajic, J. E., LeDuy, A. (1977) Pullulan, in: *Encyclopedia of Polymer Science and Technology*, Supplement Vol. 2. (Mark, H. F., Bikales, N. M., Eds.), New York: John Wiley & Sons, 643–652.

Zajic, J. E., Ho, K. K., Kosaric, N. (1979) Growth and pullulan production by *Aureobasidium pullulans* on spent sulfite liquor, in: *Developments in Industrial Microbiology* (Underkofler, L. A., Ed.) Arlington, VA: Society for Industrial Microbiology, 631–639, Vol. 20.

Zeikus, J. G., Hyun, H-H. (1988) Regulation and enhancement of enzyme production. U.S. Patent 4,737,459.

Zemek, J., Augustin, J., Kuniak, L., Kucar, S. (1980) Metabolism of α-glucans in *Aureobasidium pullulans*, *Biologia* **35**, 173–179.

12
Starch

Dr. Richard Frank Tester[1], Dr. John Karkalas[2]
[1] School of Biological and Biomedical Sciences, Glasgow Caledonian University, Cowcaddens Road Glasgow G4 0BA, UK; Tel.: +44-141-331-8514; Fax: +44-141-331-3208; E-mail: R.F.Tester@gcal.ac.uk
[2] School of Biological and Biomedical Sciences, Glasgow Caledonian University, Cowcaddens Road Glasgow G4 0BA, UK; Tel.: +44-141-331-8514; Fax: +44-141-331-3208; E-mail: JohnKarkalas@compuserve.com

1	Introduction and Historical Perspective	426
2	Occurrence, Sources and Production	426
3	Morphology and Composition of the Starch Granule	429
3.1	Morphology	429
3.2	Polysaccharides	430
3.3	Lipids	432
3.4	Proteins	432
3.5	Minerals	432
4	Biosynthesis and Granule Deposition	433
4.1	Biosynthesis	433
4.1.1	Supply and Transport of Carbon Source	433
4.1.2	Accumulation of Sucrose in the Cytoplasm	433
4.1.3	Conversion of Sucrose to Glucose-6-Phosphate in the Cytosol	433
4.1.4	Conversion of Glucose-6-Phosphate to Glucose-1-Phosphate in the Amyloplast	434
4.1.5	Conversion of Glucose-1-Phosphate to α-Glucan in the Amyloplast	434
4.2	Granule Deposition	435
5	The Structure of Starch Granules	436
5.1	Major Features of the Granule	436
5.1.1	Helical Structures	436
5.1.2	A-, B- and C-type Polymorphs	437

Polysaccharides and Polyamides in the Food Industry. Properties, Production, and Patents.
Edited by A. Steinbüchel and S. K. Rhee
Copyright © 2005 WILEY-VCH Verlag GmbH & Co. KGaA, Weinheim
ISBN: 3-527-31345-1

5.1.3	Crystalline and Noncrystalline Double Helices	438
5.1.4	Single Helices	438
5.1.5	V-type Helices	438
5.1.6	Lipid-free (FAM) and Lipid-complexed (LAM) Amylose	439
5.2	Crystallinity	439
5.3	Architecture of the Granule	439
6	**Physical Properties**	**445**
6.1	Microscopy of Starch Granules	445
6.1.1	Optical Microscopy	445
6.1.2	Electron Microscopy	445
6.1.3	Atomic Force Microscopy	446
6.2	Gelatinization and Swelling	446
6.3	Rheological Properties of Starch	448
6.3.1	Viscosity	448
6.3.2	Newtonian and Non-Newtonian Fluids	448
6.3.3	Properties of Gels	449
6.3.4	Viscometry of Gelatinized Starch Dispersions	450
6.4	Helical Complexes of Amylose	451
6.5	Retrogradation of Starch	452
6.6	Resistant Starch	453
6.6.1	Type I Resistant Starch	454
6.6.2	Type II Resistant Starch	454
6.6.3	Type III Resistant Starch	454
6.6.4	Type IV Resistant Starch	454
6.6.5	Inclusion Complexes	454
7	**Modification of Starch Properties**	**454**
7.1	The Need for Modification	454
7.1.1	Chemical Modifications	455
7.1.2	Enzymatic Modifications	456
7.1.3	Physical Modifications	456
8	**Analysis of Starch**	**457**
8.1	Early Methods of Determination of Starch	457
8.2	Quantitative Enzymatic Methods	458
8.2.1	Automated Instrumental Methods	458
8.2.2	Methods for the Characterization of Starch	459
9	**Industrial Production of Starch**	**459**
9.1	General Principles	459
9.1.1	Maize (*Zea mays*)	460
9.1.2	Wheat (*Triticum* spp.)	461
9.1.3	Rice (*Oryza sativa*)	461
9.1.4	Sorghum (*Sorghum bicolor*)	461

9.1.5	Potato (*Solanum tuberosum*)	461
9.1.6	Tapioca (*Manihot esculenta*)	461
9.1.7	Sweet potato (*Ipomoea batatas*)	462
9.1.8	Arrowroot (*Maranta arundinacea*)	462
9.1.9	Sago (*Metroxylon sagu*)	462
9.2	Uses of Starch	462
9.2.1	Food Uses	462
9.2.2	Industrial Uses	463
10	**Hydrolysis Products of Starch**	463
10.1	Introduction	463
10.1.1	Products of Acid Hydrolysis	464
10.1.2	Acid-Enzyme-produced Syrups	464
10.1.3	Enzymatically Produced Syrups	464
10.2	Other Products of Hydrolysis	466
10.2.1	Maltodextrins	466
10.2.2	Cyclodextrins	466
10.2.3	Hydrogenated Syrups and Polyols	467
10.3	Uses of Hydrolyzed Products	467
11	**Concluding Remarks**	468
12	**References**	469

ADP	adenosine diphosphate
AFM	atomic force microscopy
ATP	adenosine triphosphate
CL	chain length
Da	dalton
DBE	debranching enzyme
DE	dextrose equivalent
DP	degree of polymerization
DS	degree of substitution
DSC	differential scanning calorimetry
GBSS	granule-bound starch synthase
ΔH	gelatinization enthalpy
HFS	high-fructose syrup
HFCS	high-fructose corn syrup
IBC	iodine binding capacity
LPL	lysophospholipid
NAD	nicotinamide adenine dinucleotide (oxidized form)
NADH	nicotinamide dinucleotide (reduced form)
NADP	nicotinamide adenine dinucleotide phosphate (oxidized form)
NADPH	nicotinamide adenine dinucleotide phosphate (reduced form)
NMR	nuclear magnetic resonance

PP$_i$	inorganic pyrophosphate
SAXS	small-angle X-ray scattering
SBE	starch branching enzyme
SEM	scanning electron microscopy
SSS	soluble starch synthase
TEM	transmission electron microscopy
T$_o$	gelatinization onset temperature
T$_p$	gelatinization peak temperature
T$_c$	gelatinization conclusion temperature
UDP	uridine diphosphate
WAXS	wide-angle X-ray scattering

1
Introduction and Historical Perspective

Starch, a polymer of glucose, is an important energy reserve of plants. It is synthesized as microscopic granules in the tissues of many plant species, and has provided dietary energy for animals and man for several millennia. In addition, the useful properties of starch as an adhesive for holding fibers together in the preparation of paper (papyrus in Egypt since 3500 BC) for writing, and for stiffening fabrics were recognized by early civilizations. In fact the, word starch is derived from a cognate form of the Old High German *sterken* (to stiffen) (cf. modern German *Stärke*). The Latin word *amylum* is derived from the Greek *amylon* (non-milled) from which the widely used words amylose, amylopectin, amylase and amylolysis originate. There is considerable archaeological evidence that in ancient Egypt, Greece, Rome and China starch was utilized in both food and nonfood products. Starch research as we understand it today, however, probably originated in the early 1700s when van Leeuwenhoek (1719) observed starch granules from various sources with his newly invented microscope. Fritzsche (1834) described differences in the morphology of starch granules collected from a number of plant species. Schleiden (1849) initiated the early classification of starches, still relevant today, based on the characteristic morphology of granules from different botanical origins. Later in the same century, Nägeli (1858) further classified starch granules according to their size and shape, and this classification was later refined by Meyer (1895). By the time that Muter (1906), Winton (1906) and Kraemer (1907) described the morphology of starch granules according to their botanical origin, it was clear that much of the essential groundwork was done. From the beginning of the twentieth century onwards, many authors have written substantive accounts of the nature and properties of starch. Comprehensive reviews have been provided by Reichert (1913), McNair (1930) Radley (1968), Banks and Greenwood (1975), French (1984), Tegge (1984), Morrison and Karkalas (1990), and Zobel and Stephen (1995), to name but a few.

2
Occurrence, Sources and Production

Starch occurs in seeds, roots, tubers, fruits and stems of plants, many of which are important sources of food for the populations of temperate and tropical regions. It provides the human body with a large proportion of its energy. Rice and roots (mainly sweet potatoes, cassava and yams) are staple foods for the populations of the Far East and Africa, and in some developing countries starch may provide as much as 80% of the energy requirement. In the industrialized world a lower proportion of dietary energy is derived directly from the starch occurring in cereal foods and potatoes.

It can be calculated that, as the result of the photosynthetic process, some 1385 million tonnes of starch are produced annually by the cereals (Figure 1), roots and tubers (Figure 2) and pulses (Figure 3). Most of this starch (ca. 89%) is synthesized by cereals (Figure 4), while roots and tubers produce about 9.8% of the total. Leguminous seeds (pulses) provide a relatively small proportion (ca. 1.4%).

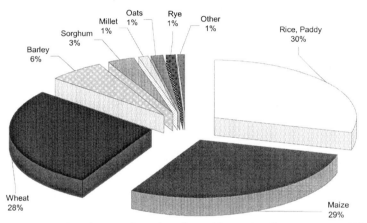

Fig. 1 Annual production of cereals in the world $= 2050 \times 10^6$ tonnes (FAO, 2000).

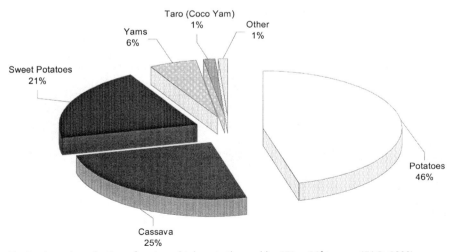

Fig. 2 Annual production of roots and tubers in the world $= 679 \times 10^6$ tonnes (FAO, 2000).

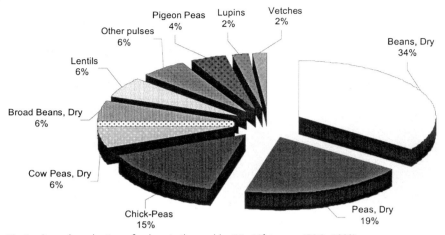

Fig. 3 Annual production of pulses in the world = 55×10^6 tonnes (FAO, 2000).

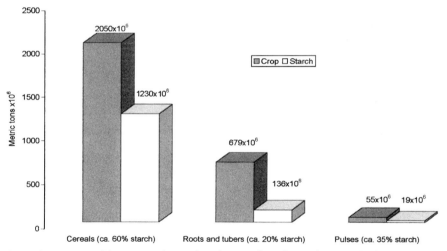

Fig. 4 Comparison of the estimated annual production of starch ($\times 10^6$ tonnes) by three principal types of crop.

Nowadays, starch is isolated in a very pure form, and in large quantities, from cereal grains (especially maize), roots and tubers in all industrialized countries. It is an important commercial commodity and finds applications in the food, pharmaceutical, textile, paper and other industries. More than half of this pure starch is hydrolyzed by acid or enzymes to glucose-containing products. The glucose is further isomerized to fructose, and is available commercially as high-fructose corn syrup. A considerable proportion of pure starch is modified by suitable chemical derivatization to meet specialized food and other industrial applications.

Large quantities of endogenous starch are also converted to soluble carbohydrate *in situ* when malted barley is extracted with hot water for the production of alcoholic beverages after fermentation of the sugars. The starch contained in other cereal grains (e.g., rice, sorghum) and in potatoes is also converted to ethanol in many parts of the world. Being a renewable resource, starch is favored as raw material for the production of fuel alcohol.

In the USA, about 47 million tonnes (1881 million bushels) of maize were used in 1999 for the manufacture of starch-derived products (Corn Refiners Association, Inc., 2000). Of this, 30.6% was used for the production of high-fructose corn syrup, 12.8% for glucose syrup and dextrose, 12.8% for starch, 29.5% for fuel alcohol, 7.0% for beverage alcohol, and 7.4% for cereals and other products. Therefore, it can be calculated that 56.2% of maize was used for the extraction of pure starch, of which 77.2% was converted to starch-based sweeteners. In the European Union, the starch trade is estimated at 3.5 million tonnes, worth $1700 million. Figures are not disclosed for starch production in the UK. However, the net supply of glucose in the UK was 825,492 tonnes to a value of £176 million (National Statistics, 2000). Manufacturers' sales of starch and starch products in 1996 amounted to £496 million in the UK (National Statistics, 1998).

3
Morphology and Composition of the Starch Granule

3.1
Morphology

Starch granules are roughly spherical (typically 1 to 100 μm in diameter), and because of their compact structure are completely insoluble in water at ambient temperature. They also exhibit considerable mechanical strength. Starch is synthesized in many parts of plants, e.g., pollen, leaves, stems, roots, tubers, bulbs, rhizomes, fruits, flowers, and seeds. The biosynthesis of starch occurs within three types of membrane-bound subcellular organelles: chloroplasts, chloro-amyloplasts, and amyloplasts. The chloroplasts produce transient starch, which is synthesized during the day and mobilized during the night. The storage depot, however, is generated by the amyloplast in familiar "starchy" tissues such as cereal endosperms and potato tubers. In many tissues there is only one granule per amyloplast ("simple" granules, e.g., wheat, barley, maize, millet, sorghum, pea, and potato), whilst in others (e.g., rice, oats, cassava, sago, and sweet potato) there are several ("compound" granules). Apart from the distinctive variations in size and shape of granules, the size distribution of granules is also indicative of origin. Further information on granule morphology is provided by Lineback (1984), Swinkels (1985), Gallant and Bouchet (1986), Blanshard (1987), Galliard and Bowler (1987a, b), Kent and Evers (1994) and Fredriksson et al. (1998) (see Table 1).

The granular form of starch is responsible for its unique properties. Although insoluble in cold water, granules swell and form a thick paste on heating above 60°C. This swelling and thickening is known as gelatinization (see Section 6.2), and has important implications with regard to the practical uses of starch, including human and animal nutrition. The physical properties of starch are discussed in Section 6.

Starch granules are composed of two major α-glucans, amylose and amylopectin, which represent approximately 98–99% of the dry weight. The ratio of the two components varies according to the botanical origin with "waxy" starches containing less than about 15% amylose (this is a relatively arbitrary figure), normal starches 20–35%

Tab. 1 Characteristics of starch granules of diverse botanical origin

Starch	Type	Character	Shape	Diameter [µm]
Barley	Cereal	Bimodal	Lenticular (A-type)	15–25
			Spherical (B-type)	2–5
Maize (waxy and normal)	Cereal	Unimodal	Spherical/ polyhedral	2–30
Amylomaize	Cereal	Unimodal	Irregular	2–30
Millet	Cereal	Unimodal	Polyhedral	4–12
Oat	Cereal	Unimodal	Polyhedral	3–10 (single)
				80 (compound)
Pea	Legume	Unimodal	Reniform (single)	5–10
Potato	Tuber	Unimodal	Lenticular	5–100
Rice	Cereal	Unimodal	Polyhedral	3–8 (Single)
				150 (compound)
Rye	Cereal	Bimodal	Lenticular (A-type)	10–40
			Spherical (B-type)	5–10
Sorghum	Cereal	Unimodal	Spherical	5–20
Tapioca	Root	Unimodal	Spherical/Lenticular	5–35
Triticale	Cereal	Unimodal	Spherical	1–30
Wheat	Cereal	Bimodal	Lenticular (A-type)	15–35
			Spherical (B-type)	2–10

and high amylose greater than about 40%. Waxy varieties of barley, maize and rice have been available for many years due to the relative ease of high amylopectin expression within these genetic backgrounds. The tetraploid/hexaploid nature of wheat has made it more difficult to generate a waxy mutant, although in the last few years waxy wheat and potatoes have become available. The moisture content of air-equilibrated cereal starches is typically 10–12% but root and tuber starches contain higher amounts (about 14–18% for potato). Cereal starches contain lipids (lysophosphilipids (LPL) and free fatty acids) and there is a positive linear relationship between the amylose and lipid content (discussed below). Waxy cereal starches contain very little lipid, while normal and cereal starches contain 1–1.5%. A clear differentiation is made between endogenous lipid and surface lipid due to contamination from other sources (Morrison, 1985, 1988, 1995). Purified starches contain <0.6% protein which is integrated into the granules and represents primarily residual biosynthetic enzymes. Minerals represent <0.4% of the granule. Phosphate ester groups attached to amylopectin and amylose are particularly abundant in potato starch.

3.2
Polysaccharides

The two major starch polysaccharides amylose and amylopectin (Figure 5) have different structures and properties, which have been discussed and reviewed in detail by several authors (French, 1972; Lineback, 1986; Galliard and Bowler, 1987a, b; Morrison and Karkalas, 1990; Seib, 1994; Buléon et al., 1998a).

Amylose is an essentially linear α-glucan containing around 99% α-$(1 \rightarrow 4)$ and 1% α-

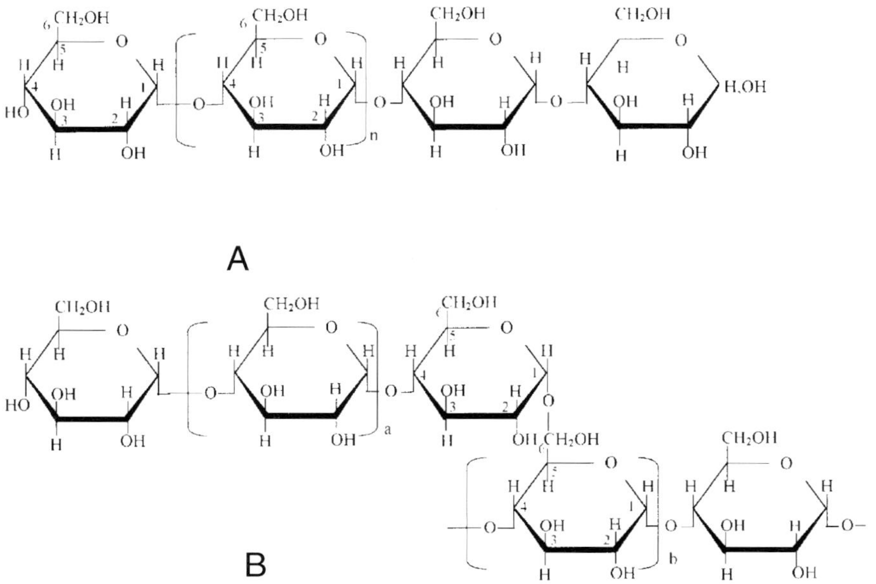

Fig. 5 The structure of (A) amylose and (B) amylopectin. In amylose, $n = \sim 1000$; in amylopectin, $a = \sim 12-23$ and $b = 20-120$. For details, see text.

$(1 \rightarrow 6)$ bonds. The molecular weight is approximately 1×10^5 to 1×10^6 (Biliaderis, 1998; Buléon et al., 1998a), with a degree of polymerization by number (DP_n) of 690 (amylomaize) to 4920 (potato) and between nine and 20 branch points per molecule (Hizukuri et al., 1981; Hizukuri, 1993). Each chain comprises in the region of 200–700 glucose residues (Morrison and Karkalas, 1990) equivalent to a molecular weight of 32,400 to 113,400 daltons. It has been calculated that there are about 1.8×10^9 amylose molecules per average starch granule (Buléon et al., 1998a). The size and structure of amylose and its polydispersity, however, varies according to botanical origin.

Amylopectin has a much higher molecular weight than amylose, typically 1×10^7 to 1×10^9 (Morrison and Karkalas, 1990; Biliaderis, 1998) and is much more heavily branched, with about 95% $\alpha\text{-}(1 \rightarrow 4)$ and about 5% $\alpha\text{-}(1 \rightarrow 6)$ bonds. In common with amylose, the size of this polymer and its polydispersity varies as a function of botanical origin, although with amylopectin there is additional variation with respect to the unit chain lengths and branching patterns. The unit chains of amylopectin are small and of the order (on average) of 20–25 units long although, depending on length and spatial orientation within the molecule, they can be subclassified (Hizukuri 1985, 1986). A- and B_1 chains are the most external (exterior) and form double helices within the native granules (see Section 5). They are approximately 12 (e.g., waxy rice) to 22 (amylomaize) glucose units long (Hizukuri, 1985, 1993), with A-type (here referring to diffraction pattern, not chain length) starches (most cereals) tending to have shorter chain lengths on average than B-type (e.g., potato). The A-chains of amylopectin are $\alpha\text{-}(1 \rightarrow 6)$ bonded to B-chains, which are themselves bonded to other B-chains or the "backbone" of the amylopectin molecule, the single C-chain. Depending on length (which correlates with clusters traversed as the chain radiates out through the native granule; see Section 5), B-chains are referred to as B_1 to B_4 (one to four clusters). Typical chain lengths (CL) for A-, B_1, B_2, B_3, and B_4 chains for four different starches are 12–16, 20–24, 42–48, 69–75, and 101–119 respectively (Hizukuri, 1986, 1988). The A- to B-chain ratio often quoted is, for example 1.26:1 for wheat (Hizukuri and Maehara, 1990), although the ratio varies according to starch type, and ranges from 1:1 to 2.6:1 for waxy maize (but not necessarily the same cultivar) depending on the literature source (Manners, 1985).

It has been proposed that certain starches (such as high-amylose maize, oat, and pea) contain an α-glucan having an intermediate molecular weight between that of amylose and amylopectin (Banks and Greenwood, 1975; Paton, 1979; Baba and Arai, 1984; Colonna and Mercier, 1984; Wang and White, 1994). However, other investigators (Tester and Karkalas, 1996) have disputed that this material exists as a discrete fraction–at least in oats. Certain maize mutants (*sugary 1, su1*) contain a soluble α-glucan which is similar to animal glycogen and has been termed phytoglycogen for this reason (Ball et al., 1996).

3.3
Lipids

The surfaces of starch granules are, if not thoroughly purified, contaminated with triglycerides, glycolipids, phospholipids, and free fatty acids derived from the amyloplast membrane and nonstarch sources. True starch (internal) lipids are composed of free fatty acids and LPL (Morrison 1985, 1988, 1993, 1995; Morrison and Karkalas, 1990). Cereal starches are unique with respect to their lipid composition, with Triticeae starches containing primarily LPL (as lysophosphatidyl-choline, -ethanolamine and -glycerol) while other cereal starches are rich in both LPL and free fatty acids. Part of the amylose fraction within starch granules exists as an amylose lipid inclusion complex which is evident by ^{13}C-cross polarization/magic-angle spinning nuclear magnetic resonance (^{13}C-CP/MAS NMR) (Morrison et al., 1993a, b, c). The presence of this fraction is also evident from iodine binding studies where nondefatted amylose from cereal starches has a lower iodine binding capacity (IBC). The amount of lipid-complexed amylose ranges from <15 to $>55\%$ in cereal starches, with oat starches being especially rich in this fraction (Morrison 1993, 1995).

3.4
Proteins

The proteins associated with starch granules may, like the lipid fraction, be derived from the nonstarchy tissues of plants, be more strongly bound to the surface of starch granules, or be integrated into the granule structure. In wheat, the starch surface protein friabilin has received much attention because of its proposed association with grain hardness (Greenwell and Schofield, 1986; Schofield, 1994; Oda and Schofield, 1997). Integral proteins have a higher molecular weight than surface proteins (50 150 kDa and 14–30 kDa, respectively), and probably represent the residues of enzymes involved in starch deposition. The starch synthases (especially soluble starch synthase) constitute the major integral protein fraction.

3.5
Minerals

Although starches contain small amounts of minerals (calcium, magnesium, phosphorus, potassium, and sodium) they are, with the exception of phosphorus, of little significance. Phosphorus is found in three major forms: phosphate monoesters, phospholipids, and inorganic phosphates. In the Triticeae starches, phosphorus content is very close to the LPL content, and there is a relatively small amount of α-glucan phosphate monoesters. Other starches contain little phosphate, with the exception of potato starch where there is essentially no lipid, but the phosphorus content (as phosphate monoester) may exceed 0.1%.

4
Biosynthesis and Granule Deposition

4.1
Biosynthesis

The biosynthesis of storage (nontransient) starch is a very complex process, and occurs in a subcellular organelle called the amyloplast (transient starch is synthesized in the chloroplast). The process involves carbon accumulation from the primary photosynthetic assimilate, sucrose, to generate the starchy polymers amylose and amylopectin. Whilst much is known about the in-vitro activity of the enzymes involved in starch deposition, the orchestration and regulation of their activity *in vivo* is still not completely understood. This represents the major challenge facing starch biochemists, and to resolve it research is focused on understanding specific mutants and transgenic plant genomes which generate alterations in the biochemical pathways associated with starch deposition.

4.1.1
Supply and Transport of Carbon Source
Sucrose is the primary photosynthetic assimilate. Upon synthesis, it is transported, via the phloem in higher plants, from photosynthetic tissues (containing chloroplasts), to cells responsible for starch biosynthesis (containing amyloplasts). There appear to be differences in the mode of sucrose uptake into different starch-synthesizing tissues. In addition, other carbon sources may be involved in the biosynthesis of the starch polysaccharides (Duffus, 1993; Smith, 1993).

4.1.2
Accumulation of Sucrose in the Cytoplasm
Little is known or understood about how the sucrose that enters the cytoplasm of cells involved in starch biosynthesis is partitioned between starch biosynthesis and other metabolic processes. However, sucrose levels do not restrict starch synthesis – at least in cereal grains where they are maintained during grain development (Duffus, 1993). The carbon flow from photosynthesis to sucrose has been discussed in detail elsewhere (ap Rees, 1993) and is beyond the scope of this review.

4.1.3
Conversion of Sucrose to Glucose-6-Phosphate in the Cytosol
A simplified representation of the process of starch synthesis (including cellular partitioning) for peas is presented in Figure 6, although it is recognized that there are plant- and tissue-specific differences (Wang et al., 1998). Using this model, sucrose within the cytosol of cells involved in starch deposition is converted to UDP-glucose and fructose (in the presence of UDP) by sucrose synthase. The UDP-glucose is then converted (in the presence of PP_i) to glucose-1-phosphate by UDP-glucose pyrophosphorylase, which is further converted to glucose-6-phosphate by phosphoglucomutase. The glucose-6-phosphate is translocated across the plastid

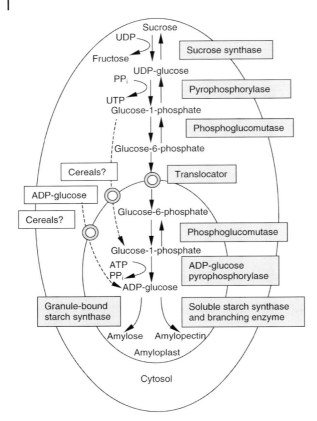

Fig. 6 Pathway for starch deposition in peas (and cereals) (adapted from Wang et al., 1998).

membrane, possibly in exchange for phosphate or triose phosphate.

In storage tissues, hexose phosphates are imported into the amyloplast as described above. There is some evidence that in cereals glucose-1-phosphate is also translocated (Bowsher et al., 1996). This must be viewed in the context that in some cereals ADP-glucose pyrophosphorylase (see Figure 6) is located outside the plastid and that ADP glucose is transported into the plastid directly (Duffus, 1993; Kleczkowski, 1996) for starch synthesis as described below.

4.1.4
Conversion of Glucose-6-Phosphate to Glucose-1-Phosphate in the Amyloplast
If glucose-6-phosphate is translocated across the amyloplast membrane (Figure 6), it is converted to glucose-1-phosphate by phosphoglucomutase. Glucose-1-phosphate and ADP-glucose feed into the biosynthetic pathway as described below.

4.1.5
Conversion of Glucose-1-Phosphate to α-Glucan in the Amyloplast
The conversion of glucose-1-phosphate to starch (within plastids) involves three major enzyme systems: (1) ADP-glucose pyrophos-

phorylase; (2) starch synthase; and (3) starch branching enzyme (see Figure 6). As mentioned previously, the specificity and kinetic activity of the individual enzymes is better understood than the mechanisms regulating the deposition of starch granules (Martin and Smith, 1995; Nelson and Pan, 1995; Smith et al., 1997; Buléon et al., 1998a; Wang et al., 1998; Smith, 1999).

ADP-glucose pyrophosphorylase catalyzes the formation of ADP-glucose from glucose-1-phosphate and ATP (Figure 6). It is a key regulatory enzyme in the biosynthesis of starch, and is inhibited by inorganic phosphate but activated by 3-phosphoglycerate (Ball et al., 1996, 1998; Smith et al., 1997; Buléon et al., 1998a; Wang et al., 1998).

Starch synthases sequentially add glucose units from ADP-glucose to the nonreducing ends of amylose and amylopectin. There are two major classes of starch synthase: "granule-bound" (GBSS, usually viewed as being responsible for amylose production); and "soluble" (SSS, sometimes abbreviated to SS when referring to different isoforms) which is responsible for amylopectin production. The assumed exclusivity with respect to GBSS and SSS activity (for amylose or amylopectin synthesis) *in vivo* may not be absolute (Wang et al., 1998; Smith, 1999). The major classes of starch synthases may be further subdivided, and SSS is currently regarded as having three major isoforms (SSI, SSII, and SSIII).

GBSSI has been shown to elongate small malto-oligosaccharides to produce amylose (Denyer et al., 1996; Ball et al., 1998; Smith, 1999). While the formation of the priming malto-oligosaccharides is poorly understood and complicated by that fact that SSS can *in vitro* elongate both amylose and amylopectin (Martin and Smith, 1995), the compartmentalization of GBSS for amylose and SSS for amylopectin is usually described as the most likely arrangement *in vivo*.

Starch branching enzyme (SBE) creates α-$(1 \rightarrow 6)$ bonds in amylopectin after transferring α-$(1 \rightarrow 4)$ glucan chains to other α-$(1 \rightarrow 4)$ glucan chains. Two major classes exist (SBEI and SBEII) which have maximal activity at different times during starch deposition (Wang et al., 1998). The chain length of amylopectin molecules may be related to the activity of different isoforms of this enzyme (Smith, 1999). In recent years it has been proposed (Ball et al., 1996; Buléon et al., 1998a; Smith, 1999) that amylopectin is pretrimmed during biosynthesis by debranching enzyme (DBE). According to this theory, when chains are long enough to become substrates for SBE, a highly branched pre-amylopectin form is created. The activity of DBE then creates small chains on which the next round of biosynthesis can work. Other models suggesting a less direct role for DBE (glucan recycling model) have also been proposed (Ball et al., 1996; Buléon et al., 1998a; Smith, 1999). Readers are referred to these papers and a similar review (Ball, 1995) for an overview with respect to current knowledge regarding the process of starch biosynthesis.

4.2
Granule Deposition

Starch granules develop in amyloplasts as previously described. In the early phases of starch deposition, the number and size of granules increase in the plant cells. Deposition of new starch occurs on the surface of these granules (often referred to as apposition) as described by Evers (1971) and discussed by Parker (1985). In wheat, the large A-type granules begin developing in the amyloplast 4–5 days after anthesis, with the final number becoming obvious after about 7 days (Parker, 1985; Bechtel et al., 1990; Morrison, 1993). The small B-type starch granules found in wheat starch develop about 10–14 days after anthesis, possibly by initiation from the A-type granules or invaginations of the amyloplast membrane (Buttrose, 1960; Morrison, 1989; Bechtel et al., 1990). The starch deposition profile in different plant species leads to the characteristic shape and distribution profile of that starch. Furthermore, there are genotypic-specific differences in a given species as in, for example, the development of maize starch granules (Boyer et al., 1977).

During the development of cereal and potato starch granules the amylose and lipid (cereal) content increases (Geddes et al., 1965; Banks et al., 1973; Asaoka et al., 1985; Kang et al., 1985; Morrison and Gadan, 1987; McDonald et al., 1991; Tester and Morrison, 1993), while in some crops (e.g., cassava and sweet potato) this increase is not obvious (Noda et al., 1992, 1997; Sriroth et al., 1999). This generates asymmetry with respect to granule composition (Morrison and Gadan, 1987; Morrison, 1989; McDonald et al., 1991). Consequently, the granule hilum (center) is relatively high with respect to amylopectin and low with respect to amylose and lipid, but with a progressive inversion of this relationship towards the granule periphery. Although the amylose content of rice starches increases during development, there is little difference in the molecular size or shape of amylose or amylopectin (Hizukuri et al., 1995). The phosphate ester content (per glucose residue) of small potato starch granules is higher than for large starch granules (Nielsen et al., 1994).

With respect to starch crystallinity and physical properties imparted during starch development, X-ray diffraction studies have indicated a constancy in the diffraction pattern and the proportion of crystalline starch for potato (Sugimoto et al., 1995) with a gradual change from C_a- to C-type for sweet potato (Noda et al., 1997). Differential scanning calorimetry (DSC) determinations have also indicated a general constancy with respect to gelatinization characteristics for barley (Tester and Morrison, 1993) and potato starches (Noda et al., 1992; Sugimoto et al., 1995), although this is most obvious at the latter stages of potato starch development (Ng et al., 1997). The gelatinization temperatures and enthalpy of sweet potato starch decrease as a function of growth temperature (Noda et al., 1997).

5
The Structure of Starch Granules

5.1
Major Features of the Granule

5.1.1
Helical Structures

The polysaccharides within starch granules assume their configuration during biosynthesis. Whilst genetic control defines the structure of the amylose and amylopectin molecules, and the key biosynthetic enzymes involved in this process are known (see Section 4.1), the mechanisms controlling their orientation and interaction which lead to the three-dimensional granular structure are poorly understood. To aid understanding of how the granule is formed, five major approaches have been adopted. The first approach involves forming association structures with the constituent α-glucans from solution and extrapolating from these structures, aspects of granular structure. The second uses nondestructive analysis of granules to try and probe the complex interactions within them. Third, the plant's genome may be modified by breeding, mutations or transgenic technology and the consequence in terms of granular architecture investigated. Fourth, the deposition profile of starch may be studied during development. Finally, granules may be chemically or physically modified after biosynthesis. The first two aspects will be considered below.

5.1.2
A-, B- and C-type Polymorphs

The formation of amylopectin double helices and subsequent registration (ordered orientation) of these double helices within starch granules has been (and this often causes confusion) modeled most effectively using amylose. Early studies by Wu and Sarko (1978a, b) on amylose double helices prepared in the A- and B-type polymorphic forms (as found in native starch granules using X-ray diffraction) indicated that the double helices comprise single helices in the 6(3.5) conformation, the notation showing that the pitch comprises six glucose residues with 3.5 Å (0.35 nm) advance per monomer. These figures can be obtained from X-ray diffraction studies (Rao et al., 1998). Hence, the single helix pitch is 21 Å (2.1 nm). Because the α-glucan chains occur as double helices, the actual repeat along the helix length is half this length 10.5 Å (1.05 nm). Although the data of Wu and Sarko (1978a, b) suggested to them that the double helices were right-handed, parallel double-stranded helices, other investigators (Imberty and Pérez, 1988; Imberty et al., 1987, 1988a, b, 1991) indicated that the helices are in a left-handed, parallel double-stranded conformation. Wu and Sarko (1978a, b) differentiated between the unit cell organization of the A- and B-type polymorphs where the A-type contains a helical core comprising a double helix which is occupied by water in the B-type (Figure 7).

Although the double helices themselves adopt a conformation (left- or right-handed), the packing of these double helices into crystalline domains also has a defined conformation where they are probably parallel (Imberty et al., 1988a, 1991) rather than antiparallel (Wu and Sarko, 1978a, b) in the unit cells (Buléon et al., 1998a).

Most cereal starches contain amylopectin which adopts the A-type crystalline polymorph, while high-amylose maize starch and tuber amylopectins adopt the B-type form, and legume, root and some fruit and stem starches the C-type form (Banks and Greenwood, 1975; French, 1984; Blanshard, 1987; Biliaderis, 1998). The previously considered distinct polymorph (C-type) is now, however, considered to be a mixture of A- and B-type (Sarko and Wu, 1978; Zobel, 1988a, b; Imberty et al., 1991; Gernat et al., 1990; Cairns et al., 1997; Buléon et al.,

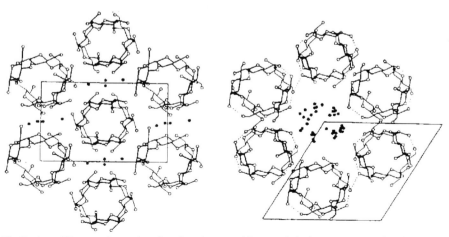

Fig. 7 A- and B-type polymorphs of amylose (reprinted from *Carbohydr. Res. 61*, Wu and Sarko, The double-helical structure of B-amylose, p. 27–40, 1978, with permission from Elsevier Science).

1998b). Although the B-type polymorph can be converted to the A-type form using for example heat moisture treatment (Sair, 1967; Lorenz and Kulp, 1982; Donovan et al., 1983; Gidley and Bociek, 1985; Kuge and Kitamura, 1985; Imberty et al., 1991; Stute, 1992; Hoover and Vasanthan, 1994; Kawabata et al., 1994; Jacobs and Delcour, 1998; Tester and Debon, 2000), it is thought that the relatively short exterior chains of amylopectins from cereal starches (DP 14–18) favor A-type crystallinity whilst B-type crystallinity is favored by longer exterior chains (DP 16–20) (Hizukuri, 1985, 1993; Gidley, 1987; Gidley and Bulpin, 1987; Whittam et al., 1990; Cheetam and Tao, 1998; Rao et al., 1998; Gérard et al., 2000). The minimum chain length to form double helices is DP 10 (Gidley and Bulpin, 1987).

It is believed by some authors that the anomalously long amylopectin chains in high-amylose starches may be responsible for the formation of the B- rather than the A-type crystalline polymorph common in cereal starches (Zobel, 1988a, b; Morrison and Karkalas, 1990) and, through their association, granule crystallinity. However, more recent studies (Shi et al., 1998; Tester et al., 2000) indicate that amylose forms double helices and registered (crystalline) associations of these helices in these starches.

5.1.3
Crystalline and Noncrystalline Double Helices
Starch crystallinity has been determined using a number of approaches, although ^{13}C-cross polarization/magic angle spinning-nuclear magnetic resonance (^{13}C-CP/MAS-NMR, for double helices) and wide-angle X-ray scattering (WAXS, crystallinity based on unit cells) are probably the most valuable. If estimates from both techniques gave the same answers, this would mean that all double helices were in crystalline arrays. However, this is not the case. The proportion of double helices to nondouble helical material for different starches (NMR) ranges from 38:62 to 53:47 (Gidley and Bociek, 1985). Starch crystallinity (primarily WAXS) ranges from 15 to 45% (Nara, 1978; Nara et al., 1978; Buléon et al., 1982; Gidley and Bociek, 1985; Zobel, 1988b; Cheetam and Tao, 1998). For native wheat starch the quantity of double helices (NMR) compared with crystalline material (WAXS) has been reported to be 46% and 36% respectively (Morrison et al., 1994). These data complement the data of others (Gidley and Bulpin, 1987; Cooke and Gidley, 1992; Gidley, 1992), and prove that double helices need not be part of crystalline arrays. Determinations of the amount of double helices (NMR) for high-amylose starches (Shi et al., 1998; Tester et al., 2000) indicate, as discussed above, that in these starches amylose double helices contribute to the crystalline fraction.

5.1.4
Single Helices
Carbon atoms quantified by ^{13}C-CP/MAS-NMR show large chemical shift differences between amorphous and crystalline materials (especially C-1 and C-4, forming the main proportion of glycosidic linkages) due to conformational differences. Amorphous starch gives a broad resonance. A large proportion of amorphous components in native starches show a C-1 and C-4 chemical shift range similar to that of amylose single helices (Gidley and Bociek 1985, 1988).

5.1.5
V-type Helices
Amylose forms helical inclusion complexes with other molecules, in the so-called V-form (Morrison and Karkalas, 1990; Zobel and Stephen, 1995; Le Bail et al., 1997; Biliaderis, 1998). These helical assemblies are left-handed with six glucose residues per pitch,

and contain a hydrophobic core. The presence of these helices in native starch granules has been proven using ^{13}C-CP/MAS-NMR (Morrison et al., 1993a, b, c)

5.1.6
Lipid-free (FAM) and Lipid-complexed (LAM) Amylose

In Section 3.3 the presence of amylose–lipid complexes in native starch granules was mentioned, and proof of their existence provided using ^{13}C-CP/MAS-NMR (Morrison et al., 1993a, c; Tester et al., 2000). It is assumed that all of the lipid in cereal starches is present as an inclusion complex with the lipid (V-form), where lipid accounts for approximately 12.5% of the complex (Morrison et al., 1993c).

5.2
Crystallinity

Starch crystallinity at the molecular level has been discussed above. There are two major ways of expressing crystallinity – absolute, and relative. These have been clearly discussed by others (Blanshard, 1987; Buléon et al., 1998a). Usually, both types are determined by X-ray diffraction. For absolute crystallinity, the integrated area of the diffraction peaks is calculated as a fraction of the diffractogram. It is generally difficult to be accurate even within a single sample. The other approach (relative crystallinity) is to compare the crystallinity of the starch sample with fully amorphous and fully crystalline references. The present authors typically use extensively acid-etched starches as the crystalline reference, and extensively ball-milled samples as the amorphous reference. Values (as cited above and reviewed by Buléon et al., 1998a) of 15–45% are typical for proportions of starch crystallinity as determined by X-ray diffraction. The amount of crystallinity may also be determined by moisture regain (as a function of crystallinity), acid hydrolysis (which initially hydrolyzes amorphous and subsequently crystalline material), spectroscopy, and NMR. Further to the previous discussion, NMR measures the number of double helices rather than crystallinity *per se* (Gidley and Bociek, 1985) and this value (38–53%) is always higher than crystallinity as not all double helices are in crystalline domains. Cooke and Gidley (1992) provided strong evidence that the (endothermic) enthalpy values obtained by DSC during gelatinization of starch granules reflect loss of double helical order rather than loss of crystallinity. The authors concluded: "This implies that the forces holding starch granules together are largely at the double helical level, and that the observed crystallinity may function as a means of achieving dense packing rather than as primary provider of structural stability".

5.3
Architecture of the Granule

Early models to describe the structure of amylopectin (Banks and Greenwood, 1975) have led to the development of the classification of the branched structure as discussed above, based on the work of Hizukuri and co-workers. This structure does not, however, indicate how the molecule is arranged within starch granules and how this three-dimensional form develops during deposition. This section focuses on this area, which still retains many unknowns.

Early microscopic work with plane-polarized light showed that birefringence patterns could be observed in native granules, and thus indicated that there was considerable internal molecular orientation (reviewed by Banks and Greenwood, 1975; Banks and Muir, 1980). The granules are positively birifringent, which implies that molecules are radially orientated. Work by the same authors (especially after using acid treatment) indicated that there were broad layers (~4–7 μm apart) with each containing sublayers of a smaller width (~1 μm). Sterling's studies (1974) using light and electron microscopy was especially valuable for visualizing the radial structure of starch granules. The progression of work on granule structure using microscopic and diffraction studies from this has led to a better understanding of the internal structure, as discussed below.

French (1972) has made a major impact on the understanding of the internal structure of starch granules by combining chemical, biochemical, microscopic, and X-ray diffraction analysis to generate a coherent form of granule structure. He incorporated the cluster model of amylopectin (crystallizing during deposition) into the three-dimensional structure of the starch granule. He suggested that the exterior chains of the amylopectin molecules as double helices within clusters form crystalline domains, which are interspersed with amorphous regions (Figure 8). French (1984) further reviewed the available data from analytical, microscopic and X-ray diffraction patterns of model systems and starch granules representing the A-, B-, and C-type polymorphs (see Section 5.1.2). This reinforced the view that starch crystals (~10–15 nm long) were present in starch granules (comprising exterior chains of amylopectin), although there was great uncertainty with respect to the role of growth rings. However, in an extension and adaptation of earlier work on the cluster/racemic model of amylopectin (French, 1972; Kainuma and French, 1972; Robin et al., 1975), French presented the structure of amylopectin molecules within growth rings (thought to represent periodicity of starch deposition), which are especially evident after acid or amylase treatment. This structure incorporated the following features. There was radial growth of amylopectin, with 16 clusters per growth ring and growth rings 120–400 nm long. The double helices forming the subgrowth ring crystalline lamellae were 5 nm long, and interspersed with amorphous branch regions 2 nm long. Different crystalline polymorphs (A- and B-type) could be accommodated in this model (Figure 9).

Blanshard (1987) further reviewed starch structure, and presented a model to describe starch structure to accommodate chemical and physical data (especially X-ray scattering) to provide spatial location for amylopectin (double helices) and amylose (free, as

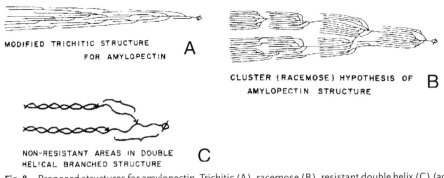

Fig. 8 Proposed structures for amylopectin. Trichitic (A), racemose (B), resistant double helix (C) (adapted from French, 1972).

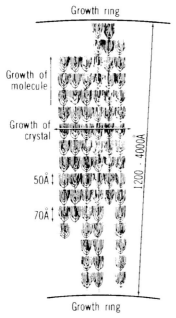

Fig. 9 Arrangement of amylopectin molecules within a growth ring (adapted from French, 1984).

V-type helices with lipids and as double helices) as shown in Figure 10. This accommodates the previously established dimensions of crystallites and amorphous material, and the associated geometry. As previously mentioned (Section 5.1.3), although not expressly discussed by Blanshard (1987), the amylopectin double helices need not all be located in starch crystallites. Eliasson et al. (1987) modeled starch structure by an analogy with quartz structure, and presented three-dimensional models of amylopectin clusters similar in some ways to more modern "blocklet" structures (discussed below).

Small-angle X-ray scattering (SAXS) (Oostergetel and van Bruggen, 1989; Cameron and Donald, 1992; Jenkins et al., 1993) and neutron scattering data (Blanshard et al., 1984; Muhr et al., 1984) contribute to the view (Donald et al., 1997) that starch granules contain alternating amorphous and crystalline lamellae, with a periodicity of ~9 nm. This dimension represents the previously mentioned single semi-crystalline lamellae which are radially aligned (16 units long) in the ordered regions of growth rings (see Figure 10), which are also apparent by atomic force microscopy (AFM), and scanning and transition electron microscopy (SEM, TEM) as discussed by others (French, 1984; Bertoft et al., 1993; Kawabata et al., 1994; Gallant et al., 1997; Smith, 1999; Baker et al., 2001).

With respect to more accurate dimensions within starch granules, according to Cameron and Donald (1992) granules contain relatively broad radial growth rings comprising semi-crystalline shells (~140 nm thick) separated by broad amorphous zones of at least the same thickness. In this model (as in Figure 11), the semi-crystalline shells themselves (140 nm) contain the 16 radiating clusters of amylopectin exterior chains (A and B_1) with the actual length of the registered double helices about 6.65 nm (equivalent to the crystalline lamellae) interspersed within amorphous lamellae of about 2.2 nm (amylopectin α-(1→6) branch points). This repeat distance of 8.85 nm is not too dissimilar to figures reported by Sterling (1962) using X-ray diffraction (9.9–10.9 nm), Oostergetel and van Bruggen (1987) using electron microscopy and diffraction, or the figures proposed by Hizukuri (1986) for amylopectin cluster dimensions (based on 22–25 glucose residues long) of 9.9–11.0 nm.

There are difficulties associated with identifying where particular components are located within starch granules. The location of amylose is particularly difficult to assign. However, there are clear indications (see Sections 5.1 and 5.2) that crystalline amylopectin, amylose–lipid complex regions and amorphous regions containing branched amylopectin and free amylose all exist (Morgan et al., 1995).

Oostergetel and van Bruggen (1993) used electron optical tomography and cryoelectron diffraction, and showed how left-handed double helices of amylopectin exterior chains could interact with each other to form super-helical structures (Figure 12). In an extension of this work, Gallant et al. (1997) proposed that crystalline (hard shell) blocklet structures exist in starch granules ranging in diameter from around 20–500 nm depending on starch type and location within the granule (Figure 13). These alternate with semi-crystalline (soft shell) small blocklets (20–50 nm). The blocklets comprise the (~9 nm) alternating crystalline and amorphous lamellae. Amorphous material exists within the semi-crystalline stacks both tangentially (predominantly α-(1→6) branch

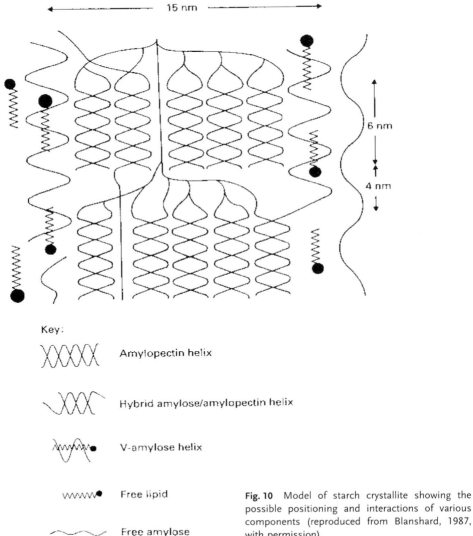

Fig. 10 Model of starch crystallite showing the possible positioning and interactions of various components (reproduced from Blanshard, 1987, with permission).

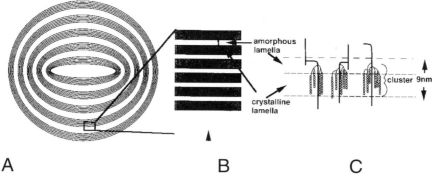

Fig. 11 Model for starch granule used to fit the SAXS data. (A) Stacks of semi-crystalline lamellae separated by amorphous growth rings. (B) Magnified view of a stack showing alternating crystalline and amorphous lamellae. (C) Double helices formed by branches of amylopectin arranged in clusters in a crystalline lamella (reproduced from Donald et al., 1997, with permission).

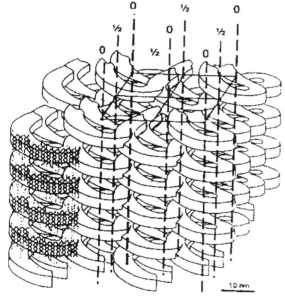

Fig. 12 Schematic model for the arrangement of amylopectin in potato starch (reprinted from *Carbohydr. Polym. 21*, Oostergetel and van Bruggen, The crystalline domains in potato starch granules are arranged in a helical fashion, p. 7–12, 1993, with permission from Elsevier Science).

points) and radially (between amylopectin clusters). Further evidence for the presence of blocklets using AFM has been reported by other authors (Ohtani et al., 2000a, b). The actual polymorphic structure within these blocklets may vary within the granule, with C-type granules themselves probably containing both A- and B-type polymorphs (Buléon et al., 1997, 1998b). Furthermore, crystalline lamellae may not all be orientated in the same direction (Waigh et al., 1999).

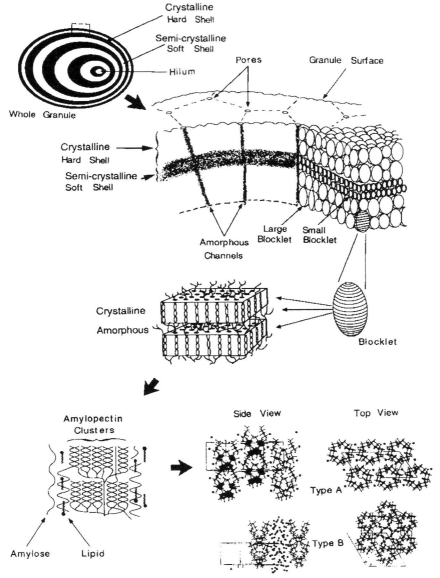

Fig. 13 Structure of a starch granule at different levels of magnification, showing the arrangement of blocklets (reprinted from *Carbohydr. Polym. 32*, Gallant et al.; Microscopy of starch: evidence of a new level of granule organisation, p. 177–191, 1997, with permission from Elsevier Science).

6
Physical Properties

6.1
Microscopy of Starch Granules

6.1.1
Optical Microscopy

The conventional optical microscope is a valuable tool for identification of the botanical origin of starch. Considerable detail can be seen with a phase-contrast microscope at a magnification of 400× to 900×. In addition, granules can be stained blue with iodine solution, which facilitates their detection in complex matrices.

An important feature of all native starches is their birefringence in polarized light. The granules show a black cross ("maltese cross") when viewed between crossed polars. This indicates that the granules have a microcrystalline structure which is ascribed to the orderly arrangement of amylopectin double-helical clusters (see Sections 5.2 and 5.3). Loss of birefringence is indicative of disorder caused by heating or by mechanical damage. The dye Congo red is often used to distinguish between intact and mechanically damaged or gelatinized granules. The dye penetrates granules that have lost their internal order.

Detailed work with mechanically damaged granules is best carried out with a petrographic microscope, which in addition to the polarizing system is fitted with a first-order red gypsum plate or a quartz wedge. In both these devices the fast ray of polarized light is in the direction NW–SE. The granules appear yellow in the direction NW–SE and blue (or blue-green) in the direction SW–NE. According to the principles of optical mineralogy, starch granules are characterized as uniaxial positive. Granules are considered as spherulites consisting of crystalline aggregates radially orientated from the hilum to the granule surface. When the microcrystalline structure is destroyed, granules appear red because they become isotropic. It should be noted that in potato and yam starch granules the hilum is not situated centrally, and the birefringence cross is not symmetrical as in wheat granules.

Before DSC became readily available, the polarizing microscope fitted with a Kofler electrically heated stage was used for the determination of gelatinization temperature of starch. The temperature of the stage was increased at a controlled slow rate, and the number of granules that lost their birefringence was counted at each particular temperature.

6.1.2
Electron Microscopy

The electron microscope is a powerful tool for the study of starch granules. The scanning electron microscope is ideal for studying the surface features of granules. Samples must be coated with an ultrathin layer of conductive material under vacuum. A resolution of 0.2 nm is possible. The internal structure of thin sections of granules is observed with the transmission electron microscope at a magnification of several thousand diameters. Fundamental studies were described by French (1984). Changes in the appearance of wheat starch granules isolated from baked products have been studied using SEM (Hoseney et al., 1977).

Gallant et al. (1992) used SEM to examine potato starch granules partially digested with pancreatic α-amylase. This treatment revealed the alternating crystalline and amorphous layers of the granule, as well as blocklets near the surface of the granule. Fannon and BeMiller (1992) studied the nature of gelatinized starch pastes and granule remnants by means of low-temperature electron microscopy, and reported that structural differences correlated strongly with rheological properties. The presence of surface pores in granules of maize, sorghum, and millet starch was reported by Fannon et al. (1992). Pores were also seen on the equatorial groove of large granules of wheat, rye, and barley starches, but not on rice, oat, potato, tapioca, arrowroot, or canna. Using SEM, Huber and BeMiller (2001) were able to observe the movement of thallium salts solutions in phosphorylated potato and sorghum starch and in hydroxypropylated waxy maize starch. In sorghum and waxy starches the flow of reagents was from the central cavity outwards and through lateral channels. In potato starch, which does not have pores, flow was by diffusion inwards. Huber and BeMiller (1997, 2000) provided evidence that maize and sorghum starch granules have radial channels connected to a central cavity. The channels appear as pores on the surface of the granules, and they remain open even under slight swelling conditions. For this experiment the authors used light, fluorescence and compositional backscattering electron microscopy.

Studies with TEM necessitate the application of various staining techniques in order to increase contrast. The PATAg reaction (periodic acid, thiosemicarbazide, silver) was used by Gallant et al. (1997) to demonstrate alternating crystalline and amorphous shells (layers) as well as blocklets in maize starch granules. The granules were treated with α-amylase to digest amorphous material and to allow visualization of resistant crystalline features.

6.1.3
Atomic Force Microscopy

AFM is a relatively new technique, which permits high resolution without the need of coating the sample. A very fine-tipped probe scans the surface of the specimen, generating a three-dimensional image. Recently, the radial organization of native maize starch granules was demonstrated with AFM (Baker et al., 2001). Blocklets of 400–500 nm corresponding to the thickness of growth rings were visible. With lintnerized samples, smaller blocklets (10–30 nm) were observed. Low-voltage SEM and AFM were used by Baldwin et al. (1998), who demonstrated that the surfaces of wheat and potato starch granules have different topographies. Protrusions on the potato granule surface were large (50–300 nm in diameter) compared with wheat (10–50 nm). In another AFM study, the existence of fine particles of ~30 nm diameter inside rice starch granules, often in chain arrangement, was attributed to blocklets or single clusters of amylopectin (Ohtani et al., 2000a).

6.2
Gelatinization and Swelling

When starch granules are heated in unlimited water the crystalline regions are progressively ruptured, leading to the loss of crystalline order, swelling and leaching of some carbohydrate (primarily amylose). DSC is the most important tool currently used to measure this progressive loss of order. Using this technique, the uniqueness of gelatinization parameters for each starch can be determined and specifically the onset (T_o), peak (T_p), and conclusion (T_c) temperatures of the endothermic event with the associated enthalpy (ΔH). In excess water, T_o is typically 45–55°C for most waxy and normal starches, T_p 55–65°C, and T_c 65–75°C. For high-amylose starches, the gelatinization endotherm is very flat and ranges from approximately 70–105°C, although this may incorporate (if not subtracted) the amylose–lipid dissociation endotherm between about 95–105°C. The gelatinization enthalpy is highly variable, ranging from about 10–20 J g^{-1} for most starches.

It is believed that amylopectin exterior chains are responsible for the gelatinization endotherm, and that amylose simply dilutes the amylopectin fraction (Tester and Morrison, 1990a, b, 1992, 1993; Tester and Karkalas, 1996; Tester, 1997b; Jane et al., 1999). The amylose–lipid fraction may also elevate the gelatinization temperature (Morrison, 1993; Morrison et al., 1993a, b, c). In high-amylose starches however, there is strong evidence that amylose chains form double helices and contribute to the broad endotherm (Tester et al., 2000). If there is insufficient water to allow unrestricted swelling in parallel with the gelatinization event, a higher temperature shoulder leading to a separate peak (if moisture is further restricted) develops (Tester and Sommerville, 2000).

The crystallinity and associated gelatinization temperatures of starches are largely genetic in origin, and are reported to reflect the exterior chain lengths of the constituent amylopectin molecules (Hizukuri, 1985). However, these are moderated by environmental effects where gelatinization temperatures are positively correlated with growth temperature (Tester et al., 1991, 1995, 1999; Tester, 1997a; Debon et al., 1998). If starch is annealed (Knutson, 1990), gelatinization temperatures can be increased in a process (comparable to growth temperature) which enhances crystallite perfection without creating more double helices (Tester et al., 1998, 2000; Debon and Tester, 2000; Tester and Debon, 2000).

Donovan (1979), Donovan and Mapes (1980) and Biliaderis et al. (1980, 1986) have discussed the excess water (G) gelatinization endotherm in terms of coupling forces between amorphous regions and crystallites (with the role of the glass transition in the last paper). Some authors (Evans and Haisman, 1982; Lund and Wirakartakusumah, 1984) and with some similarities Hoseney et al. (1986) have discussed gelatinization in terms of a range of crystallite stabilities controlling the process, while Blanshard (1987) viewed the thermal event in terms of the extent of swelling and chain mobilization in amorphous regions initiating gelatinization. Atkin et al. (1998a, b) have discussed the process in terms of the separation of amylopectin units from granular rings. Bogracheva et al. (1998) proposed that swelling of disrupted parts of granules facilitates disruption (gelatinization) of crystalline zones, and that A- and B-polymorphs within individual granules moderate the process.

At high water contents, melting of crystallites and swelling are cooperative processes (Colonna and Mercier, 1985), although cooperative forces initiated by amorphous glass transition are also relevant for intermediate moisture systems (Russell, 1987). The primary hydration of amorphous regions (and especially broad amorphous growth rings) which collectively undergo a glass–rubbery transition prior to crystallites during the gelatinization event, has been proposed and confirmed using SAXS and other X-ray scattering and DSC techniques (Zobel et al., 1988; Shi and Seib, 1992; Cameron and Donald, 1992, 1993a, b; Jenkins et al., 1994). These data have been strengthened by further studies by Donald and colleagues (Jenkins and Donald, 1996, 1998; Donald et al., 1997) and discussed in terms of the liquid crystalline characteristics of starch granules (Waigh et al., 2000).

The sequence of gelatinization has more recently been described by Tester and Debon (2000) as follows. The event involves primary hydration of amorphous regions around and above the glass transition temperature (T_g) with an associated glassy–rubbery transition. This facilitates mobility in the amorphous regions (with reversible swelling), which then provokes an irreversible molecular transition in the crystalline regions. In these regions, double helices dissociate as they hydrate and the granules swell. Double helices are responsible for the gelatinization endotherm rather than crystallinity *per se* (Cooke and Gidley, 1992; Zobel, 1992).

6.3 Rheological Properties of Starch

Starch granules suspended in water at ambient temperature make a relatively small contribution to the viscosity of the suspension, which increases as a function of concentration. At a very high concentration (>40% by weight) the suspension becomes progressively dilatant, i.e., it shows an increase in viscosity as the shear rate increases. However, the usefulness of starch as thickener, stabilizer, texture modifier or adhesive depends primarily upon its ability to form viscous solutions and pastes on heating above its gelatinization temperature. On cooling, starch pastes form gels. The rheological properties of "cooked" starches are considered below.

6.3.1 Viscosity

The viscosity of fluids is defined qualitatively as their resistance to flow. Water spreads rapidly when poured over a flat surface, while molasses will spread rather slowly under the same conditions. This resistance to flow is attributed to internal friction of the molecules. Viscous resistance can be counteracted by the application of external forces, e.g., pumping or stirring. Newton's law defines the relationship between the frictional force and the resistance to flow of a fluid:

$$F = \eta A \, dv/dr \tag{1}$$

where F is the frictional force which resists movement between two parallel planes of fluid of surface area A moving relative to each other with a velocity gradient dv/dr. The proportionality constant η is the coefficient of viscosity. When the fluid is in circular motion, the above equation is generally written as:

$$\tau = \eta \gamma \tag{2}$$

where $\tau = F/A$ = shear stress in N/m^2 or Pa, γ = shear rate in 1/s and η = viscosity in Pas.

Rheology is a branch of physics which deals with the deformation of matter. If the deformation is transient it is known as "elasticity"; if permanent it is known as "flow". Some materials have both elastic and viscous properties (e.g., gels), and these are described as viscoelastic.

6.3.2 Newtonian and Non-Newtonian Fluids

Newtonian fluids obey Newton's law (Eqn. 2). The viscosity remains constant when the shear rate is increased. Most pure liquids such as water, some vegetable oils, and solutions of small molecules (e.g., sugars or salts) in water are Newtonian. Non-Newtonian fluids do not obey this law; their viscosity depends on the shear rate, i.e., on the speed of rotation of the spindle or the cone of a rotating viscometer. There are two types of non-Newtonian fluids: pseudoplastic and dilatant. In pseudoplastic fluids (shear thinning), the viscosity decreases with increasing speed of rotation. Most polysaccharides in solution, including starch, have this property. In dilatant fluids (shear thickening), the viscosity increases with increasing speed of rotation. There are very few dilatant fluids, for example set (crystallized) honey (>80°Brix), granular starch at a high concentration (>40% w/w). Often, non-Newtonian fluids are time-dependent, i.e., their viscosity changes relative to the time of application of shear. Many pseudoplastic fluids are thixotropic when their viscosity decreases not only with increasing speed or rotation but also with the duration of application of shear. The opposite effect is observed with rheopectic fluids, i.e., time-dependent dilatant fluids.

Most non-Newtonian fluids obey the power law:

$$\tau = k\gamma^n \qquad (3)$$

where k is the consistency coefficient and n is the flow behavior index.

For Newtonian fluids, $n = 1$ and $k = \eta$; for pseudoplastic fluids $n < 1$, and for dilatant fluids $n > 1$. The power law can be written in logarithmic form:

$$\log\tau = \log k + n\log\gamma \qquad (4)$$

A plot of $\log\tau$ versus $\log\gamma$ is a straight line with slope $= n$ and an intercept $= k$. Alternatively, a plot of τ versus γ on log-log graph paper gives identical results.

Some definitions relating to the viscosity of fluids are given in Table 2.

Viscometry is widely used for the study of polymers in solution. The molecular weight and the conformation of polymer chains can be deduced by the use of appropriate techniques. The limiting viscosity number (also known as intrinsic viscosity) is obtained from a linear plot of the Huggins equation, which is applicable to very dilute solutions only:

$$\eta_{sp}/c = [\eta] + k[\eta]^2 c \qquad (5)$$

where η_{sp}/c is the y-axis and c is the x-axis. The intercept $= [\eta]$ and the slope $= k[\eta]^2 c$.

For a known limiting viscosity number it is possible to calculate the molecular weight of a polymer according to the Mark–Houwink–Sakurada equation

$$[\eta] = KM^a \qquad (6)$$

where M is the viscosity average molecular weight, and K and a are experimentally determined constants.

Capillary viscometers of the suspended level type (Ubbelohde) are used for the study of clear dilute solutions of amylose. Because the latter retrogrades rapidly in aqueous solution it is normally dissolved in dilute alkali (e.g., 0.01 M KOH) or in dimethylsulfoxide (DMSO). Methods for exploring the conformation and for determining the molecular weight of amylose in solution have been described by other authors (Banks and Greenwood, 1975). Capillary viscometry has been used by Karkalas and Raphaelides (1986) for the study of helical inclusion complexes of amylose with fatty acids.

6.3.3
Properties of Gels

Gels and very viscous solutions of hydrocolloids can behave simultaneously as vis-

Tab. 2 Some definitions relating to the viscosity of fluids

Property	Symbol	Dimensions
Viscosity	η	Pas
Relative viscosity	η/η_o	Dimensionless
Viscosity of solvent	η_o	Pas
Specific viscosity	$\eta - \eta_o/\eta_o = \eta/\eta_o - 1 = \eta_{sp}$	Dimensionless
Reduced specific viscosity or viscosity number	$1/c\,(\eta - \eta_o/\eta_o) = \eta_{sp}/c$	dL g^{-1}
Concentration	C	g dL^{-1}
Intrinsic viscosity or limiting viscosity number	$[\eta] = \lim_{c \to 0}[1/c\,(\eta - \eta_o/\eta_o)]$	dL g^{-1}
Huggins equation	$\eta_{sp}/c = [\eta] + k[\eta]^2 c$	dL g^{n1}

cous (liquid-like) and as elastic (solid-like) substances. Gels are usually self-supporting, while pastes tend to spread slowly with time. These two modes of response to a stress (viscoelasticity) can be resolved and quantified by the method of mechanical spectroscopy. The sample is placed between two parallel plates (discs), one of which oscillates at a controlled frequency and amplitude. In real systems the degree of solid-like and liquid-like character can be measured by resolving the resultant stress into its "in-" and "out-of-phase" components. The instrument used is known as a viscoelastic analyzer. The ratio of the in-phase stress to the applied strain is the elastic (or storage) modulus (G′), while the ratio of the out-of-phase stress to the applied strain is the viscous (or loss) modulus (G″). The overall response is characterized by the complex modulus G*:

$$G^* = \sqrt{(G'^2 + G''^2)} \qquad (7)$$

The frequency of oscillation (ω) is regarded as the oscillatory analogue of the shear rate (in rad s^{-1}), and a corresponding dynamic viscosity (η^*) can be expressed as:

$$\eta^* = G^*/\omega = \sqrt{(G'^2 + G''^2)}/\omega \qquad (8)$$

The units of shear stress are Newtons per square meter (N m^{-2}) or pascals (Pa); those of shear rate reciprocal seconds (s^{-1}). The units of viscosity are pascals seconds (Pas). The old unit of viscosity is the poise (P) and the more commonly used centipoise (cP). Thus, 1 cP = 0.001 Pas or 1 mPas (millipascals second).

6.3.4

Viscometry of Gelatinized Starch Dispersions

The Brabender viscoamylograph is an instrument designed to measure the development of viscosity during and after completion of the gelatinization process. The original design of the instrument consists of a rotating metal beaker in which a starch suspension (25–50 g in 500 mL water) is heated at a rate of 1.5°C min^{-1} to a temperature of 95°C. This temperature is maintained for 15–60 min, and then lowered at a rate of 1.5°C min^{-1} to 50°C and kept at this temperature for 30–60 min. The shear sensing device consists of a disc carrying several metal pins which project vertically into the starch suspension, which is contained in the beaker. As the starch gelatinizes and viscosity increases, the drag force causes a deflection of the metal pins, and this is transmitted via a torsion wire to a recording chart. With normal starches a sharp peak denotes the event of gelatinization. As the swollen granules are dispersed under shear there is a gradual decline in the viscosity (shear thinning), and when cooling begins a fairly sharp rise in viscosity is observed attributable to retrogradation (commonly known as set back) (see Section 6.5). Thus, various starches (native or modified) give characteristic curves, on the basis of which the most suitable starch for a particular application can be selected. A modern computerized version, the Micro Viscoamylograph, is now available requiring only 5–10 g of test sample.

A related device known as the Rapid Visco™ Analyser (RVA) designed by Newport Scientific requires about 2 g of sample. The instrument is computer-controlled and performs a wide range of measurements by programming the temperature/time profile as well as the shear rate.

Dilute suspensions of starch in water (<2%, w/v) give viscous solutions after complete gelatinization. Gelatinized concentrated suspensions (>4%, w/v) give a paste when hot, and normally set to a gel on cooling. Depending on the origin and concentration of starch, the gel can be soft and easily deformable, or rigid and elastic. The clarity of the gel varies considerably; potato starch gives fairly clear gels, whilst maize starch gives rather opaque gels. Gels become progressively opaque on aging because of retrogradation. Syneresis is often observed. Gelatinized starches, whether in dilute solution or in the form of a gel, consist of a mixture of exuded polysaccharide (mostly amylose) and swollen deformed granules, which retain large quantities of water. Centrifugation of the mixture gives rise to a relatively clear supernatant solution of low viscosity and an opaque semi-solid residue. Thus, gelatinized starch systems are heterogeneous. The Brabender and RVA instruments mentioned above provide empirical data which describe adequately the behavior of these mixtures. Fundamental studies of the macromolecular structure of the starch components require more sophisticated instruments based on well-recognized rheological principles, which have been effectively described by Morris (1984), Doublier et al. (1992) and Okechukwu and Rao (1998).

Starch gels can be regarded as composite systems consisting of swollen granules suspended in a continuous polymer matrix (Evans and Haisman, 1979; Eliasson and Bohlin, 1982; Eliasson 1986; Morris, 1990; Noel et al., 1993; Eliasson and Gudmundsson 1996). The viscoelastic properties of gelatinized starches have been reported by several researchers (Biliaderis and Zawistowski, 1990; Biliaderis, 1992; Doublier and Llamas, 1993). The influence of lipids on the viscoelastic behavior of starch gels has been similarly studied (Biliaderis and Tonogai, 1991; Raphaelides, 1992, 1993).

6.4
Helical Complexes of Amylose

It has long been known that amylose in solution forms insoluble helical inclusion complexes with added fatty acids, monoacyl lipids, and surfactants. The structure of the complex is similar to that formed when polyiodide ions react with amylose to give the familiar blue complex. Amylose exists as a random coil in solution. However, when an amylose chain comes into contact with the hydrocarbon moiety of, for example, monomyristoyl glycerol, a helical structure is formed, which traps the lipid molecule inside its central cavity. The helix has an outer diameter of 1.3 nm, an inside diameter of 0.6 nm and a pitch of 0.8 nm per helical turn. The continuous helix is left-handed and compact in the axial direction. The internal surface of the cavity is hydrophobic, and the guest molecule (or ligand) is held along the axis of the cavity by van der Waals forces. Each turn of the helix consists of six anhydroglucose molecules (V_6-helix), which interact with adjacent helices by H-bonding (Rappenecker and Zugenmaier, 1981). The helix owes its stability essentially to intrahelical H-bonds, and this is evident by the dissociation enthalpy of the complexes which is in the range 25–30 J g^{-1} (Raphaelides and Karkalas, 1988; Karkalas et al., 1995). Molecular modeling of amylose complexing has been proposed by Godet et al. (1993), and reviews have been published by Tomasik and Shilling (1998a, b), Biliaderis (1998) and Hoover (1998). Complexes precipitated from solution give distinct X-ray diffraction patterns (V_h form). This supermolecular structure has been studied by several groups (Bilaideris and Seneviratne, 1990; Seneviratne and Biliaderis, 1991; Godet et al., 1995), but it appears that the fundamental event during the precipitation of amylose complexes is the initial formation of a helix by H-bonding. Crystallinity, observed by X-ray diffraction, is the result of folding of the amylose–lipid helices in the form of ordered lamellar structures.

6.5 Retrogradation of Starch

As mentioned elsewhere, when gelatinized starches form double helices and then recrystallize from solution, the process is described as "retrogradation". This process has been discussed by many groups in relation to retrogradation from starch solutions and gels (Cui and Oates, 1997; Fisher and Thompson, 1997; Jacobson et al., 1997) and during the storage of food products (Zobel and Kulp, 1996). Different techniques used to follow the process (such as DSC, NMR and X-ray diffraction) have been discussed in depth elsewhere (Abd Karim et al., 2000). According to Wunderlich (1990), the mechanism of retrogradation incorporates three stages: nucleation; crystal growth; and perfection. Retrograded starches tend to adopt a B-type diffraction pattern, regardless of origin (Kitamura et al., 1984; Gidley, 1989; Leloup et al., 1992; Cairns et al., 1995; Kim et al., 1997; Abd Karim et al., 2000). However, although low-temperature storage tends to promote the formation of this B-type polymorph, higher temperatures promote the formation of A- and C-type forms (Zobel and Kulp, 1996). Short linear α-glucans may form A-, B-, or C-type crystallites from solution depending on chain length, concentration and salt concentration (Gidley, 1987, 1989; Gidley and Bulpin, 1987) where short-chain length, high concentration and higher crystallization temperatures favor the formation of the A-type polymorph over the B-type (Gidley and Bulpin, 1987).

The extent of retrogradation depends on many interrelated factors which include:

- size and structure of the α-glucans (Mua and Jackson, 1998; Klucinec and Thompson, 1999);
- length of unit chains of amylopectin and linear amylose-type molecules (Gidley et al., 1986; Gidley and Bulpin, 1989; Villareal et al., 1997; Mua and Jackson, 1998; Lai et al., 2000; Tako and Hizukuri, 2000);
- amylose content, since amylose retrogrades more readily than amylopectin (Kim et al., 1997; Klucinec and Thompson, 1999; Lai et al., 2000);
- botanical origin of the starch, where the longer average chain lengths in potato amylopectin appear to favor crystallization compared with cereal starches (Jacobson et al., 1997; Fredriksson et al., 2000);
- mutants (Fisher and Thompson, 1997; Liu and Thompson, 1998a, b; Yuan and Thompson, 1998);
- concentration and moisture content (Keetels et al., 1996a, b; Jang and Pyun, 1997);
- temperature and time where low (non-frozen) temperatures (refrigeration > room temperature > frozen) and time favor crystallization (Jang and Pyun, 1997; Kim et al., 1997; Lu et al., 1997);
- shear (Keetels et al., 1996a, b); and
- chemical modification (Perera and Hoover, 1999; Morikawa and Nishinari, 2000).

During the retrogradation of amylose, double helices of 40–70 glucose units are commonly reported (Jane and Robyt, 1984; Leloup et al., 1992; Lu et al., 1997; Klucinec and Thompson, 1999), and it is established that (linear) amylose molecules with a DP of 80–100 have the greatest tendency to retrograde (Gidley et al., 1986; Pfannemuller, 1986; Gidley and Bulpin, 1989). It is harder for amylopectin to form double helices because of the restrictions placed on the system by the branched regions. Double helices formed within retrograded systems may associate into crystallites. The crystals of amylopectin dissociate at approximately 60 °C in common with many native starches, while amylose crystallites dissociate at approximately 150 °C.

Whilst retrograded starch forms one component of resistant starch (see Section 6.6), the relevance of retrograded amylose or retrograded amylopectin with respect to digestibility is unclear (Fredriksson et al., 2000). Amylose and amylopectin may interact in retrograding systems to restrict amylose association (Klucinec and Thompson, 1999). Agents used to form inclusion complexes with amylose (e.g., monoglycerides) prevent the formation of double helices and crystallites, thus restricting staling (Knightly, 1996; Zobel and Kulp, 1996).

6.6 Resistant Starch

Resistant starch has similar effects to other forms of "dietary fiber" (nonstarch polysaccharides and lignin) in the human digestive system. Readers should, however, appreciate the similarities as well as the differences. Nonstarch polysaccharides are a diverse group of polymers comprising many monosaccharides bonded in such a way as to make them resistant to the enzymes occurring in the small intestine (amylases, lipases, and proteases). They do, therefore, pass to the large intestine where they may be fermented by the microflora generating short-chain fatty acids (SCFA) such as acetate, propionate and butyrate with hydrogen, carbon dioxide, and methane. These nonstarch polysaccharides include celluloses, pectins, alginates, glucomannans, galactomannans, and arabinoxylans. Lignin is not a polysaccharide, but is composed of organic acids rather than sugar molecules.

Englyst et al. (1992, 1996) and Englyst and Hudson (1996, 1997) have developed a useful guide to the carbohydrate fractions of the diet. They (especially Englyst and Hudson, 1996, 1997) have expressed the view that sugars, short-chain carbohydrates, starch and nonstarch polysaccharides form separate carbohydrate classes, with the starch fraction being subclassified into three major groups: rapidly digestible (RDS); slowly digestible (SDS); and resistant starch (RS). The resistant starch may be further subclassified into different types as discussed below.

Prosky and colleagues developed enzymatic gravimetric methods to determine dietary fiber in cereals and other food products which were eventually adopted by the Association of Official Analytical Chemists as official methods (Prosky et al., 1984, 1985, 1988; AOAC International, 2000). Fat-extracted test samples are gelatinized with thermostable α-amylase and digested with protease and amyloglucosidase to remove protein and nonresistant starch. Soluble fiber is precipitated in 80% ethanol. More recent publications by Prosky reinforce the value of this approach for food labeling, and that oligosaccharides (as found in many functional foods) should also be included in the fiber fraction (Lee and Prosky, 1995; Prosky, 2000a, b).

Englyst and colleagues determined the nonstarch polysaccharide fraction of foods by gas-liquid chromatography (as alditol acetate derivatives of the constituent sugars) where the resistant starch is not included (Englyst et al., 1982, 1992; Englyst and Cummings 1984, 1987, 1988). This methodology has recently been upgraded to take into account the short-chain oligosaccharides present in foods (Quigley et al., 1999).

Resistant starch refers to starch that resists digestion in the small intestine in common with the nonstarch polysaccharides and lignin as discussed above. According to Englyst et al. (1996), resistant starch may be defined as "the sum of starch and starch-degradation products that, on average, reach the human large intestine". However, there are three major types of resistant starch as discussed by Englyst et al. (1992), Englyst and Hudson (1997), and Edwards and Tester (1998).

6.6.1
Type I Resistant Starch

This form of resistant starch comprises starch granules embedded in plant tissues that are impermeable to digestive enzymes. This structure prevents starch contact with the enzymes, and hence hydrolysis. Whole grains and seeds are good examples. In this context, the starch could actually be native or gelatinized, although it is more likely that the starch is extensively ungelatinized as the integrity of the tissue is largely intact.

6.6.2
Type II Resistant Starch

Some native starches resist hydrolysis in the intestine more than others; potato starch is, for example, very resistant compared with wheat starch. Both potato and banana starches are usually cited as the prime examples of this material, and resistance is reportedly due to the more crystalline nature of these starches compared with others. The exact reason for this resistance is not fully understood, though it is the view of the present authors that it is largely due to granule size (e.g., potato compared with wheat) and the presence of amylase inhibitory components associated with the starch (e.g., banana). Large granules have a smaller surface area to volume ratio, and consequently the likelihood of surface- initiated hydrolysis by amylases is restricted. In addition, potato starch granules do not have pores.

6.6.3
Type III Resistant Starch

This form of starch comprises retrograded starch (see Section 6.5). Amylose retrogrades more easily than amylopectin, but not exclusively so. In this context, one would expect to generate resistant starch most rapidly from high-amylose, then normal, and then waxy starches. This form of starch is often produced for experimental purposes using repeated autoclaving at specific moisture contents (Berry, 1986; Ring et al., 1988; Sievert and Pomeranz, 1989; Eerlingen et al., 1993).

6.6.4
Type IV Resistant Starch

Some authors (Thomas and Atwell, 1999) report a fourth form of resistant starch, incorporating novel chemical bonds formed as a consequence of chemical or thermal processing.

6.6.5
Inclusion Complexes

The inclusion complexes naturally present in starch granules or formed during starch processing are not generally considered as resistant starch. However, studies by the present authors with the in-vitro digestion of gelatinized starch has shown that these complexes are resistant to hydrolysis by fungal α-amylase (Karkalas et al., 1992). This may be relevant to their behavior in the digestive tract.

The whole area of resistant starch is developing rapidly as more subdivisions are created and related to the chemistry of the constituent polysaccharides. More detailed discussion may be found in the following publications, where reference to industrial processing and production of the material (type III) may also be found (Baghurst et al., 1996; Escarpa and Gonzalez, 1997; Haralampu, 2000; Thompson, 2000).

7
Modification of Starch Properties

7.1
The Need for Modification

Although nature produces starches with differences in granule size, composition, structure and physical properties, limitations within this range restrict their usefulness in practical applications. Some currently available starches have been generated by mutations of the genome, e.g., waxy and high-amylose varieties. More recently, transgenic technology has enabled molecular biologists to modify further the expression of starch biosynthetic enzymes, and consequently the physical properties they regulate. However, the commercial applications of starches are more easily and cost-effectively met by modifying their structures using a raft of chemical, enzymatic and physical techniques depending on the intended use. This section is an overview of such transformations.

7.1.1
Chemical Modifications

Chemically modified starches have been reviewed in detail by Wurzburg (1995) and more briefly by Thomas and Atwell (1999). Readers are referred to these publications for additional details.

Acid-modified Starches

These are produced by treating starch slurries with mineral acids below the gelatinization point of the starch under prescribed conditions whereby hydrolysis of α-glucan is carefully regulated and controlled. The objective is to produce starches that give rise to low viscosity during gelatinization and disperse easily even at high solids content compared with their native counterparts. For these reasons they are also known as "acid-thinned" starches, and are characterized by their fluidity, which increases with the extent of treatment. They are used in the manufacture of gum confectionery. Molecular aspects of acid treatment of starches have been reviewed by Hoover (2000).

Oxidized Starches

Bleaching agents (e.g., sodium hypochlorite, peracetic acid and sodium chlorite) may be used at low concentration to facilitate removal and decolorize impurities, with little modification to the starch itself. Oxidation is achieved by treating maize starch granules with sodium hypochlorite ($<5.5\%$ available chlorine). This treatment generates keto groups at acidic or neutral pH and carbonyl groups at alkaline pH. The bulky carbonyl groups restrict association of starch chains, and hence retrogradation is also restricted.

Dextrins

These are obtained by pyroconversions in essentially dry systems (torrefaction) leading to three types: white dextrins; yellow or canary dextrins; and British gums. Dried starches are heated after the addition of a very small amount of acid (as a fine spray) or simply by heating to a high temperature (British gums). White dextrins are extensively hydrolyzed and limited branching occurs (~2.5%). Hydrolysis and transglycosylation reactions reactions in yellow dextrins, however, generate extensive branching (~25%). In British gums hydrolysis is very limited, but transglycosylation is more pronounced, resulting in even more branching (~25%). These products are used as adhesives, binders, carriers, coating, and encapsulating agents.

Crosslinked Starches

These starches resist gelatinization and hence solubilization when heated in water because of crosslinking of the α-glucan chains. They can effectively tolerate high temperatures, shear and low pH. Thus, enhanced viscosity is achieved after moderate or extreme cooking processes and results in improved paste texture. They find applications in dressings, desserts, soups, sauces and related products. Both distarch adipates and distarch phosphates are produced for this purpose. Adipates are produced by the action of adipic and acetic anhydride, and distarch phosphates with phosphorous oxychloride and sodium trimetaphosphate. Crosslinking is restricted to about one crosslink per 1000–2000 glucose units.

Stabilized (Substituted) Starches
Functional groups are bonded to the hydroxyl groups resulting in α-glucan chains that resist association and hence gelation and retrogradation. These groups may also facilitate gelatinization and swelling at lower temperatures than those found in native starches. The various reactions are carried out with starch in slurry form (acetate, octenyl succinate, or hydroxypropyl groups), or in the dry form in the case of monophosphates. The starch is subsequently washed to remove excess reagents.

The acetates are produced by esterifying starches with acetic anhydride or vinyl acetate. Acetylation amounts to $<2.5\%$ (anhydrous basis) and the degree of substitution (DS) is <0.1. The maximum DS is 3 since there are three free hydroxyl groups per anhydroglucose unit. These starches are used in many food products, especially refrigerated or frozen. Sodium octenyl succinate esters act as clouding, emulsifying and encapsulating materials, and are widely used in the food and pharmaceutical industries. Starches are treated with octenyl succinic anhydride or succinic anhydride to generate these linkages. Octenyl succinate content amounts to 3% which corresponds to a DS of 0.02 (one octenyl succinate residue for every 50 glucose units). Hydroxypropylated starch ethers are prepared by heating starch suspensions with propylene oxide in the presence of sodium hydroxide. Propylene oxide levels of $<10\%$ are permitted. Low-temperature storage stability of these products is especially valuable, although they have a broad range of industrial applications. Starch monophosphates (monoesters) are introduced into starches by heating in the presence of phosphates (ortho-, pyro-, or tripoly-). The products contain $<0.5\%$ monophosphate residues, although some of these are naturally occurring groups. These starches also gelatinize at lower temperatures than the native equivalents and have a broad range of application as thickeners in food systems. Waxy maize is the preferred starch for modification because of the relative clarity of its paste when gelatinized.

7.1.2
Enzymatic Modifications

Enzymatic modifications to starches are usually applied to produce sugar syrups rather than modify starch properties. However, the use of α-amylases to hydrolyze starch to a very low degree, as in the formation of maltodextrins, may broadly be regarded as a type of modification rather than hydrolysis. A more detailed account of enzymatic treatment is given in Section 10.2 and elsewhere (Blanshard and Katz, 1995; Kearsley and Dziedzic; 1995; Thomas and Atwell, 1999).

7.1.3
Physical Modifications

Mechanically Damaged Granules
Damaged starch is generated as a consequence of impact and attrition. Wheat flour, for example, contains damaged starch granules generated during milling. These types of modified granules contain cracks or fissures, or they may be more extensively fragmented. In commercially purified starches, small amounts of damaged granules are present which are classically identified by microscopy on the addition of Congo red, which stains damaged granules

red. The extent of damage is quantified by iodine binding of solubilized extracts, or more precisely by determination of the amount of α-glucan hydrolyzed by amylases under prescribed conditions. Damage converts the native semi-crystalline granule form into a progressively amorphous structure, which is readily hydrated, and can be detected by microscopy. There is a strong similarity between damaged starch fragments and gelatinized starch, since both are equally susceptible to fungal α-amylase. However, unlike gelatinized starch where α-glucan chains are intact, damaged starch contains cleaved amylopectin molecules and hence the molecular weight is reduced. Although these granules readily swell in (cold) water they do not form strong gels because of molecular fragmentation. More extensive description may be found elsewhere (Karkalas et al., 1992; Morrison and Tester, 1994; Morrison et al., 1994; Tester and Morrison, 1994; Tester et al., 1994). Damaged starch is not an article of commerce. However, its presence in bread flour is important because it provides a suitable substrate for the production of fermentable carbohydrate during dough development in bread making.

Pregelatinized Starches

These products, also known as physically modified starches, are obtained by the simultaneous gelatinization and drying of a starch slurry on the hot surface of a rotating drum at about 150°C. They may also be produced by spray-drying. They give a paste with cold water or milk, and form the basis of instant desserts, pastry fillings and related products when heating has to be avoided. Chemically modified starches can also be pregelatinized to avoid retrogradation after dispersion.

Annealing

According to Tester and Debon (2000), annealing represents the physical reorganization of starch granules (or appropriate polysaccharide matrices such as amylose–lipid complexes) when heated in water (or appropriate plasticizer) at a temperature between the T_g and the onset of gelatinization (T_o) of the starch (or polymeric system). Annealing leads to an elevation of the gelatinization temperature of the granules, but the gelatinization enthalpy (ΔH) remains constant or is even slightly elevated. The process does not cause the creation of new double helices, but does facilitate the registration of double helices into more perfect crystallites, with perhaps optimization of helix lengths and enhanced rigidity of amorphous regions (Jacobs and Delcour, 1998; Debon and Tester, 2000; Tester and Debon, 2000; Tester et al., 2000).

Heat-moisture Treatment

Heat-moisture treatments control molecular mobility at high temperatures (e.g., 100°C) by limiting the amount of water to less than about 35%, and hence gelatinization (Jacobs and Delcour, 1998; Tester and Debon, 2000). In common with annealing, this treatment results in physical reorganization of the granule components.

8
Analysis of Starch

8.1
Early Methods of Determination of Starch

Although starch consists entirely of glucose, which can be determined with considerable accuracy, the direct determination of starch in cereals and other food products has long challenged scientists. Early methods utilized polarimetry and acid hydrolysis (Earle and

Milner, 1944; Clegg, 1956; Lyne, 1976). Polarimetry is limited by the need for optically clear extracts, which are not easily obtained with complex matrices. Acid hydrolysis may give rise to additional glucose from nonstarch polysaccharides, and prolonged heating with acid leads to reversion products and degradation of glucose. In spite of these limitations some of these methods are still in use (AOAC International, 2000).

8.2
Quantitative Enzymatic Methods

During the past 30 years there has been a proliferation of enzymatic methods for the determination of starch. Early procedures had several drawbacks for a variety of reasons, the most important being the accessibility of starch to the added amylases used to hydrolyze the starch. For the accurate determination of starch it is necessary to ensure that: (1) all granules are thoroughly solubilized (liquefied); (2) retrogradation is avoided at the liquefaction stage; (3) starch is completely hydrolyzed to glucose; (4) the latter is specifically and accurately determined; and (5) interferences are eliminated or minimized. These criteria are essentially met by several methods, which rely on the simultaneous gelatinization and liquefaction (dextrinization) of starch in the presence of thermostable α-amylase at 85–100°C. The resulting malto-oligosaccharides are subsequently hydrolyzed quantitatively to glucose, which is determined colorimetrically after the addition of a glucose oxidase-peroxidase-chromogen reagent. A solution of pure glucose is used as an external standard (Batey, 1982; Åman and Hesselman, 1984; Karkalas, 1985; Holm et al., 1986; McCleary et al., 1994; AOAC International, 2000). If retrograded amylose (resistant starch) is present, the material is treated with 1 M NaOH (Karkalas, 1985) or with hot DMSO (McCleary et al., 1994; AOAC International, 2000) to ensure complete dissociation of double helices. Starch assay kits are also available (e.g., from Megazyme International Ireland, Ltd.).

An alternative enzymatic technique for starch quantitation relies on amyloglucosidase as the only hydrolyzing enzyme after solubilization of starch in a mixture of HCl and DMSO at 60°C. The resulting glucose is phosphorylated by adenosine-5′-triphosphate (ATP) to glucose-6-phosphate (G-6-P), catalyzed by the enzyme hexokinase. In the presence of glucose-6-phosphate dehydrogenase, G-6-P is oxidized by nicotinamide adenine dinucleotide phosphate (NADP) to gluconate-6-phosphate with the simultaneous formation of reduced nicotinamide adenine dinucleotide phosphate (NADPH) The amount of NADPH formed, measured as an increase in absorbance at 340 nm, is stoichiometric with the original amount of glucose (Beutler, 1984). There is no need for a standard solution of glucose because the millimolar absorptivity of NADPH is accurately known ($\varepsilon = 6.3$ L mmol^{-1} cm^{-1}). Suitable test kits based on this method are available (e.g., R-Biopharm GmbH, Germany).

8.2.1
Automated Instrumental Methods

These methods are useful when large numbers of samples must be routinely analyzed. Automated flow injection analysis, with colorimetric detection, for the total enzymatic determination of 30–35 samples of preliquefied starch per hour has been described by Bengtsson and Larsson (1990) and by Karkalas (1991). These user-friendly methods successfully replace those relying on the autoanalyzer and the continuous use of hot sulfuric acid/phenol or similar corrosive reagents. A totally enzymatic method for the analysis of α-glucans in eluates from gel-permeation chromatographic columns has been developed by Karkalas and Tester (1992). Continuous sequential analyzer methods can be made faster by replacing the photoelectric colorimeter with biosensors specific for glucose, or with pulsed amperometric detectors.

Near-infra red spectroscopic determination of starch can be used in conjunction with the determination of other components of cereal grains (Orman and Schumann, 1991).

8.2.2
Methods for the Characterization of Starch

Methods for laboratory-scale isolation of high-purity starches from cereals and their characterization (protein, lipid and phosphorus content, gelatinization temperature, granule size analysis, debranching of amylopectin and mechanical granule damage) have been described by Morrison (1992). Techniques for the characterization of starch granules, and of amylose and amylopectin (chain length of branches and molecular weight) by enzymatic and other procedures are described in an extensively referenced article by Hizukuri (1996). Morrison and Karkalas (1990) have also discussed related techniques.

9
Industrial Production of Starch

9.1
General Principles

Starch is produced on an industrial scale in a very pure form, and finds applications in the food, pharmaceutical, textile, paper, and other industries. Maize grain is by far the most important source because it offers distinct economic advantages. In the USA and some other countries maize is abundantly produced at low cost and high yield per unit of cultivated area. In addition, as explained below, maize affords a number of valuable byproducts, which defray a substantial part of the cost of the raw material. Thus, the final product, commonly known as "cornflour", "corn starch" or "maize starch", is available in the market at a relatively low price.

Wheat is also an important source of purified starch, particularly in the EU, Canada, and Australia. Wheat starch may be considered as a byproduct of the manufacture of wheat gluten, the latter being in considerable demand and commanding a high price. The raw material is wheat flour rather than whole-wheat grain. There are usually no other byproducts.

Potato starch has traditionally been produced in Europe, but it does not afford any important byproducts, with the exception of pulp consisting of cell-wall residues normally used as animal feed. Processing is seasonal.

Other starches such as tapioca (manioc or cassava), rice, arrowroot, sweet potato, sago, barley and sorghum are produced in small quantities in various parts of the world. Starch can also be isolated from yams (*Dioscorea* spp.), bananas, breadfruit, oats, leguminous seeds (pulses), and chestnuts, but commercial exploitation is either limited or unknown.

The technology for the isolation of starch from any source is relatively simple and can be adapted to the processing of either dry grains and seeds or fresh tubers or roots. Grains (containing normally < 16% moisture) must be steeped in water to soften the pericarp and endosperm and facilitate the separation of starch from all other components. Processing is carried out all year round. In contrast, roots and tubers (containing normally > 70% water) are shredded or crushed to disrupt the cellular structure and liberate the starch granules. In all cases cell walls and other residues are retained on sieves and the starch granules (density ~1.5 g mL^{-1}) are recovered in pure form by centrifugation. Modern processing plants are continuous in operation, and the milled raw material is suspended in water containing small quantities (~0.03%) of sulfur dioxide (SO_2) to prevent microbial proliferation and browning. Excellent accounts of the large-scale production of starch from various sources are available (Radley, 1976a; Whistler et al., 1984; Dziedzic and Kearsley, 1995).

9.1.1
Maize (*Zea mays*)

Maize grain (separated from the cobs) arriving at the plant is cleaned to remove foreign matter and washed. The grain consists approximately of 62% starch, 8% protein, 3.5% fat, 1.5% ash, 1.5% sugars, 8.5% dietary fiber, and 15% moisture. It is steeped in large vessels at about 50–52°C and in the presence of about 0.3% SO_2 for approximately 50 h. The softened grain, suspended in water, is coarsely milled to separate the germ (embryo) without damaging it. The germs, containing about 40% oil on a dry basis, are dried to a very low moisture content and maize or corn oil is obtained by expression or by solvent extraction. The oil (corn or maize oil) is refined to a high standard of purity. The residue (oil cake) is used as animal feed.

Because the process is carried out in the presence of water it is commonly known as "wet milling" in order to distinguish it from the dry milling of maize. The remaining parts of the grain (pericarp, and endosperm in slurry form) are milled further and coarsely sieved to remove the pericarp. After further milling the aqueous suspension of starch and protein is finely sieved to remove cell-wall material, and the starch is separated from the protein either by means of high-speed continuous centrifugal separators or by means of hydrocyclones. In either case, several stages of separation and counter-current washing are necessary in order to achieve satisfactory purity. The purified starch slurry emerges at a concentration of ~39% solids. It is de-watered to less than 40% moisture and dried in pneumatic or other forms of continuous drier. The final pure product normally contains ~12% moisture, 0.3% protein, 0.6% fat (mostly internal starch lipids), and <0.1% mineral matter (ash). Modern plants produce starch that consists of 99% pure carbohydrate on a dry weight basis. If starch is to be hydrolyzed for syrup production, the slurry is pumped directly to the hydrolysis plant. It should be emphasized that practically all starch-producing factories are built adjacent to hydrolysis plants as the cost of dehydrating the starch and re-suspending it in water before hydrolysis would be prohibitive. Glucose syrup, crystalline dextrose monohydrate, maltodextrins and high-fructose syrup can all be produced from maize starch at competitive prices compared with other starches.

The byproducts from "corn refining" are gluten meal, containing ~20% protein, and gluten feed, containing ~60% protein. The "steep water" from the soaking of the grain contains considerable quantities of protein, lactic acid, minerals, and some sugars. It is concentrated to ~60% total solids, and may be mixed with the feed or used as substrate for biomass in the fermentation industry for the production of antibiotics. There is very little waste generated during the refining process, and for this reason maize starch plants are described as "bottled-up". Modern automated plants are continuously operated, 24 hours a day, and a daily processing capacity of over 1250 tonnes of maize grain is common (Jackson and Shandera, 1995). The wet milling of corn has been reviewed by Singh and Eckhoff (1996).

9.1.2
Wheat (*Triticum* spp.)

Wheat grain is the source of starch, vital gluten, bran and oil-containing wheat germ (embryo). The bran and germ are separated by mechanical processing of the dry grain. The bran is used in breakfast cereals as a source of dietary fiber, and the germ is a source of oil. Wheat flour is the raw material that provides two products: vital gluten and starch.

The process of wheat starch production (commonly known as the Martin process) consists of mixing wheat flour (~70–75% starch content) with water to form a dough, which is continuously washed with sprays of water in specially designed equipment. The released starch is sieved, washed in several stages of hydrocyclones and finally de-watered in centrifugal machines. The "starch cake" is dried in a continuously operated pneumatic drier. The final starch may contain 0.25–0.5% of protein depending on the extent of purification and on the ratio of A and B granules. The latter are difficult to purify because of their slow sedimentation rate. The proteinaceous residue is similarly dried to provide "vital gluten" which is used in baking to improve the properties of bread. The soluble components of the flour constitute a disposal problem.

9.1.3
Rice (*Oryza sativa*)

The raw material for the process is broken rice grain (~80–85% starch content). The grain is mixed with 0.3–0.5% NaOH solution at ambient temperature and held for about 10 h. It is subsequently milled to form a slurry that is sieved to remove particles other than starch. The slurry is purified by repeated washing in continuous centrifugal machines or hydrocyclones. Protein in the separated residue is precipitated at the isoelectric point (pH 4.5). The starch slurry is de-watered and dried as described above for other starches.

9.1.4
Sorghum (*Sorghum bicolor*)

Grain sorghum broadly resembles maize, but the individual grains are considerably smaller. It can be processed like normal maize with suitable modifications in the steeping stage to avoid loss of the small grains. The recovery of the embryos is low, and the starch produced has a gray tint due to minute fragments of the colored pericarp. It may contain a higher protein than normal maize (>0.3%). Glucose syrup produced by hydrolysis of sorghum starch is difficult to decolorize completely.

9.1.5
Potato (*Solanum tuberosum*)

Potato tubers contain 15–20% starch depending on the origin. Due to high moisture content, storage is difficult and as a consequence processing is seasonal. The potatoes are washed to remove adhering earth and milled continuously in saw-toothed rasps or hammer mills. The slurry, to which a small quantity of SO_2 is added to prevent microbial growth, is sieved in specially designed equipment to remove cell-wall material. The latter, in the form of pulp, is used as animal feed. The starch is washed free of soluble impurities in several stages of hydrocyclones and de-watered in basket centrifugal or vacuum drum filters. The starch cake is dried in pneumatic driers. The soluble components of the potato present a serious disposal problem. Spraying on fields, or holding in lagoons, is often practiced to avoid pollution of watercourses. Potato starch contains ~0.06% protein, ~0.2% ash, 0.05% lipid, and ~16–18% moisture.

9.1.6
Tapioca (*Manihot esculenta*)

The plant known as tapioca in the East is also known as cassava in Africa and manioc or yucca in South America. There are bitter and sweet varieties. The roots contain a cyanohydrin (a β-glucoside also known as linamarin or phaseolunatin), which liberates HCN, benzaldehyde and glucose on hydrolysis in an acid medium. It can be removed after fermentation and by repeated washing of the crushed roots. Nowadays, the roots (containing ~25% starch) are processed in modern plants. High-quality starch is obtained which finds uses in a number of specialized food applications. Processing comprises rasping or hammer milling of the roots, sieving to remove fibrous and cellular material, repeated washing in centrifuges or hydrocyclones, dewatering, and drying. A thoroughly purified starch may contain 0.04% protein, <0.1% lipid, and ~0.1% ash. "Pearl tapioca" starch has been available for a long time in the form of small granules. These are formed by forcing sedimented starch through sieves (~3 mm aperture) and drying with constant stirring over heated iron sheets. The resulting spherical particles (~3 mm in diameter) have a thin film of gelatinized starch on the surface.

9.1.7
Sweet potato (*Ipomoea batatas*)

The roots of the plant are processed in a manner similar to that for tapioca. Lime water (saturated calcium hydroxide solution) has been advocated as a suspending medium to facilitate removal of impurities. China is the world's largest producer of sweet potatoes, followed by Nigeria.

9.1.8
Arrowroot (*Maranta arundinacea*)

The fleshy tubers of the plant which grows in the West Indies are sometimes locally processed in a rather elementary manner. Crushing of the roots, sieving and sedimentation of starch by the force of gravity gives rise to relatively impure starch, which can be purified in the advanced plants of industrialized countries.

9.1.9
Sago (*Metroxylon sagu*)

Sago is an aquatic palm of tropical Asia. The trunk is split and the pith (~40% starch) is scooped out. The pulp is mixed with water and sieved to retain fibrous material. The starch is separated by sedimentation under gravity. "Pearl sago" is prepared by a process similar to that for tapioca starch.

9.2
Uses of Starch

9.2.1
Food Uses

In addition to its nutritional value, starch contributes significantly to the texture of foods. The quality of wheat protein (gluten) is of paramount importance in breadmaking and in pasta manufacture. However, the quality of starch is of almost equal importance. Rain-damaged wheat contains high levels of endogenous α-amylase, which causes partial hydrolysis of starch and leads to poor quality bread. Similarly, rice and potatoes owe much of their textural organoleptic qualities to endogenous starch.

Starch is used extensively as a stabilizer or thickener because of its high viscosity in aqueous media (as already discussed). Esterified starches are used when paste clarity is required. Crosslinked waxy starches provide clarity and resistance to low pH and shear forces. Acid-modified starches are used in gum confectionery because, on heating in the presence of sugar, a high solids content can be achieved without excessive viscosity in continuously operated equipment. In the dry form starch is used in molds for the deposition and forming of gum confectionery.

Starch partially replaces oil in low-calorie mayonnaise and contributes to the desired texture. It is added to yogurt as a stabilizer and to a large number of desserts, convenience foods and ready-to-cook meals.

In the brewing industry, starch in malt and in adjuncts like maize grits or rice is converted to ethanol. The final products are beer and distilled alcoholic beverages.

9.2.2
Industrial Uses

Textile Industry
Starch finds extensive use in the textile industry. It is used in sizing yarn to improve resistance both to abrasion and to tension during weaving. Many finished textiles have an attractive appearance and feel when coated with a thin layer of gelatinized starch. In textile printing, starch is used to hold colors in the desired area of the fabric and to prevent spreading and mixing of the colors. It is also used as a finish or glaze for sewing thread.

Paper Industry
In the paper industry starch is used to increase the strength of the sheet produced in continuous machinery. As a surface coating starch is used to improve the writing and erasing characteristics of paper. Starch also provides a desirable coating to high-quality printing paper. It is also used as an adhesive for paper bags.

Miscellaneous Uses
Starches, either native of modified, find several applications in the pharmaceutical industry, particularly as excipients in the manufacture of tablets (e.g., aspirin). The food and industrial uses of starch have been described in considerable detail (Radley, 1976b; Whistler et al., 1984).

10
Hydrolysis Products of Starch

10.1
Introduction

For about 120 years starch has been an important commercial source of nutritive sweeteners, often competing with common sugar. The conversion of starch to glucose by heating with sulfuric acid is credited to the German chemist G. S. C. Kirchoff. In 1811, he obtained by this method a sweet mixture suitable for replacing sugar, which was in short supply in Europe because of a British naval blockade against Napoleon. This appears to have been the starting point for the mass production of acid-converted starch syrups that followed some 60 years later.

A second major discovery was made by the French chemist Anselme Payen, who in 1833 isolated a substance from malt extract that could convert starch to a sweet syrup. He called the substance "diastase" (Greek *diastasis* = separation), a term which is used even today. The word enzyme (Greek *en zyme* = in leaven) was coined by the German physiologist Willy Kühne some years later. In 1896, Buchner demonstrated that an extract of yeast cells (not the living cells) could ferment sugar. Thus, the potential of extracellular enzymes became apparent. However, the use of enzymes for the production of sweeteners from starch came much later, in about 1938. A breakthrough in the application of enzymes for the large-scale hydrolysis of starch occurred in the early 1970s with the production of thermostable bacterial α-amylase. By the mid-1980s, high-fructose syrup was firmly established in the market.

Nowadays, it is estimated that about 70% of the starch produced in more than 200 factories throughout the world is converted to sweeteners containing glucose (Schenck and Hebeda, 1992). The principles of starch conversion technology are discussed below.

10.1.1
Products of Acid Hydrolysis

Glucose syrup, traditionally known as "confectioners' syrup" or "corn syrup" in the USA and "starch syrup" in some European countries, contains ~80% (by weight) of dry matter, and has a dextrose equivalent (DE) of 42. DE is defined as the ratio of reducing sugars expressed as dextrose to the total carbohydrate content, multiplied by 100. On this scale, pure dextrose has a $DE = 100$ and pure starch a $DE = 0$. The word dextrose is synonymous with glucose and is commonly used in industry to describe the dry crystalline substance.

During acid hydrolysis, starch molecules are broken down in a random fashion. A 42 DE syrup invariably contains (on a dry basis) 19% dextrose, 14% maltose, 11% maltotriose, and 56% higher saccharides. Hydrolysis is effected continuously by heating starch slurry (40% DS) acidified with HCl (0.02 M, pH ~1.7) to about 140°C for ~5 min. After rapid cooling, the hydrolysate is neutralized with sodium carbonate to pH 4.8, the isoelectric point of nearly all starch protein. The product is filtered to remove residual protein and lipid (~1% in total) and decolorized with activated carbon. The liquid is concentrated in multiple effect vacuum evaporators to give a water-clear viscous syrup containing 80–82% dry solids. The syrup contains ~0.4% ash and usually <0.1% protein. It may turn yellow or brown on storage because of Maillard reaction. Color formation is also caused by hydroxymethyl furfural, which is formed under the acid conditions of hydrolysis. This compound is colorless, but it is a precursor of color formation through conjugation. In order to retard color formation, SO_2 may be added to a maximum level of 400 mg kg^{-1}. It is recognized that purification of the hydrolysate with ion-exchange resins (complete demineralization) gives rise to high-quality colorless and stable syrups (Karkalas, 1967; Howling, 1992).

Some 500,000 tonnes of syrup are produced annually in the EU, representing 45% of total production. The 42 DE syrup is the most common product on the market, although DE in the range 35–55 can also be produced (Howling, 1992). The syrup may be spray-dried for special applications.

10.1.2
Acid-Enzyme-produced Syrups

Starch is hydrolyzed with acid to a low DE (20–40) and subsequently hydrolyzed further with enzymes. The enzymatic hydrolysis is carried out with β-amylase and amyloglucosidase, or with fungal α-amylase and amyloglucosidase. The final product has a DE of 60–70 and a maltose content of 30–32%.

10.1.3
Enzymatically Produced Syrups

Liquid Dextrose and Dextrose Monohydrate
The use of enzymes produced biotechnologically has led to the development of a new range of syrups. It is impossible to deal with this topic in full, and only a brief outline is presented here. Early developments are described by Norman (1981), by Palmer (1982) and by Fullbrook (1984). Teague and Brumm (1992) have provided an account of enzymes used in industry, while processes are described by Mulvihill (1992) and Olsen (1995).

The production of pure dextrose monohydrate deserves special attention because the process leads to >95% conversion of starch to glucose. The process is essentially a two-step enzymatic conversion.

Thermostable α-amylase (from *Bacillus licheniformis*) and a small quantity of calcium salt are added to the starch slurry, adjusted to pH 6.0 and containing 30–35% dry solids.

The mixture is pumped continuously to a steam injection heater where it emerges at 105°C and is immediately flash-cooled to 100°C. Liquefaction (dextrinization) is allowed to proceed for about 2 h in sequentially operated holding tanks, whereupon the viscosity falls dramatically. The liquefied product is then flash-cooled under reduced pressure to 60°C, adjusted to pH 4.5, followed by the addition of amyloglucosidase (e.g., from *Aspergillus niger*). Saccharification is complete in about 2 days, and the DE is 98%. The hydrolysate, containing 95% dextrose on a dry basis, is filtered to remove suspended matter, concentrated by evaporation to 50% solids, decolorized with activated carbon, and deionized by passing through columns of cation- and anion-exchange resins. The effluent is further concentrated by evaporation to 71% dry solids. The clear liquor can be sold as liquid dextrose, or allowed to crystallize to yield dextrose monohydrate. Alternatively, it is sent to the isomerization plant for the production of high-fructose syrup.

Dual Enzyme Syrups

Liquefaction of starch is carried out with bacterial α-amylase. Conversion to maltose is effected with β-amylase or with fungal α-amylase. The final products are normally supplied as high-maltose syrups, and contain 50–56% maltose. Extra-high-maltose syrups contain a minimum of 70% maltose. The debranching enzyme pullulanase (from *Bacillus acidopullulyticus* or *Klebsiella planticola*) is often used, in conjunction with other enzymes, in the production of high-maltose syrups. It is known that α- and β-amylases do not hydrolyze α-$(1 \rightarrow 6)$ branch points.

Isomerized Glucose Syrups

Glucose isomerase is an enzyme that converts glucose to fructose. It is produced by a number of microorganisms, e.g., *Bacillus coagulans, Actinoplanes missouriensis, Flavobacterium arborescens*, and some *Streptomyces* spp. The optimum pH is 7–8 and the optimum temperature 55–65°C. The enzyme is used in immobilized form on a solid support. Pure dextrose syrup, described above, is pumped into large fixed-bed reactors where isomerization takes place continuously (White, 1992). Isomerization of glucose is an equilibrium reaction, and the product leaving the reactor contains ~53% glucose and 42% fructose on a dry basis. After isomerization, the product is decolorized and demineralized by ion exchange. Enrichment of fructose is achieved by a large-scale chromatographic process utilizing special resins (in the Ca form). Fructose is selectively retained, and glucose emerges from the vessel containing the resin. Several vessels are linked in series to allow efficient separation of the two sugars. Fractionation gives rise to a product containing 90% fructose (dry basis). This syrup is blended with the product containing 42% fructose to give a 55% high-fructose syrup. The latter product contains 41% glucose and 55% fructose (dry basis). These products are also known as high-fructose corn syrups (HFCS).

The two commercial high-fructose syrups designated HFS-42 (71% dry solids) and HFS-55 (77% dry solids) respectively find extensive applications, primarily in the soft drinks industry. Because of their sweetness they can compete with sucrose. However, the success of high-fructose products in the USA and their steady increase in consumption is mainly due to the low cost of maize which is the raw material. The efficient operation and management of large-scale production plants is an additional contributing factor.

10.2
Other Products of Hydrolysis

10.2.1
Maltodextrins

Maltodextrins are practically nonsweet hydrolysis products of starch of DE less than 20, consisting primarily of α-$(1 \rightarrow 4)$ bonds. They are usually available as water-soluble white powders of low bulk density.

Several patents have been issued for their preparation. There are two main processes: (1) acid-enzyme hydrolysis; and (2) enzyme hydrolysis. In the first process, starch is hydrolyzed to a very low DE in the presence of acid, neutralized, and then treated with bacterial α-amylase for a sufficient length of time to reach the desired DE. The enzyme is inactivated by heating. Subsequently the hydrolyzate is treated with activated carbon to remove color, filtered, and concentrated by vacuum evaporation. Although the product can be sold in liquid form, it is normally spray-dried. The enzymatic process involves gelatinizing the starch in the presence of α-amylase to reduce viscosity. This is usually followed by heating to a high temperature ($\sim 130°C$) to disrupt any nondegraded starch and dissociate the double helices which form easily during the gelatinization process because of the high solids content of the reaction mixture. The product is further hydrolyzed by α-amylase to the desired DE and treated as described for the previous product.

A variety of maltodextrin types is produced globally, ranging in DE from 1 up to 20. Low-DE products have limited solubility and practically no sweetness. Solubility and sweetness increase with increasing DE, while viscosity decreases. In a comprehensive review (Alexander, 1992) it is reported that about 91,000 tonnes of this product were produced in the USA. In the UK, the net supply in 1999 was 13,940 tonnes (National Statistics, 2000).

Maltodextrins find numerous applications in the food and pharmaceutical industries. They are used as a spray-drying aid together with gum arabic (acacia gum) in the encapsulation of flavors (lemon, orange) and oils. Alternately, flavors can be encapsulated in a mixture of maltodextrin and sugar and extruded into isopropanol. Maltodextrins are also used as carriers for non-nutritive sweeteners, and many instant desserts, flavored drinks and soups are based on them. Kennedy et al. (1995) have reviewed analytical methods for maltodextrins.

10.2.2
Cyclodextrins

Cyclodextrins consist of α-$(1 \rightarrow 4)$-linked glucose units in a circular form. There are three main types consisting of six, seven, or eight glucose molecules known as α-, β-, and γ-cyclodextrin, respectively. The respective solubilities in water are 12.8, 1.8, and 25.6% at 25°C. They are also known as Shardinger dextrins.

These products are obtained from partially hydrolyzed starch of low DE (< 10) by means of the enzyme cyclodextrin glycosyltransferase (CDGTase) produced by the organisms *Bacillus amylobacter* and *Bacillus macerans*. The enzyme exhibits cyclation, transglycosylation and hydrolysis simultaneously, and in aqueous media an equilibrium is established between the cyclodextrin and the short dextrins cleaved from starch. The reaction is best carried out in toluene, which forms an insoluble complex with β-cyclodextrin. If decanol is used, the α-form is favored. The γ-form is obtained in a mixture of α-naphthol and methylethyl ketone. At the end of the reaction process the insoluble complex is separated by centrifugation, and after removal of the organic solvent the product is purified and separated by crystallization.

Cyclodextrin molecules are in the form of a hollow cylinder which has one end slightly larger than the other. The inner surface of the cylinder is hydrophobic, and it attracts hydrophobic molecules known as "guests". The outer surface is hydrophilic. Thus, many essential oils and related organic compounds will complex with cyclodextrins. Complex formation with cyclodextrins can result in improved solubility, protection from light, lowering of volatility, controlled release, separation of chemicals, taste masking, catalysis and directed chemical reaction. Cyclodextrins find applications in the food, pharmaceutical, chemical, analytical, diagnostic and other areas (Hedges, 1992).

10.2.3
Hydrogenated Syrups and Polyols

Polyols or polyalcohols are reduction products of carbohydrates. D-Sorbitol (D-glucitol), D-mannitol and D-xylitol originate respectively from glucose, mannose, and xylose. All three compounds occur naturally in some fruits. Plums are particularly rich in sorbitol (about 3% fresh and 15% in prunes). Maltitol, lactitol, and isomalt (the latter is a mixture of D-glucosyl-α-(1-1)-D-mannitol, D-glucosyl-α-(1-1)-D-sorbitol and D-glucosyl-α-(1-6)-D-sorbitol) are obtained by hydrogenation of maltose, lactose, and isomaltulose respectively. Isomaltulose originates from sucrose by the action of the enzymes of *Leuconostoc mesenteroides* and *Protoaminobacter rubrum*. In addition, glucose syrups and high-maltose syrups are catalytically hydrogenated to give products with distinct properties. These products find applications mainly in foods, and in dietetic and pharmaceutical preparations.

Sorbitol and hydrogenated glucose syrups do not have carbonyl groups and do not react with amino acids (Maillard reaction). Therefore, they remain colorless on storage. They are good humectants, i.e., they do not lose moisture readily, and for this reason they are used in soft fillings in confectionery manufacture. Polyols are considered noncariogenic and are therefore used in some sweets and chewing gum. Their sweetness relative to sucrose (1 on this scale) is approximately: xylose 1, maltitol 0.9, sorbitol 0.6, and mannitol 0.4 (Le Bot and Gouy, 1995).

Polyols are slowly absorbed in the intestinal tract, and for this reason they are acceptable in diabetic foods at a low level. Intakes larger than ~20 g per day may cause flatulence. Although polyols have an energy value of 4 kcal g^{-1} like any other sugar, for labeling purposes a value of 2.4 kcal g^{-1} is accepted. This is because these compounds are only partly absorbed, and when they reach the colon they are fermented by the endogenous microflora giving rise to volatile fatty acids. However, maltitol and hydrogenated high-maltose syrups are hydrolyzed in the digestive tract to glucose and sorbitol.

10.3
Uses of Hydrolyzed Products

All hydrolysis products of starch find extensive use in the food industry, including confectionery and baking, in soft drinks, in brewing, in the pharmaceutical industry, and in various dietetic products.

Acid-converted 42 DE syrup is used in the manufacture of hard-boiled sweets (candies), which normally consist of almost equal proportions of sucrose and glucose syrup solids and have a moisture content of less than 2%. The main function of glucose syrup solids is the prevention of crystallization of sucrose by forming a clear glass-like and hard product. The high viscosity of this syrup also imparts desirable textural characteristics to toffee, caramel, and fudge.

In the UK, glucose syrups and high-maltose syrups are used in brewing to provide fermentable carbohydrate. Liquid glucose (~95 DE) is used as a substrate for the production of the widely used stabilizer xanthan gum from *Xanthomonas campestris* and for the growth of mycoprotein from *Fusarium graminearum*, which provides a textured meat analogue. Lactic, citric, and ascorbic (vitamin C) acids are also produced from glucose.

Spray-dried glucose syrups find extensive applications as an ingredient of a variety of convenience foods available in dry form for mixing with milk or with water.

Pure crystalline dextrose monohydrate ($C_6H_{12}O_6 \cdot H_2O$) has advantages because it is easily transported and stored in bags. It has numerous uses in baking, confectionery and health food products, e.g., high-energy beverages.

The consumption of nutritive sweeteners from starch differs from country to country. In the USA in 1999, the per capita consumption of sucrose was 31.1 kg, HFCS 29.9 kg, glucose syrup (dry basis) 8.4 kg, dextrose 1.6 kg, and honey 0.7 kg (Corn Refiners Association, 2000). Thus, 39.9 kg of starch-based sweeteners were used compared with 31.1 kg of sucrose.

11
Concluding Remarks

As a component of cereals and roots, starch has provided sustenance for mankind for several millennia, and more recently for domesticated animals. It remains an indispensable aid in the food, paper, textile, pharmaceutical and other industries, and is the major substrate for the production of beer and distilled alcoholic beverages. As a renewable resource it provides fuel alcohol – an area that is likely to expand in the future. Starch has had a remarkable success as a source of nutritive sweeteners for at least 120 years. The isomerization of glucose for the production of high-fructose syrups has made a considerable impact, especially in the soft drinks industry. The manufacture of specialized products in the class of maltodextrins is likely to increase. Novel, physically modified starches may appear in the future to meet specialized needs.

Nonetheless, much remains to be explored in starch research, and the differences in the properties of starches of diverse origins are only just beginning to be understood. The elucidation of the internal structure of granules, their porosity, and permeability to solutes presents a considerable challenge. The surface properties of granules and their enzymatic susceptibility also deserve attention. The nature of resistant starch and its potential merits and demerits should be explored further.

The controversial subject of genetically modified plants–and the starches that they could produce–is another area that deserves a considered approach. The day when custom-made modified starches, in seeds and roots, will be harvested directly from the field may not be very near, but the chances are that it will arrive.

12 References

Abd Karim, A., Norziah, M. H., Seow, C. C. (2000) Methods for the study of starch retrogradation, *Food Chem.* **71**, 9–36.

Alexander (1992) Maltodextrins: production, properties and applications, in: *Starch Hydrolysis Products, Worldwide Technology, Production, and Applications* (Schenck, F. W., Hebeda, R. E., Eds.), New York: VCH, 233–275.

Åman, P., Hesselman, K. (1984) Analysis of starch and other main constituents of cereal grains, *Swed. J. Agric. Res.* **14**, 135–139.

AOAC International (2000) *Official Methods of Analysis of AOAC International*, 17th edition, (Horwitz, W., Ed.), Gaithersburg: AOAC International.

ap Rees, T. (1993) Control of carbon partitioning in plants, in: *Seed Storage Compounds – Biosynthesis, Interactions and Manipulation* (Shewry, P. R., Stobart, K., Eds.), Oxford: Oxford University Press, 241–261.

Asaoka, M., Okuno, K., Sugimoto, Y., Fuwa, H. (1985) Developmental changes in the structure of endosperm starch of rice (*Oryza sativa* L.), *Agric. Biol. Chem.* **49**, 1973–1978.

Atkin, N. J., Abeysekera, R. M., Cheng, S. L., Robards, A. W. (1998a) An experimentally-based predictive model for the separation of amylopectin subunits during starch gelatinisation, *Carbohydr. Polym.* **36**, 173–192.

Atkin, N. J., Abeysekera, R. M., Robards, A. W. (1998b) The events leading to the formation of ghost remnants from the starch granule surface and the contribution of the granule surface to the gelatinisation endotherm, *Carbohydr. Polym.* **36**, 193–204.

Baba, T., Arai, Y. (1984) Structural characterisation of amylopectin and intermediate material in amylomaize starch granules, *Agric. Biol. Chem.* **48**, 1763–1775.

Baghurst, P. A., Baghurst, K. I., Record, S. J. (1996) Dietary fibre, non-starch polysaccharides and resistant starch – A review, *Food Aust.* **48**, S3–S35.

Baker, A. A., Miles, M. J., Helbert, W. (2001) Internal structure of the starch granule revealed by AFM, *Carbohydr. Res.* **330**, 249–256.

Baldwin, P. M., Adler, J., Davies, M. C., Melia, D. C. (1998) High resolution imaging of starch granule surface by atomic force microscopy, *J. Cereal Sci.* **27**, 255–265.

Ball, S. G. (1995) Recent reviews on the biosynthesis of the plant starch granule, *Trends Glycosci. Glycotechnol.* **7**, 405–415.

Ball, S., Guan, H-P, James, M., Myers, A., Keeling, P., Mouille, G., Buléon, A., Colonna, P., Preiss, J. (1996) From glycogen to amylopectin: a model for the biogenesis of the plant starch granule, *Cell* **86**, 349–352.

Ball, S. G., van de Wal, H. B. J., Visser, R. G. F. (1998) Progress in understanding the biosynthesis of amylose, *Trends Plant Sci.* **3**, 462–467.

Banks, W. and Greenwood, C. T. (1975) *Starch and its Components*. Edinburgh: Edinburgh University Press.

Banks, W., Muir, D. D. (1980) Structure and chemistry of the starch granule, in: *The Biochemistry of Plants: A Comprehensive Treatise, Volume 3, Carbohydrates: Structure and Function* (Preiss, J., Ed.), New York: Academic Press, 321–369.

Banks, W., Greenwood, C. T., Muir, D. D. (1973) Studies on the biosynthesis of starch granules. Part 5. Properties of the starch components of normal barley and barley with starch of high-amylose content during growth, *Starch/Stärke* **25**, 153–157.

Batey, I. L. (1982) Starch analysis using thermostable alpha-amylases, *Starch/Stärke* **34**, 125–128.

Bechtel, D. B., Zayas, I., Kaleikau, L., Pomeranz, Y. (1990) Size-distribution of wheat starch granules during endosperm development, *Cereal Chem.* **67**, 59–63.

Bengtsson, S., Larsson, K. (1990) Determination of starch in a flow-injection system with immobilized enzymes, *Swed. J. Agric. Res.* **20**, 27–29.

Berry, C. S. (1986) Resistant starch: Formation and measurement of starch that survives exhaustive digestion with amylolytic enzymes during the determination of dietary fibre, *J. Cereal Sci.* **4**, 301–314.

Bertoft, E., Manelius, R., Zhu, Q. (1993) Studies on the structure of pea starches. 1. Initial-stages in alpha-amylolysis of granular smooth pea starch, *Starch/Stärke* **45**, 215–220.

Beutler, H.-O. (1984) Starch, in: *Methods of Enzymatic Analysis*, 3rd edition, (Bergmeyer, H. U., Bergmeyer, J., Graßl, M., Eds.), Weinheim: Verlag Chemie, vol. VI, 2–10.

Biliaderis, C. G. (1992) Characterization of starch networks by small strain dynamic rheometry, in: *Developments in Carbohydrate Chemistry* (Alexander, R. J., Zobel, F., Eds.), St. Paul: AACC, 87–135.

Biliaderis, C. G. (1998) Structures and phase transitions of starch polymers, in: *Polysaccharide Association Structures in Foods* (Walter, R. H., Ed.), New York: Marcel Dekker, 57–168.

Biliaderis, C. G., Seneviratne, H. D. (1990) On the supermolecular structure and metastability of glycerol-monostearate-amylose complex, *Carbohydr. Polym.* **13**, 185–206.

Biliaderis, C. G., Tonogai, J. R. (1991) Influence of lipids on the thermal and mechanical properties of concentrated starch gels, *J. Agric. Food Chem.* **39**, 838–840.

Biliaderis, C. G., Zawistowski, J. (1990) Viscoelastic behaviour of aging starch gels: effects of concentration, temperature and starch hydrolysates on network properties, *Cereal Chem.* **67**, 240–246.

Biliaderis, C. G., Maurice, T. J., Vose, J. R. (1980) Starch gelatinisation phenomena studied by differential scanning calorimetry, *J. Food Sci.* **45**, 1669–1674, 1680.

Biliaderis, C. G., Page, C. M., Maurice, T. J., Juliano, B. O. (1986) Thermal characterisation of rice starches – a polymeric approach to phase-transitions of antigranulocytes starch, *J. Agric. Food Chem.* **34**, 6–14.

Blanshard, J. M. V. (1987) Starch granule structure and function: a physicochemical approach, in: *Starch: Properties and Potential* (Galliard, T., Ed.), Chichester: John Wiley & Sons, 16–54.

Blanshard, J. M. V., Bates, D. R., Muhr, A. H., Worcester, D. L., Higgins, J. (1984) Small-angle neutron-scattering studies of starch granule structure. *Carbohydr. Polym.* **4**, 427–442.

Blanshard, P. H., Katz, F. R. (1995) Starch hydrolysates, in: *Food Polysaccharides and their Applications* (Stephen, A. M., Ed.), New York: Marcel Dekker, 99–122.

Bogracheva, T. Y, Morris, V. J., Ring, S. G., Hedley, C. L. (1998) The granular structure of C-type pea starch and its role in gelatinisation, *Biopolymers* **45**, 323–332.

Bowsher, C. G., Tetlow, I. J., Lacey, A. E., Hanke, G. T., Emes, M. J. (1996) Integration of metabolism in non-photosynthetic plastids of higher plants. *CR ACAD SCI, Paris* **319**, 853–860.

Boyer, C. D., Daniels, R. R., Shannon, J. C. (1977) Starch granule (amyloplast) development in endosperm of several *Zea mays* L. genotypes affecting kernel polysaccharides, *Am. J. Bot.* **64**, 50–56.

Buléon, A., Bizot, H., Delage, M. M., Multon, J. L. (1982) Evolution of crystallinity and specific-gravity of potato starch versus water-adsorption and desorption, *Starch/Stärke* **34**, 361–366.

Buléon, A., Colonna, P., Planchot, V., Ball, S. (1998a) Starch granules: structure and biosynthesis, *Int. J. Biol. Macromol.* **23**, 85–112.

Buléon, A, Gérard, C., Riekel, C., Vuong, R., Chanzy, H. (1998b) Details of the crystalline ultrastructure of C-starch revealed by synchrotron microfocus mapping. *Macromolecules* **31**, 6605–6610.

Buléon, A., Pontoire, B., Riekel, C., Chanzy, H., Helbert, W., Vuong, R. (1997) Crystalline ultra-structure of starch granules revealed by synchrotron radiation microdiffraction mapping, *Macromolecules* **30**, 3952–3954.

Buttrose, M. S. (1960) Submicrosopic development and structure of starch granules in cereal endosperms, *J. Ultrastruct. Res.* **4**, 231–257.

Cairns, P., Sun, L., Morris, V. J., Ring, S. G. (1995) Physicochemical studies using amylose as an *in vitro* model for resistant starch, *J. Cereal Sci.* **21**, 37–47.

Cairns, P., Bogracheva, T. Y., Ring, S. G., Hedley, C. L., Morris, V. J. (1997) Determination of the polymorphic composition of smooth pea starch, *Carbohydr. Polym.* **32**, 275–282.

Cameron, R. E., Donald, A. M. (1992) A small-angle X-ray scattering study of the annealing and gelatinisation of starch, *Polymer* **33**, 2628–2635.

Cameron, R. E., Donald, R. E. (1993a) A small-angle X-ray scattering study of the absorption of water into the starch granule, *Carbohydr. Res.* **244**, 225–236.

Cameron, R. E., Donald, R. E. (1993b) A small-angle X-ray scattering study of starch gelatinisation in excess and limiting water, *J. Polym. Sci. Part B Polym. Phys.* **31**, 1197–1203.

Cheetam, N. W. H., Tao, L. (1998) Variation in crystalline type with amylose content in maize starch granules: an X-ray powder diffraction study, *Carbohydr. Polym.* **36**, 277–284.

Clegg, K. M. (1956) The application of the anthrone reagent to the estimation of starch in cereals, *J. Sci. Food Agric.* **7**, 40–44.

Colonna, P., Mercier, C. (1984) *Pisum sativum* and *Vicia faba* carbohydrates. 5. Macromolecular structure of wrinkled pea and smooth pea starch components, *Carbohydr. Res.* **126**, 233–247.

Colonna, P., Mercier, C. (1985) Gelatinisation and melting of maize and pea starches with normal and high-amylose genotypes, *Phytochemistry* **24**, 1667–1674.

Cooke, D., Gidley, M. J. (1992) Loss of crystalline and molecular order during starch gelatinisation – origin of the enthalpic transition, *Carbohydr. Res.* **227**, 103–112.

Corn Refiners Association (2000) http://www.corn.org.

Cui, R., Oates, C. G. (1997) The effect of retrogradation on enzyme susceptibility of sago starch, *Carbohydr. Polym.* **32**, 65–72.

Debon, S. J. J., Tester, R. F. (2000) In vivo and *in vitro* annealing of starches, in: *Gums and Stabilisers for the Food Industry* (Williams, P. A., Phillips, G. O., Eds.), Cambridge: The Royal Society of Chemistry, 270–276.

Debon, S. J. J., Tester, R. F., Millam, S., Davies, H. V. (1998) Effect of temperature on the synthesis, composition and physical properties of potato microtuber starch, *J. Sci. Food Agric.* **76**, 599–607.

Denyer, K., Clarke, B., Hylton, C., Tatge, H., Smith, A. M. (1996) The elongation of amylose and amylopectin chains in isolated starch granules, *Plant J.* **10**, 1135–1143.

Donald, A. M., Waigh, T. A., Jenkins, P. J., Gidley, M. J., Debet, M., Smith, A. (1997) Internal structure of starch granules revealed by scattering studies, in: *Starch Structure and Functionality* (Frazier, P. J., Donald, A. M., Richmond, P., Eds.), Cambridge: The Royal Society of Chemistry, 172–179.

Donovan, J. T., Mapes, C. J. (1980) Multiple phase transitions of starch and Nägeli amylodextrins, *Starch/Stärke* **32**, 190–193.

Donovan, J. W. (1979) Phase transitions of the starch-water system, *Biopolymers* **18**, 263–275.

Donovan, J. W., Lorenz, K., Kulp, K. (1983) Differential scanning calorimetry of heat-moisture treated wheat and potato starches, *Cereal Chem.* **60**, 381–387.

Doublier, J. L., Llamas, G. A. (1993) A rheological description of amylose-amylopectin mixtures, in: *Food Colloids and Polymers* (Dickinson, E., Walstra, P., Eds.), Cambridge: Royal Society of Chemistry, 138–146.

Doublier, J. L., Launay, B., Cuvelier, G. (1992) Viscoelastic properties of food gels, in: *Viscoelastic Properties of Foods* (Rao, M. A., Steffe, J. F., Eds.), London: Elsevier Applied Science, 371–434.

Duffus, C. M. (1993) Starch synthesis and deposition in developing cereal endosperms, in: *Seed Storage Compounds – Biosynthesis, Interactions and Manipulation* (Shewry, P. R., Stobart, K., Eds.), Oxford: Oxford University Press, 191–209.

Dziedzic, S. Z., Kearsley, M. W. (1995) The technology of starch production, in: *Handbook of Starch Hydrolysis Products and their Derivatives* (Kearsley, M. W., Dziedzic, S. Z., Eds.), London: Blackie Academic and Professional, 1–25.

Earle, F. R., Milner, R. T. (1944) Improvements in the determination of starch in corn and wheat, *Cereal Chem.* **21**, 567–575.

Edwards, C., Tester, R. F. (1998) Carbohydrates, (d) Resistant Starch and Oligosaccharides, in: *Encyclopedia of Human Nutrition* (Caballero, B., Strain, J. J., Eds.), London: Academic Press, 289–295.

Eerlingen, R. C., Crombez, M., Delcour, J. A. (1993) Enzyme-resistant starch. I. Quantitative and qualitative influence of incubation time and temperature of autoclaved starch on resistant starch formation, *Cereal Chem.* **70**, 339–344.

Eliasson, A-C. (1986) Viscoelastic behaviour during the gelatinization of starch. I. Comparison of wheat, maize, potato and waxy-barley starches, *J. Text. Stud.*, **17**, 253–255.

Eliasson, A.-C., Bohlin, L. (1982) Rheological properties of concentrated wheat starch gels, *Starch/Stärke*, **34**, 267–271.

Eliasson, A-C., Gudmundsson, M. (1996) Starch: physicochemical and functional aspects, in: *Carbohydrates in Food* (Eliasson, A.-C., Ed.), New York: Marcel Dekker, 431–503.

Eliasson, A-C., Larsson, K., Andersson, S., Hyde, S. T., Nesper, R., von Schnering, H-G. (1987) On the structure of native starch – an analogue to the quartz structure, *Starch/Stärke* **39**, 147–152.

Englyst, H. N., Cummings, J. H. (1984) Simplified method for the measurement of total non-starch polysaccharides by gas-liquid-chromatography of constituent sugars as alditol acetates, *Analyst* **109**, 937–942.

Englyst, H. N., Cummings, J. H. (1987) Digestion of polysaccharides of potato in the small intestine of man, *Am. J. Clin. Nutr.* **45**, 423–431.

Englyst, H. N., Cummings, J. H. (1988) Improved method for measurement of dietary fibre as non-starch polysaccharides in plant foods, *J. Assoc. Off. Anal. Chem.* **71**, 808–814.

Englyst, H. N., Hudson, G. J. (1996) The classification of dietary carbohydrates, *Food Chem.* **57**, 15–21.

Englyst, H. N., Hudson, G. J. (1997) Starch and health, in: *Starch Structure and Functionality* (Frazier, P. J., Donald, A. M., Richmond, P., Eds.), Cambridge: Royal Society of Chemistry, 9–21.

Englyst, H. N., Wiggins, H. S., Cummings, J. H. (1982) Determination of the non-starch polysaccharides in plant foods by gas-liquid-chromatography of constituent sugars as alditol acetates, *Analyst* **107**, 307–318.

Englyst, H. N., Kingman, S. M., Cummings, J. H. (1992) Classification and measurement of nutritionally important starch fractions, *Eur. J. Clin. Nutr.* **46**, S33–S50.

Englyst, H. N., Kingman, S. M., Hudson, G. J., Cummings, J. H. (1996) Measurement of resistant starch *in vitro* and *in vivo*, *Br. J. Nutr.* **75**, 749–755.

Escarpa, A., Gonzalez, M. C. (1997) Technology of resistant starch, *Food Sci. Technol. Int.* **3**, 149–161.

Evans, I. D., Haisman, D. R. (1979) Rheology of gelatinised starch suspensions, *J. Text. Stud.* **10**, 347–370.

Evans, I. D., Haisman, D. R. (1982) The effect of solutes on the gelatinisation temperature-range of potato starch, *Starch/Stärke* **34**, 224–231.

Evers, A. D. (1971) Scanning electron microscopy of wheat starch. III. Granule development in the endosperm, *Starch/Stärke* **23**, 157–162.

Fannon, J. E., BeMiller, J. N. (1992) Structure of corn starch paste and granule remnants revealed by low-temperature scanning electron-microscopy after cryopreparation, *Cereal Chem.* **69**, 456–460.

Fannon, J. E., Huber, R. J., BeMiller, J. N. (1992) Surface pores of starch granules, *Cereal Chem.* **69**, 284–288.

FAO (2000) http://www.apps.fao.org/default.htm.

Fisher, D. K., Thompson, D. B. (1997) Retrogradation of maize starch after thermal treatment within and above the gelatinisation range, *Cereal Chem.* **74**, 344–351.

Fredriksson, H., Silverio, J., Andersson, R., Eliasson, A-C., Åman, P. (1998) The influence of amylose and amylopectin characteristics on gelatinisation and retrogradation properties of different starches, *Carbohydr. Polym.* **35**, 119–134.

Fredriksson, H., Björck, I., Andersson, R., Liljeberg, H., Silverio, J., Eliasson, Åman, P. (2000) Studies on α-amylase degradation of retrograded starch gels from waxy maize and high-amylopectin potato, *Carbohydr. Polym.* **43**, 81–87.

French, D. (1972) Fine structure of starch and its relationship to the organisation of starch granules, *Denpun Kaguku* **19**, 8–25.

French, D. (1984) Organisation of starch granules, in: *Starch: Chemistry and Technology* (Whistler, R. L., BeMiller, J. N., Pashall, E. F., Eds.), Orlando, London: Academic Press, 183–247.

Fritzsche, C. J. (1834) *Ann. d. Physik. u. Chemie.* Vol 32.

Fullbrook, P. D. (1984) The enzymic production of glucose syrup, in: *Glucose Syrups: Science and Technology* (Dziedzic, S. Z., Kearsley, M. W., Eds.), London: Elsevier Applied Science Publishers, 65–115.

Gallant, D. J., Bouchet, B. (1986) Ultrastructure of maize starch granules – a review. *J. Food Microstruct.* **5**, 141–155.

Gallant, D. J., Bouchet, B., Buléon, A., Perez, S. (1992) Physical characteristics of starch granules and susceptibility to enzymatic degradation, *Eur. J. Clin. Nutr.* **46**, S3–S16.

Gallant, D. J., Bouchet, B., Baldwin, P. M. (1997) Microscopy of starch: evidence of a new level of granule organisation, *Carbohydr. Polym.* **32**, 177–191.

Galliard, T., Bowler, P. (1987a) Morphology and composition of starch, in: *Starch: Properties and Potential* (Galliard, T., Ed.), Chichester: John Wiley & Sons, 16–24.

Galliard, T., Bowler, P. (1987b) Morphology and composition of starch, *Crit. Rep. Appl. Chem.* **13**, 55–78.

Geddes, R., Greenwood, C. T., MacKenzie, S. (1965) Studies on the biosynthesis of starch granules. Part III. The properties of the components of starches from the growing potato tuber, *Carbohydr. Res.* **1**, 71–82.

Gérard, C., Planchot, V., Colonna, P., Bertoft, E. (2000) Relationship between branching density and crystalline structure of A- and B-type maize mutant starches, *Carbohydr. Res.* **326**, 130–144.

Gernat, C., Radosta, S., Damaschun, G., Schierbaum, F. (1990) Supramolecular structure of legume starches revealed by X-ray scattering, *Starch/Stärke* **42**, 175–178.

Gidley, M. J. (1987) Factors affecting the crystalline type (A-C) of native starches and model compounds: a rationalisation of observed effects in terms of polymorphic structures, *Carbohydr. Res.* **161**, 301–304.

Gidley, M. J. (1989) Molecular mechanisms underlying amylose aggregation and gelation. *Macromolecules*, **22**, 351–358.

Gidley, M. J. (1992) High resolution solid-state NMR of food materials, *Trends Food Sci. Technol.* **3**, 231–236.

Gidley, M. J., Bociek, S. M. (1985) Molecular organisation in starches: a ^{13}C CP/MAS NMR study, *J. Am. Chem. Soc.* **107**, 7040–7044.

Gidley, M. J., Bociek, S. M. (1988) ^{13}C CP/MAS NMR studies on amylose inclusion complexes, cyclodextrins and the amorphous phase of starch granules: relationship between glycosidic linkage conformation and solid state ^{13}C chemical shifts, *J. Am. Chem. Soc.* **110**, 3820–3829.

Gidley, M. J., Bulpin, P. V. (1987) Crystallisation of malto-oligosaccharides as models of the crystalline forms of starch: minimum chain-length requirement for the formation of double helices, *Carbohydr. Res.* **161**, 291–300.

Gidley, M. J., Bulpin, P. V. (1989) Aggregation of amylose in aqueous systems; the effect of chain length on phase behaviour and aggregation kinetics, *Macromolecules* **22**, 341–346.

Gidley, M. J., Bulpin, P. V., Kay, S. (1986) Effect of chain length on amylose retrogradation, in: *Gums and Stabilisers for the Food Industry* (Phillips, G. O., Wedlock, D. J., Williams, P. A., Eds.), London: Elsevier Science, 67–176.

Godet, M. C., Tran, V., Delage, M. M., Buleon, A. (1993) Molecular modelling of the specific interactions involved in the amylose complexation by fatty acids, *Int. J. Biol. Macromol.* **15**, 11–16.

Godet, M. C., Bizot, H., Buléon, A. (1995) Crystallization of amylose-fatty acid complexes prepared with different amylose chain lengths, *Carbohydr. Polym.* **27**, 47–52.

Greenwell, P., Schofield, J. D. (1986) A starch granule protein associated with endosperm softness in wheat, *Cereal Chem.* **63**, 379–380.

Haralampu, S. G. (2000) Resistant starch – a review of the physical properties and biological impact of RS3, *Carbohydr. Polym.* **41**, 285–292.

Hedges, A. R. (1992) Cyclodextrin: production, properties and applications, in: *Starch Hydrolysis Products, Worldwide Technology, Production, and Applications* (Schenck, F. W., Hebeda, R. E., Eds.), New York: VCH, 319–333.

Hizukuri, S. (1985) Relationship between the distribution of the chain length of amylopectin and the crystalline structure of starch granules, *Carbohydr. Res.* **141**, 295–305.

Hizukuri, S. (1986) Polymodal distribution of the chain lengths of amylopectin and its significance, *Carbohydr. Res.* **147**, 342–347.

Hizukuri, S. (1988) Recent advances in molecular structures of starch, *Denpun Kagaku* **35**, 185–198.

Hizukuri, S. (1993) Towards an understanding of the fine structure of starch molecules, *Denpun Kagaku* **40**, 133–147.

Hizukuri, S. (1996) Starch: analytical aspects, in: *Carbohydrates in Food* (Eliasson, A.-C., Ed.), New York: Marcel Dekker, 347–429.

Hizukuri, S., Maehara, Y. (1990) Fine structure of wheat amylopectin: the mode of A to B chain binding, *Carbohydr. Res.* **206**, 145–159.

Hizukuri, S., Takeda, Y., Yasuda, M., Suzuki, A. (1981) Multi branched nature of amylose and the action of debranching enzymes, *Carbohydr. Res.* **94**, 205–213.

Hizukuri, S., Takeda, Y., Juliano, B. O. (1995) Structural changes of non-waxy starch during development of rice grains, in: *Progress in Plant Polymeric Carbohydrate Research* (Meuser, F., Manners, D. J., Seibel, W., Eds.), Hamburg: B. Behr's Verlag, 38–41.

Holm, J., Björk, I, Drew, A., Asp, N.-G. (1986) A rapid method for the analysis of starch, *Starch/Stärke* **38**, 224–226.

Hoover, R. (1998) Starch–lipid interactions, in: *Polysaccharide Association Structures in Food*, (Walter, R. H., Ed.), New York: Marcel Dekker, 227–256.

Hoover, R. (2000) Acid-treated starches, *Food Rev. Int.* **16**, 369–392.

Hoover, R., Vasanthan, T. (1994) Effect of heat-moisture treatment on the structure and physicochemical properties of cereal, legume and tuber starches, *Carbohydr. Res.* **252**, 33–53.

Hoseney, R. C., Atwell, W. A., Lineback, D. R. (1977) Scanning electron microscopy of starch isolated from baked products, *Cereal Foods World* **22**, 56–61.

Hoseney, R. C., Zeleznak, K. J., Yost, D. A. (1986) A note on the gelatinisation of starch, *Starch/Stärke* **38**, 407–409.

Howling, D. (1992) Glucose syrup: production, properties and applications, in: *Starch Hydrolysis Products, Worldwide Technology, Production, and Applications* (Schenck, F. W., Hebeda, R. E., Eds.), New York: VCH, 277–317.

Huber, K. C., BeMiller, J. N. (1997) Visualization of channels and cavities of corn and sorghum granules, *Cereal Chem.* **74**, 537–541.

Huber, K. C., BeMiller, J. N. (2000) Channels of maize and sorghum starch granules, *Carbohydr. Polym.* **41**, 269–276.

Huber, K. C., BeMiller, J. N. (2001) Location of sites of reaction within starch granules, *Cereal Chem.* **78**, 173–180.

Imberty, A. and Pérez, S. (1988) A revisit to the 3-dimensional structure of B-type starch, *Biopolymers* **27**, 1205–1221.

Imberty, A., Chanzy, H., Pérez, S., Buléon, A., Tran, V. J. (1987) Three-dimensional structure analysis of the crystalline moiety of A-starch, *Food Hydrocolloid.* **1**, 455–459.

Imberty, A., Chanzy, H., Pérez, S., Buléon, A., Tran, V. J. (1988a) The double-helical nature of the crystalline part of A-starch, *J. Mol. Biol.* **201**, 365–378.

Imberty, A., Chanzy, H., Pérez, S., Buléon, A., Tran, V. J. (1988b) New 3-dimensional structure for A-type starch, *Macromolecules* **20**, 2634–2636.

Imberty, A., Buléon, A., Tran, V., Pérez, S. (1991) Recent advances in knowledge of starch structure, *Starch/Stärke* **43**, 375–384.

Jackson, D. S., Shandera, D. L. (1995) Corn wet milling: separation chemistry and technology, in: *Advances in Food and Nutrition Research* (Kinsella, J. E., Taylor, S. L., Eds.), San Diego, CA: Academic Press, Vol. 38, 271–300.

Jacobs, H., Delcour, J. A. (1998) Hydrothermal modifications of granular starch, with retention of the granular structure: a review, *J. Agric. Food Chem.* **46**, 2895–2905.

Jacobson, M. R., Obanni, M., BeMiller, J. N. (1997) Retrogradation of starches from different botanical sources, *Cereal Chem.* **74**, 511–518.

Jane, J.-L., Robyt, J. F. (1984) structure studies of amylose V-complexes and retrograded amylose by action of alpha amylases, and a new method for preparing amylodextrins, *Carbohydr. Res.* **132**, 105–118.

Jane, J., Chen, Y. Y., Lee, L. F., McPherson, A. E., Wong, K. S., Radosavljevic, M., Kasemsuwan, T. (1999) Effects of amylopectin branch chain length and amylose content on the gelatinisation and pasting properties of starch, *Cereal Chem.* **76**, 629–637.

Jang, J. K., Pyun, Y. R. (1997) Effect of moisture level on the crystallinity of wheat starch aged at different temperatures, *Starch/Stärke* **49**, 272–277.

Jenkins, P.J., Donald, A. M. (1996) Application of small-angle neutron scattering to the study of the structure of starch granules, *Polymer* **37**, 5559–5568.

Jenkins, P.J., Donald, A. M. (1998) Gelatinisation of starch: a combined SAXS/ WAXS/ DSC and SANS study, *Carbohydr. Res.* **308**, 133–147.

Jenkins, P. J., Cameron, R. E., Donald, A. M. (1993) A universal feature in the structure of starch granules from different botanical sources, *Starch/Stärke* **45**, 417–420.

Jenkins, P. J., Cameron, R. E., Donald, A. M., Bras, W., Derbyshire, G. E., Mont, G. R., Ryan, A. J. (1994) In-situ simultaneous small and wide angle X-ray scattering – a new technique to study starch gelatinisation, *J. Polym. Sci. Part B Polym. Phys.* **32**, 1579–1583.

Kainuma, K., French, D. (1972) Naegeli amylodextrin and its relationship to starch granule structure. II. Role of water in crystallisation of B-starch, *Biopolymers* **11**, 2241–2250.

Kang, M. Y., Sugimoto, Y., Sakamoto, S., Fuwa, H. (1985) Developmental changes in the amylose content of endosperm starch of barley (*Hordeum vulgare* L.) during the grain filling period after anthesis, *Agric. Biol. Chem.* **49**, 3463–3466.

Karkalas, J. (1967) Modern methods of purification of starch hydrolysates, *Starch/Stärke* **19**, 338–335.

Karkalas, J. (1985) An improved enzymic method for the determination of native and modified starch, *J. Sci. Food Agric.* **36**, 1019–1027.

Karkalas, J. (1991) Automated enzymic determination of starch by flow injection analysis, *J. Cereal Sci.* **14**, 279–286.

Karkalas, J., Raphaelides, S. (1986) Quantitative aspects of amylose–lipid interactions, *Carbohydr. Res.* **157**, 215–234.

Karkalas, J., Tester, R. F. (1992) Continuous enzymic determination of α-glucans in eluates from gel-chromatographic columns, *J. Cereal. Sci.* **15**, 175–180.

Karkalas, J., Tester, R. F., Morrison, W. R. (1992) Properties of damaged starch granules. I. Comparison of a new micromethod for the enzymic determination of damaged starch with the standard AACC and Farrand methods, *J. Cereal Sci.* **16**, 237–251.

Karkalas, J., Song, M., Morrison, W. R., Pethrick, R. A. (1995) Some factors determining the thermal properties of amylose inclusion complexes with fatty acids, *Carbohydr. Res.* **268**, 233–247.

Kawabata, A., Takase, N., Miyoshi, E., Sawayama, S., Kimura, T., Kudo, K. (1994) Microscopic observation and X-ray diffractometry of heat-moisture treated starch granules, *Starch/Stärke* **46**, 463–469.

Kearsley, M. W., Dziedzic, S. Z. (Eds.) (1995) *Handbook of Starch Hydrolysis Products and their Derivatives.* London/New York: Blackie Academic and Professional.

Keetels, C. J. A. M., van Vliet, T., Walstra, P. (1996a) Gelation and retrogradation of concentrated starch systems: 1. Gelation, *Food Hydrocolloid.* **10**, 343–353.

Keetels, C. J. A. M., van Vliet, T., Walstra, P. (1996b) Gelation and retrogradation of concentrated starch systems: 1. Retrogradation, *Food Hydrocolloid.* **10**, 355–362.

Kennedy, J. F., Knill, C. J., Taylor, D. W. (1995) Maltodextrins, in: *Handbook of Starch Hydrolysis Products and their Derivatives* (Kearsley, M. W., Dziedzic, S. Z., Eds.), London: Blackie Academic and Professional, 65–82.

Kent, N. L., Evers, A. D. (1994) *Kent's Technology of Cereals*, 4th edition. Oxford: Pergamon Press.

Kim, J.-O., Kim, W.-S., Shin, M.-S. (1997) A comparative study on retrogradation of rice starch gels by DSC, X-ray and α-amylase methods, *Starch/Stärke* **49**, 71–75.

Kitamura, S., Yoneda, S., Kuge, T. (1984) Study on the retrogradation of starch. I. Particle size and its distribution of amylose retrograded from aqueous solutions, *Carbohydr. Polym.* **4**, 127–136.

Kleczkowski, L. A. (1996) Back to the drawing board: redefining starch synthesis in cereals, *Trends Plant Sci.* **1**, 363–364.

Klucinec, J. D., Thompson, D. B. (1999) Amylose and amylopectin interact in retrogradation of dispersed high-amylose starches, *Cereal Chem.* **76**, 282–291.

Knightly, W. H. (1996) Surfactants, in: *Baked Goods Freshness: Technology, Evaluation and Inhibition of Staling* (Hebeda, R. E., Zobel, H. F., Eds.), New York: Marcel Dekker, 65–103.

Knutson, C. A. (1990) Annealing of maize starches at elevated temperatures, *Cereal Chem.* **67**, 376–384.

Kraemer, H. (1907) *A Textbook of Botany and Pharmacognosy.* London, Philadelphia, PA: Lippincott.

Kuge, T., Kitamura, A. (1985) Annealing of starch granules – warm water treatment and heat-moisture treatment, *Denpun Kagaku* **32**, 65–83.

Lai, V. M.-F., Lu, S., Lii, C.-Y. (2000) Molecular characteristics influencing retrogradation kinetics of rice amylopectins, *Cereal Chem.* **77**, 272–278.

Le Bail, P., Buléon, A, Colonna, P., Bizot, H. (1997) Structural and polymorphic transitions of amylose induced by water and temperature changes, in: *Starch Structure and Functionality* (Frazier, P. J., Donald, A. M., Richmond, P., Eds.), Cambridge: Royal Society of Chemistry, 51–58.

Le Bot, Y., Gouy, P. A. (1995) Polyols from starch, in: *Handbook of Starch Hydrolysis Products and their Derivatives* (Kearsley, M. W., Dziedzic, S. Z., Eds.), London: Blackie Academic and Professional, 155–177.

Lee, S. C., Prosky, L. (1995) International survey on dietary fibre – definition, analysis and reference materials, *J. AOAC Int.* **78**, 22–36.

Leloup, V. M., Colonna, P., Ring, S. G., Roberts, K., Wells, B. (1992) Microstructure of amylose gels, *Carbohydr. Polym.* **18**, 189–197.

Lineback, D. R. (1984) The starch granule – organisation and properties. *Baker's Dig.* **58**, 16–21.

Lineback, D. L. (1986) Current concepts of starch structure and its impact on properties, *Denpun Kagaku* **33**, 80–88.

Liu, Q., Thompson, D. B. (1998a) Retrogradation of $du\ wx$ and $su2\ wx$ maize starches after different gelatinisation heat treatments, *Cereal Chem.* **75**, 868–874.

Liu, Q., Thompson, D. B. (1998b) Effects of moisture content and different gelatinisation heating temperatures on retrogradation of waxy-type maize starches, *Carbohydr. Res.* **314**, 221–235.

Lorenz, K., Kulp, K. (1982) Cereal and root starch modification by heat-moisture treatment I. Physico-chemical properties, *Starch/Stärke* **34**, 50–54.

Lu, T.-J., Jane, J.-L., Keeling, P. L. (1997) Temperature effect on retrogradation rate and crystalline structure of amylose, *Carbohydr. Polym.* **33**, 19–26.

Lund, D. B., Wirakartakusumah, M. (1984) A model for starch gelatinisation phenomena, in: *Engineering and Food, Volume 1, Engineering Sciences for the Food Industry* (McKenna, B. M., Ed.), London/New York: Elsevier Applied Science Publishers, 425–434.

Lyne, F. A. (1976) Determination of starch in various products, in: *Examination and Analysis of Starch and Starch Products* (Radley, J. A., Ed.), London: Applied Science Publishers, 167–188.

Manners, D. J. (1985) Structural analysis of starch components by debranching enzymes, in: *New Approaches to Research on Cereal Carbohydrates* (Hill, R. D., Munck, L., Eds.), Amsterdam: Elsevier Science Publishers, 45–54.

Martin, C., Smith, A. (1995) Starch biosynthesis, *Plant Cell* **7**, 971–985.

McCleary, B. V., Solah, V., Gibson, T. S. (1994) Quantitative measurement of total starch in cereal flours and products, *J. Cereal Sci.* **20**, 51–58.

McDonald, A. M. L., Stark, J. R., Morrison, W. R., Ellis, R. P. (1991) The composition of starch granules from developing barley genotypes, *J. Cereal Sci.* **13**, 93–112.

McNair, J. B. (1930) *The differential analysis of starches.* Chicago Field Museum Of Natural History, Botanical Series 9.

Meyer, A. (1895) *Untersuchungen uber die Die Stärkekörner. Wesen und Lebensgeschichte der Stärkekörner der hoheren Pflanzen.* Jena: Gustav Fischer.

Morgan, K. R., Furneaux, R. H., Larsen, N. G. (1995) Solid-state NMR studies on the structure of starch granules, *Carbohydr. Res.* **276**, 387–399.

Morikawa, K., Nishinari, K. (2000) Effects of concentration dependence of retrogradation behaviour of dispersions for native and chemically modified potato starch, *Food Hydrocolloid.* **14**, 395–401.

Morris, E. R. (1984) Rheology of hydrocolloids, in: *Gums and Stabilisers for the Food Industry 2 – Application of Hydrocolloids* (Phillips, G. O., Wedlock, D. J., Williams, P. A., Eds.), Oxford: Pergamon Press, 57–78.

Morris, V. G. (1990) Starch gelation and retrogradation, *Trends Food Sci. Technol.*, 2–6.

Morrison, W. R. (1985) Lipids in cereal starches, in: *New Approaches to Research on Cereal Carbohydrates* (Hill, R. D., Munck, L., Eds.), Amsterdam: Elsevier Science Publishers, 61–70.

Morrison, W. R. (1988) Lipids in cereal starches: a review, *J. Cereal Sci.* **8**, 1–15.

Morrison, W. R. (1989) Uniqueness of wheat starch, in: *Wheat is Unique* (Pomeranz, Y., Ed.), St. Paul: AACC, 193–214.

Morrison, W. R. (1992) Analysis of cereal starches, in: *Seed Analysis* (Linskens, H.-F., Jackson, J. F., Eds.), Berlin: Springer-Verlag, 199–215.

Morrison, W. R. (1993) Cereal starch granule development and composition, in: *Seed Storage Compounds: Biosynthesis, Interactions and Manipulation* (Shewry, P. R., Stobart, K., Eds.), Oxford: Oxford Science Publications, 175–190.

Morrison, W. R. (1995) Starch lipids and how they relate to starch granule structure and functionality, *Cereal Food. World* **40**, 437–446.

Morrison, W. R., Gadan, H. (1987) The amylose and lipid contents of starch granules in developing wheat endosperm, *J. Cereal Sci.* **5**, 263–275.

Morrison, W. R., Karkalas, J. (1990) Starch, in: *Methods in Plant Biochemistry* (Dey, P. M., Ed.), London: Academic Press, Vol. 2, 323–352.

Morrison, W. R., Tester, R. F. (1994) Properties of damaged starch granules. IV. Composition of ball-milled wheat starches and of fractions obtained on hydration, *J. Cereal Sci.* **20**, 69–77.

Morrison, W. R., Law, R. V., Snape, C. E. (1993a) Evidence for inclusion complexes of lipids with V-amylose in maize, rice and oat starches, *J. Cereal Sci.* **18**, 107–109.

Morrison, W. R., Tester, R. F., Gidley, M. J., Karkalas, J. (1993b) Resistance to acid hydrolysis of lipid complexed amylose and free amylose in lintnerised waxy and non-waxy barley starches, *Carbohydr. Res.* **245**, 289–302.

Morrison, W. R., Tester, R. F., Snape, C. W., Law, R., Gidley, M. J. (1993c) Swelling and gelatinisation of cereal starches. IV. Effects of lipid-complexed amylose and free amylose in waxy and normal barley starches, *Cereal Chem.* **70**, 385–391.

Morrison, W. R., Tester, R. F., Gidley, M. J. (1994) Properties of damaged starch granules. II. Crystallisation, molecular order and gelatinisation of ball-milled starches, *J. Cereal Sci.* **19**, 209–217.

Mua, J. P., Jackson, D. S. (1998) Retrogradation and gel textural attributes of corn starch amylose and amylopectin fractions, *J. Cereal Sci.* **27**, 157–166.

Muhr, A. H., Blanshard, J. M. V., Bates, D. R. (1984) The effect of lintnerisation on wheat and potato starch granules, *Carbohydr. Polym.* **4**, 399–425.

Mulvihill, P. J. (1992) Crystalline and liquid dextrose products: production, properties and applications, in: *Starch Hydrolysis Products, Worldwide Technology, Production, and Applications* (Schenck, F. W., Hebeda, R. E., Eds.), New York: VCH, 23–44.

Muter, J. (1906) Organic Materia Medica, in: *Select Methods in Food Analysis* (Leffman, H., Beam, W., Eds.) Philadelphia, PA: P. Blakiston.

Nägeli, C. W. (1858) *Die Stärkekörner, Morphologische, Physiologische, Chemisch-physicalisch und systematisch-botanische Monographie.* Zürich: F. Schuithess.

Nara, S. (1978) On the relationship between regain and crystallinity of starch, *Starch/Stärke* **30**, 183–186.

Nara, S., Mori, A., Komiya, T. (1978) Study on relative crystallinity of moist potato starch, *Starch/Stärke* **30**, 111–114.

National Statistics (1998) *Annual Abstract of Statistics*, no. 134, London: The Stationery Office.

National Statistics (2000) *Products Sales and Trade*, PRA 15620, Starch and Starch Products. London: The Stationery Office (also http://www.statistics.gov.uk).

Nelson, O., Pan, D. (1995) Starch synthesis in maize endosperms, *Annu. Rev. Plant Mol. Biol.* **46**, 475–496.

Ng, K-Y., Duvick, S. A., White, P. J. (1997) Thermal properties of starch from selected maize (*Zea mays* L.) mutants during development, *Cereal Chem.* **74**, 288–292.

Nielsen, T. H., Wischmann, B., Enevoldsen, K., Moller, B. L. (1994) Starch phosphorylation in potato-tubers proceeds concurrently with de-novo biosynthesis of starch, *Plant Physiol.* **105**, 111–117.

Noda, T., Takahata, Y., Nagata, T. (1992) Developmental changes in properties of sweet potato starches, *Starch/Stärke* **44**, 405–409.

Noda, T., Takahata, Y., Sato, T., Ikoma, H., Mochida, H. (1997) Combined effects of planting and harvesting dates on starch properties of sweet potato roots, *Carbohydr. Polym.* **33**, 169–176.

Noel, T. R., Ring, S. G., Whittam, M. A. (1993) Physical properties of starch products. Structure and function, in: *Food Colloids and Polymers* (Dickinson, E., Walstra, P., Eds.), Cambridge: Royal Society of Chemistry, 126–135.

Norman, B. E. (1981) New developments in starch syrup technology, in: *Enzymes and Food Processing* (Birch, G. G., Blakebrough, N., Parker, K. J., Eds.), London: Applied Science Publishers, 15–50.

Oda, S., Schofield, J. D. (1997) Characterisation of friabilin polypeptides, *J. Cereal Sci.* **26**, 29–36.

Ohtani, T., Yoshino, T., Hagiwara, S., Maekawa, T. (2000a) High-resolution imaging of starch granule structure using atomic force microscopy, *Starch/Stärke* **52**, 150–153.

Ohtani, T., Yoshino, T., Ushiki, T., Hagiwara, S., Maekawa, T. (2000b) Structure of rice starch granules in nanometer scale as revealed by atomic force microscopy, *J. Electron Microsc.* **49**, 487–489.

Okechukwu, P. E., Rao, A. M. (1998) Rheology of structured polysaccharide food systems: starch and pectin, in: *Polysaccharide Association Structures in Food* (Walter, R. H., Ed.), New York: Marcel Dekker, 289–328.

Olsen, H. S. (1995) Enzymatic production of glucose syrups, in: *Handbook of Starch Hydrolysis Products and their Derivatives* (Kearsley, M. W., Dziedzic, S. Z., Eds.), London: Blackie Academic & Professional, 26–64.

Oostergetel, G. T., van Bruggen, E. F. J. (1987) The structure of starch: electron microscopy and electron diffraction, *Food Hydrocolloid.* **1**, 527–528.

Oostergetel, G. T., van Bruggen, E. F. J. (1989) On the origin of a low angle spacing in starch, *Starch/Stärke* **41**, 331–335.

Oostergetel, G. T., van Bruggen, E. F. J. (1993) The crystalline domains in potato starch granules are arranged in a helical fashion, *Carbohydr. Polym.* **21**, 7–12.

Orman, B. A., Schumann, R. A. (1991) Comparison of near infra red spectroscopy calibration methods for the prediction of protein, oil and starch in maize grain, *J. Agric. Food Chem.* **39**, 883–886.

Palmer, T. J. (1982) Nutritive sweeteners from starch, in: *Nutritive Sweeteners* (Birch, G. G., Parker, K. J., Eds.), London: Applied Science Publishers, 83–108.

Parker, M. L. (1985) The relationship between A-type and B-type starch granules in developing endosperm of wheat, *J. Cereal Sci.* **3**, 271–278.

Paton, D. (1979) Oat starches: some recent developments, *Starch/Stärke* **31**, 184–187.

Perera, C., Hoover, R. (1999) Influence of hydroxypropylation on retrogradation properties of native, defatted and heat-moisture treated potato starches, *Food Chem.* **64**, 361–375.

Pfannemuller, B. (1986) Models for the structure and properties of starch, *Starch/Stärke* **38**, 401–407.

Prosky, L. (2000a) What is dietary fiber? *J. AOAC Int.* **83**, 985–987.

Prosky, L. (2000b) When is dietary fiber considered a functional food? *Biofactors* **12**, 289–297.

Prosky, L., Asp, N.-G., Furda, I., DeVries, J. W., Schweizer, T. F., Harland, B. F. (1984) Determination of total dietary fiber in foods, food-products, and total diets: interlaboratory study, *J. Assoc. Off. Anal. Chem.* **67**, 1044–1052.

Prosky, L., Asp, N.-G., Furda, I., DeVries, J. W., Schweizer, T. F., Harland, B. F. (1985) Determination of total dietary fiber in foods and food-products: interlaboratory study, *J. Assoc. Off. Anal. Chem.* **68**, 677–679.

Prosky, L., Asp, N.-G., Furda, I., DeVries, J. W., Schweizer, T. F. (1988) Determination of insoluble, soluble and total dietary fiber in foods and food-products: interlaboratory study, *J. Assoc. Off. Anal. Chem.* **71**, 1017–1023.

Quigley, M. E., Hudson, G. J., Englyst, H. N. (1999) Determination of resistant short-chain carbohydrates (non-digestible oligosaccharides) using gas-liquid chromatography, *Food Chem.* **65**, 381–390.

Radley, J. A. (Ed.) (1968) *Starch and its Derivatives*. London: Chapman & Hall.

Radley, J. A. (Ed.) (1976a) *Starch Production Technology*. London: Applied Science Publishers.

Radley, J. A. (Ed.) (1976b) *Industrial Uses of Starch and its Derivatives*. London: Applied Science Publishers.

Rao, V. S. R., Qasba, P. K., Balaji, P. V., Chandrasekaran, R. (1998) *Conformation of Carbohydrates*, Amsterdam: Harwood Academic Publishers, 191–221.

Raphaelides, S. N. (1992) Flow behaviour of starch-fatty acid systems in solution, *Lebensmittelwiss. Technol.* **25**, 95–101.

Raphaelides, S. N. (1993) Rheological studies of starch-fatty acid gels, *Food Hydrocolloid.* **7**, 479–495.

Raphaelides, S., Karkalas, J. (1988) Thermal dissociation of amylose–fatty acid complexes, *Carbohydr. Res.* **172**, 65–82.

Rappenecker, G., Zugenmaier, P. (1981) Detailed refinement of the crystal structure of V_h-amylose, *Carbohydr. Res.* **89**, 11–19.

Reichert, E. T. (1913) The Differentiation and Specificity of Starches in Relation to Genera, Species, Etc. Washington, DC: Carniegie Institute.

Ring, S. G., Gee, J. M., Whittam, M., Orford, P., Johnson, I. T. (1988) Resistant starch: its chemical form in foodstuffs and effect on digestibility *in vitro*, *Food Chem.* **28**, 97–109.

Robin, J. P., Mercier, D., Duprat, F., Charbonniere, R., Guilbot, A. (1975) Amidons Lintnérisés. Etudes chromatographiques et enzymatiques des résidus insolubles provenant de l'hydrolyse chlorhydrique d'amidons de céréales, en particulier de mais cireux, *Starch/Stärke* **27**, 36–45.

Russell, P. L. (1987) Gelatinisation of starches of different amylose/amylopectin content. A study by differential scanning calorimetry, *J. Cereal Sci.* **6**, 133–145.

Sair, L. (1967) Heat-moisture treatment of starch, *Cereal Chem.* **44**, 8–26.

Sarko, A., Wu, H-C. H. (1978) The crystal structures of A-, B- and C-polymorphs of amylose and starch, *Starch/Stärke* **30**, 73–78.

Schenck, F. W., Hebeda, R. E. (1992) Starch hydrolysis products: an introduction and history, in: *Starch Hydrolysis Products, Worldwide Technology, Production, and Applications* (Schenck, F. W., Hebeda, R. E., Eds.), New York: VCH, 1–21.

Schleiden, M. J. (1849) *Principles of Scientific Botany.* London: Longman.

Schofield, J. D. (1994) Wheat proteins: structure and functionality in milling and breadmaking, in: *Wheat Production, Composition and Utilisation* (Bushuk, W., Rasper, V., Eds.), Glasgow: Blackie Academic and Professional, 73–106.

Seib, P. A. (1994) Wheat starch: Isolation, structure and properties, *Oyo Tshitsu Kagaku* **41**, 49–69.

Seneviratne, H. D., Biliaderis, C. G. (1991) Action of α-amylases on amylose-lipid superstructures, *J. Cereal Sci.* **13**, 129–143.

Shi, Y.-C., Seib, P. A. (1992) The structure of 4 waxy starches related to gelatinisation and retrogradation, *Carbohydr. Res.* **227**, 131–145.

Shi, Y.-C., Capitani, T., Trzasko, P., Jeffcoat, R. (1998) Molecular structure of a low amylopectin starch and other high amylose maize starches, *J. Cereal Sci.* **27**, 289–299.

Sievert, D., Pomeranz, Y. (1989) Enzyme resistant starch. I. Characterisation by enzymatic, thermoanalytical and microscopic methods, *Cereal Chem.* **66**, 342–347.

Singh, N., Eckhoff, S. R. (1996) Wet milling of corn – a review of laboratory scale and pilot plant scale procedures, *Cereal Chem.* **73**, 659–667.

Smith, A. M. (1993) Starch synthesis in peas, in: *Seed Storage Compounds – Biosynthesis, Interactions and Manipulation* (Shewry, P. R., Stobart, K., Eds.), Oxford: Oxford University Press, 210–223.

Smith, A. M. (1999) Making starch. *Curr. Opin. Plant Biol.* **2**, 223–229.

Smith, A. M., Denyer, K., Martin, C. (1997) The synthesis of the starch granule. *Annu. Rev. Plant Mol. Biol.* **48**, 65–87.

Sriroth, K., Santisopasri, V., Petchalanuwat, C., Kurotjanawong, K., Piyachomkwan, K. Oates, C. G. (1999) Cassava starch granule structure–function properties: influence of time and conditions of harvest on four cultivars of cassava starch, *Carbohydr. Polym.* **38**, 161–170.

Sterling, C. (1962) A low angle spacing in starch, *J. Polym. Sci.* **56**, S10–S12.

Sterling, C. (1974) Fibrillar structure of starch, *Starch/Stärke* **26**, 105–110.

Stute, R. (1992) Hydrothermal modification of starches: the difference between annealing and heat-moisture treatment, *Starch/Stärke* **44**, 205–214.

Sugimoto, Y., Yamashita, Y., Hori, I., Abe, K., Fuwa, H. (1995) Developmental changes in the properties of potato *(Solanum tuberosum* L.) starches, *Oyo Toshitsu Kagaku* **42**, 345–353.

Swinkels, J. J. M. (1985) Composition and properties of commercial native starches, *Starch/Stärke* **37**, 1–5.

Tako, M., Hizukuri, S. (2000) Retrogradation mechanism of rice starch, *Cereal Chem.* **77**, 473–477.

Teague, W. M., Brumm, P. J. (1992) Commercial enzymes for starch hydrolysis products, in: *Starch Hydrolysis Products, Worldwide Technology, Production, and Applications* (Schenck, F. W., Hebeda, R. E., Eds.), New York: VCH, 45–77.

Tegge, G. (1984) Glucose syrups – the raw material, in: *Glucose Syrups: Science and Technology* (Dziedzic, S. Z., Kearsley, M. W., Eds.), London: Elsevier Applied Science Publishers, 9–64.

Tester, R. F. (1997a) Influence of growth conditions on barley starch properties, *Int. J. Biol. Macromol.* 21, 37–45.

Tester, R. F. (1997b) Properties of damaged starch granules: composition and swelling properties of maize, rice, pea and potato starch fractions in water at various temperatures, *Food Hydrocolloid.* 11, 293–301.

Tester, R. F., Debon, S. J. J. (2000) Annealing of starch – a review. *Int. J. Biol. Macromol.* 27, 1–12.

Tester, R. F., Karkalas, J. (1996) Swelling and gelatinisation of oat starches, *Cereal Chem.* 73, 271–277.

Tester, R. F., Morrison, W. R. (1990a) Swelling and gelatinisation of cereal starches. I. Effects of amylopectin, amylose and lipids, *Cereal Chem.* 67, 551–557.

Tester, R. F., Morrison, W. R. (1990b) Swelling and gelatinisation of cereal starches. II. Waxy rice starches, *Cereal Chem.* 67, 558–563.

Tester, R. F., Morrison, W. R. (1992) Swelling and gelatinisation of cereal starches. III. Some properties of waxy and normal non-waxy barley starches, *Cereal Chem.* 69, 645–658.

Tester, R. F., Morrison, W. R. (1993) Swelling and gelatinisation of cereal starches. VI. Starches from Waxy Hector and Hector barleys at four stages of grain development, *J. Cereal Sci.* 17, 11–18.

Tester, R. F., Morrison, W. R. (1994) Properties of damaged starch granules. V. Composition and swelling of fractions of wheat starch in water at various temperatures, *J. Cereal Sci.* 20, 175–181.

Tester, R. F., Sommerville, M. D. (2000) Swelling and enzymatic hydrolysis of starch in low water systems, *J. Cereal Sci.* 33, 193–203.

Tester, R. F., South, J. B., Morrison, W. R., Ellis, R. P. (1991) The effects of ambient temperature during the grain-filling period on the composition and properties of starch from four barley genotypes, *J. Cereal Sci.* 13, 113–127.

Tester, R. F., Morrison, W. R., Gidley, M. J., Kirkland, M., Karkalas, J. (1994) Properties of damaged starch granules. III. Particle size analysis and microscopy, *J. Cereal Sci.* 20, 59–67.

Tester, R. F., Morrison, W. R., Ellis, R. P., Piggott, J. R., Batts, G. R., Wheeler, T. R., Morison, J. I. L., Hadley, P., Ledward, D. A. (1995) The effect of elevated growth temperature and carbon dioxide levels on the physico-chemical properties of wheat starch, *J. Cereal Sci.* 22, 63–71.

Tester, R. F., Debon, S. J. J., Karkalas, J. (1998) Annealing of wheat starch. *J. Cereal Sci.* 28, 259–272.

Tester, R. F., Debon, S. J. J., Davies, H. V., Gidley, M. J. (1999) Effect of temperature on the synthesis, composition and physical properties of potato starch, *J. Sci. Food Agric.* 79, 2045–2051.

Tester, R. F., Debon, S. J. J., Sommerville, M. D. (2000) Annealing of maize starch, *Carbohydr. Polym.* 42, 287–299.

Thomas, D. J., Atwell, W. A. (1999) *Starches*. St Paul: AACC.

Thompson, D. B. (2000) Strategies for the manufacture of resistant starch, *Trends Food Sci. Technol.* 11, 245–253.

Tomasik, P., Schilling, C. H. (1998a) Complexes of amylose with inorganic guests, in: *Advances in Carbohydrate Chemistry and Biochemistry*, (Horton, D., Ed.), San Diego, CA: Academic Press, Vol. 53, 263–343.

Tomasik, P., Schilling, C. H. (1998b) Complexes of amylose with organic guests, in: *Advances in Carbohydrate Chemistry and Biochemistry*, (Horton, D., Ed.), San Diego, CA: Academic Press, Vol. 53, 345–426.

Van Leeuwenhoek, A. (1719) *Epistolae physiologicae super compluribus naturae arcanis: ubi variorum animalium atque plantarum fabrica, conformatio, proprietates atque operationes, novis & hactenus inobservatis experimentis illustrantur & oculis exhibentur; item peculiares & hactenus incognitae rerum quarumdam qualitates explicantur: ut sequens pagina docet: hactenus numquam editae. Cum figuris aeneis, & indice locupletissimo*. Delphis: Apud Adrianum Beman.

Villareal, C. P., Hizukuri, S., Juliano, B. O. (1997) Amylopectin staling of cooked milled rices and properties of amylopectin and amylose, *Cereal Chem.* 74, 163–167.

Waigh, T. A., Donald, A. M., Heidelbach, F., Riekel, C., Gidley, M. J. (1999) Analysis of the native structure of starch granules with small angle X-ray microfocus scattering, *Biopolymers* 49, 91–105.

Waigh, T. A., Gidley, M. J., Komanshek, B. U., Donald, A. M. (2000) The phase transformations in starch during gelatinisation: a liquid crystalline approach, *Carbohydr. Res.* 328, 165–176.

Wang, L. Z., White, P. J. (1994) Structure and properties of amylose, amylopectin and intermediate material of oat starches, *Cereal Chem.* 71, 263–268.

Wang, T. L., Bogracheva, T. Y., Hedley, C. L. (1998) Starch: as simple as A, B, C? *J. Exp. Bot.* 49, 481–502.

Whistler, R. L., BeMiller, J. N., Pashall, E. F. (Eds.) (1984) *Starch: Chemistry and Technology*. Orlando/London: Academic Press.

White, J. S. (1992) Fructose syrup: production, properties and applications, in: *Starch Hydrolysis Products, Worldwide Technology, Production, and Applications* (Schenck, F. W., Hebeda, R. E., Eds.), New York: VCH, 177–199.

Whittam, M. A., Noel, T. R., Ring, S. G. (1990) Melting behaviour of A- and B-type starches, *Int. J. Biol. Macromol.* **12**, 359–362.

Winton, A. L. (1906) *The Microscopy of Vegetable Foods*. New York: John Wiley & Sons.

Wu, H. C. H., Sarko, A. (1978a) The double-helical molecular structure of crystalline B-amylose, *Carbohydr. Res.* **61**, 7–25.

Wu, H. C. H., Sarko, A. (1978b) The double-helical molecular structure of crystalline A-amylose, *Carbohydr. Res.* **61**, 27–40.

Wunderlich, B. (1990) The basis of thermal analysis, in: *Thermal Analysis*. New York: Academic Press, 37–78.

Wurzburg, O. B. (1995) Modified starches, in: *Food Polysaccharides and their Applications* (Stephen, A. M., Ed.), New York: Marcel Dekker, 67–97.

Yuan, R. C., Thompson, D. B. (1998) Rheological and thermal properties of aged starch pastes from three waxy maize genotypes, *Cereal Chem.* **75**, 117–123.

Zobel, H. F. (1988a) Starch crystal transformations and their industrial importance, *Starch/Stärke* **40**, 1–7.

Zobel, H. F. (1988b) Molecules to granules: A comprehensive starch review, *Starch/Stärke* **40**, 44–50.

Zobel, H. F. (1992) Starch granule structure, in: *Developments in Carbohydrate Chemistry* (Alexander, R. J., Zobel, H. F., Eds.), St Paul: AACC, 261–292.

Zobel, H. F., Kulp, K. (1996) The staling mechanism, in: *Baked Goods Freshness: Technology, Evaluation and Inhibition of Staling* (Hebeda, R. E., Zobel, H. F., Eds.), New York: Marcel Dekker, 1–64.

Zobel, H. F., Stephen, A. M. (1995) Starch: structure, analysis and application, in: *Food Polysaccharides and their Applications* (Stephen, A. M., Ed.), New York: Marcel Dekker, 19–66.

Zobel, H. F., Young, S. N., Rocca, L. A. (1988) Starch gelatinisation: an X-ray diffraction study, *Cereal Chem.* **65**, 443–446.

13
Xanthan

Dr. Karin Born[1], Dr. Virginie Langendorff[2], Dr. Patrick Boulenguer[3]
[1] DEGUSSA Texturant Systems France SAS, Research Center,
   F-50500 Baupte, France; Tel.: +33-2-33-713433; Fax: +33-2-33-713492;
   E-mail: karin.born@degussa.com
[2] DEGUSSA Texturant Systems France SAS, Research Center,
   F-50500 Baupte, France; Tel.: +33-2-33-713433; Fax: +33-2-33-713492;
   E-mail: virginie.langendorff@degussa.com
[3] DEGUSSA Texturant Systems France SAS, Research Center,
   F-50500 Baupte, France; Tel.: +33-2-33-713433; Fax: +33-2-33-713492;
   E-mail: patrick.boulenguer@degussa.com

1	Introduction	483
2	Historical Outline	483
3	Structure	484
3.1	Chemical Structure	484
3.2	Superstructure/Secondary Structure	484
4	Occurrence	485
5	Physiological Function	486
6	Analysis and Detection	486
6.1	Chemical Characterization	486
6.1.1	Sugar Composition	486
6.1.2	Pyruvic Acid Determination	487
6.1.3	Acetate Determination	487
6.2	Physical Characterization	487
6.2.1	Molecular Weight	487
6.2.2	Secondary Structure	488
6.2.3	Rheology	489

Polysaccharides and Polyamides in the Food Industry. Properties, Production, and Patents.
Edited by A. Steinbüchel and S. K. Rhee
Copyright © 2005 WILEY-VCH Verlag GmbH & Co. KGaA, Weinheim
ISBN: 3-527-31345-1

13 Xanthan

7	**Biosynthesis**	490
8	**Degradation**	492
9	**Biotechnological Production**	493
9.1	General Description of the Process	494
9.2	Process Improvement	495
9.2.1	General Improvement	495
9.2.2	Oxygen Supply	496
9.2.3	Nutrients	496
9.3	Modeling the Fermentation Process	497
9.4	Post-fermentation Treatment	498
10	**Properties**	499
10.1	Viscosity	499
10.2	Flow Behavior	500
10.3	Weak Network Formation	500
10.4	Gelation	500
10.5	Interaction of Xanthan with Other Macromolecules	501
11	**Applications**	501
11.1	Food Applications	501
11.2	Non-food Applications	503
12	**Relevant Patents**	503
13	**Current Problems and Limitations**	510
14	**Outlook and Perspectives**	511
15	**References**	513

ADI	acceptable daily intake
AFFF	asymmetric flow field fractionation
AFM	atomic force microscopy
EPS	exopolysaccharide
GM	genetically modified
GPC	gel-permeation chromatography
LALLS	low-angle laser light scattering
LBG	locust bean gum
MALLS	multi-angle laser light scattering
M_w	molecular weight
O/W	oil in water
SEC	size-exclusion chromatography
W/O	water in oil

1
Introduction

During the second half of the 20th century, many new and useful polysaccharides of scientific and commercial interest have been discovered which can be obtained from microbial fermentations. Microorganisms such as bacteria and fungi produce three distinct types of carbohydrate polymers: (1) extracellular polysaccharides, which can be found either as a capsule that envelops the microbial cell or as an amorphous mass secreted into the surrounding medium; (2) structural polysaccharides, which can be part of the cell wall; or (3) intracellular storage polysaccharides. The scientific and industrial success of polysaccharides of microbial origin is due to several factors. First, they can be produced under controlled conditions with selected species; second, they usually present a high structural regularity; and third, different microorganisms can synthesize a wide range of very specific ionic and neutral polysaccharides with widely varying compositions and properties. Such variety is not found among plant polysaccharides and, perhaps more importantly, it cannot be imitated by means of synthetic chemistry.

The usefulness of water-soluble carbohydrate polymers in industry relies on their wide range of functional properties. The most important characteristic is their ability to modify the properties of aqueous environments, that is their capacity to thicken, emulsify, stabilize, flocculate, swell and suspend or to form gels, films and membranes. Another very important aspect is that polysaccharides obtained from natural, renewable sources are both biocompatible and biodegradable.

Xanthan, a microbial biopolymer produced by the *Xanthomonas* bacterium, has provoked great scientific and industrial interest since its discovery in the late 1950s. In 1999 alone, more than 300 references of articles or patents dealing with xanthan are listed in *Chemical Abstracts*. Since 1990, more than 2000 patents have been listed in *Derwent World Patents Index*. This interest is due to the extraordinary properties of xanthan as well as to the successful establishment of an industrial process for its production.

2
Historical Outline

Xanthan gum was discovered in the 1950s by scientists of the Northern Regional Research Laboratory of the U.S. Department of Agriculture in the course of a screening which aimed at identifying microorganisms that produced water-soluble gums of commercial interest. The first industrial production of xanthan was carried out in 1960, and the product first became available commercially in 1964. Toxicology and safety studies showed that xanthan caused no acute toxicity, had no growth-inhibiting activity, and did not alter organ weights, hematological values or tumors when fed to rats or dogs, neither in short-term, nor in long-term feeding studies. The approval for food use was given by the U.S. Food and Drug administration in 1969, and the FAO/OMS specification followed in 1974. Authorization in France was given in March 1978, and approval in Europe was obtained in 1982, under the E number E415. The official definition of the EU food regulations for E415 is: "Xanthan gum is a high molecular-weight polysaccharide gum produced by a pure culture fermentation of a carbohydrate with natural strains of *Xanthomonas campestris*, purified by recovery with ethanol or propane-2-ol, dried and milled. It contains D-glucose and D-mannose as the dominant hexose units, along with D-glucuronic acid and pyruvic acid, and is prepared as the sodium, potassium or calcium salt. Its solutions are neutral. The molecular weight must be approximately 1 MDa and the color must be cream". Xanthan gum is approved as food additive with an acceptable daily intake (ADI) "not specified"; that is, no limit for ADI is defined and the gum may be used *quantum satis*, which means with just the quantity useful for the application. Today, xanthan is produced commercially by several companies, such as Monsanto/Kelco, Rhodia, Jungbunzlauer, Archer Daniels Midland, and SKW Biosystems. For the past few years xanthan gum has also been produced in China. Annual volumes worldwide are estimated to be about 35,000 tons in 2001.

3
Structure

3.1
Chemical Structure

Xanthan is a heteropolysaccharide with a very high molecular weight, consisting of repeating units (Figure 1). The sugars present in xanthan are D-glucose, D-mannose, and D-glucuronic acid. The glucoses are linked to form a β-1,4-D-glucan cellulosic backbone, and alternate glucoses have a short branch consisting of a glucuronic acid sandwiched between two mannose units. The side chain consists therefore of β-D-mannose-(1,4)-β-D-glucuronic acid-(1,2)-α-D-mannose. The terminal mannose moiety may carry pyruvate residues linked to the 4- and 6-positions. The internal mannose unit is acetylated at C-6. Acetyl and pyruvate substituents are linked in variable amounts to the side chains, depending upon which *X. campestris* strain the xanthan is isolated from. The pyruvic acid content also varies with the fermentation conditions. On average, about half of the terminal mannoses carry a pyruvate, with the number and positioning of the pyruvate and acetate residues conferring a certain irregularity to the otherwise very regular structure. Usually, the degree of substitution for pyruvate varies between 30 and 40%, whereas for acetate the degree of substitution is as high as 60–70%. Some of the repeating units may be devoid of the trisaccharide side chain.

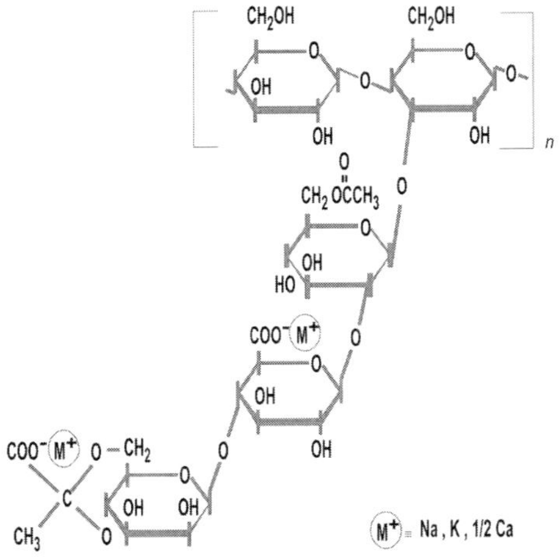

3.2
Superstructure/Secondary Structure

The secondary structure of xanthan depends on the conditions under which the molecule is characterized. The molecule may be in an ordered or in a disordered conformation. The ordered confirmation can be either native or renatured; in the native form the conformation is present at temperatures below the melting point of the molecule, a temperature which depends on the ionic strength of the medium in which xanthan is dissolved. The secondary structure of xanthan and the methods to analyze it have been recently reviewed in detail by Stokke et al. (1998). X-ray scattering results indicate that

Fig. 1 Chemical structure of xanthan.

native xanthan in the ordered conformation exists as a right-hand helix with five-fold symmetry with a pitch of 4.7 nm and a diameter of 1.9 nm (Moorhouse, 1977). Two models, a single-strand helix and a double-strand or multi-strand helix, have been proposed, though most authors currently support the idea of a double helix. The helix is stabilized by noncovalent bonds, such as hydrogen bonds, electrostatic interaction, and steric effects; its structure can be described as rigid rod. In aqueous solution, the molecule may undergo a conformational transition which can be driven by changes in temperature and ionic strength, and which depends on the degree of ionization of the carboxyl groups and acetyl contents. The temperature-induced transition from an ordered to a disordered conformation is generally attributed to a complete or partial separation of the double-strand form. Renaturation may occur under favorable conditions, which means temperatures below the transition temperature and high salt concentrations. The transition from the denatured to the renatured state is reversible, whereas that from the native to the denatured state is irreversible. The model of a double-strand structure has been supported by several studies. Capron et al. (1997) demonstrated that upon heating to temperatures above the order–disorder transition temperature, denaturation of the native ordered conformation occurred, together with a reduction in molecular weight. The molecular weight is roughly halved, which supports the model of a double-strand conformation for the native form. The molecular weight was found to be invariant after renaturation on cooling. The renaturation probably occurs as an intramolecular process, which means that the restoration of the ordered form of xanthan seems to take place within a single molecule. Most likely, the conformation for the renatured form of xanthan is that of an anti-parallel, double-stranded structure consisting of one chain folded as a hairpin loop (Liu et al., 1987). The viscosity of renatured xanthan is higher than the viscosity of native xanthan however, thereby supporting the hypothesis that single-stranded xanthan molecules associate during renaturation to form supramolecular structures.

4
Occurrence

Xanthan is produced by *X. campestris*, a plant-associated bacterium that is generally pathogenic for plants belonging to the family Brassicaceae. *Xanthomonas* causes a variety of disease symptoms such as necrosis, gummosis and vascular parenchymatous diseases on leaves, stems or fruits; an example is "black rot" of crucifers such as cabbage, cauliflower or broccoli. *Xanthomonas* does not form spores, but it is very resistant to desiccation during relatively long periods. Survival at room temperature for 25 years has been reported by Leach et al. (1957). The resistance is usually due to the protective effect of the xanthan gum produced and exuded by the bacteria. Xanthan also protects the bacteria from the effects of light, and generally causes wilting of the leaves by blocking water movements (Leach et al., 1957). Exopolysaccharides, like xanthan, are also known to provide protection against bacteriophages by building a physical barrier against the phage attack (MacNeely, 1973). The polysaccharide is not a reserve energy source because in general the bacterium is not able to catabolize its own extracellular polysaccharide.

5
Physiological Function

The physiological function of the exopolysaccharide xanthan has received little attention compared with investigations into the molecule's production, properties, and applications. The bacteria (*Xanthomonas* sp.) that produce xanthan gum as a secondary metabolite are usually phytopathogenic, or may live in asymptomatic association with plant tissues or epiphytes. *Xanthomonas* infections have been observed in over 120 monocotyledonous and over 150 dicotyledonous plant species.

Xanthan is the predominant component of the bacterial slime. The physiological function of xanthan has been deduced as being analogous to the functions of other exopolysaccharides (Yang and Tseng, 1988; De Crecy et al., 1990; Daniels and Leach, 1993; Chan and Goodwin, 1999).

Enclosure of bacterial cells in the exopolysaccharide (EPS) results in prolonged survival, and increased resistance to both temperature and ultraviolet (UV) light (Leach et al., 1957). In rice, the wilting induced by EPS seems to play a role in pathogenesis (Kuo et al., 1970). EPS may increase cell membrane leakage, which in turn leads to wilting (Vidhyasekaran et al., 1989). In general, bacterial EPS induce water-soaking of the intercellular space which is important for bacterial colonization. It is also possible that xanthan forms a gel-like slime in the intercellular space as a result of synergy with other plant polysaccharides. This gel may then promote bacterial colonization of plant tissue by retarding the desiccation of the bacterial colony, by protecting the bacteria from bacteriostatic compounds, and by preventing close morphological contact of the colony with the cell wall, thus preventing the triggering of plant defense reactions. The amount of EPS produced by *Xanthomonas* is correlated to the organism's virulence; strains with attenuated virulence usually produce less EPS (Ramirez et al., 1988) and the distribution of the polysaccharide in infected leafs coincides with that of bacteria. This was seen in a study in rice, where EPS and bacteria were distributed in both the xylem and transverse veins (Watabe et al., 1993).

6
Analysis and Detection

In order to characterize xanthan, different parameters must be taken into consideration, such as chemical structure, acetate and pyruvate contents, molecular weight, secondary structure, and rheological behavior.

6.1
Chemical Characterization

By using chemical analysis, the sugar composition of the molecule as well as the nature and the degree of substituent content can be ascertained.

6.1.1
Sugar Composition

The sugar composition of xanthan is difficult to obtain as the cellulosic backbone is highly resistant to hydrolysis. Moreover, in the side chains the presence of uronic acid prevents complete hydrolysis of the aldobiouronic (β-D-GlcAp-(1→2)-D-Manp) acid without degradation of the glucuronic acid. Well-documented reports of this situation have been made (Tait and Sutherland, 1989; Tait et al., 1990) in which the most suitable conditions to determine the neutral sugars, the aldobiouronic (β-D-GlcAp-(1→2)-D-Manp) acid and the substituents are described. A single condition with one form of hydrolysis is insufficient to characterize all the constituents of xanthan quantitatively. Due to these problems, the official description of xanthan, e.g., by the JECFA (Joint Expert Committee for Food Additives), does not refer to its chemical composition, but only to its ability to gellify in the presence of locust bean gum (LBG). In the official description, there is no reference to acetyl groups, only to pyruvic acid.

6.1.2
Pyruvic Acid Determination
After hydrolysis (Cheetham and Punruckvong, 1985; Tait et al., 1990), the pyruvic acid content of xanthan can be determined in several ways. The oldest method described is a colorimetric procedure using 2,4-dinitrophenylhydrazine (DNPH) (Slonecker and Orentas, 1962), and this is still the reference method for the JECFA. Duckworth and Yaphe (1970), have developed an enzymatic method using lactate dehydrogenase (LDH), the reaction being as follows:

$$\text{Pyruvate} + \text{NADH} + \text{H}^+ \xrightarrow{\text{LDH/LD}} \text{lactate} + \text{NAD}^+ \quad (1)$$

The amount of NAD released is measured at 340 nm. More recent determinations use high-pressure liquid chromatography (HPLC) (Cheetham and Punruckvong, 1985; Tait et al., 1990) or nuclear magnetic resonance (NMR) methods, both of which permit the simultaneous detection of both pyruvate and acetate.

6.1.3
Acetate Determination
In 1949, Hestrin published a method for the determination of acetate which uses hydroxamic acid, but today (see Section 6.1.2) NMR and HPLC methods are more often used (Cheetham and Punruckvong, 1985; Tait et al., 1990).

6.2
Physical Characterization

Besides the chemical composition of xanthan, its physical characteristics such as molecular weight, secondary structure and rheological properties are the most important determinants of the behavior of this molecule in its final application.

6.2.1
Molecular Weight
Values reported in the literature for the molecular weight (M_w) of xanthan are usually between 4 and 12×10^6 g mol^{-1}. Accurate determination of the M_w of xanthan is difficult for several reasons, including: (1) the very high molecular weight; (2) the stiffness of the molecule; and (3) the presence of aggregates. Nonetheless, several techniques have been used to determine the molecular weight of xanthan, including GPC-MALLS (gel- permeation chromatography with multi-angle laser light scattering), AFFF-MALLS (asymmetrical flow field fractionation combined with multiangle laser light scattering) and electron microscopy.

GPC-MALLS
GPC-MALLS is a technique that permits the estimation of absolute molecular weight and gyration radius of polysaccharides, without the need for column calibration methods or standards. Often used in the field of polymer analysis, this technique is constituted by a GPC system which allows molecules to be separated as a function of their molecular size, and also by MALLS, which allows information to be obtained on the molecular weight of the fraction eluted from the column. With xanthan, the GPC technique presents several difficulties: first, the high molecular weight of the xanthan combined with its rod-like structure causes the xanthan

molecules to have a very high hydrodynamic volume. The columns which are used today are unable to separate molecules with such high hydrodynamic volumes, and so xanthan appears as a monodisperse molecule. Second, for the same reason – and also due to the tight stationary phase of the column – the xanthan is submitted to a very high shear rate when eluted across the column, and this can degrade the molecule. In addition, the MALLS detection for xanthan analysis is problematic as, due to its high molecular weight, extrapolation to zero angle is not easy.

There are two classical methods to determine the molecular weight: the Zimm method; and the Debye method. In the Zimm method, we express $Kc/\Delta R_\theta = f(\sin \theta/2)$, whereby K is an optical parameter and R is the Rayleigh ratio. Since $1/M$ is obtained by extrapolation to zero angle, this method is not suitable for xanthan. The values obtained for $1/M$ are in the order of 10^{-7}, which leads to a significant variation of the value calculated for the molecular weight. In the Debye method, we express $\Delta R_\theta/Kc = f(\sin \theta/2)$; hence, by extrapolation to zero angle, M can be determined directly. This method is preferable for xanthan, but even in this case the very great angular dependence prevents the linear extrapolation and a polynomial of 3rd or 4th order is needed in order to obtain reproducible results. Some authors (Capron et al., 1997) prefer to use LALLS (low-angle laser light scattering), which provides a measure at a very low angle (5 °) and avoids this problem, but this type of apparatus does not provide any information on the gyration radius. Another problem is the presence of aggregates in the xanthan solution, and these are probably responsible for the very high molecular weight values reported in the literature. Such aggregates, even when present in very low quantities only, give very important signals in light scattering.

AFFF-MALLS

In order to avoid the problems which occur with GPC, recent investigations have used AFFF (Janca, 1988). AFFF uses a narrow channel in which a solvent flows, and a field is applied perpendicularly to this channel. Usually, the perpendicular field is created by a perpendicular flow. The sample is injected into the inlet of the channel and eluted by the solvent. At the same time, the field applied across the channel presses the sample against the wall (accumulation wall) of the channel. Due to the gradient of velocity across the channel coming from the laminar flow, the particles – depending on their distances from the wall – have different velocities. In AFFF, the separation is governed by the diffusion coefficient, and so this technique allows the separation of molecules on the basis of their hydrodynamic radius up to a size of several micrometers. An advantage of this method is that no packing material is needed, and this also avoids the problem of shear. The problem is that the different parameters – the two flows, the injected volume, and the solvent – must be carefully adjusted in order to obtain good results.

Electron Microscopy

This technique allows direct measurements of the xanthan molecule to be made after vacuum drying in the presence of glycerol and covering the molecule with a platinum film (Stokke et al., 1998). The contour length L of individual xanthan chains can be visualized by electron microscopy, and the average value of L reflects the molecular weight. Electron microscopy can also be used to detect the formation of microgels.

6.2.2
Secondary Structure

The physico-chemical properties of xanthan in aqueous solutions can be studied by means of various experimental techniques, such as light scattering measurements, hydrodynamic measurements, thermodynamic properties such as ion activity, dependence of the transition temperature on the ionic strength and calorimetric measurements. A relatively new method for studying the superstructure of xanthan is atomic force microscopy (AFM) (Capron et al., 1998a; Morris et al., 1999). AFM allows visualization of the surface of a sample, with a resolution close to the atomic scale. The mechanical properties of both native and denatured xanthan can be measured on the molecular scale (Li et al., 1999; Morris et al., 1999). AFM measures the interaction between the sample and the tip of the measuring device. A force exists between the atom of the tip of the microscope and those of the sample, separated by only a few Angstroms. By moving the tip, it is possible to follow the variation of this force on the surface of the sample and so obtain an image of the sample and estimate its shape. No modification of the sample is needed for the measurement, and this technique can be applied to both conducting and nonconducting samples. AFM avoids the drying step of the sample, and can provide images of individual xanthan molecules as well as of aggregated molecules. Molecules and molecular interactions can be imaged in the liquid environment.

Capron et al. (1997) have studied the size and conformation changes associated with the temperature-induced denaturation and renaturation of native xanthan under different salt conditions. The different methods used were LALLS, size- exclusion chromatography coupled with multi-angle light scattering (SEC-MALLS), low shear intrinsic viscosity measurement, and circular dichroism. The conformational transition can be monitored using NMR, optical rotation or calorimetric measurements. The specific optical rotation changes suddenly at the melting temperature of the molecule, from about −120 to almost zero. Circular dichroism studies near 200 nm show a decrease in overall ellipticity when passing through the transition region.

6.2.3
Rheology

Depending on the medium conditions, that is polymer concentration, salts or addition of other hydrocolloids, xanthan systems can be a Newtonian or pseudo-plastic solution or a gel. The rheological behavior can be determined using viscometers by applying shear rate and measuring shear stress and viscosity or using controlled shear stress or deformation rheometers to perform dynamic viscoelastic or flow measurements. Xanthan solutions can be characterized by classical rheological parameters, such as intrinsic viscosity $[\eta]$. The intrinsic viscosity corresponds to the hydrodynamic volume of the polymer chain in a given solvent. Classically, it can be obtained by measuring the viscosity at different low concentrations in the Newtonian domain, where the overlap between hydrodynamic volume of individual polymer chains is negligible and applying the equations of Huggins (Eq. 2) and Kraemer (Eq. 3) (Launay et al., 1986):

$$\eta_{sp}/C = [\eta] + \lambda_H [\eta]^2 C \qquad (2)$$

$$\ln(\eta_r)/C = [\eta] + \lambda_k [\eta]^2 C \text{ with} \qquad (3)$$

$$\eta_r = \eta/\eta_0 \qquad (4)$$

$$\eta_{sp} = (\eta - \eta_0)/\eta_0 \qquad (5)$$

where η is the solution viscosity, η_0 the solvent viscosity, η_r the relative viscosity, η_{sp} the specific viscosity, $[\eta]$ the intrinsic viscosity and λ_H and λ_k are constants which are functions of the hydrodynamic interaction between molecules. In dynamic measurement, the sample is submitted to a deformation $\gamma^*(t)$ or a stress $\sigma^*(t)$ which are sinusoidal function of time. When the system is viscoelastic linear, the stress $\sigma^*(t)$ is a sinusoidal function of time with the same frequency ω as $\gamma^*(t)$ and a phase angle gap δ.

$$\gamma^*(t) = \gamma_0(\cos \omega t + i \cdot \sin \omega t) \quad (6)$$

$$\sigma^*(t) = \sigma_0(\cos (\omega t + \delta) + i \cdot \sin (\omega t + \delta)) \quad (7)$$

If $\delta = 0$, stress and deformation are proportional at every moment, which means that the behavior is elastic linear. If $\delta = \pi/2$, stress and deformation speed are proportional at every moment, the behavior is Newtonian. If $0 < \delta < \pi/2$ the behavior is viscoelastic. The complex modulus G^* can be defined as:

$$G^* = \sigma^*/\gamma^* = G' + i\,G'' \text{ with} \quad (8)$$

$$G' = (\sigma_0/\gamma_0) \cdot \cos \delta \quad (9)$$

$$G'' = (\sigma_0/\gamma_0) \cdot \sin \delta \quad (10)$$

$$\tan \delta = G''/G' \quad (11)$$

G' is the storage modulus which corresponds to the elastic component of the system, and G'' is the loss modulus which corresponds to the viscous component of the system. We can also define the complex viscosity:

$$\eta^* = \sigma^*/\dot{\gamma}^* = \eta' + i\,\eta'' \text{ with} \quad (12)$$

$$\eta' = G''/\omega \quad (13)$$

$$\eta'' = G'/\omega \quad (14)$$

The gel-like structure of xanthan solutions can be characterized by deformation measurements. Small deformation measurements characterize the intact network, whilst large deformation measurements destroy the network and therefore give lower values. It is possible to measure the static yield point of a xanthan solution by applying an increasing deformation at constant shear rate and measuring the shear stress. Initially, the stress generated in resistance to the deformation increases linearly with the deformation, as in an elastic solid. Ultimately, the resistance reaches a maximum corresponding to the breaking point of the network (i.e., the yield point) and then drops again, settling down at a constant value which defines the steady shear viscosity. Xanthan gum solutions can also be characterized using dynamic light scattering. This technique allows characterization of the boundary between dilute and semi-dilute solutions. The degree of dilution is important because in a truly diluted state, the xanthan coils occupy a defined hydrodynamic volume, whereas above a critical concentration the molecules interact. Dynamic light scattering experiments can demonstrate the onset of molecular interaction and the onset of anisotropic aggregation (Rodd et al., 2000).

7
Biosynthesis

The path of xanthan biosynthesis has been described and reviewed by several authors (Leigh and Coplin, 1992; Becker et al., 1998). Xanthan synthesis starts with the assembly of the pentasaccharide repeating units, and these are then polymerized to produce the macromolecule. The oligosaccharide repeating units of xanthan are formed by the sequential addition of monosaccharides from energy-rich sugar nucleotides, involving acetyl-CoA and phosphoenolpyruvate. A polyisoprenol phosphate from the inner membrane functions as an acceptor (Ielpi et al., 1993). The first step of the pentasaccharide assembly is the transfer of glycosyl-1-phosphate from UDP-glucose to polyisoprenol phosphate. This transfer is followed by sequential transfer of the other sugar residues, D-mannose and D-glucuronic acid from GDP-mannose and UDP-glucuronic acid, respectively, which gives the complete lipid-linked repeating pentasaccharide unit. Acetyl and pyruvyl residues are added at this lipid-linked pentasaccharide stage; these are donated by acetyl-CoA and phosphoenolpyruvate, respectively. Depending on the strain and on the fermentation conditions, O-acetyl groups are attached in varying quantities to the internal mannose residue, and pyruvate is added to the terminal mannose. The xanthan chains grow at the reducing end (Figure 2). The final steps of the biosynthesis, which means the secretion from the cytoplasmic membrane, the passage across the periplasm and outer membrane and the excretion into the extracellular environment has not yet been entirely elucidated. The process requires energy, and it is probably accomplished via a specific transport system, which ensures the release of the polymer from the lipid carrier and the transport across the outer membrane (Daniels and Leach, 1993).

Many genes which are involved in xanthan biosynthesis have been identified, isolated and characterized (Figure 3). In *Xanthomonas campestris* pv. *campestris*, the biosynthesis is directed by a cluster of 12 genes, *gumB* to *gumM* (Vanderslice et al., 1989; Vojnov et al., 1998). Seven gene products are required for the transfer of the monosaccharides and for the acylation at the lipid intermediate level to form the complete acylated repeating unit. This gene cluster is not linked to the genes which are required for the synthesis of the sugar nucleotide precursors. The 12 genes of the cluster are

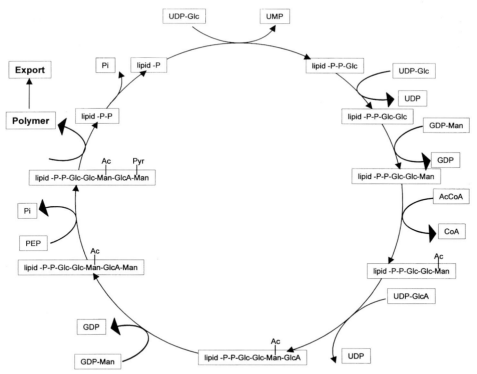

Fig. 2 Biosynthesis of xanthan. Ac: acetyl; Glc: glucose; Man: mannose; P: phosphate; PEP: phosphoenolpyruvate; Pyr: pyruvate (Adapted from Sutherland, 1993).

Fig. 3 Genetic map of the *Xanthomonas campestris* pv. *campestris* gum operon (adapted from Becker et al., 1998).

expressed as a single operon from a promoter which is located upstream of the first gene. Xanthan synthesis does not seem to be specifically controlled but is activated by the gene products of a cluster of at least five genes (Tang et al., 1990). The initial stages of xanthan biosynthesis appear to be regulated as part of a pathogenicity regulon (Coplin and Cook, 1990), and some promotors and transcription start sites have been characterized (Knoop et al., 1991; Kingsley et al., 1993; Katzen et al., 1998). It has been shown that a transposon insertion, located 15 bp upstream of the translational start site of the *gumB* gene leads to a defect in xanthan synthesis (Vojnov et al., 1998). Cells of the mutant strain were able to synthesize the lipid-linked pentasaccharide repeating unit of xanthan from the three nucleotide sugar donors (UDP-glucose, GDP-mannose, and UDP-glucuronic acid), but were unable to polymerize the pentasaccharide into mature xanthan. A subclone of the gum gene cluster carrying *gumB* and *gumC* restored xanthan production of the mutant strain to levels of almost 30% of the wild type. In contrast, subclones carrying *gumB* or *gumC* alone were not effective. These results indicate that *gumB* and *gumC* are both involved in the translocation of xanthan across the bacterial membranes. Tseng et al. (1999) have constructed a physical map of the *Xanthomonas campestris* pv. *campestris* chromosome, and the locations of eight loci involved in xanthan gum synthesis were determined. Furthermore, four loci were determined which may be involved in gum polymerization, secretion and regulation. Rodrigues and Aguilar (1997) have identified several mutants of *Xanthomonas campestris* showing increased viscosity and/or gum production after UV treatment. Xanthan solutions of the different mutants showed different intrinsic viscosity values, and no relationship was found between pyruvate or acetate contents and viscosifying ability of the xanthan. It is also possible to produce xanthan using genetically modified (GM) bacteria other than *Xanthomonas*. Pollock et al. (1997) have cloned 12 genes coding for assembly, acetylation, pyruvylation, polymerization and secretion of the polysaccharide xanthan gum in *Sphingomonas*, and these genes were sufficient for synthesis of xanthan gum. The polysaccharide from the recombinant microorganism was largely indistinguishable, both structurally and functionally, from the native xanthan gum.

8
Degradation

The xanthan molecule is very stable as long as it is in the ordered, double-stranded conformation. The double helix confers a good resistance to the molecule against degradation by free radicals, acid, enzymes or repeated freeze–thaw cycles. The stability of xanthan is significantly lowered if the molecule is in a disordered conformation, which is the case at temperatures above the transition temperature or at low ionic strength, for example in salt-free solutions at temperatures around 40–50 °C. Xanthan in solution will not degrade in a pH range from about 2 to 11, but at pH 9 xanthan gum starts to deacetylate. Temperatures up to

90 °C do not affect xanthan, and even after heat treatment at 120 °C for over 30 min the viscosity remained at about 90% of the original value. Chemical degradation of xanthan can be achieved through free radicals which are generated by strong oxidants such as hypochlorite and persulfate, at high temperature, or by a combination of H_2O_2 and Fe^{2+}. Oxidative–reductive depolymerization leads to a cleavage of the polymeric backbone (Herp, 1980). Acid hydrolysis leads preferentially to the hydrolysis of the terminal β-D-mannose in the side chain. Mechanical degradation is possible using high shear or ultrasound; however, the shear to be applied must be very high, in the order of $10^6 \, s^{-1}$. Enzymatic degradation of xanthan is usually not very efficient, but if the molecules are in the disordered state they can be degraded by a class of enzymes called xanthanases. This group of enzymes comprises endo-1,4-β-glucanases and xanthan lyases. The glucanases can partially degrade the cellulosic backbone of xanthan (Rinaudo and Milas, 1980), and they are also able to degrade different celluloses. Xanthan lyases are able to cleave the β-D-mannosyl-D-glucuronic acid linkage of the trisaccharide side chain. The presence of acetate and pyruvate hampers the action of the cellulases, whereas the activity of xanthan lyase does not seem to be affected by either pyruvate or acetyl groups. It is likely that xanthan contains cellulosic regions that do not carry side chains and are thus preferred regions for an attack by cellulases. Christensen and Smidsrød (1996) have shown that acid hydrolysis prior to enzymatic degradation increases the depolymerization of xanthan by several orders of magnitude. Removal of the side chains from xanthan with a limited degree of backbone degradation can be obtained by partial hydrolysis. The terminal β-mannose residue linked to O-4 of the glucuronic acid is hydrolyzed quite rapidly (Christensen and Smidsrød, 1991; Christensen et al., 1993a,b). The α-mannose linked to the backbone is hydrolyzed about 10 times more slowly, the hydrolysis obeying pseudo first-order kinetics. Ruijssenaars et al. (1999) have identified several xanthan-degrading enzymes in a *Paenibacillus* strain; this strain is able to degrade about one-third of the xanthan molecule without attacking the backbone of the molecule. Nankai et al. (1999) have reported in detail an enzymatic route for the depolymerization of xanthan by a *Bacillus* strain. The enzymes which are necessary for the degradation of xanthan are xanthan lysase, glucanase, glucosidase, glucuronyl hydrolase and mannosidase. The degradation starts with the cleavage of the glycosidic bond between pyruvylated mannosyl and glucuronyl residues in xanthan side chains due to the action of an extracellular xanthan lyase. The modified xanthan can then be attacked by an extracellular β-D-glucanase, which produces a tetrasaccharide, representing the repeating unit of xanthan without the terminal mannosyl residue. The tetrasaccharide is then taken into the bacterial cell and is subsequently converted by β-D-glucosidase to yield the trisaccharide unsaturated glucuronyl-acetylated mannosyl-glucose. Afterwards, a glucuronyl hydrolase generates an unsaturated glucuronic acid and the disaccharide mannosyl-glucose. The last step in the degradation is the hydrolysis of this disaccharide to mannose and glucose by α-D-mannosidase. For more details, see Chapter 23.

9
Biotechnological Production

Xanthan gum is produced by fermentation; this is a very efficient and reliable process, but it does present some intrinsic problems.

9.1 General Description of the Process

Today, xanthan gum is the most successful industrial biopolymer produced by fermentation. Xanthan gum is produced by the aerobic fermentation of *Xanthomonas campestris*, and many different strains of this bacterium have been screened for their ability to produce xanthan gum (Gupte and Kamat, 1997; De Andrade Lima et al., 1997). Xanthan fermentation can be either batch-wise, semi-batch-wise, or continuous. Industrial production is usually carried out by a batch-wise, submersed fermentation in an aerated and agitated fermenter. The different steps of an industrial process are batch fermentation, alcoholic precipitation, first drying, rinsing with an alcohol/water mixture, final drying, grinding, quality control of the batch, and packaging (Figure 4), with the production strain usually preserved in a freeze-dried state. Galindo et al. (1994) and Salcedo et al. (1992) have reported the preservation of the production strain on agar slopes as well as preservation in sterile seeds. The strain is activated by inoculation into a nutrient medium containing a carbohydrate source, a nitrogen source and mineral salts. After growth, the culture can be used to inoculate successive fermenters through to the industrial scale. Throughout the fermentation process, pH, aeration, temperature and agitation are monitored and controlled. The optimal temperature for cell growth is between 24 and 27 °C, and the best temperature for xanthan production is 30–33 °C. A pH of 6–7.5 is most suitable for growth, while for xanthan production the pH optimum range is 7–8. During the fermentation, pH decreases; however, in an industrial fermentation process the pH is maintained close to neutral in order to allow the process to continue until complete exhaustion of the carbohydrate substrate. Xanthan is produced during the bacterial growth phase as well as during the stationary phase, though maximum production is seen during the exponential growth phase. The pyruvate and acetate contents of xanthan are highest in polymers synthesized immediately after the end of exponential growth (Sutherland, 1993). The quantity of polymer increases until about 30 h after inoculation, after which time a steady state is reached with termination of bacterial growth and polymer production, at which point the fermentation is stopped. The achievable productivity and concentrations reported in the literature range from 15 to 30 g L^{-1} (sometimes higher) and up to 0.7 g L^{-1} · h) (mostly lower). Productivity in industrial fermentations may be significantly higher, however. At the end of the fermentation, the fermenter is emptied, cleaned and sterilized before the next fermentation takes place. The post-fermentation steps include a pasteurization by thermal treatment, and sometimes also

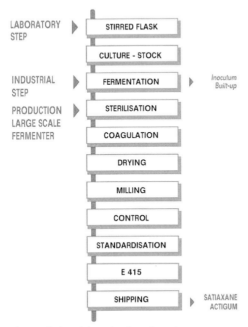

Fig. 4 Industrial production of xanthan.

purification. The heat treatment serves to kill the bacteria. Since *Xanthomonas* do not form spores, the fermentation can be stopped by heat treatment. Pasteurization also improves the properties of the gum, as it leads to a (partial) transformation of the native to the renatured state. Xanthan gum is recovered by precipitation in isopropyl alcohol or ethanol. Usually, the whole broth including biomass is precipitated, but for specific applications, purification may be necessary (see Section 8.4). In the different growth phases, xanthan molecules with different molecular weight and different degree of substitution are produced; hence, the xanthan recovered at the end of a batch fermentation is a mixture of structurally different molecules. The coagulum obtained is isolated by solid–liquid separation, rinsed with alcohol, pressed and dried. The dried xanthan is ground to obtain a white or cream-colored powder with the desired particle size (usually 80–250 µm). Xanthan powder may also be granulated in order to obtain products with specific dispersion and dissolution properties. The final steps in the industrial production chain are quality control, including chemical, physical, rheological and microbiological analysis, packaging and shipping.

9.2
Process Improvement

Although the process for xanthan production is well established, numerous and also recent publications deal with the improvement of xanthan production. Due to the rheological properties of the xanthan molecule, the production of xanthan by fermentation is technically challenging. During the course of the fermentation, the excretion of the polysaccharide results in a highly viscous and shear-thinning broth. As the fermentation broth becomes non-Newtonian, the rheology of the broth causes serious problems of mixing, oxygen supply and heat transfer, thereby limiting the maximum gum concentration achievable, as well as the product quality. Changes in viscosity during culture exceed four orders of magnitude. At the beginning of the fermentation the broth is more or less like water, whereas at the end of the process it is highly viscous and pseudoplastic, exhibiting yield stress. Besides the technical difficulties which always demand improvement, economic aspects of the fermentation process are also important, and improvement of the fermentation costs has a high priority for industrial xanthan producers.

9.2.1
General Improvement

Instead of batch fermentation, other approaches such as continuous fermentation, fed-batch or emulsion fermentation are possible. Continuous fermentation allows high productivity to be obtained, but the process is not easy to set up, the equipment needed is not standard, and stirring in the continuous process is often insufficient. Furthermore, the sterility of the process is difficult to maintain (Silman and Rogovin, 1972). Hence, continuous processes are usually not suitable for industrial production.

Fed-batch fermentation is more successful, and towards the end of the production nutrients are recharged and very good productivity and process efficiency can be obtained. Fermentation in emulsions or in water/oil dispersions can result in high yield in the water phase, but low yields in the total volume.

Emulsion fermentation can reduce the viscosity of the broth by dispersing it into an organic liquid, such as isoparaffin, with the help of an emulsifier, for example fatty acid alkanolamide. The water-in-oil emulsions obtained develop less viscosity and show less pseudoplastic behavior than normal fermentation broth. High oxygen transfer rates can be achieved, and mixing is efficient even at low power input (Schumpe et al., 1991). However, downstream processing is much more difficult than in one-phase submerged cultivation, and the gum may be contaminated with the emulsifier. The separation of cell growth and biopolymer production in a two-stage process can improve productivity (Banks and Browning, 1983; DeVuyst et al., 1987). Xanthan production by immobilized cells has also been described (Lebrun et al., 1994, Yang et al., 1996).

9.2.2
Oxygen Supply

Xanthomonas campestris is a strictly aerobic bacterium, and so oxygen is required for growth and for xanthan production. Sufficient oxygen supply is a prerequisite for efficient polymer production and good productivity. Oxygen is provided by agitation of the culture broth, but this becomes increasingly difficult during the process as the viscosity of the broth increases. Agitation is much affected by vessel geometry, liquid volume, the method of agitation and impeller properties. The productivity of *Xanthomonas* can be improved through the use of impeller systems and suitable reactor configurations which lead to increased oxygen supply for the bacteria. The various reactor configurations which have been proposed to increase oxygen transfer into the medium include stirred tank reactors, external circulation loop reactors, bubble column or air lift reactors. The impeller types which have been proposed are helical stirrers, which cause axial motion, and anchor and multiple rod stirrers which have the tangential action of flat-blade turbines (Galindo, 1994). Amanullah et al. (1998b) have investigated the agitator speed and dissolved oxygen effects in xanthan fermentation. Changes of shear stress in the vicinity of the microorganism which are caused by changes of agitation speed did not influence xanthan production up to a xanthan concentration of 20 g L^{-1}. At higher gum concentrations, xanthan production was enhanced at the higher agitation speed, due to higher microbial oxygen uptake rates.

9.2.3
Nutrients

A standard medium for xanthan production contains a nitrogen source, a carbon source, phosphate and magnesium ions, and trace elements. A typical composition of a medium is 0.06% ammonium nitrate, 0.5% potassium dihydrogen phosphate, 0.01% magnesium sulfate heptahydrate, 2.25% glucose and 97.18% water (Daniel et al., 1994). As carbon substrates, *Xanthomonas* can use different sources such as carbohydrates, amino acids and intermediates of the citric acid cycle. The most commonly used carbon source is glucose, with concentrations around 40 g L^{-1}. In fact, glucose concentrations above 50 g L^{-1} usually lead to poorer conversion into the polysaccharide. Nitrogen sources which allow growth and efficient xanthan production are ammonium salts or amino acids. The degree of substitution with pyruvate increases if $(NH_4)_2HPO_4$ is used as nitrogen source (US Secretary of Agriculture, 1978). The polymer production is promoted by a high ratio of carbon to nitrogen in the culture medium. The mode of glucose feeding can influence the productivity of the fermentation (Amanullah et al., 1998a). An improved performance cannot be achieved by increasing the initial glucose concentration above 50 g L^{-1}, nor by a single 10 g L^{-1} pulse addition with an initial glucose concentration of 40 g L^{-1}, while significant nitrogen is still present. However, a simple pulse and continuous feeding strategy, after nitrogen has been essentially exhausted and under conditions of nonlimiting dissolved oxygen, can result in a greatly enhanced performance compared with batch fermentations. A xanthan gum concentration > 60 g L^{-1} and a productivity of 0.72 g $L^{-1} \cdot$ h) were obtained, which is higher than the values usually reported in the literature for conventional stirred bioreactors.

Between 20–30% of the total process cost can be attributed to the culture medium. In order to reduce the costs of xanthan production, several cheap substrates – many of which are agricultural waste or byproducts of the agronomic industry – have been proposed for xanthan fermentation. Cost reduction is only one aspect in the selection of a nutrient, however; another important aspect is that cheap nutrients must allow both satisfactory productivity and good product quality to be obtained. Among the proposed alternative substrates are whey, corn-steep liquor, molasses, glucose syrup, and olive oil waste waters (De Vuyst and Vermeire, 1994). The cultivation of *Xanthomonas campestris* on solid substrates, such as spent malt grains, apple pomace, grape pomace and citrus peels is also possible. With most of these solid substrates, xanthan yields were comparable with those obtained in conventional submerged cultivation, the yield achieved ranging from 33 to 57 g L^{-1} (Stredansky and Conti, 1999; Stredansky et al., 1999). The biopolymers obtained were chemically and structurally similar to commercial xanthan. Moreno et al. (1998) have reported successful fermentations using different acid-hydrolyzed wastes – from melon, watermelon, cucumber, and tomato – in the culture medium for xanthan production. Melon acid-hydrolyzed waste was the best substrate for xanthan production, allowing an exopolysaccharide to be obtained the chemical composition of which was very similar to that of commercial xanthan. Sugar beet pulp added as a supplement to the xanthan culture medium can increase the yield of xanthan fermentation and allows the production of nonfood-grade xanthan gum (Yoo and Harcum, 1999). Chestnut flour can also be used as a nutrient in xanthan fermentation (Liakopoulou Kyriakides et al., 1999), with a maximum concentration of 33 g L^{-1} xanthan being obtained after 45 h of fermentation. Albergaria et al. (1999) have used LBG extracts as a growth medium in the fermentation; the maximum xanthan concentration obtained was 14 g L^{-1}, and the specific xanthan yield was 7 g g^{-1} cells. This substrate is more of academic interest however, since the yield is rather low and LBG extracts are not a cheap raw material source.

9.3
Modeling the Fermentation Process

Controlling a difficult fermentation process, especially of a highly viscous biopolymer such as xanthan gum, carries several intrinsic difficulties. The process is complex, varies over time, and is nonlinear; moreover, on-line measurements of many important variables is often not possible or is too difficult for industrial routine. A helpful approach for controlling the process can be mathematical modeling, and models for the control of a rheologically complex fermentation have been developed by several groups (Abraham et al., 1995; Garcia Ochoa et al., 1995). The proposed models are able to describe the process parameters including growth, polymer production, consumption of nutrients and evolution of dissolved oxygen, and can be used to improve significantly the production efficiency. Garcia Ochoa et al. (1998) have developed a kinetic model describing xanthan production by *Xanthomonas campestris* NRRL B-1459 in a batch stirred-tank. Parameters described include biomass production, carbon, nitrogen and oxygen consumption, xanthan production and temperature effects. The model can accurately describe fermentations over the temperature range 25–34 °C. Kuttuva et al. (1998) have simulated W/O xanthan fermentation by integrating the microbial kinetic behavior and the multiple-phase process characteristics. Two different models are proposed, one of which assumes a uniform redistribution of cells, substrates and product by frequent droplet breakup and coalescence, and another model which simulates the system of viscous aqueous phase with minimal droplet breakup and component redistribution. The evaluated parameters were the xanthan concentration in the aqueous phase and the volumetric productivity achieved at 200 h. According to the model, W/O fermentation can lead to very high xanthan concentrations in the aqueous phase of >200 g L^{-1}, and a much increased volumetric productivity of >0.8 g $L^{-1} \cdot$ h compared with conventional fermentations where the xanthan concentration reaches ~ 50 g L^{-1} and a volumetric productivity of ~ 0.5 g $L^{-1} \cdot$ h). Serrano Carreon et al. (1998) have modeled xanthan gum fermentation through the interactions between the kinetics of growth and product formation and mixing, which is related to the rheological behavior of the broth. The mixing was linked to the power drawn, as a function of the agitation rate and xanthan content.

9.4
Post-fermentation Treatment

Isolation and purification of xanthan are costly procedures, with post-fermentation treatments accounting for up to 50% of the total production costs. Industrial xanthan is usually not purified. Following pasteurization, which takes place at the end of the fermentation process, the entire fermentation broth is precipitated with isopropanol or a similar alcohol – which means that the bacterial cells remain in the product after processing. The final product contains about 85% pure xanthan gum, 5% biomass, and 10% moisture. The alcohol is recovered by distillation. The biomass imparts turbidity of the final product, and so removal of the bacterial cells is required when high transparency or clarity of the product is required. The fermentation broth can be clarified by filtration, though Yang et al. (1998) removed the cells present in xanthan fermentation broth by adsorption of the bacterial cells onto fibers. These authors were able to show that rough-surfaced cotton fibers were better than smooth-surfaced polyester fibers in the cell adsorption process. Cell adsorption was facilitated by the anionic xanthan gum present in the solution, as in the absence of xanthan gum the cell adsorption to fibers was poor. Alternatively, proteolytic enzymes which degrade the cellular debris from *Xanthomonas* can be used for clarification; this is possible for example using extracellular enzymes secreted by *Trichoderma koningii* (Triveni and Shamala, 1999).

The thermal denaturation and renaturation of a fermentation broth of xanthan can affect the rheological properties of xanthan (Capron et al., 1998b). Depending on the concentration and degree of purification, thermal treatment will have different effects. Changes in both the viscoelastic properties and molecular weight are observed after heating above the melting order–disorder temperature, these being related to the order–disorder conformational transition of the xanthan molecules. At concentrations below 10 g L^{-1}, heat denaturation occurs with dissociation of the native double-stranded structure into two single strands. Upon cooling, the single strands will fold back on themselves to form renatured helices, but at higher concentrations no complete dissociation occurs. Xanthan, which is renatured in concentrated conditions (>10 g L^{-1}) has a higher viscosity than that of the native sample, and displays more gel-like properties. In order to optimize the economy of the post-fermentation treatment of the fermentation broth, Lo et al. (1997) proposed that ultrafiltration be carried out before the alcoholic precipitation. In this way, the xanthan solution could be concentrated to up to 15% (w/v). Mixing modes can also influence the recovery of xanthan gum by precipitation. Zhao et al. (1997) measured the rheological properties of xanthan solution at varying alcohol concentration, thereby introducing the concepts of mixing alcohol concentration and precipitation alcohol concentration. The best design for the precipitation was a continuous precipitation system, as this gave the highest yield and the least amount of impurities. Parlin (1997) proposed a scalable vibrating membrane filter system designed for the separation of solid–liquid foods. Instead of recovering the polymer by alcohol precipitation it is also possible to recover the gum by direct drying with drum dryers or spray dryers (Harrison et al., 1999). Alternative agents for precipitation which have been studied include quaternary ammonium salts or calcium salts. Commercial xanthan is usually dried until it reaches a moisture content of 8–12%, after which it is ground to a particle size usually between 80 and 250 µm. Easily-dispersible commercial products may also be granulated.

10
Properties

Xanthan has very interesting rheological properties. Generally speaking, xanthan gum is a thickener, and produces extremely viscous solutions in water, even at very low concentrations, though under certain conditions xanthan may also gellify.

The properties of pure xanthan gum and xanthan gum blends will be described in this section. As detailed above, xanthan molecules can exist in either an ordered or a disordered conformation. The most interesting properties of xanthan are due to the ordered form, in which the macromolecules adopt a helical conformation whereby the secondary structure resembles rigid rods that do not have any tendency to associate. The form and the rigidity of the xanthan macromolecules determine the rheology of xanthan solutions. The organized state is stabilized in the presence of electrolytes. Parker et al. (1999) determined the dissolution kinetics of xanthan powder, and showed the solubilization of xanthan powder to depend on the size and distribution of the particles; for example, a polydisperse powder will have tendency to form lumps. The dissolution behavior is different in pure water and salt solutions, however. In pure water, dissolution begins with a burst of aggregates, whereas the addition of salt diminishes the aggregate population. The presence of anionic side chains on xanthan gum molecules enhances hydration and renders it soluble in cold water.

10.1
Viscosity

At rest, xanthan solutions develop high viscosity values at concentrations as low as 1%, this being due to the high M_w of the molecules and to their secondary structure. The rigid double-stranded helix confers stability to the molecule; this is because the rigid rods cannot gyrate freely and, at rest, the macromolecules adopt an equilibrium position, stabilized by low interactions. Xanthan solutions are quite resistant to hydrolysis, temperature, and electrolytes.

Salts usually do not reduce the viscosity, though xanthan might precipitate in the presence of polyvalent cations under alkaline conditions. The nature of the counterion in single counterion xanthan influences xanthan viscosity. Solutions of xanthan gum are generally not affected by changes in pH value, and xanthan gum dissolves in most acids or bases. The viscosity of xanthan gum is stable at pH values ranging from 2 to 11 for long periods of time, whereas other hydrocolloids lose their viscosity under the same conditions. The viscosity varies only slightly with temperature, and xanthan solutions remain viscous at temperatures even exceeding 100 °C as long as the ordered conformation is maintained. Stability is enhanced by high gum concentrations.

10.2
Flow Behavior

At very low shear rate, xanthan solutions may present a Newtonian behavior. Milas et al. (1990) showed that transition from the Newtonian regime to viscoelastic behavior was characterized by a critical value of the shear rate, $ý_r$, i.e., a relaxation time $ý_r^{-1}$. Above a certain low polymer concentration, this critical shear rate depends on the concentration: the higher the concentration, the lower the shear rate, which can be explained by the change in the entanglement degree of the system. For $ý > ý_r$, the xanthan solutions present a shear-thinning behavior which, together with a high yield value, give xanthan solutions exceptional suspension properties. The shear-thinning effect is due to the orientation of the polymer. At low shear, the molecules are randomly oriented, and the flowing properties are bad. This state is relatively stable until the shear-stress exceeds a certain value, called the yield-value. At high shear, the molecules align in the direction of the applied force and the xanthan solution flows easily. Above the yield value, the shear-thinning behavior is more pronounced than with other gums with a random coil conformation. It has been reported (Stokke et al., 1998) that the dynamic viscosity $\eta^*(\omega)$ of a xanthan solution is higher than the steady flow viscosity $\eta(ý)$ at a shear rate equal to ω, which means that xanthan does not follow the empirical Cox–Mertz rule, except at low concentration ($< 1 \text{ g L}^{-1}$), in the absence of salt and in low shear rate and frequency regions (Milas et al., 1990).

10.3
Weak Network Formation

Xanthan chains in solution form a continuous three-dimensional network, with weak cross-links. Therefore, xanthan solutions may also be characterized as weak gels. The chains separate easily under shearing, which allows the xanthan solution to flow and also accentuates the shear-thinning behavior. The mechanical spectra of xanthan solutions resembles the spectra of gels with $G' > G''$, with little frequency dependence in either modulus. Association of the negatively charged xanthan is promoted by metal ions; the order of effectiveness in inducing weak gels is $Ca^{2+} > K^+ > Na^+$.

10.4
Gelation

Under certain conditions, xanthan may gellify. This may occur in the presence of certain metal ions, or as a synergistic effect with other polymers. Heavy metal ions, such as Cr^{3+}, Al^{3+} or Fe^{3+} cross-link xanthan; in the case of Cr^{3+}, the cross-link occurs through a ligand exchange reaction, where the water molecule which is bound to the Cr^{3+} is exchanged with a carboxylic group on the xanthan molecule.

10.5
Interaction of Xanthan with Other Macromolecules

Xanthan gum can develop additive, synergistic and/or antagonistic effects with other molecules, such as galactomannans and glucomannans. Xanthan molecules in their rigid rod helical form can be cross-linked via the smooth zones of galactomannans. Xanthan shows a synergistic viscosity increase with the galactomannan guar, and it can form a strong thermoreversible gel in the presence of the galactomannan LBG, the glucomannans konjac mannan, the galactomannan tara, or modified guar (Morris, 1990). With LBG, the maximum gel strength is obtained at a xanthan:LBG ratio of 1:1. The formation of thermoreversible gels by synergistic mixtures of xanthan with certain galacto- and glucomannans has been ascribed to intermolecular binding through cocrystallization of denatured xanthan chains within the mannan crystallite (Morris, 1990). The cellulosic backbone of xanthan and the stereochemically similar mannan backbone both form ribbon-like structures. The mixed crystallites probably act as strong junction zones to consolidate the weak xanthan network. The xanthan–galactomannan system is interesting because each polysaccharide alone does not form a gel, but mixing results in a synergistic gelation. The mannose on galactose (M:G) ratio in galactomannans controls the mechanism and the temperature at which gelation occurs.

11
Applications

Xanthan is soluble in both hot and cold water, it develops high viscosity at low concentrations, is compatible with many salts even at elevated concentrations, and is stable at both acid and alkaline pH. Xanthan solutions are also quite resistant to high temperatures, even above 100 °C. These properties, in addition to the concept of a yield stress, and along with the shear-thinning behavior and water-binding capacity of xanthan gum, make it a highly valuable texturizing agent, and consequently it is used in a wide variety of applications. In the food industry, xanthan can act as thickener, as a suspending agent, as an emulsion stabilizer, and as a foam enhancer. Technical applications of xanthan make use of the same properties which are also important for the food industry, and include oil drilling emulsions, paints and glues, as well as firefighting formulations. Xanthan applications have been reviewed recently by Kang and Pettitt (1993) and Nussinovitch (1997). From the total volume of xanthan produced worldwide, about 65% is used in food applications, 15% in the petroleum industry, and about 20% in technical applications other than oil drilling.

11.1
Food Applications

Some examples of the multitude of potential food applications include:

- Use in dry mixes for products such as sauces, dressing, gravies, or desserts. Xanthan is used in these applications because it dissolves quite easily in hot or cold water, and provides a rapid build-up of viscosity. In dressings, xanthan is particularly suitable for the suspension of herbs, spices and vegetables, of which it assures an even distribution throughout the bottle over a long period of time (Figure 5). Xanthan provides the necessary high yield value and strong pseudoplasticity, and its properties are not hampered by acid pH, high salt concentra-

Fig. 5 Application of xanthan for the stabilization of herbs in salad dressing. Left: dressing without xanthan. Right: dressing containing 0.125% (w/v) xanthan.

tions, or heat treatment. Due to the shear-thinning behavior, the dressing flows easily from the bottle; however, once poured onto the salad, the sauce clings to the food, which is important for visual presentation.

- In syrup or chocolate toppings, xanthan confers a good consistency and flow properties due to the high yield value and high at-rest viscosity.
- In beverages, xanthan gives body and a good mouthfeel to the liquid and can also stabilize pulps, especially in combination with other polysaccharides such as carboxymethylcellulose.
- Applications in dairy products are often in combination with LBG or guar gum. These combinations stabilize cream emulsions, prevent whey-off and improve the physical and organoleptic properties of pasteurized products. Xanthan stabilizes air cells as well as particles, which makes it useful for mousses, whipped creams and pourable aerated desserts (Sanofi, 1993).
- In frozen desserts, xanthan gum confers heat-shock resistance, and it can also protect foods from freeze–thaw instability (MacNeely and Kang, 1973; Vafiadis, 1999). Xanthan gum is able to increase perceived creaminess and minimize the effects of temperature variation during storage. Freeze–thaw cycles often destroy the delicate texture of frozen desserts, as under unfavorable storing conditions small ice crystals will migrate towards large ice crystals during the thaw stage. During refreezing they will attach to large ice crystals, causing them to grow. Xanthan gum, together with other ingredients such as starch, slows down the recrystallization of smaller ice crystals and prevents the formation of large ice crystals.
- The heat stability and the stabilizing and suspending properties of xanthan are also used widely in canned foods.
- In baking, xanthan is helpful in dough preparation by preventing lump formation during kneading and improving dough homogeneity and volume (Collar et al., 1999). Xanthan also facilitates pumping of the dough during production. In baking, xanthan is also used to suspend larger solid particles such as fruits or nuts.

Other applications, as described by Tilly (1991) include:

- the stabilization of chocolate milk and yogurt-based beverages;
- in combination with other polysaccharides for low-fat spreads;
- the production of fat-reduced biscuits (Conforti et al., 1997); and
- reduced-calorie grape juice jellies which will show similar texture characteristics, for example gel hardness, cohesiveness and springiness, as a reduced-calorie grape juice jelly texturized with low methoxyl pectin (Gaspar et al., 1998).

The application of xanthan in combination with LBG the production of "melt-in-the-mouth" polysaccharide gelling systems for foods has also been described (Marrs, 1997).

11.2
Non-food Applications

The most important technical application for xanthan gum is its use oil drilling. In this application, xanthan is particularly useful due to its pseudoplastic behavior, temperature stability, and salt tolerance. The use of xanthan in petroleum production has been reviewed by Kang and Pettitt (1993). During oil drilling, a low viscosity is required at the drill bit, whereas a high viscosity is required in the annulus. The pseudoplasticity of xanthan solutions meets these requirements as xanthan develops low viscosity at the drill bit where the shear is high, and high viscosity in the annulus where shear is low. Therefore, drilling fluids containing xanthan allow a rapid penetration at the bit, and a suspension of cuttings in the annulus.

In textile printing, xanthan is used to provide the specific rheological properties needed in the production of sharp and clean patterns by preventing migration of the dye. Xanthan is compatible with most components of printing pastes, and it is also removed easily by washing. Xanthan is also used in ceramic glazes where it prevents agglomeration of the different components. In cleaning liquids, xanthan can be useful by providing a high viscosity to the solution at low shear, thereby allowing the cleaner to cling to inclined surfaces. Other technical applications described for xanthan gum include paint and ink, where it can stabilize and suspend pigments, wallpaper adhesives, formulation of pesticides and toothpastes, and industrial emulsions. The use of xanthan gum as a support for enzyme and cell immobilization has been described by Dumitriu and Chornet (1998). Xanthan gum can also be useful as controlled-release agent for pharmaceuticals (Philipon, 1997), with microspheres of gellified xanthan encapsulating active ingredients. The spheres are swallowed in a dry state and swell in the stomach, thereby gradually releasing the active ingredient. The active molecule can also be linked covalently to the polysaccharide and then released in the body by enzymatic hydrolysis.

12
Relevant Patents

As long ago as 1959, Esso Research patented the use of xanthan gum for the displacement of oil from partially depleted reservoirs. Later, several patents followed concerning processes for fermentation (1963, 1966), recovery (1963) and application (1975) of xanthan. In 1960, several patents of Jersey Production Research Co. appeared, covering a thickening agent and the process for its production, a substituted heteropolysaccharide, and a process for synthesizing polysaccharides. Kelco Biospecialities Ltd. has patented processes for xanthan production and recovery as well as xanthan application, since the early 1960s. This includes several patents concerning processes for producing the polysaccharides deposited in 1966, a process for xanthan gum production (1981), an inoculation procedure for xanthan gum production (1981), or the production of xanthan gum by emulsion fermentation (1981). Patents concerning the post-fermentation treatment include patents regarding the treatment of a *Xanthomonas* hydrophilic colloid and the resulting product (1964, 1968) or the precipitation of xanthan gum (1981). Special xanthans such as etherified *Xanthomonas* hydrophilic colloids (1965), polymeric derivatives of a cationic *Xanthomonas* colloid (1964), cationic ethers of *Xanthomonas* hydrophilic colloids (1963) or xanthan having a low pyruvate content (1981) have also been patented by Kelco. Applications patented by Kelco include edible compositions comprising oil-in-water emulsions (1960), joint-filling compositions (1963), and a dehydrated food product (1971). Patents of Rhone-Poulenc appeared mainly in the 1970s and 1980s; those of the 1970s covered improvements of the fermentation process (1973, 1976, 1978) and applications as a thickening agent (1975) and in oil drilling (1977). During the 1980s, other patents relating to the fermentation process were deposited (1984, 1987, 1988). Merck began working on xanthan in the 1970s, and this resulted in several patents concerning improvements in the process to obtain xanthan solutions with increased viscosity (1975), the preparation of a xanthan copolymer (1975), a deacetylated borate-biosynthetic gum composition (1979), a process for producing low-calcium xanthan gum by fermentation (1978), a process to prepare a xanthan gum which does not contain any cellulases (1979), the production of low-calcium, smooth-flow xanthan gum (1979), and a process for improved recovery of xanthan gum with high viscosity (1979). In the 1980s, the Merck patents covered a dispersible xanthan gum composite (1980) and a heteropolysaccharide and its production and use (1985). In 1986, Merck patented a recombinant DNA plasmid for xanthan gum synthesis containing some essential genetic material for xanthan gum synthesis and in 1991, a patent about low-ash xanthan gum ($<2\%$) appeared which is claimed to be particularly useful in the preparation of ceramics. Patents of Standard Oil Co. about xanthan gum appeared mainly in 1980/1981, and referred to different strains and the corresponding processes, such as *Xanthomonas campestris* ATCC 31600 (1980), *Xanthomonas campestris* NRRL B-12075 and NRRL B-12074 (1980), *Xanthomonas campestris* ATCC 31602 (1980) and *Xanthomonas campestris* ATCC 31601 (1980). Patents relating to the production process appeared in 1980; these included the production of xanthan gum from a chemically defined medium, a method for improving xanthan yield, and also for improving specific xanthan productivity during continuous fermentation. Standard Oil Co. has also patented a semicontinuous method for the production of xanthan gum using different *Xanthomonas* strains (1981). A method of producing a low-viscosity xanthan gum was patented in 1981. Sanofi has patented production processes (1986, 1987) and the application of xanthan (1993), as well as a mutant strain which produces a xanthan that does not develop any viscosity (1990). Oil companies other than Esso which have worked on xanthan development include Shell, which has patented the treatment of polysaccharide solutions (1980), a process for preparing *Xanthomonas* heteropolysaccharide, the heteropolysaccharide as prepared by this process and its use (1983), as well as a method for improving the filterability of a microbial broth and its use (1984). Mobil Oil Corporation, with patents relating to a waterflood process employing thickened water (1966) and a method for clarifying polysaccharide solutions (1973), and Phillips Petroleum Co., with a patent about recovery of a microbial polysaccharide (1971). Other companies which have patented xanthan-related processes or applications include: Archer Daniels Midland Co., who claim the biochemical synthesis of industrial gums (1962); Pfizer, who have patented a batch process (1979) and a continuous production of *Xanthomonas* biopolymer (1982); Stauffer Chemical Co., who have patented the fermentation of whey to produce a thickening polymer (1981); and Hoechst AG, who claim the lowering of viscosity of fermentation broths (1983). In addition, Celanese Corp. patented concentrated xanthan gum solutions obtained by ultrafiltration (1980), and Jungbunzlauer AG claimed an improved fermentation process (1982) and a process for obtaining polysaccharides as grains (1989). Several xanthan-related patents have been deposited by General Mills Chemicals Inc. about *Xanthomonas* gum amine salts (1974), about flash-drying of xanthan gum (1976), or about a cationic polysaccharide obtained by reacting a xanthan with isopropanol and NaOH (1986). Hercules Inc. has patented a xanthan recovery process (1976), and Henkel KGaA has patented some biopolymers obtained from *Xanthomonas* (1981) as well as a process for the preparation of exocellular biopolymers (1986). The Akademie der Wissenschaften der DDR of the former East Germany has worked on a method for microbial production of xanthan (1986). Getty Scientific Development Co. has patented a polysaccharide polymer made by *Xanthomonas* in 1985, and the recombinant-DNA mediated production of xanthan gum in 1986. The latter patent claims a gene cluster encoding enzymes necessary for the biosynthesis of xanthan gum which was isolated from *Xanthomonas campestris*. The U.S. Secretary of Agriculture has patented a method for producing an atypically salt-responsive alkali-deacetylated polysaccharide (1959), a method of recovering microbial polysaccharides from their fermentation broths (1962), a continuous process for producing *Xanthomonas* heteropolysaccharide (1967) as well as a nitrogen source for improved production of microbial polysaccharides (1967). The Institut Français du Petrole has patented the clarification of xanthan in 1980, and a process for the production of an improved xanthan for the oil drilling application in 1993. Cerestar Holding B.V. patented a fermentation feedstock containing simple sugars including mannose and glucose (1993). One of the most recent patents has been from Rhodia Inc. (1997), and claims the use of a liquid carbohydrate fermentation product in food. The claimed product can be either food grade or pharmaceutical grade, and it is delivered in a carbohydrate medium other than dairy whey, without any drying steps prior to use (Table 1).

Tab. 1 Relevant patents relating to xanthan products

Publication date	Patent holder	Inventors	Patent No.	Patent title
1961	U.S. Secretary of Agriculture	Jeanes, A. R., Sloneker, J. H.	US 3 000 790	Method of producing an atypically salt-responsive alkali-deacetylated polysaccharide.
1962	Jersey Production Research Co.	Patton, J. T., Lindblom, G. P.	US 3 020 206	Process for synthesizing polysaccharides.
1962	Jersey Production Research Co.	Patton, J. T.	US 3 020 207	Thickening agent and process for producing same.
1962	Kelco Co.	O'Connell, J. J.	US 3 067 038	Edible compositions comprising oil in water emulsions.
1964	Jersey Production Research Co.	Lindblom, G. P., Patton, J. T.	US 3 163 602	Substituted heteropolysaccharide.
1964	U.S. Secretary of Agriculture	Rogovin, S. P., Albrecht, W. J.	US 3 119 812	Method of recovering microbial polysaccharides from their fermentation broths.

Tab. 1 (cont.)

Publication date	Patent holder	Inventors	Patent No.	Patent title
1966	Archer Daniels Midland Co.	Weber, R. O., Horan, F. E.	US 3 271 267	Biochemical synthesis of industrial gums.
1966	Esso Production Research Co.	Lipps, B. J.	US 3 251 749	Fermentation process for preparing polysaccharides.
1966	Esso Production Research Co.	Lipps, B. J.	US 3 281 329	Fermentation process for producing a heteropolysaccharide.
1966	Kelco Co.	Schweiger, R. G.	US 3 244 695	Cationic ethers of *Xanthomonas* hydrophilic colloids.
1966	Kelco Co.	Schuppner, H. R.	US 3 279 934	Joint filling composition.
1967	Esso Production Research Co.	Lindblom, G. P., Ortloff, G. D., Patton, J. T.	US 3 305 016	Displacement of oil from partially depleted reservoirs.
1967	Esso Production Research Co.	Lindblom, G. P., Patton, J. T.	US 3 328 262	Heteropolysaccharide fermentation process.
1967	Kelco Co.	O'Connell, J. J.	US 3 355 447	Treatment of *Xanthomonas* hydrophilic colloid and resulting product.
1967	Kelco Co.	Schweiger, R. G.	US 3 349 077	Etherified *Xanthomonas* hydrophilic colloids and process of preparation.
1968	Esso Production Research Co.	Patton, J. T., Holman, W. E.	US 3 382 229	Polysaccharide recovery process.
1968	Kelco Co.	Schweiger, R. G.	US 3 376 282	Polymeric derivatives of cationic *Xanthomonas* colloid derivatives.
1968	Kelco Co.	MacNeely, W. H.	US 3 391 060	Process for producing polysaccharides.
1968	Kelco Co.	MacNeely, W. H.	US 3 391 061	Process for producing polysaccharides.
1968	Mobil Oil Corp.	Sherrod, A. W.	US 3 373 810	Waterflood process employing thickened water.
1969	General Mills Inc.	Nordgren, R., Wittcoff, H. A.	DE 1 919 790	Polysaccharide B1459 cationique.
1969	Kelco Co.	Macneely, W. H.	US 3 427 226	Process for preparing polysaccharide.
1969	U.S. Secretary of Agriculture	Rogovin, S. P.	US 3 485 719	Continuous process for producing *Xanthomonas* heteropolysaccharide.
1970	Kelco Co.	Colegrove, G. T.	US 3 516 983	Treatment of *Xanthomonas* hydrophilic colloid and resulting product.
1970	Mobil Oil Corp.	Abdo, M. K.	US 3 711 462	Method of clarifying polysaccharide solutions.
1972	Kelco Co.	Edlin, R. L.	US 3 694 236	Method of producing a dehydrated food product.
1973	Philips Petroleum Co.	Buchanan, B. B., Cottle, J. E.	US 3 773 752	Recovery of microbial polysaccharide.
1975	General Mills Chemicals Inc.	Jordan, W. A., Carter, W. H.	US 3 928 316	*Xanthomonas* gum amine salts.
1975	Société des Usines Chimiques Rhone-Poulenc		FR 2 251 620	Perfectionnement à la production de polysaccharides par fermentation.
1976	Rhone-Poulenc Industries	Falcoz, P., Celle, P., Campagne, J. C.	FR 2 299 366	Nouvelles compositions épaississantes à base d'hétéropolysaccharides.

Tab. 1 (cont.)

Publication date	Patent holder	Inventors	Patent No.	Patent title
1977	Exxon Research & Engineering Co.	Naslund, L. A., Laskin, A. I.	FR 2 330 697	Procédé de traitement d'un hétéropolysaccharide et son application.
1977	General Mills Chemicals Inc.	Cahalan, P. T., Peterson, J. A., Arndt, D. A.	US 4 053 699	Flash drying of xanthan gum and product produced thereby.
1977	Hercules Inc.	Towle, G. A.	US 4 051 317	Xanthan recovery process.
1977	Merck & Co. Inc.	Kang, K. S., Burnett, D. B.	FR 2 318 926	Perfectionnement aux procédés pour accroitre la viscosité de solutions aqueuses de gomme xanthane et aux compositions obtenues.
1977	Merck & Co. Inc.	Cottrell, I. W.	FR 2 325 666	Copolymère greffe d'un colloide hydrophile de *Xanthomonas* et procédé pour sa préparation.
1977	Rhone-Poulenc Industries	Campagne, J. C.	FR 2 342 339	Procédé de production de polysaccharides par fermentation.
1979	Institut Français du Pétrole des Carburants & Lubrifiants - Rhone-Poulenc Industries	Ballerini, D., Claude, O., Chauveteau, G., Kohler, N., Vandecasteele, J. P.	FR 2 398 874	Utilisation de mouts de fermentation pour la récupération assistée du pétrole.
1979	Rhone-Poulenc Industries	Contat, F., Lartigau, G., Nocolas, O.	FR 2 414 555	Procédé de production de polysaccharides par fermentation.
1980	Merck & Co. Inc.	Empey, R., Dominik, J. G.	EP 20 097	Production of low-calcium smooth-flow xanthan gum.
1980	Merck & Co. Inc.	Roche, R. E.	FR 2 457 322	Procédé de récupération pour améliorer la viscosité de la gomme xanthane.
1980	Merck & Co. Inc.	Racciato, J. S., Cottrell, I. W.	US 4 214 912	Deacetylated borate-biosynthetic gum composition.
1981	Celanese Corp.	Lee, H. L.	US 4 299 825	Concentrated xanthan gum solutions.
1981	Merck & Co. Inc.	Kang, K. S.	FR 2 458 589	Procédé de préparation de gomme xanthane exempte de cellulase.
1981	Pfizer Inc.	Wernau, W. C.	US 4 282 321	Fermentation process for production of xanthan.
1981	Standard Oil Co.	Weisrock, W. P.	US 4 301 247	Method for improving xanthan yield.
1982	Kelco Biospecialities Ltd.	Jarman, T. R.	EP 66 957	Inoculation procedure for xanthan gum production.
1982	Henkel KGaA	Bahn, M., Engelskirchen, K., Schieferstein, L., Schindler, J., Schmid, R.	EP 58 364	Biopolymères à partir de *Xanthomonas*.

Tab. 1 (cont.)

Publication date	Patent holder	Inventors	Patent No.	Patent title
1982	Institut Français du Pétrole	Rinaudo, M., Milas, M., Kohler, N.	FR 2 491 494	Procédé enzymatique de clarification de gommes de xanthane permettant d'améliorer leur injectivité et leur filtrabilité.
1982	Kelco Biospecialities Ltd.	Jarman, T. R., Pace, G. W.	EP 66 961	Production of xanthan having a low pyruvate content.
1982	Kelco Biospecialties Ltd.	Jarman, T. R.	EP 66 377	Process for xanthan gum production.
1982	Kelco Biospecialties Ltd.	Maury, L. G.	US 4 352 882	Production of xanthan gum by emulsion fermentation.
1982	Merck & Co. Inc.	Sandford, P. A., Baird, J. K.	US 4 357 260	Dispersible xanthan gum composite.
1982	Shell Internationale Research Maatschappij B.V.	Van Lookeren Campagne, C. J., Roest, J. B.	EP 49 012	Treatment of polysaccharide solutions.
1982	Standard Oil Co.	Weisrock, W. P., MacCarthy, E. F.	EP 46 007	*Xanthomonas campestris* ATCC 31601 and process for use.
1982	Standard Oil Co.	Weisrock, W. P.	US 4 311 796	Method for improving specific xanthan productivity during continuous fermentation.
1982	Standard Oil Co.	Weisrock, W. P.	US 4 328 310	Semi-continuous method for production of xanthan gum using *Xanthomonas campestris* ATCC 31601.
1983	Kelco Biospecialties Ltd.	Smith, I. H.	EP 68 706	Precipitation of xanthan gum.
1983	Merck & Co. Inc.	Richmon, J. B.	US 4 375 512	Process for producing low calcium xanthan gum by fermentation.
1983	Standard Oil Co.	Weisrock, W. P., Klein, H. S.	US 4 374 929	Production of xanthan gum from a chemically defined medium introduction.
1983	Standard Oil Co.	Bauer, K. A., Khosrovi, B.	US 4 400 467	Process for using *Xanthomonas campestris* NRRL B-12075 and NRRL B-12074 for making heteropolysaccharide.
1983	Standard Oil Co.	Weisrock, W. P., MacCarthy, E. F.	US 4 407 950	*Xanthomonas campestris* ATCC 31602 and process for use.
1983	Standard Oil Co.	Weisrock, W. P., MacCarthy, E. F.	US 4 407 951	*Xanthomonas campestris* ATCC 31600 and process for use.
1983	Standard Oil Co.	Weisrock, W. P.	US 4 377 637	Method of producing a low viscosity xanthan gum.
1983	U.S. Secretary of Agriculture	Cadmus, M. C., Knutson, C. A.	US 4 394 447	Production of high-pyruvate xanthan gum on synthetic medium.
1984	Jungbunzlauer A.G.	Kirkovits, A. E., Waltenberger, I.	AT 373 916	Xanthan.

Tab. 1 (cont.)

Publication date	Patent holder	Inventors	Patent No.	Patent title
1984	Pfizer Inc.	Young, T. B.	EP 115 154	Continuous production of *Xanthomonas* biopolymer.
1984	Stauffer Chemical Co.	Schwartz, R. D., Bodie, E. A.	US 4 444 792	Fermentation of whey to produce a thickening polymer.
1985	Hoechst A.G.	Voelskow, H., Keller, R., Schlingmann, M.	DE 3 330 328	Lowering the viscosity of fermentation broths.
1985	Shell International Research Maatschappij B.V.	Downs, J. D.	EP 130 647	Process for preparing *Xanthomonas* heteropolysaccharide; heteropolysaccharide as prepared by the latter process and its use.
1986	Rhone-Poulenc Specialités Chimiques	Leproux, V., Peignier, M., Cros, P., Beucherie, J., Kennel, Y.	EP 187 092	Procédé de production de polysaccharides de type xanthane.
1986	Shell International Research Maatschappij B.V.	Drozd, J. W., Rye, A. J.	EP 184 882	Method for improving the filtrability broth and the use.
1987	Akademie der Wissenschaften der DDR	Behrens, U., Stottmeister, U.	DD 250720	Method for microbial synthesis of polysaccharides.
1987	Getty Scientific Development Co.	Vanderslice, R. W., Shanon, P.	EP 211 288	A polysaccharide polymer made by *Xanthomonas*.
1987	Getty Scientific Development Co.	Capage, M. A., Doherty, D. H., Betlach, M. R., Vanderslice, R. W.	WO 87/05 938	Recombinant-DNA mediated production of xanthan gum.
1987	Merck & Co. Inc.	Peik, J. A., Steenbergen, S. M., Veeder, G. T.	EP 209 277	Heteropolysaccharide and its production and use.
1987	Merck & Co. Inc.	Cleary, J. M., Rosen, I. G., Harding, N. E., Cabanas, D. K.	EP 233 019	Recombinant DNA plasmid for xanthan gum synthesis.
1988	Sanofi - Méro Rousselot Satia S.A.	Eyssautier, B.	EP 296 965	Procédé de fermentation pour l'obtention d'une polysaccharide de type xanthane.
1988	Sanofi Elf Bio-Industries	Eyssautier, B.	FR 2 606 423	Procédé d'obtention d'un xanthane à fort pourvoir épaississant et application de ce xanthane.
1989	Henkel KGaA	Viehweg, H.	US 4 871 665	Process for the preparation of exocellular biopolymers.
1989	Rhone-Poulenc Chimie	Tavernier, C.	FR 2 624 135	Procédé de production de polysaccharides.

Tab. 1 (cont.)

Publication date	Patent holder	Inventors	Patent No.	Patent title
1990	Jungbunzlauer A.G.	Westermayer, R., Stojan, O., Eder, J.	FR 2 646 857	Procédé pour obtenir sous forme grenue des polysaccharides formés par les bactéries du genre *Xanthomonas* ou *Arthrobacter*.
1990	Rhone-Poulenc chimie	Nicolas, O.	EP 365 390	Procédé de production de polysaccharides par fermentation d'une source carbonée à l'aide de microorganismes.
1992	Merck & Co. Inc.	Talashek, T., Cleary, J. M.	EP 511 784	Low-ash xanthan gum.
1992	Sanofi - Société Nationale Elf Aquitaine	Salome, M.	FR 2 671 097	Souche mutante de *Xanthomonas campestris*. Procédé d'obtention de xanthane et xanthane non visqueux.
1994	Cerestar Holding B.V.	De Troostemberghm J. C., Beck, R. H. F., De Wannemaeker, B. L. T.	EP 609 995	Fermentation feedstock.
1994	Institut Français du Pétrole	Monot, F., Noik, C., Ballerini, D.	FR 2 701 490	Procédé de production d'un mout de xanthane ayant une propriété améliorée. Composition obtenue et application de la composition dans une boue de forage de puits.
1995	Sanofi	Tilly, G.	EP 649 599	Stabilizer composition enabling the production of a pourable aerated dairy dessert.
1999	Rhodia Inc.	Hoppe, C. A., Lawrence, J., Shaheed, A.	WO 99/25 208	Use of liquid carbohydrate fermentation product in foods.

13
Current Problems and Limitations

As a food additive which is produced by fermentation, xanthan is affected by the current debate regarding the use and danger of GM organisms. Even though the xanthan production strains used today are not GM, the consumer is skeptical about a product which is obtained via a biotechnological process. Indeed, the consumer demands that the whole process from the very beginning is accomplished without using any substrate that may have a GM source. The culture medium used for xanthan production must contain carbon and nitrogen sources; in order for the production to be cost-competitive, these sources must be cheap and efficient, easy to handle during the fermentation, and lead to a product of good quality. These requirements are fulfilled by sources such as corn-steep water as a carbon source and soy protein as a nitrogen source. However, a significant part of the world production of soy and corn today is obtained from GM plants. Although it seems very unlikely that GM carbon and nitrogen sources, after having been metabolized by the bacterium to yield the biopolymer, will confer any risk to the final xanthan product, the customer – especially the European customer – is not accepting GM sources in the culture medium. One of the reasons is certainly that standard commercial xanthans are not 100% pure, but contain about 5% of biomass. The components of the culture medium – especially glucose and nitrogen – should be used completely at the end of the production and no GM material should be left in the final product, though no guarantee can be given for this. Hence, the culture medium must be adapted to contain only non-GM sources, yet costs, productivity and product quality should remain unchanged. This is a major challenge for today's xanthan producers.

The current refusal of genetic modification for food additives by the consumer also limits innovation concerning xanthan. GM *Xanthomonas* strains could be developed in order to increase productivity, to improve the use of nutrients, to enable the metabolism of cheap substrates and their conversion into xanthan gum. It has been proved in the past that this is possible. Fu and Tseng (1990) have constructed a *Xanthomonas campestris* strain which carries a recombinant β-galactosidase-encoding gene and which is able to convert whey into xanthan. GM strains could also be modified in order to produce biopolymers with different substitution patterns, different properties in terms of rheology, dissolution and dispersion behavior, or to provide new or different synergetic interactions with other molecules. For the oil drilling industry, xanthan with increased temperature stability would be valuable. However, the current GM debate makes such development highly unlikely in the near future.

14
Outlook and Perspectives

Today, xanthan gum is a very successful biopolymer, and this success is likely to continue as no other biopolymer with similar properties is available commercially at a similar price. The world market for xanthan gum is growing and new markets are emerging for xanthan consumption as well as for xanthan production; for example, China is currently estimated to produce 5–10% of the world's xanthan. As discussed above, it appears highly unlikely that any future development of xanthan gum will be based on genetic modification, and developments will rather focus on specialty xanthans with improved handling properties. However, some perspectives based on genetic modification and improvement will clearly be required, and these are discussed in the following section.

Common genetic methods such as conjugation, electroporation, chemical and transposon mutagenesis and site-directed mutagenesis can be used with *Xanthomonas campestris*. By modifying the biosynthetic pathway for xanthan production, the carbohydrate structure and substitution pattern of the polymer can be genetically controlled to produce polysaccharides with quite different properties (Betlach et al., 1987). Mutants that lack glucuronic acid and pyruvate residues have been constructed, as well as strains producing xanthan with an increased pyruvate content. Tait and Sutherland (1989) have constructed a strain producing truncated xanthan. Until today, strains for xanthan production have been selected and improved by conventional methods. Attempts to construct strains with improved xanthan yield have been unsuccessful in the past, and are not very likely in the future since the conversion rate of carbon to xanthan is very high (Linton, 1990). Improvement of yield by genetic methods seems less promising than improvement by a more efficient fermenter design and better culture media. Another perspective for the development of xanthan is the improvement of the molecule by chemical modification. Potential modifications might include oxidation, reduction, changing of the substitution pattern, altering the side chains, or grafting other molecules onto the xanthan molecule. However, any such modification would lead to products that today are not food-approved. Controlled degradation of xanthan might lead to oligosaccharides with bioactive properties.

Acknowledgements

The authors thank Annick Bourdais and Patricia Poutrel, without whom this chapter would never have been accomplished.

15
References

Abraham, N. H., Kent, C. A., Satti, S. M. (1995) Modeling for control of poorly-mixed bioreactors, *I. Chem E. Research Event* **2**, 1049–1051.

Akademie der Wissenschaften der DDR (1986) Method for microbial synthesis of polysaccharides, DD 250720.

Albergaria, H., Roseiro, J. C., Amaral Collaco, M. T. (1999) Technological aspects and kinetics analysis of microbial gum production in carob, *Agro-Food-Ind. Hi-Tech* **10**, 24–26.

Amanullah, A., Satti, S., Nienow, A. W. (1998a) Enhancing xanthan fermentations by different modes of glucose feeding, *Biotechnology* **14**, 265–269.

Amanullah, A., Tuttiet, B., Nienow, A. W. (1998b) Agitator speed and dissolved oxygen effects in xanthan fermentation, *Biotechnol. Bioeng.* **57**, 198–210.

Archer-Daniels-Midland Co. (1962) Biochemical synthesis of industrial gums, U.S. Patent No. 3 271 267.

Banks, G., Browning, F. (1983) The development of a two stage xanthan gum fermentation, *Process Biochem. Suppl. Proc. Conf. Adv. Ferment.* 163–170.

Becker, A., Katzen, F., Puhler, A., Ielpi, L. (1998) Xanthan gum biosynthesis and application: a biochemical/genetic perspective, *Appl. Microbiol. Biotechnol.* **50**, 145–152.

Betlach, M. R., Capage, M. A., Doherty, D. H., Hassler, R. A., Henderson, N. M., Vanderslice, R. W., Marreli, J. D., Ward, M. B. (1987) Genetically engineered polymers: manipulation of xanthan biosynthesis, in: *Industrial Polysaccharides: Genetic Engineering, Structure/Property Relations and Applications* (Yalpani, M., Ed.), Amsterdam: Elsevier, 35–50.

Bih-Ying, Y., Tseng, Y. H. (1988) Production of exopolysaccharide and levels of protease and pectinase activity in pathogenic and non-pathogenic strains of *Xanthomonas campestris* pv. *campestris*, *Bot. Bull. Acad. Sinica* **29**, 93–99.

Capron, I., Brigand, G., Muller, G. (1997) About the native and renatured conformation of xanthan exopolysaccharide, *Polymer* **38**, 5289–5295.

Capron, I., Alexandre, S., Muller, G. (1998a) An atomic force microscopy study of the molecular organisation of xanthan, *Polymer* **39**, 5725–5730.

Capron, I., Brigand, G., Muller, G. (1998b) Thermal denaturation and renaturation of a fermentation broth of xanthan: rheological consequences, *Int. J. Biol. Macromol.* **23**, 215–225.

Celanese Corp. (1980) Concentrated xanthan gum solutions, U.S. Patent No. 4 299 825.

Cerestar Holding B.V. (1993) Fermentation feedstock, EP 609 995.

Chan, J. W. Y. F., Goodwin, P. H. (1999) The molecular genetics of virulence of *Xanthomonas campestris*, *Biotechnol. Adv.* **17**, 489–508.

Cheetham, N. W. H., Punruckvong, A. (1985) An HPLC method for the determination of acetyl and pyruvyl groups in polysaccharides, *Carbohydr. Polym.* **5**, 399–406.

Christensen, B. E., Smidsrød, O. (1991) Hydrolysis of xanthan in dilute acid: effects on chemical composition, conformation, and intrinsic viscosity, *Carbohydr. Res.* **214**, 55–69.

Christensen, B. E., Smidsrød, O. (1996) Dependence of the content of unsubstituted (cellulosic) regions in prehydrolysed xanthans on the rate of hydrolysis by *Trichoderma reesei* endoglucanase, *Int. J. Biol. Macromol.* **18**, 93–99.

Christensen, B. E., Smidsrød, O., Elgsaeter, A., Stokke, B. T. (1993a) Depolymerization of double-stranded xanthan by acid hydrolysis: characterization of partially degraded double strands and single-stranded oligomers released from the ordered structures, *Macromolecules* **26**, 6111–6120.

Christensen, B. E., Smidsrød, O., Stokke, B. T. (1993b) Xanthans with partially hydrolysed side chains: conformation and transitions, in: *Carbohydrates and Carbohydrate Polymers, Analysis, Biotechnology, Modification, Antiviral, Biomedical and Other Applications* (Yalpani, M., Ed.), ATL Press, 166–173.

Collar, C., Andreu, P., Martinez, J. C., Armero, E. (1999) Optimisation of hydrocolloid addition to improve wheat bread dough functionality: a response surface methodology study, *Food Hydrocolloids* **13**, 467–475.

Conforti, F. D., Charles, S. A., Duncan, S. E. (1997) Evaluation of a carbohydrate-based fat replacer in a fat-reduced baking powder biscuit, *J. Food Qual.* **20**, 247–256.

Coplin, D. L., Cook, D. (1990) Molecular genetics of extracellular polysaccharide biosynthesis in vascular phytopathogenic bacteria, *Mol. Plant-Microbe Interact.* **3**, 271–279.

Daniel, J. R., Whistler, R. L., Voragen, A. C. J., Pilnik W. (1994) Starch and other polysaccharides, in: *Ullmann's Encyclopedia of Industrial Chemistry* (Elvers, B., Hawkins, S., Russey, W., Eds.), Weinheim: VCH, 1–62.

Daniels, M. J., Leach, J. E. (1993) Genetics of *Xanthomonas* in: *Xanthomonas* (Swings, J. G., Civerolo, E. L., Eds.), London: Chapman & Hall, 301–339.

De Andrade Lima, M. A. G., De Araujo, J. M., De Franca, F. P. (1997) The evaluation of different parameters to characterize xanthan gum-producing strains of *Xanthomonas* pv. *campestris*, *Arq. Biol. Tecnol.* **40**, 179–187.

De Crecy Lagard, V., Glaser, P., Lejeune, P., Sismeiro, O., Barber, C. E., Daniels, M. J., Danchin, A. (1990) A *Xanthomonas campestris* pv. *campestris* protein similar to catabolite activation factor is involved in regulation of phytopathogenicity, *J. Bacteriol.* **172**, 5877–5883.

De Vuyst, L., Vermeire, A. (1994) Use of industrial medium components for xanthan production by *Xanthomonas campestris* NRRL-B-1459, *Appl. Microbiol. Biotechnol.* **42**, 187–191.

De Vuyst, L., Van Loo, J., Vandamme, E. J. (1987) Two stage fermentation process for improved xanthan production by *Xanthomonas campestris* NRRL B-1459, *J. Chem. Technol. Biotechnol.* **39**, 263–273.

Duckworth, M., Yaphe, W. (1970) Definitive assay for pyruvic acid in agar and other algal polysaccharide, *Chem. Ind.* **23**, 747–748.

Dumitriu, S., Chornet, E. (1998) Polysaccharides as support for enzyme and cell immobilisation, in: *Polysaccharides. Structural Diversity and Functional Versatility* (Dumitriu, S., Ed.), New York: Marcel Dekker, 629–748.

Esso Production Research Co. (1959) Displacement of oil from partially depleted reservoirs, U.S. Patent No. 3 305 016.

Esso Production Research Co. (1963) Fermentation process for preparing polysaccharides, U.S. Patent No. 3 251 749.

Esso Production Research Co. (1963) Fermentation process for producing a heteropolysaccharide, U.S. Patent No. 3 281 329.

Esso Production Research Co. (1963) Polysaccharide recovery process, U.S. Patent No. 3 382 229.

Esso Production Research Co. (1966) Heteropolysaccharide fermentation process, U.S. Patent No. 3 328 262.

Exxon Research & Engineering Co. (1975) Procédé de traitement d'un hétéropolysaccharide et son application, FR 2 330 697.

Fu, J. F., Tseng, Y. H. (1990) Construction of lactose-utilizing *Xanthomonas campestris* and production of xanthan gum, *Appl. Environ. Microbiol.* **56**, 919–923.

Galindo, E. (1994) Aspects of the process for xanthan production, *Trans. I. Chem. E* **72**, 227–237.

Galindo, E., Salcedo, G., Ramirez, M. E. (1994) Preservation of *Xanthomonas campestris* on agar slopes: effects on xanthan production, *Appl. Microbiol. Biotechnol.* **40**, 634–637.

Garcia Ochoa, F., Santos, V. E., Alcon, A. (1995) Xanthan gum production: an unstructured kinetic model, *Enzyme Microb. Technol.* **17**, 206–217.

Garcia Ochoa, F., Santos, V. E., Alcon, A. (1998) Metabolic structured kinetic model for xanthan production, *Enzyme Microb. Technol.* **23**, 75–82.

Gaspar, C., Laureano, O., Sousa, I. (1998) Production of reduced-calorie grape juice jelly with gellan, xanthan and locust bean gums: sensory and objective analysis of texture, *Z. Lebensm. Unters. Forsch.* **206**, 169–174.

General Mills Chemicals Inc. (1974) *Xanthomonas* gum amine salts, U.S. Patent No. 3 928 316.

General Mills Chemicals Inc. (1976) Flash drying of xanthan gum and product produced thereby, U.S. Patent No. 4 053 699.

General Mills Inc. (1968) Polysaccharide B1459 cationique, DE 1 919 790.

Getty Scientific Development Co. (1985) A polysaccharide polymer made by *Xanthomonas*, EP 211 288.

Getty Scientific Development Co. (1986) Recombinant-DNA mediated production of xanthan gum, WO 87/05 938.

Gupte, M. D., Kamat, M. Y. (1997) Isolation of wild *Xanthomonas* strains from agricultural produce, their characterisation and potential related to polysaccharide production, *Folia Microbiol.* **42**, 621–628.

Harris, P. J., Fergusson, L. R., (1999) Dietary fibres may protect or enhance carcinogenesis, *Mutat. Res.* **443**, 95–110.

Harrison, G. M., Mun, R., Cooper, G., Boger, D. V. (1999) A note on the effect of polymer rigidity and concentration on spray atomisation, *J. Non-Newtonian Fluid Mech.* **85**, 93–104.

Henkel KGaA (1981) Biopolymères à partir de *Xanthomonas*, EP 58 364.

Henkel KGaA (1986) Process for the preparation of exocellular biopolymers, U.S. Patent No. 4 871 665.

Hercules Inc. (1976) Xanthan recovery process, U.S. Patent No. 4 051 317.

Herp, A. (1980) Oxidative-reductive depolymerization of polysaccharides, in: *The Carbohydrates*, Vol. Ib (Pigman, W., Horton, D., Eds.), New York: Academic Press, 1276–1297.

Hestrin, S. (1949) Reaction of acetylcholine and other carboxylic acid derivatives with hydroxylamine, and its analytical application, *J. Biol. Chem.* **180**, 249–261.

Hoechst A.G. (1983) Lowering the viscosity of fermentation broths, DE 3 330 328.

Ielpi, L., Couso, R. O., Dankert, M. A. (1993) Sequential assembly and polymerization of the polyprenol linked pentasaccharide repeating unit of the xanthan polysaccharide in *Xanthomonas campestris*, *J. Bacteriol.* **175**, 2490–2500.

Institut Français du Pétrole des Carburants & Lubrifiants - Rhone-Poulenc Industries (1977) Utilisation de mouts de fermentation pour la récupération assistée du pétrole, FR 2 398 874.

Institut Français du Pétrole (1980) Procédé enzymatique de clarification de gommes de xanthane permettant d'améliorer leur injectivité et leur filtrabilité, FR 2 491 494.

Institut Français du Pétrole (1993) Procédé de production d'un mout de xanthane ayant une propriété améliorée. Composition obtenue et application de la composition dans une boue de forage de puits, FR 2 701 490.

Janca, J. (Ed.) (1988) Field-Flow Fractionation: analysis of macromolecules and particles, New York: Marcel Dekker.

Jarman, T. R., Pace, G. W. (1984) Energy requirement for microbial expolysaccharide synthesis, *Arch. Microbiol.* **137**, 231–235.

Jersey Production Research Co. (1960) Process for synthesizing polysaccharides, U.S. Patent No. 3 020 206.

Jersey Production Research Co. (1960) Thickening agent and process for producing same, U.S. Patent No. 3 020 207.

Jersey Production Research Co. (1960) Substituted heteropolysaccharide, U.S. Patent No. 3 163 602.

Jungbunzlauer A.G. (1982) Xanthan, AT 373 916.

Jungbunzlauer A.G. (1989) Procédé pour obtenir sous forme grenue des polysaccharides formés par les bactéries du genre *Xanthomonas* ou *Arthrobacter*, FR 2 646 857.

Kang, K. S., Pettitt, D. L. (1993) Xanthan, gellan, welan and rhamsan, in: *Industrial Gums*, 3rd edn (Whistler, R. L., BeMiller, J. N., Eds.), San Diego, CA: Academic Press, 341–397.

Katzen, F., Ferreiro, D. U., Oddo, C. G., Ielmini, M. V., Becker, A., Puhler, A., Ielpi, L. (1998) *Xanthomonas campestris* pv. *campestris* gum mutants: effects on xanthan biosynthesis and plant virulence, *J. Bacteriol.* **180**, 1607–1617.

Kelco Biospecialities Ltd. (1981) Inoculation procedure for xanthan gum production, EP 66 957.

Kelco Biospecialities Ltd. (1981) Production of xanthan having a low pyruvate content, EP 66 961.

Kelco Biospecialties Ltd. (1981) Process for xanthan gum production, EP 66 377.

Kelco Biospecialties Ltd. (1981) Precipitation of xanthan gum, EP 68706.

Kelco Biospecialties Ltd. (1981) Production of xanthan gum by emulsion fermentation, U.S. Patent No. 4 352 882.

Kelco Co. (1960) Edible compositions comprising oil in water emulsions, U.S. Patent No. 3 067 038.

Kelco Co. (1963) Cationic ethers of *Xanthomonas* hydrophilic colloids, U.S. Patent No. 3 244 695.

Kelco Co. (1963) Joint filling composition, U.S. Patent No. 3 279 934.

Kelco Co. (1964) Treatment of *Xanthomonas* hydrophilic colloid and resulting product, U.S. Patent No. 3 355 447.

Kelco Co. (1964) Polymeric derivatives of cationic *Xanthomonas* colloid derivatives, U.S. Patent No. 3 376 282.

Kelco Co. (1965) Etherified *Xanthomonas* hydrophilic colloids and process of preparation, U.S. Patent No. 3 349 077.

Kelco Co. (1966) Process for producing polysaccharides, U.S. Patent No. 3 391 060.

Kelco Co. (1966) Process for producing polysaccharides, U.S. Patent No. 3 391 061.

Kelco Co. (1966) Process for preparing polysaccharide, U.S. Patent No. 3 427 226.

Kelco Co. (1968) Treatment of *Xanthomonas* hydrophilic colloid and resulting product, U.S. Patent No. 3 516 983.

Kelco Co. (1971) Method of producing a dehydrated food product, U.S. Patent No. 3 694 236.

Kingsley, M., Gabriel, D., Marlow, G., Roberts, P. (1993) The *opsX* locus of *Xanthomonas campestris* affects host range and biosynthesis of lipopolysaccharide and extracellular polysaccharide, *J. Bacteriol.* **175**, 5839–5850.

Knoop, V., Staskawicz, B., Bonas, U. (1991) Expression of the avirulence gene *avrBs3* from *Xanthomonas campestris* pv. *vesicatoria* is not under the control of *hrp* genes and is independent of plant factors, *J. Bacteriol.* **173**, 7142–7150.

Kuo, T. T., Lin, B. C., Li, C. C. (1970) Bacterial leaf blight of rice plant. III – Phytotoxic polysaccharides produced by *Xanthomonas oryzae*, *Bot. Bull. Acad. Sinica* **11**, 46–54.

Kuttuva, S. G., Sundararajan, A., Ju, L. K. (1998) Model simulation of water-in-oil xanthan fermentation, *J. Dispersion Sci. Technol.* **19**, 1003–1029.

Launay, B., Doublier, J.L., Cuvelier, G. (1986) Flow properties of aqueous solutions and dispersion of polysaccharides, in: *Functional Properties of Food Macromolecules* (Mitchell, J. R., Ledward, D. A., Eds.), London, New York: Elsevier Applied Science Publisher, 1–78.

Leach, J. G., Lilly, V. G., Wilson, H. A., Purvis, M. R. (1957) Bacterial polysaccharides: the nature and function of the exudate produced by *Xanthomonas phaseoli*, *Phytopathology* **47**, 113–120.

Lebrun, L., Junter, G. A., Jouenne, T., Mignot, L. (1994) Exopolysaccharide production by free and immobilized microbial cultures, *Enzyme Microb. Technol.* **16**, 1048–1054.

Leigh, J. A., Coplin, D. L. (1992) Exopolysaccharides in plant–bacterial interactions, *Annu. Rev. Microbiol.* **46**, 307–346, 1048–1054.

Li, H., Rief, M., Oesterhelt, F., Gaub, H. E. (1999) Force spectroscopy on single xanthan molecules, *Appl. Physics A* **68**, 407–410.

Liakopoulou Kyriakides, M., Psomas, S. K., Kyriakidis, D. A. (1999) Xanthan gum production by *Xanthomonas campestris* w.t. fermentation from chestnut extract, *Appl. Biochem. Biotechnol.* **82**, 175–183.

Linton, J. D. (1990) The relationship between metabolite production and the growth efficiency of the producing organism, *FEMS Microbiol. Rev.* **75**, 1–18.

Liu, W., Sato, T., Norisuye, T., Fujita, H. (1987) Thermally induced conformational change of xanthan in 0.01M aqueous sodium chloride, *Carbohydr. Res.* **160**, 267–281.

Lo, Y. M., Yang, S. T., Min, D. B. (1997) Ultrafiltration of xanthan gum fermentation broth: process and economic analyses, *J. Food Eng.* **31**, 219–236.

Marrs, M. (1997) Melt-in-mouth gels, *World Ingr.* June, 39–40.

MacNeely, W. H., Kang, K. S. (1973) Xanthan and some other biosynthetic gums, in: *Industrial Gums*, 2nd edn (Whistler, R. L., BeMiller, J. N., Eds.), New York: Academic Press, 473–497.

Merck & Co. Inc. (1975) Copolymère greffe d'un colloide hydrophile de *Xanthomonas* et procédé pour sa préparation, FR 2 325 666.

Merck & Co. Inc. (1975) Perfectionnement aux procédés pour accroitre la viscosité de solutions aqueuses de gomme xanthane et aux compositions obtenues, FR 2 318 926.

Merck & Co. Inc. (1978) Process for producing low calcium xanthan gum by fermentation, U.S. Patent No. 4 375 512.

Merck & Co. Inc. (1979) Production of low-calcium smooth-flow xanthan gum, EP 20 097.

Merck & Co. Inc. (1979) Procédé de préparation de gomme xanthane exempte de cellulase, FR 2 458 589.

Merck & Co. Inc. (1979) Deacetylated borate-biosynthetic gum composition, U.S. Patent No. 4 214 912.

Merck & Co. Inc. (1980) Dispersible xanthan gum composite, U.S. Patent No. 4 357 260.

Merck & Co. Inc. (1985) heteropolysaccharide and its production and use, EP 209 277.

Merck & Co. Inc. (1986) Recombinant DNA plasmid for xanthan gum synthesis, EP 233 019.

Merck & Co. Inc. (1991) Low-ash xanthan gum, EP 511 784.

Milas, M., Rinaudo, M., Knipper, M., Schuppiser, J.L. (1990) Flow and viscoelastic properties of xanthan gum solutions, *Macromolecules* **23**, 2506–2511.

Mobil Oil Corp. (1966) Waterflood process employing thickened water, U.S. Patent No. 3 373 810.

Mobil Oil Corp. (1973) Method of clarifying polysaccharide solutions, U.S. Patent No. 3 711 462.

Moorhouse, R., Walkinshaw, M. D., Arnott, S. (1977) Xanthan gum. Molecular conformation and interactions, in: *ACS Symposium Series 45, Extracellular Microbial Polysaccharides* (Sandford, P. A., Laskin, A., Eds.), Washington, DC: American Chemical Society, 90–102.

Moreno, J., Lopez, M. J., Vargas Garcia, C., Vasquez, R. (1998) Use of agricultural wastes for xanthan production by *Xanthomonas campestris*, *J. Ind. Microbiol. Biotechnol.* **21**, 242–246.

Morris, V. J. (1990) Science, structure and applications of microbial polysaccharides, in: *Gums and Stabilisers for the Food Industry 5* (Phillips, G. O., Wedlock, D. J., Williams, P. A., Eds.), New York: IRL Press, 315–328.

Morris, V. J., Kirby, A. R., Gunning, A. P. (1999) Using atomic force microscopy (AFM) to probe food biopolymer functionality, *Scanning* **21**, 287–292.

Nankai, H., Hashimoto, W., Miki, H., Kawai, S., Murata, K. (1999) Microbial system for polysaccharide depolymerisation: enzymatic route for xanthan depolymerisation by *Bacillus* sp. strain GL1, *Appl. Environ. Microbiol.* **65**, 2520–2526.

Nussinovitch, A. (Ed.) (1997) Xanthan gum, in: *Hydrocolloids Applications: Gum Technology in the Food and Other Industries*, London: Blackie Academic & Professional, 154–168.

Parker, A., Michel, R., Vigouroux, F., Reed, W. F. (1999) Dissolution kinetics of polymer powders, *Polym. Prep. Amer. Chem. Soc. Div. Polym. Chem.* **40**, 685–686.

Parlin, S. (1997) Good vibrations. New scalable vibrating membrane filter system separates liquids, solids, *Food Process* **58**, 106–107.

Pfizer Inc. (1979) Fermentation process for production of xanthan, U.S. Patent No. 4 282 321.

Pfizer Inc. (1982) Continuous production of *Xanthomonas* biopolymer, EP 115 154.

Philipon, P. (1997) Des médicaments libérés sur commande, *Biofutur* **171**, 25–27.

Philips Petroleum Co. (1971) Recovery of microbial polysaccharide, U.S. Patent No. 3 773 752.

Pollock, T. J., Mikolajczak, M., Yamazaki, M., Thorme, L., Armentrout, R. W. (1997) Production of xanthan gum by *Sphingomonas* bacteria carrying genes from *Xanthomonas campestris*, *J. Ind. Microbiol. Biotechnol.* **19**, 92–97.

Ramirez, M. E., Fucikovsky, L., Garcia-Jimenez, F., Quintero, R., Galindo, E. (1988) Xanthan gum production by altered pathogenicity variants of *Xanthomonas campestris*, *Appl. Microbiol. Biotechnol.* **29**, 5–10.

Rhodia Inc. (1997) Use of liquid carbohydrate fermentation product in foods, WO 99/25 208.

Rhone-Poulenc Chimie (1987) Procédé de production de polysaccharides, FR 2 624 135.

Rhone-Poulenc Chimie (1988) Procédé de production de polysaccharides par fermentation d'une source carbonée à l'aide de microorganismes, EP 365 390.

Rhone-Poulenc Industries (1975) Nouvelles compositions épaississantes à base d'hétéropolysaccharides, FR 2 299 366.

Rhone-Poulenc Industries (1976) Procédé de production de polysaccharides par fermentation, FR 2 342 339.

Rhone-Poulenc Industries (1978) Procédé de production de polysaccharides par fermentation, FR 2 414 555.

Rhone-Poulenc Specialités Chimiques (1984) Procédé de production de polysaccharides de type xanthane, EP 187 092.

Rinaudo, M., Milas, M. (1980) Enzymic hydrolysis of the bacterial polysaccharide xanthan by cellulase, *Int. J. Biol. Macromol.* **2**, 45–48.

Rodd, A. B., Dunstan, D. E., Boger, D. V. (2000) Characterisation of xanthan gum solutions using dynamic light scattering and rheology, *Carbohydr. Polym.* **42**, 159–174.

Rodriguez, H., Aguilar, L. (1997) Detection of *Xanthomonas campestris* mutants with increased xanthan production, *J. Ind. Microbiol. Biotechnol.* **18**, 232–234.

Ruijssenaars, H. J., De Bont, J. A. M., Hartmans, S. (1999) A pyruvated mannose-specific xanthan lyase involved in xanthan degradation by *Paenibacillus alginolyticus* XL-1, *Appl. Environ. Microbiol.* **65**, 2446–2452.

Salcedo, G., Ramirez, M. E., Flores, C., Galindo, E. (1992) Preservation of *Xanthomonas campestris* in *Brassica oleracea* seeds, *Appl. Microbiol. Biotechnol.* **37**, 723–727.

Sanofi Elf Bio-Industries (1986) Procédé d'obtention d'un xanthane à fort pourvoir épaississant et application de ce xanthane, FR 2 606 423.

Sanofi (1993) Stabilizer composition enabling the production of a pourable aerated dairy dessert, EP 649 599.

Sanofi, Méro Rousselot Satia S.A. (1987) Procédé de fermentation pour l'obtention d'une polysaccharide de type xanthane, EP 296 965.

Sanofi, Société Nationale Elf Aquitaine (1990) Souche mutante de *Xanthomonas campestris*, procédé d'obtention de xanthane et xanthane non visqueux, FR 2 671 097.

Schumpe, A., Diedrichs, S., Hesselink, P. G. M., Nene, S., Deckwer, W. D. (1991) Xanthan production in emulsions, *Proceedings of the Second*

International Symposium on Biochemical Engineering, Stuttgart, 196–199.

Serrano Carreon, L., Corona, R. M., Sanchez, A., Galindo, E. (1998) Prediction of xanthan fermentation development by a model linking kinetics, power drawn and mixing, *Proc. Biochem.* **33**, 133–146.

Shell International Research Maatschappij B.V. (1983) Process for preparing *Xanthomonas* heteropolysaccharide, heteropolysaccharide as prepared by the latter process and its use, EP 130 647.

Shell International Research Maatschappij B.V. (1984) Method for improving the filterability broth and the use, EP 184 882.

Shell International Research Maatschappij B.V. (1980) Treatment of polysaccharide solutions, EP 49 012.

Silman, R. W., Rogovin, P. (1972) Continuous fermentation to produce xanthan biopolymer: effect of dilution rate, *Biotechnol. Bioeng.* **14**, 23–31.

Sloneker, J. H., Orentas, D. G. (1962) Exocellular bacterial polysaccharide from *Xanthomonas campestris* NRRL B61459. II – Linkage of the pyruvic acid, *Can. J. Chem.* **40**, 2188–2189.

Société des Usines Chimiques Rhone-Poulenc (1973) Perfectionnement à la production de polysaccharides par fermentation, FR 2 251 620.

Standard Oil Co. (1980) *Xanthomonas campestris* ATCC 31601 and process for use, EP 46 007.

Standard Oil Co. (1980) Method for improving xanthan yield, U.S. Patent No. 4 301 247.

Standard Oil Co. (1980) Method for improving specific xanthan productivity during continuous fermentation, U.S. Patent No. 4 311 796.

Standard Oil Co. (1980) Production of xanthan gum from a chemically defined medium introduction, U.S. Patent No. 4 374 929.

Standard Oil Co. (1980) Process for using *Xanthomonas campestris* NRRL B-12075 and NRRL B-12074 for making heteropolysaccharide, U.S. Patent No. 4 400 467.

Standard Oil Co. (1980) *Xanthomonas campestris* ATCC 31602 and process for use, U.S. Patent No. 4 407 950.

Standard Oil Co. (1980) *Xanthomonas campestris* ATCC 31600 and process for use, U.S. Patent No. 4 407 951.

Standard Oil Co. (1981) Semi-continuous method for production of xanthan gum using *Xanthomonas campestris* ATCC 31601, U.S. Patent No. 4 328 310.

Standard Oil Co. (1981) Method of producing a low viscosity xanthan gum, U.S. Patent No. 4 377 637.

Stauffer Chemical Co. (1981) Fermentation of whey to produce a thickening polymer, U.S. Patent No. 4 444 792.

Stokke, B. J., Christensen, B. E., Smidsrød, O. (1998) Macromolecular properties of xanthan, in: *Polysaccharides. Structural, Diversity and Functional Versatility* (Dumitriu, S., Ed.), New York: Marcel Dekker, 433–472.

Stredansky, M., Conti, E. (1999) Xanthan production by solid state fermentation, *Process Biochem.* **34**, 581–587.

Stredansky, M., Conti, E., Navarini, L., Bertocchi, C. (1999) Production of bacterial exopolysaccharides by solid substrate fermentation, *Process Biochem.* **34**, 11–16.

Sutherland, I. W. (1993) Xanthan, in: *Xanthomonas* (Swings, J. G., Civerolo, E. L., Eds.), London: Chapman & Hall, 363–388.

Tait, M. I., Sutherland, I. W. (1989) Synthesis and properties of a mutant type of xanthan, *J. Appl. Bacteriol.* **66**, 457–460.

Tait, M. I., Sutherland, I. W., Clarke-Sturman, A. J. (1990) Acid hydrolysis and high-performance liquid chromatography of xanthan, *Carbohydr. Polym.* **13**, 133–148.

Tang, J. L., Gough, C. L., Daniels, M. J. (1990) Cloning of genes involved in negative regulation of production of extracellular enzymes and polysaccharide of *Xanthomonas campestris* pathovar *campestris*, *Mol. Gen. Genet.* **222**, 157–160.

Tilly, G. (1991) Stabilization of dairy products by hydrocolloids, in: *Food Ingredients Europe: Conference Proceedings* (Van Zeijst, R., Ed.) Maarsen: Expoconsult Publishers, 105–121.

Triveni, R., Shamala, T. R., (1999) Clarification of xanthan gum with extracellular enzymes secreted by *Trichoderma koningii*, *Process Biochem.* **34**, 49–53.

Tseng, Y. H., Choy, K. T., Hung, C. H., Lin, N. T., Liu, J. Y., Lou, C. H., Yang, B. Y., Wen, F. S., Wu, J. R. (1999) Chromosome map of *Xanthomonas campestris* pv. *campestris* 17 with locations of genes involved in xanthan gum synthesis and yellow pigmentation, *J. Bacteriol.* **181**, 117–125.

U. S. Secretary Agriculture (1959) Method of producing an atypically salt-responsive alkali-deacetylated polysaccharide, U.S. Patent No. 3 000 790.

U. S. Secretary Agriculture (1962) Method of recovering microbial polysaccharides from their fermentation broths, U.S. Patent No. 3 119 812.

U. S. Secretary Agriculture (1967) Continuous process for producing *Xanthomonas* heteropolysaccharide, U.S. Patent No. 3 485 719.

U. S. Secretary Agriculture (1978) Production of high-pyruvate xanthan gum on synthetic medium, U.S. Patent No. 4 394 447.

Vafiadis, D. (1999) Anti-shock treatment (frozen dairy products formulation), *Dairy Field* **182**, 85–88.

Vanderslice, R. W., Doherty, D. H., Capage, M. A., Betlach, M. R., Hassler, R. A., Henderson, N. M., Ryan-Graniero, J., Tecklenburg, M. (1989) Genetic engineering of polysaccharide structure in *Xanthomonas campestris*, in: *Biomedical and Biotechnological Advances in Industrial Polysaccharides* (Crescenzi, V., Dea, I. C. M., Paoletti, S., Stivala, S. S., Sutherland, I. W., Eds.), New York: Gordon and Breach, 145–156.

Vidhyasekaran, P., Alvenda, M. E., Mew, T. W. (1989) Physiological changes in rice seedlings induced by extracellular polysaccharide produced by *Xanthomonas campestris* pv. *oryzae*, *Physiol. Mol. Plant Pathol.* **35**, 391–402.

Vojnov, A. A., Zorreguieta, A., Dow, J. M., Daniels, M. J., Dankert, M. A. (1998) Evidence for a role for the gumB and gumC gene products in the formation of xanthan from its pentasaccharide repeating unit by *Xanthomonas campestris*, *Microbiology* **144**, 1487–1493.

Watabe, M., Yamaguchi, M., Kitamura, S., Horino, O. (1993) Immunohistochemical studies on localization of the extracellular polysaccharide produced by *Xanthomonas oryzae* pv. *oryzae* in infected rice leaves, *Can. J. Microbiol.* **39**, 1120–1126.

Yang, B. Y., Tseng, Y. H. (1988) Production of exopolysaccharide and levels of protease and pectinase activity in pathogenic and non-pathogenic strains of *Xanthomonas campestris* pv. *campestris*, *Bot. Bull. Academia Sinica* **29**, 93–99.

Yang, S. T., Lo, Y. M., Min, D. B. (1996) Xanthan gum fermentation by *Xanthomonas campestris* immobilized in a novel centrifugal fibrous-bed bioreactor, *Biotechnol. Prog.* **12**, 630–637.

Yang, S. T., Lo, Y. M., Chattopadhyay, D. (1998) Production of cell-free xanthan fermentation broth by cell adsorption on fibers, *Biotechnol. Prog.* **14**, 259–264.

Yoo, S. D., Harcum, S. W. (1999) Xanthan gum production from paste waste sugar beet pulp, *Bioresource Technol.* **70**, 105–109.

Zhao, X. M., Li, X. H., Ban, R., Zhu, Y. (1997) The effects of mixing modes on recovery of xanthan gum by precipitation, *BHR Group Conf. Ser. Publ.* **25**, 3–8.